Other Titles in This Series

(Continued in the back of this publication)

Mischa Cotlar

Harmonic Analysis
and Operator Theory

CONTEMPORARY MATHEMATICS

189

Harmonic Analysis and Operator Theory

A Conference in Honor of Mischa Cotlar
January 3–8, 1994
Caracas, Venezuela

S. A. M. Marcantognini
G. A. Mendoza
M. D. Morán
A. Octavio
W. O. Urbina
Editors

American Mathematical Society
Providence, Rhode Island

The International Conference on Harmonic Analysis and Operator Theory in Honor of Mischa Cotlar was held in Caracas, Venezuela, from January 3–8, 1994, with support from CONICIT (Venezuela), Fundación Polar, Congreso Nacional de Venezuela, Universidad Central de Venezuela, Universidad Simón Bolívar, Instituto Venezolano de Investigaciones Científicas, and the National Science Foundation, Grant No. NSF/INT 9309850.

1991 *Mathematics Subject Classification*. Primary 43–XX, 47–XX; Secondary 44–XX, 45–XX, 46–XX.

Library of Congress Cataloging-in-Publication Data

Harmonic analysis and operator theory : a conference in honor of Mischa Cotlar, January 3–8, 1994, Caracas, Venezuela / S. A. M. Marcantognini ... [et al.].

p. cm. — (Contemporary mathematics, ISSN 0271-4132; v. 189)

Includes bibliographical references.

ISBN 0-8218-0304-2

1. Harmonic analysis—Congresses. 2. Operator theory—Congresses. I. Cotlar, Mischa II. Marcantognini, S. A. M. (Stefania A. M.), 1960– . III. Series: Contemporary mathematics (American Mathematical Society); v. 189.

QA403.H224   1995

515′.724–dc20   95-23965

                                                       CIP

Contents

CONTENTS

Preface

Professor Mischa Cotlar arrived in Venezuela for the first time in 1971, and we have been fortunate to have him with us permanently since 1974. Throughout the years since, he has had a decisive influence in the development here of the areas of harmonic analysis and operator theory. In 1993, when Mischa celebrated his 80th birthday, his friends in Caracas, students and colleagues, decided to pay homage to this extraordinary person. The celebration crystalized as a conference on Harmonic Analysis and Operator Theory held in Caracas, January 3–8, 1994, and this volume, comprising papers submitted by a number of participants.

The organizers of the conference wish to take this opportunity to acknowledge the financial support received from CONICIT (Venezuela), Fundación Polar, Congreso Nacional de Venezuela, Universidad Central de Venezuela, Universidad Simón Bolívar, Instituto Venezolano de Investigaciones Científicas, and the U.S. National Science Foundation, and to thank the close to one hundred mathematicians who through their participation contributed to the success of the conference.

The editors of this volume wish to thank the authors for their contribution to this celebration of Mischa Cotlar, the referees for their careful scrutiny of the papers, and the AMS for providing this outlet for the volume.

Caracas, March of 1995

Contemporary Mathematics
Volume **189**, 1995

Nuclear Hankel Matrices and Orthogonal Trigonometric Polynomials

Adamyan V.M. and Nechayev S.E.

ABSTRACT. This article is a study of the Generalized Schur Problem (GSP) in the case when the corresponding infinite Hankel matrix generates a trace class operator. A multiplicative representation in terms of Schur parameters is proposed for the matrix of the linear fractional transformation in the known formula [2],[3] furnishing all solutions of the GSP. In connection with this representation a recurrent system defining Szego orthogonal trigonometric polynomials with respect to the absolutely continuous measure on the unit circle is investigated.

Introduction

This article contains some partial additions to the investigations [1],[2],[3],[5] dealing with infinite Hankel matrices and approximation problems connected with them. All problems mentioned in [3] are anyhow directly related to the following Generalized Schur problem (GSP):

Given a sequence $\gamma_1, \ldots, \gamma_n, \ldots$ of complex numbers. To find all bounded functions on the unit circle

$$f(\zeta) = \sum_{-\infty}^{\infty} c_n(f)\zeta^n, \qquad |\zeta| = 1,$$

such that

$$1) \; c_{-j}(f) = \gamma_j, \qquad j = 1, 2, \ldots$$

$$2) \; \|f\|_\infty = ess\,sup|f(\zeta)| \le 1.$$

In particular classical Caratheodory-Fejer and Nevanlinna-Pick problems are special cases of GSP [3].

The approach to GSP developed in [1],[2],[3],[5] uses as a main tool a Hankel operator Γ in l^2 generated by a matrix $\left(\gamma_{j+k-1}\right)_1^\infty$. In this article we consider special cases of GSP when Γ is a trace class operator and $\|\Gamma\| < 1$. Under this conditions we reveal here a new content of the known formula [2],[3] furnishing

1991 Mathematics Subject Classification. Primary 47B10, 47B35; Secondary 47Bxx

all solutions of GSP in the indefinite cases. The fixed entries $p(\zeta), q(\zeta)$ in this formula turn out to be connected with the solution of the recurrent system

(1)
$$u_{k+1}(\zeta) = \zeta u_k(\zeta) - \alpha_{k+1} v_k(\zeta),$$
$$v_{k+1}(\zeta) = -\zeta \overline{\alpha}_{k+1} u_k(\zeta) + v_k(\zeta),$$

with the condition at infinity

$$\lim_{k \to \infty} \zeta^{-k} \begin{pmatrix} u_k(\zeta) \\ v_k(\zeta) \end{pmatrix} = \begin{pmatrix} 1 \\ 0 \end{pmatrix},$$

by the relation

$$\begin{pmatrix} u_0(\zeta) \\ v_o(\zeta) \end{pmatrix} = \begin{pmatrix} p(\zeta) \\ q(\zeta) \end{pmatrix}.$$

The coefficients $(\alpha_k)_1^\infty$ in (1) called Shur parameters are uniquely determined by the initial GSP data $(\gamma_j)_1^\infty$.

Remind that in general $p, q, [p-q]^{-1} \in H^2$ when $\|\Gamma\| < 1$ [2],[3]. Another solution of the system (1) that is singled out by the condition

$$\begin{pmatrix} u_0(\zeta) \\ v_0(\zeta) \end{pmatrix} = \begin{pmatrix} 1 \\ 0 \end{pmatrix},$$

defines the sequence $(u_k(\zeta))_0^\infty$ of the Szego type orthogonal polynomials on the unit circle with respect to the absolutely continuous measure

$$d\sigma(\vartheta) = \frac{a}{2\pi} \frac{1}{|p(\zeta)-q(\zeta)|^2} d\vartheta, \qquad d\vartheta = |d\zeta|, \, a > 0.$$

At the same time the function

$$S(\zeta) = -\frac{\overline{p}(\zeta) - \overline{q}(\zeta)}{p(\zeta) - q(\zeta)}$$

is a solution of GSP.
Establishing of these connections is the main purpose of this work.

1. Truncated GSP and Shur Parameters.

Remined that GSP is solvable if $\|\Gamma\| \le 1$, this is the essence of the Nehari Theorem [6].

If $\|\Gamma\| = 1$ and is attained, i.e. there is a Γ-extremal vector ξ in l_2:

$$\overline{\Gamma}\Gamma\xi = \xi, \; \xi \ne 0 \; (\overline{\Gamma} = \Gamma^* = (\overline{\gamma}_{j+k-1})_1^\infty).$$

then the unique solution is connected with *any* Schmidt pair:

$$\xi = (\xi_i)_1^\infty, \quad \eta = (\eta_-)_1^\infty, \quad \Gamma\xi = \eta, \quad \overline{\Gamma}\eta = \xi$$

by the formula

$$f(\zeta) = \eta_-(\zeta)/\xi_+(\zeta), \quad \xi_+(\zeta) = \sum_{j=1}^\infty \xi_j \zeta^{j-1}, \quad \eta_-(\zeta) = \sum_{j=1}^\infty \eta_j \zeta^{-j}.$$

Let T be the shift operator in l_2:

$$T(\xi_1, \xi_2, \dots) = (0, \xi_1, \xi_2, \dots).$$

Hankel operators satisfy the following characteristic property:

$$T^*\Gamma = \Gamma T.$$

This property implies $\left|\xi_+(\zeta)\right| = \left|\eta_-(\zeta)\right|$ a.e. and thus $\left|f(\zeta)\right| = 1$ a.e.

Now following [3] we provide the description of all solutions in the case $\|\Gamma\| < 1$.
Put

$$e = (1, 0, \dots), \quad R = (I - \overline{\Gamma}\Gamma)^{-1}, \quad a = (Re, e),$$

$$p = Re = (p_j)_1^\infty \qquad p(\zeta) = p_+(\zeta) = \sum_{j=1}^\infty \zeta^{j-1} p_j,$$

$$q = T\overline{\Gamma}Re = (q_j)_1^\infty \quad q(\zeta) = q_+(\zeta) = \sum_{j=2}^\infty \zeta^{j-1} q_j.$$

Observe that

$$p(0) = a, \quad q(0) = 0.$$

The set of vectors $\psi_\varepsilon = p + \varepsilon q$ with the parameter ε running through the unit circle $|\varepsilon| = 1$ satisfies the condition:

$$\Gamma\psi_\varepsilon = T^*\left(T\Gamma Re + \varepsilon T \overline{\Gamma} Re\right) = T^*(\overline{q} + \varepsilon \overline{p}), \quad \text{i.e.}$$

(2)
$$\Gamma\psi_\varepsilon = \varepsilon T^* \overline{\psi}_\varepsilon.$$

Now we can consider the matrix $\Gamma' = (\gamma'_{j+k-1})_1^\infty$, such that $\gamma'_k = \gamma_{k-1}$ for $k > 1$, and $\gamma'_1 = \gamma_0$ is an arbitrary complex number. This matrix generates the corresponding Hankel operator in l_2:

$$\Gamma' = (\bullet, h)e + T\Gamma, \quad h = \overline{\gamma}_0 e + T\overline{\Gamma}e,$$

(3)
$$\overline{\Gamma'}\Gamma' = \overline{\Gamma}\Gamma + (\bullet, h)h.$$

Select among all one-step extensions $(\Gamma \to \Gamma')$ only those which satisfy $\|\Gamma'\| = 1$.

If we take

$$\gamma_0 = -\frac{1}{a}\left(p, T\overline{\Gamma}e\right) + \frac{1}{a}\varepsilon, \quad |\varepsilon| = 1,$$

then from (2) and (3) it follows that vectors ψ_ε and $\varepsilon\overline{\psi}_\varepsilon$ form a Schmidt pair of Γ'_ε corresponding to

$$\Gamma'_\varepsilon\psi_\varepsilon = 1\psi^+_\varepsilon, \quad \overline{\Gamma}'_\varepsilon\psi^+_\varepsilon = \psi_\varepsilon, \quad \psi^+_\varepsilon = \varepsilon\overline{\psi}_\varepsilon.$$

Indeed,

$$\Gamma'\psi_\varepsilon = (\psi_\varepsilon, h)e + T\Gamma\psi_\varepsilon = \left\langle(\psi_\varepsilon, h), \Gamma\psi_\varepsilon\right\rangle = \left\langle(\psi_\varepsilon, h), T^*\varepsilon\overline{\psi}_\varepsilon\right\rangle = \left\langle(\psi_\varepsilon, h), T^*\left(\overline{q} + \varepsilon\overline{p}\right)\right\rangle$$

where:

$$(\psi_\varepsilon, h) = \left(\psi_\varepsilon, \overline{\gamma}_0 e + T\overline{\Gamma}e,\right) = a\gamma_0 + \left(p + \varepsilon q, T\overline{\Gamma}e\right) = a\gamma_0 + \left(p, T\overline{\Gamma}e\right) + \varepsilon\left(q, T\overline{\Gamma}e\right)$$

$$= -\left(p, T\overline{\Gamma}e\right) + \varepsilon + \left(p, T\overline{\Gamma}e\right) + \varepsilon\left(q, T\overline{\Gamma}e\right) = \varepsilon + \varepsilon\left(\Gamma T^*q, e\right)$$

$$= \varepsilon + \varepsilon\left(\Gamma T^*T\overline{\Gamma}Re, e\right) = \varepsilon + \varepsilon\left(\Gamma\overline{\Gamma}Re, e\right)$$

$$= \varepsilon + \varepsilon\left(\overline{p} - e, e\right) = \varepsilon + \varepsilon(a - 1) = \varepsilon a = \left(\overline{q} + \varepsilon\overline{p}, e\right).$$

From this relation and (3) we can conclude that $\|\Gamma'\| = 1$ and the certain family of solutions of GSP admits the following representation:

$$f_\varepsilon(\zeta) = \frac{\overline{p}(\zeta)\varepsilon + \overline{q}(\zeta)}{p(\zeta) + q(\zeta)\varepsilon}, \quad |\varepsilon| = 1.$$

Description of all solutions was obtained in [1]:

$$f_E(\zeta) = \frac{\overline{p}(\zeta)E(\zeta) + \overline{q}(\zeta)}{p(\zeta) + q(\zeta)E(\zeta)},$$

where the parameter E runs over the unit ball $\left(\|E\|_\infty \leq 1\right)$ in the Hardy space H^∞.

Remind that for any nuclear operator A (i.e. $A \in \gamma_1$) in the Hilbert space the value of the functional

$$\text{trace}\,T = \sum_{n \geq 0}\left(Te_n, e_n\right),$$

where $\{e_n\}_{n \geq 0}$ is an (any) orthonormal basis independent of the choice of the basis and more over this serie is absolutely convergent. It is easy to check that if $\Gamma \in \gamma_1$ then

$$\sum_j |\gamma_j| < \infty.$$

Indeed:

$$\sum |\gamma_i| = \sum |(e_i, \mathrm{T}\Gamma e_i)| + \sum |(e_i, \Gamma e_i)| < \infty$$

It is interesting to show that $p, q \in l^1$ in this case and therefore $p(\zeta), q(\zeta) \in W_+$ (W denotes the Wiener algebra of continuous functions $f \in C$ on the unit circle with

Fourier coefficients $c_k(f)$ for which $\sum_{-\infty}^{\infty} |c_k(f)| < \infty$, W_+ and $W_-(\overline{W_+})$ its complex

conjugate subalgebras: $W_+ = \{f : f \in W, c_k(f) = 0, k < 0\}$).

In fact, if $\sum |\gamma_i| < \infty$ then the matrix Γ and $\overline{\Gamma}$ with it generate completely continuous operators in l_1. Consequently a completely continuous operator will be generated by the matrix $\overline{\Gamma}\Gamma$ in l_2. Thus for $\|\Gamma\| < 1$ and $\xi \in l_1$, the vectors $= (I - \overline{\Gamma}\Gamma)^{-1}\xi$ and $\Gamma\eta$ also belong to l_1. Setting $\xi = e_1$, we get $p(\zeta), q(\zeta) \in W_+$. It is clear now that f_E will belong to the algebra W if and only if $E \in B \bigcap W_+$.

Now we shall consider Hankel operator $\Gamma_k = (\gamma_{j+i+k-1})_1^\infty$ for the truncated sequence $(\gamma_{k+1}, \gamma_{k+2}, \ldots)$.

Set as before

$$R_k = (I - \overline{\Gamma}_k \Gamma_k)^{-1}, \qquad P_k = R_k e, \quad q_k = \mathrm{T}\overline{\Gamma}_k \overline{R}_k e,$$

$$p_k(\zeta) = \sum_{j=1}^\infty p_k \zeta^{j-1}, \quad q_k(\zeta) = \sum_{j=i}^\infty q_k \zeta^j,$$

$$a_k = (R_k e, e).$$

Observe that in accordance with (3):

$$\|\Gamma_k\| \le \|\Gamma_{k-1}\| \le \ldots \|\Gamma\|, \text{ and } \overline{\Gamma}_{k+1}\Gamma_{k+1} \le \overline{\Gamma}_k \Gamma_k., \text{ and}$$

(4)
$$\lim_{k \to \infty} \max |p_k(\zeta) - 1| \to 0, \quad \lim_{k \to \infty} \max |q_k(\zeta)| \to 0,$$

$$\lim_{k \to \infty} a_k = 1.$$

Indeed:

$$\max |p_k(\zeta) - 1| \le \|p_k - e\|_{l_1} = \|(R_k - I)e\|_{l_1} \le \|R_k \overline{\Gamma}\Gamma_k\|_1 \le \frac{\|\Gamma_k\|_1^2}{1 - \|\Gamma_k\|_1^2},$$

$$\max\left|q_k(\zeta)\right| \le \left\|q_k\right\|_{l_1} \le \frac{\left\|\Gamma_k\right\|_1}{1-\left\|\Gamma_k\right\|_1^2}, \qquad\qquad a_k \le 1 + \frac{\left\|\Gamma_k\right\|_1^2}{1-\left\|\Gamma_k\right\|_1^2},$$

where

$$\left\|\Gamma_k\right\|_1 = \sup_{\left\|\xi\right\|_{l_1}=1}\left\|\Gamma_k\xi\right\|_{l_1}.$$

Since

$$\left\|\Gamma_k\xi\right\|_{l_1} \le \left\{\left|\xi_1\right|\sum_{j=k+1}^{\infty}\left|\gamma_j\right| + \left|\xi_2\right|\sum_{j=k+2}^{\infty}\left|\gamma_j\right| + \ldots\right\} \le \sum_{k+1}^{\infty}\left|\gamma_j\right|\left(\sum_{1}^{\infty}\left|\xi_j\right|\right) \le \sum_{j=k+1}^{\infty}\left|\gamma_j\right|\left\|\xi\right\|_{l_1},$$

$$\left\|\Gamma_k e\right\|_{l_1} = \sum_{j=k+1}\left|\gamma_j\right| \ge \left\|\Gamma_k\right\|_1,$$

then

$$\left\|\Gamma_k\right\|_1 = \sum_{j=k+1}^{\infty}\left|\gamma_j\right|, \qquad\qquad \lim_{k\to\infty}\left\|\Gamma_k\right\|_1 = 0,$$

and, therefore, relations (4) hold.
If we set

(5)
$$C_k = \left\{f_{E,k}(\zeta): f_{E,k} = \frac{\overline{p}_k(\zeta)\mathrm{E}(\zeta)+\overline{q}_k(\zeta)}{q_k(\zeta)\mathrm{E}(\zeta)+p_k(\zeta)}, \mathrm{E}\in H^{\infty}, \left\|\mathrm{E}\right\|_{\infty} \le 1\right\},$$

then for all $f(\zeta)\in C_k$ we have $f\in C_{k+1}$ and therefore the inclusion $C_k \subset C_{k+1}$ holds. Now consider the following questions: What is the condition that might be imposed on the function $\mathrm{E}(\zeta)$ in order to satisfy the inclusion $\overline{\zeta}\,f_{E,k+1}\in C_k$?

The answer follows from the lemma.

LEMMA 1. *The function* $\overline{\zeta}\,f_{E,k+1}$ *belongs to* C_k *if and only if*

(6)
$$\mathrm{E}(0) = \alpha_k = a_k\gamma_k + \left(p_k, T\overline{\Gamma}_k e\right).$$

Proof.

Let us put $\mathrm{E}(\zeta) = 0$ in $f_{E,k+1}$, then $f_{0,k+1} = \dfrac{\overline{q}_{k+1}}{p_{k+1}}$.

If $\overline{\zeta}\,f_{E,k+1}\in C_k$, then $\left(f_{E,k+1}\right)_0$ (0-Fourier coefficient of $f_{E,k+1}$) satisfies the condition:

$$\left(f_{E,k+1}\right)_0 = \frac{1}{2\pi}\int_0^{2\pi}f_{E,k+1}(\zeta)d\vartheta = \gamma_{k+1},$$

and

$$\gamma_{k+1} - (f_{0,k+1})_0 = \frac{1}{2\pi} \int_0^{2\pi} \left(\frac{\overline{q}_{k+1} + E\overline{p}_{k+1}}{p_{k+1} + Eq_{k+1}} - \frac{\overline{q}_{k+1}}{p_{k+1}} \right) d\vartheta = \frac{a_{k+1}}{2\pi} \int_0^{2\pi} \frac{E(\zeta)}{(p_{k+1} + Eq_{k+1})p_{k+1}} d\vartheta$$

$$(7) \qquad = \frac{a_{k+1}E(0)}{[p_{k+1}(0) + E(0)q_{k+1}(0)]p_{k+1}(0)} = \frac{E(0)}{a_{k+1}}.$$

We applied the "Mean Value" Theorem for the H^∞-functions in the unit circle.

This implies $(f_{E,k+1})_0 = \gamma_{k+1}$ if

$$E(0) = a_{k+1}\left[\gamma_{k+1} - (f_{0,k+1})_0 \right].$$

Setting, $E(\zeta) = \varepsilon = e^{i\varphi}$ we can conclude:

$$(f_{\varepsilon,k+1})_0 = -\frac{1}{a_{k+1}}\left(p_{k+1}, T\overline{\Gamma}_{k+1}e \right) + \frac{1}{a_{k+1}}\varepsilon,$$

and hence:

$$(f_{0,k+1})_0 = \frac{1}{2\pi} \int_0^{2\pi} \frac{\overline{q}_{k+1}}{p_{k+1}} d\vartheta = \frac{1}{4\pi^2} \int_0^{2\pi} d\vartheta \int_0^{2\pi} \frac{\overline{q}_{k+1} + e^{i\varphi}\overline{p}_{k+1}}{p_{k+1} + e^{i\varphi}q_{k+1}} d\varphi =$$

$$\frac{1}{4\pi^2} \int_0^{2\pi} d\varphi \int_0^{2\pi} \frac{\overline{q}_{k+1} + \varepsilon\overline{p}_{k+1}}{p_{k+1} + \varepsilon q_{k+1}} d\vartheta = \frac{-1}{2\pi a_{k+1}} \int_0^{2\pi} \left[(p_{k+1}, T\overline{\Gamma}_{k+1}e) - \varepsilon \right] d\varphi = \frac{-1}{a_{k+1}}(p_{k+1}, T\overline{\Gamma}e)$$

(8)

The Fubiny and "Mean-value" theorems were used in the last chain of transformations. After the substitution (8) in (7) we get:

$$(9) \quad E(0) = a_{k+1}\left[\gamma_{k+1} + \frac{1}{a_{k+1}}(p_{k+1}, T\overline{\Gamma}_{k+1}e) \right] = \left[a_{k+1}\gamma_{k+1} + (p_{k+1}, T\overline{\Gamma}_{k+1}e) \right] = \alpha_{k+1}.$$

The numbers α_k defined by formula (6) are the so called "Shur Parameters".

One can consider (9) as Shur's problem with one element:

$$1) E(z) = \alpha_k + a_1 z + a_2 z^2 + \dots$$
$$2) \sup_{z \in d} |E(z)| \le 1$$

It is known that all the solutions of this problem admit the representation

$$(9') \qquad E(z) = \frac{z\tilde{E}(z) + \alpha_k}{1 + \overline{\alpha}_k\tilde{E}(z)z}, \quad \tilde{E}(z) \in H^\infty, \|\tilde{E}\| \le 1.$$

which maps the unit ball of H^∞ onto itself.

Finally, summarizing (9), (8), (7) we can conclude that $\overline{\zeta} f_{E,k+1} \in C_k$, if $E(\zeta)$ is chosen out of (9').

This result accepts alternative interpretation, setting:

$$(10) \qquad W_k(\zeta) = \begin{pmatrix} p_k(\zeta) & q_k(\zeta) \\ \overline{q_k(\zeta)} & \overline{p_k(\zeta)} \end{pmatrix}, \quad \det W_k(\zeta) = a_k.$$

It is easy to behold that the linear fractional transformations of the unit ball in H^∞ defined by $W_k(\zeta)$ and

$$\tilde{W}_k(\zeta) = \begin{pmatrix} 1 & 0 \\ 0 & \overline{\zeta} \end{pmatrix} \begin{pmatrix} p_{k+1} & q_{k+1} \\ \overline{q_{k+1}} & \overline{p_{k+1}} \end{pmatrix} \begin{pmatrix} 1 & \zeta\overline{\alpha}_1 \\ \alpha_1 & \zeta \end{pmatrix}$$

coincide.

Therefore,

$$W_k^{-1}(\zeta)\tilde{W}_k(\zeta)$$

maps the unit ball in H^∞ onto itself.

Using properties $p_k(0) = a_k$, $q_k(0) = 0$ we can carry out the following formula:

$$(11) \qquad W_k(\zeta) = \frac{a_k}{a_{k+1}} \begin{pmatrix} 1 & 0 \\ 0 & \overline{\zeta} \end{pmatrix} W_{k+1}(\zeta) \begin{pmatrix} 1 & \zeta\overline{\alpha}_k \\ \alpha_k & \zeta \end{pmatrix},$$

where

$$a_k = (R_k e, e), \quad \alpha_k = a_k \gamma_k + (p_k, T\overline{\Gamma}_k e) \quad .$$

The constituents of the Smidt pairs p, q, p_n, q_n are connected by the multiplicative formula which follows from the:

LEMMA 2.

$$(12) \qquad \begin{pmatrix} p & q \\ \overline{q} & \overline{p} \end{pmatrix} = \frac{a}{a_n} \begin{pmatrix} p_n & q_n\zeta^n \\ \overline{\zeta}^n \overline{q}_n & \overline{p}_n \end{pmatrix} \prod_{k=1}^{n} \begin{pmatrix} 1 & \zeta^k \overline{\alpha}_k \\ \overline{\zeta}^k \alpha_k & 1 \end{pmatrix}$$

Using induction in (11) it is easy to prove (12). Now we are going to affirm the important statement:

LEMMA 3. *If the Hankel operator* Γ *is of the trace-class then the sequence of the Shur parameters* $\{\alpha_k\}$ *is absolutely summable:*

$$\sum_1^\infty |\alpha_k| < \infty.$$

Proof.

$$\sum_{k=1}^{\infty}|\alpha_k| = \sum_{k=1}^{\infty}\left|a_k\gamma_k + \left(p_k, T\overline{\Gamma}_k e\right)\right| \le \sum_{k=1}^{\infty}a_k\gamma_k + \sum_{k=1}^{\infty}\left|\left(p_k, T\overline{\Gamma}_k e\right)\right|.$$

Convergence of $\sum|\gamma_k|$ and $\lim_{k\to\infty}a_k = 1$ implies the absolute convergence of $\sum a_k\gamma_k$.

Further

$$\sum\left|\left(p_k, T\overline{\Gamma}_k e\right)\right| = \sum\left|\left(R_k e, T\overline{\Gamma}_k e\right)\right| = \sum\left|\left(\left(I + R\overline{\Gamma}_k\Gamma_k\right)e, T\overline{\Gamma}_k e\right)\right|$$
$$\le \sum\left|\left(e, T\overline{\Gamma}_k e\right)\right| + \sum\left|\left(R_k\overline{\Gamma}_k\Gamma_k e, T\overline{\Gamma}_k e\right)\right| \le \sum\|R_k\|\,\|\Gamma_k\|\,\|\Gamma_k e\|^2 < \infty$$

The last inequality is a simple consequence of the proven fact:

$$\lim_{n\to\infty}\|R_n\|\,\|\Gamma_n\| = 0,$$

and the relation

$$\sum\|\Gamma_k e\|^2 = \sum k|\gamma_k|^2 < \infty,$$

which is valid for any Hilbert-Smidt Hankel operator and, therefore, Γ is a trace class operator.

It is a well-known fact that convergence of the serie $\sum_1^{\infty}\|A_k(\zeta)\|$, implies the uniformal convergence of the product $\prod_1^{\infty}\left(I + A_k(\zeta)\right)$, where $A_k(\zeta)$ is the finite order matrix continuously depending on ζ.

In our case

$$A_k(\zeta) = \begin{pmatrix} 0 & \zeta^k\overline{\alpha}_k \\ \overline{\zeta}^k\alpha_k & 0 \end{pmatrix} \text{ and } \|A_k\| \le |\alpha_k|.$$

Hence we can conclude that $\sum|\alpha_k| < \infty$ implies the uniformal convergence of the infinite product on the unit circle:

(13)
$$\prod_{k=1}^{\infty}\begin{pmatrix} 1 & \zeta^k\overline{\alpha}_k \\ \overline{\zeta}^k\alpha_k & 1 \end{pmatrix} = \begin{pmatrix} p(\zeta) & q(\zeta) \\ \overline{q}(\zeta) & \overline{p}(\zeta) \end{pmatrix}.$$

Combining all the results obtained above, we can formulate:

THEOREM 1. *If* Γ *is a nuclear Hankel operator then the representation (13) holds; the convergence of the infinite product in (13) is uniform on the unit circle.*

2. First order Discrete System Connected with GSP

Recalling (10) it is possible to rewrite (11) in the form of recurrence equations:

$$\frac{\overline{p}_{k+1}(z)}{a_{k+1}} = \frac{1}{1-|\alpha_{k+1}|^2}\left[\frac{\overline{p}_k(z)}{a_k} - \overline{\alpha}_{k+1}z\frac{\overline{q}_k(z)}{a_k}\right], \qquad k=0,1,2\ldots,$$

(14)
$$\frac{\overline{q}_{k+1}(z)}{a_{k+1}} = \frac{1}{1-|\alpha_{k+1}|^2}\left[-\alpha_{k+1}\frac{\overline{p}_k(z)}{a_k} + z\frac{\overline{q}_k(z)}{a_k}\right].$$

If we make the substitution:

$$u_k(z) = \frac{\overline{q}_k(z)}{a_k}\prod_{i=1}^{k}\left(1-|\alpha_i|^2\right),$$

$$v_k(z) = \frac{\overline{p}_k(z)}{a_k}\prod_{i=1}^{n}\left(1-|\alpha_i|^2\right), \qquad z=\zeta=e^{i\vartheta},$$

then (11) may be rewritten as a recurrent system:

(15)
$$\begin{cases} u_{k+1} = \zeta u_k - \alpha_{k+1}v_k \\ v_{k+1} = -\zeta\overline{\alpha}_{k+1}u_k + v_k \end{cases}.$$

This is recursive formulas for Szego polynomials on the unit circle [4].
Let us consider more general system:

(16)
$$\begin{cases} u_{k+1} = A_k\zeta u_k + B_k v_k \\ v_{k+1} = \overline{B}_k\zeta u_k + A_k v_k \end{cases},$$

where

$$A_k = \frac{1}{1-|\alpha_{k+1}|^2}, \qquad B_k = -\alpha_{k+1}a_k.$$

It is easy to check that the following pairs:

$$\begin{pmatrix}\zeta^k P_k(\zeta) \\ \zeta^k Q_k(\zeta)\end{pmatrix}, \quad \begin{pmatrix}\overline{Q}_k(\zeta) \\ \overline{P}_k(\zeta)\end{pmatrix},$$

where

$$P_k(\zeta) = \frac{p_k(\zeta)}{a_k},$$

$$Q_k(\zeta) = \frac{q_k(\zeta)}{a_k}$$

satisfy the system (15).

This solutions are singled out by conditions:

$$\lim_{k \to \infty} \zeta^{-k}\begin{pmatrix} \hat{u}_k \\ \hat{v}_k \end{pmatrix} = \begin{pmatrix} 1 \\ 0 \end{pmatrix}, \lim_{k \to \infty}\begin{pmatrix} \tilde{u}_k \\ \tilde{v}_k \end{pmatrix} = \begin{pmatrix} 0 \\ 1 \end{pmatrix}.$$

In the case of the initial conditions

(17) $$u_0 = u_0(\zeta) = 1, \quad v_0 = v_0(\zeta) = 1$$

the first components $u_n(\zeta)$ are orthogonal polynomials on the unit circle

$$\int_0^{2\pi} u_r(\zeta)\bar{u}_s(\zeta)d\tau(\vartheta) = \frac{a_r}{a}\delta_{rs}, \quad r,s = 0,1,2,\ldots$$

for the non-negative measure $d\tau(\vartheta)$ [4].

If $r > l, l \geq 1$ then

$$\int_0^{2\pi} \zeta^r d\tau(\vartheta) = \frac{1}{2\pi}\frac{a_r}{a}\int_0^{2\pi} \zeta^r \frac{1}{|u_e(\zeta)|^2}d\vartheta,$$

and for continuous f we have

$$\lim_{l \to \infty} \frac{a_r}{2\pi a}\int_0^{2\pi} f(\zeta)\frac{1}{|u_l(\zeta)|^2}d\vartheta = \int_0^{2\pi} f(\zeta)d\tau.$$

We shall denote by

$$u_{nl}(\zeta), v_{nl}(\zeta), \quad n = 0,1,2,\ldots$$

a sequence of polynomials which are solutions of (9) with different initial conditions:

$$u_{01} = u_{01}(\zeta) = 1, \quad v_{01} = v_{01}(\zeta) = -1.$$

Now the Finite Boundary Problem may be posed.
For fixed $\alpha, 0 \leq \alpha \leq 2\pi$, we ask for ζ-values for which (15) has a nontrivial solution

such that

(18) $$u_0 = v_0, \quad u_m = \exp(i\alpha)v_m.$$

It is possible to show that the eigenvalues, the roots of (18), will be exactly m in number, and they lie on the unit circle [4].

If we define the characteristic function

(19)
$$f_m(u) = \frac{u_{ml}(z) - e^{i\alpha} v_{ml}(z)}{u_m(z) - e^{i\alpha} v_m(z)}, \qquad |z| < 1,$$

then the function (19) has the representation

$$f_m(z) = \int_0^{2\pi} \left(z + e^{i\varphi}\right)\left(z - e^{i\varphi}\right)^{-1} d\tau_{m,\alpha}(\varphi).$$

Where $\tau_{m,\alpha}(\varphi)$ is a nondecreasing step function on the interval $[0, 2\pi)$ with jumps at roots of (17) and such that

$$\int_0^{2\pi} \zeta^l d\tau_{m,\alpha}(\varphi) = \int_0^{2\pi} \zeta^l d\tau(\varphi), \qquad 0 \le l < m.$$

Tending m to infinity one can derive the following representation for the characteristic function corresponding to the infinite boundary problem:

(20)
$$f(z) = \int_0^{2\pi} \left(z + e^{i\varphi}\right)\left(z - e^{i\varphi}\right)^{-1} d\tau(\varphi), \qquad |z| < 1.$$

The representation of the characteristic function (20) in terms of Hankel Operator follows from the

LEMMA 4. *For the recurrent system (15) generated by nuclear Hankel operator Γ the sequence of finite characteristic functions $f_m(z)$ converges uniformly on sets inside the unit circle $|z| < 1$. The infinite characteristic function accepts the representation:*

$$f(\zeta) = -\frac{p(z) + q(z)}{p(z) - q(z)}.$$

Proof.

Representing the polynomials in (19) in terms of two independent solutions of (15):

$$\begin{pmatrix} z^k p_k \\ z^k q_k \end{pmatrix}, \qquad \begin{pmatrix} \overline{q}_k \\ \overline{p}_k \end{pmatrix}$$

one can get

$$f_{n,\exp i\alpha}(z) = \frac{2z^n}{p(z)-q(z)} \frac{p_n(z)-e^{i\alpha}q_n(z)}{u_n(z)-e^{i\alpha}v_n(z)} - \frac{p(z)+q(z)}{p(z)-q(z)}.$$

Observing that $\dfrac{2z^n}{p(z)-q(z)} \to 0$ inside the unit circle, we can derive that

$$f(z) = -\frac{p(z)+q(z)}{p(z)-q(z)}.$$

Relying on this result we can conclude that the spectral function also may be represented in terms of generating the Hankel operator

THEOREM 2. *The spectral function of the infinite boundary problem accepts the representation:*

$$d\tau(\vartheta) = \frac{1}{2\pi} a \frac{1}{|p(z)-q(z)|^2} d\vartheta.$$

Proof.

If $f(z)$ is a holomorphic function inside the unite circle $|z|<1$,then according to Schwartz formula we can establish for $|z|<r<1$:

$$-f(z) = i\mathrm{I}_m f(0) + \frac{1}{2\pi} \int_0^{2\pi} \frac{re^{i\vartheta}+z}{re^{i\vartheta}-z} u(r,\vartheta)d\vartheta, \text{ and}$$

$$-f(rz) = \frac{1}{2\pi} \int_0^{2\pi} \frac{e^{i\vartheta}+z}{e^{i\vartheta}-z} d\tau_r(\vartheta),$$

$$(0<r<1, \; |z|<1).$$

Denote by $\tau_r(\vartheta)$ the nondecreasing function:

$$\tau_r(\vartheta) = \int_{-\pi}^{\vartheta} u(r,\psi)d\psi.$$

The functions $\tau_r(\vartheta)$ arrange the uniformly bounded family, since:

$$0 < \tau_r(\vartheta) \le \tau_r(2\pi) = 2\pi \, Re[f(0)] = 2\pi.$$

Using the Helly's Theorems and the

$$-f(z) = \frac{1}{2\pi} \int_0^{2\pi} \frac{e^{i\vartheta} + z}{e^{i\vartheta} - z} d\tau,$$

which coincides with Ries-Herghlotz representation. And, at last using the identity

$$|p(\zeta)|^2 - |q(\zeta)|^2 \equiv a, \qquad |\zeta| = 1$$

we have

$$\tau(\vartheta) = -\frac{1}{2\pi} \left[Re \, f(\zeta) \right] d\vartheta = \frac{1}{2\pi} \int_0^{\vartheta} \frac{a}{|p(\zeta') - q(\zeta')|^2} d\psi, \qquad \zeta' = e^{i\psi} \qquad .$$

It is obvious that the spectral density

$$\frac{d\tau}{d\vartheta} = \frac{a}{|p(\zeta) - q(\zeta)|^2} \in W \text{ (Wiener's algebra)}.$$

The above results are summarized in the following theorem.

MAIN THEOREM 1. *The trace-class Hankel operator* Γ *generates the infinite boundary problem (15), (17) with the spectral function*

$$d\tau(\zeta) = w(\zeta) d\vartheta, \qquad \zeta = e^{i\vartheta}$$

and

$$w(\zeta) = |p(\zeta) - q(\zeta)|^{-2} \in B_1 \text{ (Besov class)}.$$

2. *The absolutely continuous measure* $d\tau(\zeta) = w(\zeta) d\vartheta,$ $\zeta = e^{i\vartheta}$ *where* $w(\zeta) \in B_1,$
and

$$w^{-1} \in L_1(T), \quad \log w^{-1} \in L_1,$$

is the spectral function for the infinite boundary problem generated by nuclear Hankel operator Γ *generated by negative Fourier coefficients* $\gamma_j = c_{-j} \left(\overline{h} / h \right),$ *where* $w^{-1} = |h|^2;$ $h \in H^\infty \cap B_1;$ *h is outer function, i.e.*

$$h(z) = \exp \left(\frac{1}{2} \int_T \frac{\zeta + z}{\zeta - z} \log w^{-1}(\zeta) d\vartheta \right).$$

The proof of the last part of the Theorem follows from the reversability of our consideration and the famous result of Peller [7], [6].
Let us mention that the function $S(\zeta)$:

$$S(\zeta) = -\frac{\overline{p}(\zeta) - \overline{q}(\zeta)}{p(\zeta) - q(\zeta)}$$

is a GSP solution.
It is clear that such Inverse problems have been solved:
a) Reconstruction of the difference system (set of Schur parameters α_k) by $d\tau(\vartheta)$;
b) Reconstruction of the difference system by $S(\zeta)$.

References

1. Adamyan V.M., Arov D.Z. and Krein M.G., *Infinite Hankel matrices and generalized problems of Caratheodory-Fejer and F. Riecz.*, Functional Anal. Appl. **2** (1968), 1-19.

2. Adamyan V.M., Arov D.Z. and Krein M.G., *Infinite Hankel matrices and generalized problems of Caratheodory-Fejer and I.Schur*, Functional Anal. Appl. **2** (1968), 269-281.

3. Adamyan V.M., Arov D.Z. and Krein M.G., *Analytic properties of Schmidt pairs for a Hankel operator and the generalized Schur-Takagi problem*, Math USSR Sbornik, **15-1**, (1971), 31-73.

4. Atkinson F., *Discrete and continuous boundary problems*, Acad. Press, New York, 1964, 222-232

5. Nechayev S., *Functions of class on unit circle and Hankel Kernel operators*, Math. Notes, **37-4**, (1985), 293-294.

6. Nehari Z., *On bounded bilinear forms*, Ann. Of Math. **2**, (1957), 153-162.

7. Peller V.V. and Khrushchev S.V., Hankel operators, best approximations and stationary Gaussian processes, Usp. Mat. Nauk **1** (1982), 53-124.

DEPARTMENT OF THEORETICAL PHYSICS, ODESSA STATE UNIVERSITY, ODESSA, UKRAINE 270045
E-mail address: vma @ dtp odessa.ua

DEPARTMENT OF MATHEMATICS, POLYTECHNIC UNIVERSITY, BROOKLYN, NY 11201
E-mail address: nechayev @ magnus.poly.edu

Contemporary Mathematics
Volume **189**, 1995

An Invariant Submean Value Property and Hyponormal Toeplitz Operators

PATRICK AHERN AND ŽELJKO ČUČKOVIĆ

Introduction

For a positive integer n, let V denote the Lebesgue volume measure on $\mathbb{C}^n$. If $B_n = \{z \in \mathbb{C}^n \mid |z| < 1\}$, then for each $a \in B_n$ we have the automorphism of B_n

$$\varphi_a(z) = \frac{a - Pz - (1 - |a|^2)^{1/2} Qz}{1- <z, a>},$$

where $Pz = <z, a> \frac{a}{|a|^2}$, $Qz = z - Pz$, $<z, a> = \sum_{i=1}^{n} z_i \bar{a}_i$.

For $u \in L^1(B_n)$, we can define the Berezin transform

$$Tu(z) = \frac{1}{V(B_n)} \int_{B_n} u \circ \varphi_z dV.$$

The Bergman space $L^2_a(B_n)$ is the set of analytic functions on B_n such that

$$\int_{B_n} |f|^2 dV < \infty.$$

The Bergman space is a closed subspace of $L^2(B_n)$, and so there is an orthogonal projection $P : L^2(B_n) \to L^2_a(B_n)$. If $\varphi \in L^\infty(B_n)$, we define the Toeplitz operator $T_\varphi : L^2_a(B_n) \to L^2_a(B_n)$ by $T_\varphi f = P(\varphi f)$. For $a \in B_n$, we have the Bergman kernel k_a of the form

$$k_a(z) = \frac{1}{(1- <a, z>)^{n+1}}.$$

The invariant Laplacian is defined for $f \in C^2(B_n)$ by $(\tilde{\Delta} f)(z) = \Delta(f \circ \varphi_z)(0)$, where Δ is the ordinary Laplacian. A function f is called harmonic if $\Delta f = 0$, and M-harmonic if $\tilde{\Delta} f = 0$.

1991 *Mathematics Subject Classification.* Primary 31B05, 32A37, 47B20.

A bounded linear operator A on a Hilbert space H is normal if it commutes with its adjoint. It is called hyponormal if $A^*A \geq AA^*$, i.e. $\|Ax\| \geq \|A^*x\|$ for all $x \in H$. Here $\|\cdot\|$ is the norm on H.

In (1), the authors studied functions satisfying a conformally invariant mean value inequality $Tu \geq u$ on the unit disk $\mathbb{D}$ and then gave necessary condition for hyponormality of T_φ acting on $L_n^2(\mathbb{D})$, if φ is a bounded harmonic function on $\mathbb{D}$. In this paper we generalize those results and show that $Tu \geq u$ on B_n implies

$$\limsup_{z \to \zeta} \frac{\tilde{\Delta}u}{(1-|z|^2)^{n+1}} \geq 0 \quad \text{for all} \quad \zeta \in S = \partial B_n.$$

We say that u is almost M-subharmonic. Then we give another version of almost M-subharmonicity. If σ denotes the surface area measure on S, and if $\tilde{\Delta}u(r\zeta)d\sigma$ has a weak-star limit as $r \to 1$ on an open set $S_0 \subset S$, then the limit must be a positive measure on S. An application of these results gives us necessary conditions for hyponormality of $T_{f+\bar{g}}$, where f and g are bounded analytic functions on B_n. We show that if $T_{f+\bar{g}}$ is hyponormal, then

$$\limsup_{z \to \zeta}(|\nabla_T f|^2 - |\nabla_T g|^2) \geq 0 \quad \text{for every} \quad \zeta \in S.$$

Here ∇_T denotes the tangential gradient defined below. If, in addition, all partial derivatives $D_i f, D_i g \in H^2(S_0)$, the second version of almost M-subharmonicity will give

$$|\nabla_T f(\zeta)| \geq |\nabla_T g(\zeta)| \quad \text{a.e. on} \quad S_0.$$

Submean value property

In this section we study functions satisfying the inequality $Tu \geq u$ in B_n. We say that such functions have the invariant submean value property.

Let $0 < r < 1$. We have the following analogue of Green's identity for the invariant Laplacian $\tilde{\Delta}$:

If u and v are C^2-functions on a neighborhood of $\{z \in \mathbb{C}^n \mid |z| \leq r\}$, then

$$(1) \qquad \int_{|z|<r} (u\tilde{\Delta}v - v\tilde{\Delta}u)d\mu = \frac{(1-r^2)^{-n+1}}{2r} \int_{|z|=r} (uNv - vNu)d\sigma$$

where

$$N = \frac{1}{2}\left(\sum_{j=1}^n z_j D_j + \sum_{j=1}^n \bar{z}_j \bar{D}_j\right) \quad \text{and} \quad d\mu = (1-|z|^2)^{-n-1}dV.$$

If u and v are C^2 on a neighborhood of $\{z \mid \epsilon < |z| \leq r\}$, then we can extend u and v to be C^2 on a neighborhood of $\{|z| \leq r\}$. From (1) we obtain:

$$\int_{\epsilon<|z|<r} (u\tilde{\Delta}v - v\tilde{\Delta}u)d\mu = \frac{(1-r^2)^{-n+1}}{2r} \int_{|z|=r} (uNv - vNu)d\sigma -$$

(2) $$\frac{(1-\epsilon^2)^{-n+1}}{2\epsilon}\int_{|z|=\epsilon}(uNv-vNu)d\sigma.$$

Let $G(z)=\frac{2}{\sigma(S)}\int_{|z|^2}^1\frac{(1-t)^{n-1}}{t^n}dt$. It is well known that $\tilde{\Delta}G\equiv 0$ on $B_n\backslash\{0\}$. Applying (2) to $v=G-G(r)$ and letting $\epsilon\to 0$ gives

(3) $$u(0)=\frac{1}{\sigma(S_r)}\int_{S_r}ud\sigma-\frac{2}{\sigma(S)}\int_{|z|<r}\tilde{\Delta}u(z)\left[\int_{|z|^2}^{r^2}\frac{(1-t)^{n-1}}{t^n}dt\right]d\mu.$$

We multiply both sides of (3) by $2nr^{2n-1}$ and integrate on r from 0 to 1. We obtain

$$u(0)=\int_{B_n}u(z)\frac{dV(z)}{V(B_n)}-\frac{4n}{\sigma(S)}\int_0^1 r^{2n-1}\int_{B_n}\tilde{\Delta}u(z)\int_{|z|^2}^{r^2}\frac{(1-t)^{n-1}}{t^n}dtd\mu(z)dr.$$

After changing the order of integration, we get

(4) $$u(0)=\int_{B_n}u(z)dV(z)-\frac{2}{\sigma(S)}\int_{B_n}\tilde{\Delta}u(z)K(|z|^2)d\mu(z),$$

where $K(x)=\int_x^1\frac{(1-t)^{n-1}(1-t^n)}{t^n}dt$. Notice that $K(x)\geq 0$ if $0<x<1$. If we differentiate $K(x)$ we get $K'(x)=-(1-x)^{n-1}\cdot\frac{1-x^n}{x^n}$, or $K'(x)=-(1-x)^n h(x)$, where

$$h(x)=\frac{1+x+x^2+\cdots+x^{n-1}}{x^n}.$$

Since $h(1)\neq 0$, we can conclude that $K'(1)=K''(1)=\cdots=K^{(n)}(1)=0$, but $K^{(n+1)}(1)\neq 0$. An application of Taylor's formula shows that

$$K(x)=\frac{K^{(n+1)}(t)}{(n+1)!}(x-1)^{n+1},\quad 0<x<t<1.$$

The formula for $K^{(n+1)}(t)$ shows that there are constants $C_1,C_2>0$ such that

(5) $$K(x)\geq C_1(1-x)^{n+1}\quad\text{for}\quad 0<x<1\quad\text{and}$$

(6) $$K(x)\leq C_2(1-x)^{n+1}\quad\text{for}\quad\epsilon<x<1,\quad\text{for any}\quad\epsilon>0.$$

Thus (4) holds if $u\in C^2(B_n)\cap L^1(B_n)$ and $\int_{B_n}|\tilde{\Delta}u(z)|(1-|z|^2)^{n+1}d\mu<\infty$. If we apply (4) to $u\circ\varphi_a$ instead of u, we obtain

$$Tu(a)-u(a)=\frac{2}{\sigma(S)}\int_{B_n}\tilde{\Delta}(u\circ\varphi_a)(z)K(|z|^2)d\mu(z).$$

Since $\tilde{\Delta}(u\circ\varphi_a)(z)=(\tilde{\Delta}u)(\varphi_a(z))$, the change of variables $w=\varphi_a(z)$ will give

$$Tu(a)-u(a)=\frac{2}{\sigma(S)}\int_{B_n}\tilde{\Delta}u(w)K(|\varphi_a(w)|^2)d\mu(w),$$

because $d\mu$ is the invariant measure. Thus we have proved the following theorem.

Theorem 1. *Suppose that $u \in C^2(B_n) \cap L^1(B_n)$ and that $\int_{B_n} |\tilde{\Delta}u(z)| dV(z) < \infty$. Then*

(7)
$$Tu(z) - u(z) = \int_{B_n} \tilde{\Delta}u(w) K(z,w) d\mu(w),$$

where $K(z,w) = \frac{2}{\sigma(S)} \int_{|\varphi_z(w)|^2}^1 \frac{(1-t)^{n-1}(1-t^n)}{t^n} dt$. Moreover, K satisfies:

(8)
$$K(z,w) \geq C_1 \left[\frac{(1-|z|^2)(1-|w|^2)}{|1-\langle z,w \rangle|^2} \right]^{n+1} \quad \text{for all} \quad z,w \in B_n$$

and

(9)
$$K(z,w) \leq C_2 \left[\frac{1-|z|^2)(1-|w|^2)}{|1-\langle z,w \rangle|^2} \right]^{n+1}$$

whenever $|\varphi_z(w)|^2 > \epsilon$, where $C_1, C_2, \epsilon > 0$ are constants not depending on z and w, and C_2 depends on ϵ.

PROOF. Inequalities (8) and (9) follow from (5) and (6) and the identity

$$1 - |\varphi_z(w)|^2 = \frac{(1-|z|^2)(1-|w|^2)}{|1-\langle z,w \rangle|^2}.$$

We will use (7) to study functions satisfying the invariant mean value inequality $Tu \geq u$ on B_n. We show that u must be 'almost' M-subharmonic. Here is our result.

THEOREM 2. *Suppose that $u \in C^2(B_n) \cap L^1(B_n)$ and that $\int_{B_n} |\tilde{\Delta}(z)| dV(z) < \infty$. If $\limsup_{z \to \zeta_0} \frac{\tilde{\Delta}u(z)}{(1-|z|^2)^{n+1}} < 0$ for some $\zeta_0 \in S$, then there exists $\delta > 0$ such that $Tu(z) < u(z)$ for all $z \in B_n$ such that $|z - \zeta_0| < \delta$.*

PROOF. Without loss of generality, we can assume that $\zeta_0 = 1 = (1, 0, 0, \cdots, 0)$. By our assumption, there exists $a > 0$ and $\epsilon > 0$ such that

$$\tilde{\Delta}u(z) \leq -a(1-|z|^2)^{n+1}$$

if $z \in B_n \cap \{z \in \mathbb{C}^n : |z-1| < \epsilon\} = B(1, \epsilon)$. Then

(10)
$$\int_{B_n} \tilde{\Delta}u(w) K(z,w) d\mu(w) = \int_{B(1,\epsilon)} \tilde{\Delta}u(w) K(z,w) d\mu(w)$$

$$+ \int_{B(1,\epsilon)^c} \tilde{\Delta}u(w) K(z,w) d\mu(w).$$

If $|z - 1| < \frac{\epsilon}{2}$ and $w \in B(1, \epsilon)^c$, then

$$\frac{(1-|z|^2)(1-|w|^2)}{|1-\langle z,w \rangle|^2} \leq C(1-|z|^2) < \epsilon$$

if $1 - |z|^2$ is sufficiently small. This means that $|\varphi_z(w)|^2 > \epsilon$, so by (9), we have $K(z,w) \leq C(1-|z|^2)^{n+1}(1-|w|^2)^{n+1}$. Hence

$$|\int_{B(1,\epsilon)^c} \tilde{\Delta}u(w)K(z,w)d\mu(w)| \leq C(1-|z|^2)^{n+1}\int_{B(1,\epsilon)^c} |\tilde{\Delta}u(w)|dV.$$

Thus the second integral in (10) is $O[(1-|z|^2)^{n+1}]$. The first integral is

$$\int_{B(1,\epsilon)} \tilde{\Delta}u(w)K(z,w)d\mu(w) \leq -a\int_{B_n}(1-|w|^2)^{n+1}K(z,w)d\mu(w)$$

$$+ a\int_{B(1,\epsilon)^c} K(z,w)d\mu(w)$$

$$\leq -C_1 a\int_{B_n}(1-|z|^2)^{n+1}\frac{(1-|w|^{2(n+1)})}{|1-\langle z,w\rangle|^{2n+2}}d\mu(w) + O[(1-|z|^2)^{n+1}]$$

$$= -C_1 a(1-|z|^2)^{n+1}\int_{B_n}\frac{(1-|w|^2)^{n+1}}{|1-\langle z,w\rangle|^{2n+2}}dV(w) + O[(1-|z|^2)^{n+1}].$$

Combining these estimates we have

$$Tu - u \leq -C_1 a(1-|z|^2)^{n+1}\int_{B_n}\frac{(1-|w|^2)^{n+1}}{|1-\langle z,w\rangle|^{2n+2}}dV(w) + O[(1-|z|^2)^{n+1}].$$

The integral in the above expression $\to \infty$ by [2], pg 18, hence $Tu - u < 0$.

An immediate application of Theorem 2 is a necessary condition for hyponormality of Toeplitz operators. Suppose that f and g are bounded analytic functions in B_n. As in [1], hyponormality of $T_{f+\bar{g}}$ implies that the function $u(z) = |f(z)|^2 - |g(z)|^2$ satisfies $Tu \geq u$ in B_n. Hence, we have

COROLLARY 3. *Suppose that f and g are bounded analytic function on B_n and that $T_{f+\bar{g}}$ is hyponormal. Then*

$$\limsup_{z\to\zeta}(|\nabla_T f(z)|^2 - |\nabla_T g(z)|^2) \geq 0 \quad \text{for every} \quad \zeta \in S,$$

where $|\nabla_T f|^2 = \sum_{i-1}^n |D_i f|^2 - |\sum_{i=1}^n z_i D_i f|^2$. In particular, if $D_i f$ and $D_i g$ are continuous at $\zeta \in S$ for all $i = 1, \cdots, n$, then $|\nabla_T f(\zeta)| \geq |\nabla_T g(\zeta)|$.

PROOF. Notice that $\tilde{\Delta}|f|^2 = (1-|z|^2)|\nabla_T f|^2$.

Submean value property 2

We will continue the study of functions that have the invariant submean value property. We saw that they have to be almost M-subharmonic. We want to give another version of almost M-subharmonicity. For that we need the following definition and lemma.

Let U_n denotes the group of the unitary transformations of $\mathbb{C}^n$. Haar measure is a Borel probability measure on U_n such that

$$\int_{U_n} f(UV)dV = \int_{U_n} f(V)dV,$$

for all $V \in U_n$ and $f \in C(U_n)$. Haar measure has the property

$$\int_{U_n} f(U\zeta)dV = \int_S f d\sigma \quad \text{for} \quad \zeta \in S, f \in C(S).$$

LEMMA 4. *Suppose $u \in L^1(B_n)$ satisfies $Tu \geq u$ in B_n. Suppose $w \geq 0$ is a bounded measurable function on U. For $z \in B_n$, we define*

$$U(z) = \int_{U_n} u(Vz)w(V)dV.$$

Then $U \in L^1(B_n)$ and $TU \geq U$ in B_n.

PROOF. By hypothesis,

$$U(z) = \int_{U_n} u(Vz)w(V)dV \leq \int_{U_n} (Tu)(Vz)w(V)dV$$

$$= \frac{1}{V(B_n)} \int_{U_n} \int_{B_n} u(\zeta) \frac{(1-|z|^2)^{n+1}}{|1-\langle Vz,\zeta\rangle|^{2(n+1)}} dV(\zeta)w(V)dV$$

If we let $\zeta = V\tau$, then

$$U(z) \leq \frac{1}{V(B_n)} \int_{U_n} \int_{B_n} u(V\tau) \frac{(1-|z|^2)^{n+1}}{|1-\langle Vz,V\tau\rangle|^{2(n+1)}} dV(\tau)w(V)dV$$

$$= \frac{1}{V(B_n)} \int_{B_n} \frac{(1-|z|^2)^{n+1}}{|1-\langle z,\tau\rangle|^{2(n+1)}} [\int_{U_n} u(V\tau)w(V)dV]dV(\tau)$$

$$= TU(z)$$

The next theorem gives a weak-* version of almost M-subharmonicity.

THEOREM 5. *Suppose that $u \in C^2(B_n) \cap L^1(B_n)$ and that $\int_{B_n} |\tilde{\Delta}u(w)|dV(w) < \infty$. Suppose further that $Tu \geq u$ in B_n and that there is an open subset $I \subseteq S$ and a finite Borel measure ν on I, and an a, such that $0 \leq a \leq n+1$, and that for all continuous functions φ with compact support on I we have*

$$\lim_{r \to 1} \int_I \frac{\tilde{\Delta}u(r\zeta)}{(1-r^2)^a} \varphi(\zeta)d\sigma(\zeta) = \int_I \varphi d\nu.$$

Then $\nu \geq 0$ on I.

PROOF. Without loss of generality, we can assume $1 = (1,0,\cdots,0) \in I$. Let φ be a continuous non-negative function with compact support in I and we define $w(V) = \varphi(V1)$. Let

$$U(z) = \int_{U_n} u(Vz)w(V)dV.$$

From Lemma 4, we know that $TU \geq U$ in B_n. Since $\tilde{\Delta}U(z) = \int_{U_n} (\tilde{\Delta}u)(Vz)w(V)dV$, it follows that $\int_{B_n} |\tilde{\Delta}U(z)|dV(z) < \infty$.

Theorem 2 implies that there is a sequence $z_k = r_k\zeta_k \to 1$ such that

$$\limsup_{k \to \infty} \frac{\tilde{\Delta}U(z_k)}{(1-|z_k|^2)^a} \geq 0.$$

Now we have

$$\int_I \varphi d\nu - \frac{\tilde{\Delta}U(z_k)}{(1-|z_k|^2)^a}$$

$$= [\int_I \varphi d\nu - \int_I \varphi(\zeta)\frac{\tilde{\Delta}u(r_k\zeta)}{(1-r_k^2)^a}d\sigma(\zeta)]$$

$$+ [\int_I \varphi(\zeta)\frac{\tilde{\Delta}u(r_k\zeta)}{(1-r_k^2)^a}d\sigma(\zeta) - \int_{U_n}\frac{(\tilde{\Delta}u)(Vz_k)}{(1-r_k^2)^a}w(V)dV].$$

Since $z_k \to 1$, it follows that $r_k \to 1$, so that the first difference goes to 0. The second difference is equal to

$$(11) \qquad \int_{U_n}\frac{\tilde{\Delta}u(r_kV1)}{(1-r_k^2)^a}\varphi(V1)dV - \int_{U_n}\frac{(\tilde{\Delta}u)(Vz_k)}{(1-r_k^2)^a}\varphi(V1)dV$$

Let $V_k \in U$ such that $V_kz_k = r_k \cdot 1$ and let $W = VV_k^*$. Then $V = WV_k$ in the second integral of (11) gives

$$\frac{1}{(1-r_k^2)^a}[\int_{U_n}\tilde{\Delta}u(r_kV1)\varphi(V1)dV - \int_{U_n}\tilde{\Delta}u(Wr_k1)\varphi(WV_k1)dV].$$

If we replace W by V again, we obtain

$$\int_{U_n}\frac{\tilde{\Delta}(r_kV1)}{(1-r_k^2)^a}[\varphi(V1) - (\varphi(VV_k1)]dV.$$

Now since $V_k \to id$ it follows that there is a compact subset of K of I such that the above integral is bounded by

$$[\int_K \frac{|\tilde{\Delta}u(r_k\zeta)|}{(1-r_k^2)^a}d\sigma(\zeta)] \cdot [\sup_{k,V\in U_n}|\varphi(V1) - \varphi(VV_k1)|].$$

The first factor is bounded, by the principle of uniform boundedness, and the second goes to 0 by uniform continuity of φ. This shows that $\int \varphi d\nu \geq 0$ for all such φ and hence that $\nu \geq 0$.

As an application of Theorem 5 we could give a result about hyponormal Toeplitz operators in the ball that is completely analogous to Theorem 5 of [1]. We omit the details.

We want to say a few words about the significance of the fact that there is a certain flexibility in the choice of a in Theorem 5 above. We want to give a concrete application of this. Suppose that u is smooth up to the boundary of B_n and that $Tu \geq u$ in B_n. Let $Lu(z) = \sum(\delta_{ij} - z_i\bar{z}_j)\frac{\partial^2 u}{\partial z_i\partial\bar{z}_j}$, then $\tilde{\Delta}u(z) = (1-|z|^2)Lu(z)$. From Theorem 5 we may conclude that $Lu(\zeta) \geq 0$ for all $\zeta \in S$. Suppose that there is a $\zeta_0 \in S$ and $Lu(z) \leq 0$ in a neighborhood Δ of S in B_n. It follows that $\lim_{r\to 1}\frac{Lu(r\zeta)}{1-r^2}$ exists for $\zeta \in \Delta$, because u is smooth. This limit is $\frac{-\partial Lu}{\partial r}(\zeta)$. On the other hand, by Theorem 5 with $a = 2$ this limit is ≥ 0. So we see that $\frac{\partial Lu}{\partial r} \leq 0$ in $\Delta \cap S$. If there were some $\zeta \in \Delta$ with $\frac{\partial Lu}{\partial r}(\zeta) < 0$, then we would have $Lu(r\zeta) > 0$ for all r close to 1. This contradicts our assumption, hence $\frac{\partial Lu}{\partial r} = 0$ in $\Delta \cap S$. It now follows that $\frac{Lu(r\zeta)}{(1-r^2)^2}$ has a limit as $r \to 1$, for

$\zeta \in \Delta \cap S$ and from Theorem 5 this limit is ≥ 0. Again if this limit is > 0 at some point in $\Delta \cap S$ we get a contradiction. We may continue to do this n times. We arrive at the following conclusion.

COROLLARY 6. *If $Tu \geq u$ and u is smooth out to the boundary of B_n and if there is a $\zeta_0 \in S$ and a neighborhood Δ of ζ_0 such that $Lu \leq 0$ in $\Delta \cap B_n$ then*

$$Lu = \frac{\partial}{\partial r} Lu = \cdots = \frac{\partial^n Lu}{\partial r^n} = 0 \quad on \quad \Delta \cap S.$$

It can be shown by example that the corollary is sharp.

REFERENCES

[1] P. Ahern and Ž. Čučković, *A mean value inequality with applications to Bergman space operators*, Pacific J. Math. (to appear).
[2] W. Rudin, *Function theory in the Unit Ball of $\mathbb{C}^n$*, Springer-Verlag, New York, 1980.

DEPARTMENT OF MATHEMATICS, UNIVERSITY OF WISCONSIN, MADISON, WI 53706
E-mail address: ahern@math.wisc.edu

DEPARTMENT OF MATHEMATICS, UNIVERSITY OF TOLEDO, TOLEDO, OH 43606
E-mail address: zcuckovi@math.utoledo.edu

Contemporary Mathematics
Volume 189, 1995

The Commutator of the Ergodic Hilbert Transform

A. MICHAEL ALPHONSE AND SHOBHA MADAN

ABSTRACT. In this paper we study the commutator of the ergodic Hilbert transform and give a characterization of the ergodic BMO space. We use known results on the commutator of the Hilbert transform on **R**, and to these inequalities we apply the Calderón-Coifman-Weiss transference principle.

1. Introduction.

In this paper we study the commutator of the ergodic Hilbert Transform $\tilde{H}$ and the operator of pointwise multiplication by a function b. Let $(X, \mathbf{B}, m)$ be a probability space and U an invertible measure preserving transformation on X. Then the ergodic Hilbert transform is defined as

$$\tilde{H}f(x) = \lim_{N \to \infty} \sum_{k=-N}^{N}{}' \frac{f(U^k x)}{k}$$

where the prime in the summation means exclusion of the term k=0. For results on the almost everywhere convergence of this series for functions in $L^p(X), 1 < p < \infty$ and the boundedness of the operator on $L^p(X), 1 < p < \infty$ see [9]. In [15] Petersen proved the weak (1,1) inequality using a transference argument.

The commutator of the operator of pointwise multiplication by a function $b \in L^1(X)$ and the ergodic Hilbert transform is given by

$$[b, \tilde{H}]f(x) = b(x)\tilde{H}f(x) - \tilde{H}(bf)(x).$$

We use the transference argument due to Calderón-Coifman-Weiss ([4], [7]) to prove that if b is a function in the ergodic BMO space (see definition in §3), then the above operator is bounded on $L^p(X)$, for $1 < p < \infty$. This technique also gives the L^p inequalities for the associated maximal operator.

1991 *Mathematics Subject Classification.* Primary 42B20; Secondary 28D05, 47A35.

The transference method has been applied twice for the final result. First, in §2 we study the boundedness of the commutator of the discrete Hilbert transform on the sequence space ℓ^p, $1 < p < \infty$, by transferring the results from **R**. For the commutator theorem on **R** we refer to [5] and for the maximal commutator inequalities see [16]. We remark here that the real variable methods, as in [16], can be used to prove Theorem 2.1 directly, but this is very technical. The details are in [13]. The proof given below using transference methods is much simpler.

The results in §2 are then transferred to the ergodic case in §3. The last result, Theorem 3.6 is a converse, namely that if [b,H] is bounded on $L^p(X)$ for some p, $1 < p < \infty$ and if in addition U is ergodic then $b \in \mathrm{BMO}(X)$.

As is the usual practice $C, C_p \dots$ denote constants which may change from one line to the next. The authors would like to thank Prof. Guido Weiss; and Prof. Aline Bonami for some useful suggestions.

2. The commutator on sequence spaces.

The discrete Hilbert transform, defined initially on finite sequences $\{a(n)\}_{n \in Z}$ is given by

$$Ha(n) = \sum_{k \neq 0} \frac{a(n-k)}{k}$$

It is well known that this operator extends to a bounded operator on the sequence spaces $\ell^p, 1 < p < \infty$ and is weak type (1,1).

Let $b = b(n)$ and consider the commutator given by

$$\begin{aligned} [b, H]a(n) &= b(n)Ha(n) - H(ba)(n) \\ &= \sum_{-\infty}^{\infty}{}' \frac{b(n) - b(n-k)}{k} a(n-k) \end{aligned}$$

For $1 < p < \infty$, we ask the question: For which sequences b is $[b, H]$ a bounded operator on ℓ^p? It turns out that the answer is given by the space BMO(**Z**), which is defined as

$$\mathrm{BMO}(\mathbf{Z}) = \left\{ b = \{b(n)\} : \sup_I \frac{1}{Card\,I} \sum_{k \in I} |b(k) - b_I| = \|b\|_{*,\mathbf{Z}} < \infty \right\}$$

Here the supremum is taken over all finite intervals in **Z**, and

$$b_I = \frac{1}{Card\,I} \sum_{j \in I} b(j).$$

We recall that a locally integrable function b defined on **R** is said to be in BMO(**R**) if

$$\sup_I \frac{1}{|I|} \int_I |b(x) - b_I| dx = \|b\|_{*,\mathbf{R}} < \infty,$$

where $b_I = \frac{1}{|I|} \int_I b(x) dx$ and the supremum is taken over all finite intervals in **R**. $\| \ \|_{*,\mathbf{R}}$ is defined as the norm on BMO(**R**). With this norm BMO(**R**) is a Banach space. For standard results on BMO($\mathbf{R}$) see [17]. Most results on BMO($\mathbf{R}$) carry over to BMO($\mathbf{Z}$) and the proofs are more or less the same. For details we refer to [13].

In [5] Coifman, Rochberg and Weiss proved that if $b \in$ BMO($\mathbf{R}$) then $[b, H]$ is a bounded operator on $L^p(\mathbf{R})$ for $1 < p < \infty$. Conversely, if $[b, H]$ is bounded on $L^p(\mathbf{R})$ for some p, $1 < p < \infty$, then $b \in$ BMO($\mathbf{R}$).

For the transference in §3 we will need inequalities for the associated maximal operator. Let

$$H_\epsilon f(x) = \int_{|x-y|>\epsilon} \frac{f(x)}{x - y} dy$$

Then the maximal commutator is given by

$$[b, H]^* f(x) = \sup_{\epsilon > 0} |[b, H_\epsilon] f(x)|$$

Segovia and Torrea [16] proved that if $b \in$ BMO($\mathbf{R}$), then $[b, H]^*$ is bounded on $L^p(\mathbf{R})$, $1 < p < \infty$. We will use this result in the following theorem.

THEOREM 2.1. *Let* $b \in BMO(\mathbf{Z})$ *and* $1 < p < \infty$. *Then there exists a constant* $C_p > 0$ *such that*

$$\|[b, H]^* a\|_p \leq C_p \|b\|_{*,\mathbf{Z}} \|a\|_p \quad \forall\, a \in \ell^p.$$

where $[b, H]^* a(n) = \sup_N | \sum_{k=-N}^{N}{}' \frac{b(n) - b(n-k)}{k} a(n - k)|$

For the proof of this theorem, we will need the following lemma.

LEMMA 2.2. *Let* $b \in BMO(\mathbf{Z})$; *define a function* $\tilde{b}$ *on* $\mathbf{R}$ *as*

$$\tilde{b}(x) = \sum_{k \in Z} b(k) \chi_{[k-1/2, \, k+1/2)}(x).$$

Then
 (i) $\tilde{b} \in BMO(\mathbf{R})$, $\|\tilde{b}\|_{*,\mathbf{R}} \leq 6 \|b\|_{*,\mathbf{Z}}$ *and*
 (ii) $|\tilde{b}(x) - \tilde{b}(y)| \leq C \|b\|_{*,\mathbf{Z}} \log(2 + |x - y|)$.

PROOF. (i) Let I be an interval in $\mathbf{R}$. If $I \subset [k - 1/2, k + 1/2)$ for some k, then $\frac{1}{|I|} \int_I |\tilde{b}(x) - \tilde{b}_I| dx = 0$. Suppose $I = [j - 1/2 - \delta_1, k + 1/2 + \delta_2] \subset \mathbf{R}$ with $0 < \delta_1 < 1$ and $0 \leq \delta_2 < 1$. Then $|I| = k - j + 1 + \delta_1 + \delta_2$. Let J be the integer interval $[j - 1, k + 1] \subset \mathbf{Z}$. Then $Card J = k - j + 3$, so that $(1/3) Card J \leq |I| \leq Card J$. We have,

$$\int_I |\tilde{b}(x) - \tilde{b}_I| dx = |b(j - 1) - \tilde{b}_I| \delta_1 + \sum_{l=j}^{k} |b(l) - \tilde{b}_I| + |b(k + 1) - \tilde{b}_I| \delta_2$$

$$\leq \sum_{l=j-1}^{k+1} |b(l) - b_I| + (k - j + 1 + \delta_1 + \delta_2)|b_I - \tilde{b}_I|$$

$$\leq \; Card\, J\|b\|_{*,\mathbf{Z}}$$
$$+ \||I|b_I - (b(j-1)\delta_1 + b(j) + ... + b(k) + b(k+1)\delta_2)|$$

$$\leq \; Card\, J\|b\|_{*,\mathbf{Z}}$$
$$+\delta_1|b(j-1) - b_I| + \sum_{l=j}^{k} |b(l) - b_I| + \delta_2|b(k+1) - b_I|$$

$$\leq \; 2Card\, J\|b\|_{*,\mathbf{Z}} \leq 6|I|\|b\|_{*,\mathbf{Z}}.$$

Hence $\|\tilde{b}\|_{*,\mathbf{R}} \leq 6\|b\|_{*,\mathbf{Z}}$.

(ii) Let $I_j = [j-1/2, j+1/2]$ and suppose $x\epsilon I_j, y\epsilon I_k$. Then $\tilde{b}(x) = \tilde{b}_{I_j}, \tilde{b}(y) = \tilde{b}_{I_k}$ and so by standard results on BMO($\mathbf{R}$) [10] we have,

$$|\tilde{b}_{I_j} - \tilde{b}_{I_k}| \; \leq \; c\,log(2 + dist(I_j, I_k))\|\tilde{b}\|_{*,\mathbf{R}}$$
$$\leq \; 6c\,log(2 + |x - y|)\|b\|_{*,\mathbf{Z}}$$

since $dist(I_j, I_k) \leq |x - y|$. $\square$

PROOF OF THEOREM 2.1. For $a = a(k)\,\epsilon\,\ell^p$ and $b\,\epsilon\,$BMO($\mathbf{Z}$), define

$$f(x) \; = \; \sum_{k=-\infty}^{\infty} a(k)\chi_{I'_k}(x) \quad \text{and}$$

$$\tilde{b}(x) \; = \; \sum_{k=-\infty}^{\infty} b(k)\chi_{I_k}(x)$$

where $I'_k = [k - 1/4, k + 1/4]$ and $I_k = [k - 1/2, k + 1/2]$. Then $f\epsilon L^p(\mathbf{R})$ with $\|f\|_p = 2^{-1/p}\|a\|_p$, and by Lemma 2.2, $\tilde{b}\epsilon$BMO($\mathbf{R}$) with $\|\tilde{b}\|_{*,\mathbf{R}} \leq 6\|b\|_{*,\mathbf{Z}}$.

Fix $N \geq 0$. Then for $x\epsilon I'_j$, we have

$$\left| \sum_{k:|k-j|>N} \frac{[b(j) - b(k)]a(k)}{(j - k)} - 2\int_{|x-y|>N+1/2} \frac{[\tilde{b}(x) - \tilde{b}(y)]f(y)}{(x - y)}dy \right|$$

$$= \; \left|2 \sum_{k:|k-j|>N} \int_{I'_k} [\frac{1}{(j - k)} - \frac{1}{(x - y)}][\tilde{b}(x) - \tilde{b}(y)]f(y)dy\right|$$

$$\leq \; C \sum_{k:|k-j|>N} \int_{I'_k} \frac{|\tilde{b}(x) - \tilde{b}(y)|}{(x - y)^2}|f(y)|dy$$

$$\leq \; C \sum_{k:|k-j|>N} \|b\|_{*,\mathbf{Z}} \int_{I'_k} \frac{log(2 + |x - y|)}{(x - y)^2}|f(y)|dy$$

$$\leq \; C\|b\|_{*,\mathbf{Z}} \int_{|x-y|>1/2} \frac{log(2 + |x - y|)}{(x - y)^2}|f(y)|dy \leq C\|b\|_{*,\mathbf{Z}}Mf(x)$$

where Mf is the Hardy-Littlewood maximal function. The last inequality follows since

$$\int_{|x-y|>1/2} \frac{\log(2+|x-y|)}{(x-y)^2}|f(y)|dy = \sum_{k=0}^{\infty} \int_{2^{k-1}<|x-y|\leq 2^k} \frac{\log(2+|x-y|)}{(x-y)^2}dy$$

$$\leq \sum_{k=0}^{\infty} \frac{k+2}{2^{2(k-1)}} \int_{|x-y|\leq 2^k} |f(y)|dy \leq CMf(x)\sum_{k=0}^{\infty}\frac{k}{2^k}.$$

In particular, putting N $= 0$, we have

$$|[b,H]a(j)| \leq [\tilde{b},H]^*f(x) + C\|b\|_{*,\mathbf{Z}}Mf(x) \quad \text{for } x \epsilon I_j'.$$

Finally

$$[b,H]^*a(j) = \sup_N | \sum_{|k-j|\leq N} \frac{[b(j)-b(k)]a(k)}{(j-k)}|$$

$$\leq |[b,H]a(j)| + \sup_N | \sum_{|k-j|>N} \frac{[b(j)-b(k)]a(k)}{(j-k)}|$$

$$\leq |[b,H]a(j)| + [\tilde{b},H]^*f(x) + C\|b\|_{*,\mathbf{Z}}Mf(x), \quad \text{for } x \epsilon I_j'$$

$$\leq C\{[\tilde{b},H]^*f(x) + \|b\|_{*,\mathbf{Z}}Mf(x)\}, \quad \text{for } x \epsilon I_j'$$

Therefore

$$\|[b,H]^*a\|_p \leq C_p\|b\|_{*,\mathbf{Z}}\|f\|_p \leq C_p\|b\|_{*,\mathbf{Z}}\|a\|_p.$$

using the result on the boundedness of the maximal commutator on $L^p(\mathbf{R})$, $1 < p < \infty$ [**16**].

In the next theorem we prove a converse to this result

THEOREM 2.3. *Let $1 \leq p < \infty$. Suppose there exists a constant $C > 0$ such that $\|[b,H]a\|_p \leq C\|a\|_p \forall a \epsilon \ell^p$. Then $b \epsilon BMO(\mathbf{Z})$.*

PROOF. Let I be any interval in $\mathbf{Z}$ and n_0 the centre of I. Then for $n \epsilon I$, we have

$$|b(n) - b_I| = |\frac{1}{Card\,I}\sum_{k\epsilon I}[b(n)-b(k)]|$$

$$= |\sum_{k\epsilon I} \frac{[b(n)-b(k)]}{(n-k)} \cdot \frac{(n-n_0+n_0-k)}{Card\,I}|$$

$$\leq |\sum_{k=-\infty}^{\infty} \frac{[b(n)-b(k)]}{(n-k)}\chi_I(k)|$$

$$+ |\sum_{k=-\infty}^{\infty} \frac{[b(n)-b(k)]}{(n-k)} \cdot \frac{(n_0-k)}{Card\,I}\chi_I(k)|$$

$$= |[b,H]a_1(n)| + |[b,H]a_2(n)|$$

where $a_1(k) = \chi_I(k)$ and $a_2(k) = \frac{(n_o-k)}{Card\,I}\chi_I(k)$.

Clearly $\|a_j\|_p \leq (CardI)^{1/p}, j = 1, 2$ and we have

$$
\begin{aligned}
\frac{1}{Card\,I} \sum_{n\epsilon I} |b(n) - b_I| &\leq (\frac{1}{Card\,I} \sum_{n\epsilon I} |b(n) - b_I|^p)^{1/p} \\
&\leq (\frac{1}{Card\,I} \sum_{n\epsilon I} |[b, H]a_1(n)|^p)^{1/p} \\
&\quad +(\frac{1}{Card\,I} \sum_{n\epsilon I} |[b, H]a_2(n)|^p)^{1/p} \\
&\leq C\frac{1}{(Card\,I)^{1/p}}[\|a_1\|_p + \|a_2\|_p] \leq C.
\end{aligned}
$$

Hence $b\epsilon$BMO($\mathbf{Z}$). $\quad\square$

3. Commutator of the ergodic Hilbert Transform.

For a probability space $(X, \mathbf{B}, m)$ and an invertible measure preserving transformation on X, the space BMO(X) is defined as the space of those functions $b \epsilon L^1(X)$ satisfying

$$
ess.sup_{x\epsilon X} [\sup_{N\geq 1} \frac{1}{2N+1} \sum_{k=-N}^{N} |b(U^k x) - \frac{1}{2N+1} \sum_{j=-N}^{N} b(U^j x)|] = \|b\|_* < \infty
$$

In [8] Coifman and Weiss showed that this space is the dual space of the ergodic Hardy space $H^1(X) = \{f\epsilon L^1(X) : \tilde{H}f\epsilon L^1(X)\}$. In [14] Petersen showed that if X is a Lebesgue space then BMO(X) contains $L^\infty(X)$ as a proper subspace. As in the classical case the ergodic Hardy space $H^1(X)$ has an atomic decomposition [3]. The ergodic atoms are defined using ergodic rectangles and we need some results about these in section 3.

DEFINITION 3.1. *Let E be a subset of X with positive measure and let $K \geq 1$ be such that $U^i E \bigcap U^j E = \phi$ if $i \neq j$ and $-K \leq i, j \leq K$. Then the set $R = \bigcup_{i=-K}^{K} U^i E$ is called an ergodic rectangle of length $2K + 1$ with base E.*

For the following results we refer to [1].

PROPOSITION 3.2 ([1]). *Let $(X, \mathbf{B}, m)$ be a probability space, U an ergodic invertible measure perserving transformation on X and K a positive integer.*
  (i) *If $F \subseteq X$ is a set of positive measure then there exists a subset $E \subseteq F$ of positive measure such that E is the base of an ergodic rectangle of length $2K + 1$.*
  (ii) *There exists a countable family $\{E_j\}$ of bases of ergodic rectangles of length $2K + 1$ such that $X = \bigcup_j E_j$.*

Let $(X, \mathbf{B}, m)$ be a probability space and U an invertible measure preserving transformation on X. In this section we study the commutator of the operator of pointwise multiplication by a function $b \in L^1(X)$ and the ergodic Hilbert transform. This is defined as

$$\begin{aligned} [b, \tilde{H}]f(x) &= b(x)\tilde{H}f(x) - \tilde{H}(bf)(x) \\ &= \sum_{k=-\infty}^{\infty}{}' \frac{[b(x) - b(U^{-k}x)]f(U^{-k}x)}{k}. \end{aligned}$$

We first show that this operator is well defined on a dense subset of $L^p(X)$, $1 \le p \le 2$.

LEMMA 3.3. *Let* $1 \le p \le 2$ *and*

$$D = \{f \epsilon L^p(X) : f = foU \ a.e.\} + \{f \epsilon L^p(X) : f = g - goU \ a.e., g \epsilon L^\infty(X)\}.$$

Then for each f *in* D, $[b, \tilde{H}]f$ *exists a.e.*

PROOF. If $f \epsilon L^p(X)$ with $f = foU$ a.e., then

$$[b, \tilde{H}]f(x) = b(x)\tilde{H}f(x) - f(x)\tilde{H}b(x)$$

which exists a.e. $x \epsilon X$, since f and b belong to $L^1(X)$.

If $f = g - goU$ a.e., $g \epsilon L^\infty(X)$ then f and bf are in $L^1(X)$ so that $[b, \tilde{H}]f(x)$ exists a.e. $x \epsilon X$. For the density of the set D see [4]. $\square$

Consider now the associated maximal operator,

$$[b, \tilde{H}]^* f(x) = \sup_{N \ge 1} | \sum_{k=-N}^{N} \frac{[b(x) - b(U^{-k}x)]f(U^{-k}x)}{k}|.$$

The following theorem is proved by a transference argument.

THEOREM 3.4. *Let* $1 < p < \infty$ *and* $b \epsilon BMO(X)$. *Then there exists a constant* $C_p > 0$ *such that*

$$\|[b, \tilde{H}]^* f\|_p \le C_p \|f\|_p, \quad \forall f \epsilon L^p(X).$$

PROOF. Observe that if $b \epsilon BMO(X)$ then $b \epsilon L^1(X)$ and for a.e. x, the sequence b_x given by $b_x(n) = b(U^n x)$ is in BMO($\mathbf{Z}$) with $\|b_x\|_{*, \mathbf{Z}} \le C$, where C is independent of x.

For $J \ge 1$, let

$$[b, \tilde{H}]^*_J f(x) = \sup_{N \le J} | \sum_{k=-N}^{N}{}' \frac{[b(x) - b(U^{-k}x)]f(U^{-k}x)}{k}|.$$

We will prove that

$$\|[b, \tilde{H}]^*_J f\|_p \le C \|f\|_p, \quad \forall f \epsilon L^p(X).$$

where the constant C is independent of f and J. Then the theorem will follow by monotone convergence theorem.

For a.e. $x \epsilon X$, let

$$[b_x, H]^*_J a(n) = \sup_{N \leq J} | \sum_{k=-N}^{N}{}' \frac{[b_x(n) - b_x(n - k)]a(n - k)}{k}|.$$

It is easy to see that $[b_x, H]^*_J$ is sublinear. Further observe that if supp $a \subseteq \{n : |n| > L\}$, then supp $[b, H]^*_J a \subseteq \{n : |n| > L - J\}$. For a.e. $x \epsilon X$, let $f_x(n) = f(U^n x)$. For $K \epsilon \mathbf{Z}_+$, define

$$f_x^K(n) = \begin{cases} f_x(n) & \text{if } |n| \leq K \\ O & \text{if } |n| > K. \end{cases}$$

Since $[b_x, H]^*_J$ is sublinear, for $K, L \epsilon \mathbf{Z}_+$, and $n \epsilon \mathbf{Z}$, we have

$$[b_x, H]^*_J f_x(n) \leq [b_x, H]^*_J f_x^{K+L}(n) + [b_x, H]^*_J (f_x - f_x^{K+L})(n).$$

We can choose L large enough so that $supp[b_x, H]^*_J(f_x - f_x^{K+L}) \subseteq \{n : |n| \leq K\}$. since supp $(f_x - f_x^{K+L}) \subseteq \{n : |n| > K + L\}$.

Note that L depends only on J and not on K. Therefore

$$[b_x, H]^*_J f_x(n) \leq [b_x, H]^*_J f_x^{K+L}(n) \quad \text{for } |n| \leq K.$$

Also for a.e. $x \epsilon X$ and $j \epsilon Z$, we have

$$
\begin{aligned}
[b, \tilde{H}]^*_J f(U^j x) &= \sup_{N \leq J} | \sum_{k=-N}^{N}{}' \frac{[b(U^j x) - b(U^{j-k}x)]f(U^{j-k}x)}{k}| \\
&= \sup_{N \leq J} | \sum_{k=-N}^{N}{}' \frac{[b_x(j) - b_x(j - k)]f_x(j - k)}{k}| \\
&= [b_x, H]^*_J f_x(j).
\end{aligned}
$$

Then

$$
\begin{aligned}
\int_X ([b,\tilde{H}]^*_J f(x))^p dm &= \frac{1}{2K+1} \sum_{j=-K}^{K} \int_X ([b,\tilde{H}]^*_J f(U^j x))^p dm \\
&= \frac{1}{2K+1} \sum_{j=-K}^{K} \int_X ([b_x,H]^*_J f_x(j))^p dm \\
&\leq \frac{1}{2K+1} \sum_{j=-K}^{K} \int_X ([b_x,H]^*_J f_x^{K+L}(j))^p dm \\
&\leq \frac{C}{2K+1} \int_X \sum_{j=-(K+L)}^{(K+L)} |f_x(j)|^p dm \\
&\leq \frac{C}{2K+1} \sum_{j=-(K+L)}^{(K+L)} \int_X |f(U^j x)|^p dm \\
&\leq \frac{C[2(K+L)+1]}{2K+1} \|f\|^p
\end{aligned}
$$

Choosing K sufficiently large, we get

$$
\|[b,\tilde{H}]^*_J f\|_p \leq C_p \|f\|_p.
$$

□

As a consequence of Lemma 3.3 and standard results in harmonic analysis we have the following Corollary.

COROLLARY 3.5. *Let $1 < p < \infty$ and $b \epsilon BMO(X)$. Then the commutator of the ergodic Hilbert transform $[b,\tilde{H}]f$ exists a.e., $\forall f \epsilon L^p(X)$.*

We now prove a converse of Theorem 3.4 with an additional hypothesis that U be ergodic.

THEOREM 3.6. *Let $(X, \mathbf{B}, m)$ be a non-atomic probability space, U an invertible, ergodic measure preserving transformation on X and $b \epsilon L^1(X)$. If $[b,\tilde{H}]$ is bounded on $L^p(X)$, for some p, $1 < p < \infty$, then $b \epsilon BMO(X)$.*

PROOF. Let $\{a(k)\}$ be any sequence in ℓ^p. We will prove that

$$
\sup_N \sum_{j=-N}^{N} |\sum_{k=-N}^{N} \frac{[b_x(j)-b_x(k)]a(k)}{j-k}|^p \leq C\|a\|_p^p, \ a.e. \ x \epsilon X \tag{3.7}
$$

where for a.e. x, the sequence b_x is defined as $\{b_x(k)\} = \{b(U^k x)\}$. If inequality (3.7) is proved, then by the same argument as in the proof of Theorem 2.3, it follows that $b_x \epsilon BMO(\mathbf{Z})$ a.e. x and $\|b_x\|_{*,\mathbf{Z}} \leq C$, where C is independent of x so that $b \epsilon BMO(X)$.

For inequality (3.7), fix $N\epsilon\mathbf{Z}_+$ and take an ergodic rectangle $R = \bigcup_{j=-N}^{N} U^j B$, of length $2N+1$ with base B. Define

$$f(x) = \sum_{k=-N}^{N} a(k)\chi_{U^k B}(x).$$

Let $\epsilon > 0$. For $n \geq 1$, let A_n be the set

$$\{x\epsilon B : |\sum_{|k|>r}' \frac{[b(U^j x) - b(U^{k+j}x)]f(U^{k+j}x)}{k}|$$

$$< \frac{\epsilon}{(2N+1)^{1/p}} \; ; \forall r \geq n, -N \leq j \leq N\}$$

Clearly the A_ns are increasing. Since $f \epsilon L^\infty(X)$, f and bf belong to $L^1(X)$, hence $[b, \tilde{H}]f(x)$ exists a.e. $x\epsilon X$, which in turn implies that $B = \bigcup_{n=1}^{\infty} A_n$.

Now fix $n \geq 1$ and let $J = max\{n, N\}$. Take an ergodic rectangle $S = \bigcup_{k=-2J}^{2J} U^k E$, of length $4J+1$ with base $E \subseteq A_n$. Let F be any measurable subset of E and call $S' = \bigcup_{k=-N}^{N} U^k F$. Define a function g on X by

$$g(x) = \sum_{k=-N}^{N} a(k)\chi_{U^k F}(x).$$

Since F is the base of an ergodic rectangle of length of $4J+1$, for $x\epsilon S'$ we have

$$|\sum_{k=-2N}^{2N}' \frac{[b(x) - b(U^k x)]g(U^k x)}{k}| \leq |[b, \tilde{H}]g(x)| + |\sum_{|k|>2J} \frac{[b(x) - b(U^k x)]g(U^k x)}{k}|.$$

Now,

$$[\int_{S'} |\sum_{k=-2N}^{2N} \frac{[b(x) - b(U^k x)]g(U^k x)}{k}|^p \, dm]^{1/p}$$

$$\leq \quad \|[b, \tilde{H}]g\|_p + [\int_{S'} |\sum_{|k|>2J}' \frac{[b(x) - b(U^k x)]g(U^k x)}{k}|^p \, dm]^{1/p}$$

$$\leq \quad C\|g\|_p + \frac{\epsilon}{(2N+1)^{1/p}}m(S')^{1/p} \leq [C\|a\|_p + \epsilon]m(F)^{1/p}$$

since $[b, \tilde{H}]$ is bounded on $L^p(X)$ and $g = f|_S$.

Now,

$$[\int_{S'} |\sum_{k=-2N}^{2N}' \frac{[b(x) - b(U^k x)]g(U^k x)}{k}|^p \, dm]^{1/p}$$

$$= \left[\sum_{j=-N}^{N} \int_{U^j F} \Big| \sideset{}{'}\sum_{k=-2N}^{2N} \frac{[b(x) - b(U^k x)]g(U^k x)}{k} \Big|^p dm \right]^{1/p}$$

$$= \left[\int_F \sum_{j=-N}^{N} \Big| \sideset{}{'}\sum_{k=-2N}^{2N} \frac{[b(U^j x) - b(U^{k+j} x)]g(U^{k+j} x)}{k} \Big|^p dm \right]^{1/p}$$

$$= \left[\int_F \sum_{j=-N}^{N} \Big| \sideset{}{'}\sum_{k=-N}^{N} \frac{[b(U^j x) - b(U^k x)]g(U^k x)}{k - j} \Big|^p dm \right]^{1/p}$$

The last equality follows since supp $g \subseteq S'$. Hence

$$\frac{1}{m(F)} \int_F \sum_{j=-N}^{N} \Big| \sum_{k=-2N}^{2N} \frac{[b(U^j x) - b(U^k x)]g(U^k x)}{(k - j)} \Big|^p dm \le [C\|a\|_p + \epsilon]^p.$$

Since F was an arbitrary subset of E, we conclude that

$$\sum_{j=-N}^{N} \Big| \sum_{k=-2N}^{2N} \frac{[b(U^j x) - b(U^k x)]a(k)}{(k - j)} \Big|^p dm \le [C\|a\|_p + \epsilon]^p \qquad (3.8)$$

for a.e. $x \epsilon E$. Since A_n can be written as a countable union of bases of ergodic rectangles of length $4J + 1$, inequality (3.8) holds for a.e. $x \epsilon A_n$ for each n. But $B = \bigcup_{n=1}^{\infty} A_n$ so that (3.8) holds for a.e. $x \epsilon B$. Lastly $X = \bigcup_{j=1}^{\infty} B_j$, where each B_j is the base of an ergodic rectangle of length $2N + 1$, hence inequality (3.8) holds for a.e. $x \epsilon X$. Since ϵ and N are arbitrary (3.7) follows. $\square$

REFERENCES

1. E. Attencia and A. de la Torre. A dominated ergodic estimate for L^p spaces with weights. Studia Math. 74 (1982), 35-47.

2. E. Berkson, T. A. Gillespie and P. S. Muhly. Abstract spectral decomposition guaranteed by the Hilbert transform. Proc. London Math. Soc. 53 (1986), 489-517.

3. R. Caballero and A. de la Torre. An atomic theory of ergodic H^p spaces. Studia Math. 82 (1985), 39-59.

4. A. P. Calderón. Ergodic theory and translation invariant operators. Proc. Nat. Acad. Sci. USA. 59 (1968), 349-353.

5. R. R. Coifman, R. Rochberg and G. Weiss. Factorization of Hardy spaces. Ann. Math. 103 (1976), 611-635.

6. R. R. Coifman and G. Weiss. Operators associated with representations of amenable groups, Singular integrals induced by ergodic flows, the rotation method and multipliers. Studia Math. 47 (1973), 285-303.

7. R. R. Coifman and G. Weiss. Transference methods in Analysis. CBMS Regional Conference series in Math 31, Amer. Math. Soc, 1977.

8. R. R. Coifman and G. Weiss. Maximal functions and H^p spaces defined by ergodic transfomations. Proc. Nat. Acad. Sci. USA. 70 (1973), 1761-1763.

9. M. Cotlar. A unified theory of Hilbert transform and ergodic theory. Rev. Mat. Cuyana. 1 (1955), 105-107.

10. J.B.Garnett. Bounded Analytic Functions. Academic Press. 1981.
11. R. Hunt, B. Muckenhoupt and R. Wheeden. Weighted norm inequalities for the conjugate function and Hilbert transform. Trans. Amer. Math. Soc. 176 (1973), 227-251.
12. S. Janson, J. Peetre and S. Semmes. On the action of Hankel and Toeplitz operators on some function spaces. Duke Math. 51 (1984), 937-958.
13. A. Michael Alphonse. A class of operators on sequence spaces and Transference. Ph.D. Thesis. Indian Institute of Technology, Kanpur, 1992.
14. K. Petersen. A construction of ergodic BMO functions. Proc. Amer. Math. Soc. 79 (1980) 549-555.
15. K. Petersen Another proof of the existence of the ergodic Hilbert transform. Proc. Amer. Math. Soc. 88 (1983), 39-43.
16. C. Segovia and J. L. Torrea. Vector valued commutators and applications. Indiana Univ. Math. J. 38 (1989), 959-971.
17. A. Torchinsky. Real variable methods in Harmonic Analysis, Academic Press 1986.

INDIAN STATISTICAL INSTITUTE CALCUTTA, AND INDIAN INSTITUTE OF TECHNOLOGY, KANPUR, INDIA
E-mail address: madan@iitk.ernet.in

Contemporary Mathematics
Volume **189**, 1995

THE HOMOGENEOUS SPACE
OF REPRESENTATIONS
OF A NUCLEAR C*-ALGEBRA

E. ANDRUCHOW, G. CORACH AND D. STOJANOFF

ABSTRACT. Let $R = R(\mathcal{A}, \mathcal{N})$ be the space of bounded representations of the C*-algebra $\mathcal{A}$ onto the von Neumann algebra $\mathcal{N}$. In [ACS] we proved that when $\mathcal{A}$ is nuclear and $\mathcal{N}$ is injective, the space R is a Banach homogeneous reductive space. In this paper we focus on the applications of that result. Namely, geometric descriptions of the spaces of representations of finite dimensional C*-algebras [AS2], spectral measures [ARS] and representations of locally compact groups G such that $C^*(G)$ is amenable. This latter example generalizes [MR] and [M], where norm continuous representations of compact groups are considered.

0. *Introduction.*

The space of representations of a nuclear C*-algebra $\mathcal{A}$ on an injective von Neumann algebra $\mathcal{N}$ with separable predual can be regarded as a general example of a homogeneous reductive space ([ACS]). The existence of such a structure on this space implies several results of the kind in different contexts: spectral measures ([ARS]), joint similarity orbits of operators on a C*-algebra ([AS2], [AFHPS]) , representations of amenable groups (see below).

The general result provides a model of a homogeneous structure for the space of SOT-continuous representations of a locally compact amenable group (or more generally, the similarity orbits of representations of a locally compact group G such that $C^*(G)$ is amenable). In [M] and [MR] it was shown that the space

1991 *Mathematics Subject Classification.* 46L99, 58B20.
Partially supported by CONICET (Argentina)

$S_\infty(G, \mathcal{B})$ of norm continuous representations of G on a C*-algebra $\mathcal{B}$ admits a homogeneous structure as a submanifold of $C(G, \mathcal{B})$ (= space of continuous maps from G to $\mathcal{B}$ with supremum norm) when the group G is compact.

Our method requires that the C*-algebra must be an injective von Neumann algebra, but allows one to consider more general groups and, most important, SOT-continuous representations.

In this paper we also view the ways to construct, given a bounded algebra representation, a positive operator intertwining it with a selfadjoint representation. There are several methods to obtain this positive elements. Most remarkable are Haagerup's solution of the similarity problem [H1] and Porta and Recht's decomposition of invertible elements in the domain of a conditional expectation [PR]. It is an interesting question to decide whether this two methods coincide.

Once it is established that a homogeneous space has a reductive structure, via a conditional expectation, one may like to know how the geometry depends on the particular choice of the conditional expectation, or if there is a canonical choice. We have only partial results, but the evidence suggests that in most cases there exists no canonical choice, and that different conditional expectations give rise to spaces with different geodesics.

This paper can be regarded as a continuation of [ACS]. Here we focus on the applications of the general results obtained there.

The paper is divided in four sections. In Section 1 we describe the general facts about homogeneous spaces modelled on Banach spaces which admit a reductive structure, as well as the results of [ACS] concerning the space of representations of a nuclear C*-algebra. In Section 2 we consider the case of finite dimensional C*-algebras, and the relationship between the space of representations of these algebras and joint similarity orbits of n-tuples of elements of C*-algebras (see [AS2] and [AFHPS]). In Section 3 the case of commutative von Neumann algebras is considered. The representations of such algebras can be put in a one-to-one correspondence with spectral measures valued on the algebra (see [ARS]). Sections 4 contains the description of the model of a differential geometry for the space of SOT-continuous representations of a locally compact amenable group G.

1. Homogeneous reductive spaces

(1.1) General facts and definitions

Let $\mathcal{A}$ be a C*-algebra , $\mathcal{E}$ a Banach space such that the group $G_\mathcal{A}$ of invertible elements of $\mathcal{A}$ acts continuously on $\mathcal{E}$ and let $\mathcal{R}$ be a subset of $\mathcal{E}$ invariant under the action of $G_\mathcal{A}$. We say that $\mathcal{R}$ is an analytic (resp C^∞) homogeneous space under the action of $G_\mathcal{A}$ iff

- $\mathcal{R}$ is an analytic (resp. C^∞) submanifold of $\mathcal{E}$
- for every $x_0 \in \mathcal{R}$, the map

$$\tau_{x_0} : G_\mathcal{A} \to \mathcal{R} , \quad \tau_{x_0}(g) = g \bullet x_0$$

is an analytic (resp. C^∞) submersion.

For basic facts and definition about differentiable manifolds modelled on Banach spaces see [L] or [Ra]. A completely analogous definition can be made for the action of the unitary group $\mathcal{U}_A$ of $\mathcal{A}$.

Note that an analytic submersion, by definition, has analytic local cross sections, i.e., for each $x \in \mathcal{R}$ there exist a neighborhood $x \in \mathcal{V} \subset \mathcal{R}$ and an analytic map $s : \mathcal{V} \to G_A$ such that $s(y) \bullet x_0 = y$ for all $y \in \mathcal{V}$. It turns out that the existence of merely continuous local cross sections is one of the main tasks in proving that a given G_A-invariant set is a homogeneous space (and in some examples it is equivalent to it, see [AS1]). For submanifolds of infinite dimensional Banach spaces there are no simple conditions ensuring that a given action of G_A defines a homogeneous structure. For example, in the case of similarity orbits of operators ([AFHV], [AS1]), the orbits that admit a homogeneous structure are very specific.

Suppose that $\mathcal{R}$ is a homogeneous space. For each $x_0 \in \mathcal{R}$, denote by $\mathcal{T}_{x_0}$ the derivative of τ_{x_0} at $1 \in G_A$. Since the Lie algebra of G_A may be identified with $\mathcal{A}$, $\mathcal{T}_{x_0}$ can be regarded as a map from $\mathcal{A}$ to $(T\mathcal{R})_{x_0}$ the tangent space of $\mathcal{R}$ at x_0. Therefore $(T\mathcal{R})_{x_0} = \mathcal{T}_{x_0}(\mathcal{A})$.

Let us denote by I_{x_0} the isotropy group of x_0, i.e. the subgroup of G_A consisting of the elements which leave x_0 fixed. Let Υ_{x_0} be the Lie algebra of I_{x_0}.

Given a homogeneous space $\mathcal{R}$ under the action of G_A, we say that $\mathcal{R}$ admits a reductive structure if and only if

1) For each $x_0 \in \mathcal{R}$ there exists a closed linear subspace $\mathcal{H}_{x_0} \subset \mathcal{A}$, invariant under inner automorphisms by elements of I_{x_0}, such that

$$\mathcal{H}_{x_0} \oplus \Upsilon_{x_0} = \mathcal{A}$$

2) The distribution $x \mapsto \mathcal{H}_x$ is analytic.

Statement 2) means that if P_x is the projection onto $\mathcal{H}_x$ given by the decomposition $\mathcal{A} = \mathcal{H}_x \oplus \Upsilon_x$, then the map

$$\mathcal{R} \ni x \mapsto P_x \in L(\mathcal{A})$$

is analytic.

The existence of a reductive structure for a given homogeneous space is not a trivial fact. One example is the case of similarity orbits of operators in a separable Hilbert space, when the orbit admits a reductive structure if and only if the operator is normal and has finite spectrum [ARS]. In another example, when $\mathcal{R}$ is the space of representations of a fixed C*-algebra on an also fixed Hilbert space, the existence of a reductive structure implies that the C*-algebra must be nuclear [ACS].

Note that Υ_{x_0} is the kernel of $\mathcal{T}_{x_0} : \mathcal{A} \to T(\mathcal{R})_{x_0}$, and therefore $\mathcal{T}_{x_0}$ restricted to $\mathcal{H}_{x_0}$ is a bounded linear isomorphism between $\mathcal{H}_{x_0}$ and $T(\mathcal{R})_{x_0}$. Let us denote by K_{x_0} the inverse of this isomorphism. Then $P_{x_0} = K_{x_0} \circ \mathcal{T}_{x_0}$.

A reductive structure induces a linear connection in the tangent bundle of $\mathcal{R}$. We shall not go into the details (see [KN] and [MR]). Let us just describe the geodesics of this connection. They are of the form

$$\mathbb{R} \ni t \mapsto e^{tz} \bullet x_0$$

for $x_0 \in \mathcal{R}$ and $z \in \mathcal{H}_{x_0}$.

(1.2) The space of representations of a C*-algebra

Let $\mathcal{A}$ be a C*-algebra, $\mathcal{M}$ a von Neumann algebra and $L(H)$ the algebra of all bounded linear operators on a separable Hilbert space H. Denote by $R = R(\mathcal{A})$ the set of bounded non-degenerate homomorphisms (shortly, representations) of $\mathcal{A}$ on $L(H)$, $R^w = R^w(\mathcal{M}) \subset R(\mathcal{M})$ the ultraweakly continuous representations of $\mathcal{M}$. Denote by $R_0 = R_0(\mathcal{A})$ and $R_0^w = R_0^w(\mathcal{M})$ the respective subsets of *-representations. We regard R and R_0 (resp. R^w and R_0^w) as subsets of the Banach space $L(\mathcal{A}, L(H))$ of bounded linear operators from $\mathcal{A}$ to $L(H)$ with its usual norm (resp. $L(\mathcal{M}, L(H))$). In other words, using notations of (1.1), $\mathcal{A} = L(H)$, $\mathcal{E} = L(\mathcal{A}, L(H))$ (or $L(\mathcal{M}, L(H))$) and $\mathcal{R} = R$ or R_0 (R^w or R_0^w). The action is $G \bullet \pi = Ad(G) \circ \pi$, that is, $G \bullet \pi(x) = G\pi(x)G^{-1}$, for $x \in \mathcal{A}$ (or $\mathcal{M}$) and G invertible or unitary in $L(H)$.

In [ACS] we proved that if $\mathcal{A}$ is nuclear, then R and R_0 are discrete unions of homogeneous spaces. The same is true for R^w and R_0^w if $\mathcal{M}$ is injective.

Moreover, R_0 is a real analytic submanifold of R, and since every bounded representation π of $\mathcal{A}$ is similar to a selfadjoint representation π_0 (for $\mathcal{A}$ is nuclear, see [H1]), the connected components of R consist of the similarity orbits of elements of R_0.

If $\pi \in R$, the isotropy group I_π is the group of invertible elements of the Banach algebra $\pi(\mathcal{A})'$, and therefore $\Upsilon_\pi = \pi(\mathcal{A})'$. Also $\mathcal{T}_\pi$ can be explicitly computed,

$$\mathcal{T}_\pi(X) = \delta_\pi(X) = X\pi - \pi X \in L(\mathcal{A}, L(H)).$$

Fix $\pi_0 \in R_0 \subset R$. Since $\pi_0(\mathcal{A})'$ is an injective von Neumann algebra, let us also fix a conditional expectation

$$E = E_{\pi_0} : L(H) \to \pi_0(\mathcal{A})'.$$

(E_{π_0} may not be unique). Using this conditional expectation we can introduce a reductive structure on R (and R_0). Let $E_\pi = Ad(G) \circ E_{\pi_0} \circ Ad(G^{-1})$, if $\pi = G\pi_0 G^{-1}$ (it is easy to see that E_π does not depend on the particular choice of G). E_π may not be a conditional expectation, unless for instance G is unitary; anyway it is a $\pi(\mathcal{A})'$-bimodule idempotent map from $L(H)$ onto $\pi(\mathcal{A})'$. Therefore if we put

$$\mathcal{H}_\pi = \ker E_\pi$$

it is straightforward to verify that it defines a reductive structure for R. When G varies in the unitary group of H, we obtain a reductive structure for R_0. Also, in an analogous way, we can introduce reductive structures to R^w and R_0^w.

In [ACS] it was proven that if the space R of an arbitrary C*-algebra admits a reductive structure, then the algebra must be nuclear (or injective in the von Neumann case).

All the results described in this paragraph remain valid if one replaces $L(H)$ by an injective von Neumann algebra $\mathcal{N}$ with separable predual. This fact will be used through the rest of this paper.

(1.3) Geodesics of R

By simply applying the general theory to the example, it turns out that geodesics in R are of the form

$$\mathbb{R} \ni t \mapsto e^{tX} \, \pi \, e^{-tX}$$

for $\pi \in R$ and $E_\pi(X) = 0$. ¿From the definition of $\{\mathcal{H}_\pi\}_{\pi \in R}$ it is apparent that the reductive structure depends on the choice of E_{π_0}. Therefore one may like to know how the invariants of the induced connection vary if one changes E_{π_0}.

It turns out that different conditional expectation produce different geodesics.

(1.3.1) PROPOSITION *With the above notations, let $Z \in L(H)$. The curve $\alpha(t) = e^{tZ} \pi e^{-tZ}$ is a geodesic if and only if Z commutes with $E_\pi(Z)$.*

Proof. It follows from the general theory of homogeneous reductive spaces (see [MR]) that a curve δ is a geodesic if and only if

$$K_{\delta(t)}(\dot{\delta}(t))\dot{} + [K_{\delta(t)}(\dot{\delta}(t)), K_{\delta(0)}(\dot{\delta}(0))] = 0$$

where $\cdot$ =derivative and the bracket $[A, B] = AB - BA$ for $A, B \in L(H)$. Checking this condition for the curve $\alpha(t) = e^{tZ} \pi e^{-tZ}$, and using that $K_{\alpha(t)} = Ad(e^{tZ}) \circ K_\pi \circ Ad(e^{-tZ})$, one gets the equation:

$$
\begin{aligned}
0 &= \{Ad(e^{tZ} \circ K_\pi([Z,\pi]))\dot{}\} + \left[Ad(e^{tZ} \circ K_\pi([Z,\pi])), K_\pi([Z,\pi])\right] \\
&= e^{tZ}\{[K_\pi([Z,\pi]), Ad(e^{-tZ}) \circ K_\pi([Z,\pi])] + \\
&\quad + [K_\pi([Z,\pi]), Ad(e^{-tZ}) \circ K_\pi([Z,\pi])]\}e^{-tZ}.
\end{aligned}
$$

Noting that $K_\pi([Z,\pi]) = K_\pi \circ \delta_\pi(Z) = (I - E_\pi)(Z)$, one obtains

$$0 = [Z - E_\pi(Z), e^{-tZ} E_\pi(Z) e^{tZ}].$$

If Z commutes with $E_\pi(Z)$, then the equation above reduces to

$$[Z - E_\pi(Z), E_\pi(Z)] = 0,$$

which holds. Conversely, putting $t = 0$ in the equation, it follows that Z and $E_\pi(Z)$ commute.

Let us exhibit an elementary example of two conditional expectations onto the same algebra, with different geodesics. Let H be $\mathbb{C}^4$ and $\mathcal{A} = \mathbb{C}^{2 \times 2}$, and embed $\mathbb{C}^{2 \times 2}$ into $\mathbb{C}^{4 \times 4}$ in the following manner:

$$\pi(A) = \begin{pmatrix} A & 0 \\ 0 & A \end{pmatrix}.$$

Let us consider the conditional expectations E_1 and E_2 from $\mathbb{C}^{4\times 4}$ onto $\pi(\mathbb{C}^{2\times 2})$ given by

$$E_1\left(\begin{pmatrix} A & B \\ C & D \end{pmatrix}\right) = \begin{pmatrix} A & 0 \\ 0 & A \end{pmatrix} \quad \text{and} \quad E_2\left(\begin{pmatrix} A & B \\ C & D \end{pmatrix}\right) = \begin{pmatrix} \frac{A+D}{2} & 0 \\ 0 & \frac{A+D}{2} \end{pmatrix}$$

where A, B, C and $D \in \mathbb{C}^{2\times 2}$. Consider $X, Y \in \mathbb{C}^{2\times 2}$ such that $[X, Y] \neq 0$. It is clear that if

$$Z = \begin{pmatrix} 0 & X \\ 0 & Y \end{pmatrix}$$

then

- $Z \in \ker E_1$;
- $E_2(Z)$ and Z do not commute.

(1.4) Horizontal cross sections

In [ACS] we introduced the continuous (and afterwards analytic) local cross sections for τ_π:

$$R \ni G\pi G^{-1} \mapsto s_\pi(G\pi G^{-1}) = GE_\pi(G)^{-1} \in \mathcal{G}l(H).$$

Fix π_0 a selfadjoint representation of $\mathcal{A}$. π_0 extends to a selfadjoint normal representation of $\mathcal{A}^{**}$ (also denoted π_0). Let p be a projection in the centre of $\mathcal{A}^{**}$ such that $\ker \pi_0 = p\mathcal{A}^{**}$. Since $\mathcal{A}$ is nuclear and H is separable, $(1-p)\mathcal{A}^{**}$ is hyperfinite. Let $\mathcal{U}_0$ be an amenable WOT-dense subgroup of the unitary group of $(1-p)\mathcal{A}^{**}$ and let m be an invariant mean on $\mathcal{U}_0$. A conditional expectation E_{π_0} is obtained via m by means of

$$E_{\pi_0}(X) = \int_{\mathcal{U}_0} \pi_0(u) X \pi_0(u^*) dm.$$

Using this conditional expectation, one obtains the following expression for the local cross sections. Let $G, T \in \mathcal{G}l(H)$ such that $\pi = T\pi_0 T^{-1}$ and $\pi_1 = G\pi G^{-1}$. Note that all these representations have the same kernel as π_0. Then

$$\begin{aligned}
s_\pi(\pi_1) &= G\{Ad(T) \circ E_{\pi_0} \circ Ad(T^{-1})\,(G)\}^{-1} \\
&= G\{T[\int_{\mathcal{U}_0} \pi_0(u)T^{-1}GT\pi_0(u^*)dm]T^{-1}\}^{-1} \\
&= \{\int_{\mathcal{U}_0} \pi(u)\pi_1(u^*)dm\}^{-1}.
\end{aligned}$$

Observe that this expression of the local cross section has the same appearence as the local cross section for the spaces of systems of projections ([CPR2]) and spectral measures ([ARS]).

Another property of this cross section is that it is "horizontal". That is, for each tangent vector $V \in T(R)_\pi$,

$$d(s_\pi)_\pi(V) \in \mathcal{H}_\pi = \ker E_\pi.$$

Indeed, let $\nu(t)$ be a smooth curve in R such that $\nu(0) = \pi$ and $\dot{\nu}(0) = V$. Then there is a smooth lifting of ν, i.e., a curve $G(t) \in \mathcal{G}l(H)$ such that $\tau_\pi(G(t)) = \nu(t)$. Then

$$d(s_\pi)_\pi(V) = \frac{d}{dt} s_\pi(\nu(t))\big|_{t=0} = \frac{d}{dt}\{G(t)[E_\pi(G(t))]^{-1}\}\big|_{t=0}$$
$$= \dot{G}(0) - E_\pi(\dot{G}(0)) \in \ker E_\pi.$$

Therefore, $d(s_\pi)_\pi$ coincides with the map K_π and a differentiable description of E_π can be obtained:

$$E_\pi = Id - K_\pi \circ \mathcal{T}_\pi = Id - d(s_\pi)_\pi \circ d(\tau_\pi)_1 = Id - d(s_\pi \circ \tau_\pi)_1.$$

(1.5) The similarity problem

In [H1] (Theorem 1.10), Haagerup proved that if π is a bounded representation of a C*-algebra $\mathcal{A}$ on a Hilbert space H, then the following conditions are equivalent:

(i) π is completely bounded;

(ii) π is similar to a selfadjoint representation.

Moreover, if these conditions are fulfilled, then there exists a positive operator $T \in \mathcal{G}l(H)$ such that $T \pi T^{-1}$ is a selfadjoint representation and

$$\|T\|\|T^{-1}\| = \sup_{n \in \mathbb{N}} \|\pi^{(n)}\|$$

where $\pi^{(n)}$ denotes the representation of $\mathcal{A} \otimes \mathbb{C}^{n \times n}$ on H^n induced by π.

In particular, if $\mathcal{A}$ is nuclear, Haagerup [H1] observed that every representation of $\mathcal{A}$ is similar to a selfadjoint representation (see also [Ch3]).

Let $E : \mathcal{B} \to \mathcal{C}$ be a conditional expectation of arbitrary C*-algebras. Porta and Recht [PR] proved that for every invertible element $b \in \mathcal{B}$ there exist unique elements u, x and c such that u is unitary in $\mathcal{B}$, x is selfadjoint and $E(x) = 0$, c is positive and invertible in $\mathcal{C}$ and

$$b = u \, e^x \, c.$$

Using this decomposition in the case $E_\rho : L(H) \to \rho(\mathcal{A})'$ for a suitable ρ in R_0 one obtains the following result: for every $\pi_0 \in R$ there exists a unique $\rho \in R_0$ and a unique positive operator e^X with $X^* = X \in \ker E_\rho$ such that $\pi_0 = e^X \rho \, e^{-X}$.

It is an open question whether this latter positive intertwiner is one of the operators T of Haagerup, or in other words, whether $\|e^X\|\|e^{-X}\|$ coincides with the completely bounded norm of π_0.

2. *Finite dimensional C*-algebras*

(2.1) Let $\mathcal{A}$ be a finite dimensional C*-algebra. Since every finite dimensional C*-algebra is nuclear, the space $R(\mathcal{A}, \mathcal{N})$ is a homogeneous reductive space under the action of $G_{\mathcal{N}}$, provided that $\mathcal{N}$ is a injective von Neuman algebra with separable predual. But it is easy to check in the proof of 1.1 that it is also the case for any C*-algebra $\mathcal{N}$.

The algebra $\mathcal{A}$ can be seen as $\mathcal{A} = M_{n_1}(\mathbb{C}) \oplus \cdots \oplus M_{n_k}(\mathbb{C})$. Let q_{n_i} be the nilpotent Jordan block in $M_{n_i}(\mathbb{C})$ and $a = (1+q_{n_1}) \oplus (2+q_{n_2}) \oplus \cdots \oplus (k+q_{n_k}) \in \mathcal{A}$. It is easy to see that $\mathcal{A} = C^*(a)$. Let $\pi_0 \in R(\mathcal{A}, \mathcal{N})$ be the inclusion $\pi_0 : \mathcal{A} \hookrightarrow \mathcal{N}$. Then for every $g \in G_{\mathcal{N}}$, the representation $g\pi_0 g^{-1}$ is determined by the elements gag^{-1} and ga^*g^{-1}.

Let $S(a, a^*) = \{(gag^{-1}, ga^*g^{-1}) : g \in G_{\mathcal{N}}\} \subset \mathcal{N} \times \mathcal{N}$ be the joint similarity orbit of the pair (a, a^*). Using the previous comments we deduce that there exists a one to one map $\Phi : S_{\mathcal{N}}(\pi_0) \to S(a, a^*)$ which is onto and such that the following diagram is commutative:

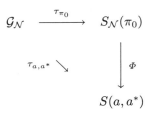

In [AFHPS] and [AS2] the orbit $S(a, a^*)$ is described as a homogeneous reductive space and submanifold of $\mathcal{N} \times \mathcal{N}$, in the case when $C^*(a)$ is finite dimensional. By the existence of continuous local cross sections for both τ_{π_0} and τ_{a,a^*} (1.1 and [AS2]), Φ is a homeomorphism. Moreover, the differential and reductive structures coincide with those of the natural model $G_{\mathcal{N}}/G_{\mathcal{A}' \cap \mathcal{N}}$.

One obtains the same geometrical structure by considering the orbit of $\pi_0|_{\mathcal{U}_{\mathcal{A}}}$ as a representation of the compact group $\mathcal{U}_{\mathcal{A}}$ (unitary group of $\mathcal{A}$) in $\mathcal{N}$, following [M], [MR] and [MS].

(2.2) <u>Weaker topologies on $R(\mathcal{A}, \mathcal{N})$</u>:

Given $a \in \mathcal{N}$, consider the joint similarity orbit $S(a, a^*) \subset \mathcal{N} \times \mathcal{N}$, $\mathcal{A} = C^*(a)$ and $\pi_0 \in R(\mathcal{A}, \mathcal{N})$ the inclusion. On the similarity orbit $S_{\mathcal{N}}(\pi_0)$ we consider the following three topologies:

1) $\pi_i \to \pi$ as elements of $S(a, a^*)$ if $\|\pi_i(a) - \pi(a)\| \to 0$ and $\|\pi_i(a^*) - \pi(a^*)\| \to 0$.
2) $\pi_i \to \pi$ pointwise in norm if $\|\pi_i(b) - \pi(b)\| \to 0$ for all $b \in \mathcal{A}$.
3) $\pi_i \to \pi$ uniformly if $\|\pi_i - \pi\| \to 0$ (i.e. as elements of $L(\mathcal{A}, \mathcal{N})$).

Remark 4.6 of [ACS] states that if $\mathcal{A}$ is nuclear and $\mathcal{N}$ is injective with separable predual, then τ_{π_0} has local cross sections which are continuous in the third topology. If $\mathcal{A}$ is finite dimensional the three topologies coincide. Moreover, in [F] and [AS1] it was proven that if $\mathcal{N}$ is $L(H)$ then the existence of local cross sections continuous in the first topology implies that $\mathcal{A}$ is finite dimensional. We conjecture that if $\mathcal{N}$ is a factor, then there exist similarity cross sections for π_0 continuous in the topology (2) iff $\mathcal{A}$ is finite dimensional.

(2.3) <u>A canonical conditional expectation</u>

In what follows $\mathcal{A}$ will be a finite dimensional C*-algebra. We shall construct a conditional expectation from $\mathcal{N}$ onto $\mathcal{A}' \cap \mathcal{N}$. It is characterized as the only conditional expectation invariant under inner automorphisms of $\mathcal{N}$ determined by unitaries of $\mathcal{A}$:

$$E : \mathcal{N} \to \mathcal{A}' \cap \mathcal{N} \quad \text{such that} \quad E(uxu^*) = E(x),$$

for all u unitary in $\mathcal{A}$ and x in $\mathcal{N}$.

We may suppose that $\mathcal{A} = M_{n_1}(\mathbb{C}) \oplus \cdots \oplus M_{n_k}(\mathbb{C})$.

<u>Case $k = 1$</u>

$\mathcal{A}$ is a finite type I factor, then $\mathcal{N}$ is isomorphic to $\mathcal{A} \otimes (\mathcal{A}' \cap \mathcal{N})$. Define $E = tr \otimes Id_{\mathcal{A}' \cap \mathcal{N}}$ where tr is the normalized trace of $\mathcal{A}$. It is apparent that this is the unique invariant conditional expectation onto $\mathcal{A}' \cap \mathcal{N}$ because for any other such expectation E', its restriction to $\mathcal{A}$ is a tracial state, and then $E' = E$.

<u>Case $k > 1$</u>

If E is an expectation from $\mathcal{N}$ onto $\mathcal{A}' \cap \mathcal{N}$ and $p_1, ..., p_k$ are the minimal central projections of $\mathcal{A}$, then for $x \in \mathcal{N}$

$$E(x) = \sum_{i=1}^{k} p_i E(x) p_i = E(\sum_{i=1}^{k} p_i x p_i) = E(E_0(x))$$

where $E_0 : \mathcal{N} \to \mathcal{A} \cap \mathcal{A}'$ is the unique conditional expectation onto the center of $\mathcal{A}$. Then E can be factorized as

$$E = (\oplus_{i=1}^{k} E_i) \circ E_0$$

where $E_i : p_i \mathcal{N} p_i \to p_i(\mathcal{A}' \cap \mathcal{N})$ are conditional expectations. E is invariant if and only if every E_i is. Then to obtain an invariant conditional expectation E, one must put E_i as in the previous case:

$$E = \{\oplus_{i=1}^{k} (tr_{p_i \mathcal{A}} \otimes Id)\} \circ E_0.$$

(2.4) <u>Local cross sections</u>

Let $p_1, ..., p_k$ as in (2.3) and a a generator of $\mathcal{A}$ as in (2.1)). Then there exist polynomials $E_{j,m}^i(X, Y)$ in two non commuting variables X and Y, such that for each $1 \leq i \leq k$, $E^i_{j,m} = E^i_{j,m}(a, a^*)$, $1 \leq j, m \leq n_i$ form a system of matrix units for the factor $p_i \mathcal{A}$.

It is clear that an element $x \in \mathcal{N}$ is in the kernel of E if and only if

$$\sum_{j=1}^{n_i} E^i_{1,j} x E^i_{j,1} = 0 \quad \text{for } i = 1, ..., k.$$

In section 1 we defined the local continuous cross sections compatible with the reductive structure

$$G \pi_0 G^{-1} \to G E(G)^{-1}.$$

Using the polynomials $E^i_{j,m}(X, Y)$, one obtains these cross sections for the joint similarity orbit of the pair (a, a^*) in the following manner.

Let $(b, c) = (GaG^{-1}, Ga^*G^{-1})$ close enough to (a, a^*), then

$$s_{(a,a^*)}(b, c) = G\{\sum_{i=1}^{k} \frac{1}{n_i} \sum_{k,m=1}^{n_i} E^i_{k,m} \, G \, E^i_{m,k}\}^{-1}$$

$$= \{\sum_{i=1}^{k} \frac{1}{n_i} \sum_{k,m=1}^{n_i} E^i_{k,m} \, G \, E^i_{m,k} \, G^{-1}\}^{-1}$$

$$= \{\sum_{i=1}^{k} \frac{1}{n_i} \sum_{k,m=1}^{n_i} E^i_{k,m} E^i_{m,k}(b, c)\}^{-1}.$$

Note that $G E^i_{m,k} G^{-1} = G E^i_{m,k}(a, a^*) G^{-1} = E^i_{m,k}(GaG^{-1}, Ga^*G^{-1})$.

3. *Commutative von Neumann algebras*

(3.1) <u>Spectral measures</u>

Commutative C*-algebras are nuclear, and therefore the results of section 1 apply to these algebras. Let $\mathcal{M}$ be a commutative von Neumann algebra acting on a separable Hilbert space. It is well known such an algebra is *-isomorphic to a maximal abelian algebra $L^\infty_\mu(\Omega)$ acting on $L^2_\mu(\Omega)$, where Ω is a second countable Haussdorff space and μ is a probability measure on Ω. Let $\Phi : \mathcal{M} \to L^\infty_\mu(\Omega)$ be such an isomorphism.

For each Borel set $\Delta \subset \Omega$ let us put $\sigma(\Delta) = \Phi^{-1}(\chi_\Delta)$. Then σ defines what in [ARS] is called a spectral measure in Ω with values in $\mathcal{M}$. That is, σ is a map defined in the Borel sets of Ω which verifies

1) $\sigma(\Delta)$ is a selfadjoint projection in $\mathcal{M}$.

2) $\sigma(\Delta \cup \Delta') = \sigma(\Delta) + \sigma(\Delta')$ if $\Delta \cap \Delta' = \emptyset$.

3) $\sigma(\Delta \cap \Delta') = \sigma(\Delta)\,\sigma(\Delta')$.

4) $\sigma(\emptyset) = 0$ and $\sigma(\Omega) = 1$.

Let $M(\Omega)$ be the space of spectral measures. Consider the space $X(\Omega)$ of $\mathcal{M}$-valued measures on Ω, i.e. functions η defined on the Borel sets of Ω which satisfy condition 2) and such that $\sup\{\|\eta(\Delta)\| : \Delta$ Borel subset of $\Omega\}$ is finite. Clearly, $M(\Omega) \subset X(\Omega)$. Also it is easy to verify that $X(\Omega)$ is a Banach space with the norm $\|\eta\| = \sup \|\eta(\Delta)\|$.

In other words, the inclusion $\pi_0 : \mathcal{M} \hookrightarrow L(H)$ gives rise to a spectral measure σ such that the projections in the image of σ generate $\mathcal{M}$. Conversely, if $\mathcal{B}(\Omega)$ is the algebra of Borel subsets of Ω, and $\sigma : \mathcal{B}(\Omega) \to L(H)$ is a spectral measure, one obtains a commutative von Neumann algebra by considering the algebra generated by $\{\sigma(\Delta) : \Delta \in \mathcal{B}(\Omega)\}$. We can construct the following (commutative) diagram:

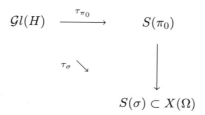

Here τ_σ is the map induced by the action of $\mathcal{G}l(H)$ on the set $X(\Omega)$, which is $G \bullet \sigma = G\sigma G^{-1}$. In [ARS] it was proven that with this action the similarity orbits of measures are analytic homogeneous spaces. The vertical arrow of the diagram is an analytic diffeomorphism, which maps $G\pi_0 G^{-1}$ to $G\sigma G^{-1}$.

An analogous diagram describes the action of the unitary group. The unitary orbits of spectral of spectral measures, which correspond to selfadjoint representations , are the connected components of $M(\Omega)$ in $X(\Omega)$. In particular they are submanifolds of the Banach space $X(\Omega)$. Many of the results in [ARS] (reductive structure, properties of the unitary orbit relative to the similarity orbit) can be obtained from the properties of the more general example of section 1.

(3.2) <u>Positive intertwiners</u>

In section 1 we considered different ways to construct a positive invertible operator in $L(H)$ (or $\mathcal{N}$) intertwining a given nonselfadjoint representation of a C*-algebra with a selfadoint one. The first way using Haagerup's technique. Given π a completely bounded representation of $\mathcal{M}$, one obtains an invertible operator $T > 0$ such that $T\pi T^{-1}$ is selfadjoint and $\|T\|\|T^{-1}\| = \sup_n \|\pi^{(n)}\|$, the completely bounded norm of π (see [H1]).

Another way to construct a positive operator is by considering Porta and Recht's factorization of positive invertible elements by means of a given conditional expectation. This procedure was described in section 1, let us denote by e^X the positive operator obtained in this way (which depends strongly on the conditional expectation).

There is a third and more explicit way to obtain a positive operator in the case of a commutative von Neumann algebra. Let γ be a bounded spectral measure, and let $\mathcal{P}$ be a finite partition of Ω. Consider the net of positive invertible elements

$$A_{\gamma,\mathcal{P}} = \sum_{\Delta \in \mathcal{P}} \gamma(\Delta)^* \gamma(\Delta)$$

indexed by the set of all finite partitions of Ω, order = refinement. In [ARS] it was proven that this net is bounded and each limit of a WOT-convergent subnet is a positive invertible operator. Denote by A_γ one of these limits. Then it is apparent that $A_\gamma \gamma = \gamma^* A_\gamma$. Therefore $A_\gamma^{1/2} \gamma A_\gamma^{-1/2}$ is a spectral measure.

In general, e^X and A_γ do not coincide. Take $\Omega = \{0,1\}$ and γ given by

$$\gamma(0) = \begin{pmatrix} 1 & 1 \\ 0 & 0 \end{pmatrix} \text{ and } \gamma(1) = \begin{pmatrix} 0 & -1 \\ 0 & 1 \end{pmatrix}.$$

Then

$$A_\gamma^2 = \begin{pmatrix} 1 & 1 \\ 1 & 2 \end{pmatrix}.$$

Let E be the conditional expectation in $\mathbb{C}^{2\times 2}$ onto the diagonal matrices and let $X = X^* \in \ker E$. Then

$$X = \begin{pmatrix} 0 & z \\ \bar{z} & 0 \end{pmatrix} \quad \text{and} \quad e^X = \begin{pmatrix} \cosh(|z|) & \frac{z}{|z|}\sinh(|z|) \\ \frac{\bar{z}}{|z|}\sinh(|z|) & \cosh(|z|) \end{pmatrix}.$$

It is clear that $A_\gamma \neq e^X$.

(3.3) Geodesics in $S(\sigma) \subset X(\Omega)$: an example

Let us consider the case of the spectral measure σ of the operator M_t, multiplication by t, in $L^2(0,2\pi)$. Clearly $\sigma(\Delta) = M_{\chi_\Delta}$, i.e. multiplication by the characteristic function of Δ. In [S] Suarez constructed an example of two different conditional expectations E_1, E_2 which are limits of von Neumann's operation in $L^2(0,2\pi)$. That is, conditional expectations

$$E_i : L(L^2(0,2\pi)) \to L^\infty(0,2\pi) , \quad i=1,2$$

which are cluster points of the net $E_\mathcal{P}$, $\mathcal{P}$ finite partitions of Ω, for the topology of WOT-convergence at each $X \in L(L^2(0,2\pi))$, where $E_\mathcal{P}$ is defined as follows:

$$E_\mathcal{P}(X) = \sum_{\Delta \in \mathcal{P}} M_{\chi_\Delta} X M_{\chi_\Delta}.$$

Moreover, Suarez also exhibited a projection $P \in L(L^2(0, 2\pi))$, with range equal to the span of a certain infinite subset of the Fourier basis of $L^2(0, 2\pi)$ such that

$$E_1(P) = I \quad \text{and} \quad E_2(P) = 0.$$

Using this example, consider the operator $X = M_t P$. Clearly, $E_2(X) = 0$, and therefore $\delta(t) = e^{tX}\sigma e^{-tX}$ is a geodesic for the reductive structure of $S(\sigma)$ given by E_2. But δ is not a geodesic of the reductive structure given by E_1. Indeed, using (1.3.1.) if δ is a geodesic for E_1 then X commutes with $E_1(X) = E_1(M_t P) = M_t$, and therefore commutes with all $L^\infty(0, 2\pi)$. Therefore $X \in L^\infty(0, 2\pi)$, which leads to a contradiction, since it would imply that $E_2(X) = X \neq 0$.

In other words, using Suarez' example we have exhibited two different connections for the tangent bundle of $S(\sigma)$ (or equivalently, of the similarity orbit of the natural representation of $L^\infty(0, 2\pi)$ in $L^2(0, 2\pi)$) which have different geodesic curves.

4. *Groups G such that $C^*(G)$ is amenable*

The same results which are valid for the space of unital representations of nuclear C*-algebras with unit hold for the space of non degenerate representations of nuclear C*-algebras without unit. Indeed, by [H1, Lemma 1.2], a non degenerate representation of a C*-algebra extends uniquely and with the same norm to a unital representation of the unitization. On the other hand, the unitization of a nuclear C*-algebra is nuclear.

Let us also denote by R (resp. R_0) the space of bounded (resp. selfadjoint) non degenerate representations of $\mathcal{A}$. In what follows, every algebra representation considered is non degenerate.

(4.1) Let G be a compact group and $\mathcal{B}$ a unital C*-algebra. Let $S_\infty(G, \mathcal{B})$ denote the set of all norm continuous homomorphisms from G to $\mathcal{B}$ which preserve the identity. The differential structure of the space $S_\infty(G, \mathcal{B})$ has been studied in [M], [MR] and [MS]. In these papers it is shown that $S_\infty(G, \mathcal{B})$ has a homogeneous reductive structure and several of its geometrical invariants are determined. However, the methods developed there do not seem to allow the study of spaces of strongly continuous representations of non compact groups. In what follows we shall describe how we can study the differential geometry of the space $S(G, \mathcal{N})$ of strongly continuous homomorphisms of an amenable group G into an injective von Neumann algebra $\mathcal{N}$. Recall that a locally compact group G is amenable if there exists an invariant mean m on G, i.e. an $m \in L^\infty(G)^*$ such that $m(1) = 1 = \|m\|$ and $m(\Phi_g) = m(\Phi)$ for all $g \in G$ and $\Phi \in L^\infty(G)$, where $\Phi_g(x) = \Phi(gx)$ $(x \in G)$. We refer the reader to [Pa] for the results on amenability (of groups and algebras) needed here. Abelian locally compact groups and solvable locally compact groups are amenable, just to mention two important classes. The notion of amenability for groups has been first considered by von Neumann in his study of the Banach-Tarski paradox. Forty years later B.E. Johnson [Jo] introduced the notion of amenable Banach algebra and proved that the group G is amenable if and only if the Banach algebra $L^1(G)$ is amenable (recall that a Banach algebra B is amenable if every norm continuous derivation $D : B \to X$ is inner, i.e. there exists $\xi \in X$ such that $D(f) = f\xi - \xi f$ $\forall f \in B$, where X is a dual Banach B-module). With these definitions it can be proved that $C^*(G)$ is amenable if G is amenable [J], [B] but the converse is not true in general. The class of all groups G such that $C^*(G)$ is amenable is not well understood yet. However it is known that it contains, besides the amenable groups, the type I groups (i.e. groups G such that $C^*(G)$ is a type I C*-algebra) [J] and the almost connected groups (i.e. groups G such that G/G_e is compact, where G_e denotes the connected component of the neutral element e) [LP].

Amenable groups have the following property, which is crucial in order to apply our results to the differential geometry of $S(G, \mathcal{N})$: every bounded strongly continuous representation of G in $L(H)$ is similar to a unitary representation (see [Gr], Theorem 3.4.1). All the results from now on are valid for groups G with the property above and such that $C^*(G)$ is nuclear.

(4.2) <u>Representations with values in an injective von Neumann algebra</u>

Let $\mathcal{N}$ be an injective von Neumann algebra and let $E_\mathcal{N} : L(H) \to \mathcal{N}$ be a conditional expectation onto $\mathcal{N}$.

If $\Pi : G \to L(H)$ is a unitary representation, and $\pi : C^*(G) \to L(H)$ is the non degenerate selfadjoint representation associated to it, then the range of Π is contained in the von Neumann algebra $\mathcal{N} \subset L(H)$ if and only if the range of π is. Indeed, first suppose that $\Pi(G) \subset \mathcal{N}$. Then clearly

$$\pi(f) = \int_G \Pi(g)f(g)d\ell g \in \mathcal{N}$$

for all $f \in L^1(G)$. Therefore $\pi(L^1(G)) \subset \mathcal{N}$, and then the image under π of the completion of $L^1(G)$ (in any C*-algebra norm) lies in $\mathcal{N}$. Conversely, suppose that $\pi(C^*(G)) \subset \mathcal{N}$. In particular $\pi(L^1(G)) \subset \mathcal{N}$. Let $\phi_\alpha \in L^1(G)$ be an approximation of the identity; then for $\psi, \eta \in H$

$$< \Pi(g)\psi, \eta > = \lim_\alpha < \pi(\phi_\alpha^{g^{-1}})\psi, \eta >$$

where $f^g(h) = f(gh)$ for $f \in L^1(G)$ and $g, h \in G$. Then $\Pi(g) \in \mathcal{N}$ for all $g \in G$.

Let $\Pi : G \to \mathcal{N}$ be a not necessarily unitary representation and $T \in \mathcal{G}l(H)$ positive, such that $T\Pi T^{-1}$ is a unitary representation. Then, for $g \in G$,

$$T^2\Pi(g^{-1}) = \Pi(g)^* T^2.$$

Let $S = E_\mathcal{N}(T^2)^{1/2} \in \mathcal{N}$. Then , for $g \in G$, $S^2\Pi(g^{-1}) = \Pi(g)^* S^2$ and $\Pi_0 = S\Pi S^{-1}$ is a unitary representation of G in $\mathcal{N}$. In other words, if the range of Π lies in $\mathcal{N}$ and Π is similar to a unitary representation, then Π is similar to a unitary representation with range in $\mathcal{N}$ and the positive intertwiner operator can also be chosen in $\mathcal{N}$. The analogous statement is true for algebra representations.

In the case of an amenable group G, one can use the reduced C*-algebra $C_r^*(G)$ instead of $C^*(G)$. Then it is clear that bounded representations of G into $\mathcal{N}$ are in one to one correspondence with bounded representations of $C_r^*(G)$ into $\mathcal{N}$, and with ultraweak continuous representations of the von Neumann algebra of G into $\mathcal{N}$ (this latter bijection is a well known result [KR, Lemma 10.1.10 and Th. 10.1.12] and [Pe,Th. 3.7.7]).

(4.3) <u>Norm continuous representations of compact groups</u>

Martin [M] and Mata Lorenzo and Recht [MR] studied the geometry of the space of norm continuous representations of a compact group G on a fixed C*-algebra $\mathcal{B}$ (let us denote this space by $S_\infty(G,\mathcal{B})$). They proved that the space $S_\infty(G,\mathcal{B})$ is a homogeneous space under the action of the group $\mathcal{G}_\mathcal{B}$ of invertibles of $\mathcal{B}$, and a submanifold of $C(G,\mathcal{B})$, the Banach space of continuous maps from G to $\mathcal{B}$ with the supremum norm. In particular, the maps

$$\tau_\Pi : \mathcal{G}_\mathcal{B} \to S_\infty(G,\mathcal{B})$$

have continuous local cross sections, for all $\Pi \in S_\infty(G,\mathcal{B})$. Therefore the connected component $S_\mathcal{B}(\Pi)$ of Π in $S_\infty(G,\mathcal{B})$ is homeomorphic to $\mathcal{G}_\mathcal{B}/\Pi(G)' \cap \mathcal{B}$.

Let π be the non degenerate representation of $C^*(G)$ associated to Π. Taking $\mathcal{B} = \mathcal{N}$ an injective von Neumann algebra, our results show that the connected component $S_\mathcal{N}(\pi)$ of π in $R = R(C^*(G),\mathcal{N})$ is also homeomorphic to the quotient $\mathcal{G}_\mathcal{N}/\Pi(G)' \cap \mathcal{N}$ (note that $\Pi(G)' \cap \mathcal{N} = \pi(C^*(G))' \cap \mathcal{N}$).

Therefore $S_\mathcal{N}(\Pi)$ and $S_\mathcal{N}(\pi)$ are homeomoprphic. Moreover since the diagram

$$
\begin{array}{ccccc}
\mathcal{G}_\mathcal{N} & \xrightarrow{\tau_\pi} & S_\mathcal{N}(\pi) & \subset & R(C^*(G),\mathcal{N}) \\
& \tau_\Pi \searrow & \downarrow \varrho & & \\
& & S_\mathcal{N}(\Pi) & \subset & S_\infty(G,\mathcal{N})
\end{array}
$$

is commutative, the vertical arrow ϱ, which denotes the bijection between representations of G and representations of $C^*(G)$, is a homeomorphism.

In other words, uniform convergence in the space of representations of a compact group corresponds, via the natural bijection, to the operator norm topology of $R \subset L(C^*(G),\mathcal{N})$.

It would be interesting to characterize which algebra representations induce norm continuous representations of G.

(4.4) <u>Representations of amenable groups</u>

One can construct the same commutative diagram as in the previous paragraph for a general locally compact group such that $C^*(G)$ is amenable, replacing

$S_\infty(G, \mathcal{N})$ by the similarity orbit of a given unitary representation of G. When G is amenable, this orbit coincides with the connected component of the representation in the space $S(G, \mathcal{N})$, for we have already noted that every bounded representation of G is similar to a unitary representation (see [B] or [CH3]). Let Π_0 be a unitary representation of G and π_0 the associated selfadjoint representation of $C^*(G)$. If ϱ denotes the bijection of the previous paragraph extended to the similarity orbit in the natural way, we have the following commutative diagram:

$$
\begin{array}{ccc}
\mathcal{G}_\mathcal{N} & \xrightarrow{\ \tau_{\pi_0}\ } & S_\mathcal{N}(\pi_0) \quad \subset \quad R(C^*(G), \mathcal{N}) \\[2em]
{\scriptstyle \tau_{\Pi_0}} \searrow & & \Big\downarrow {\scriptstyle \varrho} \\[2em]
& & S_\mathcal{N}(\Pi_0) \quad \subset \quad S(G, \mathcal{N})
\end{array}
$$

where $S_\mathcal{N}(\Pi_0)$ denotes the similarity orbit of Π_0 under the action of $\mathcal{G}_\mathcal{N}$.

The bijection ϱ, which in the compact case is a homeomorphism, induces a topology and a homogeneous reductive structure for the orbits $S_\mathcal{N}(\Pi_0)$.

It is an interesting question to describe this topology of $S(G, \mathcal{N})$, and to answer if it coincides with the uniform convergence (or uniform convergence on compact subsets) topology of $S(G, \mathcal{N}) \subset C(G, \mathcal{N})$, as it is the case when G is compact.

References

[ACS]   E. Andruchow, G. Corach and D. Stojanoff, *A geometric characterization of nuclearity and injectivity.* Preprint.

[AFHPS] E. Andruchow, L. Fialkow, D. A. Herrero, M. P. de Herrero and D. Stojanoff, *Joint similarity orbits with local cross sections*, Integral Equations Oper. Theory **13** (1990), 1-48.

[ARS]   E. Andruchow, L. Recht and D. Stojanoff, *The space of spectral measures is a homogeneous reductive space*, Integral Equations Oper. Theory **16** (1993), 1-14.

[AS1]   E. Andruchow and D. Stojanoff, *Differentiable structure of similarity orbits*, J. Oper. Theory **21** (1989), 349-366.

[AS2]   E. Andruchow and D. Stojanoff, *Geometry of unitary orbits*, J. Oper. Theory **26** (1991), 25-41.

[A]   W. Arveson, *Subalgebras of C^*-algebras*, Acta Math. **123** (1969), 141-224.

[B]   J. W. Bunce, *The similarity problem for representations of C^*-algebras*, Proc. Amer. Math. Soc. **81** (1981), 409-414.

[BP]   J. W. Bunce and W. L. Paschke, *Quasi-expectations and amenable von Neumann algebras*, Proc. Amer. Math. Soc. **71** (1978), 232-236.

[Ch1]   E. Christensen, *Derivations and their relation to perturbations of operator algebras*, Proc. Symp. Pure. Math. **38** (1982, Part 2), 261-273.

[Ch2]   E. Christensen, *Near inclusions of C^*-algebras*, Acta Math. **144** (1980), 249-265.

[Ch3]   E. Christensen, *On non selfadjoint representations of C^*-algebras*, Amer. J. Math. **103** (1981), 817-833.

[C1]   A. Connes, *Classification of injective factors*, Ann. Math. **104** (1976), 73-115.

[C2]   A. Connes, *On the cohomology of operator algebras*, J. Funct. Anal. **28** (1978), 248-253.

[CPR1]   G. Corach, H. Porta and L. Recht, *The geometry of spaces of projections in C*-algebras*, Adv. Math. (to appear).

[CPR2]   G. Corach, H. Porta and L. Recht, *Differential geometry of systems of projections in Banach algebras*, Pacific J. Math. **140** (1990), 209-228.

[D]   J. Dixmier, *Les C*-algebres et leurs representations*, Gauthier-Villars, Paris, 1964.

[EL]   E.G. Effros and E.C. Lance, *Tensor products of operator algebras*, Advances in Math. **25** (1977), 1-34.

[G]   F.P. Greenleaf, *Invariant means on topological groups and their applications*, Van Nostrand-Reinhold, New York, 1969.

[H1]   U. Haagerup, *Solution of the similarity problem for cyclic representations of C*-algebras*, Ann. Math. **118** (1983), 215-240.

[H2]   U. Haagerup, *All nuclear C*-algebras are amenable*, Invent. Math. **74** (1983), 305-319.

[J]   B.E. Johnson, *Cohomology in Banach algebras*, Mem. Amer. Math. Soc. **127** (1972).

[K]   R. V. Kadison, *On the orthogonalization of operator representations*, Amer. J. Math. **82** (1978), 1-62.

[KR]   R.V. Kadison and J.R. Ringrose, *Fundamentals of the theory of operator algebras I, II*, Academic Press, New York, 1986.

[LP]   A.T. Lau and A.L.T. Paterson, *Amenability for twisted covariance algebras and group C*-algebras*, J. Funct. Anal. **100** (1991), 59-86.

[LR]   A. Larotonda, L. Recht, *Orbits of conditional expectations*. Preprint.

[M]   M. Martin, Projective representations of compact groups in C*-algebras, Operator Theory: Advances and Applications 43, Birkhauser-Verlag, Basel, 1990, 237-253.

[MS]   M. Martin and N. Salinas, *The canonical complex structure of flag manifolds in a C*-algebra*. Preprint (1993).

[MR]   L. Mata Lorenzo, L. Recht, *Infinite dimensional homogeneous reductive spaces*, Acta Cient. Venezol. **43** (1992), 76-90.

[Pa]   A.T. Paterson, *Amenability*, Mathematical Surveys and Monographs 29, American Mathematical Society, Providence, RI, 1988.

[Pe]   G.K. Pedersen, *C*-algebras and their automorphism groups*, Academic Press, London, 1979.

[PR]   H. Porta, L. Recht, *Conditional expectations and operator decompositions*. Preprint (1992).

[Ra]   I. Raeburn, *The relationship between a commutative Banach algebra and its maximal ideal space*, J. Funct. Anal. **25** (1977), 366-390.

[St]   S. Stratila, *Modular theory in operator algebras*, Abacus Press, Kent, 1981.

[S]   D. Suarez, *Von Neumann's operation in $L^2(0, 2\pi)$*. Preprint.

INSTITUTO ARGENTINO DE MATEMÁTICA, VIAMONTE 1636 1ER PISO, 1055 BUENOS AIRES ARGENTINA AND DEPARTAMENTO DE MATEMÁTICA, FACULTAD DE CIENCIAS EXACTAS, UNIVERSIDAD DE BUENOS AIRES, CIUDAD UNIVERSITARIA, 1428 BUENOS AIRES ARGENTINA

E-mail address: demetrio@iamba.edu.ar, esteban@iamba.edu.ar, gustavo@iamba.edu.ar

Contemporary Mathematics
Volume **189**, 1995

On the Nagy-Foias-Parrott
commutant lifting theorem.

RODRIGO AROCENA

Para Mischa, testimonio pequeño
del afecto grande
de todos los matemáticos uruguayos
y del agradecimiento al maestro
de uno de sus muchos alumnos.

ABSTRACT. A new proof and two-dimensional generalizations of Parrott's refinement of the commutant lifting are given.

Introduction

The Nagy-Foias commutant lifting theorem is an abstract generalization of Sarason's generalized interpolation theorem with a wide range of applications. An extension of that theorem due to Parrott yields interpolation results for analytic functions from the unit disc to a von Newmann algebra.

In the first section of this paper a proof of the Nagy-Foias-Parrott theorem is given by means of the method of unitary extensions of isometrics. In section II that result is extended to the minimal regular unitary dilations of bicommutative couples of contractions. In section III a two-dimensional version of the Nagy-Foias theorem is stated. In section IV it is extended to another two-dimensional version of the Nagy-Foias-Parrott theorem.

For basic definitions and results we refer to [**NF**] and [**FF**].

I. The one dimensional case

Let us fix the notation. Unless otherwise stated, every space is a Hilbert space. $\mathcal{L}(E)$ is the set of bounded operators in the space E; if G is a closed subspace of E, $P_G^E = P_G$ is the orthogonal projection onto G and $i_G^E \equiv i_G$ is the injection of G in E; if $E = G \oplus F$ then $F = E \ominus G$. $\bigvee\{...\}$ denotes the closed

---

1991 *Mathematics Subject Classification.* 47A20.

linear span of $\{...\}$. If $X \in \mathcal{L}(E)$ is a contraction, $D_X = (I - X^*X)^{\frac{1}{2}}$ and $\mathcal{D}_X$ is the closure of the range of D_X.

In this section $\{T_1, T_2, X\}$ denotes a given **commutant**, i.e., for $j = 1, 2$, $T_j \in \mathcal{L}(E_j)$ is a contraction, $U_j \in \mathcal{L}(F_j)$ is its minimal unitary dilation and $X \in \mathcal{L}(E_1, E_2)$ is such that $XT_1 = T_2X$.

The proof of the following result extends the one given in [**Pau**] of the well known fact that, when $T \in \mathcal{L}(E)$ is a contraction, the function $T(\cdot) : \mathbb{Z} \to \mathcal{L}(E)$ defined by $T(m) = T^{-m}$ if $m \geq 0$ and $T(m) = T^{*-m}$ if $m < 0$ is of positive type. This fact is also proved in (I.8) of [**NF**].

(I.1) PROPOSITION. *If* $\|X\| \leq 1$, *the following conditions are equivalent:*

(a) $\exists \hat{X} \in \mathcal{L}(F_1, F_2)$ *that extends* X, *such that* $\hat{X}U_1 = U_2 \hat{X}$ *and* $\|\hat{X}\| \leq 1$;

(b) $I - T_1^*T_1 - X^*X + (T_2X)^*T_2X \geq 0$;

(c) *The inequality*

$$\sum_{m,n\in\mathbb{Z}} \langle T_2(m-n)Xh(m), Xh(n)\rangle \leq \sum_{m,n\in\mathbb{Z}} \langle T_1(m-n)h(m), h(n)\rangle$$

holds for every finitely supported function $h : \mathbb{Z} \to E$.

SKETCH OF THE PROOF. We must have $\hat{X}(U_1^n e) = U_2^n Xe$, $\forall n \in \mathbb{Z}$, $e \in E_1$, so (a) holds iff $\|\sum_m U_2^m Xh(m)\|_{F_2}^2 \leq \|\sum_m U_1^m h(m)\|_{F_1}^2$ is always true, i.e: iff (c) holds. If v is a positive integer such that supp $h := \{n \in \mathbb{Z} : h(n) \neq 0\} \subset [-v, v]$ we may assume that $h \in E_1^{2v+1}$; let $R = [R_{jk}]_{-v \leq j,k \leq v} \in \mathcal{L}(E_1^{2v+1})$ be given by $R_{j,j+1} = T_1$ and $R_{jk} = 0$ if $k - j \neq 1$. Then, for $u = (I - R)^{-1}h \in E_1^{2v+1}$, we have $\sum \langle T_1(m-n)h(m), h(n)\rangle = \|u\|^2 - \|Ru\|^2 = \sum \|u(m)\|^2 - \sum \|T_1 u(m)\|^2$ and $\sum \langle T_2(m-n)Xh(m), Xh(n)\rangle = \sum \|Xu(m)\|^2 - \sum \|T_2 Xu(m)\|^2$; thus, (c) and (b) are equivalent.

Condition (b) above holds in the following particular cases: i) X is an isometry, ii) T_2 is an isometry, iii) $\|T_1x\|^2 + \|Xx\|^2 \leq \|x\|^2$, $\forall x \in E_1$; iv) $X^*T_2 = T_1X^*$ (in this case it follows that $D_XT_1 = T_1D_X$). The last shows that, if $T \in \mathcal{L}(E)$ is a contraction with minimal unitary dilation $U \in \mathcal{L}(F)$, every $A \in \mathcal{L}(E)$ that bicommutes with T (i.e., that commutes with T and T^*) has a (unique) extension $\hat{A} \in \mathcal{L}(F)$ that commutes with U; it is such that $\|\hat{A}\| = \|A\|$ and $\hat{A}^*$ extends A^*.

Let $L(X)$ be the set of **liftings** of X, i.e.,

$$L(X) = \{\tau \in \mathcal{L}(F_1, F_2) : \tau U_1 = U_2\tau, \; P_{E_2}\tau \mathbin{\uparrow}_{E_1} = X, \; \|\tau\| = \|X\|\}.$$

From now on we assume that $\|X\| = 1$. Then, conditions (I.b) and (I.c) are equivalent to the existence of an extension $\hat{X}$ of X that belongs to $L(X)$. In particular, $X^*T_2 = T_1X^*$ iff $\exists \hat{X} \in L(X)$ such that $P_{E_2}\hat{X} = XP_{E_1}$.

The Nagy-Foias commutant lifting theorem states that $L(X)$ is never empty ([**NF**], [**FF**]). An extension due to Parrott of that result is closely connected with the following

(I.2) THEOREM. *Let the commutant $\{T_1, T_2, X\}$ be given. Set $(A_1, A_2) \in \mathcal{A}$ if $A_j \in \mathcal{L}(E_j)$ bicommutes with T_j, $j = 1, 2$ and $XA_1 = A_2X$, $XA_1^* = A_2^*X$; let $\hat{A}_j \in \mathcal{L}(F_j)$ be the extension of A_j that commutes with U_j and is such that $\|\hat{A}_j\| = \|A_j\|$, $j = 1, 2$. Then there exists $\tau \in L(X)$ such that $\tau\hat{A}_1 = \hat{A}_2\tau$, $\forall (A_1, A_2) \in \mathcal{A}$.*

Our proof is based on the following

(I.3) LEMMA. *Let V be an isometry with domain D and range R, both closed subspaces of the Hilbert space H, and Δ the set of operators $\delta \in \mathcal{L}(H)$ such that D and R are invariant under δ and δ^*, and $V\delta \restriction_D = \delta V$. Then the minimal unitary dilation $U \in \mathcal{L}(F)$ of $VP_D \in \mathcal{L}(H)$ is a unitary extension of V to $F \supset H$ such that every $\delta \in \Delta$ has a (unique) extension $\hat{\delta} \in \mathcal{L}(F)$ that commutes with U; moreover, $\|\hat{\delta}\| = \|\delta\|$ and $\hat{\delta}^*$ extends δ^*.*

Let N and M be the orthogonal complements of D and R in H, respectively; then $F = (\bigoplus_{n<0} U^n M) \oplus H \oplus (\bigoplus_{n>0} U^n N)$ and $\hat{\delta} \in \mathcal{L}(F)$ is defined by $\hat{\delta}(U^n v) = U^n(\delta v)$ for $v \in M$, $n < 0$ and for $v \in N$, $n > 0$. The assertion follows.

We now sketch the proof of Theorem (I.2), extending the one given in [**A2**] of the commutant lifting theorem.

Set $M_1 = \bigvee\{U_1^n E_1 : n \geq 0\}$ and $M_2' = \bigvee\{U_2^n E_2 : n \leq 0\}$. Let H be a Hilbert space such that $H = M_1 \vee M_2'$ and $P_{M_2'}^H \restriction_{M_1} = X' := XP_{E_1}^{M_1}$. Every $(A_1, A_2) \in \mathcal{A}$ defines an operator $A \in \mathcal{L}(H)$ by $A(g_2' + g_1) = \hat{A}_2 g_2' + \hat{A}_1 g_1$, $\forall g_2' \in M_2'$ and $g_1 \in M_1$; in fact, $X' \hat{A}_1 \restriction_{M_1} = A_2 X'$ so $D_{X'} \hat{A}_1 \restriction_{M_1} = \hat{A}_1 D_{X'}$ and consequently

$$
\begin{aligned}
\|\hat{A}_2 g_2' + \hat{A}_1 g_1'\|_H^2 &= \|D_{X'} \hat{A}_1 g_1\|_{M_1}^2 + \|X'\hat{A}_1 g_1 + \hat{A}_2 g_2'\|_{M_2'}^2 \\
&= \|\hat{A}_1 D_{X'} g_1\|_{M_1}^2 + \|\hat{A}_2 (X' g_1 + g_2')\|_{M_2'}^2 \\
&\leq \max\{\|A_1\|^2 \ \|A_2\|^2\}\{\|D_{X'} g_1\|_{M_1}^2 + \|X' g_1 + g_2'\|_{M_2'}^2\} \\
&= \max\{\|A_1\|^2, \|A_2\|^2\}\|g_2' + g_1'\|_H^2.
\end{aligned}
$$

Set $D = U_2^* M_2' \vee M_1$; let V be the isometry given by $V(U_2^* g_2' + g_1) = g_2' + U_1 g_1$ and apply (I.3). We may assume that $F = F_1 \vee F_2$ and that $U \restriction_{F_j} = U_j$. Since $A \in \Delta$ it extends to $\hat{A} \in \mathcal{L}(F)$ such that $\hat{A}U = U\hat{A}$, so $\hat{A} \restriction_{F_j} = \hat{A}_j$. Set $\tau = P_{F_2}^F \restriction_{F_1}$; then $\tau \in L(X)$ and $\tau\hat{A}_1 = \hat{A}_2\tau$.

Note that, in $H \subset F$, M_1 is orthogonal to $(M_2' \ominus E_2)$, since $F_2 = (M_2' \ominus E_2) \oplus M_2$, it follows that $\tau(M_1) \subset M_2$.

Parrott's theorem [**Parr**] says that:

(I.4) THEOREM. *Let $T \in \mathcal{L}(E)$ be a contraction, $W \in \mathcal{L}(M)$ its minimal isometric dilation and $X \in \mathcal{L}(E)$ such that $TX = XT$. Let $\mathcal{U}$ be the algebra of all operator in $\mathcal{L}(E)$ that bicommute with T and X; let $A' \in \mathcal{L}(M)$ be the unique extension of $A \in \mathcal{U}$ that commutes with W. There exists $X' \in \mathcal{L}(M)$ such that: $X'A' = A'X'$, $\forall A \in \mathcal{U}$; $P_E A' W^m X'^n \restriction_E = AT^m X^n$ for every $m, n \geq 0$ and $A \in \mathcal{U}$; $\|X'\| = \|X\|$.*

PROOF. Assume $\|X\| = 1$ and let $\tau \in L(X)$ be given by theorem (I.2), with $T_1 = T_2 = T$, $M_1 = M$ and $U_1 \restriction_M = W$. If $A \in \mathcal{U}$, $(A, A) \in \mathcal{A}$ and $A' = \hat{A} \restriction_M$. Since $\tau(M) \subset M$, $X' := \tau \restriction_M \in \mathcal{L}(M)$ commutes with W and A', and is such that $P_E X' = X P_E$. The result follows.

II. On minimal regular unitary dilations of commuting contractions

Let $(T', T'') \subset \mathcal{L}(E)$ be a commutative couple of contractions; set

$$T(m, n) = (T'^{m_-} T''^{n_-})^* (T'^{m_+} T''^{n_+}),$$

with $m_- = \sup(-m, 0)$ and $m_+ = \sup(m, 0)$. It is said that $(U', U'') \subset \mathcal{L}(F)$ is the minimal regular unitary dilation (m.r.u.d., from now on) of $(T', T'') \subset \mathcal{L}(E)$ if (U', U'') is a commutative couple of unitary operators, E is a closed subspace of F, $P_E U'^m U''^n \restriction_E = T(m, n)$, $\forall m, n \in \mathbb{Z}$, and $F = \bigvee \{U'^n U''^m E : m, n \in \mathbb{Z}\}$. It is known (see [**NF**]) that

(II.1) PROPOSITION. *Given the commutative couple of contractions*

$$(T', T'') \subset \mathcal{L}(E)$$

the following conditions are equivalent:

(a) *there exists a m.r.u.d. $(U', U'') \subset \mathcal{L}(F)$ of (T', T'');*
(b) $I - T'^* T' - T''^* T'' + (T'T'')^* (T'T'') \geq 0$;
(c) $T : \mathbb{Z} \to E$ *is a fuction of positive type;*
(d) $\sum_{m,n\in\mathbb{Z}} \langle T'(m-n) T'' h(m), T'' h(n) \rangle \leq \sum_{m,n\in\mathbb{Z}} \langle T'(m-n) h(m), h(n) \rangle$ *holds for every finetely supported fuction $h : \mathbb{Z} \to E$.*

In fact, the equivalence of (a) and (c) follows from Naimark's dilation theorem, and the one of (b) and (d) from (I.1). Assume (a): set $F' = \bigvee \{U'^n E : n \in \mathbb{Z}\}$ and $\hat{T}'' = P_{F'} U'' \restriction_{F'}$; then (a) shows that $\hat{T}''$ extends T'' and that $\hat{T}'' U'^n e = U'^n T'' e$ holds for every $n \in \mathbb{Z}$ and $e \in E$, so (d) follows. Conversely, assume the last: if $W' \in \mathcal{L}(F')$ is the minimal unitary dilation of T', (I.1) shows that there exists a norm preserving extension of T'', $\hat{T}'' \in \mathcal{L}(F')$, that commutes with W'; let $U'' \in \mathcal{L}(F)$ be the minimal unitary dilation of $\hat{T}''$ and $U' \in \mathcal{L}(F)$ defined by $U'(U''^m f) = U''^m W' f$, $m \in \mathbb{Z}$ and $f \in F'$; (a) follows. Moreover, if $A \in \mathcal{L}(E)$ bicommutes with T' and T'', there exists an extension $A' \in \mathcal{L}(F')$ of A that commutes with W', bicommutes with $\hat{T}''$ and is such that $\|A'\| = \|A\|$ and $A'^* = A^{*'}$; consequently there exists an extension $\hat{A} \in \mathcal{L}(F)$ of A' that commutes with U'' and is such that $\|\hat{A}\| = \|A'\|$ and $\hat{A}^*$ extends A'^*; $\hat{A}$ also commutes with U'. Thus:

(II.2) COROLLARY. *Let $(T', T'') \subset \mathcal{L}(E)$ be a contractive couple of contractions such that its m.r.u.d. $(U', U'') \subset \mathcal{L}(F)$ exists. Every $A \in \mathcal{L}(E)$ that bicommutes with T' and T'' can be uniquely extended to $\hat{A} \in \mathcal{L}(F)$ that commutes with U' and U'' and is such that $\|\hat{A}\| = \|A\|$ and $\hat{A}^* = (A^*)\hat{}$.*

Now we can state a bidimensional extension of Theorem (I.2).

(II.3) THEOREM. *Let $(T_j', T_j'') \subset \mathcal{L}(E_j)$ be a bicommutative couple of contractions and $(U_j', U_j'') \subset \mathcal{L}(F_j)$ its minimal regular unitary dilation, $j = 1, 2$. Let $X \in \mathcal{L}(E_1, E_2)$ be such that $XT_1' = T_2'X$, $XT_1'' = T_2''X$ and $XT_1''^* = T_2''^*X$. Set*

$$\mathcal{A} = \{(A_1, A_2) : A_j \in \mathcal{L}(E_j) \text{ bicommutes with } T_j' \text{ and } T_j'', \ j = 1, 2,$$
$$XA_1 = A_2X, \ XA_1^* = A_2^*X\}.$$

Then there exists $\tau \in \mathcal{L}(F_1, F_2)$ such that $\tau U_1' = U_2'\tau$, $\tau U_1'' = U_2''\tau$, $P_{E_2}\tau \restriction_{E_1} = X$, $\|\tau\| = \|X\|$ and $\tau \hat{A}_1 = \hat{A}_2\tau$, $\forall (A_1, A_2) \in \mathcal{A}$.

PROOF. We assume $\|X\| = 1$ and keep the above notations, with obvious complements. Set $F_j' = \bigvee\{U_j'^n E_j : n \in \mathbb{Z}\}$ and let $W_j' \in \mathcal{L}(F_j')$ be the minimal unitary dilation of T_j', $\hat{T}_j'' \in \mathcal{L}(F_j')$ the extension of $T''j$ that commutes with W_j' and $A_j' = \hat{A}_j \restriction_{F_j'}$ if $(A_1, A_2) \in \mathcal{A}$, $j = 1, 2$. Theorem (I.2) says that there exists $\tau' \in \mathcal{L}(F_1', F_2')$ such that $\tau'W_1' = W_2'\tau'$, $P_{E_2}\tau' \restriction_{E_1} = X$, $\|\tau'\| = \|X\|$, $\tau'\hat{T}_1'' = \hat{T}_2''\tau'$, $\tau'\hat{T}_1''^* = \hat{T}_2''^*\tau'$ and $\tau'A_1' = A_2'\tau'$, $\forall (A_1, A_2) \in \mathcal{A}$. We may assume that $U_j'' \in \mathcal{L}(F_j)$ is the minimal unitary dilation of $\hat{T}_j''$ and that U_j' is the extension of W_j' to F_j that commutes with U_j''. We know that τ' can be extended to $\tau \in \mathcal{L}(F_1, F_2)$ that intertwines U_1'' with U_2'' and is such that $\|\tau\| = \|\tau'\|$, $P_{F_2'}\tau = \tau'P_{F_1'}$; consequently, $\|\tau\| = \|X\|$ and $P_{E_2}\tau \restriction_{E_1} = X$. For any $m, n \in \mathbb{Z}$ and $e \in E_1$,

$$(\tau U_1')(U_1'^n U_1''^m e) = \tau U_1''^m W_1'^{n+1}e = U_2''^m \tau' W_1'^{n+1}e = U_2''^m W_2'\tau'W_1'^n e$$
$$= U_2'U_2''^m \tau'W_1'^n e = U_2'\tau U_1''^m W_1'^n e = (U_2'\tau)(U_1'^n U_1''^m e);$$

also,

$$(\tau \hat{A}_1)(U_1'^n U_1''^m e) = U_2''^m \tau \hat{A}_1 W_1'^n e$$
$$= U_2''^m \tau' A_1' W_1'^n e = U_2''^m A_2' \tau' W_1'^n e = (\hat{A}_2\tau)(U_1'^n U_1''^m e).$$

The proof is complete.

III. A two-dimensional version of the Nagy-Foias theorem

For $j = 1, 2$ let $(T_j', T_j'') \subset \mathcal{L}(E_j)$ be any pair of commuting contractions and (U_j', U_j'') a commuting pair of unitary operators in $F_j = \bigvee\{U_j'^m U_j''^n E_j : (m, n) \in \mathbb{Z}^2\}$ such that $P_{E_j}U_j'^m U_j''^n \restriction_{E_j} = T_j'^m T_j''^n$, $\forall \ m, n \geq 0$. The existence of such (U_j', U_j'') was proved by Ando (see [**NF**]). For $X \in \mathcal{L}(E_1, E_2)$ such that $XT_1' = T_2'X$ and $XT_1'' = T_2''X$, set

$$L(X) = \{\tau \in \mathcal{L}(F_1, F_2) : \tau U_1' = U_2'\tau, \ \tau U_1'' = U_2''\tau, \ P_{E_2}\tau \restriction_{E_1} = X, \ \|\tau\| = \|X\|\}.$$

In order to give conditions that ensure that $L(X)$ is non void we may be assumed that $\|X\| = 1$. Then ([**A1**], [**A3**]):

(III.1) THEOREM. *Set*

$$M_1 = \bigvee_{m,n \geq 0} U_1'^m U_1''^n E_1 \text{ and } \tilde{M}_2 = \bigvee_{m,n \leq 0} U_2'^m U_2''^n E_2;$$

let H be the Hilbert space such that $H = M_1 \vee \tilde{M}_2$ and $P^H_{\tilde{M}_2} 1_{M_1} = X P^{M_1}_{E_1}$; set $D' = U_2'^* \tilde{M}_2 \vee M_1$, $D'' = U_2''^* \tilde{M}_2 \vee M_1$ and let V', V'' be the isometries with domains D', D'', respectively, given by $V'(U_2'^* v + u) = v + U_1' u$, $V''(U_2''^* v + u) = v + U_1'' u$, $\forall u \in \tilde{M}_2$, $v \in M_1$. The following assertions hold.

(a) *There exists $\tau \in L(X)$ iff there exists a commutative pair of unitary operators U', U'' that extend V', V'', respectively, to a space F that contains H.*

(b) *When (a) is satisfied, it may assume that: $F = F_1 \vee F_2$; $U' 1_{F_j} = U_j'$ and $U'' 1_{F_j} = U_j''$, $j = 1, 2$; $\tau = P^F_{F_2} 1_{F_1}$.*

For convenience of the reader we shall sketch the proof of the above facts.

Assume that $\exists \tau \in L(X)$; let F be a Hilbert space such that $F = F_1 \vee F_2$ and $\tau = P^F_{F_2} 1_{F_1}$; define U' in F by $U'(f_1 + f_2) = U_1' f_1 + U_2' f_2$, $\forall (f_1, f_2) \in F_1 \times F_2$; define U'' analogously; then $U', U'' \in \mathcal{L}(F)$ are commuting unitary operators that extend V', V'', respectively, to $F \supset H$. Conversely, if such (U', U'') exist, we may assume that $F = \bigvee \{U'^m U''^n H : m, n \in \mathbb{Z}\}$, so $F = \bigvee (\{U'^m U''^n E_j : m, n \in \mathbb{Z}\} : j = 1, 2)$; thus, we may also assume that $F = F_1 \vee F_2$, $U' 1_{F_j} = U_j'$ and $U'' 1_{F_j} = U_j''$, $j = 1, 2$. Set $\tau = P^F_{F_2} 1_{F_1}$; let $v_1 \in E_1$, $v_2 \in E_2$ and $s', s'' \geq 0$; then

$$\langle (P_{E_2} \tau 1_{M_1}) U_1'^{s'} U_1''^{s''} v_1, v_2 \rangle_{E_2} = \langle U'^{s'} U''^{s''} v_1, v_2 \rangle_F = \langle X P_{E_1} U_1'^{s'} U_1''^{s''} v_1, v_2 \rangle_{E_2},$$

so $P_{E_2} \tau 1_{M_1} = X P_{E_1}$ and $\|\tau\| = \|X\| = 1$. If $s', s'', m', m'', n', n'' \in \mathbb{Z}$, then

$$\langle \tau U_1'^{s'} U_1''^{s''} U_1'^{m'} U_1''^{m''} v_1, U_2'^{n'} U_2''^{n''} v_2 \rangle_{F_2}$$
$$= \langle U'^{m'} U''^{m''} v_1, U'^{n'-s'} U''^{n''-s''} v_2 \rangle_{F_2}$$
$$= \langle P_{F_2} U'^{m'} U''^{m''} v_1, U'^{n'-s'} U''^{n''-s''} v_2 \rangle_{F_2},$$

so $\tau U_1'^{s'} U_1''^{s''} = U_2'^{s'} U_2''^{s''} \tau$. Assertions (III.1.a) and (III.1.b) follow.

In order to apply the above result our main tool is the following.

(III.2) LEMMA. *Let V', V'' be isometries with domains D', D'' and ranges R', R'', respectively, all of them closed subspaces of the Hilbert space H, such that one of the following conditions holds:*

(i) $P_{R''}(V' P_{D'})^n V'' P_{D''} = V'' P_{D''}(V' P_{D'})^n P_{D''}$, $n = 1, 2, ...;$

(ii) $P_{D''}(V'^{-1} P_{R'})^n V''^{-1} P_{R''} = V''^{-1} P_{R''}(V'^{-1} P_{R'})^n P_{R''}$, $n = 1, 2, ...$

Let Δ be the set of operators $\delta \in \mathcal{L}(H)$ such that D', D'', R' and R'' are invariant under δ and δ^*, and $V'\delta 1_{D'} = \delta V'$, $V''\delta 1_{D''} = \delta V''$. Then there exist U', U'', commutative unitary operators that extend V', V'', respectively, to a space F that contains H, such that $F = \bigvee \{U'^m U''^n H : m, n \in \mathbb{Z}\}$ and that every $\delta \in \Delta$ has a (unique) extension $\hat{\delta} \in \mathcal{L}(F)$ that commutes with U' and U''; moreover, $\|\hat{\delta}\| = \|\delta\|$ and $\hat{\delta}^*$ extends δ^*.

PROOF. Assume (i). Let $V_1' \in \mathcal{L}(H_1)$ be the minimal unitary dilation of $(V'P_{D'}) \in \mathcal{L}(H)$ and $D_1'' = \bigvee\{V_1'^m D'' : m \in \mathbb{Z}\}$. By (i), setting $V_1''(V_1'^m d) = V_1'^m V''d$ for every $m \in \mathbb{Z}$ and $d \in D''$, an isometry V_1'' with domain D_1'' and range $R_1'' = \bigvee\{V_1'^m R'' : m \in \mathbb{Z}\}$ is defined in such a way that it extends V''. Also, $V_1'D_1'' = D_1''$, $V_1'R_1'' = R_1''$ and $V_1''V_1' \restriction_{D_1''} = V_1'V_1''$, so $(V_1''P_{D_1''})^n V_1' = V_1'(V_1''P_{D_1''})^n$, $n = 1, 2, ...$ Lemma (I.3) shows that every $\delta \in \Delta$ extends to $\delta_1 \in \mathcal{L}(H_1)$ that commutes with V_1' and is such that $\|\delta_1\| = \|\delta\|$ and δ_1^* extends δ^*. Moreover, D_1'' and R_1'' are invariant under δ_1 and δ_1^*, and $V_1''\delta_1 \restriction_{D_1''} = \delta_1 V_1''$. Now we repeat the same argument. Let $U'' \in \mathcal{L}(F)$ be the minimal unitary dilation of $(V_1''P_{D_1''}) \in \mathcal{L}(H_1)$: V_1' has a unique unitary extension $U' \in \mathcal{L}(F)$ that commutes with U''. Lemma (I.3) shows that, for every $\delta \in \Delta$, δ_1 has a unique extension $\hat{\delta} \in \mathcal{L}(F)$ that commutes with U'' and is such that $\|\hat{\delta}\| = \|\delta_1\|$ and $\hat{\delta}^*$ extends δ_1^*. If $m \in \mathbb{Z}$ and $h \in H_1$, $(U'\hat{\delta})(U''^m h) = U''^m U'\delta_1 h = U''^m V_1'\delta_1 h = U''^m \delta_1 V_1' h = \hat{\delta}U''^m U' h = (\hat{\delta}U')(U''^m h)$. The result follows.

Let V' and V'' be given by (III.1); we shall give some conditions that ensure that (III.2.i) holds.

Set $D = U_2''^* U_2'^* \tilde{M}_2 \vee M_1$; note that $V'D \subset D''$, $V''D \subset D'$, and $V'V'' \restriction_D = V''V' \restriction_D$. If $D = D''$, $V'P_{D'}V''P_{D''} = (V'V'' \restriction_D)P_{D''} = V''P_{D''}(V'P_{D'})P_{D''}$ and (III.2.i) follows.

If $E_2 \subset U_2'^* \tilde{M}_2$, then $U_2'^m U_2''^n E_2 \subset U_2'^* \tilde{M}_2$ for any $m, n \le 0$, so $\tilde{M}_2 = U_2'^* \tilde{M}_2$ and $D = D''$.

If T_2' is an isometry, $U_2'^* T_2'$ is the identity in E_2, so $E_2 \subset U_2'^* \tilde{M}_2$.

The following formulas are not difficult to prove:

$$H = \tilde{M}_2 \oplus \{(I - X)a : a \in E_1\}^- \oplus (M_1 \ominus E_1) = \tilde{M}_2 \oplus \mathcal{D}_X \oplus (M_1 \ominus E_1);$$
$$D'' = U_2''^* \tilde{M}_2 \oplus \{(I - U_2''^* XT_1'')a : a \in E_1\}^- \oplus (M_1 \ominus E_1);$$
$$D = U_2''^* U_2'^* \tilde{M}_2 \oplus \{(I - U_2''^* U_2'^* XT_1''T_1')a : a \in E_1\}^- \oplus (M_1 \ominus E_1);$$
$$E_2 \vee U_2'^* \tilde{M}_2 = \{(I - U_2'^* T_2')b : b \in E_2\}^- \oplus U_2'^* \tilde{M}_2.$$

Assume $\tilde{M}_2 = E_2 \vee U_2'^* \tilde{M}_2$. Then $\tilde{M}_2 \ominus U_2'^* \tilde{M}_2 = \{(I - U_2'^* T_2')b : b \in E_2\}^-$, so $(D'' \ominus D) \oplus \{(I - U_2''^* U_2'^* XT_1''T_1')a : a \in E_1\}^- = \{U_2''^*(I - U_2'^* T_2')b : b \in E_2\}^- \oplus \{(I - U_2''^* XT_1'')a : a \in E_1\}^-$ and, consequently

$$D'' \ominus D \approx [\mathcal{D}_{T_2'} \oplus \mathcal{D}_{T_2''X}] \ominus \{D_{T_2'}T_2'Xc \oplus D_{T_2''}Xc : c \in E_1\}^-.$$

Recall that if $A \in \mathcal{L}(G_2, G_3)$ and $B \in \mathcal{L}(G_1, G_2)$ are contractions and $C = AB$ it is said that AB is a regular factorization of C if $\mathcal{D}_A \oplus \mathcal{D}_B = \{D_A Bv \oplus D_B v : v \in G_1\}^-$.

Thus, if $\tilde{M}_2 = E_2 \vee U_2'^* \tilde{M}_2$ and the factorization $T_2'(T_2''X)$ of the contraction $XT_1'T_1'' = T_2''T_2'X$ is regular, $D = D''$.

Set $R = \tilde{M}_2 \vee U_1''U_1'M_1$; if $R = R''$ then (II.2.ii) holds.

If $E_1 \subset U_1'M_1$, then $U_1'^m U_1'^n E_2 \subset U_1'M_1$ for any $m, n \ge 0$, so $M_1 = U_1'M_1$ and $R = R''$.

Now we consider the following formulas:

$$R'' = \tilde{M}_2 \oplus \{(I - XP_{E_1}^{M_1})U_1''a : a \in E_1\}^- \oplus U_1''(M_1 \ominus E_1);$$
$$R = \tilde{M}_2 \oplus \{(I - XP_{E_1}^{M_1})U_1'U_1''a : a \in E_1\}^- \oplus U_1'U_1''(M_1 \ominus E_1);$$
$$E_1 \vee U_1'M_1 = E_1 \oplus \{(U_1' - T_1')a : a \in E_1\}^- \oplus U_1'(M_1 \ominus E_1).$$

Assume $M_1 = E_1 \vee U_1'M_1'$. Then $M_1 \ominus E_1 = \{(U_1' - T_1')a : a \in E_1\}^- \oplus U_1'(M_1 \ominus E_1)$, so

$$(R'' \ominus R) \oplus \{(I - XP_{E_1}^{M_1})U_1'U_1''a : a \in E_1\}$$
$$= (I - XP_{E_1}^{M_1})U_1''a : a \in E_1\}^- \oplus \{U_1''(U_1' - T_1')a : a \in E_1\}^-$$

and consequently, $R'' \ominus R \approx [\mathcal{D}_{XT_1''} \oplus \mathcal{D}_{XT_1'}] \ominus \{D_{XT_1''}T_1'c \oplus D_{T_1'}c : c \in E_1\}^-$. Thus, if $M_1 = E_1 \vee U_1'M_1$ and the factorization $(XT_1'')T_1'$ of the contraction $XT_1'T_1'' = T_2''T_2'X$ is regular, $R = R''$.

Summing up:

(III.3) THEOREM. *With the same conditions as in Theorem (III.1), $L(X)$ is non void if one of the following conditions holds:*

 a) $E_2 \subset U_2'^* \tilde{M}_2$;
 b) $E_1 \subset U_1'M_1$;
 c) $\tilde{M}_2 = E_2 \vee U_2'^* \tilde{M}_2$ *and the factorization $T_2'(T_2''X)$ of the contraction $XT_1'T_1'' = T_2''T_2'X$ is regular;*
 d) $M_1 = E_1 \vee U_1'M_1$ *and the factorization $(XT_1'')T_1'$ of the contraction $XT_1'T_1'' = T_2''T_2'X$ is regular.*

In particular, $L(X)$ is non void if one of the operator $T_1'^*$, $T_1''^*$, T_2', T_2'' is an isometry.

IV. A two-dimensional version of the Nagy-Foias-Parrott theorem

Here we shall prove the following

(IV.1) THEOREM. *For $j = 1, 2$ let $(T_j', T_j'') \subset \mathcal{L}(E_j)$ be a pair of commuting contractions with minimal regular unitary dilation $(U_j', U_j'') \subset \mathcal{L}(F_j)$; let $X \in L(E_1, E_2)$ be such that $XT_1' = T_2'X$ and $XT_1'' = T_2''X$. Set*

$$L(X) = \{\tau \in \mathcal{L}(F_1, F_2) : \tau U_1' = U_2'\tau, \tau U_1'' = U_2''\tau, P_{E_2}\tau \restriction_{E_1} = X, \|\tau\| = \|X\|\};$$

set $(A_1, A_2) \in \mathcal{A}$ if $A_j \in \mathcal{L}(E_j)$ bicommutes with T_j' and T_j'', $j = 1, 2$, and $XA_1 = A_2X$, $XA_1^* = A_2^*X$, and let $\hat{A}_j \in \mathcal{L}(F_j)$ be the extension of A_j that commutes with U_j' and U_j'', and is such that $\|\hat{A}_j\| = \|A_j\|$, $j = 1, 2$. Set

$$M_1 = \bigvee_{m,n \geq 0} U_1'^m U_1''^n E_1 \text{ and } \tilde{M}_2 = \bigvee_{m,n \leq 0} U_2'^m U_2''^n E_2.$$

Assume that one of the following conditions holds:

 a) $E_2 \subset U_2'^* \tilde{M}_2$;

b) $E_1 \subset U'_1 M_1$;

c) $\tilde{M}_2 = E_2 \vee U'^*_2 \tilde{M}_2$ and the factorization $T'_2(T''_2 X)$ of the contraction $X T'_1 T''_1 = T''_2 T'_1 X$ is regular;

d) $M_1 = E_1 \vee U'_1 M_1$ and the factorization $(X T''_1) T'_1$ of the contraction

$$X T'_1 T''_1 = T''_2 T'_1 X$$

is regular.

Then there exists $\tau \in L(X)$ such that $\tau \hat{A}_1 = \hat{A}_2 \tau$, $\forall (\hat{A}_1, \hat{A}_2) \in \mathcal{A}$.

PROOF. Let the Hilbert space H and the isometries V', V'' be as in Theorem (III.1); we know that conditions (a), (b), (c) or (d) ensure that the assertions of Lemma (III.2) are true.

For any $(A_1, A_2) \in \mathcal{A}$ we shall see that an operator $A \in \mathcal{L}(H)$ is defined by $A(v_2 + v_1) = \hat{A}_2 v_2 + \hat{A}_1 v_1, \forall \, v_1 \in M_1, \, v_2 \in \tilde{M}_2$. Set $X' = X P_{E_1}^{M_1}$; then $X' \hat{A}_1 \upharpoonright_{M_1} = A_2 X'$; since $(A_1^*, A_2^*) \in \mathcal{A}$, $X'^* X' \hat{A}_1 \upharpoonright_{M_1} = X'^* \hat{A}_2 X' = \hat{A}_1 X'^* X'$ so $D_{X'} \hat{A}_1 \upharpoonright_{M_1} = \hat{A}_1 D_{X'}$. Thus,

$$\begin{aligned}
\|\hat{A}_2 v_2 + \hat{A}_1 v_1\|_H^2 &= \|D_{X'} \hat{A}_1 v_1\|_{M_1}^2 + \|X' \hat{A} v_1 + \hat{A}_2 v_2\|_{\tilde{M}_2}^2 \\
&= \|\hat{A}_1 D_{X'} v_1\|_{M_1}^2 + \|\hat{A}_2(X' v_1 + v_2)\|_{\tilde{M}_2}^2 \\
&\leq \max(\|\hat{A}_1\|^2, \|\hat{A}_2\|^2)[\|D_{X'} v_1\|_{M_1}^2 + \|X' v_1 + v_2\|_{\tilde{M}_2}^2] \\
&= \max(\|A_1\|^2, \|\hat{A}_2\|^2) \|v_1 + v_2\|_H^2.
\end{aligned}$$

Now, D', D'', R' and R'' are invariant under A, and $V' A \upharpoonright_{D'} = AV'$, $V'' A \upharpoonright_{D''} = AV''$. Moreover, A^* is obtained in the same way from $(A_1^*, A_2^*) \in \mathcal{A}$. Thus, with notations as in Lemma (III.2), $A \in \Delta$, so it can be extended to $\hat{A} \in \mathcal{L}(F)$ that commutes with U' and U'', and is such that $\hat{A}^* = (A^*)\hat{\,}$. We may assume that $F = F_1 \vee F_2$, $U'_j \upharpoonright_{F_j} = U'_j$ and $U''_j \upharpoonright_{F_j} = U''_j$, $j = 1, 2$; moreover $\tau := P_{F_2}^H \upharpoonright_{F_1} \in L(X)$. For $m, n \in \mathbb{Z}$, $u_j \in E_j$ and $j = 1, 2$, $\hat{A} U'^m_j U''^n_j u_j = \hat{A} U'^m U''^n u_j = U'^m U''^n A_j u_j = U'^m_j U''^n_j A_j u_j = \hat{A}'_j U'^m_j U''^n_j u_j$, so $\hat{A} \upharpoonright_{F_j} = \hat{A}_j$; if also $s, t \in \mathbb{Z}$, $\langle \tau \hat{A}_1 U'^m_1 U''^n_1 u_1, U'^s_1 U''^t_1 u_2 \rangle_{F_2} = \langle \hat{A} U'^m U''^n u_1, U'^s U''^t u_2 \rangle_F = \langle U'^m U''^n u_1, \hat{A}^* U'^s U''^t u_2 \rangle_F = \langle \tau U'^m_1 U''^n_1 u_1, \hat{A}_2^* U'^s_1 U''^t_1 u_2 \rangle_{F_2}$, so $\tau \hat{A}_1 = \hat{A}_2 \tau$.

The proof of Theorem (IV.1) is complete.

In fact, the following extension of Theorem (IV.1) has also been proved.

(IV.2) THEOREM. *In the same conditions as in Theorem (IV.1), let V', V'' be defined as in Theorem (III.1) and such that one of the following conditions holds:*

i) $P_{R''}(V' P_{D'})^n V'' P_{D''} = V'' P_{D''}(V' P_{D'})^n P_{D''}$, $n = 1, 2, ...$;

ii) $P_{D''}(V'^{-1} P_{R'})^n V''^{-1} P_{R''} = V''^{-1} P_{R''}(V'^{-1} P_{R'})^n P_{R''}$, $n = 1, 2, ...$

Then there exists $\tau \in L(X)$ such that $\tau \hat{A}_1 = \hat{A}_2 \tau$, $\forall \, (A_1, A_2) \in \mathcal{A}$.

REFERENCES

[A1]   R. Arocena, *On the extension problem for a class of translation invariant forms*, J. Operator Theory **21** (1989), 232–347.

[A2]   ―――, *Unitary extensions of isometries and contractive intertwining dilatations*, Operator Theory: Adv. and Appl. **41** (1989), 13–23.

[A3]   ―――, *On some extensions of the commutant lifting theorem*, Publicaciones Matemáticas del Uruguay **5** (1992), 61–76.

[FF]   C. Foias and A. E. Frazho, *The Commutant Lifting Approach to interpolation Problems*, Birkhäuser, 1990.

[NF]   B. Sz.-Nagy and C. Foias, *Harmonic Analysis of operator on Hilbert Space*, North-Holland, 1970.

[Parr] S. Parrott, *On a Quotient Norm and the Sz.-Nagy-Foias Lifting Theorem*, J. Func. An. **30** (1978), 311–328.

[Pau]   V. I. Paulsen, *Completely Bounded maps and Dilations*, Longman, 1986.

CENTRO DE MATEMÁTICA, FACULTAD DE CIENCIAS, URUGUAY. POSTAL ADDRESS: JOSÉ M. MONTERO 3006, AP. 503-MONTEVIDEO, URUGUAY

E-mail address: rarocena@cmat.edu.uy

Contemporary Mathematics
Volume **189**, 1995

Weighted Bergman projections in domains of finite type in $\mathbb{C}^2$

ALINE BONAMI AND SANDRINE GRELLIER

ABSTRACT. Let Ω be a smooth bounded pseudo-convex domain of finite type in $\mathbb{C}^2$. We study the weighted Bergman projections for weights which are a power of the distance to the boundary. We define a class of operators of Bergman type for which we develop a functional calculus. Sobolev estimates, both of isotropic and non isotropic type, are then deduced.

1. Introduction

Let $\Omega \subset \mathbb{C}^2$ be a bounded, smooth domain, given by

$$\Omega = \{z \in \mathbb{C}^2, \quad r(z) < 0\}$$

with r a $\mathcal{C}^\infty$-function such that $|\nabla r| = 1$ whenever $r(z) = 0$.

We shall assume that Ω is of finite type m, and is pseudo-convex. The behavior of the Bergman kernel of Ω is now well known (see [**McN**] and [**NRSW**]). Let dV denote the Lebesgue measure in $\mathbb{C}^2$, and dV_q, for $q > -1$, denote the weighted measure

$$dV_q(z) = (-r(z))^q dV(z).$$

We shall call $\mathbf{B}_q$ the weighted Bergman projection for the measure dV_q, that is the orthogonal projection in $L^2(\Omega, dV_q)$ onto the subspace of holomorphic functions. It is given by the kernel $B_q(z, w)$:

$$\mathbf{B}_q f(z) = \int_\Omega B_q(z, w) f(w) dV_q(w).$$

Our aim is to show that, for q a positive integer, B_q satisfies the same estimates as $B = B_0$, except for the fact that, now, volume dV_q is employed instead

1991 *Mathematics Subject Classification.* 32A10 32A25.

Key words and phrases. Several Complex variables, Bergman Projections, pseudo-convex domains, finite type.

of ordinary volume. Moreover, $\mathbf{B}_q$ and $\mathbf{B}_{q-1}$ are related by the two following relations:

(1) There exists a differential operator of order 1 with $\mathcal{C}^\infty$-coefficients, D, so that

$$B_q(z,w) = D_w B_{q-1}(z,w) + \text{ error term};$$

(2) $\mathbf{B}_q = \mathbf{B}_{q-1} + \mathbf{B}_{q-1}(<\omega, \overline{\partial}>) + \text{ error term }$,
where ω is a $(0,1)$ form with $\mathcal{C}^\infty$-coefficients which are $\mathcal{O}((-r))$.

When $q = 0$, *(1)* and *(2)* are still valid if $\mathbf{B}_{-1}$ is replaced by the Szegö projection $\mathbf{S}$. Such equalities *(1)* and *(2)* are valid for ellipsoïds without error terms as it follows from elementary integrations by parts (see for instance [**B&C**] and [**S**]). For strictly pseudo-convex domains in $\mathbb{C}^n$, such comparison results have been developed using Henkin-Ramirez formulas (see [**Cu**] and [**Co**]).

Let us remark also that the weighted Bergman kernel depends on the choice of the defining function r but the difference between the weighted Bergman kernels corresponding to two different choices of defining function is again a smoothing error term.

One way to study weighted Bergman projections is to increase the dimension as in [**L1**]. We shall use her results which give Sobolev estimates and so, $\mathcal{C}^\infty$-estimates up to the boundary but, for precise estimates, we restrict ourselves to q integer and we rely only on the estimates on the Bergman kernel given by [**NRSW**] and [**McN**] to obtain $\mathbf{B}_{q+1}$ from $\mathbf{B}_q$ via an integration by parts. This idea is already contained in [**NRSW**] where the Szegö projection is obtained from the Bergman kernel via an integration along the integral curves of a vector field (which is more or less *(1)* for $q = 0$). We did not say yet what we meant by an error term. To do it, we have to develop a functional calculus which is adapted to the Bergman projections. This is the aim of §2, and it should be compared, of course, to the properties of NIS operators in [**NRSW**].

We should emphasize that our methods would as well work for strictly pseudo-convex domains in $\mathbb{C}^n$ (see [**L2**] and [**P**] for some results in the strictly pseudo-convex case).

Before writing the estimates satisfied by the weighted Bergman kernel, let us give a few definitions.

Denote by L any $(1,0)$-vector field which is $\mathcal{C}^\infty(\overline{\Omega})$ and tangential on $\partial\Omega$ *i.e.* $Lr = 0$ on $\partial\Omega$. Denote by $R = N + iT$ a complex normal vector field, *i.e.*, R is a $(1,0)$-vector field which is $\mathcal{C}^\infty(\overline{\Omega})$ and satisfies $R(r) = N(r) > 0$ on $\partial\Omega = \{r(z) = 0\}$. It means that N goes outwards of Ω, and T is transverse to the complex tangent space at the boundary.

As in [**NRSW**], we consider the distance d_c on $\partial\Omega$ defined by:
$d_c(z,\zeta)$ *is the infimum of length of piecewise smooth curves Φ joining z and ζ such that Φ' may be written*

$$\Phi'(t) = \alpha(t)X_1 + \beta(t)X_2,$$

where X_1 and X_2 are the real and imaginary parts of L.

The corresponding balls $B(z, \delta)$ are anisotropic and are of size $\Lambda(z, \delta)$ in the T direction, where $\Lambda(z, \delta)$ is defined as in [**NRSW**]. It is why we shall prefer to use the pseudo-distance:

$$d(z, \zeta) = \Lambda(z, d_c(z, \zeta)).$$

For this pseudo-distance, the balls $\mathcal{B}(z, \delta)$ have size $\mu(z, \delta)$ in the complex tangent space, where $\delta \to \mu(z, \delta)$ is the inverse function of $\delta \to \Lambda(z, \delta)$.

For z, ζ in a neighborhood of $\partial\Omega$ in $\overline{\Omega}$, define

$$D(z, \zeta) = d(\pi(z), \pi(\zeta)) + \delta(z) + \delta(\zeta)$$

where π denotes the projection to the boundary and $\delta(z)$ the distance from z to $\partial\Omega$.

For any ζ in a neighborhood of $\partial\Omega$ in $\overline{\Omega}$, we define the *"tent"* over $\mathcal{B}(\pi(\zeta), \varepsilon) = \{z \in \mathbb{C}^2, \quad d(z, \pi(\zeta)) \le \varepsilon\}$ as the set

$$\mathcal{B}^{\#}(\zeta, \varepsilon) = \overline{\Omega} \cap \{z; \quad D(z, \zeta) \le \varepsilon\}.$$

Then $Vol_q[\mathcal{B}^{\#}(\zeta, \varepsilon)] \simeq \mu(\zeta, \varepsilon)^2 \times \varepsilon^{2+q}$.

The following theorem gives the estimates for the weighted Bergman kernel.

THEOREM. *Let q be a non-negative integer. For each l, there exists a constant C_l so that if $Y_1, ..., Y_l$ are vector fields, each of which is one of $\{L_z, \overline{L}_\zeta, R_z, \overline{R}_\zeta\}$, with α either L_z or $\overline{L}_\zeta$ and $\beta = l - \alpha$ either R_z or $\overline{R}_\zeta$, then*

$$|Y_1...Y_l B_q(z, \zeta)| \le C_l \frac{D(z, \zeta)^{-\beta} \mu(z, D(z, \zeta))^{-\alpha}}{Vol_q(\mathcal{B}^{\#}(z, D(z, \zeta)))} \text{ for any } (z, \zeta) \in \overline{\Omega} \times \overline{\Omega}.$$

Once this behavior of the weighted Bergman projection is known, one can obtain for $\mathbf{B}_q$ commutation identities with the vector fields L and T (see [**BCG**] for the case $q = 0$). Sobolev estimates of isotropic and anisotropic type are then easy to write. We shall also prove estimates for the difference $\mathbf{B}_q - \mathbf{B}_{q-1}$.

2. The class of weighted Bergman type operators

Before defining our class of weighted Bergman type operators, we need to define a class of bump functions.

DEFINITION 2.1. Let $N \in \mathbb{N}$. A function φ of class $\mathcal{C}^N$ in $\overline{\Omega}$ is said to be bump of order N if φ is supported in some $\mathcal{B}^{\#}(w, \delta)$ and if, for any $l \in \mathbb{N}, l \le N$,

$$\sup_{z \in \mathcal{B}^{\#}(w, \delta)} \delta^\beta \mu(w, \delta)^\alpha |Y_1...Y_l \varphi(z)| \le 1$$

whenever α of the Y_j's are $\{L, \overline{L}\}$ and $\beta = l - \alpha$ are $\{R, \overline{R}\}$.

REMARK. Let φ be a bump function supported in some $\mathcal{B}^{\#}(w, \delta)$. Let $s(z, \zeta)$ be a $\mathcal{C}^\infty(\overline{\Omega} \times \overline{\Omega})$-function satisfying

$$s(z, \zeta) = \mathcal{O}(\mu(z, D(z, \zeta)))$$

then, there exist some constants C_1, C_2 and C_3 so that, for any $z \in \mathcal{B}^{\#}(w, \delta)$, the functions

$$C_1 \frac{(-r)}{\delta} \varphi, \quad C_2 \frac{s(z,.)}{\mu(w,\delta)} \varphi \text{ and } C_3(-r)\nabla\varphi$$

are also bump in $\mathcal{B}^{\#}(w, \delta)$.

Define the function h from $\mathbb{R}$ to $\{2, m\}$ as $h(t) = 2\chi_{t \geq 0} + m\chi_{t < 0}$ where χ_E is the characteristic function of the set E.

DEFINITION 2.2. An operator $\mathbf{K}$ belongs to the class of weighted Bergman type operators $\mathcal{B}_{(q,\mathbf{r})}$ of order $\mathbf{r} = (r_1, r_2)$ and weight $q > -1$, for $r_1 + (r_2 - 2)/h(2 - r_2) < q + 2$, if the following properties hold:

• $\mathbf{K}$ is given by a kernel $K(z, \zeta)$ with respect to the measure $dV_q(\zeta)$ **i.e.**

$$\text{if } f \in L^1(dV_q) \quad \mathbf{K}f(z) = \int_\Omega K(z,\zeta)f(\zeta)dV_q(\zeta), \quad z \in \Omega.$$

• The kernel $K(z, \zeta)$ belongs to $\mathcal{C}^\infty(\overline{\Omega} \times \overline{\Omega} \setminus \Delta_{\partial\Omega})$ (where $\Delta_{\partial\Omega}$ denotes the diagonal of $\partial\Omega$) and satisfies the following estimates.

For each l, there exists a constant C_l so that if $Y_1, ..., Y_l$ are vector fields, each of which is one of $\{L, \overline{L}, R, \overline{R}\}$, with α either L or $\overline{L}$ and $\beta = l - \alpha$ either R or $\overline{R}$, then

$$|Y_1...Y_l K(z,\zeta)| \leq C_l \frac{D(z,\zeta)^{r_1-\beta} \mu(z, D(z,\zeta))^{r_2-\alpha}}{Vol_q(\mathcal{B}^{\#}(z, D(z,\zeta)))} \quad (z,\zeta) \in \overline{\Omega} \times \overline{\Omega}.$$

Each Y_j acts either in z or ζ.

• $\mathbf{K}$ is restrictedly regular in the following sense:

for any $l \in \mathbb{N}$, there exist a constant C_l and a positive integer N_l such that, whenever φ is a bump function of order $\geq N_l$ supported in $\mathcal{B}^{\#}(w, \delta)$, then

$$\sup_{z \in \mathcal{B}^{\#}(w,\delta)} |Y_1...Y_l \mathbf{K}(\varphi)(z)| \leq C_l \delta^{r_1-\beta} \mu(w,\delta)^{r_2-\alpha}$$

whenever α of the Y_j's are $\{L, \overline{L}\}$ and $\beta = l - \alpha$ are $\{R, \overline{R}\}$.

• The adjoint $\mathbf{K}^*$ in $L^2(dV_q)$ is restrictedly regular.

REMARK. • It has been proved independently in [**McN**] and in [**NRSW**] that the Bergman projection is in $\mathcal{B}_{(0,0)}$.

• It follows from the definition that $\mathcal{B}_{(q+1,\mathbf{r})} \subset \mathcal{B}_{(q,\mathbf{r})}$ and that, for instance, $\mathcal{B}_{(q,(r_1,r_2))} \subset \mathcal{B}_{(q,(r_1+\frac{r_2}{h(-r_2)},0))} \cap \mathcal{B}_{(q,(0,r_2+r_1h(r_1)))}$.

• the condition on $\mathbf{r} = (r_1, r_2)$ and q allows the right hand side of the pointwise estimates to tend to ∞ as z and ζ go to the diagonal $\Delta_{\partial\Omega}$.

PROPOSITION 2.1. *Let $q > -1$. Let $\mathbf{T}_1$ and $\mathbf{T}_2$ be two operators in $\mathcal{B}_{(q,\mathbf{r})}$ and in $\mathcal{B}_{(q,\mathbf{s})}$ respectively. Then $\mathbf{T}_1 \mathbf{T}_2$ is in $\mathcal{B}_{(q,\mathbf{r}+\mathbf{s})}$ if the following condition holds:*

$$\varrho = r_1 + s_1 + (r_2 + s_2 - 2)/h(r_2 + s_2 - 2) - q - 2 < 0.$$

PROOF. The proof of the proposition follows the same line as the proof given in [**NRSW**] for composition of NIS operators. For completeness, we give some details of the proof. It is sufficient to consider the operator $\mathbf{T}_1\mathbf{T}_2$, the conditions for its adjoint will follow.

First, we show that, under the assumption of the proposition, the kernel of $\mathbf{T}_1\mathbf{T}_2$ and its derivatives satisfy the right pointwise estimates. For $(z,w) \in \overline{\Omega} \times \overline{\Omega}$, let $\delta = D(z,w)$. We want to show that

$$|T_1T_2(z,w)| \leq \frac{C\delta^{r_1+s_1}\mu(z,\delta)^{r_2+s_2}}{Vol_q(\mathcal{B}^{\#}(z,\delta))}.$$

The estimates for the derivatives will follow the same lines.

Let φ_1 be some bump function supported in $\mathcal{B}^{\#}(z,\frac{\delta}{C})$ so that $\varphi_1(z) = 1$ on $\mathcal{B}^{\#}(z,\frac{\delta}{2C})$ for some $C > 1$ depending on the pseudo-distance D. We also assume that φ_2 is a bump function supported in $\mathcal{B}^{\#}(w,\frac{\delta}{C})$ and $\varphi_2(z) = 1$ in $\mathcal{B}^{\#}(w,\frac{\delta}{2C})$. Define

$$\psi = 1 - \varphi_1 - \varphi_2.$$

By definition

$$T_1T_2(z,w) = \int_{\Omega} T_1(z,\zeta)T_2(\zeta,w)dV_q(\zeta)$$

So

$$T_1T_2(z,w) = \int_{\partial\Omega} T_1(z,\zeta)T_2(\zeta,w)\varphi_1(\zeta)dV_q(\zeta)$$

$$+ \int_{\partial\Omega} T_1(z,\zeta)T_2(\zeta,w)\varphi_2(\zeta)dV_q(\zeta)$$

$$+ \int_{\partial\Omega} T_1(z,\zeta)T_2(\zeta,w)\psi(\zeta)dV_q(\zeta)$$

$$= I + II + III.$$

But

$$|I| \leq |\mathbf{T}_1(\varphi_1)(z)| \times \sup_{\zeta \in \text{supp}\varphi_1} |T_2(\zeta,w)|$$

$$\leq |\mathbf{T}_1(\varphi_1)(z)| \times \sup_{\zeta \in \text{supp}\varphi_1} \frac{D(\zeta,w)^{s_1}\mu(w,D(\zeta,w))^{s_2}}{Vol_q(\mathcal{B}^{\#}(z,D(z,w)))}$$

$$\leq \frac{C\delta^{r_1+s_1}\mu(z,\delta)^{r_2+s_2}}{Vol_q(\mathcal{B}^{\#}(z,\delta))},$$

since, on $\text{supp}\varphi_1$, $D(\zeta,w) \simeq \delta$.

To handle II, the argument is similar (it uses the assumption on the adjoint of $\mathbf{T}_2$). It remains to estimate the last term III. Denote by E the set of $\zeta \in \Omega$ where $D(z,\zeta) \geq \frac{\delta}{2C}$ and $D(w,\zeta) \geq \frac{\delta}{2C}$.

$$|III| = \left| \int_{\Omega} T_1(z,\zeta)T_2(\zeta,w)\psi(\zeta)dV_q(\zeta)) \right|$$

$$\leq C \int_E |T_1(z,\zeta)T_2(\zeta,w)|dV_q(\zeta))$$

$$\leq \int_{E\cap\{D(z,\zeta)\leq 2\delta\}} + \int_{E\cap\{D(z,\zeta)\geq 2\delta\}}$$

$$= III_1 + III_2.$$

In the region of the integral III_1, we have $D(z,\zeta) \simeq \delta$ and $D(w,\zeta) \simeq \delta$. Therefore,

$$III_1 \leq \frac{C\delta^{r_1+s_1}\mu(z,\delta)^{r_2+s_2}}{Vol_q(\mathcal{B}^\#(z,\delta))^2} \int_{D(z,\zeta)\leq 2\delta} dV_q(\zeta)$$

$$\leq \frac{C\delta^{r_1+s_1}\mu(z,\delta)^{r_2+s_2}}{Vol_q(\mathcal{B}^\#(z,\delta))}.$$

In the region of the integral III_2, we have $D(\zeta,w) \simeq D(z,\zeta)$. Hence

$$III_2 \leq C\int_{D(z,\zeta)\geq 2\delta} \frac{D(z,\zeta)^{r_1+s_1}\mu(z,D(z,\zeta))^{r_2+s_2}}{Vol_q(\mathcal{B}^\#(z,D(z,\zeta)))^2} dV_q(\zeta)$$

$$= \sum_{j=1}^{\infty} \int_{2^j\delta \leq D(z,\zeta) < 2^{j+1}\delta}$$

$$\leq C\sum_{j=1}^{\infty} \frac{(2^j\delta)^{r_1+s_1}\mu(z,2^j\delta)^{r_2+s_2}}{Vol_q(\mathcal{B}^\#(z,2^j\delta))}$$

$$\leq \frac{C\delta^{r_1+s_1}\mu(z,\delta)^{r_2+s_2}}{Vol_q(\mathcal{B}^\#(z,\delta))} \times \sum_{j=1}^{\infty}(2^j)^\varrho.$$

It remains to show that $\mathbf{T}_1\mathbf{T}_2$ is restrictedly regular.

Let ϕ be a bump function supported in some $\mathcal{B}^\#(w,\delta)$ of order sufficiently large.

As in [**NRSW**], it is possible to construct a partition of unity $1 = \sum_{j\geq 0}\psi_j$ with ψ_j a bump function supported in $\mathcal{B}^\#(w,c2^j\delta) \setminus \mathcal{B}^\#(w,2^j\delta)$, $j \geq 1$, where c is some fixed constant, and ψ_0 is a function supported in $\mathcal{B}^\#(w,c\delta)$.

We have

$$\mathbf{T}_2(\phi) = \sum_{j\geq 0}\psi_j\mathbf{T}_2(\phi).$$

But, using the properties of $\mathbf{T}_2$, we have

$$\psi_j\mathbf{T}_2(\phi)(z) = \psi_j(z)\int_\Omega T_2(z,\zeta)\phi(\zeta)dV_q(\zeta)$$

$$= \lambda_j\phi_j(z)$$

where ϕ_j is a bump function supported in $\mathcal{B}^\#(w,2^j\delta)$ and

$$\lambda_j = C\frac{Vol_q(\mathcal{B}^\#(w,\delta))}{Vol_q(\mathcal{B}^\#(w,2^j\delta))} \times (2^j\delta)^{s_1}\mu(z,2^j\delta)^{s_2}.$$

For $j = 0$, this follows immediately from the estimates on $\mathbf{T}_2(\phi)$. For $j \geq 1$, we use only the size estimates of the kernel $T_2(z,\zeta)$ and its derivatives which give (with the notations of Definition 2.1)

$$|Y_1...Y_l T_2(z,\zeta)| \leq C\frac{(2^j\delta)^{s_1-\beta}\mu(z,2^j\delta)^{s_2-\alpha}}{Vol_q(\mathcal{B}^\#(z,2^j\delta))}$$

when $\zeta \in \mathcal{B}^{\#}(w,\delta)$ and $z \in \mathcal{B}^{\#}(w,c2^j\delta) \setminus \mathcal{B}^{\#}(w,2^j\delta)$ (since $D(z,\zeta) \simeq 2^j\delta$). So

$$\mathbf{T}_1\mathbf{T}_2\phi(z) = \int_\Omega T_1(z,\zeta)\mathbf{T}_2(\phi)(\zeta)dV_q(\zeta)$$
$$= \sum \lambda_j \mathbf{T}_1(\phi_j)(z)$$

so that

$$\sup_{z\in\mathcal{B}^{\#}(w,\delta)} |\mathbf{T}_1\mathbf{T}_2(\phi)(z)| \le \sum \lambda_j \sup_{z\in\mathcal{B}^{\#}(w,\delta)} |\mathbf{T}_1(\phi_j)(z)|$$
$$\le C\delta^{r_1+s_1}\mu(w,\delta)^{r_2+s_2}\sum_{j=1}^{\infty}(2^j)^\varrho$$
$$\le C\delta^{r_1+s_1}\mu(w,\delta)^{r_2+s_2}.$$

This gives the result.

The next two propositions give simple rules of functional calculus in the operators of Bergman type.

PROPOSITION 2.2. *Let $q > -1$ and $\mathbf{K}$ be an operator in $\mathcal{B}_{(q,\mathbf{r})}$. Let us consider the operator $\tilde{\mathbf{K}}$ defined by*

$$\tilde{\mathbf{K}}f = \mathbf{K}((-r)f).$$

Then, $\tilde{\mathbf{K}} \in \mathcal{B}_{(q+1,\mathbf{r}+(1,0))}$.

PROOF. It is obvious that the kernel of $\tilde{\mathbf{K}}$ is equal to $K(z,\zeta)$ and that it satisfies the right pointwise estimate since

$$Vol_q(\mathcal{B}^{\#}(z,\zeta)) \times D(z,\zeta) \simeq Vol_{q+1}(\mathcal{B}^{\#}(z,\zeta)).$$

Let us consider the action of $\tilde{\mathbf{K}}$ on bump functions. Let φ be any bump function supported in $\mathcal{B}^{\#}(w,\delta)$ of order sufficiently large. With the notations of Definition 2.2,

$$\sup_{z\in\mathcal{B}^{\#}(w,\delta)} |Y_1...Y_l\tilde{\mathbf{K}}(\varphi)|(z)$$
$$= \sup_{z\in\mathcal{B}^{\#}(w,\delta)} |Y_1...Y_l\mathbf{K}((-r)\varphi)|(z) \le C\delta^{r_1+1-\beta}\mu(w,\delta)^{r_2-\alpha}$$

since $C\frac{(-r)}{\delta}\varphi$ is a bump function. The same holds for the adjoint in $L^2(dV_q)$ since $\tilde{\mathbf{K}}^*(\varphi) = \mathbf{K}^*((-r)\varphi)$.

PROPOSITION 2.3. *Let $q > -1$ and $\mathbf{K}$ be an operator in $\mathcal{B}_{(q,\mathbf{r})}$. Let $s(z,\zeta)$ be a $\mathcal{C}^\infty$-function in $\overline{\Omega} \times \overline{\Omega}$ satisfying*

$$s(z,\zeta) = \mathcal{O}(\mu(z,D(z,\zeta))^\theta))$$

for $\theta \in \{0,1\}$. Let us consider the operator $\tilde{\mathbf{K}}$ defined by

$$\tilde{\mathbf{K}}f(z) = T\mathbf{K}(s(z,.)f)(z).$$

Then, $\tilde{\mathbf{K}} \in \mathcal{B}_{(q,\mathbf{r}+(-1,\theta))}$.

We first prove the following lemma which will be useful in the following:

LEMMA 2.1. *For any $q > -1$, there exists a $\mathcal{C}^\infty$-function a_q in $\overline{\Omega}$ so that, for every couple of $\mathcal{C}^\infty$-functions φ and ψ, the following holds:*

$$\int_\Omega T\varphi(\zeta)\psi(\zeta)dV_q(\zeta) = -\int_\Omega \varphi(\zeta)[T\psi(\zeta) + a_q(\zeta)\psi(\zeta)]dV_q(\zeta).$$

PROOF. It follows from integration by parts since T is tangential and since there exists a $\mathcal{C}^\infty$-function b in $\overline{\Omega}$ so that $Tr(\zeta) = b(\zeta)r(\zeta)$.

PROOF OF PROPOSITION 2.3. Assume $\theta = 1$. The other case is similar. It is obvious that the kernel of $\tilde{\mathbf{K}}$ is equal to $T_z K(z, \zeta)s(z, \zeta)$ and that it satisfies the right pointwise estimate. So is the kernel of its adjoint.

Let us consider the action of $\tilde{\mathbf{K}}$ and of its adjoint on bump functions. Let φ be any bump function supported in $\mathcal{B}^\#(w, \delta)$ of order sufficiently large.

$$\sup_{z \in \mathcal{B}^\#(w,\delta)} |\tilde{\mathbf{K}}(\varphi)|(z) = \sup_{z \in \mathcal{B}^\#(w,\delta)} |T\mathbf{K}(s(z,.)\varphi)|(z) \leq C\delta^{r_1-1}\mu(w,\delta)^{r_2+1}$$

since $C\frac{s(z,\zeta)}{\mu(z,\delta)}\varphi$ is bump for any $z \in \mathcal{B}^\#(w, \delta)$.

$$\tilde{\mathbf{K}}^*(\varphi(z)) = \int_\Omega T_\zeta \overline{K}(\zeta, z)\overline{s}(\zeta, z)\varphi(\zeta)dV_q(\zeta)$$
$$= \mathbf{K}^*(\overline{s}(., z)T\varphi + a_q(z, .)\varphi)(z)$$

where a_q is a smooth function, by Lemma 2.1. So, it gives

$$\sup_{z \in \mathcal{B}^\#(w,\delta)} |\tilde{\mathbf{K}}^*(\varphi)(z)| \leq C(\delta^{r_1-1}\mu(w,\delta)^{r_2+1} + \delta^{r_1}\mu(w,\delta)^{r_2})$$

$$\leq C\delta^{r_1-1}\mu(w,\delta)^{r_2+1}$$

since $C\frac{\delta}{\mu(w,\delta)}\overline{s}(., z)T\varphi$ is a bump function for any $z \in \mathcal{B}^\#(w, \delta)$ and $\delta\mu(w,\delta)^{-1} \leq \delta^{1/2} \leq C$. The estimates for the derivatives are similar.

3. The weighted Bergman projection

The aim of this paragraph is to prove the following result which has been roughly stated in the introduction.

THEOREM 3.1. *The weighted Bergman projection $\mathbf{B}_q$ belongs to the class $\mathcal{B}_{(q,0)}$ for any $q \in \mathbb{N}$.*

PROOF. We are going to prove this result by induction on q. As we noticed before, the result is well known when $q = 0$ (see [McN] and [NRSW]). Now, assume that $\mathbf{B}_{q-1} \in \mathcal{B}_{(q-1,0)}$ for some $q \geq 1$. We want to prove that $\mathbf{B}_q \in \mathcal{B}_{(q,0)}$.

We first prove the following proposition.

PROPOSITION 3.1. *There exist an operator $\mathbf{H}_q \in \mathcal{B}_{(q,0)}$ and an operator $\mathbf{E}_q \in \mathcal{B}_{(q,(0,1))}$ so that*

$$\mathbf{B}_q = \mathbf{H}_q^* + \mathbf{E}_q\mathbf{B}_q = \mathbf{H}_q - \mathbf{B}_q\mathbf{E}_q.$$

Here again the adjoint is taken in $L^2(dV_q)$.

REMARK. The idea is to construct $\mathbf{H}_q$ as an oblique projection in $L^2(dV_q)$. Such a construction has been done in [**S**].

PROOF OF PROPOSITION 3.1. First, since by assumption $\overline{R}r = 1$ on $\partial\Omega$, there exist two functions b_1, b_2 in $\mathcal{C}^\infty(\overline{\Omega})$, b_1 real-valued on $\partial\Omega$, so that

$$1 = b_1(\zeta)\overline{R}r(\zeta) + b_2(\zeta)r(\zeta).$$

Let f be a holomorphic function in $\mathcal{C}^\infty(\overline{\Omega})$. By the reproducing property, we have, for any $z \in \Omega$,

$$
\begin{aligned}
f(z) &= \int_\Omega B_{q-1}(z,\zeta)f(\zeta)dV_{q-1}(\zeta) \\
&= \int_\Omega B_{q-1}(z,\zeta)f(\zeta)[b_1(\zeta)\overline{R}r(\zeta) + b_2(\zeta)r(\zeta)]dV_{q-1}(\zeta) \\
&= -\int_\Omega B_{q-1}(z,\zeta)f(\zeta)b_1(\zeta)\frac{\overline{R}(-r(\zeta))^q}{q}dV(\zeta) \\
&\quad + \int_\Omega B_{q-1}(z,\zeta)f(\zeta)b_2(\zeta)r(\zeta)dV_{q-1}(\zeta) \\
&= \int_\Omega [\frac{b_1(\zeta)}{q}\overline{R}_\zeta B_{q-1}(z,\zeta) + b_3(\zeta)B_{q-1}(z,\zeta)]f(\zeta)dV_q(\zeta) \\
&= \int_\Omega [-\frac{2i}{q}b_1(\zeta)T_\zeta B_{q-1}(z,\zeta) + b_3(\zeta)B_{q-1}(z,\zeta)]f(\zeta)dV_q(\zeta)
\end{aligned}
$$

for some smooth function b_3, by integration by parts ($r = 0$ on $\partial\Omega$ and f and $\overline{B}(z,.)$ are holomorphic).

So, there exist two $\mathcal{C}^\infty(\overline{\Omega})$-functions A_1 and A_2, A_1 real valued on $\partial\Omega$, so that

$$H_q(z,\zeta) = -iA_1(\zeta)T_\zeta B_{q-1}(z,\zeta) + A_2(\zeta)B_{q-1}(z,\zeta)$$

is a reproducing kernel for the holomorphic functions in $\mathcal{C}^\infty(\overline{\Omega})$ with respect to the measure dV_q. So, if $f \in \mathcal{C}^\infty(\overline{\Omega})$, then, since by [**L1**], $\mathbf{B}_q f \in \mathcal{C}^\infty(\overline{\Omega})$, we get

$$\mathbf{B}_q f = \mathbf{H}_q \mathbf{B}_q f.$$

Furthermore, by construction, $\mathbf{H}_q^*(f - \mathbf{B}_q f) = 0$ since the kernel of $\mathbf{H}_q^*$:

$$H_q^*(z,\zeta) = i\overline{A_1}(z)T_z B_{q-1}(z,\zeta) + \overline{A_2}(z)B_{q-1}(z,\zeta)$$

is antiholomorphic in ζ.

We get

$$\mathbf{B}_q f = \mathbf{H}_q^* f + (\mathbf{H}_q - \mathbf{H}_q^*)\mathbf{B}_q f = \mathbf{H}_q f - \mathbf{B}_q(\mathbf{H}_q - \mathbf{H}_q^*)f$$

for $f \in \mathcal{C}^\infty(\overline{\Omega})$ and hence, for $f \in L^2(dV_q)$ by density.

It remains to show that $\mathbf{H}_q \in \mathcal{B}_{(q,0)}$ and that $\mathbf{E}_q = \mathbf{H}_q - \mathbf{H}_q^* \in \mathcal{B}_{(q,(0,1))}$.

First, let us consider $\mathbf{H}_q$. By definition,

$$\mathbf{H}_q^*(f)(z) = i\overline{A_1}(T\mathbf{B}_{q-1})((-r)f)(z) + \overline{A_2}\mathbf{B}_{q-1}((-r)f)(z).$$

By Propositions 2.2 and 2.3, since, by assumption $\mathbf{B}_{q-1} \in \mathcal{B}_{(q-1;\mathbf{0})}$, we get that $\mathbf{H}_q \in \mathcal{B}_{q,\mathbf{0}}$.

Let us consider now the operator $\mathbf{E}_q$. We need first the following lemma.

LEMMA 3.1. *Let q be an integer greater than 1. Up to operators in $\mathcal{B}_{q,(0,1)}$,* $\mathbf{E}_q(f)$ *is equal to*

$$i\left(\mathbf{B}_{q-1}T(A_1(-r)f) - T\mathbf{B}_{q-1}(A_1(-r)f)\right).$$

PROOF. Let us consider first the term corresponding to the kernel

$$-iA_1(\zeta)T_\zeta B_{q-1}(z,\zeta).$$

We have

$$-i\int_\Omega T_\zeta B_{q-1}(z,\zeta)A_1(\zeta)f(\zeta)dV_q(\zeta)$$

$$=i\int_\Omega B_{q-1}(z,\zeta)[T((-r)A_1f)(\zeta) + A_3(\zeta)(-r(\zeta))f(\zeta)]dV_{q-1}(\zeta)$$

$$=i\mathbf{B}_{q-1}T(A_1(-r)f) + \mathbf{B}_{q-1}(A_3(-r)f)(z).$$

So,

$$\begin{aligned}
\mathbf{E}_q(f)(z) &= \mathbf{B}_{q-1}((A_2+A_3)(-r)f)(z) - \overline{A_2}(z)\mathbf{B}_{q-1}((-r)f)(z) \\
&\quad + i\mathbf{B}_{q-1}T(A_1(-r)f)(z) - i\overline{A_1}(z)T\mathbf{B}_{q-1}((-r)f)(z) \\
&= \mathbf{B}_{q-1}((A_2+A_3)(-r)f)(z) - \overline{A_2}(z)\mathbf{B}_{q-1}((-r)f)(z) \\
&\quad + i\mathbf{B}_{q-1}T(A_1(-r)f)(z) - iT\mathbf{B}_{q-1}(A_1(-r)f)(z) \\
&\quad + i[A_1(z) - \overline{A_1}(z)]T\mathbf{B}_{q-1}((-r)f)(z) \\
&\quad + iT\mathbf{B}_{q-1}([A_1 - A_1(z)](-r)f)(z).
\end{aligned}$$

Remark that $A_1(\zeta) - A_1(z) = \mathcal{O}(\mu(z, D(z,\zeta)))$ and that, since A_1 is real valued on $\partial\Omega$, $A_1(z) - \overline{A_1}(z) = \mathcal{O}(\mu(z, D(z,\zeta)))$. So, by Propositions 2.2 and 2.3, the corresponding operators are in $\mathcal{B}_{q,(0,1)}$. Similarly, by Proposition 2.2, the two first terms are in $\mathcal{B}_{q,(1,0)}$, so in $\mathcal{B}_{q,(0,1)}$.

It remains to prove that the remaining term is in the good class. To do it, we prove the following lemma.

LEMMA 3.2. *There exist two $\mathcal{C}^\infty$-functions c_q and d_q in $\overline{\Omega} \times \overline{\Omega}$ so that*

$$\mathbf{B}_{q-1}T((-r)f) - T\mathbf{B}_{q-1}((-r)f) =$$

$$= \mathbf{B}_{q-1}[d_q(-r)f] + \mathbf{B}_{q-1}(\overline{c_q}\mathbf{B}_{q-1}((-r)f)) + [\mathbf{R}_1^* + \mathbf{R}_1]((-r)f)$$

where

$$\mathbf{R}_1 f(z) = \sum_j [I - \mathbf{B}_{q-1}]\left[\frac{\partial}{\partial z_j}\mathbf{B}_{q-1}\right](s_j(z,.)(-r)f)(z)$$

with $s_j(z,\zeta) = \mathcal{O}(\mu(z, D(z,\zeta))$ and $\mathbf{R}_1^*$ is the adjoint of $\mathbf{R}_1$ in $L^2(dV_{q-1})$.

REMARK. We are going to show that the right hand side is smoothing. Close commutation properties can be found in the context of NIS operators (see [**CNS**]).

END OF THE PROOF OF PROPOSITION 3.1. Assume Lemma 3.2 proved. By Propositions 2.2, 2.3, we get that the two first operators appearing in the preceding lemma are in $\mathcal{B}_{(q,(1,0))}$. It remains to consider $\mathbf{R}_1$. Let us prove that the operator defined by $\tilde{\mathbf{K}}(f)(z) = \sum_j \frac{\partial}{\partial z_j} \mathbf{B}_{q-1}(s_j(z,.)(-r)f)(z)$ is in $\mathcal{B}_{(q,(0,1))}$. It is first obvious that the kernels of $\tilde{K}$ and its adjoint satisfy the right estimates. Assume now that φ is a bump function of order sufficiently large supported in some $\mathcal{B}^{\#}(w,\delta)$. It follows immediately from the definition of $\mathcal{B}_{(q-1,0)}$ that $\tilde{K}(\varphi)$ satisfies the right property. Let us consider the action of its adjoint.

$$\tilde{\mathbf{K}}^*(\varphi)(z) = \sum_j \int_\Omega \frac{\partial}{\partial \zeta_j} B_{q-1}(z,\zeta)\overline{s_j}(\zeta,z)\varphi(\zeta)dV_q(\zeta)$$

$$= \sum_j \int_\Omega B_{q-1}(z,\zeta)\left[\overline{s_j}(\zeta,z)\frac{\partial \varphi}{\partial \zeta_j}(\zeta)(-r)(\zeta) + a_q(z,\zeta)\varphi(\zeta)\overline{s_j}(\zeta,z)+\right.$$

$$\left.+ b_q(z,\zeta)\varphi(\zeta)(-r)(\zeta)\right]dV_{q-1}(\zeta)$$

by integration by parts, there is no term on $\partial\Omega$ since $q \geq 1$,

$$= \sum_j \mathbf{B}_{q-1}[\overline{s_j}(.,z)\frac{\partial \varphi}{\partial \zeta_j}(-r) + a_q(z,.)\varphi\overline{s_j}(.,z) + b_q(z,.)\varphi(-r)]$$

and the result follows from the assumption on $\mathbf{B}_{q-1}$ and from the fact that

$$C\frac{\overline{s_j}(.,z)}{\mu(w,\delta)}(-r)\nabla\varphi$$

is a bump function for any $z \in \mathcal{B}^{\#}(w,\delta)$.

PROOF OF LEMMA 3.2. Denote by $<,>$ the scalar product in $L^2(dV_{q-1})$. Write $T_z B_{q-1}(z,\zeta)$ as $\sum_{j=1}^2 b_j(z)\frac{\partial}{\partial z_j} B_{q-1}(z,\zeta)$ (by holomorphy in z). We have

$$< T\mathbf{B}_{q-1}((-r)f)(z), g(z) - \mathbf{B}_{q-1}g(z) >$$

$$= \sum_j < \frac{\partial}{\partial z_j}\mathbf{B}_{q-1}(b_j(z) - b_j)((-r)f)(z), g(z) - \mathbf{B}_{q-1}g(z) >$$

$$+ \sum_j < \frac{\partial}{\partial z_j}\mathbf{B}_{q-1}(b_j(-r)f)(z), g(z) - \mathbf{B}_{q-1}g(z) >$$

$$= \sum_j < [I - \mathbf{B}_{q-1}]\left[\frac{\partial}{\partial z_j}\mathbf{B}_{q-1}\right]((b_j(z) - b_j)(-r)f)(z), g(z) >$$

$$= < \mathbf{R}_1((-r)f), g > .$$

Similarly

$$< [I - \mathbf{B}_{q-1}]((-r)f)(z), T\mathbf{B}_{q-1}(g)(z) > = < \mathbf{R}_1^*((-r)f), g > .$$

So,

$$< T\mathbf{B}_{q-1}((-r)f), g >=< T\mathbf{B}_{q-1}((-r)f), \mathbf{B}_{q-1}g > + < \mathbf{R}_1((-r)f), g >$$
$$= - < \mathbf{B}_{q-1}((-r)f), T\mathbf{B}_{q-1}g > + < \mathbf{B}_{q-1}((-r)f), c_q\mathbf{B}_{q-1}g >$$
$$+ < \mathbf{R}_1((-r)f), g >$$

by integration by parts, Lemma 2.1

$$= - < (-r)f, T\mathbf{B}_{q-1}g > + < [\mathbf{R}_1^* + \mathbf{R}_1]((-r)f), g >$$
$$+ < \mathbf{B}_{q-1}((-r)f), c_q\mathbf{B}_{q-1}g >$$
$$=< T((-r)f), \mathbf{B}_{q-1}g > + < d_q(-r)f), \mathbf{B}_{q-1}g > +$$
$$+ < [R_1^* + R_1]((-r)f), g > + < \mathbf{B}_{q-1}((-r)f), c_q\mathbf{B}_{q-1}g >$$

by integration by parts, Lemma 2.1

$$=< \mathbf{B}_{q-1}T((-r)f), g > + < \mathbf{B}_{q-1}[d_q(-r)f]), g > +$$
$$+ < [\mathbf{R}_1^* + \mathbf{R}_1]((-r)f), g > + < \mathbf{B}_{q-1}(\overline{c_q}\mathbf{B}_{q-1}((-r)f)), g > .$$

So, it gives that the remaining term is equal to

$$\mathbf{B}_{q-1}[d_q(-r)f] + \mathbf{B}_{q-1}(\overline{c_q}\mathbf{B}_{q-1}((-r)f)) + [\mathbf{R}_1^* + \mathbf{R}_1]((-r)f).$$

END OF THE PROOF OF THEOREM 3.1. By Proposition 3.1, we have

$$\mathbf{B}_q = \mathbf{H}_q^* + \mathbf{E}_q\mathbf{B}_q.$$

So, by iteration,

$$\mathbf{B}_q = \sum_{j=0}^{N-1} \mathbf{E}_q^j\mathbf{H}_q^* + \mathbf{E}_q^N\mathbf{B}_q.$$

Taking the adjoint in $L^2(dV_q)$, we have also

$$\mathbf{B}_q = \sum_{j=0}^{N-1} \mathbf{H}_q(\mathbf{E}_q^*)^j + \mathbf{B}_q(\mathbf{E}_q^*)^N.$$

By substitution in the above, we get

$$\mathbf{B}_q = \sum_{j=0}^{N-1} \mathbf{E}_q^j\mathbf{H}_q^* + \mathbf{E}_q^N \sum_{j=0}^{N-1} \mathbf{H}_q(\mathbf{E}_q^*)^j + \mathbf{E}_q^N\mathbf{B}_q(\mathbf{E}_q^*)^N$$
$$= \mathbf{B}_{q,N} + \mathbf{E}_q^N\mathbf{B}_q(\mathbf{E}_q^*)^N$$

where $\mathbf{B}_{q,N}$ is, by Proposition 2.1, an operator in $\mathcal{B}_{(q,0)}$ for each N. Therefore it suffices to show that, for each fixed l, N can be chosen large enough so that $Y_1...Y_l\mathbf{E}_q^N\mathbf{B}_q(\mathbf{E}_q^*)^N(z,\zeta)$ is bounded. This kernel corresponds to an operator of the form $Y_1...Y_{l_1}\mathbf{E}_q^N\mathbf{B}_q(\mathbf{E}_q^*)^N(Y_1...Y_{l_2})^*$ for some l_1, l_2 so that $l_1 + l_2 = l$. So, it suffices to prove that for N large enough, this operator sends $L^1(dV_q)$ into L^∞. Since $\mathbf{B}_q$ maps $L^2(dV_q)$ into itself, it remains to show that for N large enough, $Y_1...Y_{l_1}\mathbf{E}_q^N$ maps $L^2(dV_q)$ into L^∞ and that $(\mathbf{E}_q^*)^N(Y_1...Y_{l_2})^*$ maps $L^1(dV_q)$ into $L^2(dV_q)$. By Proposition 2.1, for M large enough, we may assume that $Y_1...Y_{l_1}\mathbf{E}_q^M$ is in $\mathcal{B}_{(q,0)}$. So, we can see $Y_1...Y_{l_1}\mathbf{E}_q^N$ as a product of $N - M$ operators in $\mathcal{B}_{(q,(0,1))}$. Now, let $\mathbf{K}$ be an operator in $\mathcal{B}_{(q,(0,1))}$. From the pointwise estimates of the kernel of $\mathbf{K}$, it is easy to see that

$$\int_\Omega |K(z,\zeta)|^r dV_q(\zeta) \leq C$$

independently of z as long as

$$1 \leq r \leq \frac{2m + mq + 2}{2m + mq + 1}.$$

The same estimate holds for the adjoint of $\mathbf{K}$.

Hence such an operator maps $L^1(dV_q)$ into $L^r(dV_q)$ and $L^{r'}(dV_q)$ into L^∞. By interpolation, we conclude that such an operator maps $L^{p_1}(dV_q)$ into $L^{p_2}(dV_q)$ if $1/p_2 = 1/p_1 - 1/r'$. It gives the result.

¿From the explicit form of $\mathbf{H}_q$ and from the smoothing property of $\mathbf{E}_q$, we obtain the following (which gives (1) and (2) in the introduction).

THEOREM 3.2. *Let $q \geq 1$. There exists a differential operator D with $\mathcal{C}^\infty(\overline{\Omega})$-coefficients so that*

$$B_q(z,\zeta) = D_\zeta B_{q-1}(z,\zeta) + F_q(z,\zeta)$$

where F_q is the kernel of a smoothing operator $\mathbf{F}_q$ in $\mathcal{B}_{(q,(0,1))}$.

PROOF. Simply take $D_\zeta B_{q-1}(z,\zeta) = H_q(z,\zeta)$. Then $\mathbf{F}_q$ is the operator $-\mathbf{B}_q\mathbf{E}_q$. Use Theorem 3.1, Propositions 3.1 and 2.1.

COROLLARY 3.1. *Let $q \geq 1$. There exist two $\mathcal{C}^\infty(\overline{\Omega})$-functions a_1 and a_2 so that*

$$\mathbf{B}_q = \mathbf{B}_{q-1} + \mathbf{B}_{q-1}\left((-r)\sum_{j=1}^2 a_j \frac{\partial}{\partial \overline{z_j}}\right) + \mathbf{F}_q.$$

PROOF. In the beginning of the proof of Proposition 3.1, we wrote $\mathbf{B}_{q-1}f$ for a holomorphic function. From the same integration by parts, if f is not holomorphic, we get

$$\mathbf{B}_{q-1}f(z) = \mathbf{H}_q f(z) - \int_\Omega B_{q-1}(z,\zeta)[\overline{R}f(\zeta)]b_1(\zeta)dV_q(\zeta).$$

So,

$$\mathbf{B}_{q-1} = \mathbf{B}_q - \mathbf{B}_{q-1}\left((-r)\sum_j a_j \frac{\partial}{\partial \overline{z_j}}\right) - \mathbf{F}_q.$$

4. Sobolev inequalities for the weighted Bergman projections

To prove Sobolev inequalities, we rely on commutation properties for $\mathbf{B}_q$. We refer to [**BCG**] where the proof is given for $q = 0$. The general case needs little changes.

PROPOSITION 4.1. *Let* $k \in \mathbb{N}$.

(a) *There exist operators* $\{\mathbf{B}_{q,j}\}_{j=0}^k$ *in* $\mathcal{B}_{(q,0)}$ *so that*

$$\nabla^k \mathbf{B}_q = \sum_{j=0}^k \mathbf{B}_{q,j} T^j.$$

(b) *Let* $\{Y_j\}_{j=0}^k$ *be* k *vector fields, each of which is* L *or* $\overline{L}$, *then there exist operators* $\{\mathbf{B}_{q,j}\}_{j=0}^k$ *in* $\mathcal{B}_{(q,0)}$ *so that*

$$Y_1...Y_k \mathbf{B}_q = \sum_{j=0}^k \mathbf{B}_{q,j} L^j.$$

The next proposition is as in [**BCG**] a consequence of Schur's Lemma which is well known to give L^p estimates (see [**FR**]).

PROPOSITION 4.2. *Let* $\mathbf{K}$ *in* $\mathcal{B}_{(q,0)}$. *Then* $\mathbf{K}$ *maps* $L^p(dV_\alpha)$ *into itself for* $1 < p < \infty$ *and* $-1 < \alpha < qp + p - 1$.

Let k be an integer.
Let $W^{p,k,\alpha}(\Omega)$ be the usual Sobolev space related to the measure dV_α.
Define

$$W_T^{p,k,\alpha}(\Omega) = \{f \in L^p(\Omega, dV_\alpha); \quad T^j f \in L^p(\Omega, dV_\alpha), \quad 1 \le j \le k\},$$
$$W_{anis}^{p,k,\alpha}(\Omega) = \{f \in L^p(\Omega, dV_\alpha); \ Y_1...Y_j f \in L^p(\Omega, dV_\alpha), \ 1 \le j \le k, \ Y_j \in \{L, \overline{L}\}\},$$

and

$$W_L^{p,k,\alpha}(\Omega) = \{f \in L^p(\Omega, dV_\alpha); \quad L^j f \in L^p(\Omega, dV_\alpha), \quad 1 \le j \le k\}.$$

The anisotropic spaces are the spaces defined in [**CNS**].

THEOREM 4.1. *The weighted Bergman projection maps the space* $W_T^{p,k,\alpha}(\Omega)$ *into* $W^{p,k,\alpha}(\Omega)$ *and* $W_L^{p,k,\alpha}(\Omega)$ *into* $W_{anis}^{p,k,\alpha}(\Omega)$ *for* $1 < p < \infty$ *and* $-1 < \alpha < qp + p - 1$. *In particular, the isotropic and anisotropic spaces,* $W^{p,k,\alpha}(\Omega)$ *and* $W_{anis}^{p,k,\alpha}(\Omega)$, *are left invariant by* $\mathbf{B}_q$.

One can as well write Lipschitz estimates for $\mathbf{B}_q$, following [**NRSW**] and [**McNS**] or more precisely [**BCG**] where the case $q = 0$ is considered.

REMARK. We have considered $\mathbf{F}_q$ as a smoothing operator. This is justified by the following proposition.

PROPOSITION 4.3. *The operator* $\mathbf{F}_q$ *maps* $W_L^{p,k,\alpha}(\Omega)$ *into* $W_{anis}^{p,k+1,\alpha}(\Omega)$ *for* $1 < p < \infty$ *and* $-1 < \alpha < qp + p - 1$.

SKETCH OF THE PROOF. For $k = 0$, the result follows from the fact that $L\mathbf{F}_q \in \mathcal{B}_{q,0}$.

For simplicity, we will just consider the case $k = 1$. Since $\mathbf{F}_q = -\mathbf{B}_q \mathbf{E}_q$, by Proposition 4.1, it is sufficient to prove $L^2 \mathbf{E}_q f \in L^p$ if f and Lf are in L^p.

Coming back to the proof of Proposition 3.1, one sees that $\mathbf{E}_q f$ involves terms as

$$\mathbf{B}_{q-1}(\alpha_1(-r)f), \quad \mathbf{B}_{q-1}(\alpha_2\mathbf{B}_{q-1}((-r)f)), \quad \mathcal{O}(\mu)\nabla\mathbf{B}_{q-1}((-r)f),$$
$$\nabla\mathbf{B}_{q-1}(\mathcal{O}(\mu)(-r)f), \quad \mathbf{B}_{q-1}(\mathcal{O}(\mu)\nabla B_{q-1}((-r)f)$$

for some smooth functions α_j, $j = 1, 2$.

Using Proposition 4.1 for $\mathbf{B}_{q-1}$, we obtain that

$$L\mathbf{E}_q f = \mathbf{E}_q^1 Lf + \mathbf{E}_q^0 f$$

where $\mathbf{E}_q^j \in \mathcal{B}_{q,(0,1)}$, $j = 0, 1$, from which the result follows.

As a corollary, we obtain the following result on $\mathbf{B}_q - \mathbf{B}_{q-1}$ which can be compared to the results of [**Cu**] in the strictly pseudo-convex case.

COROLLARY 4.4.

$$||(\mathbf{B}_q - \mathbf{B}_{q-1})u||_{W_{anis}^{p,2,0}(\Omega)} \leq C||\bar{\partial}u||_{L^p}.$$

PROOF. We may assume that u is the minimal solution of $\bar{\partial}v = f$ where $f = \bar{\partial}u \in L^p$. Then, by [**CNS**],

$$||u||_{W_{anis}^{p,1,0}(\Omega)} \leq C||\bar{\partial}u||_{L^p},$$

so

$$||\mathbf{F}_q u||_{W_{anis}^{p,2,0}(\Omega)} \leq C||\bar{\partial}u||_{L^p}$$

while

$$\left\|\mathbf{B}_{q-1}\left((-r)\sum_j a_j\frac{\partial u}{\partial\bar{z}_j}\right)\right\|_{W_{anis}^{p,2,0}(\Omega)} \leq C\left\|\mathbf{B}_{q-1}\left((-r)\sum_j a_j\frac{\partial u}{\partial\bar{z}_j}\right)\right\|_{W^{p,1,0}(\Omega)}$$
$$\leq C||\bar{\partial}u||_{L^p}$$

since $W^{p,1,0}(\Omega) \cap \mathcal{H}(\Omega) \subset W_{anis}^{p,2,0}(\Omega) \cap \mathcal{H}(\Omega)$.

It follows in particular that the minimal solution u, for the ordinary measure dV, and the solution $u - \mathbf{B}_q u$ are, in some sense, close.

REFERENCES

[BC]   A. Bonami & Ph. Charpentier, *Comparing the Bergman and Szegö projections*, Math. Zeit. **204** (1990), 225-233.

[BCG]   A. Bonami, D.C. Chang & S. Grellier, *Commutation Properties and Lipschitz estimates for the Bergman and Szegö projections*, Preprint (1994).

[CNS]   D.C. Chang, A. Nagel, & E.M. Stein, *Estimates for the $\bar{\partial}$-Neumann problem in pseudoconvex domains of finite type in $\mathbb{C}^n$*, Acta Math **169** (1992), 153-228.

[Co]   W. Cohn, *Weighted Bergman projections and tangential area integrals*, Studia Math. **106 (1)** (1993), 59-76.

[Cu]   A. Cumenge, *Comparaison des projecteurs de Bergman et Szegö et applications*, Arkiv för Mat. **28** (1990), 23-47.

[FR]   F. Forelli & W. Rudin, *Projections on spaces of holomorphic functions in balls*, Indiana Univ. Math. J. **24** (1974), 593-602.

[L1]   E. Ligocka, *Forelli-Rudin Constructions and weighted Bergman projections*, Studia Math. **94** (1989), 257-272.

[L2]   E. Ligocka, *The regularity of the weighted Bergman projection, Lecture Notes in Math: Seminar of Deformation Theory 1982/1984* **1165** (1985), Springer, 197-203.

[McN]   J. McNeal, *Boundary Behavior of the Bergman kernel function in* $\mathbb{C}^2$, Duke Math J **58** (1989), 499-512.

[McNS]   J. McNeal & E.M. Stein, *Mapping properties of the Bergman projection on convex domains of finite type*, Duke Math J **73** (1994), 177-199.

[NRSW]   A. Nagel, J.-P. Rosay, E.M. Stein, & S. Wainger, *Estimates for the Bergman and Szegö kernels in* $\mathbb{C}^2$, Ann. Math **129** (1989), 113-149.

[P]   M. Peloso, *Sobolev regularity of the weighted Bergman Projections*, Preprint (1993).

[S]   F. Symesak, *Décomposition atomique des espaces de Bergman*, Prépublication et Thèse de l'université d'Orléans (1994), France.

URA 1803, MATHÉMATIQUES, UNIVERSITÉ D'ORLÉANS, BP 6759, 45067 ORLÉANS CEDEX 2, FRANCE

E-mail address: bonami@anh.matups.fr

DÉPARTMENT DE MATHÉMATIQUES, UNIVERSITÉ PARIS SUD, BÂT. N° 425, 91405 ORSAY CEDEX, FRANCE

E-mail address: grellier@anh.matups.fr

Contemporary Mathematics
Volume **189**, 1995

COMMUTATORS OF SINGULAR INTEGRALS AND FRACTIONAL INTEGRALS ON HOMOGENEOUS SPACES

MARCO BRAMANTI AND M. CRISTINA CERUTTI

ABSTRACT. $L^p - L^q$ estimates are proved for commutators of singular and fractional integral operators with multiplication with a *BMO* function on homogeneous spaces. For both types of operators the exponents p and q have the same values as in the corresponding theorems in Euclidean space. A characterization of *BMO* functions is also proved.

1. INTRODUCTION

This paper deals with commutators of integral operators with multiplication by a *BMO* function on a homogeneous space. The results we obtain confirm and sometimes even improve the results that have been proved to hold in R^n. More precisely, let (X, d, μ) be a homogeneous space (X a set, d a quasidistance, μ a Borel measure: see beginning of sect. 2 for the precise definition) and let T be either a singular integral operator or a fractional integral operator and $a \in BMO(X)$, (definition in sect. 2). In this paper we prove that the commutator $[a, T]$ defined as $[a, T]f = [a \cdot Tf - T(af)]$ is a bounded continuous operator $[a, T] : L^p(X) \to L^q(X)$ satisfying

$$(1.1) \qquad \|[a, T]f\|_q \le C \|a\|_* \|f\|_p$$

where $\|a\|_*$ is the *BMO* semi-norm of a, $p = q$ with $1 < p < \infty$ when T is a singular integral operator, and $\dfrac{1}{q} = \dfrac{1}{p} - \alpha$ with $1 < p < \dfrac{1}{\alpha}$ when T is a fractional integral operator ($\alpha \in (0, 1)$ is the number appearing in the fractional integral kernel, see sect.2). For the precise statement of these results see theorems 2.9, 2.10

1991 *Mathematics Subject Classification.* Primary 43A85; Secondary 42B20 .

Key words and phrases. Singular integrals, fractional integrals, commutators, homogenous spaces.

This work was supported in part by the Italian CNR-GNAFA

81

and 2.12. These results were known to hold for Eucledean spaces (see [6] for the result for singular integrals and [4] for the results for fractional integrals) and were first proved for homogeneous spaces in [1] and [2]. In the case of fractional integrals we actually prove a result which is stronger than (1.1) and therefore stronger than the result proved by Chanillo in [4] for Euclidean spaces. Our result is contained in theorem 2.11 and in particular implies that the estimate (1.1) holds when we replace the fractional integral kernel with an equivalent kernel. Here we also prove a "converse theorem" which is completely new, i.e. we prove that the validity of theorem 2.11 for any triple p, q, α characterizes functions in $BMO(X)$. Chanillo in [4] has a result in this direction, where he proves, in the Euclidean case, that if an estimate of the kind (1.1) holds for a function a and a triple p, q, α with $n \cdot \alpha$ an even number, then $a \in BMO(R^n)$.

The proofs we present here of estimate (1.1) in the two cases, follow the main lines of the proofs in [1] and [2], though they are somewhat improved and simplified. The methods used in the two cases are very different. Here we would also like to discuss the applicability of those methods and to point out why, we believe, it is in the nature of each of these problems to be treated differently.

The method used to prove the result for singular integrals follows an idea of Torchinsky (see [13]) and makes use of powerful harmonic analysis tools and results that have been proved to hold in the context of homogeneous spaces, such as maximal function and sharp function estimates, as well as a John-Nirenberg lemma. A main point is the assumption in the definition of a singular integral operator that the kernel satisfies a pointwise Hörmander condition (see (2.4)). We point out that we prove the above estimates for singular integrals also in the case X is a finite measure space, under a few extra natural geometric conditions on X and the kernel of T.

In the case of fractional integral operators we could not use the above proof because the fractional integral kernel does not, in general, satisfy a Hörmander condition. It seemed instead more natural to follow an idea contained in [6] as a suggestion for an alternative proof for their L^p continuity of commutators of singular integrals on Euclidean spaces. As mentioned above, we prove a stronger result in this case. We point out that if we replace a kernel with an equivalent one in commutators of singular integral operators, in general, L^p estimates are no longer true, since in this case the cancellation properties of the kernels play a major role. The idea of the proof is to show that it is possible to reduce the problem of getting estimates (1.1) to that of obtaining certain *weighted* estimates for the operator T. The reduction is done by showing that a certain operator is a holomorphic function in a neighborhood of the origin of the complex plane and then using classic complex analysis tools. We need to assume a natural geometric condition on the space X (see property (**P**), (2.8) below) in order to assure that the proof of weighted estimates by Sawyer and Wheeden (see [12]) carries through to our case. Condition (2.8) is quite general, since it allows for X being a finite measure space as well as for existence of atoms in X.

The reason why we did not apply this proof in the singular integral case is that, to our knowledge, there are no general weighted estimates of the kind we would need for singular integrals. Moreover, while it seemed that these estimates may follow quite straightforwardly from the proofs in R^n in the case $\mu(X) = \infty$, it seemed that in the finite measure case it would have been harder to prove those estimates than to use Torchinsky's approach to obtain the commutator estimates directly.

2. DEFINITIONS, PRELIMINARY RESULTS AND STATEMENT OF MAIN RESULTS

Let X be a set. A *quasi-distance* on X is a function $d : X \times X \to [0, +\infty)$ such that

(i) $d(x, y) = 0$ if and only if $x = y$;

(ii) $d(x, y) = d(y, x)$ for every $x, y \in X$;

(iii) $d(x, y) \leq c_d (d(x, z) + d(z, y))$ for some $c_d \geq 1$ and for every $x, y, z \in X$.

The balls $B_r(x) = \{y \in X : d(x, y) < r\}$, $r > 0$, form a base for a complete system of neighborhoods of $x \in X$. The space X endowed with this topology (which we will refer to as the d-topology of X) is a Hausdorff space. We'll say that two quasi-distances $d(x, y)$ and $d'(x, y)$ on X are equivalent if there exists two positive constants a_1 and a_2 such that $a_1 d(x, y) \leq d'(x, y) \leq a_2 d(x, y)$ for every x and y in X. In particular equivalent quasi-distances induce the same topology on X.

Observe that the balls are not in general open sets in the d-topology. Let X be a set endowed with a quasi-distance $d(x, y)$ such that the balls are open sets in the d-topology and let μ be a positive Borel measure defined on a σ-algebra of subsets of X which contains the d-balls, for which there exists a constant, $c_\mu \geq 1$ such that the following *doubling condition* is satisfied

$$(2.1) \qquad 0 < \mu(B_{2r}(x)) \leq c_\mu \cdot \mu(B_r(x)) < \infty$$

for every $x \in X$ and $r > 0$. We will call (X, d, μ) a *homogeneous space*.

We'll say that a homogeneous space (X, d, μ) is *normal* if there exist two positive constants c_1 and c_2, such that

$$(2.2) \qquad c_1 r \leq \mu(B_r(x)) \leq c_2 r$$

for every $x \in X$ and every $r \in (\mu\{x\}, \mu(X))$.

The following three theorems, due to Macías and Segovia (see [10]), contain very useful results about quasi-distances and homogeneous spaces.

Theorem 2.1. *Let X be a homogeneous space. Then the set M of atoms of X (i. e. of points $x \in X$ with $\mu\{x\} > 0$) is countable and for each $x \in M$ there exists $r > 0$ such that $B_r(x) = \{x\}$.*

Theorem 2.2. *Let $d(x, y)$ be a quasi-distance on a set X. Then there exists a quasi-distance $d'(x, y)$ on X equivalent to d and a number $0 < \alpha < 1$, such that for every x, y and z in X and $r > 0$*

$$|d'(x, z) - d'(y, z)| \leq C [d'(x, z) + d'(y, z)]^{1-\alpha} (d'(x, y))^\alpha$$

whenever are both smaller than r. Observe that balls corresponding to the quasi-distance d' are open in the d'-topology.

Theorem 2.3. *Let (X, d, μ) be a homogeneous space. Then the function $\delta(x, y)$ defined as follows*

$$(2.3) \qquad \delta(x, y) = \begin{cases} \inf\{\mu(B) | B \text{ ball containing } x, y\} & \text{if } x \neq y \\ 0 & \text{if } x = y \end{cases}$$

is a quasi-distance on X. Moreover (X, δ, μ) is a normal space and the topologies induced by d and by δ coincide.

Another interesting known result, about homogeneous spaces is the following:

Theorem 2.4. *Let X be a homogeneous space. Then $\mu(X) < \infty$ if and only if X is bounded, i.e. if there exists a ball $B = B_r(x)$ such that $X = B$.*

We will adapt the definitions of some spaces of functions on R^n to the context of homogeneous spaces.

The space $C_0^\alpha(X)$ of Hölder continuous functions on X, is the space of functions f with bounded support in X and such that

$$\|f\|_\alpha = \sup_{x \neq y} \frac{|f(x) - f(y)|}{d(x, y)^\alpha} < \infty.$$

The spaces $L^p(X)$, $1 \leq p \leq \infty$ are defined as usual. The space $\mathcal{D}(X)$ of test functions, is the space of functions $f \in L^\infty(X)$ with bounded support.

Let $\fint_\Omega f = \frac{1}{|\Omega|} \int_\Omega f d\mu = f_\Omega$. In what follows B will denote a ball in X. Given $f : X \to R$, we will call *maximal function* of f, the function

$$Mf(x) \equiv \sup_{B \ni x} \fint_B |f(y)| d\mu(y)$$

and *sharp function* of f, the function

$$f^\sharp(x) \equiv \sup_{B \ni x} \fint_B |f(y) - f_B| d\mu(y)$$

We will say that $f \in BMO(X)$, if $f \in L^1_{loc}(X)$ and $f^\sharp \in L^\infty$. If $f \in BMO(X)$ we will call *BMO seminorm* of f the quantity

$$\|f\|_* \equiv \sup_x f^\sharp(x) = \sup_B \fint_B |f(y) - f_B| d\mu(y)$$

The following two lemmas contain important properties of BMO functions (proofs may be found e.g. in [13]).

Lemma 2.5. *Let $f \in BMO(X)$ and let*

$$f_n(x) = \begin{cases} f(x) & \text{if } |f(x)| < n \\ n & \text{if } f(x) \geq n \\ -n & \text{if } f(x) \leq -n \end{cases}$$

Then $f_n \in BMO(X)$ and $\|f_n\|_* \leq c \|f\|_*$.

Lemma 2.6. *Let* $f \in BMO$ *and* $M > 1$. *Then for every ball* $B_r(x)$

$$\left| f_{B_{M^j r}} - f_{B r} \right| \leq c \cdot j \left\| f \right\|$$

for every positive integer j, *where* c *depends only on* M *and* c_μ.

Let's now define singular integral and fractional integral operators on homogeneous spaces.

Definition 2.7. We say that K is a Calderòn-Zygmund operator (CZ operator) on the homogeneous space (X, d, μ) if:

(i) $K : L^p(X) \to L^p(X)$ is linear continuous for every $p \in (1, \infty)$;
(ii) there exists a measurable function $k : X \times X \to R$ such that for every $f \in \mathcal{D}$ and a.e. $x \notin sprtf$, $Kf(x) = \displaystyle\int_X k(y, x) f(y) d\mu(y)$;
(iii) k satisfies a *pointwise Hörmander condition* i.e. there exist $c > 0$, $\beta > 0$ and $h > 1$ such that for every $x_0 \in X$, $r > 0$, $x \in B_r(x_0)$, $y \notin B_{hr}(x_0)$

$$(2.4) \qquad |k(x_0, y) - k(x, y)| \leq \frac{c}{\mu\left(B(x_0; y)\right)} \cdot \frac{d(x_0, x)^\beta}{d(x_0, y)^\beta}$$

(where $B(x; y) = B_{d(x,y)}(x)$).

Incidentally observe that condition (2.4) in definition 2.7 implies an integral Hörmander condition (for a proof see [1]); this condition is assumed in [5] and [7] to prove L^p continuity of a CZ operator which is L^2 continuous.

Definition 2.8. For $\alpha \in (0, 1)$ we call *fractional integral kernel* the function $k_\alpha(x, y) = \mu\left(B(x; y)\right)^{\alpha - 1}$ and *fractional integral operator* the operator I_α defined by

$$(2.5) \qquad I_\alpha f(x) = \int_{X \backslash \{x\}} k_\alpha(x, y) f(y) d\mu(y)$$

We are now ready to state the main results in this paper. Theorems 2.9 and 2.10 contain the results for singular integrals; theorems 2.11 and 2.12 contain the results for fractional integrals and theorem 2.13 the "converse fractional integral" result.

Theorem 2.9. *Let* (X, d, μ) *be an infinite measure homogeneous space,* K *a CZ operator on* X *and* $a \in BMO(X)$. *Then the commutator* $[a, K]$ *satisfies*

$$(2.6) \qquad \left\| [a, K] f \right\|_p \leq C \left\| a \right\|_* \left\| f \right\|_p$$

for every $p \in (1, \infty)$

Theorem 2.10. *Let* (X, d, μ) *be a finite measure homogeneous space,* K *a CZ operator on* X *and* $a \in BMO(X)$. *Moreover let the following conditions hold:*

(i) X *is separable and* μ *is regular;*
(ii) *the adjoint kernel* k^* $\left(\text{defined as } k^*(x, y) = k(y, x)\right)$ *satisfies the pointwise Hörmander condition* (2.4);

(iii) k *satisfies the* growth condition:

$$(2.7) \qquad |k(x,y)| \leq \frac{c}{\mu\big(B(x;y)\big)}, \quad \text{for every } x, \ y \in X.$$

Then (2.6) *holds.*

We need a few more definitions in order to state the precise result for fractional integrals. For $x \in X$ let

$$R_x = \begin{cases} \inf\{R|B_R(x) = X\} & \text{if } \mu(X) < \infty \\ \infty & \text{otherwise} \end{cases}$$

and

$$r_x = \begin{cases} \sup\{r|B_r(x) = x\} & \text{if } \mu\{x\} > 0 \\ 0 & \text{otherwise} \end{cases}.$$

We'll say the space X satisfies property **(P)** if:

(2.8) *there exists* $m \in (0,1)$ *such that* $\forall\, r \in (r_x, R_x)$, $B_r(x) \setminus B_{mr}(x) \neq \emptyset$.

Finally let the operator C_α^a be defined as $C_\alpha^a f(x) = \displaystyle\int_X |a(x) - a(y)|\, k_\alpha(x,y)\, f(y)\, dy.$

Theorem 2.11. *Let* (X, d, μ) *be a homogeneous space, satisfying property* **(P)** *and* $a \in BMO(X)$. *Then the operator* C_α^a *satisfies*

$$(2.9) \qquad \|C_\alpha^a f\|_q \leq C \|a\|_* \|f\|_p$$

for every $\alpha \in (0,1)$, $p \in \left(1, \frac{1}{\alpha}\right)$ *and* $\frac{1}{q} = \frac{1}{p} - \alpha$.

Theorem 2.12. *Let* (X, d, μ) *be a homogeneous space, satisfying property* **(P)** *and* $a \in BMO(X)$. *Then the commutator* $[a, I_\alpha]$ *satisfies*

$$(2.10) \qquad \|[a, I_\alpha]f\|_q \leq C \|a\|_* \|f\|_p$$

for every $\alpha \in (0,1)$, $p \in \left(1, \frac{1}{\alpha}\right)$ *and* $\frac{1}{q} = \frac{1}{p} - \alpha$.
Moreover inequality (2.10) *holds if the kernel* k_α *is replaced with any equivalent kernel* $\tilde{k}_\alpha$.

Theorem 2.13. *Assume that for* $a \in L^1_{loc}(X)$ *the operator* C_α^a *satisfies inequality* (2.9) *for some* $\alpha \in (0,1)$, *some* $p \in \left(1, \dfrac{1}{\alpha}\right)$ *and* $\dfrac{1}{q} = \dfrac{1}{p} - \alpha$. *Then* $a \in BMO(X)$.

3. COMMUTATORS OF SINGULAR INTEGRALS: PROOFS OF THMS. 2.9 AND 2.10

In the next theorem we'll recall known results about thr L^p-continuity of maximal function and the sharp function as well as a John-Nirenberg type lemma that have been proved to hold in homogeneous spaces. For the proofs see [7] and [3]

Theorem 3.1. (i) (Maximal Inequality) *The operator* $M : f \longmapsto Mf$ *maps* L^p *in* L^p *for every* $p \in (1, \infty]$ *and*

$$(3.1) \qquad \|Mf\|_p \leq c \|f\|_p.$$

(ii) (Sharp Inequality) *For every* $f \in L^p$, $p \in (1, \infty]$ *if* $\mu(X) = \infty$:

$$(3.2) \qquad \|f\|_p \leq c \|f^\sharp\|_p;$$

if $\mu(X) < \infty$:

$$(3.3) \qquad \left\| f - \fint_X f \right\|_p \leq c \|f^\sharp\|_p.$$

(iii) (John-Nirenberg Lemma) *Let* $p \in [1, \infty)$ *and* $B \subset X$ *a ball; then* $f \in BMO(X)$ *with norm* $\|f\|_*$ *if and only if*

$$(3.4) \qquad \left(\fint_B |f(x) - f_B|^p d\mu(x) \right)^{\frac{1}{p}} \leq \|f\|_*.$$

Inequality (3.4) *also implies that there exists* $\eta > 0$ *such that* $f^\eta \in A_2$ (A_2 *is the Muckenhoupt class*).

In view of John-Nirenberg lemma (theorem 3.1, equation (3.4)), lemma 2.5 and the density of $\mathcal{D}$ in L^p, it is enough to prove theorems 2.9 and 2.10 for $a \in L^\infty$ and $f \in \mathcal{D}$. The first main result in the proof of estimates for commutators of singular integrals is the following.

Theorem 3.2. *Let* K *be a CZ operator and* $p \in (1, \infty)$; *then there exists* $c = c(p, K, X)$ *such that for every* $a \in L^\infty$ *and every* $f \in \mathcal{D}$

$$(3.5) \qquad \|[a, K]^\sharp\| \leq c \|a\|_* \|f\|_p$$

Proof. In view of the L^p-continuity of K and of the maximal inequality (theorem 3.1 (i)) in order to prove (3.5) it is enough to prove that for every $r \in (1, \infty)$ there exists $c = c(r, K, X)$ such that the following inequality

$$(3.6) \qquad |[a, K]f^\sharp(x)| \leq c \cdot \|a\|_* \left\{ (M|Kf|^r(x))^{\frac{1}{r}} + (M|f|^r(x))^{\frac{1}{r}} \right\}$$

holds for every $x \in X$.

The proof follows Torchinsky ([13], p.418). Let's fix $B_\delta = B_\delta(x_0)$ containing x.

Then for $y \in B_\delta$

$$[a, K]f(y) =$$

$$= K\left((a - a_{B_\delta})f \cdot \chi_{B_{h\delta}}\right)(y) + K\left((a - a_{B_\delta})f \cdot \chi_{B \setminus X_{h\delta}}\right)(y) - (a(y) - a_{B_\delta}) \cdot Kf(y) =$$

$$= A(y) + B(y) + C(y).$$

Now by the John-Nirenberg lemma (theorem 3.1 (iii)), lemma 2.6 and the L^p continuity of K, we immediately get

$$\fint_{B_\delta} |A(y) - A_{B_\delta}| d\mu(y) \le c \|a\|_* \left[M\left(|f|^r\right)(x)\right]^{\frac{1}{r}}$$

and

$$\fint_{B_\delta} |C(y) - C_{B_\delta}| d\mu(y) \le c \|a\|_* \left[M\left(|Kf|^r\right)(x)\right]^{\frac{1}{r}}.$$

Next observe that from Hörmander, we obtain

$$|B(y) - B(x_0)| \le$$

$$(3.7) \qquad \le C \cdot \delta^\beta \left(\int_{X \setminus B_{h\delta}} \frac{|a(z) - a_{B_\delta}|^{r'}}{\mu\left(B(x_0; z)\right) \cdot d(x_0, z)^\beta} d\mu(z)\right)^{\frac{1}{r'}} \cdot$$

$$\cdot \left(\int_{X \setminus B_{h\delta}} \frac{|f(z)|^r}{\mu\left(B(x_0; z)\right) \cdot d(x_0, z)^\beta} d\mu(z)\right)^{\frac{1}{r}}.$$

The last term can be written as

$$\sum_{j=1}^\infty \int_{h^j \delta \le d(x_0, z) < h^{j+1}\delta} \frac{|f(z)|^r}{\mu\left(B(x_0; z)\right) \cdot d(x_0, z)^\beta} d\mu(z)$$

and by the doubling condition (2.1) it is $\le C \cdot \dfrac{1}{\delta^\beta} \cdot M\left(|f|^r\right)(x)$. Using the analogous expansion together with lemma 2.6 for the first term in (3.7) we get that it is

$$\le \frac{1}{\delta^\beta} \sum_{j=1}^\infty c \cdot \|a\|_*^{r'} \left(\frac{1 + j^{r'}}{M^{\beta j}}\right) = c \cdot \frac{1}{\delta^\beta} \|a\|_*^{r'}.$$

Observing finally that $\fint_{B_\delta} |B(y) - B_{B_\delta}| \, d\mu(y) \le 2 \fint_{B_\delta} |B(y) - B(x_0)| \, d\mu(y)$, the theorem is proved. $\square$

From the above theorem, combined with the sharp inequality (theorem 3.1 (ii)) we obtain, respectively, theorem 2.9 when $\mu(X) = \infty$ and

(3.8)
$$\left\| [a, K]f - \fint_X [a, K]f \right\|_p \le c \, \|a\|_* \, \|f\|_p \,,$$

when $\mu(X) < \infty$.

Proof (of theorem 2.10). From now on we'll assume $\mu(X) < \infty$ together with hypothesis (i)-(iii) of theorem 2.10. We may also assume d to be Hölder continuous, since if we replace d with d' as defined in theorem 2.2 all the conditions in theorem 2.10 still hold. Accordingly the number α will be the Hölder exponent of d. Let's now define the truncating function

$$\psi_\varepsilon(t) = \begin{cases} 0 & t \le \varepsilon \\ \dfrac{t}{\varepsilon} - 1 & \varepsilon < t < 2\varepsilon \\ 1 & t \ge 2\varepsilon \end{cases}$$

and the truncated kernels

$$k_\varepsilon(x, y) = k(x, y) \cdot \psi_\varepsilon\left(\delta(x, y)\right),$$

for $\varepsilon \in (0, 1)$.

The first result we need is

Lemma 3.3. *The truncated kernels k_ε satisfy Hörmander condition (2.4) with constants independent of ε and in particular with exponent $\gamma = min\,(\alpha, \beta)$.*

(For a proof see [2].)

The following L^p continuity result may now be obtained adapting to our context the proof of the well known result by M. Cotlar's (see e. g. [11]).

Theorem 3.4. *Let K be a CZ operator satisfying the hypothesis of theorem 2.10 and let the operators K_ε be defined as $K_\varepsilon f(x) = \displaystyle\int_X k_\varepsilon(x, y) \, f(y) \, d\mu(y)$. Then, for every $p \in (1, \infty)$, there exists a constant c independent of ε such that*

(3.9)
$$|K_\varepsilon f(x)| \le c \left\{ M(Kf)(x) + \left(M(|f|^p)(x)\right)^{\frac{1}{p}} + Mf(x) \right\},$$

for every $f \in L^p$. Moreover the operator K^* defined by $K^* f(x) = \sup\limits_{\varepsilon > 0} |K_\varepsilon f(x)|$ is continuous from L^p into L^p for every $p \in (1, \infty)$.

The next step is now to prove that the commutators of K_ε are L^p continuous. This will be a consequence of theorem 3.4, lemma 3.3 and the following theorem.

Theorem 3.5. *Let K be an operator as in the hypothesis of theorem 2.10. Assume moreover that for $f \in \mathcal{D}$ the kernels k and k^* are such that $\displaystyle\int_X |k(x,y)| \, f(y) \, d\mu(y)$ and $\displaystyle\int_X |k^*(x,y)| \, f(y) \, d\mu(y)$ are bounded. Then the commutator $[a, K]$ is a continuous operator on L^p and equation (2.6) holds with constant C depending on K only through the norms of K and K^* as operators on L^p and the constants in (2.4) and (2.7).*

Proof. Recalling equation (3.8) we have that

(3.10)

$$\|[a,K]f\|_p \le c\,\|a\|_* \,\|f\|_p + \mu(X)^{\frac{1}{p}-1}\left|\int_X\left(\int_X k(x,y)[a(x)-a(y)]\,d\mu(y)\right)d\mu(x)\right|.$$

By Fubini's theorem the term in absoloute value equals

(3.11) $$\left|\int_X f(y)\left(\int_X k(x,y)[a(y)-a(x)]\,d\mu(x)\right)d\mu(y)\right| \le \|f\|_p\,\|[K^*,a]\,1\|_{p'}$$

Recalling that K^* satisfies the same hypothesis as K, and applying (3.10) and (3.11) with K^* in place of K we obtain

(3.12)

$$\|[K^*,a]\,1\|_{p'} \le c\,\|a\|_*\,\mu(X)^{\frac{1}{p'}} + \mu(X)^{-\frac{1}{p}}\left|\int_X Ka(y)\,d\mu(y) - \int_X K^*a(x)\,d\mu(x)\right|.$$

Finally, using Hölder's inequality, the L^p estimates and the John-Nirenberg lemma (see theorem 3.1 (iii)) we get

(3.13) $$\left|\int_X Ka(y)\,d\mu(y)\right| \le c\,\|a\|_p\cdot\mu(X)^{\frac{1}{p'}} \le c\,\|a\|_*\cdot\mu(X),$$

and the analogous inequality for K^*. The thesis of the theorem can now be obtained by combining inequalities (3.10) to (3.13). □

As a final step we need to show that $[a,K_\varepsilon] \to [a,K]$ in some sense. In general it is not true that the truncated operators K_ε converge to K, but we'll show in the following theorem that they converge to an operator T (in a suitable weak sense) and that T and K have the same commutator. The analogous result has been proved for Calderòn-Zygmund operators on R^n; in this last case the proof uses tools, such as Fourier transforms, which are characteristic of R^n; our proof instead relies entirely on topological and measure theoretic arguments.

Theorem 3.6. *In the hypothesis of theorem 2.10 there exists a sequence* $\varepsilon_n \to 0$ *and a function* $b \in L^\infty$ *such that*

(3.14) $$\int_X f\,K_{\varepsilon_n}g\,d\mu \to \int_X f\,Kg\,d\mu + \int_X f\,b\,g\,d\mu$$

for every $f,\,g \in \mathcal{D}$.

Proof. Because of the hypothesis we have made on X and on the measure μ, the space $L^p(X)$ in separable for every $p \in [1,\infty)$. Therefore theorem 3.4 implies that there exists a sequence $\varepsilon_n \to 0$ such that $K_{\varepsilon_n}f$ converges weakly in L^p for f in a dense subset of L^p. Again theorem 3.4 implies that

(3.15) $$\int_X g\,K_{\varepsilon_n}f\,d\mu \to \int_X g\,\tilde{K}f\,d\mu$$

for every $f \in L^p$ and $g \in L^{p'}$. Our goal is to show that $K_1 = K - \tilde{K}$ is multiplication times an L^∞ function. Observe first of all that if f, $g \in \mathcal{D}$ and $d(sprt\, f, sprt\, g) \geq \varepsilon_0 > 0$ then

$$\int_X g\, K f\, d\mu \to \int_X g\, \tilde{K} f\, d\mu.$$

Taking $f = \chi_B$ with E a Borel set whose closure is contained in $X \setminus sprt\, f$ this implies that $Kg = \tilde{K}g$ a. e. in $X \setminus sprt\, f$. Now for E a Borel set let C_n be an increasing sequence of closed sets and A_n be an decreasing sequence of open sets such that $\mu(A_n \setminus C_n) \to 0$ and $C_n \subset E \subset A_n$. Let now

$$b(x) = (K_1 1)(x) = \left(K_1 \chi_{C_n}\right)(x) + \left(K_1 \chi_{A_{h\delta} \setminus C_n}\right)(x) + \left(K_1 \chi_{A_n^c}\right)(x).$$

Taking limits in n we get that $K_1 \chi_E(x) = b(x)$ for a.e. $x \in E$. Analogously $K_1 \chi_E(x) = 0$ for a.e. $x \notin E$, i. e. $K_1 \chi_E(x) = b(x) \cdot \chi_E(x)$. By linearity this proves the assertion. To see that $b \in L^\infty$ observe that for $f \in \mathcal{D}$ from the linearity of K_1 we have that

$$\left| \int_X b\, h\, d\mu \right| = \left| \int_X h^{\frac{1}{p}} K_1\, h^{\frac{1}{p'}}\, d\mu \right| \leq c \|h\|_1. \qquad \square$$

The proof of theorem 2.10 is now complete. $\square$

4. COMMUTATORS OF FRACTIONAL INTEGRALS: PROOF OF THM. 2.11

First of all let's recall the following $L^p - L^q$ continuity result that holds for the operator I_α.

Theorem 4.1. *Let (X, d, μ) be a homogeneous space; then the fractional integral operator I_α satisfies*

(4.1) $$\|I_\alpha f\|_q \leq C \|f\|_p$$

for every $\alpha \in (0, 1)$, $p \in \left(1, \frac{1}{\alpha}\right)$ and $\dfrac{1}{q} = \dfrac{1}{p} - \alpha$.

Observe that no extra hypothesis are needed on X. The proof of this result can be obtained by the results in [8] combined with theorem 2.1.

Let's now make a few observations and remarks on property **(P)** (see (2.8)).

(i) By the definitions of R_x and r_x (2.8) is equivalent to the following

(4.2) $\exists\, m \in (0, 1)$ *such that* $\forall\, r \in \left(r_x, \dfrac{R_x}{m}\right)$, $B_r(x) \setminus B_{mr}(x) \neq \emptyset$.

(ii) (2.8) is quite general since it allows for the existence of athoms in X as well as for X to be bounded (and therefore finite measure). What it actually requires is that the ratio between outer and inner radii of empty anulli is bounded. In [2], though, we construct an example of a (very pathological) homogeneous space for which condition (2.8) does not hold.

(iii) It is trivial that if (2.8) holds for a certain p than it holds for $p' < p$. Because of (i) the same is true for (4.2).

(iv) (2.8) implies the following *reverse doubling condition* (RDC) there exists $\delta > 0$, $K > 1$ and k_1, $k_2 \in (0,1)$ such that

$$(4.3) \quad \mu\left(B_{Kr}(x)\right) \geq (1+\delta) \cdot \mu\left(B_r(x)\right), \ \forall \, x \in X, \ r \in (k_1 r_x, k_2 R_x).$$

In order to prove theorem 2.11 observe first of all that, because of lemma 2.5 and Fatou's theorem, it is enough to prove the theorem for $f \geq 0$ and $a \in L^\infty$. The first main result is the following theorem.

Theorem 4.2. *For $a \in BMO$ and $z \in C$ (i.e. z complex), let $b(x) = e^{\left(Re \, z \cdot a(x)\right)}$. Assume there exists $\varepsilon > 0$ such that the following weighted estimate*

$$(4.4) \qquad \qquad \|I_\alpha f\|_{L^q(b^q \, d\mu)} \leq c \cdot \|f\|_{L^p(b^p \, d\mu)}$$

holds for every z with $|z| < \varepsilon$. Then theorem 2.11 holds.

Proof. We follow an idea contained in [6] as a suggestion for an alternative proof to their result on the commutator of classical Calderón-Zygmund operators on euclidean space. For $z \in C$ define

$$T_z f(x) = \int_{X \setminus \{x\}} e^{\left(a(x) - a(y)\right) \cdot z} k_\alpha(x,y) f(y) \, sgn \, (a(x) - a(y)) \, d\mu(y);$$

then formally

$$\frac{d}{dz} T_z \Big|_{z=0} z = 0 f(x) = \int_X |a(x) - a(y)| \, k_\alpha(x,y) \, f(y) \, dy = C_\alpha^a f(x).$$

Therefore our first goal is to show that the map $z \to T_z$ is analytic in a neighborhood of the origin of C. From the definition of T we get

$$|T_z f(x)| \leq b(x) \int_{X \setminus \{x\}} k_\alpha(x,y) |f(y)| \, b^{-1}(y) \, d\mu(y)$$

and from the weighted estimates

$$\|T_z\|_q \leq \left\|I_\alpha(b^{-1}f)\right\|_{L^q(b^q \, d\mu)} \leq c \left\|b^{-1}f\right\|_{L^p(b^p \, d\mu)} = c \, \|f\|_p.$$

Therefore T maps the disk $|z| < \varepsilon$ into the linear continuous operators from L^p to L^q. By taking difference quotients and using the John-Nirenberg lemma together with the above reasoning it is easy to see that T_z is anaytic in the same disk. We can now use Cauchy's integral formula and get $C_\alpha^a f(x) = \dfrac{1}{2\pi i} \displaystyle\int_{|w| = \frac{\varepsilon}{2}} \dfrac{T_w f(x)}{w^2} \, dw.$ Finally using the continuity of T_z the above gives $\|C_\alpha^a f\|_q \leq c \|f\|_p$. The linear dependence of the constant c on $\|a\|_*$ follows by the homogeneity of the estimate. The proof of theorem 4.2 is complete. $\square$

Theorem 4.3. *Let (X, d, μ) be a homogeneous space satisfying property (2.8) and let K the operator acting on functions on X defined by*

$$Kf(x) = \int_X k(x,y) f(y) \, d\mu(y)$$

where the kernel $k : X \times X \to R^+$ is a measurable function. Let p, q, α be as in theorem 2.11 and assume further:

(i) *the kernel k is such that the quantity defined by $\Phi(B) = \sup\{k(x,y)| \ x,y \in B, \ d(x,y) \geq hr\}$ is equivalent to $\mu(B)^{1-\alpha}$ for every ball $B = B_r(x)$ with $r \in (r_x, R_x)$, i.e. $c_1 \, \mu(B)^{1-\alpha} \leq \Phi(B) \leq c_2 \, \mu(B)^{1-\alpha}$.*

(ii) *v and w are two nonnegative measurable function such that for some $s > 1$*
$$v^{(1-p)s} \in A_{t_1} \ \text{ and } \ w^s \in A_{t_2} \ \text{ with } \ t_1 = 1 + \frac{p'}{q} \ \text{ and } \ t_2 = 1 + \frac{q}{p'}.$$

Then

$$(4.5) \qquad \|I_\alpha f\|_{L^q(w \, d\mu)} \leq c \cdot \|f\|_{L^p(v \, d\mu)}$$

The above theorem is a consequence of the weighted estimates proved by Sawyer and Wheeden in [12]. Actually the hypothesis on X in [12] are more restrictive since they require X to have no empty anulli. To see how to adapt their proof to our case as well as how theorem 4.3 follows from Sawyer and Wheeden's estimates see [2].

Proof. (of Theorem 2.11) To conclude the proof of theorem 2.11 we only need to show that our kernel k_α satisfies hypothesis (i) and our weights $v = b^p$ and $w = b^q$ satisfy hypothesis (ii) in the above theorem. Observe that the RDC (4.3) implies that for every $\eta \in (0,1)$ there exists two positive constants c, γ depending on X and η such that for every $x \in X$ we have $\mu(B_R(x)) \geq c \left(\frac{R}{r}\right)^\gamma \mu(B_r(x))$, for $\eta \, r_x \leq r \leq R \leq \frac{1}{\eta} R_x$. Now (i) follows form the above property, triangle inequality and the doubling condition. As for condition (ii), since $a \in L^\infty$, by the John-Nirenberg Lemma for every $t \in (1,\infty)$ there exists ε such that $e^{(\varepsilon \cdot a(x))} \in A_t$ with A_t constant bounded by $c(t, \|a\|_*)$. Therefore, for every $t \in (1,\infty)$, $\beta > 0$ there exists ε such that for any $z \in \mathbf{C}$ with $|z| < \varepsilon$ the function $b^\beta \in A_p$. The proof is now complete. $\square$

5. A CHARACTERIZATION OF BMO FUNCTIONS: PROOF OF THM. 2.13

We will now prove theorem 2.13. Let $B = B_r(z)$ be a fixed ball in X and $x \in B$; then

$$(5.1) \qquad \left| a(x) - \fint_B a(y) dy \right| \leq \fint_B |a(x) - a(y)| \, dy.$$

Moreover, observing that if $y \in B$ then $B_{d(x,y)}(x) \subset B_{3c_d r}(z)$, we have

$$(5.2) \quad C_\alpha^a \left(\chi_B \right)(x) = \int_B |a(x) - a(y)| \frac{1}{\mu\left(B_{d(x,y)}(x)\right)^{1-\alpha}} \, dy \geq$$
$$\geq \frac{1}{\mu\left(B_{3c_d^2 r}(z)\right)^{1-\alpha}} \int_B |a(x) - a(y)| \, dy \geq \frac{C}{\mu(B)^{1-\alpha}} \int_B |a(x) - a(y)| \, dy$$

where we used the doubling condition and where $C = C\left(c_d, c_\mu\right)$. From (5.1) and (5.2) we get

$$|a(x) - a_B| \leq C \, \mu(B)^{-\alpha} C_\alpha^a \left(\chi_B \right)(x).$$

From the continuity of C_α we obtain

(5.3) $$\left\{ \int_B |a(x) - a_B|^q \; dx \right\}^{\frac{1}{q}} \leq C \; \mu(B)^{\frac{1}{p} - \alpha} = C \; \mu(B)^{\frac{1}{q}}$$

In view of the John-Nierenberg Lemma (see (3.4) (i)) (5.3) proves the theorem. □

REFERENCES

1. Bramanti, M., and Cerutti, M.C., *Commutators of Singular Integrals on Homogeneous Spaces*, to appear.
2. Bramanti, M., and Cerutti, M.C., *Commutators of Fractional Integrals on Homogeneous Spaces*, to appear.
3. Burger, N., *Espace de Fonctions à Variation Moyennne Bornée sur un Espace de Nature Homogène*, C.R.Acad.Sc.Paris, t. **286** (1978), 139-142.
4. Chanillo, *A note on Commutators*, Indiana Univ. Math, Jour. **31** n. 1 (1982), 7-16.
5. Coifman, R., and de Guzmán, M., *Singular Integrals and Multipliers on Homogeneous Spaces*, Revista de la Union Matematica Argentina, **25** (1970), 137-143.
6. Coifman, R., Rochberg, R., and Weiss, G., *Factorization Theorems on Hardy Spaces*, Annals of Math., **103** (1976), 611-635.
7. Coifman, R., and Weiss, G., *Analys Harmonique Non-commutative sur Certain Espaces Homogènes*, Lecture Notes in Mathematics, n. **242** (1968), Springer-Verlag, Berlin-New York-Heidelberg 1971.
8. Gatto, A.E., and Vàgi, S., *Fractional Intregrals on spaces of Homogeneous Type*, Analysis and Partial Differential Equations, ed Cora Sadosky, Lecture Notes in Pure and Applied Math., vol. **122** (1990) 171-216.
9. Gatto, A.E., and Vàgi, S., *On Molecules and Fractional Intregrals on spaces of Homogeneous Type with Finite Measure*, Studia Math., **103** (1) (1992) 25-39.
10. Macìas, R. A., and Segovia, C., *Lipschitz Functions on spaces of Homogeneous Type*, Adv. in Math. **33** (1979) 257-270.
11. Meyer, Y., *Ondelettes et Opérateurs II: Opérateur de Calderón-Zygmund*, Hermann, Paris 1990.
12. Sawyer, E., and Wheeden, R. L., *Weighted Inequalities for Fractional Integrals on Euclidean and Homogeneous Spaces*, Amer. Jour. of Math. **114** (1992) n.4
13. Torchinsky, A., *Real Variable Methods in Harmonic Analysis*, Academic Press, New York 1986.

DIPARTIMENTO DI MATEMATICA, POLITECNICO DI MILANO, PIAZZA LEONARDO DA VINCI, 32, 20133 MILANO ITALY
E-mail address, M. Bramanti: marbra@ipmma1.polimi.it

E-mail address, M.C. Cerutti: cricer@ipmma1.polimi.it

Contemporary Mathematics
Volume **189**, 1995

A Note on the Besov-Lipschitz And Triebel-Lizorkin Spaces

H.-Q. BUI, M. PALUSZYŃSKI AND M. TAIBLESON

Let ϕ be a function in $\mathcal{S}$ with the property that supp $\hat{\phi} \subset \{1/2 \leq |\xi| \leq 2\}$ and $|\phi(\xi)| \geq c$ for $3/5 \leq |\xi| \leq 5/3$ for some $c > 0$. The homogeneous Besov-Lipschitz and Triebel-Lizorkin spaces are then defined by their norms

$$\|f\|_{\dot{F}_p^{\alpha,q}} = \left\| \left(\int_0^\infty \left(t^{-\alpha} |\phi_t * f| \right)^q \frac{dt}{t} \right)^{1/q} \right\|_p,$$

$$\|f\|_{\dot{B}_p^{\alpha,q}} = \left(\int_0^\infty \left(t^{-\alpha} \|\phi_t * f\|_p \right)^q \frac{dt}{t} \right)^{1/q},$$

where $\phi_t(x) = t^{-n}\phi(x/t)$. These norms are not the usual norms for these spaces, however, they are equivalent to the standard "discrete" ones, and the equivalence can be shown using the same techniques as in the proof of their independence of the particular ϕ (see [2]). One of the advantages of the above norms is that they are immediately seen to be homogeneous:

$$\|f_t\|_{\dot{F}_p^{\alpha,q}} = t^{n(1/p-1)-\alpha} \|f\|_{\dot{F}_p^{\alpha,q}} ;$$

with the same result for $\|f\|_{\dot{B}_p^{\alpha,q}}$. Suppose we have two finite measures on $\mathbf{R}^n$, μ and ν. Suppose μ is either compactly supported or is absolutely continuous with density in $\mathcal{S}$ ($\mu \in \mathcal{S}$), and suppose it satisfies the standard Tauberian condition:

$$\forall \xi \in \mathbf{R}^n,\ \xi \neq 0 \exists t > 0 \text{ such that } \hat{\mu}(t\xi) \neq 0.$$

Let us suppose that $\nu \in \mathcal{S}$ and has m_0 vanishing moments:

$$\int_{\mathbf{R}^n} x^\gamma d\nu = 0 \qquad \forall \gamma,\ |\gamma| \leq m_0.$$

Let us adopt the convention that this last condition is empty if $m_0 < 0$. We then have the following theorem, which is valid for *all* $\alpha \in \mathbf{R}$:

1991 *Mathematics Subject Classification.* Primary 42B25; Secondary 26B35.

THEOREM 1. *Suppose the above assumptions on μ and ν hold. Let $[\alpha] \leq m_0$. Then, for $1 \leq p, q \leq \infty$ there exists a constant c, such that*

$$\frac{1}{c}\left(\int_0^\infty \left(s^{-\alpha}\|\nu_s * f\|_p\right)^q \frac{ds}{s}\right)^{1/q} \leq \|f\|_{\dot{B}_p^{\alpha,q}} \leq c\left(\int_0^\infty \left(s^{-\alpha}\|\mu_s * f\|_p\right)^q \frac{ds}{s}\right)^{1/q},$$

where the dilation μ_s should be interpreted appropriately if μ is not a function.

REMARKS. 1.) When $\alpha > 0$ this result is well known (see [4]). The observation that the proof actually carries over for $\alpha \leq 0$ (when ν is a function in $\mathcal{S}$) and the Triebel-Lizorkin space counterpart (Theorem 2) appear to be new.
2.) If $\mu = \nu$, then the above theorem provides a characterization of the Besov-Lipschitz spaces, while Theorem 2 below provides a characterization of the Triebel-Lizorkin spaces. 3.) Using size estimates of the Heideman type, which are different from Lemmas 1 and 2, and the Peetre-Triebel maximal function, we have also proved the above theorem for the weighted space case (with A_∞ weights), and that proof includes the range of indices $0 < p, q \leq 1$ (see [1]). It would be interesting to see if the techniques used here can be applied to this more general case.

The proof relies on the following lemma (see [3]):

LEMMA 1. *Suppose ν is a finite measure on $\mathbf{R}^n$ either compactly supported, or with density in $\mathcal{S}$. Suppose ν has vanishing moments up to order m_1. Let $\eta \in \mathcal{S}$. Then*

$$\|\nu_s * \eta_t\|_1 \leq c\left(\frac{s}{t}\right)^m, \qquad m = 0, \ldots, m_1 + 1, \quad t, s > 0,$$

where c is a constant depending on ν and η, but not on s or t. If, in addition, we assume that $\nu \in \mathcal{S}$, and η has vanishing moments up to order m_2, we have

$$\|\nu_s * \eta_t\|_1 \leq c\left(\frac{t}{s}\right)^m, \qquad m = 0, \ldots, m_2 + 1, \quad t, s > 0.$$

PROOF. The first inequality can be obtained as in [3], and the second is the immediate consequence of the first, simply by interchanging the roles of ν and η. □

The proof of Theorem 1 now follows along the lines of [4].

PROOF OF THEOREM 1. Let us fix $f \in \mathcal{S}'$. Let k be the order of f as a distribution; that is, let k be an integer such that

$$|\langle f, \psi \rangle| \leq c \sum_{|\gamma| \leq k} \sup_{x \in \mathbf{R}^n} (1 + |x|)^k |D^\gamma \psi(x)|, \quad \forall \psi \in \mathcal{S}.$$

It has been shown in [4] that if μ satisfies the Tauberian condition, then there exists an $\eta \in \mathcal{S}$ with $\hat{\eta}$ supported on an annulus, such that

$$f = \int_0^\infty \mu_s * \eta_s * f \frac{ds}{s},$$

where the integral converges in $\mathcal{S}'/\mathcal{P}_{k+1}$ (tempered distributions modulo polynomials of degree at most $k + 1$. Convolution with ϕ is a bounded operator on $\mathcal{S}'/\mathcal{P}_{k+1}$, and thus

$$\phi_t * f = \int_0^\infty \phi_t * \eta_s * \mu_s * f \frac{ds}{s} \quad \text{in} \quad \mathcal{S}'/\mathcal{P}_{k+1}.$$

The integral in the above expression actually converges in $\mathcal{S}'$, since the convolution with ϕ_t annihilates the polynomials. Let us assume that

$$\left(\int_0^\infty \left(s^{-\alpha} \| \mu_s * f \|_p \right)^q \frac{ds}{s} \right)^{1/q} < \infty.$$

We will show that $t^{-\alpha} \phi_t * f$ exists as an element of $l^q(L^p)$, and the right-hand inequality in our theorem holds. By our lemma,

$$t^{-\alpha} \| \phi_t * f \|_p \le ct^{-\alpha} \int_0^\infty \| \mu_s * f \|_p \min \left\{ \left(\frac{t}{s} \right)^{m_1}, \left(\frac{s}{t} \right)^{m_2} \right\} \frac{ds}{s}, \quad m_1, m_2 \ge 0,$$

since both η and ϕ have all moments vanishing. Let us choose m_1 so that $m_1 > \alpha$ (0 if $\alpha < 0$) and m_2 so that $m_2 > -\alpha$ (again, $m_2 = 0$ if $\alpha > 0$). Thus,

$$\left(\int_0^\infty \left(t^{-\alpha} \| \phi_t * f \|_p \right)^q \frac{dt}{t} \right)^{1/q}$$

$$\le c \left(\int_0^\infty \left(t^{-\alpha} \int_0^\infty \| \mu_s * f \|_p \min \left\{ \left(\frac{t}{s} \right)^{m_1}, \left(\frac{s}{t} \right)^{m_2} \right\} \frac{ds}{s} \right)^q \frac{dt}{t} \right)^{1/q}$$

$$\le c \left(\int_0^\infty \left(t^{-\alpha} \int_0^t \| \mu_s * f \|_p \left(\frac{s}{t} \right)^{m_2} \frac{ds}{s} \right)^q \frac{dt}{t} \right)^{1/q} +$$

$$+ c \left(\int_0^\infty \left(t^{-\alpha} \int_t^\infty \| \mu_s * f \|_p \left(\frac{t}{s} \right)^{m_1} \frac{ds}{s} \right)^q \frac{dt}{t} \right)^{1/q}.$$

By Hardy's inequalities,

$$\left(\int_0^\infty \left(t^{-\alpha} \| \phi_t * f \|_p \right)^q \frac{dt}{t} \right)^{1/q}$$

$$\le c \left(\frac{1}{m_2 + \alpha} + \frac{1}{m_1 - \alpha} \right) \left(\int_0^\infty \left(s^{-\alpha} \| \mu_s * f \|_p \right)^q \frac{ds}{s} \right)^{1/q}.$$

The right-hand inequality in our theorem is proved. The left-hand inequality can be proved in the same way. $\square$

We now state the Triebel-Lizorkin counterpart of Theorem 1:

THEOREM 2. *Under the assumptions of Theorem 1, and if* $1 < p, q \le \infty$, $p < \infty$, *there is a constant* c *such that*

$$\frac{1}{c} \left\| \left(\int_0^\infty \left(s^{-\alpha} | \nu_s * f | \right)^q \frac{ds}{s} \right)^{1/q} \right\|_p \le \| f \|_{\dot{F}_p^{\alpha,q}} \le c \left\| \left(\int_0^\infty \left(s^{-\alpha} | \mu_s * f | \right)^q \frac{ds}{s} \right)^{1/q} \right\|_p.$$

PROOF. The proof relies on the following variant of Lemma 1:

LEMMA 2. *Suppose ν and η are as in Lemma 1. Let $f \in L^1_{loc}$ and $M(f)$ be the Hardy-Littlewood maximal function. Then*

$$|f * \nu_s * \eta_t(x)| \le cM(f)(x) \left(\frac{s}{t}\right)^m, \qquad m = 0, \ldots, m_1 + 1, \quad 0 < s \le t,$$

where the constant c only depends on ν, η and n (but not on f, s, t or x). If, in addition, we assume that $\nu \in \mathcal{S}$, and η has vanishing moments up to order m_2, we have

$$|f * \nu_s * \eta_t(x)| \le cM(f)(x) \left(\frac{t}{s}\right)^m, \qquad m = 0, \ldots, m_2 + 1, \quad 0 < t \le s.$$

PROOF. As before, the second part of the lemma is an immediate consequence of the first. Let us prove the first part of the lemma. It is enough to let $t = 1$. To see this, suppose we have our estimate for $t = 1$. Pick any $t \ge s$. Then

$$\begin{aligned}
|f * \nu_s * \eta_t(x)| &= \left|f * \left(\nu_{s/t} * \eta\right)_t (x)\right| \\
&= \frac{1}{t^n} \left|f_{1/t} * \nu_{s/t} * \eta\left(\frac{x}{t}\right)\right| \\
&\le \frac{c}{t^n} M(f_{1/t})\left(\frac{x}{t}\right)\left(\frac{s}{t}\right)^m,
\end{aligned}$$

since $s/t \le 1$. Clearly,

$$\frac{1}{t^n} M(f_{1/t})\left(\frac{x}{t}\right) = M(f)(x),$$

so, indeed, it is enough to take $t = 1$. Let Q_k be a cube centered at the origin, with side length 2^k, and let $R_k = Q_k \setminus Q_{k-1}$.

$$\begin{aligned}
|f * \eta * \nu_s(x)| &\le \sum_{k=1}^{\infty} \int_{R_k} |f(x-y)||\eta * \nu_s(y)| \, dy + \\
&\quad + \int_{Q_0} |f(x-y)||\eta * \nu_s(y)| \, dy \\
&\le \sum_{k=1}^{\infty} \int_{R_k} |f(x-y)| \, dy \sup_{y \in R_k} |\eta * \nu_s(y)| + \\
&\quad + \int_{Q_0} |f(x-y)| \, dy \sup_{y \in Q_0} |\eta * \nu_s(y)| \\
&\le \sum_{k=1}^{\infty} \frac{1}{|Q_k|} \int_{Q_k} |f(x-y)| \, dy \, 2^{kn} \sup_{|y| \ge 2^{k-2}} |\eta * \nu_s(y)| + \\
&\quad + \int_{Q_0} |f(x-y)| \, dy \sup_{y \in \mathbf{R}^n} |\eta * \nu_s(y)| \\
&\le M(f)(x) \left(\sum_{k=1}^{\infty} 2^{kn} \sup_{|y| \ge 2^{k-2}} |\eta * \nu_s(y)| + \sup_{y \in \mathbf{R}^n} |\eta * \nu_s(y)|\right).
\end{aligned}$$

We will show that in the present setting

$$|\boldsymbol{\eta} * \boldsymbol{\nu}_s(y)| \le cs^m \min\left\{\frac{1}{|y|^{n+1}}, 1\right\},$$

with c appropriately independent, which will complete the proof of the lemma. We will use the Taylor formula with the remainder in the integral form:

$$\boldsymbol{\eta}(y-h) = \sum_{|\gamma|\le m} \frac{(-h)^\gamma \boldsymbol{\eta}^{(\gamma)}(y)}{\gamma!} + (m+1)\sum_{|\gamma|=m+1} \frac{(-h)^\gamma}{\gamma!}\int_0^1 (1-t)^m \boldsymbol{\eta}^{(\gamma)}(x-ht)dt.$$

It is enough to consider $m = m_1 + 1$. Using the fact that $\boldsymbol{\nu}$ has vanishing moments, we get

$$\begin{aligned}
|\boldsymbol{\eta} * \boldsymbol{\nu}_s(y)| &= \left|\int_{\mathbf{R}^n} \boldsymbol{\nu}_s(h)\boldsymbol{\eta}(y-h)dh\right| \\
&= \left|\int_{\mathbf{R}^n} \boldsymbol{\nu}_s(h)m\sum_{|\gamma|=m}\frac{h^\gamma}{\gamma!}\int_0^1 (1-t)^{m-1}\boldsymbol{\eta}^{(\gamma)}(y-ht)dt\,dh\right| \\
&\le m\sum_{|\gamma|=m}\int_{\mathbf{R}^n}|\boldsymbol{\nu}_s(h)||h|^m\left|\int_0^1 (1-t)^{m-1}\boldsymbol{\eta}^{(\gamma)}(y-ht)dt\right|dh \\
&\le c\sum_{|\gamma|=m}\int_{\mathbf{R}^n}|\boldsymbol{\nu}_s(h)||h|^m\int_0^1\left|\boldsymbol{\eta}^{(\gamma)}(y-ht)\right|dt\,dh.
\end{aligned}$$

Observe, that this last integral:

$$\int_0^1 \left|\boldsymbol{\eta}^{(\gamma)}(y-ht)\right|dt \le c,$$

with c independent of y and h in $\mathbf{R}^n$. Thus,

$$|\boldsymbol{\eta} * \boldsymbol{\nu}_s(y)| \le c\int_{\mathbf{R}^n}|\boldsymbol{\nu}_s(h)||h|^m dh = cs^m.$$

This is one estimate we need. To get the other one, we fix $y \in \mathbf{R}^n$ and define a region $A(y) \subset \mathbf{R}^n$:

$$A(y) = \{h \in \mathbf{R}^n : |y-ht| \ge |y|/2 \quad \forall t, 0 \le t \le 1\}.$$

From the computation we have just completed, we have

$$\begin{aligned}
|\boldsymbol{\eta} * \boldsymbol{\nu}_s(y)| &\le c\sum_{|\gamma|=m}\int_{\mathbf{R}^n}|\boldsymbol{\nu}_s(h)||h|^m\int_0^1 |\boldsymbol{\eta}^{(\gamma)}(y-ht)|dt\,dh \\
&= c\sum_{|\gamma|=m}\left(\int_{A(y)}\cdots + \int_{A(y)^c}\cdots\right) = I + II.
\end{aligned}$$

Since $\boldsymbol{\eta}^{(\gamma)} \in \mathcal{S}$, we have

$$\left|\boldsymbol{\eta}^{(\gamma)}(x)\right| \le \frac{c}{|x|^{n+1}}.$$

Hence,

$$I \leq c \int_{A(y)} |\boldsymbol{\nu}_s(h)||h|^m \sup_{0 \leq t \leq 1} \frac{1}{|y - ht|^{n+1}} dh$$

$$\leq c \int_{A(y)} |\boldsymbol{\nu}_s(h)||h|^m dh \frac{1}{|y|^{n+1}}$$

$$= cs^m \frac{1}{|y|^{n+1}} \int_{\mathbf{R}^n} |h|^m d|\boldsymbol{\nu}|(h)$$

$$= cs^m \frac{1}{|y|^{n+1}}.$$

To estimate II we need

$$\left|\boldsymbol{\eta}^{(\gamma)}(x)\right| \leq c,$$

which is a consequence of $\boldsymbol{\eta}^{(\gamma)} \in \mathcal{S}$.

$$II \leq c \int_{A(y)^c} |\boldsymbol{\nu}_s(h)||h|^{m+n+1} \frac{1}{|h|^{n+1}} dh$$

$$\leq c \int_{A(y)^c} |\boldsymbol{\nu}_s(h)||h|^{m+n+1} dh \frac{1}{|y|^{n+1}}$$

$$\leq cs^{m+n+1} \frac{1}{|y|^{n+1}}.$$

The second inequality follows from the easy observation that if $h \in A(y)^c$ then $|h| \geq |y|/2$. Recall, that $s \leq 1$, so $s^{m+n+1} \leq s^m$. Thus, we obtained the second part of our estimate

$$|\boldsymbol{\eta} * \boldsymbol{\nu}_s(y)| \leq cs^m \frac{1}{|y|^{n+1}}.$$

Our lemma is proved. □

To conclude the proof of Theorem 2 we proceed as in Theorem 1 with Lemma 2 in place of Lemma 1, and use a vector-valued maximal inequality (see [5], Theorem 1). (The ordinary maximal inequality suffices if $q = \infty$.) □

REMARK. Observe that the elements of $\dot{F}_p^{\alpha,q}$ or $\dot{B}_p^{\alpha,q}$ we obtain in the proof of the right-hand inequalities of the theorems are defined via the reproducing formula, and are equal to f only modulo a polynomial of degree no higher than $k + 1$. This is not relevant since the spaces in question are defined modulo all polynomials. It becomes relevant in the proof of the left-hand inequalities, since for many particular measures ν the respective spaces are defined modulo polynomials of a fixed degree. The left-hand inequalities in Theorems 1 and 2 should be interpreted accordingly: if f is in $\dot{B}_p^{\alpha,q}$ or $\dot{F}_p^{\alpha,q}$, then there exists a representative of f modulo polynomials which is in the respective space defined by ν, and the inequality holds.

ACKNOWLEDGEMENT. The research in this paper was done when the first two authors were visiting Washington University in St. Louis. They would like to express their gratitude to colleagues in the Department of Mathematics for their warm hospitality and support.

REFERENCES

1. H.-Q. Bui, M. Paluszyński and M. Taibleson, *A maximal function characterization of weighted Besov-Lipschitz and Triebel-Lizorkin spaces*, In preparation.
2. M. Frazier and B. Jawerth, *A discrete transform and decompositions of distribution spaces*, J. Func. Anal. **93** (1990), 34–170.
3. N. J. H. Heideman, *Duality and fractional integration in Lipschitz spaces*, Studia Math. **50** (1974), 65–85.
4. S. Janson and M. Taibleson, *I teoremi di rappresentazione di Calderón*, Rend. Sem. Mat. Univ. Politecn. Torino **39** (1981), 27–35.
5. V. M. Kokilasvili, *Maximal inequalities and multipliers in weighted Lizorkin-Triebel spaces*, (English translation), Soviet Math. Dokl. **19** (1978), 272–276.

DEPARTMENT OF MATHEMATICS & STATISTICS, UNIVERSITY OF CANTERBURY, CHRISTCHURCH 1, NEW ZEALAND

E-mail address: hqb@math.canterbury.ac.nz

INSTITUTE OF MATHEMATICS, UNIVERSITY OF WROCŁAW, 50-384 WROCŁAW, POLAND

E-mail address: mpal@math.uni.wroc.pl

DEPARTMENT OF MATHEMATICS, WASHINGTON UNIVERSITY, ST. LOUIS, MO 63-130, USA

E-mail address: mitch@math.wustl.edu

Contemporary Mathematics
Volume 189, 1995

Nonlinear Superposition on spaces of analytic functions

G.A. CÁMERA

ABSTRACT. In this article we discuss the action and continuity of the nonlinear superposition operator in Hardy and Bergman spaces of analytic functions in the unit disc. The situation for the Nevanlinna space and the Smirnov class is also considered.

1. Introduction

Let $H(\Omega)$ be the space of analytic functions in a domain $\Omega \subseteq \mathbb{C}$, with the topology of the uniform convergence in compact sets of Ω. Let $f : \mathbb{C} \to \mathbb{C}$ and define

$$F_f(u)(z) = f(u(z)), \quad \forall u \in H(\Omega), \quad z \in \Omega.$$

The operator F_f is called *The Autonomous Nonlinear Superposition Operator*. This is the name given in [1] to this operator acting on various spaces of real-valued functions. If H, G are two subspaces of $H(\Omega)$ and $F_f(u) \in G$ for all $u \in H$ we shall say that F_f acts from H to G. It is easy to see that the operator F_f acts from $H(\Omega)$ to $H(\Omega)$ if and only if f is an entire function. In this case F_f is continuous, bounded and compact.

Let Δ be the unit disc in $\mathbb{C}$. In this paper we review some results concerning the action of this operator on the clasical Hardy Spaces H^p, $0 < p \le \infty$ and Bergman spaces B_p (Section 2) and prove some new results. We will see that when the action takes place it is necessarily locally-lipschitz continuous (Section 3). This has been proved for L^p spaces [7] and for Sobolev spaces [8]. We also study the action on the Nevanlinna space of analytic functions $u \in H(\Delta)$ such that $\log |u|$ has a harmonic majorant. Necessary and sufficient conditions are given in order that F_f acts from $\bigcup_{p<q} H^q$ to N (Section 4). We pay special attention to the action between H^p and B_q. Again, in this case we prove that

---

1991 *Mathematics Subject Classification.* 31A05, 30D15, 30D20.

mere action implies continuity (Section 5). In Section 6 we prove that F_f acts from N to N if and only if f is a polynomial. Some other results are included for the action between H^p and N and $\bigcup_{p<q} H^q$ and the Smirnov class N_*. We end up this article with a short section of open questions (Section 7).

2.   The Action in H^p and B_p, $0 < p \le \infty$

For $0 < p < \infty$ we write

$$||u||_p = \lim_{r \to 1} \left(\frac{1}{2\pi} \int_0^{2\pi} |u(re^{i\theta})|^p \, d\theta \right)^{1/p}, \quad u \in H(\Delta).$$

The classical Hardy Space H^p is defined as $\{u \in H(\Delta) : ||u||_p < +\infty\}$. We shall denote by H^∞ the Banach Space of bounded analytic functions in Δ with the uniform norm.

The Bergman space is defined by $B_p = H(\Delta) \cap L^p(\Delta, dxdy)$, $0 < p \le +\infty$ with the metric induced by

$$||u||_{B_p} = \left(\frac{1}{\pi} \iint_\Delta |u(z)|^p dxdy \right)^{\frac{1}{p}}.$$

The symbol BN (it stands for Bergman-Nevanlinna) shall denote the set of functions u in $H(\Delta)$ such that

$$\iint_\Delta \log^+ |u(z)| dxdy < \infty.$$

Let $f : \mathbb{C} \to \mathbb{C}$ be a function. In this section we would like to consider the action:

$$F_f : H^p \to H^q$$
$$u \mapsto f \circ u, \quad 0 < p, q \le \infty.$$

It is not difficult to see that if this action takes place then f is an entire function.

If f is an entire function then clearly F_f acts from H^∞ to H^∞, and therefore to H^q for all q, $0 < q \le \infty$. Moreover, this action is continuous. However, if $p < \infty$ and f is an entire function it is not necessarily true that F_f acts from H^p to H^q. It is easy, for instance, to give an example of a function $u \in H^p$ such that $u^2 \notin H^p$. The problem that we consider in this section is what kind of entire function is f if F_f acts from H^p to H^q. The answer is given in the next theorem. In what follows the symbol $[s]$ denotes the integer part of s.

THEOREM 1. *Let f be an entire function and $0 < p, q < \infty$. Then F_f acts from H^p to H^q or from B_p to B_q if and only if f is a polynomial of degree n, with $n \le [p/q]$.*

COROLLARY 1. *If f is an entire function and F_f acts from H^p to H^∞, $0 < p < \infty$ or from N to H^q, for some q, $0 < q \le \infty$, then f is constant.*

PROOF OF COROLLARY 1   The action from H^p to H^∞ implies that F_f acts from H^p to H^q for any q. The conclusion follows by applying Theorem 1 and letting q tend to infinity. The action from N to H^q implies that F_f acts from H^p to H^q for all $p > 0$. We obtain the conclusion by applying Theorem 1 and letting p tend to zero.

Remarks: Theorem 1 says, in particular, that the composition operator does not improve functions in H^p unless f is constant. The corollary can be viewed as an extension of the classical Liouville theorem.

To prove Theorem 1 we shall need the following lemma which is normally used to prove that the injection $H^p \to H(\Delta)$ is continuous, bounded and compact. It also shows that evaluation at a point of Δ is a continuous linear functional on H^p, and that bounded subsets of H^p are normal families. Better versions of it can be given but for our purposes the following is enough.

LEMMA 1.   *Let $u \in H^q$ $(0 < q < \infty)$. Then*

$$|u(z)| \le \frac{2^{1/q}\|u\|_q}{(1-|z|)^{1/q}}.$$

A proof of this result can be found for instance in [4]. We proceed to prove Theorem 1 for the Hardy case. For the Bergman case one argues in a similar way using the analogue of Lemma 1 for Bergman functions and appropriate examples (see [3]).

PROOF OF THEOREM 1   Let $\epsilon > 0$ and

$$u_\epsilon(z) = \left(\frac{1}{1-z} - \frac{1}{2}\right)^{1/(p+\epsilon)}.$$

Then $u_\epsilon \in H^p$, $\forall \epsilon > 0$. Therefore $f \circ u_\epsilon \in H^q$ and by Lemma 1

$$|f(u_\epsilon(z))| \le \frac{c}{(1-|z|)^{1/q}}.$$

Let $w_1 = u_\epsilon(z)$. Then $|z| = \left|\frac{w_1^{p+\epsilon}-1/2}{w_1^{p+\epsilon}+1/2}\right|$. Thus

$$|f(w_1)| \le \frac{c}{\left(1 - \left|\frac{w_1^{p+\epsilon}-1/2}{w_1^{p+\epsilon}+1/2}\right|\right)^{1/q}} = \frac{c\left|w_1^{p+\epsilon}+1/2\right|^{1/q}}{\left(\left|w_1^{p+\epsilon}+1/2\right| - \left|w_1^{p+\epsilon}-1/2\right|\right)^{1/q}}.$$

If we take w_1 in the sector $S_1 = \left\{\frac{-\pi}{4(p+\epsilon)} \le \arg w_1 \le \frac{\pi}{4(p+\epsilon)}\right\}$ and outside the ball $B = B(0, 1/2^{\frac{1}{p+\epsilon}})$ we obtain

$$|f(w_1)| \le c_1 \left|w_1^{p+\epsilon}+1/2\right|^{1/q}, \quad w \in S_1 \cap B^c,$$

with an appropiate constant c_1.

Now we argue in the same way starting with the function $w_2 = e^{\frac{2\pi i}{4(p+\epsilon)}} u_\epsilon(z)$. Thus

$$|z| = \left| \frac{w_2^{p+\epsilon} e^{-\frac{\pi i}{2}} - \frac{1}{2}}{w_2^{p+\epsilon} e^{-\frac{\pi i}{2}} + \frac{1}{2}} \right|$$

and

$$|f(w_2)| \leq \frac{c}{\left(1 - \left| \frac{w_2^{p+\epsilon} e^{-\frac{\pi i}{2}} - \frac{1}{2}}{w_2^{p+\epsilon} e^{-\frac{\pi i}{2}} + \frac{1}{2}} \right|\right)^{1/q}} = \frac{c \left| w_2^{p+\epsilon} e^{-\frac{\pi i}{2}} + \frac{1}{2} \right|^{1/q}}{\left(\left| w_2^{p+\epsilon} e^{-\frac{\pi i}{2}} + \frac{1}{2} \right| - \left| w_2^{p+\epsilon} e^{-\frac{\pi i}{2}} - \frac{1}{2} \right| \right)^{1/q}}.$$

If we take $w_2 \in S_2 = \left\{ \frac{\pi}{4(p+\epsilon)} \leq \arg w_2 \leq \frac{3\pi}{4(p+\epsilon)} \right\}$ we obtain

$$(2.1) \qquad |f(w_2)| \leq c_2 \left| w_2^{p+\epsilon} e^{-\frac{\pi i}{2}} + \frac{1}{2} \right|^{1/q}, \quad w_2 \in S_2 \cap B^c,$$

for a suitable constant c_2.

We keep on repeating this process a finite number of times (depending on p and ϵ) until we fill $\mathbb{C} \setminus B$ with sectors of the form

$$S_n = \left\{ \frac{(2n-3)\pi}{4(p+\epsilon)} \leq \arg w \leq \frac{(2n-1)\pi}{4(p+\epsilon)} \right\}, \quad n = 1, 2, \ldots l.$$

In dealing with the sector S_n we consider the function $w_n = e^{\frac{2(n-1)\pi i}{4(p+\epsilon)}} u_\epsilon(z)$, which is also in H^p for every $\epsilon > 0$, and obtain, in the same way as we got (2.1), the inequality

$$(2.2) \qquad |f(w_n)| \leq c_n \left| w_n^{p+\epsilon} e^{-\frac{(n-1)\pi}{2} i} + \frac{1}{2} \right|^{1/q}, \quad w_n \in S_n \cap B^c,$$

for some constant c_n. The inequality (2.2) and the fact that $\mathbb{C} \setminus \cup_{n=1}^l S_n$ is bounded prove that f is a polynomial of degree at most $(p+\epsilon)/q$. By letting ϵ tend to zero we obtain the desired result.

3. Continuity of F_f

Let $0 < p, q < +\infty$ and $F_f : H^p \to H^q$. We shall prove in this section that F_f is necessarily continuous. If $q > p$ then, in view of Theorem 1, F_f is constant and so continuous. Let us assume then that $q \leq p$. We shall need the following lemma.

LEMMA 2. *If $u_k \to u$ in H^p, $n \in \mathbb{N}$ and $n \leq [p/q]$, then $u_k^n \to u^n$ in H^q.*

PROOF   The case $n = 1$ is trivial. So let us assume that $n > 1$. In what follows the functions in the integrands should be interpreted as the non-tangentially boundary values of the functions involved. First of all we prove that $u_k^n, u^n \in H^q$.

In fact, since $nq \leq p$

(3.1)
$$\left(\frac{1}{2\pi} \int_0^{2\pi} |u_k^n|^q \, d\theta \right)^{\frac{1}{nq}} \leq \left(\frac{1}{2\pi} \int_0^{2\pi} |u_k|^p \, d\theta \right)^{\frac{1}{p}}.$$

Thus, in general, if $u \in H^p$ then $u^n \in H^q$ and

$$\|u^n\|_q \leq \|u\|_p^n, \quad \forall n \in \mathbb{N}, \quad n \leq [p/q].$$

To prove that $u_k^n \to u^n$ in H^q we argue as follows:

$$\|u_k^n - u^n\|_q = \left(\frac{1}{2\pi} \int_0^{2\pi} |u_k^n - u^n|^q \, d\theta \right)^{1/q}$$

$$= \left(\left(\frac{1}{2\pi} \int_0^{2\pi} \left(|u_k^n - u^n|^{1/n} \right)^{nq} \, d\theta \right)^{1/nq} \right)^n$$

$$\leq \left(\frac{1}{2\pi} \int_0^{2\pi} |u_k^n - u^n|^{p/n} \, d\theta \right)^{n/p}$$

by means of (3.1). Thus

$$\|u_k^n - u^n\|_q \leq \left(\frac{1}{2\pi} \int_0^{2\pi} |u_k - u|^{p/n} \left| u_k^{n-1} + u_k^{n-2} u + \cdots + u^{n-1} \right|^{p/n} \, d\theta \right)^{n/p}.$$

If we use Hölder's inequality

$$\left| \int f \cdot g \right| \leq \left(\int f^r \right)^{1/r} \left(\int g^s \right)^{1/s},$$

with $f = |u_k - u|^{p/n}$, $g = \left| u_k^{n-1} + \cdots + u^{n-1} \right|^{p/n}$, $r = n > 1$ and $\frac{1}{n} + \frac{1}{s} = 1$, (thus $s = \frac{n}{n-1}$) we obtain

(3.2) $\|u_k^n - u^n\|_q$

$$\leq \left(\frac{1}{2\pi} \int_0^{2\pi} |u_k - u|^p \, d\theta \right)^{1/p} \left(\frac{1}{2\pi} \int_0^{2\pi} \left| u_k^{n-1} + \cdots + u^{n-1} \right|^{p/(n-1)} \, d\theta \right)^{\frac{n-1}{p}}$$

$$\leq \|u_k - u\|_p \sum_{l=0}^{n-1} \left\| u_k^{n-1-l} u^l \right\|_{p/(n-1)}.$$

Since $u_k, u \in H^p$ then $u_k^{n-1-l} \in H^{p/(n-1-l)}$ and $u^l \in H^{p/l}$. Now we use Hölder's inequality in the following form:

$$f_i \in L^{p_i}, \ i = 1, 2, \ldots, m, \quad f = \prod_{i=1}^{m} f_i \quad \text{and} \quad \frac{1}{p} = \sum_{i=1}^{m} \frac{1}{p_i} \implies$$

$$f \in L^p \quad \text{and} \quad \|f\|_p \leq \prod_{i=1}^{m} \|f_i\|_{p_i}.$$

And so we get that $u_k^{n-1-l} \cdot u^l \in H^{p/(n-1)}$, $\forall l = 0, 1, \ldots, n-1$ and

$$\|u_k^{n-1-l} u^l\|_{p/(n-1)} \leq \|u_k^{n-1-l}\|_{p/(n-1-l)} \|u^l\|_{p/l} = \|u_k\|_p^{n-1-l} \|u\|_p^l.$$

This inequality implies that all summands on the right hand side of (3.2) are bounded above as $k \to \infty$. Therefore we get from (3.2) that $u_k^n \to u^n$ in H^q as required.

Now we are ready to prove the following theorem.

THEOREM 2. *Let f be an entire function and suppose that F_f acts from H^p to H^q or from B_p to B_q, $0 < p, q < \infty$. Then F_f is locally-lipschitz continuous.*

PROOF Again we prove the result for the Hardy case. By Theorem 1, $f(z) = a_n z^n + \cdots + a_0$ with $n \leq [p/q]$. Let $u_k \to u$ in H^p. Thus

$$F_f(u_k) - F_f(u) = a_n (u_k^n - u^n) + \cdots + a_1 (u_k - u).$$

Next we write

$$|F_f(u_k) - F_f(u)|^q \leq c(f, q, n) \left(|u_k^n - u^n|^q + \cdots + |u_k - u|^q \right).$$

Now integrating both sides of this inequality and using Lemma 2 we get the desired result.

Remark: The cases $F_f : H^p \to H^\infty$ and $F_f : H^\infty \to H^q$ are trivial.

4. The action from $\bigcup_{p<q} H^q$ to N

We shall need the following lemma.

LEMMA 3. *If $u \in N$ then*

$$\log |u(z)| \leq \frac{C}{1 - |z|}$$

for some constant C.

PROOF If $|u(z)| \leq 1$ there is nothing to prove. Let us suppose that $|u(z)| > 1$. Since $\log |u(z)|$ is majorized by the Poisson integral of a measure μ of finite total variation $\|\mu\|$ then

$$\log |u(z)| \leq \frac{1}{2\pi} \int_{\partial \Delta} \frac{1 - |z|^2}{|e^{i\theta} - z|^2} d\mu(\theta)$$

$$\leq \frac{1 + |z|}{2\pi(1 - |z|)} \|\mu\| \leq \frac{C}{1 - |z|}$$

as required.

THEOREM 3. *Let f be a function from $\mathbb{C}$ to $\mathbb{C}$. Then F_f acts from $\bigcup_{p<q} H^q$ to N ($0 < p < \infty$) if and only if f is an entire function of order at most p.*

Remark: The action $F_f : H^\infty \to N$ holds for any entire function f. This theorem is also true for the action between $\bigcup_{p<q} B_q$ and BN (see [3]). With similar arguments one can also tackle the action between $\bigcup_{p<q} H^q$ and BN. The result in this case is that f must be of order at most $2p$ ([3]).

PROOF   Let us assume first that F_f acts from $\bigcup_{p<q} H^q$ to N, for some p, $0 < p < \infty$. It is clear that f must be entire. For any $\epsilon > 0$ the functions

$$\omega_1 = u_\epsilon(z) = \left\{ \frac{1}{1-z} - \frac{1}{2} \right\}^{\frac{1}{p+\epsilon}}$$

belong to $\bigcup_{p<q} H^q$. Then, by hypothesis, $f \circ u_\epsilon \in N$ and so, by means of lemma 3, for some positive constant C we have

$$|f(\omega_1)| \leq e^{\frac{C}{1-|z|}}.$$

Next, we argue as we did to prove Theorem 1 to obtain that f is of order at most p.

Conversely, let f be of order at most p and $u \in \bigcup_{p<q} H^q$. There exits $\epsilon > 0$ so that $u \in H^{p+\epsilon}$ and $c = c(\epsilon) > 0$ such that $\log^+ M(r,f) \leq r^{p+\epsilon} + c$, $\forall r \geq 0$, where $M(r,f)$ denotes the maximun modulus of f on $\{|z| = r\}$. Thus

$$\int_0^{2\pi} \log^+ |f(u(re^{i\theta}))| d\theta \leq \int_0^{2\pi} \log^+ M(|u(re^{i\theta})|, f) d\theta$$
$$\leq \int_0^{2\pi} |u(re^{i\theta})|^{p+\epsilon} d\theta + 2\pi c,$$

which proves the desired conclusion.

As regard the action of F_f from H^p to N we have the following result.

THEOREM 4.   *Let f be an entire function of order less that p ($0 < p < \infty$) or of order p and finite type. Then F_f acts from H^p to N and from B_p to BN.*

PROOF   It is enough to assume that f is of order p and finite type $\sigma - \delta > 0$. Thus there exists a constant C such that

$$\log^+ M(r,f) \leq \sigma r^p + C, \quad r \geq 0.$$

If $u \in H^p$ then

$$\sup_{0 \le r < 1} \frac{1}{2\pi} \int_0^{2\pi} \log^+ |f(u(re^{i\theta}))| d\theta \le \sup_{0 \le r < 1} \frac{1}{2\pi} \int_0^{2\pi} \log^+ M(|u(re^{i\theta})|, f) d\theta$$

$$\le \sigma \sup_{0 \le r < 1} \frac{1}{2\pi} \int_0^{2\pi} |u(re^{i\theta})|^p d\theta + C < \infty.$$

Thus $f \circ u \in N$. The proof for the Bergman case is similar.

Before passing onto the next theorem we shall recall the definition of the Smirnov class N_* and some notation. If $u \in H(\Delta)$ and $0 \le r < 1$, let $u_r(z) = u(rz)$. In what follows the symbol u^* will denote the non-tangential limits of u on $\partial \Delta$. An analytic function u in Δ is said to belong to N_* if

$$\lim_{r \to 1} \int_{\partial \Delta} \log(1 + |u_r(\zeta)|) d\sigma(\zeta) = \int_{\partial \Delta} \log(1 + |u^*(\zeta)|) d\sigma(\zeta),$$

where σ denotes the Lebesgue measure on Δ normalized to have total mass 1. It should be said that this is not the usual definition of N_*. The reader is refered to ([10], Prop. 1.2) for the equivalence.

THEOREM 5.   *Let $0 < p < \infty$ and f be an entire function of order less than p or of order p and finite type. Then F_f acts from $\bigcup_{p < q} H^q$ to N_*.*

PROOF   The hypothesis implies that there are constants C_1 and C_2 such that

$$(4.1) \qquad \log^+ M(r, f) \le C_1 r^p + C_2, \quad \forall r \ge 0.$$

Let $u \in \bigcup_{p < q} H^q$. To see that $f \circ u \in N_*$ one has to prove that

$$(4.2) \quad \lim_{r \to 1} \int_{\partial \Delta} \log(1 + |(f \circ u)_r(\zeta)|) d\sigma(\zeta) = \int_{\partial \Delta} \log(1 + |(f(u^*(\zeta))|) d\sigma(\zeta).$$

Since f is of finite type, by Theorem 4 we have that $f \circ u \in N$ and so the integral on the right hand side of (4.2) is finite (see [3], Theorem 5.3). Therefore, all we have to prove is that

$$(4.3) \qquad \lim_{r \to 1} \int_{\partial \Delta} \log \frac{1 + |f(u(r\zeta))|}{1 + |f(u^*(\zeta))|} d\sigma(\zeta) = 0.$$

On the other hand, using (4.1) and the inequality $\log(1 + t) \le 1 + \log^+ t$, $t \ge 0$ we can write

$$(4.4)$$
$$\left| \log \frac{1 + |f(u(r\zeta))|}{1 + |f(u^*(\zeta))|} \right| \le 2 + \log^+ M(|u(r\zeta)|, f) + \log^+ M(|u^*(\zeta)|, f)$$
$$\le C_1(|u(r\zeta)|^p + |u^*(\zeta)|^p) + 2(C_2 + 1).$$

The estimate in (4.4) and a refined version of the Lebesgue dominated convergence Theorem allow us to take the limit in (4.3) inside the integral. This concludes the proof of Theorem 5.

5. The action between Hardy and Bergman spaces

We consider in this section the case of transforming Hardy functions into Bergman functions by means of superposition. As regard the other direction we have the following result

THEOREM 6. *Let f be an entire function. If $p \neq \infty$ then F_f acts from B_p to H^q if and only if f is constant.*

The proof of this theorem can be carried out arguing as in Theorem 1. We shall use the following notation.

$$M_p(r,u) = \left(\frac{1}{2\pi} \int_0^{2\pi} |u(re^{i\theta})|^p d\theta \right)^{\frac{1}{p}}, \quad 0 < p < \infty,$$
$$M_\infty(r,u) = \sup_{0 \le \theta < \pi} |u(re^{i\theta})|.$$

The next result (Theorem 7) depends on some inequalities of Hardy-Littlewood ([6]) about comparative growth of means of analytic functions. They are contained in the following two theorems.

THEOREM A. *(a) Let u be analytic in Δ, $1 < p < \infty$, $1 < a < \infty$ and $-1 < b < \infty$. Then*

$$\int_0^1 (1-r)^b M_p^a(r,u)dr \le C\left\{ |u(0)|^a + \int_0^1 (1-r)^{a+b} M_p^a(r,u')dr \right\},$$

where C is a constant independent of u.

(b) If

$$M_p(r,u) \le \frac{C}{(1-r)^\beta}, \quad 0 < p < \infty, \quad \beta \ge 0,$$

then there exists a constant $K = K(p,\beta)$ such that

$$M_q(r,u) \le \frac{KC}{(1-r)^{\beta+\frac{1}{p}-\frac{1}{q}}}, \quad p < q \le \infty.$$

A proof of this theorem can be found in [4, pp. 81-84].

As a consequence of Theorem A we have the following version of two inequalities also by Hardy and Littlewood. We state it in a way which keeps track of the H^p norm, a fact that will be needed in dealing with the continuity.

THEOREM B. *(a) Let $u \in H^p$, $0 < p < \infty$, $p < q \le \infty$. Then*

$$\int_0^1 (1-r)^{-\frac{p}{q}} M_q^p(r,u)dr \le C\|u\|_p^p,$$

where C is a constant depending on p and q.

(b) Let $0 < p \le q \le \infty$ and $u \in H^p$. Then

$$\int_0^1 (1-r)^{1-\frac{2p}{q}} M_q^{2p}(r,u)dr \le C\|u\|_p^p.$$

THEOREM 7. *Let f be an entire function. F_f acts from H^p to B_q if and only if f is a polynomial of degree at most $\left[\frac{2p}{q}\right]$. Furthermore, the action is locally Lipschitz continuous.*

PROOF   The proof of the first part is in [3]. We procceed to prove the continuity (this is not in [3]). We shall also need the following elementary lemmas.

LEMMA 4. *If $u \in H^p$ then $u^2 \in B_p$ and*

$$\|u^2\|_{B_p} \le C\|u\|_p^2.$$

PROOF   In fact,

$$
\begin{aligned}
\|u^2\|_{B_p}^p &= \frac{1}{\pi}\iint_\Delta |u^2|^p dx dy = \frac{1}{\pi}\int_0^1 r dr \int_0^{2\pi} |u|^p |u|^p d\theta \\
&\le \frac{1}{\pi}\int_0^1 r dr \int_0^{2\pi} |u|^p M_\infty(r,u)^p d\theta \le 2\|u\|_p^p \int_0^1 M_\infty^p(r,u) dr \\
&\le C\|u\|_p^{2p}
\end{aligned}
$$

in view of Theorem B(a) (setting there $q = \infty$).

Another way of stating this result is by saying that $H^p \subset B_{2p}$ and the injection is continuous. There is a short proof of this for $p = \frac{1}{2}$ in [9] based on inner-outer factorization.

LEMMA 5. *If $u, v \in B_{H^p}(0, R)$, $0 < p \le \infty$ then*

$$\|u^2 - v^2\|_{B_p} \le C\|u - v\|_p,$$

where C is a constant depending on p and R.

PROOF   We have

$$
\begin{aligned}
\|u^2 - v^2\|_{B_p}^p &= \frac{1}{\pi}\iint_\Delta |u^2 - v^2|^p dx dy = \frac{1}{\pi}\int_0^1 r dr \int_0^{2\pi} |u^2 - v^2|^p d\theta \\
&\quad \frac{1}{\pi}\int_0^1 r dr \int_0^{2\pi} |u-v|^p |u+v|^p d\theta \\
&\le \frac{C(p)}{\pi}\int_0^1 r dr \int_0^{2\pi} |u-v|^p (|u|^p + |v|^p) d\theta \\
&\le \frac{C(p)}{\pi}\int_0^1 r dr \int_0^{2\pi} |u-v|^p \left(M_\infty^p(r,u) + M_\infty^p(r,v)\right) d\theta \\
&\le 2C(p)\left\{\int_0^1 r M_\infty^p(r,u) dr + \int_0^1 r M_\infty^p(r,v) dr\right\} \|u-v\|_p^p \\
&\le C(p,R)\|u-v\|_p^p,
\end{aligned}
$$

in view of Theorem B(a).

LEMMA 6. *If $u, v \in B_{H^p}(0, R)$, $0 < p, q \leq \infty$ and $1 \leq n < \left[\frac{2p}{q}\right]$ then $u^n, v^n \in B_q$ and*

$$\|u^n - v^n\|_{B_q} \leq C\|u - v\|_{H_p},$$

where $C = C(p, q, n, R)$.

PROOF   The case $n = 1$ is obvious. If $n = 2$ we are in the situation of Lemma 5. Let us assume that $n > 2$. Then

$$\|u^n - v^n\|_{B_q} \leq \left(\frac{1}{\pi} \iint_\Delta |u^n - v^n|^{\frac{2p}{n}} dx dy\right)^{\frac{n}{2p}}$$

$$= \left(\frac{1}{\pi} \iint_\Delta |u - v|^{\frac{2p}{n}} |u^{n-1} + \cdots + v^{n-1}|^{\frac{2p}{n}} dx dy\right)^{\frac{n}{2p}}.$$

Next, we use Hölder's inequality $\int fg \leq \left(\int f^r\right)^{\frac{1}{r}} \left(\int g^s\right)^{\frac{1}{s}}$ with

$$f = |u - v|^{\frac{2p}{n}}, \ g = |u^{n-1} + \cdots + v^{n-1}|^{\frac{2p}{n}}, \ r = \frac{n}{2}, \ s = \frac{n}{n-2}$$

and obtain

$$\|u^n - v^n\|_{B_q}$$

$$\leq \left(\frac{1}{\pi} \iint_\Delta |u - v|^p dx dy\right)^{\frac{1}{p}} \left(\frac{1}{\pi} \iint_\Delta |u^{n-1} + \cdots + v^{n-1}|^{\frac{2p}{n-2}} dx dy\right)^{\frac{n-2}{2p}}$$

$$\leq C\|u - v\|_{B_p} \left(\sum_{l=0}^{n-1} \|u^{n-1-l} v^l\|_{B_{\frac{2p}{n-2}}}\right).$$

But again by Hölder's inequality

$$\|u^{n-1-l} v^l\|_{B_{\frac{2p}{n-2}}} \leq \|u^{n-1-l}\|_{B_{\frac{2p}{n-1-l}}} \|v^l\|_{B_{\frac{2p}{l}}}.$$

Therefore

$$\|u^n - v^n\|_{B_q} \leq C\|u - v\|_{B_p} \left(\sum_{l=0}^{n-1} \|u^{n-1-l}\|_{B_{\frac{2p}{n-1-l}}} \|v^l\|_{B_{\frac{2p}{l}}}\right)$$

$$= C\|u - v\|_{B_p} \left(\sum_{l=0}^{n-1} \|u^2\|_{B_p}^{n-1-l} \|v^2\|_{B_p}^l\right)$$

$$\leq C_1 \|u - v\|_p \left(\sum_{l=0}^{n-1} \|u\|_p^{2(n-1-l)} \|v\|_p^{2l}\right),$$

in view of Lemma 4.

Now we are ready to prove the locally Lipschitz continuity. We know that f has the form

$$f(z) = a_m z^m + \cdots + a_0, \quad m \leq \left[\frac{2p}{q}\right].$$

Thus if $u, v \in B_{H^p}(0, R)$, then

$$\|F_f(u) - F_f(v)\|_{B_q} \le C_1 \left(\|u^m - v^m\|_{B_q} + \cdots + \|u - v\|_{B_q} \right)$$
$$\le C\|u - v\|_p,$$

by means of Lemma 6, where C depends on p, q, f and R.

An easy application of this gives

THEOREM 8. *If* $\log^+ M(r, f) = O(r^{2p})$ *as* r *tends to* ∞ *then* F_f *acts from* H^p *to* BN.

6.   The action from N to N

Our next problem consists in characterizing all entire functions f for which F_f acts from N to N. If F_f acts from N to N then, in particular, it acts from $\bigcup_{p<q} H^q$ to N for all positive real number p. Thus, by Theorem 3, order of f is less than or equal to p, $\forall p > 0$. Therefore f is an entire function of order zero. As a matter of fact one can prove the following result.

THEOREM 9.   *Let* $f : \mathbb{C} \to \mathbb{C}$. *Then* F_f *acts from* N *to* N *if and only if* f *is a polynomial.*

PROOF   If $\log M(r, f) = O(\log r)$ then take C_1 and C_2 such that

$$\log^+ M(r, f) \le C_1 \log^+ r + C_2, \quad \forall r \ge 0.$$

Hence

$$\int_0^{2\pi} \log^+ |f(u(re^{i\theta}))| d\theta \le \int_0^{2\pi} \log^+ M(|u(re^{i\theta})|, f) d\theta$$
$$\le C_1 \int_0^{2\pi} \log^+ |u(re^{i\theta})| d\theta + 2\pi C_2.$$

Thus $f \circ u \in N$, whenever $u \in N$. Next, let us assume that f is an entire function such that F_f acts from N to N. The function $\omega = u(z) = exp\left\{ \frac{1+z}{1-z} \right\}$ belongs to N. Therefore, $f(u(z)) \in N$ and by Lemma 3

$$(6.1) \qquad\qquad \log |f(\omega)| \le \frac{C}{1 - |z|}.$$

We shall confine ourselves to z in the Stolz angle $S = \Delta \bigcap \{z : |1 - z| \le C_1(1 - |z|), \; Re\; z > c_0\}$. If S is big enough then $u(S) \supseteq \{\omega : |\omega| > R\}$ for some R. Since $|\omega| = exp\left\{ \frac{1-|z|^2}{|1-z|^2} \right\}$ we can write

$$\log |\omega| = \frac{1}{1 - |z|} \frac{(1 - |z|)^2 (1 + |z|)}{|1 - z|^2}.$$

Therefore

(6.2)
$$\frac{1}{1-|z|} = \frac{|1-z|^2}{(1-|z|)^2(1+|z|)} \log |\omega|$$

$$\leq \frac{|1-z|^2}{(1-|z|)^2} \log |\omega| \leq C_1^2 \log |\omega|, \quad z \in S.$$

From (6.1) and (6.2) one gets

$$\log |f(\omega)| \leq C_2 \log |\omega|,$$

for all ω outside $B(0,R)$. This proves Theorem 9.

7. Some questions

We end this article with a series of questions concerning the autonomous nonlinear superposition operator F_f. (1) Is the converse to Theorem 4 true? (2) Does the converse to Theorem 5 hold? (3) All seems to indicate that if F_f acts from N to N_* then f is constant. (4) When does F_f act from H^p to N_* or from N_* to N_*? (5) Does the action of F_f from N_* to N_* imply continuity? (6) Does the action of F_f from $\bigcup_{p<q} H^q$ to N_* imply continuity? (7) Is Theorem 9 true when we replace N by BN?

In this set of questions we have omitted the problem of the action between N and BN. In this case it is possible to prove the following result.

THEOREM 10. *Let f be an entire function. Then F_f acts from N to BN if and only if*

$$\int_1^\infty \frac{\log \ M(e^t, f)}{t^3} dt < \infty.$$

To prove this theorem we need a sharpening of the obvious analog of Lemma 1 for BN functions. A kind of Hardy-Littlewood inequality for Hardy harmonic functions is the other ingredient necessary to prove Theorem 10 (for details we refer the reader to [2]).

REFERENCES

1. Appel, J., Zabrejko, P.P., *Nonlinear Superposition Operators*, Cambridge University Press, 1990.
2. Cámera, G.A., *A sharp inequality for Bergman-Nevanlinna functions*, unpublished.
3. Cámera, G.A. and Gimenez, J., *The non-linear superposition operator acting on Bergman spaces*, to appear in Compositio Mathematica.
4. Duren, P., *Theory of H^p Spaces*, Academic Press, New York., 1970.
5. Garnett, J., *Bounded analytic functions*, Academic Press, New York., 1981.
6. Hardy, G.H. and Littlewood, J.E., *Some properties of fractional integrals, II.*, Math. Z. **28** (1928), 612-634.
7. Krasnoselskij, M.A., *On the continuity of the operator $Fu(x) = f(x, u(x))$ (Russian)*, Doklady Akad. Nauk SSSR **77,2** (1951), 185-188.
8. Marcus,M., Mizel, V.J.,, *Every superposition operator mapping one Sobolev space into another is continuous*, J. Funct. Anal. **33** (1979), 217-229.

9. Shapiro, J.H., *Remarks on F-spaces of analytic functions*, in Banach spaces of analytic functions, J. Baker, C. Cleaver and J. Distel, editors, Springer Lecture Notes in Mathematics **604** (1977).

10. Shapiro, J.H. and Shields, A.L., *Unusual topological properties of the Nevanlinna class.*, Amer. J. Math. **97 no 4** (1976), 915-936.

IVIC-MATEMÁTICAS, APARTADO 21827, CARACAS 1020-A, VENEZUELA

Contemporary Mathematics
Volume **189**, 1995

A version of cotlar's lemma for L^p spaces and some applications

A. CARBERY

Dedicated to Professor M. Cotlar on the occasion of his eightieth birthday.

ABSTRACT. A variant of the Cotlar-Stein Lemma which is valid for L^p spaces is discussed. Some applications —in particular to Hilbert transforms along flat variable plane curves —are also briefly considered.

The very well-known "almost-orthogonality" lemma of Cotlar and Stein is as follows:

COTLAR'S LEMMA. *Suppose that* $\{T_j\}_{j\in\mathbb{Z}}$ *is a sequence of bounded linear operators on a Hilbert space such that*

$$\|T_j T_k^*\| \leq \alpha(j-k)$$
$$and \quad \|T_j^* T_k\| \leq \alpha(j-k).$$

Then

$$\|\sum_j T_j\| \leq \sum_j \alpha(j)^{1/2}.$$

See [**S**], Ch. 7 for the elegant proof, and a discussion of the lemma. This reference also contains a discusssion of many of the successes of the Cotlar-Stein lemma: L^2 estimates for pseudodifferential operators in the $S_{\delta,\delta}^0$ classes (Calderón -Vaillancourt theorems); Singular Integral Operators on homogeneous groups; the David-Journé $T1$ theorem on L^2 boundedness of Singular Integral Operators of non-convolution type. It has also played a pivotal role in the theory of Singular Integrals and Maximal Functions along curves in homogeneous groups (and indeed in wider settings). In this connection see also the application below.

1991 *Mathematics Subject Classification.* 42B25.
Research partially supported by ICMS, Edinburgh, Scotland.

There is a standard way to apply Cotlar's lemma to maximal functions of the form $f \longmapsto \sup_k |S_k f|$. Suppose we know already that for some "closely associated" maximal function $f \longmapsto \sup_k |P_k f|$ we have L^2 boundedness. (One should think of $\sup_k |P_k f|$ as being some variant of the Poisson or Hardy-Littlewood maximal function.) We can then dominate $\sup_k |S_k f|$ by

$$\sup_k |S_k f| \leq \left(\sum_k |(S_k - P_k)f|^2 \right)^{1/2} + \sup_k |P_k f|,$$

and hope that $\sum_k \pm(S_k - P_k)$ satisfies the hypothesis of Cotlar's lemma (independently of the choice of $\pm$ if at all!)

One of the chief virtues of Cotlar's lemma is its universality – that is, as an abstract principle it is valid in any Hilbert space , and that the quantitative conclusion depends upon nothing other than the quantification in the hypothesis. At the same time, this leads to one of its chief drawbacks– as a Hilbert space phenomenon it is valid only on L^2, not on L^p, where, of course, substantial interest for the harmonic analyst lies. In order to obtain L^p estimates, one traditionally does some Calderón-Zygmund theory and then interpolates (to oversimplify matters). But doing this introduces dependence on the underlying geometry necessary for Calderón-Zygmund theory–and in particular upon the dimension of the underlying $\mathbb{R}^n$ (if indeed that is where we are working). Thus it would seem to be reasonable to develop an *abstract* almost-orthogonality theory for operators on L^p spaces , $1 < p < \infty$. Of course such a theory cannot be as universal as the Hilbert space one is, as simple examples show; the price we have to pay is having to incorporate the Calderón-Zygmund theory – in the guise of Littlewood-Paley theory –into the formulation of the principle. This is not unreasonable as the previous discussion of the application of Cotlar's lemma to maximal functions demonstrates: there we need some "abstract" control over the Poisson-like maximal function to retain the abstract nature of the procedure.

We are now lead to the theory by the following observation.

OBSERVATION. *Suppose we have a decomposition of the identity*

(1) $$\sum_{j \in \mathbb{Z}} Q_j = I$$

(in, say, the strong operator toplogy on a Hilbert space.) Suppose the Littlewood-Paley operators $\{Q_j\}$ satisfy

(2) $$\|Q_j^* Q_k\| + \|Q_j Q_k^*\| \leq C 2^{-\epsilon |j-k|}$$

for some $\epsilon > 0$, (so that $\{Q_j\}$ satisfies the hypothesis of Cotlar's Lemma.) Suppose $\{T_j\}_{j \in \mathbb{Z}}$ satisfies

(3) $$\|T_j\| \leq C$$

and

(4) $$\|T_j Q_{j+k}^*\| + \|T_j^* Q_{j+k}\| \le C2^{-\epsilon|k|}$$

for some $\epsilon > 0$. Then $\{T_j\}$ also satisfies the hypothesis of Cotlar's Lemma, and so $\sum_j T_j$ is a bounded operator on the Hilbert space.

PROOF. $T_j T_k^* = \sum_{l,m} T_j Q_l^* Q_m T_k^*$, and each term in the sum can be estimated in norm by $\|T_j Q_l^*\| \|Q_m T_k^*\|$ as well as by $C\|Q_l^* Q_m\|$. Thus

$$\|T_j T_k^*\| \le C \sum_{l,m} 2^{-\epsilon'|j-l|} 2^{-\epsilon'|m-k|} 2^{-\epsilon'|l-m|}$$

$$\le C2^{-\epsilon''|j-k|}.$$

Similarly for $T_j^* T_k$. $\square$

This observation suggests the following formulation of an L^p almost orthogonality principle for operators.

TENTATIVE ALMOST ORTHOGONALITY PRINCIPLE. *Suppose we have Littlewood-Paley operators on L^2 satisfying (1) and (2), and linear operators T_j on L^2 satisfying (3) and (4). Given that we accept the need for Littlewood-Paley theory, let us suppose*

(5) $$\left\| \sum \pm Q_j \right\|_{p_0-p_0} + \left\| \sum \pm Q_j^* \right\|_{p_0-p_0} \le C$$

uniformly in the choices of $\pm$, for some $p_0 \ne 2$, with $1 < p_0 < \infty$. Bearing in mind the nature of the conclusion we desire, we suppose

(6) $$\|T_j\|_{p_0-p_0} \le C.$$

Can we then conclude that

$$\left\| \sum T_j \right\|_{p-p} \le A(C, \epsilon, p, p_0)$$

for p between p_0 and 2 ?

Unfortunately, the answer is no, as a classical counterexample due to Littman, McCarthy and Rivière shows. (See[**LMcCR**].) Indeed, in that example the Q_j are standard Littlewood-Paley operators in $\mathbb{R}$ associated to the intervals $[2^{-j}, 2^{-j+1}]$, the T_j are Fourier multiplier operators with multipliers supported in $[2^{-j}, 2^{-j+1}]$ satisfying $\|T_j\|_{1-1} \le C$; but $\sum T_j$ is bounded only on L^2. So clearly some extra ingredient is needed to make our principle work.

Undeterred by the failure of our tentative almost orthogonality principle, let us proceed as if it were valid, and see what it would imply about maximal functions.

If $\{P_j\}$ are the natural Poisson -like averages to associate to a problem – in so far as the the $P_j's$ form an approximation of the identity with $P_j \to I$ as $j \to -\infty$ and $P_j \to 0$ as $j \to \infty$ – then the natural Littlewood-Paley difference operators Q_j are given by

$$Q_j = P_j - P_{j+1}$$

so that

$$P_j = \sum_{l=0}^{\infty} Q_{j+l}$$

PROPOSITION 1. *(Almost Orthogonality Principle for maximal functions.)*
Let P_j and Q_j be related as above. Suppose that $\{Q_j\}$ satisfies (1), (2) and (5),
and that $\{S_j - P_j\}$ satisfies (3), (4) and (6). Suppose also that

$$(7) \qquad\qquad \| \sup_j |P_j f| \|_r \leq C_r \|f\|_r$$

for $r \in [p_0, 2]$ or $[2, p_0]$. Then

$$\| \sup_j |S_j f| \|_p \leq A(C, \epsilon, p, p_0) \|f\|_p$$

for p between 2 and p_0 , if either $p_0 > 2$ or $p_0 < 2$ and S_j and P_j are positive
(in the sense that $f \geq 0 \Rightarrow S_j f$ and $P_j f \geq 0$.)

LEMMA 1. *Suppose that $\{Q_j\}$ satisfies (1),(2) and (5). Then there is an $N =$*
$N(C, p, \epsilon, p_0)$ such that $D = \sum_{j \in \mathbb{Z}, |k| \leq N} Q_j^* Q_{j+k}$ is invertible on $L^p, p \in (p_0, p_0')$.

PROOF. For a fixed k, $\sum_j Q_j^* Q_{j+k}$ is bounded on L^{p_0} uniformly in k since
for $f \in L^{p_0}$, $g \in L^{p_0'}$,

$$\left\langle \sum_j Q_j^* Q_{j+k} f, g \right\rangle = \sum_j \langle Q_{j+k} f, Q_j g \rangle$$

$$\leq \int \left(\sum_j |Q_{j+k} f|^2 \right)^{1/2} \left(\sum_j |Q_j g|^2 \right)^{1/2}$$

$$\leq \| \left(\sum |Q_j f|^2 \right)^{1/2} \|_{p_0} \| \left(\sum |Q_j g|^2 \right)^{1/2} \|_{p_0'}$$

$$\leq C \|f\|_{p_0} \|g\|_{p_0'} \quad \text{by (5)}.$$

On the other hand, by Cotlar's lemma (or the Observation) , $\sum_j Q_j^* Q_{j+k}$
is bounded on L^2 with a constant dominated by $2^{-\epsilon|k|}$. Hence $\sum_j Q_j^* Q_{j+k}$ is
bounded on L^p with a constant dominated by $2^{-\epsilon'|k|}$. Thus $\sum_j \sum_{|k| \geq N} Q_j^* Q_{j+k}$
is bounded on L^p with a constant dominated by $2^{-\epsilon' N}$. But $I = \sum_j Q_j^* \sum_k Q_{j+k}$
$= \sum_j \sum_{|k| \leq N} + \sum_j \sum_{|k| > N} = D + B$ where $\|B\|_{p-p} \leq C 2^{-\epsilon' N} < 1/2$ if N is
chosen sufficiently large. Thus

$$\sum_{j \in \mathbb{Z}, |k| \leq N} Q_j^* Q_{j+k} = D = I - B$$

is invertible on L^p (by considering the Neumann series $I + B + B^2 + \dots$) $\quad \square$

REMARK 1. The proof of the lemma shows that $\sum_j \pm Q_j^* Q_{j+k}$ is bounded
on L^{p_0} independently of k , and on L^p with exponential decay in $|k|$. Using
Rademacher functions we see that the same is true of the quadratic expression
$f \longmapsto (\sum_j |Q_j^* Q_{j+k} f|^2)^{1/2}$.

PROOF OF PROPOSITION 1.

Case(i) $p_0 > 2$

By Lemma 1 it suffices to show that

$$\| \sup_j |S_j \sum_{l \in \mathbb{Z}} Q^*_{j+l} Q_{j+l+k} f| \|_p \leq A \|f\|_p$$

with bounds independent of k. Now

$$\sup_j |S_j (\sum_l Q^*_{j+l} Q_{j+l+k} f)| \leq \left(\sum_j |(S_j - P_j)(\sum_l Q^*_{j+l} Q_{j+l+k} f)|^p \right)^{1/p}$$

$$+ \sup_j |P_j \sum_l Q^*_l Q_{l+k} f|.$$

The second term here is controlled by (7) and the fact that $\sum_l Q^*_l Q_{l+k}$ is bounded on L^p (see Remark 1.) For the first term we integrate, interchange the order of integration and summation, and pull out the l-sum to obtain

$$\sum_l \left(\sum_j \|(S_j - P_j)Q^*_{j+l} Q_{j+l+k} f\|_p^p \right)^{1/p} \leq \sum_l C 2^{-\epsilon'|l|} \left(\sum_j \|Q_{j+l+k} f\|_p^p \right)^{1/p}$$

(since $\|(S_j - P_j)Q^*_{j+l}\|_{2-2} \leq C 2^{-\epsilon|l|}$ and $\|(S_j - P_j)Q^*_{j+l}\|_{p_0 - p_0} \leq C$,)

$$= C' \|(\sum_j |Q_j f|^p)^{1/p}\|_p$$

$$\leq C'' \|(\sum_j |Q_j f|^2)^{1/2}\|_p \text{(since } p > 2 \text{)}$$

$$\leq A \|f\|_p \quad \text{(by (5) and a Rademacher}$$
$$\text{function argument).}$$

Case(ii) $p_0 < 2$.

We establish this case by means of several steps. Fix p with $p_0 < p < 2$.

1.Because of (7) it suffices to show that

(8) $$\| \sup_j |(S_j - P_j)f| \|_p \leq A \|f\|_p$$

2.Assume temporarily that we are only considering j with $|j| \leq M$. Then since the P_j and S_j are uniformly bounded on L^p, the version of (8) corresponding to this setting holds; that is we can assert that

$$\| \sup_{|j| \leq M} |(S_j - P_j)f| \|_p \leq A_p(M) \|f\|_p$$

where $A_p(M)$ is , by definition, the best constant such that the inequality holds. (Obviously $A_p(M) = O(M)$.)

3.(Still in the case that $|j| \leq M$.) Consider inequalities of the form

(*) $$\| \|(S_j - P_j)g_j\|_{l^s} \|_{L^r} \leq Const. \| \|g_j\|_{l^s} \|_{L^r}.$$

When $s = r = p_0$, (*) holds since the $S_j - P_j$ are uniformly bounded on L^{p_0} , by (6). When $s = \infty$ and $r = p$, letting $G = \sup_j |g_j| \in L^p$, we see that positivity of S_j and P_j gives

$$\sup_j |(S_j - P_j)g_j| \le \sup_j |S_j G| + \sup_j |P_j G|$$

$$\le \sup_j |(S_j - P_j)G| + 2\sup_j |P_j G|$$

whose L^p norm is dominated by

$$(A_p(M) + 2C)\,\|G\|_p \le C'A_p(M)\|\,\|g_j\|_{l^\infty}\|_{L^p}$$

(by (7).) Interpolating, we see that (*) holds for $s = 2$ and some $r = p_1 \in (p_0, p)$ with constant dominated by $A_p(M)^\theta$ for some $\theta \in (0, 1)$; that is

$$(9) \qquad \|\left(\sum_j |(S_j - P_j)g_j|^2\right)^{1/2}\|_{p_1} \le CA_p(M)^\theta\|\left(\sum_j |g_j|^2\right)^{1/2}\|_{p_1}$$

4.(Still in the case that $|j| \le M$.) With p_1 as above,

$$\|\sup_j |(S_j - P_j)Q^*_{j+l}Q_{j+l+k}f|\|_{p_1} \le \|\left(\sum_j |(S_j - P_j)Q^*_{j+l}Q_{j+l+k}f|^2\right)^{1/2}\|_{p_1}$$

$$\le CA_p(M)^\theta\|\left(\sum_j |Q^*_{j+l}Q_{j+l+k}f|^2\right)^{1/2}\|_{p_1}$$

$$\text{(by (9))}$$

$$\le CA_p(M)^\theta\|f\|_{p_1}.$$

On the other hand, an application of either Cotlar's Lemma or the Observation shows that

$$\|\sum_j \pm(S_j - P_j)Q^*_{j+l}Q_{j+l+k}f\|_2 \le C2^{-\epsilon|l|}\|f\|_2$$

which, upon dominating $\|.\|_{l^\infty}$ by $\|.\|_{l^2}$ and using Rademacher functions implies

$$\|\sup_j |(S_j - P_j)Q^*_{j+l}Q_{j+l+k}f|\|_2 \le C2^{-\epsilon|l|}\|f\|_2.$$

Interpolating yields

$$\|\sup_j |(S_j - P_j)Q^*_{j+l}Q_{j+l+k}f|\|_p \le C2^{-\epsilon'|l|} < A_p(M)^{\theta'}\|f\|_p$$

where $0 < \theta' < 1$ and $\epsilon' > 0$. Hence

$$\|\sup_j |(S_j - P_j)\{\sum_{m\in\mathbb{Z},|k|\le N} Q^*_m Q_{m+k}\}f|\|_p \le C_N A_p(M)^{\theta'}\|f\|_p.$$

The lemma now implies

$$\| \sup_j |(S_j - P_j)f| \|_p \leq C A_p(M)^{\theta'} \|f\|_p.$$

5. Returning to step 2, we recall that $A_p(M)$ was defined to be the best constant in the preceding inequality. So $A_p(M) \leq A_p(M)^{\theta'}$, which, since $\theta' < 1$, implies that $A_p(M)$ is uniformly bounded as $M \longrightarrow \infty$. Thus (8) holds and the proof is concluded. $\square$

Thus, in spite of the failure of the tentative almost orthogonality principle, we are nevertheless able to act as if it were true in terms of its application to maximal functions, so long as we insist in addition that P_j and S_j be positive when $p_0 < 2$. Can we now turn matters round, and use the almost orthogonality principle for maximal functions to obtain one for sums of operators? The answer is yes, and we now address this matter. We first give a preliminary lemma and proposition.

LEMMA 2. *Suppose $\{Q_j\}$ satisfies (1), (2) and (5). Then for $p_0 < p < p_0'$,*

$$\|g\|_{p'} \leq C\|(\sum_j |Q_j g|^2)^{1/2}\|_{p'}.$$

PROOF. By Lemma 1, $\|g\|_{p'} \approx \|\sum_{j \in \mathbb{Z}, |k| \leq N} Q_j^* Q_{j+k} g\|_{p'}$ and to estimate this we let $\|f\|_p = 1$ and examine (for fixed $|k| \leq N$)

$$< \sum_j Q_j^* Q_{j+k} g, f > = \sum_j < Q_{j+k} g, Q_j f >$$
$$\leq \|(\sum_j |Q_j g|^2)^{1/2}\|_{p'} \|(\sum_j |Q_j f|^2)^{1/2}\|_p$$
$$\leq C\|(\sum_j |Q_j g|^2)^{1/2}\|_{p'} \text{ by (5)}. \qquad \square$$

A first version of the almost orthogonality principle for sums of operators is contained in Proposition 2 below. It does not depend on maximal functions, but instead we have to assume the vector-valued inequality (6')-which in itself is rather strong.

PROPOSITION 2. *Let $p_0 < p < 2$. Suppose (1),(2),(3),(4),(5) hold, that*

$$(10) \qquad \|\left(\sum_k |Q_k g_k|^2\right)^{1/2}\|_{p_0'} \leq C\|\left(\sum_k |g_k|^2\right)^{1/2}\|_{p_0'},$$

and that

$$(6') \qquad \|\left(\sum_j |T_j g_j|^2\right)^{1/2}\|_{p_0} \leq C\|\left(\sum_j |g_j|^2\right)^{1/2}\|_{p_0}.$$

Then $\sum_j T_j$ is bounded on L^p.

PROOF. We argue by duality. By Lemma 2,

$$\| \sum_j T_j^* f \|_{p'} \le C \| \left(\sum_k |Q_k \sum_j T_j^* f|^2 \right)^{1/2} \|_{p'}$$

$$= C \| \left(\sum_k |Q_k \sum_j T_{j+k}^* f|^2 \right)^{1/2} \|_{p'}$$

$$\le \sum_j \| \left(\sum_k |Q_k T_{j+k}^* f|^2 \right)^{1/2} \|_{p'}.$$

Now, for j fixed,

$$\| \left(\sum_k |Q_k T_{j+k}^* f|^2 \right)^{1/2} \|_{p'} = \| \left(\sum_k |Q_k T_{j+k}^* \sum_l Q_{k+l} f|^2 \right)^{1/2} \|_{p'}$$

$$\le \sum_l \| \left(\sum_k |Q_k T_{j+k}^* Q_{k+l} f|^2 \right)^{1/2} \|_{p'},$$

(since $I = \sum_k Q_k$ in $L^{p'}$ (weakly)). Now, for j and l fixed,

$$\| \left(\sum_k |Q_k T_{j+k}^* Q_{k+l} f|^2 \right)^{1/2} \|_{p_0'} \le C \| \left(\sum_k |T_{j+k}^* Q_{k+l} f|^2 \right)^{1/2} \|_{p_0'} \quad \text{(by (10))}$$

$$\le C \| \left(\sum_k |Q_{k+l} f|^2 \right)^{1/2} \|_{p_0'} \quad \text{(by (6'))}$$

$$(\dagger) \qquad\qquad\qquad\qquad \le C \| f \|_{p_0'} \quad \text{(by (5).)}$$

On the other hand, for j and l fixed,

$$\| \sum_k \pm Q_k T_{j+k}^* Q_{k+l} \|_{2-2} \le C 2^{-\epsilon|j|} 2^{-\epsilon|l|},$$

and hence

$$(\ddagger) \qquad \| \left(\sum_k |Q_k T_{j+k}^* Q_{k+l} f|^2 \right)^{1/2} \|_2 \le C 2^{-\epsilon|j|} 2^{-\epsilon|l|} \| f \|_2,$$

as hypotheses (2),(3) and (4) and the Observation show.

Thus interpolating ($\dagger$) with ($\ddagger$) gives

$$\| \left(\sum_k |Q_k T_{j+k}^* Q_{k+l} f|^2 \right)^{1/2} \|_{p'} \le C 2^{-\epsilon'|j|} 2^{-\epsilon'|l|} \| f \|_{p'},$$

for $p_0 < p < 2$. Summing in j and l shows $\| \sum_j T_j^* f \|_{p'} \le C \| f \|_{p'}$; that is, $\sum_j T_j$ is bounded on L^p. $\square$

REMARK 2. a) As the proof of Proposition 2 shows, we can replace hypotheses (10) and (6') by their analogues on L^{p_1}, $p_1 > p_0$ but arbitrarily close to p_0, without affecting the conclusion.

b) Since we know that the Q_k^*'s are uniformly bounded on L^{p_0} (by hypothesis (5) and a Rademacher function argument), we have that

$$\| \, \|Q_j^* g_j\|_{l^{p_0}} \|_{L^{p_0}} \leq C\| \, \|g_j\|_{l^{p_0}} \|_{L^{p_0}}.$$

Were we also to have

(11) $$\| \, \|Q_j^* g_j\|_{l^{\infty}} \|_{L^{p_0}} \leq C\| \, \|g_j\|_{l^{\infty}} \|_{L^{p_0}}$$

we could interpolate to obtain (10). If $Q_j = P_j - P_{j+1}$ with $P_j \geq 0$, (11) would follow from

(10') $$\| \sup_j |P_j^* f| \|_{p_0} \leq C\|f\|_{p_0}$$

which in many cases can be established directly. On the other hand, inequality (10) will usually follow directly from the theory of vector-valued singular integrals – in establishing hypothesis (5), inequality (10) often comes almost for free.

c) The main drawback with Proposition 2 is its reliance upon the vector-valued inequality (6') for the $\{T_j\}$. However, taking our lead from the discussion in b) above and bearing in mind a), we note that

(**) $$\| \, \|T_j g_j\|_{l^s} \|_{L^r} \leq C\| \, \|g_j\|_{l^s} \|_{L^r}$$

for $s = r = p_0$, and for $s = \infty$, all $p \in (r_0, 2)$ would imply (6') with $p_1 > p_0$ in place of p_0, with p_1 arbitrarily close to p_0. But (**) with $s = r = p_0$ follows from (6), and

$$\| \sup_j |T_j g_j| \|_p \leq C\| \sup_j |g_j| \|_p$$

would follow if we had

$$|T_j g| \leq S_j |g|$$

with $S_j \geq 0$ and satisfying

$$\| \sup_j |S_j f| \|_p \leq C\|f\|_p.$$

But this is precisely the conclusion of the almost orthogonality principle for maximal operators! We have thus proved the following:

PROPOSITION 3. *(Almost-Orthogonality Principle for Sums of Operators)* *Suppose that* $\{Q_j\}$ *satisfies*

(1) $$\sum_{j \in \mathbb{Z}} Q_j = I,$$

(2) $$\|Q_j^* Q_k\|_{2-2} + \|Q_j Q_k^*\|_{2-2} \leq C 2^{-\epsilon|j-k|},$$

and

(5)
$$\|\sum_j \pm Q_j\|_{p_0-p_0} + \|\sum_j \pm Q_j^*\|_{p_0-p_0} \leq C$$

for some $p_0 \in (1,2)$.

Suppose that $Q_j = P_j - P_{j+1}$ where $P_j \geq 0$ and

(7)
$$\|\sup_j |P_j f|\|_r \leq C_r \|f\|_r$$

for $p_0 \leq r \leq 2$.

Suppose also that either

(10)
$$\left\|\left(\sum_k |Q_k g_k|^2\right)^{1/2}\right\|_{p_0'} \leq C \left\|\left(\sum_k |g_k|^2\right)^{1/2}\right\|_{p_0'}$$

or

(10')
$$\|\sup_j |P_j^* f|\|_{p_0} \leq C\|f\|_{p_0}.$$

Suppose that $\{T_j\}$, $\{S_j\}$ satisfy

$$|T_j f| \leq S_j |f|$$

where $S_j \geq 0$.

Suppose

(6")
$$\|S_j\|_{r-r} \leq C$$

for $p_0 \leq r \leq 2$, that

(4')
$$\|(S_j - P_j)Q_{j+k}^*\|_{2-2} + \|(S_j - P_j)^* Q_{j+k}\|_{2-2} \leq C2^{-\epsilon|k|},$$

and that

(4)
$$\|T_j Q_{j+k}^*\|_{2-2} + \|T_j^* Q_{j+k}\|_{2-2} \leq C2^{-\epsilon|k|}.$$

Then $\sum_j T_j$ is bounded on L^p, $p_0 < p \leq 2$.

REMARK 3. Hypotheses (1),(2),(5),(7),(10) and (10') assert that we have a Calderón-Zygmund or Littlewood-Paley theory. The form of T_j usually dictates S_j via the pointwise domination hypothesis. Usually (6") will be trivial to verify. The only link between the $\{P_j, Q_j\}$ and the $\{S_j, T_j\}$ comes in (4) and (4') - the same inequalities one needs in the original Observation - and it is these inequalities which express the compatibility of the Littlewood-Paley theory with the operator $T = \sum_j T_j$ one wants to control. It is in verification of (4) and (4') that one needs to do some hard work. Although we have stated Proposition 3 for $p_0 < 2$, duality implies a similar proposition for $p_0 > 2$.

A REMARK ABOUT HISTORY. None of the ideas in what has been presented above is new; all of them can be traced to work of Nagel, Stein and Wainger [NSW], Christ [C], Duoandikoetxea and Rubio de Francia [DRdeF], David,

Journé and Semmes [**DJS**], and Seeger and the present author [**C**],[**CS**]. The formulation of Proposition 3, while implicit in [**CS**], is new.

Application to high-dimensional analysis. Because of their abstract nature, the almost-orthogonality principles formulated above can be used to obtain quasi-abstract results which would be difficult to obtain by more traditional methods. As an example to illustrate this we consider the Hardy-Littlewood maximal operator associated to an arbitrary convex body B in $\mathbb{R}^n$, normalised to have volume 1. Let

$$Mf(x) = \sup_{k \in \mathbb{Z}} \frac{1}{2^{kn}} \int_{2^k B} |f(x-y)| dy$$

(where tB is the body concentric with B dilated by a factor t.)

THEOREM [**C**]. *If $p > 1$ then there is an A_p independent of B and n such that*

$$\|Mf\|_p \leq A_p \|f\|_p.$$

The quasi-abstract nature of this result is indicated by the independence from the dimension; the usual geometric arguments give an estimate of $3^{n/p}$ instead of A_p. The proof uses Proposition 1, where, after a suitable affine scaling, we take P_j to be the Poisson kernel of height 2^j. The relevant dimension-free estimates for the Poisson maximal function and its associated Littlewood-Paley operators had earlier been established by Stein, [**S1**].

Application to Hilbert transforms and Maximal functions along variable curves. For another application of the almost orthogonality principles, we wish to consider maximal functions and Hilbert transforms along certain (flat) variable plane curves. This is joint work with Jim Wright and Stephen Wainger. Let $\gamma : (0, \infty) \longrightarrow \mathbb{R}$ be convex and satisfy $\gamma(0) = 0 = \gamma'(0)$; extend γ to be odd on $\mathbb{R}$. We define a family of curves $\gamma : \mathbb{R}^2 \times \mathbb{R} \longrightarrow \mathbb{R}$ by $\gamma(x, t) = (x_1 - t, x_2 - x_1 \gamma(t))$. Let

$$M_\gamma f(x) = \sup_{j \in \mathbb{Z}} 2^{-j} |\int_{2^j}^{2^{j+1}} f(\gamma(x,t)) dt| = \sup_j |S_j f|$$

and

$$H_\gamma f(x) = \text{p.v.} \int_{-\infty}^{\infty} f(\gamma(x,t)) dt/t = \sum_j \int_{2^j \leq |t| \leq 2^{j+1}} \cdots = \sum_j T_j f$$

be the maximal function and Hilbert transform associated to γ.

THEOREM. *(Carbery, Wright and Wainger, [**CWW**]) Suppose the quantity $t\gamma''(t)/\gamma'(t)$ is decreasing and bounded below on $(0, \infty)$. Then M_γ and H_γ are bounded on $L^p(R^2)$ for $1 < p < \infty$.*

It is important to note that included in this theorem are flat examples such as $\gamma(t) = e^{-t^{-2}}$ to which the curvature condition considered by Christ, Nagel, Stein and Wainger in [**CNSW**] does not apply. Their curvature condition is satisfied in our case if and only if $\gamma^{(j)}(0) \neq 0$ for some $j \in N$.

We wish to employ the almost orthogonality principles above to obtain this result, so we first need to define the less singular averaging and difference operators P_j and Q_j.

To our curve γ we associate the dilation matrices $A(t) = diag\{t, t\gamma(t)\}, t > 0$. When $\gamma(t) = t^k$ these matrices have the property of making the curve $\gamma(x, t)$ homogeneous with respect to them in the sense that

$$\gamma(x, ts) = A(s)\gamma(A^{-1}(s)x, t) := [A(s)_*\gamma](x, t).$$

One hopes that in the more general case the curve will be approximately homogeneous with respect to these dilations in the sense that the $[A(s)_*^{-1}\gamma](x, ts)$, where s is regarded as a parameter, should, if not indeed identical, nevertheless satisfy the same estimates in t uniformly over the parameter s. Once one has an initial averaging operator P_0 with kernel $p_0(x, y)$ it is natural to define P_j to be the operator $A_{j*}P_0$ with kernel $det A_j^{-1}p_0(A_j^{-1}x, A_j^{-1}y)$ (where we are using the notation $A_j = A(2^j)$.)

The initial averaging operator is dictated to us by the local algebraic form of the curve. Let $M : \mathbb{R}^2 \longrightarrow \mathbb{R}^2$ be defined by $M(x, y) = (x_1 - y_1, (x_2 - y_2)/(1 + |x_1|))$, and let

$$P_0 f(x) = \int \phi(|M(x, y)|)f(y)dy / \int \phi(|M(x, u)|)du,$$

where ϕ is a nonnegative member of the Schwartz space on the real line. Notice that our matrices satisfy the crucial Rivière condition

(R) $|M(A(s)^{-1}A(t)x, A(s)^{-1}A(t)y)| \leq (t/s)^\epsilon |M(x, y)|$ for $s \geq t$.

THEOREM. ([**CVWW**]) *With the above definitions, $f \mapsto \sup_j |P_j f|$ and $f \mapsto \sum_j \pm Q_j f$ are bounded on $L^p, 1 < p < \infty$, and*

$$\|(\sum |Q_j g_j|^2)^{1/2}\|_p \leq C\|(\sum |g_j|^2)^{1/2}\|_p, \quad 1 < p < \infty.$$

The proof of this theorem is a variant of the classical Calderón-Zygmund theory, and it is then easy to verify all of the remaining conditions required for the almost orthogonality principles, with the exception of

(12) $\begin{cases} \|Q_{j+k}T_j^*\|_{2-2} + \|Q_{j+k}^*T_j\|_{2-2} \leq C2^{-\epsilon|k|} \\ \|Q_{j+k}(S_j - P_j)^*\|_{2-2} + \|Q_{j+k}^*(S_j - P_j)\|_{2-2} \leq C2^{-\epsilon|k|} \end{cases}$

For $k \geq 0$, these estimates are variants of the "standard computation" and follow from smoothness of $\{P_j\}$ together with support properties of $\{T_j, S_j, P_j\}$ and the important cancellation properties $T_j 1 = T_j^* 1 = (S_j - P_j)1 = (S_j - P_j)^* 1 = 0$. For $k \leq 0$, one has to work harder to obtain the L^2 smoothing estimates.

In the case $k \geq 0$ let us indicate how to prove, for example $\|Q_{j+k}^*T_j\|_{2-2} \leq C2^{-\epsilon k}$. This is equivalent to $\|T_j^*Q_{j+k}\|_{2-2} \leq C2^{-\epsilon k}$ which would follow from a similar estimate for $\|T_j^* P_{j+k}\|_{2-2}$. This in turn is equivalent to $\|T_{jk}^* P_0\|_{2-2} \leq C2^{-\epsilon k}$ where T_{jk} is the dilate of T_j by the matrix A_{j+k}; in particular $T_{jk}^* 1 = 0$ and T_{jk} has its distribution kernel essentially supported in

$$\{(x,y) | |M(A_j^{-1} A_{j+k} x, A_j^{-1} A_{j+k} y)| \leq C\}$$

which is contained in $\{|M(x,y)| \leq C2^{-\epsilon k}\}$ using (R). It is now easy to see that $\|T_{jk}^* P_0\|_{2-2} \leq C2^{-\epsilon k}$ for $k \geq 0$.

For $k \leq 0$, we first note that, for example, $\|Q_{j+k}(S_j - P_j)^*\|_{2-2} \leq C2^{\epsilon k}$ is equivalent to $\|Q_{j+k} S_j^*\|_{2-2} \leq C2^{\epsilon k}$ (since we already know that $\|Q_{j+k} Q_j^*\|_{2-2} \leq C2^{\epsilon k}$.) To obtain this, we first normalise using the matrix A_j; that is, it is enough to show that

(13)
$$\|A_{j*}^{-1} Q_{j+k} A_{j*}^{-1} S_j^*\|_{2-2} \leq C2^{\epsilon k}$$

Note that

$$A_{j*}^{-1} S_j f(x) = \int_1^2 f(\tilde{\gamma}_j(x,t)) dt$$

where

$$\tilde{\gamma}_j(x,t) = [A_{j*}^{-1}\gamma](x, 2^j t) = (x_1 - t, x_2 - x_1 \gamma(2^j t)/\gamma(2^j)),$$

and, as above, $A_{j*}^{-1} Q_{j+k} = [A_j^{-1} A_{j+k}]_* P_0 - [A_j^{-1} A_{j+k+1}]_* P_0$ (which is a normalised difference operator with kernel essentially supported in

$$\{(x,y) | |M(x,y)| \leq C2^{\epsilon k}\}).$$

We now observe that the hypotheses on the curve γ and the estimate (13) we are trying to establish are both invariant under transformations of the form $(t, \gamma(t)) \longmapsto (t, a\gamma(t/a)), a > 0$. Thus, in proving (13), it suffices to assume that $j = 0$, i.e. to prove that

(14)
$$\|Q_k S_0^*\|_{2-2} \leq C2^{\epsilon k}, k \leq 0,$$

under the additional hypothesis $\gamma(1) = 1$. Thus we may now assume

(15)
$$\gamma'' \geq C\gamma' \geq C'\gamma \geq C'' \text{ on } [1,2]$$

We turn to the proof of (14), which is completely different from and somewhat simpler than the one given in [**CWW**]. This proof is also due to Wainger, Wright and the present author. Somewhat ironically, (in view of the fact that Cotlar's lemma was first devised in order to handle situations where one could not employ the Fourier transform), we use the Fourier transform to establish (14), (somewhat in the spirit of [**Se**].) Indeed we use the method of Phong and Stein [**PS**] whose main thrust we now recall.

Let $K : \mathbb{R}^m \longrightarrow C$ be a kernel, and $A : \mathbb{R}^p \times \mathbb{R}^m \longrightarrow \mathbb{R}^q$ be a function. Let $Tf(x) = \int_{R^m} f(x_1 - y_1, x_2 - A(x_1, y)) K(y) dy$ where $x = (x_1, x_2) \in \mathbb{R}^p \times \mathbb{R}^q$ and $y = (y_1, y_2) \in \mathbb{R}^p \times \mathbb{R}^{m-p}$. Define $T_\lambda h(x) = \int_{\mathbb{R}^m} h(x - y_1) e^{i\lambda A(x,y)} K(y) dy$ where now $x \in \mathbb{R}^p$. Then $(TS^*)_\lambda = T_\lambda S_\lambda^*$, and Plancherel's theorem in the $x_2 \in \mathbb{R}^q$

variable shows that $\|T\|_{2-2} = \sup_\lambda \|T_\lambda\|_{2-2}$. Thus to prove (14) it suffices to prove $\|(Q_k)_\lambda (S_0)_\lambda^*\|_{2-2} \leq C2^{\epsilon k}$ or indeed

$$(16) \qquad \|(Q_k)_\lambda (S_0)_\lambda^* (S_0)_\lambda\|_{2-2} \leq C2^{\epsilon k}$$

uniformly in $\lambda \in R$.

Since $S_0 f(x) = \int f(x_1 - t, x_2 - x_1 \gamma(t)) \alpha(t) dt$ where $\alpha = \chi_{[1,2]}$ or a smoothed out version) , $S_{0\lambda} h(x) = \int h(x - t) e^{i\lambda x \gamma(t)} \alpha(t) dt$ and so the kernel of $S_{0\lambda}^* S_{0\lambda}$ is (as a function of x and y)

$$K_\lambda(x, y) = \int e^{i\lambda z[\gamma(z-y) - \gamma(z-x)]} \alpha(z - y) \overline{\alpha}(z - x) dz$$

while that of $(P_k)_\lambda$ is, (as a function of x and u)

$$2^{-k} (F_2 \phi)((x - u)2^{-k}, 2^k \gamma(2^k) \lambda(1 + 2^{-k}|x|))$$

where F_2 denotes the Fourier transform in the second variable. Let us suppose, as we may, that $F_2 \phi(x, \xi) = \alpha(x) \beta(\xi)$ where $\alpha, \beta \in C_c^\infty(R)$ and are supported in [-2,2], and that β is identically one on [-1,1]. So

$$\begin{aligned}
ker&(Q_k)_\lambda(x, u) \\
&= 2^{-k} \alpha((x - u)2^{-k}) \beta(\gamma(2^k) \lambda(2^k + |x|)) \\
&\quad - 2^{-k-1} \alpha((x - u)2^{-k-1}) \beta(\gamma(2^{k+1}) \lambda(2^{k+1} + |x|)) \\
&= [2^{-k} \alpha((x - u)/2^k) - 2^{-k-1} \alpha((x - u)/2^{k+1})] \beta(\gamma(2^k) \lambda(2^k + |x|)) \\
&\quad + [\beta(\gamma(2^k) \lambda(2^k + |x|) - \beta(\gamma(2^{k+1}) \lambda(2^{k+1} + |x|))] 2^{-k-1} \alpha((x - u)/2^{k+1}).
\end{aligned}$$

Note that the second term here is nonzero only when $\gamma(2^{k+1}) \lambda(2^{k+1} + |x|) \geq 1$.

Now to prove (16) it suffices to show that

$$\int ker(Q_k)_\lambda(x, x') K_\lambda'(x', y) dx'$$

has L^1 operator norm dominated by $2^{\epsilon k}$; that is

$$(17) \qquad \int |\int ker(Q_k)_\lambda(x, x') K_\lambda'(x', y) dx'| dx \leq C2^{\epsilon k}$$

uniformly in $y \in R, \lambda \in R$, (where K_λ' denotes $K_\lambda(x, y) \chi_{\{x \leq y\}}$,) together with a similar estimate corresponding to $\chi_{\{x \geq y\}}$.

PROPOSITION. *There exists an* $M \in \mathbb{N}$ *(M=4 will do) such that*

$$a) \qquad \int |K_\lambda'(x, y)| dx \leq C/\{\lambda(1 + |y|)\}^{1/(2M+2)}$$

$$b) \qquad \int |K_\lambda'(x + h, y) - K_\lambda'(x, y)| dx \leq C|h|^{1/(2M+2)}.$$

Once we have the proposition, the estimate (17) follows easily: given λ, y and k, either $(1 + |y|)\lambda\gamma(2^{k+1}) > 1/10$ —in which case a) implies that

$$\int |K_{\lambda'}(x, y)| dx \leq C\{\gamma(2^{k+1})\}^{1/(2M+2)}| \leq C2^{\epsilon k}$$

—or $(1 + |y|)\lambda\gamma(2^{k+1}) \leq 1/10$, which implies

$$\gamma(2^{k+1})\lambda(2^{k+1} + |x|) \leq \gamma(2^{k+1})\lambda + \gamma(2^{k+1})\lambda|x - y| + \gamma(2^{k+1})\lambda|y|$$
$$\leq 1/10 + |x - y|/10 < 1/2$$

provided $|x - y| < 4$. Now for y fixed, $K_{\lambda}'(x', y)$ is supported in $|x' - y| \leq 2$ and $ker(Q_k)_\lambda(x, x')$ is supported in $|x - x'| \leq 2$. Hence $\int ker(Q_k)_\lambda(x, x') K_\lambda'(x', y) dx'$ is supported in $|x - y| < 4$, and hence $\gamma(2^{k+1})\lambda(2^{k+1} + |x|) < 1/2$.
So in this case,

$$\int ker(Q_k)_\lambda(x, x') K_\lambda'(x', y) dx' = q_k *_1 K_\lambda'(x, y)$$

where q_k is $2^{-j}\alpha(2^{-j}.) - 2^{-j-1}\alpha(2^{-j-1})$ and $*_1$ denotes convolution with respect to the first variable. But $\int |q_k *_1 K_{\lambda'}(x, y)| dx \leq C2^{k/(2M+2)}$, by part b) of the proposition.

SKETCH OF PROOF OF PROPOSITION. Let $\phi(z) = \phi_{xy}(z) = z[\gamma(z - y) - \gamma(z - x)]$ for $x \leq y$. Consider, for fixed y, $\Delta = \{(z, x) | y + 1 \leq z \leq y + 2, y - 1 \leq x \leq y, x \geq z - 2\}$. It can be shown (see [**CWW**]) that there exist $M \in N$ (M=3 will do) and (at most) two decreasing functions μ and ν such that if

$$F = \partial\Delta \cup (\Delta \cap \{z = 0\}) \cup (\Delta \cap \{z, z - \mu(z - y)\}) \cup (\Delta \cap \{z, z - \nu(z - y)\},$$

and if B is a ball of radius δ contained in $\Delta - F$ whose distance from F is proportional to δ (a Whitney ball for F), then on B

(18) $$|\phi'(z)| \geq C\gamma'(z - x)(1 + |y|)\delta^M.$$

Using this and (15) together with the fact that

(19) $$\gamma'(t) - \gamma'(s) \geq C(t - s)\gamma'(t) \text{ for } t \geq s$$

(which itself follows from (15)) one arrives at the following:

LEMMA. *There exists an $M \in N$ (M=3 will do) such that if B is a Whitney ball for F of radius δ we have*

$$\int_B |\partial/\partial z\{z\gamma'(z - x)\alpha(x, y, z)/[(1 + |y|)\phi'(z)]\}| + |\partial/\partial z\{\alpha(x, y, z)/\phi'(z)\}| dx dz$$
$$\leq C/\delta^{2M}(1 + |y|)$$

(Here, $\alpha(x, y, z) = \alpha(z - y)\overline{\alpha}(z - x)$.)

Now for $\delta > 0$ and y and x fixed, let $S_x = \{z \in [y + 1, y + 2] | (z, x) \in \Delta, dist((z, x), F) \geq \delta\}$. Then S_x is the union of at most 4 intervals.

COROLLARY.

$$\int_{|x-y|\geq\delta} |\int_{S_x} z\gamma'(z-x)e^{i\lambda\phi(z)}\alpha(x,y,z)dz(1+|y|)^{-1}|dx$$

$$+\int_{|x-y|\geq\delta} |\int_{S_x} e^{i\lambda\phi(z)}\alpha(x,y,z)dz|dx$$

(20) $$\leq C/\lambda\delta^{2M+1}(1+|y|).$$

PROOF. In the inner integrals, one integrates by parts on each of the intervals comprising S_x (using smooth cutoff functions which are nonconstant only on δ intervals) and obtains an estimate of

$$C/\lambda \int_{S_x} |\partial/\partial z\{z\gamma'(z-x)\alpha(x,y,z)/[(1+|y|)\phi'(z)]\}|dz$$

together with a similar term not including the expression $z\gamma'(z-x)/(1+|y|)$. Thus we can dominate the left hand side of (20) by

$$C/\lambda \int_{|x-y|\leq\delta} \int_{S_x} |\partial/\partial z\{\quad\}|dzdx$$

(21) $$= C/\lambda \int\int_{\{(z,x):dist((z,x),F)\geq\delta\}} |\partial/\partial z\{\quad\}|dzdx.$$

Now we can break up $\Delta - F$ into Whitney cubes or balls of size $2^{-j}, j \geq 0$, where, because of the nature of F, there are at most 20.2^j of size 2^{-j}. Thus, by the lemma, we can dominate (21) by

$$1/\lambda \sum_{\{j:2^{-j}\geq\delta\}} 2^j/2^{-2jM}(1+|y|) \leq C/\lambda(1+|y|)\delta^{2M+1}. \quad \square$$

Finally, we turn to the estimates of the proposition.

PROOF OF A). Since $|K_\lambda'(x,y)| \leq C$, we always have $\int_{|x-y|\leq\delta} |K_\lambda'(x,y)|dx \leq C\delta$. Moreover, $|\int_{S_x^c} e^{i\lambda\phi}\alpha dz| \leq |S_x^c| \leq C\delta$. Hence

$$\int |K_\lambda'(x,y)|dx \leq C\delta + \int_{|x-y|\geq\delta} |\int_{S_x} dz|dx$$

$$\leq C\left(\delta + 1/[\lambda\delta^{2M+1}(1+|y|)]\right)$$

by (20). Optimising in δ then yields the required estimate.

PROOF OF B). We may assume that $|h|\lambda(1+|y|) < 1$ (for if not, we can apply a) directly). So fix x and y, and consider, for $|x-y| \geq \delta >> |h|$,

$$K_\lambda'(x+h,y) - K_\lambda'(x,y) = \int [e^{i\lambda\phi_{x+h,y}(z)} - e^{i\lambda\phi_{x,y}(z)}]\alpha(x+h,y,z)dz$$

$$+ \int e^{i\lambda\phi_{x,y}(z)}[\alpha(x+h,y,z) - \alpha(x,y,z)]dz.$$

The second term is clearly $O(|h|)$, while the first is dominated by

$$\delta + |\int_{S_x} \{e^{i\lambda\phi_{x+h,y}(z)} - e^{i\lambda\phi_{x,y}(z)}\}\alpha(x+h,y,z)dz|$$

$$= \delta + |\int_{S_x} \int_0^h (\partial/\partial t)e^{i\lambda\phi_{x+t,y}(z)}dt\alpha(x+h,y,z)dz|$$

$$\leq \delta + \int_0^h |\int_{S_x} \lambda(\partial/\partial x)\phi_{.,y}(z)_{[x+t]}e^{i\lambda\phi_{x+t,y}(z)}\alpha(x+h,y,z)dz|dt.$$

Hence,

$$\int_{|x-y|\geq\delta} |K'_\lambda(x+h,y) - K'_\lambda(x,y)|dx$$

$$\leq C[|h| + \delta + |h|\lambda \int_{|x-y|\geq\delta} |\int_{S_x} z\gamma'(z-.)e^{i\lambda\phi_{.,y}(z)_{[x+t]}}\alpha(x+h,y,z)dz|dx]$$

$$\leq C[|h| + \delta + |h|\lambda \int_{|x-y|\geq\delta/2} |\int_{S_x} z\gamma'(z-x)e^{i\lambda\phi_{x,y}(z)}\alpha(x+h-t,y,z)dz|dx]$$

$$\leq C[|h| + \delta + |h|/\delta^{2M+1}]$$

by (20).

Therefore

$$\int |K'_\lambda(x+h,y) - K'_\lambda(x,y)|dx \leq C(\delta + |h|/\delta^{2M+1})$$

(assuming that $|h| << \delta$) which is optimised when $\delta = |h|^{1/2M+2} >> |h|$, giving the required estimate, and finishing the proof of (17). $\square$

Similar arguments when $x \geq y$ give (14), and similar arguments give $\|Q_k^* S_0\|_{2-2} \leq C2^{\epsilon k}$, which is the last ingredient in the proof of the theorem.

REFERENCES

[C]   A.Carbery, *An almost orthogonality principle with applications to maximal functions associated to convex bodies*, B.A.M.S. **14** (1986), 269–273.

[CS]   A.Carbery and A.Seeger, *Conditionally convergent series of linear operators on L^p spaces and L^p estimates for pseudodifferential operators*, P.L.M.S. **57** (1988), 481–510.

[CVWW]   A.Carbery, J.Vance, S.Wainger and J.Wright, *A variant of the notion of a space of homogeneous type*, to appear, J.F.A..

[CWW]   A.Carbery, S.Wainger and J.Wright, *Hilbert transforms and maximal functions along flat variable plane curves*, to appear, J.F.A.A..

[CNSW]   M.Christ, A.Nagel, E.M.Stein and S.Wainger, to appear.

[DJS]   G.David, J-L.Journé and S.Semmes, *Operateurs de Calderón-Zygmund, fonctions para-accretives et interpolation*, Revista Matematica Iberoamericana **1** (1985), 1–56.

[DRdeF] J.Duoandikoetxea and J.L. Rubio de Francia, *Maximal and singular integral operators via Fourier transform estimates*, Inv.Math. **84** (1986), 541–561.

[LMcCR] W.Littman, C.McCarthy and N.Rivière, *L^p multiplier theorems*, Studia Math. **30** (1968), 193–217.

[NSW]   A.Nagel, E.M.Stein and S.Wainger, *Differentiation in lacunary directions*, Proc. Nat. Ac. Sci. **75** (1978), 1060–1062.

[PS]   D.H.Phong and E.M.Stein, *Hilbert integrals, singular integrals and Radon trans-forms,I*, Acta Mathematica **157** (1986), 99–157.

[Se]   A.Seeger, L^2 *estimates for a class of singular oscillatory integrals*, to appear Math. Res. Letters.

[S]   E.M.Stein, *Harmonic Analysis*, Princeton Univ. Press, 1993.

[S1]   E.M.Stein, *Some results in harmonic analysis in R^n for $n \to \infty$*, B.A.M.S. **9** (1983), 71–73.

DEPARTMENT OF MATHEMATICS AND STATISTICS, UNIVERSITY OF EDINBURGH, KING'S BUILDINGS, MAYFIELD RD., EDINBURGH EH9 3JZ U.K.

Contemporary Mathematics
Volume **189**, 1995

Julia Operators and the Schur Algorithm

GENE CHRISTNER AND JAMES ROVNYAK

To Mischa Cotlar, with best wishes on his eightieth birthday.

ABSTRACT. Julia operators are used to formulate the Schur algorithm for power series whose coefficients are Kreĭn space operators. The elements of the Kreĭn space Schur class are analytic functions in a neighborhood of the origin which possess certain two-sided positivity properties. They are characterized as linear fractional transformations of analytic functions on the unit disk whose values are Hilbert space contractions.

1. Introduction

The Schur class S is the set of analytic functions $f(z)$ which are bounded by one on the unit disk of the complex plane. Each such function generates a sequence of functions in the class by the formulas $f_0(z) = f(z)$ and

$$f_{n+1}(z) = \frac{1}{z} \frac{f_n(z) - f_n(0)}{1 - \bar{f}_n(0)f_n(z)}, \qquad n \geq 0.$$

If $f_r(0)$ has modulus one for some r, Schwartz's lemma implies that $f_r(z)$ is a constant, and we take $f_k(z) \equiv 0$ for all $k > r$. The construction is called the Schur algorithm, and the numbers

$$\alpha_n = f_n(0), \qquad n \geq 0,$$

are the Schur parameters of the function $f(z)$. The Schur parameters label S by sequences $\{\alpha_n\}_{n=0}^{\infty}$ of numbers in the closed unit disk such that if some number has modulus one then all subsequent numbers are zero.

1991 *Mathematics Subject Classification*. Primary 47A57, 47B50; Secondary 30E05.

This work was supported by the National Science Foundation.

We thank D. Alpay, D. Z. Arov, and M. Dritschel for helpful conversations on the material of the paper.

These ideas were applied by Schur [29] to the Carathéodory-Fejér problem. Schur showed that if $c_0, \ldots, c_n$ are given numbers, then there exists an element $f(z)$ in S of the form

$$f(z) = c_0 + c_1 z + \ldots + c_n z^n + higher\ powers$$

if and only if the matrix

$$\begin{pmatrix} c_0 & & & \\ c_1 & c_0 & & 0 \\ & \ddots & \ddots & \\ c_n & & c_1 & c_0 \end{pmatrix}$$

acts as a contraction on $\mathbf{C}^{n+1}$. This result, which today is also called the Carathéodory-Fejér theorem, is one of many beautiful theorems of the subject.

The commutant lifting theorem for contraction operators on a Hilbert space [22] and theory of Toeplitz kernels [16] incorporate far-reaching operator extensions with many applications. Two recent books [7, 21] treat matrix and operator generalizations of the ideas introduced by Schur and their applications in analysis. See [24] for a historical survey and [28] for a different overview by the second author. We note that the structural analysis of J-contractive matrix-valued functions was initiated by V. P. Potapov [27]. Potapov's student, T. Ivanchenko, studied the Schur algorithm for matrix-valued functions [25]. Some recent works extend such Schur methods to other areas of Kreĭn space operator theory. The paper [17] generalizes results of [3–6] to a Schur parametrization of unitary extensions of isometric operators on Kreĭn spaces. Kreĭn space methods based on complementation theory [9] appear in [10, 11, 30]. Methods based on reproducing kernels are given in [1, 2].

In a study of coefficients of univalent functions with Kin Y. Li [13], the authors encountered the problem of finding indefinite operator generalizations of certain of these ideas. A commutant lifting theorem for Kreĭn space operators was proved in [18] but does not directly yield a Carathéodory-Fejér theorem as in the Hilbert space case. The first author proved a Carathéodory-Fejér theorem [12] for Kreĭn space operators following a choice sequence method in [14, 15] and [7]. This paper is an account of [12] and includes additional results. The operator factorization theorems needed to justify results are taken from [19, 20]. We formulate the choice sequence method in a calculus of Julia operators. Similar constructions can be used to label solutions in the Kreĭn space commutant lifting theorem but are not included here.

Notation and Kreĭn space background.

Underlying spaces are assumed to be Kreĭn spaces unless the contrary is stated. The space of continuous linear operators from $\mathfrak{H}$ to $\mathfrak{K}$ is written $\mathfrak{L}(\mathfrak{H}, \mathfrak{K})$ or $\mathfrak{L}(\mathfrak{H})$ when the spaces are the same. An operator $T \in \mathfrak{L}(\mathfrak{H}, \mathfrak{K})$ is a **contraction** if $T^*T \leq 1$ in the partial ordering of selfadjoint operators, and a **bicontraction** if both T and T^* are contractions. We assume familiarity with basic concepts of Kreĭn space operator theory [8].

We recall some notions from [**19, 20**]. If $T \in \mathfrak{L}(\mathfrak{H}, \mathfrak{K})$, a **defect operator** for T is any operator $\tilde{D} \in \mathfrak{L}(\tilde{\mathfrak{D}}, \mathfrak{H})$ having zero kernel such that

$$(1\text{-}1) \qquad\qquad 1 - T^*T = \tilde{D}\tilde{D}^*.$$

These conditions imply that

$$\begin{pmatrix} T \\ \tilde{D}^* \end{pmatrix}$$

is an isometric extension of T. A **Julia operator** for T is any unitary operator of the form

$$(1\text{-}2) \qquad\qquad \begin{pmatrix} T & D \\ \tilde{D}^* & -L^* \end{pmatrix} \in \mathfrak{L}(\mathfrak{H} \oplus \mathfrak{D}, \mathfrak{K} \oplus \tilde{\mathfrak{D}})$$

such that $\mathfrak{D}$ and $\tilde{\mathfrak{D}}$ are Kreĭn spaces and D and $\tilde{D}$ are operators with zero kernel. Then $\tilde{D}$ is a defect operator for T, and D is a defect operator for T^*. New defect and Julia operators can be created by replacing the **defect spaces** $\tilde{\mathfrak{D}}$ and $\mathfrak{D}$ by isomorphic copies. Defect and Julia operators are said to be **essentially unique** if this is the only way in which such operators can be generated (see the discussion preceding Theorem 4.4).

Defect and Julia operators always exist. They are essentially unique for important classes of operators, including contractions. The notation for Julia operators used here conforms with a forthcoming account by M. A. Dritschel and the second author: *Operators on indefinite inner product spaces*, Lectures presented at a meeting on Applications of Operator Theory (The Fields Institute, September, 1994). It differs slightly from [**20**].

The study of contractions and bicontractions automatically brings in Hilbert spaces. If $T \in \mathfrak{L}(\mathfrak{H}, \mathfrak{K})$ has Julia operator (1–2), then T is a contraction if and only if $\tilde{\mathfrak{D}}$ is a Hilbert space. The operator T is a bicontraction if and only if both $\tilde{\mathfrak{D}}$ and $\mathfrak{D}$ are Hilbert spaces. These assertions follow from (1–1) and its counterpart in which the roles of T and T^* are reversed.

The **negative index** of a Kreĭn space $\mathfrak{H}$ is the maximum dimension ind$_-\mathfrak{H}$ of a Hilbert space whose antispace is contained isometrically in $\mathfrak{H}$. The **negative index** of a selfadjoint operator T on a Kreĭn space $\mathfrak{H}$ is the supremum ind$_-T$ of all nonnegative integers r such that there exists a nonpositive and invertible matrix of the form

$$\left[\langle Tf_j, f_k \rangle_{\mathfrak{H}}\right]_{j,k=0}^{r}, \qquad f_1, \dots, f_r \in \mathfrak{H}.$$

Positive indices ind$_+\mathfrak{H}$ and ind$_+T$ are defined in a similar way .

Kreĭn spaces which are used as generalizations of the complex numbers are called **coefficient spaces** and denoted by letters $\mathfrak{A}, \mathfrak{B}, \mathfrak{C}, \dots$, often with subscripts. For any coefficient space $\mathfrak{C}$, by $\mathfrak{C}^n$ we mean the n-fold cartesian product of $\mathfrak{C}$ in the usual inner product. The elements of $\mathfrak{C}^n$ are written as column vectors, and operators on $\mathfrak{C}^n$ are written as $n \times n$ matrices of elements of $\mathfrak{L}(\mathfrak{C})$.

2. The Schur algorithm

The main results of the paper are stated in this section and proved except for
Theorems 2.2, 2.4, and 2.5. These theorems depend on more general results in
§3 and §4, and their proofs will be given in §4.

If $\mathfrak{A}$ and $\mathfrak{B}$ are Kreĭn spaces, we define the **Schur class** $S(\mathfrak{A}, \mathfrak{B})$ to be the
set of all formal power series

(2-1) $$B(z) = B_0 + B_1 z + B_2 z^2 + \cdots$$

with coefficients in $\mathfrak{L}(\mathfrak{A}, \mathfrak{B})$ such that for each nonnegative integer r, the matrix

(2-2) $$T_r = \begin{pmatrix} B_0 & & \\ B_1 & B_0 & \\ & \ddots & \ddots \\ B_r & & B_1 & B_0 \end{pmatrix}$$

acts as a bicontraction on $\mathfrak{A}^{r+1}$ to $\mathfrak{B}^{r+1}$. Formal power series are usually not
distinguished from analytic functions which they represent.

The class $S(\mathfrak{A}, \mathfrak{B})$ is nonempty if and only if the Kreĭn spaces $\mathfrak{A}$ and $\mathfrak{B}$ have the
same negative index. For if (2–1) belongs to the class, then B_0 is a bicontraction
on $\mathfrak{A}$ to $\mathfrak{B}$, and this forces $\mathrm{ind}_- \mathfrak{A} = \mathrm{ind}_- \mathfrak{B}$ by a well known result (see [**20,
Th. 1.3.6**]). Conversely, the class $S(\mathfrak{A}, \mathfrak{B})$ contains constant elements if $\mathfrak{A}$ and
$\mathfrak{B}$ have the same negative index. In the Hilbert space case, our definition of the
Schur class agrees with standard terminology [**7, 21**]. When $\mathfrak{A}$ and $\mathfrak{B}$ coincide
with the complex numbers, $S(\mathfrak{A}, \mathfrak{B})$ reduces to the Schur class S defined in §1.

The following characterization of the Schur class was suggested by D. Alpay
and D. Z. Arov.

THEOREM 2.1. *A series (2–1) with coefficients in $\mathfrak{L}(\mathfrak{A}, \mathfrak{B})$ belongs to $S(\mathfrak{A}, \mathfrak{B})$
if and only if it represents an analytic function in a neighborhood of the origin
and*

(2-3) $$\frac{1 - B(z)B(w)^*}{1 - \bar{w}z} \quad and \quad \frac{1 - \tilde{B}(z)\tilde{B}(w)^*}{1 - \bar{w}z}$$

define positive definite kernels in a neighborhood of the origin, where

(2-4) $$\tilde{B}(z) = B_0^* + B_1^* z + B_2^* z^2 + \cdots .$$

The main structure theorem for the Schur class $S(\mathfrak{A}, \mathfrak{B})$ is a linear fractional
representation of any element of the class.

THEOREM 2.2 ([**12**]). *Let B_0 be a bicontraction in $\mathfrak{L}(\mathfrak{A}, \mathfrak{B})$ with Julia operator*

(2-5) $$\begin{pmatrix} B_0 & D_0 \\ \tilde{D}_0^* & -L_0^* \end{pmatrix} \in \mathfrak{L}(\mathfrak{A} \oplus \mathfrak{D}_0, \mathfrak{B} \oplus \tilde{\mathfrak{D}}_0).$$

If $F(z)$ belongs to the Schur class $S(\tilde{\mathfrak{D}}_0, \mathfrak{D}_0)$, then the power series

$$(2\text{-}6) \qquad B(z) = B_0 + zD_0F(z)\left(1 + zL_0^*F(z)\right)^{-1}\tilde{D}_0^*$$

belongs to $S(\mathfrak{A}, \mathfrak{B})$. Every element $B(z)$ of $S(\mathfrak{A}, \mathfrak{B})$ with $B(0) = B_0$ has this form, and the choice of $F(z)$ in $S(\tilde{\mathfrak{D}}_0, \mathfrak{D}_0)$ is unique when the Julia operator (2–5) is fixed.

Since B_0 is a bicontraction, $\tilde{\mathfrak{D}}_0$ and $\mathfrak{D}_0$ are Hilbert spaces. Thus $S(\tilde{\mathfrak{D}}_0, \mathfrak{D}_0)$ is the space of holomorphic functions on the unit disk of the complex plane whose values are contraction operators from $\tilde{\mathfrak{D}}_0$ to $\mathfrak{D}_0$.

PROOF OF THEOREM 2.1 ASSUMING THEOREM 2.2. As this argument uses mainly standard methods, a sketch of the proof will be sufficient. We only need Theorem 2.2 to know that an element (2–1) in the Schur class represents an analytic function in a neighborhood of the origin. If $B(z) = B_0 + B_1 z + B_2 z^2 + \cdots$ converges in a neighborhood of the origin, then for small $|z|$ and $|w|$,

$$\frac{1 - B(z)B(w)^*}{1 - \bar{w}z} = \sum_{m,n=0}^{\infty} C_{mn}\, z^m \bar{w}^n,$$

where $[C_{mn}]_{m,n=0}^r = 1 - T_r T_r^*$ with T_r given by (2–2) for each $r \geq 0$. The positivity of the kernel is equivalent to the positivity of $1 - T_r T_r^*$ for all $r \geq 0$ by an argument which uses a Cauchy integral representation of the kernel. A similar identity holds for the second kernel in (2–3), with $B_0, B_1, \ldots$ replaced by $B_0^*, B_1^*, \ldots$, and the result follows on combining these facts. $\square$

Theorem 2.2 allows a complete implementation of the Schur algorithm. Let $B(z)$ belong to $S(\mathfrak{A}, \mathfrak{B})$ for some Kreĭn spaces $\mathfrak{A}$ and $\mathfrak{B}$. Put $\mathfrak{A}_0 = \mathfrak{A}$ and $\mathfrak{B}_0 = \mathfrak{B}$. Define $B_0(z) \in S(\mathfrak{A}_0, \mathfrak{B}_0)$ and $\alpha_0 \in \mathfrak{L}(\mathfrak{A}_0, \mathfrak{B}_0)$ by

$$(2\text{-}7) \qquad B_0(z) = B(z) \qquad \text{and} \qquad \alpha_0 = B_0(0),$$

and choose a Julia operator

$$(2\text{-}8) \qquad \begin{pmatrix} \alpha_0 & \beta_0 \\ \gamma_0 & \delta_0 \end{pmatrix} \in \mathfrak{L}(\mathfrak{A}_0 \oplus \mathfrak{B}_1, \mathfrak{B}_0 \oplus \mathfrak{A}_1).$$

The defect spaces $\mathfrak{A}_1$ and $\mathfrak{B}_1$ are Hilbert spaces because α_0 is a bicontraction. By Theorem 2.2, there is a unique solution $B_1(z) \in S(\mathfrak{A}_1, \mathfrak{B}_1)$ of the functional equation

$$(2\text{-}9) \qquad B_0(z) = \alpha_0 + z\beta_0 B_1(z)\left(1 - z\delta_0 B_1(z)\right)^{-1}\gamma_0.$$

Set $\alpha_1 = B_1(0)$, and then choose a Julia operator for α_1 with defect Hilbert spaces $\mathfrak{A}_2$ and $\mathfrak{B}_2$ and repeat the construction. For the inductive step, suppose that for each $k = 0, \ldots, n$, we have chosen $B_k(z) \in S(\mathfrak{A}_k, \mathfrak{B}_k)$ and a Julia operator

$$(2\text{-}10) \qquad \begin{pmatrix} \alpha_k & \beta_k \\ \gamma_k & \delta_k \end{pmatrix} \in \mathfrak{L}(\mathfrak{A}_k \oplus \mathfrak{B}_{k+1}, \mathfrak{B}_k \oplus \mathfrak{A}_{k+1})$$

such that $\alpha_k = B_k(0)$ and $\mathfrak{A}_{k+1}$ and $\mathfrak{B}_{k+1}$ are Hilbert spaces. Use Theorem 2.2 to choose $B_{n+1}(z) \in S(\mathfrak{A}_{n+1}, \mathfrak{B}_{n+1})$ such that

$$(2\text{-}11) \qquad B_n(z) = \alpha_n + z\beta_n B_{n+1}(z) \left(1 - z\delta_n B_{n+1}(z)\right)^{-1} \gamma_n.$$

Set $\alpha_{n+1} = B_{n+1}(0)$ and choose a Julia operator for α_{n+1} with new defect Hilbert spaces $\mathfrak{A}_{n+2}$ and $\mathfrak{B}_{n+2}$ to complete the inductive step.

We call the Julia operators

$$(2\text{-}12) \qquad \begin{pmatrix} \alpha_n & \beta_n \\ \gamma_n & \delta_n \end{pmatrix} \in \mathfrak{L}(\mathfrak{A}_n \oplus \mathfrak{B}_{n+1}, \mathfrak{B}_n \oplus \mathfrak{A}_{n+1}), \qquad n = 0, 1, 2, \dots,$$

or more simply the operators

$$(2\text{-}13) \qquad\qquad \alpha_0, \alpha_1, \alpha_2, \dots,$$

the **Schur parameters** of $B(z)$. They are defined up to replacement of the underlying spaces by isomorphic copies. For since $\alpha_0 \in \mathfrak{L}(\mathfrak{A}_0, \mathfrak{B}_0)$ is a bicontraction and $\alpha_n \in \mathfrak{L}(\mathfrak{A}_n, \mathfrak{B}_n)$, $n \geq 1$, are Hilbert space contractions, the Julia operators which appear in the Schur algorithm are essentially unique.

The construction coincides with the Schur algorithm as it appears in [**7, 21**] in the Hilbert space case, and it reduces to the classical Schur algorithm described in §1 in the scalar case (with natural choices of Julia operators).

THEOREM 2.3. *If $B(z)$ is in $S(\mathfrak{A}, \mathfrak{B})$ and has Schur parameters $\alpha_0, \alpha_1, \dots$, then $\tilde{B}(z)$ belongs to $S(\mathfrak{B}, \mathfrak{A})$ and has Schur parameters $\alpha_0^*, \alpha_1^*, \dots$.*

PROOF. The relation (2–9) gives

$$\begin{aligned} \tilde{B}_0(z) &= \alpha_0^* + z\gamma_0^* \left(1 - z\tilde{B}_1(z)\delta_0^*\right)^{-1} \tilde{B}_1(z)\beta_0^* \\ &= \alpha_0^* + z\gamma_0^* \tilde{B}_1(z) \left(1 - z\delta_0^* \tilde{B}_1(z)\right)^{-1} \beta_0^*, \end{aligned}$$

which has the same form as (2–9) but with (2–8) replaced by its adjoint. Since $\mathfrak{A}_1$ and $\mathfrak{B}_1$ are Hilbert spaces, $\tilde{B}_1(z)$ belongs to $S(\mathfrak{B}_1, \mathfrak{A}_1)$, and hence $\tilde{B}(z)$ belongs to $S(\mathfrak{B}, \mathfrak{A})$ by Theorem 2.2. The calculation of Schur parameters follows on repeating the preceding argument. $\quad\square$

The Schur parameters label the Schur class.

THEOREM 2.4. *Let (2–12) be any Julia operators such that $\mathfrak{A}_1, \mathfrak{A}_2, \dots$ and $\mathfrak{B}_1, \mathfrak{B}_2, \dots$ are Hilbert spaces. Then there exists a unique $B(z)$ in $S(\mathfrak{A}, \mathfrak{B}) = S(\mathfrak{A}_0, \mathfrak{B}_0)$ whose Schur parameters are (2–12).*

Since the basic setup is now the same as in the Hilbert space case [**7, 21**], it is clear that numerous results from these and other sources can be duplicated in the Kreĭn space generalization. We give just one result.

THEOREM 2.5 ([**12**]). *Given operators $B_0, \dots, B_r$ in $\mathfrak{L}(\mathfrak{A}, \mathfrak{B})$, $\mathfrak{A}$ and $\mathfrak{B}$ Kreĭn spaces, occur as the initial segment of coefficients of an element $B(z)$ in $S(\mathfrak{A}, \mathfrak{B})$ if and only if (2–2) is a bicontraction.*

3. On the calculus of Julia operators

The choice sequence approach to the Schur algorithm given in Bakonyi and Constantinescu [7] is recast here in a formal calculus of Julia operators.

The assumptions in this section are weaker than those in §2. We assume given Kreĭn spaces $\mathfrak{A} = \mathfrak{A}_0, \mathfrak{A}_1, \mathfrak{A}_2, \dots$ and $\mathfrak{B} = \mathfrak{B}_0, \mathfrak{B}_1, \mathfrak{B}_2, \dots$, and Julia operators

$$(3\text{-}1) \qquad \begin{pmatrix} \alpha_k & \beta_k \\ \gamma_k & \delta_k \end{pmatrix} \in \mathfrak{L}(\mathfrak{A}_k \oplus \mathfrak{B}_{k+1}, \mathfrak{B}_k \oplus \mathfrak{A}_{k+1}),$$

$k = 0, 1, 2, \dots$. From this data, we shall construct a formal power series

$$B(z) = B_0 + B_1 z + B_2 z^2 + \cdots$$

with coefficients in $\mathfrak{L}(\mathfrak{A}, \mathfrak{B})$ in such a way that the iterative algorithm (2–7) – (2–11) formally applied to $B(z)$ produces the given sequence of Julia operators (3–1) in the role of (2–10). The series $B(z)$ need not belong to the Schur class, however, and the results in this section are of a purely algebraic nature. The special case of the Schur class is taken up again in §4.

The calculus uses Julia operators for certain row and column extensions of a given operator. Define

$$
(3\text{-}2) \quad \begin{pmatrix} R_k & D_{R_k} \\ \tilde{D}_{R_k}^* & -L_{R_k}^* \end{pmatrix}
$$

$$
= \begin{pmatrix}
\begin{pmatrix} \alpha_1 & \beta_1\alpha_2 & \beta_1\beta_2\alpha_3 & \cdots & \beta_1\beta_2\cdots\beta_{k-1}\alpha_k \end{pmatrix} & \begin{pmatrix} \beta_1\beta_2\cdots\beta_k \end{pmatrix} \\
\begin{pmatrix} \gamma_1 & \delta_1\alpha_2 & \delta_1\beta_2\alpha_3 & \cdots & \delta_1\beta_2\cdots\beta_{k-1}\alpha_k \\ 0 & \gamma_2 & \delta_2\alpha_3 & \cdots & \delta_2\beta_3\cdots\beta_{k-1}\alpha_k \\ \vdots & \vdots & \vdots & & \vdots \\ 0 & 0 & 0 & \cdots & \gamma_k \end{pmatrix} & \begin{pmatrix} \delta_1\beta_2\cdots\beta_k \\ \delta_2\beta_3\cdots\beta_k \\ \vdots \\ \delta_k \end{pmatrix}
\end{pmatrix}
$$

$$
= \begin{pmatrix} \alpha_1 & \beta_1 & & 0 \\ \gamma_1 & \delta_1 & & \\ & & 1 & \\ & & & 1 \\ 0 & & & \ddots \\ & & & & 1 \end{pmatrix} \begin{pmatrix} 1 & & & \\ & \alpha_2 & \beta_2 & & 0 \\ & \gamma_2 & \delta_2 & & \\ & & & 1 & \\ 0 & & & & \ddots \\ & & & & & 1 \end{pmatrix} \cdots \begin{pmatrix} 1 & & & \\ & 1 & & 0 \\ & & 1 & \\ & & & \ddots \\ 0 & & & & \alpha_k & \beta_k \\ & & & & \gamma_k & \delta_k \end{pmatrix}.
$$

The formula defines an operator acting on

$$(3\text{-}3) \qquad (\mathfrak{A}_1 \oplus \cdots \oplus \mathfrak{A}_k) \oplus \mathfrak{D}_{R_k} = (\mathfrak{A}_1 \oplus \cdots \oplus \mathfrak{A}_k) \oplus \mathfrak{B}_{k+1}$$

to

$$(3\text{-}4) \qquad \mathfrak{B}_1 \oplus \tilde{\mathfrak{D}}_{R_k} = \mathfrak{B}_1 \oplus (\mathfrak{A}_2 \oplus \cdots \oplus \mathfrak{A}_{k+1}).$$

By inspection, it is a Julia operator for the row extension

$$(3\text{-}5) \quad R_k = \begin{pmatrix} \alpha_1 & \beta_1\alpha_2 & \beta_1\beta_2\alpha_3 & \cdots & \beta_1\beta_2\cdots\beta_{k-1}\alpha_k \end{pmatrix} \in \mathfrak{L}(\mathfrak{A}_1 \oplus \cdots \oplus \mathfrak{A}_k, \mathfrak{B}_1)$$

of α_1.

In a similar way, the operator

$$(3\text{-}6) \quad \begin{pmatrix} C_k & D_{C_k} \\ \tilde{D}_{C_k}^* & -L_{C_k}^* \end{pmatrix}$$

$$= \left(\begin{pmatrix} \alpha_1 \\ \alpha_2\gamma_1 \\ \alpha_3\gamma_2\gamma_1 \\ \vdots \\ \alpha_k\gamma_{k-1}\cdots\gamma_1 \\ (\gamma_k\gamma_{k-1}\cdots\gamma_1) \end{pmatrix} \begin{pmatrix} \beta_1 & 0 & \cdots & 0 \\ \alpha_2\delta_1 & \beta_2 & \cdots & 0 \\ \alpha_3\gamma_2\delta_1 & \alpha_3\delta_2 & \cdots & 0 \\ \vdots & \vdots & & \vdots \\ \alpha_k\gamma_{k-1}\cdots\gamma_2\delta_1 & \alpha_k\gamma_{k-1}\cdots\gamma_3\delta_2 & \cdots & \beta_k \\ (\gamma_k\gamma_{k-1}\cdots\gamma_2\delta_1 & \gamma_k\gamma_{k-1}\cdots\gamma_3\delta_2 & \cdots & \delta_k\,) \end{pmatrix} \right)$$

$$= \begin{pmatrix} 1 & & & & \\ & 1 & & 0 & \\ & & 1 & & \\ & & & \ddots & \\ & 0 & & & \alpha_k\ \beta_k \\ & & & & \gamma_k\ \delta_k \end{pmatrix} \cdots \begin{pmatrix} 1 & & & & \\ & \alpha_2\ \beta_2 & & 0 & \\ & \gamma_2\ \delta_2 & & & \\ & & 1 & & \\ & 0 & & \ddots & \\ & & & & 1 \end{pmatrix} \begin{pmatrix} \alpha_1\ \beta_1 & & 0 & \\ \gamma_1\ \delta_1 & & & \\ & 1 & & \\ & & 1 & \\ 0 & & & \ddots \\ & & & & 1 \end{pmatrix}$$

acting on

$$(3\text{-}7) \qquad\qquad \mathfrak{A}_1 \oplus \mathfrak{D}_{C_k} = \mathfrak{A}_1 \oplus (\mathfrak{B}_2 \oplus \cdots \oplus \mathfrak{B}_{k+1})$$

to

$$(3\text{-}8) \qquad (\mathfrak{B}_1 \oplus \cdots \oplus \mathfrak{B}_k) \oplus \tilde{\mathfrak{D}}_{C_k} = (\mathfrak{B}_1 \oplus \cdots \oplus \mathfrak{B}_k) \oplus \mathfrak{A}_{k+1}.$$

defines a Julia operator for the column extension

$$(3\text{-}9) \qquad C_k = \begin{pmatrix} \alpha_1 \\ \alpha_2\gamma_1 \\ \alpha_3\gamma_2\gamma_1 \\ \vdots \\ \alpha_k\gamma_{k-1}\cdots\gamma_1 \end{pmatrix} \in \mathfrak{L}(\mathfrak{A}_1, \mathfrak{B}_1 \oplus \cdots \oplus \mathfrak{B}_k)$$

of α_1.

Our first two results of this section, Theorems 3.1 and 3.2, are stated in a form that only uses initial segments of the sequence (3–1).

THEOREM 3.1. *Assume that Julia operators of the form (3–1) are given for* $0 \le k \le r$, *and define matrices (3–2) and (3–6) for* $k = 1,\ldots,r$. *Construct Julia operators*

$$(3\text{-}10) \quad \begin{cases} \begin{pmatrix} T_k & D_{T_k} \\ \tilde{D}_{T_k}^* & -L_{T_k}^* \end{pmatrix} \in \mathfrak{L}(\mathfrak{A}^{k+1} \oplus \mathfrak{D}_{T_k}, \mathfrak{B}^{k+1} \oplus \tilde{\mathfrak{D}}_{T_k}) \\[2mm] \tilde{\mathfrak{D}}_{T_k} = \mathfrak{A}_1 \oplus \cdots \oplus \mathfrak{A}_{k+1}, \\[2mm] \mathfrak{D}_{T_k} = \mathfrak{B}_1 \oplus \cdots \oplus \mathfrak{B}_{k+1}, \end{cases}$$

for $0 \leq k \leq r$ by setting

(3-11)
$$\begin{pmatrix} T_0 & D_{T_0} \\ \tilde{D}^*_{T_0} & -L^*_{T_0} \end{pmatrix} = \begin{pmatrix} \alpha_0 & \beta_0 \\ \gamma_0 & \delta_0 \end{pmatrix}$$

and

(3-12)
$$\begin{pmatrix} T_k & D_{T_k} \\ \tilde{D}^*_{T_k} & -L^*_{T_k} \end{pmatrix}$$

$$= \begin{pmatrix} 1 & 0 & 0 & 0 \\ 0 & \alpha_0 & \beta_0 & 0 \\ 0 & \gamma_0 & \delta_0 & 0 \\ 0 & 0 & 0 & 1 \end{pmatrix} \begin{pmatrix} 1 & 0 & 0 & 0 \\ 0 & 1 & 0 & 0 \\ 0 & 0 & R_k & D_{R_k} \\ 0 & 0 & \tilde{D}^*_{R_k} & -L^*_{R_k} \end{pmatrix} \begin{pmatrix} T_{k-1} & 0 & D_{T_{k-1}} & 0 \\ 0 & 1 & 0 & 0 \\ \tilde{D}^*_{T_{k-1}} & 0 & -L^*_{T_{k-1}} & 0 \\ 0 & 0 & 0 & 1 \end{pmatrix}$$

$$= \begin{pmatrix} \begin{pmatrix} T_{k-1} & 0 \\ \beta_0 R_k \tilde{D}^*_{T_{k-1}} & \alpha_0 \end{pmatrix} & \begin{pmatrix} D_{T_{k-1}} & 0 \\ -\beta_0 R_k L^*_{T_{k-1}} & \beta_0 D_{R_k} \end{pmatrix} \\ \begin{pmatrix} \delta_0 R_k \tilde{D}^*_{T_{k-1}} & \gamma_0 \\ \tilde{D}^*_{R_k} \tilde{D}^*_{T_{k-1}} & 0 \end{pmatrix} & \begin{pmatrix} -\delta_0 R_k L^*_{T_{k-1}} & \delta_0 D_{R_k} \\ -\tilde{D}^*_{R_k} L^*_{T_{k-1}} & -L^*_{R_k} \end{pmatrix} \end{pmatrix}$$

for $1 \leq k \leq r$. Then for $1 \leq k \leq r$,

(3-13)
$$\begin{pmatrix} T_k & D_{T_k} \\ \tilde{D}^*_{T_k} & -L^*_{T_k} \end{pmatrix}$$

$$= \begin{pmatrix} 1 & 0 & 0 & 0 \\ 0 & T_{k-1} & D_{T_{k-1}} & 0 \\ 0 & \tilde{D}^*_{T_{k-1}} & -L^*_{T_{k-1}} & 0 \\ 0 & 0 & 0 & 1 \end{pmatrix} \begin{pmatrix} 1 & 0 & 0 & 0 \\ 0 & 1 & 0 & 0 \\ 0 & 0 & C_k & D_{C_k} \\ 0 & 0 & \tilde{D}^*_{C_k} & -L^*_{C_k} \end{pmatrix} \begin{pmatrix} \alpha_0 & 0 & \beta_0 & 0 \\ 0 & 1 & 0 & 0 \\ \gamma_0 & 0 & \delta_0 & 0 \\ 0 & 0 & 0 & 1 \end{pmatrix}$$

$$= \begin{pmatrix} \begin{pmatrix} \alpha_0 & 0 \\ D_{T_{k-1}} C_k \gamma_0 & T_{k-1} \end{pmatrix} & \begin{pmatrix} \beta_0 & 0 \\ D_{T_{k-1}} C_k \delta_0 & D_{T_{k-1}} D_{C_k} \end{pmatrix} \\ \begin{pmatrix} -L^*_{T_{k-1}} C_k \gamma_0 & \tilde{D}^*_{T_{k-1}} \\ \tilde{D}^*_{C_k} \gamma_0 & 0 \end{pmatrix} & \begin{pmatrix} -L^*_{T_{k-1}} C_k \delta_0 & -L^*_{T_{k-1}} D_{C_k} \\ \tilde{D}^*_{C_k} \delta_0 & -L^*_{C_k} \end{pmatrix} \end{pmatrix}.$$

The motivation for the recursive definition will become transparent in due course. The main point at this stage is the equivalence of the relations (3–12) and (3–13).

We leave it to the reader to work out the spaces on which operator matrices such as those that appear in (3–12) and (3–13) act. Such identities admit only one possible interpretation which is determined by the requirement that the expressions are compatible for the indicated operations.

PROOF. It follows from (3–11) and (3–12) that (3–10) is a Julia operator for each $k = 0, \ldots, r$. Clearly (3–13) holds for $k = 1$. We proceed by induction, assuming that (3–13) holds for some $k < r$. Then

$$
\begin{pmatrix} 1 & 0 & 0 & 0 \\ 0 & T_k & D_{T_k} & 0 \\ 0 & \tilde{D}_{T_k}^* & -L_{T_k}^* & 0 \\ 0 & 0 & 0 & 1 \end{pmatrix}
\begin{pmatrix} 1 & 0 & 0 & 0 \\ 0 & 1 & 0 & 0 \\ 0 & 0 & C_{k+1} & D_{C_{k+1}} \\ 0 & 0 & \tilde{D}_{C_{k+1}}^* & -L_{C_{k+1}}^* \end{pmatrix}
\begin{pmatrix} \alpha_0 & 0 & \beta_0 & 0 \\ 0 & 1 & 0 & 0 \\ \gamma_0 & 0 & \delta_0 & 0 \\ 0 & 0 & 0 & 1 \end{pmatrix}
$$

$$
(a) =
\begin{pmatrix} 1 & 0 & 0 & 0 & 0 & 0 \\ 0 & 1 & 0 & 0 & 0 & 0 \\ 0 & 0 & \alpha_0 & \beta_0 & 0 & 0 \\ 0 & 0 & \gamma_0 & \delta_0 & 0 & 0 \\ 0 & 0 & 0 & 0 & 1 & 0 \\ 0 & 0 & 0 & 0 & 0 & 1 \end{pmatrix}
\begin{pmatrix} 1 & 0 & 0 & 0 & 0 & 0 \\ 0 & 1 & 0 & 0 & 0 & 0 \\ 0 & 0 & 1 & 0 & 0 & 0 \\ 0 & 0 & 0 & R_k & D_{R_k} & 0 \\ 0 & 0 & 0 & \tilde{D}_{R_k}^* & -L_{R_k}^* & 0 \\ 0 & 0 & 0 & 0 & 0 & 1 \end{pmatrix}
$$

$$
\times
\begin{pmatrix} 1 & 0 & 0 & 0 & 0 & 0 \\ 0 & T_{k-1} & 0 & D_{T_{k-1}} & 0 & 0 \\ 0 & 0 & 1 & 0 & 0 & 0 \\ 0 & \tilde{D}_{T_{k-1}}^* & 0 & -L_{T_{k-1}}^* & 0 & 0 \\ 0 & 0 & 0 & 0 & 1 & 0 \\ 0 & 0 & 0 & 0 & 0 & 1 \end{pmatrix}
\begin{pmatrix} 1 & 0 & 0 & 0 & 0 & 0 \\ 0 & 1 & 0 & 0 & 0 & 0 \\ 0 & 0 & 1 & 0 & 0 & 0 \\ 0 & 0 & 0 & 1 & 0 & 0 \\ 0 & 0 & 0 & 0 & \alpha_{k+1} & \beta_{k+1} \\ 0 & 0 & 0 & 0 & \gamma_{k+1} & \delta_{k+1} \end{pmatrix}
$$

$$
\times
\begin{pmatrix} 1 & 0 & 0 & 0 & 0 & 0 \\ 0 & 1 & 0 & 0 & 0 & 0 \\ 0 & 0 & 1 & 0 & 0 & 0 \\ 0 & 0 & 0 & C_k & D_{C_k} & 0 \\ 0 & 0 & 0 & \tilde{D}_{C_k}^* & -L_{C_k}^* & 0 \\ 0 & 0 & 0 & 0 & 0 & 1 \end{pmatrix}
\begin{pmatrix} \alpha_0 & 0 & 0 & \beta_0 & 0 & 0 \\ 0 & 1 & 0 & 0 & 0 & 0 \\ 0 & 0 & 1 & 0 & 0 & 0 \\ \gamma_0 & 0 & 0 & \delta_0 & 0 & 0 \\ 0 & 0 & 0 & 0 & 1 & 0 \\ 0 & 0 & 0 & 0 & 0 & 1 \end{pmatrix}
$$

$$
(b) =
\begin{pmatrix} 1 & 0 & 0 & 0 & 0 & 0 \\ 0 & 1 & 0 & 0 & 0 & 0 \\ 0 & 0 & \alpha_0 & \beta_0 & 0 & 0 \\ 0 & 0 & \gamma_0 & \delta_0 & 0 & 0 \\ 0 & 0 & 0 & 0 & 1 & 0 \\ 0 & 0 & 0 & 0 & 0 & 1 \end{pmatrix}
\begin{pmatrix} 1 & 0 & 0 & 0 & 0 & 0 \\ 0 & 1 & 0 & 0 & 0 & 0 \\ 0 & 0 & 1 & 0 & 0 & 0 \\ 0 & 0 & 0 & R_k & D_{R_k} & 0 \\ 0 & 0 & 0 & \tilde{D}_{R_k}^* & -L_{R_k}^* & 0 \\ 0 & 0 & 0 & 0 & 0 & 1 \end{pmatrix}
$$

$$
\times
\begin{pmatrix} 1 & 0 & 0 & 0 & 0 & 0 \\ 0 & 1 & 0 & 0 & 0 & 0 \\ 0 & 0 & 1 & 0 & 0 & 0 \\ 0 & 0 & 0 & 1 & 0 & 0 \\ 0 & 0 & 0 & 0 & \alpha_{k+1} & \beta_{k+1} \\ 0 & 0 & 0 & 0 & \gamma_{k+1} & \delta_{k+1} \end{pmatrix}
\begin{pmatrix} 1 & 0 & 0 & 0 & 0 & 0 \\ 0 & T_{k-1} & 0 & D_{T_{k-1}} & 0 & 0 \\ 0 & 0 & 1 & 0 & 0 & 0 \\ 0 & \tilde{D}_{T_{k-1}}^* & 0 & -L_{T_{k-1}}^* & 0 & 0 \\ 0 & 0 & 0 & 0 & 1 & 0 \\ 0 & 0 & 0 & 0 & 0 & 1 \end{pmatrix}
$$

$$
\times
\begin{pmatrix} 1 & 0 & 0 & 0 & 0 & 0 \\ 0 & 1 & 0 & 0 & 0 & 0 \\ 0 & 0 & 1 & 0 & 0 & 0 \\ 0 & 0 & 0 & C_k & D_{C_k} & 0 \\ 0 & 0 & 0 & \tilde{D}_{C_k}^* & -L_{C_k}^* & 0 \\ 0 & 0 & 0 & 0 & 0 & 1 \end{pmatrix}
\begin{pmatrix} \alpha_0 & 0 & 0 & \beta_0 & 0 & 0 \\ 0 & 1 & 0 & 0 & 0 & 0 \\ 0 & 0 & 1 & 0 & 0 & 0 \\ \gamma_0 & 0 & 0 & \delta_0 & 0 & 0 \\ 0 & 0 & 0 & 0 & 1 & 0 \\ 0 & 0 & 0 & 0 & 0 & 1 \end{pmatrix}
$$

$$(c) = \begin{pmatrix} 1 & 0 & 0 & 0 \\ 0 & \alpha_0 & \beta_0 & 0 \\ 0 & \gamma_0 & \delta_0 & 0 \\ 0 & 0 & 0 & 1 \end{pmatrix} \begin{pmatrix} 1 & 0 & 0 & 0 \\ 0 & 1 & 0 & 0 \\ 0 & 0 & R_{k+1} & D_{R_{k+1}} \\ 0 & 0 & \tilde{D}^*_{R_{k+1}} & -L^*_{R_{k+1}} \end{pmatrix} \begin{pmatrix} T_k & 0 & D_{T_k} & 0 \\ 0 & 1 & 0 & 0 \\ \tilde{D}^*_{T_k} & 0 & -L^*_{T_k} & 0 \\ 0 & 0 & 0 & 1 \end{pmatrix}$$

$$(d) = \begin{pmatrix} T_{k+1} & D_{T_{k+1}} \\ \tilde{D}^*_{T_{k+1}} & -L^*_{T_{k+1}} \end{pmatrix},$$

which proves (3–13) with k replaced by $k+1$.

Notes on the matrix manipulations: (a) follows from (3–12) and (3–6); (b) is obtained by permuting the third and fourth factors; (c) uses (3–2) together with the identity

$$\begin{pmatrix} T_k & 0 & D_{T_k} \\ 0 & 1 & 0 \\ \tilde{D}^*_{T_k} & 0 & -L^*_{T_k} \end{pmatrix}$$

$$= \begin{pmatrix} 1 & 0 & 0 & 0 & 0 \\ 0 & T_{k-1} & 0 & D_{T_{k-1}} & 0 \\ 0 & 0 & 1 & 0 & 0 \\ 0 & \tilde{D}^*_{T_{k-1}} & 0 & -L^*_{T_{k-1}} & 0 \\ 0 & 0 & 0 & 0 & 1 \end{pmatrix} \begin{pmatrix} 1 & 0 & 0 & 0 & 0 \\ 0 & 1 & 0 & 0 & 0 \\ 0 & 0 & 1 & 0 & 0 \\ 0 & 0 & 0 & C_k & D_{C_k} \\ 0 & 0 & 0 & \tilde{D}^*_{C_k} & -L^*_{C_k} \end{pmatrix} \begin{pmatrix} \alpha_0 & 0 & 0 & \beta_0 & 0 \\ 0 & 1 & 0 & 0 & 0 \\ 0 & 0 & 1 & 0 & 0 \\ \gamma_0 & 0 & 0 & \delta_0 & 0 \\ 0 & 0 & 0 & 0 & 1 \end{pmatrix},$$

which holds by the inductive hypothesis; (d) follows from (3–12). □

The next result shows that the operators T_k which appear in Theorem 3.1 have a Toeplitz structure, and it computes the entries in a number of ways.

THEOREM 3.2. *Under the assumptions of Theorem 3.1, there exist operators $B_0, \ldots, B_r$ in $\mathfrak{L}(\mathfrak{A}, \mathfrak{B})$ such that*

$$(3\text{-}14) \qquad T_k = \begin{pmatrix} B_0 & & & \\ B_1 & B_0 & & 0 \\ & \ddots & \ddots & \\ B_k & & B_1 & B_0 \end{pmatrix}$$

for each $k = 0, \ldots, r$. They are given by

$$(3\text{-}15) \qquad B_0 = \alpha_0$$

and for each $k = 1, \ldots, r$,

$$(3\text{-}16) \qquad \begin{pmatrix} B_k & B_{k-1} & \cdots & B_1 \end{pmatrix} = \beta_0 R_k \tilde{D}^*_{T_{k-1}},$$

$$(3\text{-}17) \qquad B_k = \beta_0 R_k \begin{pmatrix} \delta_0 R_{k-1} \\ \tilde{D}^*_{R_{k-1}} \end{pmatrix} \cdots \begin{pmatrix} \delta_0 R_1 \\ \tilde{D}^*_{R_1} \end{pmatrix} \gamma_0,$$

and

$$
(3\text{-}18) \qquad \begin{pmatrix} B_1 \\ B_2 \\ \vdots \\ B_k \end{pmatrix} = D_{T_{k-1}} C_k \gamma_0,
$$

$$
(3\text{-}19) \qquad B_k = \beta_0 \left(C_1 \delta_0 \quad D_{C_1} \right) \cdots \left(C_{k-1}\delta_0 \quad D_{C_{k-1}} \right) C_k \gamma_0.
$$

Moreover

$$
(3\text{-}20) \qquad B_1 = \beta_0 \alpha_1 \gamma_0,
$$

$$
(3\text{-}21) \qquad B_k = -\beta_0 R_{k-1} L^*_{T_{k-2}} C_{k-1}\gamma_0 + \beta_0 \cdots \beta_{k-1} \alpha_k \gamma_{k-1} \cdots \gamma_0,
$$

where in (3–21) *we assume* $k \geq 2$.

For example, the intermediate spaces for the product on the right of (3–17) are given by the diagram

$$
\mathfrak{A}_0 \to \mathfrak{A}_1 \to \mathfrak{A}_1 \oplus \mathfrak{A}_2 \to \quad \cdots \quad \to \mathfrak{A}_1 \oplus \cdots \oplus \mathfrak{A}_k \to \mathfrak{B}_1 \to \mathfrak{B}_0.
$$

A similar diagram can be constructed for (3–19).

PROOF. Define $B_0, \ldots, B_r$ by (3–15), (3–20), and (3–21).

We prove (3–14) and (3–16) by induction. By (3–15) and (3–20), these identities hold for $k = 0, 1$. Assume that they hold for some $k = 1, \ldots, r - 1$. By (3–21) and the formula for $\tilde{D}^*_{T_k}$ given in (3–13),

$$
\beta_0 R_{k+1} \tilde{D}^*_{T_k} = \left(\beta_0 R_k, \quad \beta_0 \beta_1 \cdots \beta_k \alpha_{k+1} \right) \begin{pmatrix} -L^*_{T_{k-1}} C_k \gamma_0 & \tilde{D}^*_{T_{k-1}} \\ \gamma_k \cdots \gamma_0 & 0 \end{pmatrix}
$$

$$
= \left(-\beta_0 R_k L^*_{T_{k-1}} C_k \gamma_0 + \beta_0 \cdots \beta_k \alpha_{k+1} \gamma_k \cdots \gamma_0, \quad \beta_0 R_k \tilde{D}^*_{T_{k-1}} \right)
$$

$$
= \left(B_{k+1} \quad B_k \quad \ldots \quad B_1 \right),
$$

which is (3–16) with k replaced by $k + 1$. Using this result and (3–12), we get

$$
T_{k+1} = \begin{pmatrix} T_k & 0 \\ \beta_0 R_{k+1} \tilde{D}^*_{T_k} & \alpha_0 \end{pmatrix} = \begin{pmatrix} B_0 & & & \\ B_1 & B_0 & & 0 \\ & \ddots & \ddots & \\ B_{k+1} & & B_1 & B_0 \end{pmatrix},
$$

which is (3–14) with k replaced by $k + 1$. This completes the inductive proof of (3–14) and (3–16).

The identity (3–18) follows on comparing (3–14) and the formula

$$
T_k = \begin{pmatrix} \alpha_0 & 0 \\ D_{T_{k-1}} C_k \gamma_0 & T_{k-1} \end{pmatrix},
$$

which follows from (3–13).

We next prove (3–17). By (3–12),

$$\tilde{D}^*_{T_j} = \begin{pmatrix} \delta_0 R_j \tilde{D}^*_{T_{j-1}} & \gamma_0 \\ \tilde{D}^*_{R_j} \tilde{D}^*_{T_{j-1}} & 0 \end{pmatrix}, \qquad j = 1, \dots, r.$$

The domain of this operator is $\mathfrak{A}^{j+1}$, and we may embed $\mathfrak{A}$ into this space by the mapping

$$E_j = \begin{pmatrix} E_{j-1} \\ 0 \end{pmatrix} = \begin{pmatrix} 1 \\ 0 \\ \vdots \\ 0 \end{pmatrix}.$$

Then

$$\tilde{D}^*_{T_j} E_j = \begin{pmatrix} \delta_0 R_j \\ \tilde{D}^*_{R_j} \end{pmatrix} \tilde{D}^*_{T_{j-1}} E_{j-1}.$$

Applying this identity repeatedly with $j = 1, \dots, k-1$, we get

$$B_k = \begin{pmatrix} B_k & B_{k-1} & \dots & B_1 \end{pmatrix} E_{k-1} = \beta_0 R_k \tilde{D}^*_{T_{k-1}} E_{k-1}$$

$$= \beta_0 R_k \begin{pmatrix} \delta_0 R_{k-1} \\ \tilde{D}^*_{R_{k-1}} \end{pmatrix} \cdots \begin{pmatrix} \delta_0 R_1 \\ \tilde{D}^*_{R_1} \end{pmatrix} \tilde{D}^*_{T_0} E_0.$$

This reduces to (3–17) because E_0 is the identity operator and $\tilde{D}^*_{T_0} = \gamma_0$.

To obtain (3–19) in parallel fashion, start with

$$D_{T_j} = \begin{pmatrix} \beta_0 & 0 \\ D_{T_{j-1}} C_j \delta_0 & D_{T_{j-1}} D_{C_j} \end{pmatrix}, \qquad j = 1, \dots, r,$$

which we obtain from (3–13). The range of this operator is $\mathfrak{B}^{j+1}$, and the projection F_j of this space onto the last coordinate is $F_j = \begin{pmatrix} 0 & F_{j-1} \end{pmatrix} = \begin{pmatrix} 0 & \dots & 0 & 1 \end{pmatrix}$. Then

$$F_j D_{T_j} = F_{j-1} D_{T_{j-1}} \begin{pmatrix} C_j \delta_0 & D_{C_j} \end{pmatrix},$$

and repeated application of this relation gives

$$B_k = F_{k-1} \begin{pmatrix} B_1 \\ B_2 \\ \vdots \\ B_k \end{pmatrix} = F_{k-1} D_{T_{k-1}} C_k \gamma_0$$

$$= F_0 D_{T_0} \begin{pmatrix} C_1 \delta_0 & D_{C_1} \end{pmatrix} \cdots \begin{pmatrix} C_{k-1} \delta_0 & D_{C_{k-1}} \end{pmatrix} C_k \gamma_0 ,$$

which yields (3–19). □

The structure of the operators $B_0, B_1, B_2, \ldots$ is most clearly exhibited in different identities that use formal matrices of infinite order. Set

$$(3\text{-}22) \qquad \begin{pmatrix} \widehat{\alpha}_0 & \widehat{\beta}_0 \\ \widehat{\gamma}_0 & \widehat{\delta}_0 \end{pmatrix} = \begin{pmatrix} (\alpha_0) & (\beta_0 \ \ 0 \ \ 0 \ \ \ldots \ ) \\ \begin{pmatrix} \gamma_0 \\ 0 \\ 0 \\ \vdots \end{pmatrix} & \begin{pmatrix} \delta_0 & 0 & 0 & \ldots \\ 0 & 1 & 0 & \ldots \\ 0 & 0 & 1 & \ldots \\ & & & \ddots \end{pmatrix} \end{pmatrix}$$

The identity operators in the formula for $\widehat{\delta}_0$ act on the spaces $\mathfrak{A}_2, \mathfrak{A}_3, \ldots$. Set

$$(3\text{-}23) \qquad X = \begin{pmatrix} \alpha_1 & \beta_1 & & & 0 \\ \gamma_1 & \delta_1 & & & \\ & & 1 & & \\ & & & 1 & \\ 0 & & & & 1 \\ & & & & & \ddots \end{pmatrix} \begin{pmatrix} 1 & & & & \\ & \alpha_2 & \beta_2 & & 0 \\ & \gamma_2 & \delta_2 & & \\ & & & 1 & \\ & 0 & & & 1 \\ & & & & & \ddots \end{pmatrix} \cdots$$

$$= \begin{pmatrix} \alpha_1 & \beta_1\alpha_2 & \beta_1\beta_2\alpha_3 & \beta_1\beta_2\beta_3\alpha_4 & \ldots \\ \gamma_1 & \delta_1\alpha_2 & \delta_1\beta_2\alpha_3 & \delta_1\beta_2\beta_3\alpha_4 & \ldots \\ 0 & \gamma_2 & \delta_2\alpha_3 & \delta_2\beta_3\alpha_4 & \ldots \\ 0 & 0 & \gamma_3 & \delta_3\alpha_4 & \ldots \\ & & \ldots & & \end{pmatrix}.$$

The present calculations involving these matrices use only finite sections and are purely algebraic. Later in §4, where additional hypotheses are imposed, the matrices represent continuous operators on direct sum spaces.

THEOREM 3.3. *Assume that Julia operators* (3–1) *are given, and set*

$$(3\text{-}24) \qquad B(z) = B_0 + B_1 z + B_2 z^2 + \cdots ,$$

where $B_0, B_1, B_2, \ldots$ *are the operators constructed by the recursive procedure in Theorem 3.2. Then*

$$(3\text{-}25) \qquad B(z) = \widehat{\alpha}_0 + z\widehat{\beta}_0 X \left(1 - z\widehat{\delta}_0 X \right)^{-1} \widehat{\gamma}_0$$

in the sense of formal power series.

PROOF. We must show that $B_0 = \widehat{\alpha}_0$, $B_1 = \widehat{\beta}_0 X \widehat{\gamma}_0$, and

$$B_n = \widehat{\beta}_0 X \left(\widehat{\delta}_0 X \right)^{n-1} \widehat{\gamma}_0, \qquad n \geq 2.$$

The formulas for B_0 and B_1 are immediate. By the definitions of X and $\widehat{\delta}_0$,

$$(3\text{-}26) \qquad \widehat{\delta}_0 X = \begin{pmatrix} \delta_0\alpha_1 & \delta_0\beta_1\alpha_2 & \delta_0\beta_1\beta_2\alpha_3 & \delta_0\beta_1\beta_2\beta_3\alpha_4 & \ldots \\ \gamma_1 & \delta_1\alpha_2 & \delta_1\beta_2\alpha_3 & \delta_1\beta_2\beta_3\alpha_4 & \ldots \\ 0 & \gamma_2 & \delta_2\alpha_3 & \delta_2\beta_3\alpha_4 & \ldots \\ 0 & 0 & \gamma_3 & \delta_3\alpha_4 & \ldots \\ & & \ldots & & \end{pmatrix}.$$

In view of the nearly triangular form of (3–26), we get

(3-27)
$$\begin{cases} P_2\left(\widehat{\delta_0}X\right)P_1 = \left(\widehat{\delta_0}X\right)P_1, \\[2mm] P_3\left(\widehat{\delta_0}X\right)P_2\left(\widehat{\delta_0}X\right)P_1 = \left(\widehat{\delta_0}X\right)^2 P_1, \\[2mm] \cdots \end{cases}$$

where P_n has 1's (identity operators on $\mathfrak{A}_1, \ldots, \mathfrak{A}_n$) in the first n diagonal entries and all other terms zero. If we identify $P_k\left(\widehat{\delta_0}X\right)P_{k-1}$ with the finite section of the first k rows and $k-1$ columns, then by (3–2),

(3-28)
$$P_k\left(\widehat{\delta_0}X\right)P_{k-1} = \begin{pmatrix} \delta_0 R_{k-1} \\ \tilde{D}^*_{R_{k-1}} \end{pmatrix}.$$

Hence by (3–17),

$$\begin{aligned} B_n &= \beta_0 R_n \begin{pmatrix} \delta_0 R_{n-1} \\ \tilde{D}^*_{R_{n-1}} \end{pmatrix} \cdots \begin{pmatrix} \delta_0 R_1 \\ \tilde{D}^*_{R_1} \end{pmatrix} \gamma_0 \\ &= \widehat{\beta_0}X P_n\left(\widehat{\delta_0}X\right) P_{n-1}\left(\widehat{\delta_0}X\right) \cdots P_2\left(\widehat{\delta_0}X\right) P_1 \widehat{\gamma_0} \\ &= \widehat{\beta_0}X \left(\widehat{\delta_0}X\right)^{n-1} \widehat{\gamma_0}, \end{aligned}$$

as was to be shown. $\square$

THEOREM 3.4. *Starting with the Julia operators in* (3–1),

$$\begin{pmatrix} \alpha_0 & \beta_0 \\ \gamma_0 & \delta_0 \end{pmatrix}, \begin{pmatrix} \alpha_1 & \beta_1 \\ \gamma_1 & \delta_1 \end{pmatrix}, \begin{pmatrix} \alpha_2 & \beta_2 \\ \gamma_2 & \delta_2 \end{pmatrix}, \ldots,$$

construct the formal power series $B(z) = B_0 + B_1 z + B_2 z^2 + \cdots$ *in the manner described above. In a similar way, starting with*

$$\begin{pmatrix} \alpha_1 & \beta_1 \\ \gamma_1 & \delta_1 \end{pmatrix}, \begin{pmatrix} \alpha_2 & \beta_2 \\ \gamma_2 & \delta_2 \end{pmatrix}, \ldots$$

construct $B'(z) = B'_0 + B'_1 z + B'_2 z^2 + \cdots$. *Then*

(3-29)
$$B(z) = \alpha_0 + z\beta_0 B'(z)\left(1 - z\delta_0 B'(z)\right)^{-1}\gamma_0$$

in the sense of formal power series.

PROOF. Write $B(z)$ in the form (3–25), and let

(3-30)
$$B'(z) = \widehat{\alpha}_1 + z\widehat{\beta}_1 X'\left(1 - z\widehat{\delta}_1 X'\right)^{-1}\widehat{\gamma}_1$$

be the analogous representation of $B'(z)$. Observe that

$$X = \begin{pmatrix} \widehat{\alpha}_1 & \widehat{\beta}_1 \\ \widehat{\gamma}_1 & \widehat{\delta}_1 \end{pmatrix}\begin{pmatrix} 1 & 0 \\ 0 & X' \end{pmatrix} = \begin{pmatrix} \widehat{\alpha}_1 & \widehat{\beta}_1 X' \\ \widehat{\gamma}_1 & \widehat{\delta}_1 X' \end{pmatrix},$$

and so

$$1 - z\widehat{\delta_0}X = \begin{pmatrix} 1 - z\delta_0\widehat{\alpha}_1 & -z\delta_0\widehat{\beta}_1 X' \\ -z\widehat{\gamma}_1 & 1 - z\widehat{\delta}_1 X' \end{pmatrix}.$$

The operators Z_1, Z_2 in the first column of the inverse

$$\left(1 - z\widehat{\delta_0}X\right)^{-1} = \begin{pmatrix} Z_1 & * \\ Z_2 & * \end{pmatrix}$$

are therefore characterized as the unique solution of the operator equations

$$(1 - z\delta_0\widehat{\alpha}_1)\, Z_1 - z\delta_0\widehat{\beta}_1 X' Z_2 = 1,$$

$$-z\widehat{\gamma}_1 Z_1 + \left(1 - z\widehat{\delta}_1 X'\right) Z_2 = 0.$$

By direct verification, these solutions are

$$Z_1 = \left(1 - z\delta_0 \left[\widehat{\alpha}_1 + z\widehat{\beta}_1 X' \left(1 - z\widehat{\delta}_1 X'\right)^{-1} \widehat{\gamma}_1\right]\right)^{-1},$$

$$Z_2 = z\left(1 - z\widehat{\delta}_1 X'\right)^{-1} \widehat{\gamma}_1 \left(1 - z\delta_0 \left[\widehat{\alpha}_1 + z\widehat{\beta}_1 X' \left(1 - z\widehat{\delta}_1 X'\right)^{-1} \widehat{\gamma}_1\right]\right)^{-1}.$$

Thus

$$Z_1 = (1 - z\delta_0 B'(z))^{-1},$$

$$Z_2 = z\left(1 - z\widehat{\delta}_1 X'\right)^{-1} \widehat{\gamma}_1 \left(1 - z\delta_0 B'(z)\right)^{-1}.$$

Therefore by (3–25) and (3–30),

$$B(z) = \alpha_0 + z\,(\,\beta_0 \quad 0\,) \begin{pmatrix} \widehat{\alpha}_1 & \widehat{\beta}_1 X' \\ \widehat{\gamma}_1 & \widehat{\delta}_1 X' \end{pmatrix} \begin{pmatrix} Z_1 & * \\ Z_2 & * \end{pmatrix} \begin{pmatrix} \gamma_0 \\ 0 \end{pmatrix}$$

$$= \alpha_0 + z\beta_0 \left(\widehat{\alpha}_1 Z_1 + \widehat{\beta}_1 X' Z_2\right) \gamma_0$$

$$= \alpha_0 + z\beta_0 B'(z) \left(1 - z\delta_0 B'(z)\right)^{-1} \gamma_0,$$

which is (3–29). □

A uniqueness question arises:

Is it possible for two essentially different sequences of Julia operators (3–1) to yield the same formal power series $B(z) = B_0 + B_1 z + B_2 z^2 + \cdots$ by the constructions given above?

An affirmative answer cannot be expected unless the operators $\alpha_0, \alpha_1, \alpha_2, \ldots$ are of a type that have essentially unique Julia operator. We show in §4 that there is an affirmative answer for the Schur class.

4. Applications to the Schur class

We are now ready to supply the proofs of Theorems 2.2, 2.4, and 2.5. The various parts of these results will be done in somewhat different order and pulled together at the end.

The formulas of §3 immediately yield a large number of elements of the Schur class.

THEOREM 4.1. *Let*

$$(4\text{-}1) \qquad \begin{pmatrix} \alpha_n & \beta_n \\ \gamma_n & \delta_n \end{pmatrix} \in \mathfrak{L}(\mathfrak{A}_n \oplus \mathfrak{B}_{n+1}, \mathfrak{B}_n \oplus \mathfrak{A}_{n+1}), \qquad n \geq 0,$$

be Julia operators such that $\mathfrak{A}_1, \mathfrak{A}_2, \ldots$ *and* $\mathfrak{B}_1, \mathfrak{B}_2, \ldots$ *are Hilbert spaces. Define*

$$B(z) = B_0 + B_1 z + B_2 z^2 + \cdots,$$

where $B_0, B_1, B_2, \ldots$ *are constructed from the sequence of Julia operators (4–1) as in §3. Then* $B(z)$ *belongs to* $S(\mathfrak{A}, \mathfrak{B})$ *where* $\mathfrak{A} = \mathfrak{A}_0$ *and* $\mathfrak{B} = \mathfrak{B}_0$.

PROOF. Theorems 3.1 and 3.2 compute a Julia operator for (3–14) with the defect spaces

$$\tilde{\mathfrak{D}}_{T_k} = \mathfrak{A}_1 \oplus \cdots \oplus \mathfrak{A}_{k+1},$$

$$\mathfrak{D}_{T_k} = \mathfrak{B}_1 \oplus \cdots \oplus \mathfrak{B}_{k+1}.$$

The defect spaces are Hilbert spaces by our hypotheses, and therefore each operator (3–14) is a bicontraction. This is the condition for $B(z)$ to belong to the Schur class $S(\mathfrak{A}, \mathfrak{B})$. $\square$

Some factorization properties of Kreĭn space operators are needed to reverse the process. We use the following results of M. A. Dritschel:

(I) ([**20, Th. 2.2.1 (ii)**]). *Let* $X \in \mathfrak{L}(\mathfrak{H}, \mathfrak{K})$ *be a bicontraction with defect operator* $\tilde{D} \in \mathfrak{L}(\tilde{\mathfrak{D}}, \mathfrak{H})$. *A column extension*

$$\begin{pmatrix} X \\ \tilde{E}^* \end{pmatrix} \in \mathfrak{L}(\mathfrak{H}, \mathfrak{K} \oplus \tilde{\mathfrak{E}}),$$

with $\tilde{\mathfrak{E}}$ *a Hilbert space, is a bicontraction if and only if* $\tilde{E} = \tilde{D}\tilde{G}$ *where* $\tilde{G} \in \mathfrak{L}(\tilde{\mathfrak{E}}, \tilde{\mathfrak{D}})$ *is a Hilbert space contraction.*

(II) ([**20, Th. 2.3.3**]). *Let* $X \in \mathfrak{L}(\mathfrak{H}, \mathfrak{K})$ *be a contraction, and let* $D \in \mathfrak{L}(\mathfrak{D}, \mathfrak{K})$ *be a defect operator for* X^*. *A row extension*

$$(X \quad E) \in \mathfrak{L}(\mathfrak{H} \oplus \mathfrak{E}, \mathfrak{K}),$$

with $\mathfrak{E}$ *a Kreĭn space, is a bicontraction if and only if* $E = DG$ *where* $G \in \mathfrak{L}(\mathfrak{E}, \mathfrak{D})$ *is a bicontraction.*

The main problem is to determine the structure of lower triangular matrices which appear in the analysis. The form that is needed occurs in a particular case in [**13**]. This form is a consequence of the hypothesis that the lower triangular matrix is bicontractive.

LEMMA 4.2. *Let*

(4-2) $$T = \begin{pmatrix} T_1 & 0 \\ Q & T_2 \end{pmatrix} \in \mathfrak{L}(\mathfrak{H}_1 \oplus \mathfrak{H}_2, \mathfrak{K}_1 \oplus \mathfrak{K}_2)$$

be a given operator. Let $\tilde{D}_1 \in \mathfrak{L}(\tilde{\mathfrak{D}}_1, \mathfrak{H}_1)$ be a defect operator for T_1 and $D_2 \in \mathfrak{L}(\mathfrak{D}_2, \mathfrak{K}_2)$ a defect operator for T_2^*. Then T is a bicontraction if and only if T_1^* and T_2 are contractions and

(4-3) $$Q = D_2 Y \tilde{D}_1^*$$

where $Y \in \mathfrak{L}(\tilde{\mathfrak{D}}_1, \mathfrak{D}_2)$ is a bicontraction.

PROOF. Assume that T is a bicontraction. By the triangular form of (4–2), T_2 and T_1^* are contractions. Apply (II) to

$$T^* = \left(\begin{pmatrix} T_1^* \\ 0 \end{pmatrix} \begin{pmatrix} Q^* \\ T_2^* \end{pmatrix} \right)$$

using the defect operator $\begin{pmatrix} \tilde{D}_1 & 0 \\ 0 & 1 \end{pmatrix} \in \mathfrak{L}(\tilde{\mathfrak{D}}_1 \oplus \mathfrak{H}_2, \mathfrak{H}_1 \oplus \mathfrak{H}_2)$ for $(\,T_1 \quad 0\,)$. We get

(4-4) $$\begin{pmatrix} Q^* \\ T_2^* \end{pmatrix} = \begin{pmatrix} \tilde{D}_1 & 0 \\ 0 & 1 \end{pmatrix} \begin{pmatrix} G_1 \\ G_2 \end{pmatrix}$$

where

$$\begin{pmatrix} G_1 \\ G_2 \end{pmatrix} \in \mathfrak{L}(\mathfrak{K}_2, \tilde{\mathfrak{D}}_1 \oplus \mathfrak{H}_2)$$

is a bicontraction. Necessarily, $G_2 = T_2^*$. Therefore

$$(\,T_2 \quad G_1^*\,) = (\,G_2^* \quad G_1^*\,) \in \mathfrak{L}(\mathfrak{H}_2 \oplus \tilde{\mathfrak{D}}_1, \mathfrak{K}_2)$$

is a bicontraction. Again by (II), $G_1^* = D_2 Y$ where $Y \in \mathfrak{L}(\tilde{\mathfrak{D}}_1, \mathfrak{D}_2)$ is a bicontraction. By (4–4), $Q = G_1^* \tilde{D}_1^* = D_2 Y \tilde{D}_1^*$, so the conditions of the lemma are necessary. The sufficiency follows on reversing these steps. $\square$

THEOREM 4.3. *Every $B(z)$ in the Schur class $S(\mathfrak{A}, \mathfrak{B})$ has the form described in Theorem 4.1.*

PROOF. Let $B(z) = B_0 + B_1 z + B_2 z^2 + \cdots$ be a given element of $S(\mathfrak{A}, \mathfrak{B})$. By definition, this means that each of the operators (3–14) is a bicontraction. We construct Hilbert spaces $\mathfrak{A}_1, \mathfrak{A}_2, \ldots$ and $\mathfrak{B}_1, \mathfrak{B}_2, \ldots$ and Julia operators

(4-5) $$\begin{pmatrix} \alpha_j & \beta_j \\ \gamma_j & \delta_j \end{pmatrix} \in \mathfrak{L}(\mathfrak{A}_j \oplus \mathfrak{B}_{j+1}, \mathfrak{B}_j \oplus \mathfrak{A}_{j+1}), \qquad j \geq 0,$$

which yield the coefficients $B_0, B_1, B_2, \ldots$ by the construction in Theorems 3.1 and 3.2.

Put $\mathfrak{A}_0 = \mathfrak{A}$, $\mathfrak{B}_0 = \mathfrak{B}$ and $\alpha_0 = B_0$, and choose a Julia operator

$$\begin{pmatrix} \alpha_0 & B_0 \\ \gamma_0 & \delta_0 \end{pmatrix} \in \mathfrak{L}(\mathfrak{A}_0 \oplus \mathfrak{B}_1, \mathfrak{B}_0 \oplus \mathfrak{A}_1).$$

By the case $k = 0$ in (3–14), α_0 is a bicontraction, and therefore $\mathfrak{A}_1, \mathfrak{B}_1$ are Hilbert spaces. By the case $k = 1$ in (3–14) and Lemma 4.2, $B_1 = \beta_0 \alpha_1 \gamma_0$, where $\alpha_1 \in \mathcal{L}(\mathfrak{A}_1, \mathfrak{B}_1)$ is a Hilbert space contraction. Choose a Julia operator

$$\begin{pmatrix} \alpha_1 & \beta_1 \\ \gamma_1 & \delta_1 \end{pmatrix} \in \mathcal{L}(\mathfrak{A}_1 \oplus \mathfrak{B}_2, \mathfrak{B}_1 \oplus \mathfrak{A}_2).$$

Since α_1 is a Hilbert space contraction, $\mathfrak{A}_2, \mathfrak{B}_2$ are Hilbert spaces. We have produced Hilbert spaces $\mathfrak{A}_1, \mathfrak{A}_2, \mathfrak{B}_1, \mathfrak{B}_2$ and the first two Julia operators in the sequence (4–5) such that the required properties hold up to this stage.

We proceed by induction. Assume that Hilbert spaces $\mathfrak{A}_1, \dots, \mathfrak{A}_{r+1}$ and $\mathfrak{B}_1, \dots, \mathfrak{B}_{r+1}$ and $r + 1$ Julia operators in the sequence (4–5) have been constructed such that the required properties hold up to this stage, that is, the coefficients $B_0, \dots, B_r$ are represented by the equations in Theorem 3.2. By hypothesis, the matrix (3–14) defines a bicontraction in the case $k = r + 1$. Thus

$$T_{r+1} = \begin{pmatrix} \begin{pmatrix} B_0 \\ B_1 \\ \vdots \\ B_{r+1} \end{pmatrix} & \begin{matrix} 0 \\ \\ T_r \end{matrix} \end{pmatrix}$$

is a bicontraction in the form that appears in Lemma 4.2. In that result, use the defect operators

$$\tilde{D}_{B_0} = \gamma_0^*,$$

$$D_{T_r} = \begin{pmatrix} D_{T_{r-1}} & 0 \\ -\beta_0 R_r L_{T_{r-1}}^* & \beta_0 \beta_1 \cdots \beta_r \end{pmatrix},$$

for B_0 and T_r^*. By Lemma 4.2,

$$(4\text{-}6) \qquad \left(\begin{pmatrix} B_1 \\ \vdots \\ B_r \\ B_{r+1} \end{pmatrix} \right) = \begin{pmatrix} D_{T_{r-1}} & 0 \\ -\beta_0 R_r L_{T_{r-1}}^* & \beta_0 \beta_1 \cdots \beta_r \end{pmatrix} \begin{pmatrix} Y_1 \\ Y_2 \end{pmatrix} \gamma_0,$$

where $\begin{pmatrix} Y_1 \\ Y_2 \end{pmatrix} \in \mathcal{L}(\mathfrak{A}_1, \mathfrak{D}_{T_{r-1}} \oplus \mathfrak{B}_{r+1})$ is a Hilbert space contraction. Thus

$$(4\text{-}7) \qquad \begin{pmatrix} B_1 \\ \vdots \\ B_r \end{pmatrix} = D_{T_{r-1}} Y_1 \gamma_0.$$

Comparing (4–7) with (3–18), we find that $Y_1 = C_r$ (because $D_{T_{r-1}}$ has zero kernel and γ_0 has dense range). By (3–6) with $k = r$, a defect operator for $Y_1 = C_r$ is given by $\tilde{D}_{C_r} = \gamma_1^* \cdots \gamma_r^*$. Hence by (I), $Y_2 = \alpha_{r+1} \tilde{D}_{C_r}^* = \alpha_{r+1} \gamma_r \cdots \gamma_1$, where $\alpha_{r+1} \in \mathfrak{L}(\mathfrak{A}_{r+1}, \mathfrak{B}_{r+1})$ is a Hilbert space contraction. Choose a Julia operator for α_{r+1} of the form (4–5) with $j = r + 1$. By (3–9) with $k = r + 1$,

$$\begin{pmatrix} Y_1 \\ Y_2 \end{pmatrix} = \begin{pmatrix} C_r \\ \alpha_{r+1}\gamma_r \cdots \gamma_1 \end{pmatrix} = C_{r+1}.$$

Thus (4–6) implies (3–18) with $k = r + 1$.

We have constructed Hilbert spaces $\mathfrak{A}_1, \ldots, \mathfrak{A}_{r+2}$ and $\mathfrak{B}_1, \ldots, \mathfrak{B}_{r+2}$ and Julia operators (4–5) for $j = 1, \ldots, r + 1$ such that $B_0, \ldots, B_{r+1}$ are represented by the formulas in Theorem 3.2. The result follows by induction. $\square$

Recall that two Julia operators

$$\begin{pmatrix} T & D \\ \tilde{D}^* & -L^* \end{pmatrix} \in \mathfrak{L}(\mathfrak{H} \oplus \mathfrak{D}, \mathfrak{K} \oplus \tilde{\mathfrak{D}}), \qquad \begin{pmatrix} T & D' \\ \tilde{D}'^* & -L'^* \end{pmatrix} \in \mathfrak{L}(\mathfrak{H} \oplus \mathfrak{D}', \mathfrak{K} \oplus \tilde{\mathfrak{D}}')$$

for the same operator $T \in \mathfrak{L}(\mathfrak{H}, \mathfrak{K})$ are considered essentially the same if

$$\begin{pmatrix} T & D' \\ \tilde{D}'^* & -L'^* \end{pmatrix} = \begin{pmatrix} 1 & 0 \\ 0 & U \end{pmatrix} \begin{pmatrix} T & D \\ \tilde{D}^* & -L^* \end{pmatrix} \begin{pmatrix} 1 & 0 \\ 0 & V^* \end{pmatrix},$$

where $U : \tilde{\mathfrak{D}} \to \tilde{\mathfrak{D}}'$ and $V : \mathfrak{D} \to \mathfrak{D}'$ are Kreĭn space isomorphisms.

THEOREM 4.4. *Let $B(z)$ and $B'(z)$ be elements of the Schur class $S(\mathfrak{A}, \mathfrak{B})$ which are generated by sequences of Julia operators*

$$(4\text{-}8) \qquad \begin{pmatrix} \alpha_n & \beta_n \\ \gamma_n & \delta_n \end{pmatrix} \in \mathfrak{L}(\mathfrak{A}_n \oplus \mathfrak{B}_{n+1}, \mathfrak{B}_n \oplus \mathfrak{A}_{n+1}), \qquad n \geq 0,$$

$$(4\text{-}9) \qquad \begin{pmatrix} \alpha'_n & \beta'_n \\ \gamma'_n & \delta'_n \end{pmatrix} \in \mathfrak{L}(\mathfrak{A}'_n \oplus \mathfrak{B}'_{n+1}, \mathfrak{B}'_n \oplus \mathfrak{A}'_{n+1}), \qquad n \geq 0,$$

as in Theorem 4.1. Then $B(z) = B'(z)$ if and only if there exist Kreĭn space isomorphisms $\varphi_j : \mathfrak{A}_j \to \mathfrak{A}'_j$ and $\psi_j : \mathfrak{B}_j \to \mathfrak{B}'_j$, $j \geq 0$, such that

(1) *φ_0, ψ_0 are identity operators and $\alpha_0 = \alpha'_0 = B_0$, and*
(2) *for all $j = 0, 1, 2, \ldots,$*

$$(4\text{-}10) \qquad \begin{pmatrix} \alpha'_j & \beta'_j \\ \gamma'_j & \delta'_j \end{pmatrix} = \begin{pmatrix} \psi_j & 0 \\ 0 & \varphi_{j+1} \end{pmatrix} \begin{pmatrix} \alpha_j & \beta_j \\ \gamma_j & \delta_j \end{pmatrix} \begin{pmatrix} \varphi_j^* & 0 \\ 0 & \psi_{j+1}^* \end{pmatrix}.$$

PROOF. By assumption, the coefficients of $B(z) = B_0 + B_1 z + B_2 z^2 + \cdots$ are derived from Theorems 3.1 and 3.2 using (4–8). Intermediate in the calculation are the operators $R_n, C_n, T_n, \ldots$. Write $B'(z) = B'_0 + B'_1 z + B'_2 z^2 + \cdots$ and $R'_n, C'_n, T'_n, \ldots$ for the corresponding quantities associated with (4–9).

SUFFICIENCY. Assume that isomorphisms $\varphi_j : \mathfrak{A}_j \to \mathfrak{A}'_j$ and $\psi_j : \mathfrak{B}_j \to \mathfrak{B}'_j$ exist for $j \geq 0$ such that (1) and (2) hold. Write

$$\Delta(u_1, \ldots, u_k) = \begin{pmatrix} u_1 & & \\ & \ddots & 0 \\ 0 & & u_k \end{pmatrix}$$

for a diagonal matrix with diagonal entries $u_1, \ldots, u_n$. Then

$$(4\text{-}11) \quad \begin{pmatrix} R'_n & D_{R'_n} \\ \tilde{D}^*_{R'_n} & -L^*_{R'_n} \end{pmatrix} = \begin{pmatrix} \psi_1 & 0 \\ 0 & \Delta(\varphi_2, \ldots, \varphi_{n+1}) \end{pmatrix} \begin{pmatrix} R_n & D_{R_n} \\ \tilde{D}^*_{R_n} & -L^*_{R_n} \end{pmatrix}$$
$$\times \begin{pmatrix} \Delta(\varphi_1, \ldots, \varphi_n)^* & 0 \\ 0 & \psi^*_{n+1} \end{pmatrix}$$

$$(4\text{-}12) \quad \begin{pmatrix} C'_n & D_{C'_n} \\ \tilde{D}^*_{C'_n} & -L^*_{C'_n} \end{pmatrix} = \begin{pmatrix} \Delta(\psi_1, \ldots, \psi_n) & 0 \\ 0 & \varphi_{n+1} \end{pmatrix} \begin{pmatrix} C_n & D_{C_n} \\ \tilde{D}^*_{C_n} & -L^*_{C_n} \end{pmatrix}$$
$$\times \begin{pmatrix} \varphi^*_1 & 0 \\ 0 & \Delta(\psi_2, \ldots, \psi_{n+1})^* \end{pmatrix}$$

for $n \geq 1$. This follows on viewing (4–10) as information relating the factors in (3–2) and (3–6) for the systems (4–8) and (4–9). On writing

$$\Phi_n = \Delta(\varphi_1, \ldots, \varphi_n), \qquad \Psi_n = \Delta(\psi_1, \ldots, \psi_n),$$

we also get

$$(4\text{-}13) \quad \begin{pmatrix} T'_n & D_{T'_n} \\ \tilde{D}^*_{T'_n} & -L^*_{T'_n} \end{pmatrix} = \begin{pmatrix} 1 & 0 \\ 0 & \Phi_{n+1} \end{pmatrix} \begin{pmatrix} T_n & D_{T_n} \\ \tilde{D}^*_{T_n} & -L^*_{T_n} \end{pmatrix} \begin{pmatrix} 1 & 0 \\ 0 & \Psi^*_{n+1} \end{pmatrix}$$

for $n \geq 0$. The proof of (4–13) is by induction. The case $n = 0$ is easily checked. Assume that (4–13) has been established up to some stage n. Then

$$T'_{n+1} = \begin{pmatrix} T'_n & 0 \\ \beta'_0 R'_{n+1} \tilde{D}^*_{T'_n} & \alpha'_0 \end{pmatrix} = \begin{pmatrix} T_n & 0 \\ \beta_0 \psi^*_1 \psi_1 R_{n+1} \Phi^*_{n+1} \Phi_{n+1} \tilde{D}^*_{T_n} & \alpha_0 \end{pmatrix}$$
$$= \begin{pmatrix} T_n & 0 \\ \beta_0 R_{n+1} \tilde{D}^*_{T_n} & \alpha_0 \end{pmatrix} = T_{n+1}.$$

In a similar way,

$$D_{T'_{n+1}} = D_{T_{n+1}} \Psi^*_{n+2},$$
$$\tilde{D}^*_{T'_{n+1}} = \Phi_{n+2} \tilde{D}^*_{T_{n+1}},$$
$$-L^*_{T'_{n+1}} = -\Phi_{n+2} L^*_{T_{n+1}} \Psi^*_{n+2},$$

which yields (4–13) with n replaced by $n+1$. This completes the inductive proof of (4–13). Sufficiency follows, because $T_n = T'_n$ for all n by (4–13).

NECESSITY. Assume now that $B(z) = B'(z)$, that is, $B_n = B'_n$ for all $n \geq 0$. Since a bicontraction has essentially unique Julia operator, there are isomorphisms $\varphi_j : \mathfrak{A}_j \to \mathfrak{A}'_j$ and $\psi_j : \mathfrak{B}_j \to \mathfrak{B}'_j$, $j = 0, 1$, such that φ_0, ψ_0 are identity operators and

$$\begin{pmatrix} \alpha'_0 & \beta'_0 \\ \gamma'_0 & \delta'_0 \end{pmatrix} = \begin{pmatrix} \psi_0 & 0 \\ 0 & \varphi_1 \end{pmatrix} \begin{pmatrix} \alpha_0 & \beta_0 \\ \gamma_0 & \delta_0 \end{pmatrix} \begin{pmatrix} \varphi^*_0 & 0 \\ 0 & \psi^*_1 \end{pmatrix}.$$

Since $B_1 = B'_1$, by (3–20),

$$\beta'_0 \alpha'_1 \gamma'_0 = \beta_0 \alpha_1 \gamma_0.$$

In this identity, observe that $\beta_0' = \beta_0 \psi_1^*$ and $\gamma_0' = \varphi_1 \gamma_0$, β_0 has zero kernel, and γ_0 has dense range. Therefore

$$\psi_1^* \alpha_1' \varphi_1 = \alpha_1.$$

Again by the essential uniqueness of the Julia operator of a bicontraction (which is now a Hilbert space contraction), there are isomorphisms $\varphi_2 : \mathfrak{A}_2 \to \mathfrak{A}_2'$ and $\psi_2 : \mathfrak{B}_2 \to \mathfrak{B}_2'$ such that

$$\begin{pmatrix} \psi_1^* & 0 \\ 0 & 1 \end{pmatrix} \begin{pmatrix} \alpha_1' & \beta_1' \\ \gamma_1' & \delta_1' \end{pmatrix} \begin{pmatrix} \varphi_1 & 0 \\ 0 & 1 \end{pmatrix} = \begin{pmatrix} 1 & 0 \\ 0 & \varphi_2 \end{pmatrix} \begin{pmatrix} \alpha_1 & \beta_1 \\ \gamma_1 & \delta_1 \end{pmatrix} \begin{pmatrix} 1 & 0 \\ 0 & \psi_2^* \end{pmatrix}.$$

This is (4–10) with $j = 1$.

Suppose that isomorphisms $\varphi_j : \mathfrak{A}_j \to \mathfrak{A}_j'$ and $\psi_j : \mathfrak{B}_j \to \mathfrak{B}_j'$ have been found for $j = 0, \dots, r+1$ such that (1) holds and (4–10) holds for $j = 0, \dots, r$. This has been verified for $j = 0, 1$. The identities (4–11), (4–12), and (4–13) hold for $n = 1, \dots, r$ by the proof of sufficiency. We shall produce isomorphisms $\varphi_{r+2} : \mathfrak{A}_{r+2} \to \mathfrak{A}_{r+2}'$ and $\psi_{r+2} : \mathfrak{B}_{r+2} \to \mathfrak{B}_{r+2}'$ such that (4–10) holds for $j = r+1$. Starting from

$$B_{r+1} = B_{r+1}'$$

and (3–21), we obtain

$$(4\text{-}14) \qquad -\beta_0' R_r' L_{T_{r-1}'}^* C_r' \gamma_0' + \beta_0' \cdots \beta_r' \alpha_{r+1}' \gamma_r' \cdots \gamma_0'$$
$$= -\beta_0 R_r L_{T_{r-1}}^* C_r \gamma_0 + \beta_0 \cdots \beta_r \alpha_{r+1} \gamma_r \cdots \gamma_0.$$

By (4–11)–(4–13),

$$-\beta_0' R_r' L_{T_{r-1}'}^* C_r' \gamma_0' = -\beta_0 \psi_1^* \psi_1 R_r \Phi_r^* \Phi_r L_{T_{r-1}}^* \Psi_r^* \Psi_r C_r \varphi_1^* \varphi_1 \gamma_0$$
$$= -\beta_0 R_r L_{T_{r-1}}^* C_r \gamma_0.$$

Our inductive hypothesis and (4–10) imply further that

$$\beta_0' \beta_1' \cdots \beta_r' = \beta_0 \beta_1 \cdots \beta_r \psi_{r+1}^*,$$
$$\gamma_r' \gamma_{r-1}' \cdots \gamma_0' = \varphi_{r+1} \gamma_r \gamma_{r-1} \cdots \gamma_0.$$

Therefore by (4–14),

$$\psi_{r+1}^* \alpha_{r+1}' \varphi_{r+1} = \alpha_{r+1}.$$

By the essential uniqueness of the Julia operator for α_{r+1}, there exist isomorphisms $\varphi_{r+2} : \mathfrak{A}_{r+2} \to \mathfrak{A}_{r+2}'$ and $\psi_{r+2} : \mathfrak{B}_j \to \mathfrak{B}_j'$ such that

$$\begin{pmatrix} \psi_{r+1}^* & 0 \\ 0 & 1 \end{pmatrix} \begin{pmatrix} \alpha_{r+1}' & \beta_{r+1}' \\ \gamma_{r+1}' & \delta_{r+1}' \end{pmatrix} \begin{pmatrix} \varphi_{r+1} & 0 \\ 0 & 1 \end{pmatrix}$$
$$= \begin{pmatrix} 1 & 0 \\ 0 & \varphi_{r+2} \end{pmatrix} \begin{pmatrix} \alpha_{r+1} & \beta_{r+1} \\ \gamma_{r+1} & \delta_{r+1} \end{pmatrix} \begin{pmatrix} 1 & 0 \\ 0 & \psi_{r+2}^* \end{pmatrix},$$

that is, such that (4–10) holds for $j = r+1$. This completes the inductive step, and necessity follows. $\square$

We are now able to prove the main results in §2. The Schur algorithm is based on Theorem 2.2.

PROOF OF THEOREM 2.2. Every $B(z)$ in $S(\mathfrak{A}, \mathfrak{B})$ has the form (2–6) for some $F(z)$ in $S(\tilde{\mathfrak{D}}_0, \mathfrak{D}_0)$. For by Theorem 4.3, $B(z)$ has the form described in Theorem 4.1 for some Julia operators

$$
(4\text{-}15) \qquad \begin{pmatrix} \alpha_0 & \beta_0 \\ \gamma_0 & \delta_0 \end{pmatrix}, \begin{pmatrix} \alpha_1 & \beta_1 \\ \gamma_1 & \delta_1 \end{pmatrix}, \begin{pmatrix} \alpha_2 & \beta_2 \\ \gamma_2 & \delta_2 \end{pmatrix}, \ldots.
$$

We may furthermore choose

$$
(4\text{-}16) \qquad \begin{pmatrix} \alpha_0 & \beta_0 \\ \gamma_0 & \delta_0 \end{pmatrix} = \begin{pmatrix} B_0 & D_0 \\ \tilde{D}_0^* & -L_0^* \end{pmatrix}.
$$

Define $F(z) \in S(\tilde{\mathfrak{D}}_0, \mathfrak{D}_0)$ by the same prescription but using the Julia operators

$$
(4\text{-}17) \qquad \begin{pmatrix} \alpha_1 & \beta_1 \\ \gamma_1 & \delta_1 \end{pmatrix}, \begin{pmatrix} \alpha_2 & \beta_2 \\ \gamma_2 & \delta_2 \end{pmatrix}, \ldots.
$$

Then (2–6) holds by Theorem 3.4.

Conversely, every $F(z) \in S(\tilde{\mathfrak{D}}_0, \mathfrak{D}_0)$ determines an element $B(z) \in S(\mathfrak{A}, \mathfrak{B})$ with $B(0) = B_0$ by (2–6). For given $F(z) \in S(\tilde{\mathfrak{D}}_0, \mathfrak{D}_0)$, we may apply Theorem 4.3 to obtain Julia operators (4–17) which represent it as above. By adjoining (4–16), we obtain a sequence (4–15) which determines an element $B(z)$ of $S(\mathfrak{A}, \mathfrak{B})$ by Theorem 4.1. This $B(z)$ is related to $F(z)$ as in (2–6) by Theorem 3.4, and $B(0) = B_0$ by construction.

Suppose that $B(z)$ is in $S(\mathfrak{A}, \mathfrak{B})$ and of the form (2–6) for two functions $F(z)$ and $F'(z)$ in $S(\tilde{\mathfrak{D}}_0, \mathfrak{D}_0)$. This gives rise to two representations of $B(z)$, one using the Julia operators (4–15) and another

$$
(4\text{-}18) \qquad \begin{pmatrix} \alpha_0 & \beta_0 \\ \gamma_0 & \delta_0 \end{pmatrix}, \begin{pmatrix} \alpha_1' & \beta_1' \\ \gamma_1' & \delta_1' \end{pmatrix}, \begin{pmatrix} \alpha_2' & \beta_2' \\ \gamma_2' & \delta_2' \end{pmatrix}, \cdots
$$

with the same first term (4–16). By Theorem 4.4, the sequences (4–15) and (4–18) are related by isomorphisms $\varphi_j : \mathfrak{A}_j \to \mathfrak{A}_j'$ and $\psi_j : \mathfrak{B}_j \to \mathfrak{B}_j'$, $j \geq 0$ (we use our usual scheme for labeling spaces). The case $j = 0$ in (4–10) implies that φ_1 is the identity operator on $\mathfrak{A}_1 = \mathfrak{A}_1' = \tilde{\mathfrak{D}}_0$, and ψ_1 is the identity operator on $\mathfrak{B}_1 = \mathfrak{B}_1' = \mathfrak{D}_0$. Another application of Theorem 4.4 then shows that $F(z) = F'(z)$. Therefore the representation (2–6) is unique. $\square$

The Schur parameters

$$
\begin{pmatrix} \alpha_n & \beta_n \\ \gamma_n & \delta_n \end{pmatrix} \in \mathfrak{L}(\mathfrak{A}_n \oplus \mathfrak{B}_{n+1}, \mathfrak{B}_n \oplus \mathfrak{A}_{n+1}), \qquad n = 0, 1, 2, \ldots,
$$

of an element $B(z)$ of $S(\mathfrak{A}, \mathfrak{B})$ may only be altered by replacement of the defect spaces by isomorphic copies: a second sequence of Julia operators

$$
\begin{pmatrix} \alpha_n' & \beta_n' \\ \gamma_n' & \delta_n' \end{pmatrix} \in \mathfrak{L}(\mathfrak{A}_n' \oplus \mathfrak{B}_{n+1}', \mathfrak{B}_n' \oplus \mathfrak{A}_{n+1}'), \qquad n = 0, 1, 2, \ldots,
$$

can occur as the Schur parameters of the same element if and only if there exist isomorphisms $\varphi_j : \mathfrak{A}_j \to \mathfrak{A}'_j$ and $\psi_j : \mathfrak{B}_j \to \mathfrak{B}'_j$, $j \geq 0$, such that

(1) φ_0, ψ_0 are identity operators and $\alpha_0 = \alpha'_0 = B_0$, and
(2) for all $j = 0, 1, 2, \ldots,$

$$\begin{pmatrix} \alpha'_j & \beta'_j \\ \gamma'_j & \delta'_j \end{pmatrix} = \begin{pmatrix} \psi_j & 0 \\ 0 & \varphi_{j+1} \end{pmatrix} \begin{pmatrix} \alpha_j & \beta_j \\ \gamma_j & \delta_j \end{pmatrix} \begin{pmatrix} \varphi_j^* & 0 \\ 0 & \psi_{j+1}^* \end{pmatrix}.$$

We have seen this condition previously, in Theorem 4.4.

DEFINITION 4.5. *By the* **choice parameters** *of an element $B(z)$ of the Schur class $S(\mathfrak{A}, \mathfrak{B})$ we mean the sequence of Julia operators associated with $B(z)$ as in Theorems 4.1 and 4.3.*

THEOREM 4.6. *The Schur parameters of an element $B(z)$ of the Schur class $S(\mathfrak{A}, \mathfrak{B})$ coincide with the choice parameters of $B(z)$.*

PROOF. Let $B(z) = B_0(z)$ be an element of $S(\mathfrak{A}, \mathfrak{B})$ generated from choice parameters (4–1) as in Theorem 4.1. Then $B_0(0) = \alpha_0$. By Theorem 2.2, the equation (2–9) has a unique solution $B_1(z)$ in $S(\mathfrak{A}_1, \mathfrak{B}_1)$. By the proof of Theorem 2.2, this solution is associated with the sequence (4–1) but shifted so as to drop the first term. Therefore $B_1(0) = \alpha_1$. Repeating this argument, we see that $B(z)$ has the Schur parameters (4–1). $\square$

PROOF OF THEOREM 2.4. Existence follows from Theorems 4.1 and 4.6, uniqueness from Theorem 4.4. $\square$

PROOF OF THEOREM 2.5. By Lemma 4.2, if (2–2) is a bicontraction, so is

$$T_k = \begin{pmatrix} B_0 & & & \\ B_1 & B_0 & & 0 \\ & \ddots & \ddots & \\ B_k & & B_1 & B_0 \end{pmatrix}$$

for every $k = 0, \ldots, r$. The inductive construction in the proof of Theorem 4.3 can be carried out up to the r-th stage. This shows that the operators $B_0, \ldots, B_r$ have the form described in §3 relative to Julia operators

$$\begin{pmatrix} \alpha_k & \beta_k \\ \gamma_k & \delta_k \end{pmatrix} \in \mathfrak{L}(\mathfrak{A}_k \oplus \mathfrak{B}_{k+1}, \mathfrak{B}_k \oplus \mathfrak{A}_{k+1}), \qquad 0 \leq k \leq r.$$

We may complete this finite sequence to an infinite sequence of Julia operators (4-1), and then the element $B(z)$ of $S(\mathfrak{A}, \mathfrak{B})$ provided by Theorem 4.1 has the required properties. $\square$

5. Concluding Remarks

(1) The choice sequence solution to the labeling problem for the commutant lifting theorem can be recast in a calculus of Julia operators similar to that described in §3.

(2) There is a generalization of the Schur class in which the condition that the operators (2–2) are bicontractions is replaced by the hypothesis that these operators together with their adjoints have the same finite negative hermitian indices (see §1). M. Dritschel has pointed out that the key factorization theorems in the proof of Theorem 4.3 can be obtained in this generality from the results in his paper [19]. This generalization makes contact with work of Kreĭn-Langer [26] (see Part I, pp. 188–189 and Theorem 6.3, p. 231). These questions are being pursued in joint work with A. Dijksma and M. Dritschel.

(3) It is tempting to look for a one-sided theory in which the operators (2–2) are simple contractions. A large class of examples is generated by the constructions in §3 by choosing Julia operators in which $\mathfrak{A}_1, \mathfrak{A}_2, \ldots$ are Hilbert spaces and $\mathfrak{B}_1, \mathfrak{B}_2, \ldots$ are Kreĭn spaces. However, there are difficulties in reversing the construction because of the failure of an adequate counterpart to Lemma 4.2. As pointed out by M. Dritschel, extra terms emerge in the case of simple contractions. The possibility that these terms might be excluded in some way remains, but at this time we do not know how to create a satisfactory one-sided theory.

References

1. D. Alpay and H. Dym, *On applications of reproducing kernel spaces to the Schur algorithm and rational J unitary factorization*, Oper. Theory: Adv. Appl., vol. 18, Birkhäuser, Basel-Boston, 1986, pp. 89–159.

2. _____, *On reproducing kernel spaces, the Schur algorithm, and interpolation in a general class of domains*, Oper. Theory: Adv. Appl., vol. 59, Birkhäuser, Basel-Boston, 1992, pp. 30–37.

3. R. Arocena, *Schur analysis of a class of translation invariant forms*, Analysis and Partial Differential Equations (C. Sadosky, ed.), Marcel Dekker, 1990, pp. 355–369.

4. _____, *Unitary extensions of isometries and contractive intertwining dilations*, Oper. Theory: Adv. Appl., vol. 41, Birkhäuser, Basel-Boston, 1989, pp. 13–23.

5. D. Z. Arov and I. V. Grossman, *Scattering matrices in the theory of extensions of isometric operators*, Dokl. Akad. Nauk. SSSR **270** (1983), 17–20; English transl. in Soviet Math. Dokl. **27** (1983), 518–522.

6. _____, *Scattering matrices in the theory of unitary extensions of isometric operators*, Math. Nachr. **157** (1992), 105–123.

7. M. Bakonyi and T. Constantinescu, *Schur's Algorithm and Several Applications*, Pitman Research Notes, vol. 261, Longman Scientific and Technical, 1992.

8. J. Bognár, *Indefinite Inner Product Spaces*, Springer-Verlag, Berlin-New York, 1974.

9. L. de Branges, *Complementation in Krein spaces*, Trans. Amer. Math. Soc. **305** (1988), 277–291.

10. _____, *A construction of Krein spaces of analytic functions*, J. Funct. Anal. **98** (1991), 1–41.

11. _____, *Square Summable Power Series*, in preparation.

12. G. Christner, *Applications of the Extension Properties of Operators on Krein Spaces*, Dissertation, University of Virginia, 1993.

13. G. Christner, Kin Y. Li, and J. Rovnyak, *Julia operators and coefficient problems*, Oper. Theory Adv. Appl., vol. 73, Birkhäuser, Basel-Boston, 1994, pp. 138–181.

14. T. Constantinescu, *On the structure of positive Toeplitz forms*, Oper. Theory: Adv. Appl., vol. 11, Birkhäuser, Basel-Boston, 1983, pp. 127–149.

15. _____, *An algorithm for the operatorial Caratheodory-Fejer problem*, Oper. Theory: Adv. Appl., vol. 14, Birkhäuser, Basel-Boston, 1984, pp. 81–107.

16. M. Cotlar and C. Sadosky, *A lifting theorem for subordinated invariant kernels*, J. Funct. Anal. **67** (1986), 345–359.

17. A. Dijksma, S. A. M. Marcantognini, and H. S. V. de Snoo, *A Schur type analysis of the minimal unitary Hilbert space extensions of a Kreĭn space isometry whose defect subspaces are Hilbert spaces*, Z. Anal. Anwendungen **13** (1994), 233–260.

18. M. A. Dritschel, *Extension Theorems for Operators on Kreĭn Spaces*, Dissertation, University of Virginia, 1989.

19. _____, *The essential uniqueness property for linear operators in Kreĭn spaces*, J. Funct. Anal. **118** (1993), 198–248.

20. M. A. Dritschel and J. Rovnyak, *Extension theorems for contraction operators on Kreĭn spaces*, Oper. Theory: Adv. Appl., vol. 47, Birkhäuser, Basel-Boston, 1990, pp. 221–305.

21. V. K. Dubovoj, B. Fritzsche, and B. Kirstein, *Matricial Version of the Classical Schur Problem*, Teubner-Texte zur Mathematik, vol. 129, B. G. Teubner Verlagsgesellschaft, Stuttgart, 1992.

22. C. Foias and A. Frazho, *The Commutant Lifting Approach to Interpolation Problems*, Oper. Theory: Adv. Appl., vol. 44, Birkhäuser, Basel-Boston, 1990.

23. I. Gohberg (ed.), *I. Schur Methods in Operator Theory and Signal Processing*, Oper. Theory: Adv. Appl., vol. 18, Birkhäuser, Basel-Boston, 1986.

24. G. Herglotz, I. Schur, et al., *Ausgewählte Arbeiten zu den Ursprüngen der Schur-Analysis* (B. Fritzsche and B. Kirstein, eds.), B. G. Teubner, Stuttgart, 1991.

25. T. Ivanchenko, *The Schur problem for J-contractive matrix functions*, Technical Report, 31/7/79, VINITI #2868–79.

26. M. G. Kreĭn and H. Langer, *Über einige Fortsetzungsprobleme, die eng mit der Theorie hermitescher Operatoren im Raume Π_κ zusammenhangen. I. Einige Funktionenklassen und ihre Darstellungen*, Math. Nachr. **77** (1977), 187–236; II. *Verallgemeinerte Resolventen, u-Resolventen und ganze Operatoren*, J. Funct. Anal. **30** (1978), 390–447.

27. V. P. Potapov, *The multiplicative structure of J-contractive matrix functions*, Trudy Moskov. Mat. Obšč. **4**, 125–236; English transl. in Amer. Math. Soc. Transl. (2) **15** (1960), 131–243.

28. J. Rovnyak, Review of *Schur's Algorithm and Several Applications* by M. Bakonyi and T. Constantinescu, Bull. Amer. Math. Soc. **30** (1994), 270–276.

29. I. Schur, *Über Potenzreihen, die im Innern des Einheitskreises beschränkt sind* I, J. Reine Angew. Math. **147** (1917), 205–232; II, ibid. **148** (1918), Gesammelte Abhandlungen, vol. II, nos. 29, 30.

30. A. Yang, *A construction of unitary linear systems*, Integral Equations and Operator Theory **19** (1994), 477–499.

MARTIN MARIETTA, P.O. BOX 8048, PHILADELPHIA, PA 19101

DEPARTMENT OF MATHEMATICS, MATHEMATICS-ASTRONOMY BUILDING, UNIVERSITY OF VIRGINIA, CHARLOTTESVILLE, VA 22903-3199

E-mail address: rovnyak@Virginia.EDU

Contemporary Mathematics
Volume **189**, 1995

Weighted H^p Spaces
for One Sided Maximal Functions

LILIANA DE ROSA AND CARLOS SEGOVIA

ABSTRACT. The purpose of this paper is to present a maximal theory for
weighted Hardy spaces on the real line with weights belonging to the class
A_q^+ introduced by E. Sawyer in [8]. Since we work on the real line we use
the method developed by R. Coifman in [1] in order to obtain decompositions into atoms for the non-weighted case. García-Cuerva, also using
R. Coifman's method, obtained in [2] decompositions into atoms for the
weighted Hardy spaces with weights belonging to the class A_q of B. Muckenhoupt. We work on the real line for two reasons: the first one is that
no-characterizations are known for the weights that preserve weighted L^q
spaces for the possible generalizations of the one sided Hardy-Littlewood
maximal function to higher dimension; the second one is that there are
difficulties of a geometric nature that we could overcome in dimension one
only. It is crucial in the theory of the Hardy spaces to be able to approximate distributions by functions. In order to get such approximation we
modify an argument due to R.A. Macías and C. Segovia, see [6].

1. Notations and Definitions

As usual, $C_0^\infty(\mathbb{R})$ is the set of all functions with compact support having
derivatives of all orders. Let r, $-\infty \le r < \infty$, we shall denote by $\mathcal{D}(r, \infty)$ the
space of all $C_0^\infty(\mathbb{R})$-functions with support contained in (r, ∞) equipped with
the usual topology and by $\mathcal{D}'(r, \infty)$ the space of distributions on (r, ∞).

Given an integer $\gamma \ge 1$ and $x \in \mathbb{R}$, we shall say that a $C_0^\infty(\mathbb{R})$-function $\psi(t)$
belongs to the class $\Phi_\gamma(x)$ if there exists a bounded interval $I_\psi = [x, \beta]$ containing
the support of $\psi(t)$ such that $D^\gamma \psi(t)$ satisfies

$$(1.1) \qquad |I_\psi|^{\gamma+1} \|D^\gamma \psi\|_\infty \le 1.$$

1991 *Mathematics Subject Classification*. Primary 42B30, Secondary 42B25.

The authors were supported by a grant from Consejo Nacional de Investigaciones Científicas
y Técnicas, Argentina

Let F be a distribution in $\mathcal{D}'(r,\infty)$. We define the one-sided maximal function $F_{+,\gamma}^*(x)$ as

$$(1.2) \qquad F_{+,\gamma}^*(x) = \sup\{|\langle F,\psi\rangle| : \psi \in \Phi_\gamma(x)\},$$

for every $x > r$. We observe that $F_{+,\gamma}^*(x)$ is a lower semicontinuous function.

Given a Lebesgue measurable set $E \subseteq \mathbb{R}$, we denote its Lebesgue measure by $|E|$ and the characteristic function of E by $\chi_E(x)$.

Let $f(x)$ be a measurable function defined on $\mathbb{R}$. The one-sided Hardy- Littlewood maximal functions $M^-f(x)$ and $M^+f(x)$ are

$$(1.3) \qquad M^-f(x) = \sup_{h>0} \frac{1}{h} \int_{x-h}^{x} |f(t)|dt \text{ and } M^+f(x) = \sup_{h>0} \frac{1}{h} \int_{x}^{x+h} |f(t)|dt.$$

As usual, a weight $w(x)$ is a measurable and non-negative function. If $E \subseteq \mathbb{R}$ is a measurable set, we denote its w-measure by $w(E) = \int_E w(t)dt$. A function $f(x)$ belongs to $L^s(w)$, $0 < s \leq \infty$, if $\|f\|_{L^s(w)} = \left(\int_{-\infty}^{\infty} |f(t)|^s w(t)dt\right)^{1/s}$ is finite. If $w(x) \equiv 1$, we simply write L^s.

A weight $w(x)$ belongs to the class A_q^+, $1 \leq q < \infty$, defined by E. Sawyer in [8], if there exists a constant c such that

$$\sup_{h>0}\left(\frac{1}{h}\int_{x-h}^{x} w(t)dt\right)\left(\frac{1}{h}\int_{x}^{x+h} w(t)^{-\frac{1}{q-1}}dt\right)^{q-1} \leq c\ ,$$

for all real number x. We observe that $w(x)$ belongs to A_1^+ if $M^-w(x) \leq cw(x)$ holds for almost every x.

Given $w(x)$ belonging to A_q^+, $1 \leq q < \infty$, we can define $x_{-\infty} \geq -\infty$ and $x_\infty \leq \infty$, such that

(i) $w(x) \equiv 0$ \qquad in $(-\infty, x_{-\infty})$,

(ii) $w(x) \equiv \infty$ \qquad in (x_∞, ∞) and

(iii) $0 < w(x) < \infty$ for almost every $x \in (x_{-\infty}, x_\infty)$.

We always have $x_{-\infty} \leq x_\infty$. In order to avoid the non-interesting case of $x_{-\infty} = x_\infty$, it is necessary and sufficient to assume that there exists a measurable set E satisfying $0 < w(E) < \infty$.

Let $w(x) \in A_q^+$ and $0 < p \leq 1$. For every integer $\gamma \geq 1$ satisfying $(\gamma + 1)p \geq q > 1$ or $(\gamma+1)p > 1$ if $q = 1$, we shall say that the distribution F in $\mathcal{D}'(x_{-\infty}, \infty)$, belongs to $H_{+,\gamma}^p(w)$ if the "p-norm"

$$\|F\|_{H_{+,\gamma}^p(w)} = \left(\int_{x_{-\infty}}^{\infty} F_{+,\gamma}^*(x)^p w(x)dx\right)^{1/p},$$

is finite.

A function $a(x)$ defined on $\mathbb{R}$ is called a p-atom with respect to $w(x)$ if there exists an interval I (not necessarily bounded) containing the support of $a(x)$, such that

(1.4)

    (i) I is contained in $(x_{-\infty}, \infty)$ and $w(I) < \infty$,

    (ii) $\|a\|_\infty \leq w(I)^{-1/p}$ and

    (iii) if the length of I is less than the distance $d(x_{-\infty}, I)$ from $x_{-\infty}$ to I, then

$$\int_I a(x) x^k dx = 0,$$

holds for every integer k, $0 \leq k < \gamma$.

We shall say that I is the interval associated to the atom $a(x)$.

2. Statement of the main results

With the notations and definitions given in 1. we can state the main results of this paper.

THEOREM 2.1. (Approximation of distributions in $H^p_{+,\gamma}(w)$ by functions). *Let $w(x) \in A^+_q$, $\gamma \geq 1$ be an integer and $0 < p \leq 1$ such that $(\gamma+1)p \geq q > 1$ or $(\gamma+1)p > 1$ if $q = 1$. Then, if $F \in H^p_{+,\gamma}(w)$, for any given $\varepsilon > 0$, there exists a function $g(x)$ locally integrable in $(x_{-\infty}, \infty)$ such that*

$$\int_{x_{-\infty}}^\infty (F-g)^*_{+,\gamma}(x)^p w(x) dx < \varepsilon .$$

THEOREM 2.2. (Decomposition into atoms). *Let $w(x) \in A^+_q$, $\gamma \geq 1$ be an integer and $0 < p \leq 1$ such that $(\gamma+1)p \geq q > 1$ or $(\gamma+1)p > 1$ if $q = 1$. Then, if F belongs to $H^p_{+,\gamma}(w)$, there exists a sequence $\{a_k(x)\}$ of p-atoms with respect to $w(x)$ and a sequence $\{\lambda_k\}$ of real numbers such that*

$$F = \sum \lambda_k a_k(x) \text{ in } \mathcal{D}'(x_{-\infty}, \infty) ,$$

and

(2.3) $$c'_p \|F\|^p_{H^p_{+,\gamma}(w)} \leq \sum |\lambda_k|^p \leq c_p \|F\|^p_{H^p_{+,\gamma}(w)} ,$$

holds.

We observe that in the case of $x_{-\infty} = -\infty$, if the interval I involved in (1.4) is bounded then, by part (iii) of (1.4), it follows that the p-atom has null moments up to the order $\gamma - 1$.

THEOREM 2.4. *Under the hypotheses of Theorem (2.2) and if, in addition, we assume that $x_{-\infty} = -\infty$, then the p-atoms $\{a_k(x)\}$ in the decomposition can be taken in such that way that the corresponding intervals I_k in (1.4) are bounded* *and therefore all the p-atoms in the decomposition have null moments up to the order $\gamma - 1$.*

In the sequel, we shall assume that γ is given and we shall omit γ in the notation of $\Phi_\gamma(x)$, $F^*_{+,\gamma}(x)$ and $H^p_{+,\gamma}(w)$.

3. Basic lemmas

The next lemma contains the results on weights that shall be needed.

LEMMA 3.1.

(i) *Let $1 \leq q_1 < q_2 < \infty$. If the weight $w(x)$ belongs to the class $A^+_{q_1}$, then it also belongs to $A^+_{q_2}$.*

(ii) *Let $1 < q < \infty$. The one-sided Hardy-Littlewood maximal M^+ is bounded on $L^q(w)$ if and only if $w(x)$ belongs to A^+_q.*

(iii) *Given $w(x) \in A^+_q$, $1 \leq q < \infty$, for every $a \in \mathbb{R}$, the w-measure of the interval (a, ∞) is infinite.*

The proof of parts (i) and (iii) are very simple and shall be omitted. Part (ii) of this lemma is contained in Theorem 1, page 54 of [8].

LEMMA 3.2. *Let $w(x) \in A^+_q$, $\gamma \geq 1$ an integer and $0 < p \leq 1$ such that $(\gamma + 1)p \geq q > 1$ or $(\gamma + 1)p > 1$ if $q = 1$. Let $f(x) \in L^\infty$ with support contained in an interval $I \subset (x_{-\infty}, \infty)$. If $|I| < d(x_{-\infty}, I)$ we assume that $\int_{-\infty}^{\infty} f(x) x^k dx = 0$ holds for every k, $0 \leq k < \gamma$. Then for any $x > x_{-\infty}$, we have*

$$(3.3) \qquad f^*_+(x) \leq c_\gamma \|f\|_\infty [M^+ \chi_I(x)]^{\gamma+1}.$$

Moreover,

$$(3.4) \qquad \|f\|_{H^p_+(w)} \leq c_{\gamma,w} \|f\|_\infty w(I)^{1/p}.$$

The constants c_γ and $c_{\gamma,w}$ do not depend on $f(x)$.

PROOF. Let $\alpha < \beta$ be the end points of I. If $\max(x_{-\infty}, \alpha - |I|) \leq x$, by (1.2) and (1.3) we have that $f^*_+(x) \leq M^+ f(x) \leq \|f\|_\infty M^+ \chi_I(x)$. Since $\frac{1}{2} \leq M^+ \chi_I(x)$, we get (3.3). If there exists x such that $x_{-\infty} < x < \alpha - |I|$, then $|I| < d(x_{-\infty}, I)$. Let $\psi(t) \in \Phi(x)$ and I_ψ the interval associated with $\psi(t)$ in this class. We have

$$|\langle f, \psi \rangle| = \left| \int_I f(t) \left[\psi(t) - \sum_{s=0}^{\gamma-1} \frac{D^s \psi(\alpha)}{s!} (t - \alpha)^s \right] dt \right|$$

$$\leq \frac{1}{\gamma!} \|D^\gamma \psi\|_\infty |I|^\gamma \int_I |f(t)| dt.$$

We may assume $I \cap I_\psi \neq \emptyset$, then $\alpha - x \leq |I_\psi|$ and applying (1.1), we get

$$|\langle f, \psi \rangle| \leq c_\gamma \|f\|_\infty \left(\frac{|I|}{\alpha - x} \right)^{\gamma+1} \leq 2^\gamma c_\gamma \|f\|_\infty [M^+ \chi_I(x)]^{\gamma+1},$$

showing that (3.3) holds. The inequality (3.4) is an immediate consequence of parts (i) and (ii) of Lemma (3.1). ∎

COROLLARY 3.5. *Let $a(x)$ be a p-atom with respect to $w(x)$. Then*

$$\|a\|_{H_+^p(w)} \le c_{\gamma,w},$$

where the constant $c_{\gamma,w}$ is finite and independent of $a(x)$.

PROOF. It follows by (3.4) of Lemma (3.2) and (1.4)(ii). ∎

LEMMA 3.6. *Let $\gamma \ge 1$. If $\psi(t)$ belongs to $\Phi_\gamma(x)$ and I_ψ is the interval associated with $\psi(t)$ in this class, then for every s, $0 \le s \le \gamma$, we have*

$$|I_\psi|^{s+1}\|D^s\psi\|_\infty \le 1 .$$

PROOF. This is an immediate consequence of (1.1) and the formula

$$D^s\psi(y) = \int_x^y \frac{(y-t)^{\gamma-s-1}}{(\gamma-s-1)!} D^\gamma\psi(t)\, dt ,$$

that holds for all $y \ge x$ and $0 \le s \le \gamma-1$. ∎

LEMMA 3.7. (Completeness of $H_{+,\gamma}^p(w)$). *Let $w(x) \in A_q^+$, $\gamma \ge 1$ be an integer and $0 < p \le 1$ such that $(\gamma+1)p \ge q > 1$ or $(\gamma+1)p > 1$ if $q = 1$. Let $\{F_n\}$ be a sequence in $H_+^p(w)$ such that for every $\varepsilon > 0$ there exists N_ε satisfying*

$$\|F_n - F_m\|_{H_+^p(w)}^p < \varepsilon ,$$

for every $n, m \ge N_\varepsilon$. Then, there exists $F \in H_+^p(w)$ such that

$$\lim_{n\to\infty} \|F_n - F\|_{H_+^p(w)}^p = 0 .$$

The proof is simple and shall be omitted. ∎

LEMMA 3.8. (Fefferman-Stein type lemma). *Let $w(x) \ge 0$. For every s, $1 < s \le \infty$, there exists a constant c_s such that*

$$\int_{-\infty}^\infty M^+f(x)^s\, w(x)dx \le c_s \int_{-\infty}^\infty |f(x)|^s M^- w(x)dx ,$$

holds.

The proof is essentially that of the (1,1)-weak type for the maximal operator M^+ (see [5]).

LEMMA 3.9. *Let $1 < q < \infty$ and $w(x) \in A_q^+$. If $\{I_k\}$ is a sequence of pairwise disjoint intervals then, for $1 < s < \infty$,*

$$\int \left(\sum [M^+\chi_{I_k}(x)]^s\right)^q w(x)dx \le c_{s,q,w} w(\cup I_k) ,$$

where $c_{s,q,w}$ is finite and does not depend on the sequence $\{I_k\}$.

The proof of this lemma depends on Lemma (3.8) and it is essentially that of Theorem 6.1, page 519 of [3].

LEMMA 3.10. *Let $\gamma \geq 1$ and let G belong to $\mathcal{D}'(r, \infty)$, where $-\infty \leq r < \infty$. If $G_+^*(x)$ is locally integrable on (r, ∞) then there exists a locally integrable function $g(x)$ such that*

$$\langle G, \psi \rangle = \int g(x)\psi(x)dx \; ,$$

holds for every $\psi(x) \in \mathcal{D}(r, \infty)$.

PROOF. Let $\rho(x) \geq 0$ be a $C_0^\infty(\mathbb{R})$-function with support contained in $[0, 1]$ and $\int \rho(x)dx = 1$. We define $\varphi(x) = \int_{x-1}^x \rho(t)dt$. Then for $\varepsilon > 0$ the functions $\varphi_{\varepsilon,k}(x) = \varphi\left(\frac{x-\varepsilon k}{\varepsilon}\right)$, $k \in \mathbb{Z}$, are a partition of the unit in $\mathbb{R}$, i.e. $\sum_k \varphi_{\varepsilon,k}(x) \equiv 1$. We observe that the support of $\varphi_{\varepsilon,k}(x)$ is contained in the interval $I_{\varepsilon,k} = [\varepsilon k, \varepsilon(k+2)]$ for every $k \in \mathbb{Z}$.

Let $\psi(x) \in D(r, \infty)$ with support contained in $[a, b]$, where $r < a$. If ε is small enough, then $[a, b]$ is included in a union of a finite number of $I_{k,\varepsilon} \subset (r, \infty)$. Then

$$(3.11) \qquad \langle G, \psi \rangle = \sum_k \langle G, \psi \varphi_{\varepsilon,k} \rangle \; ,$$

where the sum has a finite number of terms different from zero. Let us estimate $|\langle G, \psi \varphi_{\varepsilon,k} \rangle|$. If $y \in I_{\varepsilon,k-2}$ and $I = [y, \varepsilon(k+2)]$ then $\text{supp}\,(\psi \varphi_{\varepsilon,k}) \subset I$. We observe that $|I| \leq 4\varepsilon$. By Leibnitz formula $\|D^\gamma(\psi \varphi_{\varepsilon,k})\|_\infty$ is less than or equal to a constant times a sum of terms of the form $\|D^\alpha \psi D^\beta \varphi_{\varepsilon,k}\|_\infty$, $\alpha + \beta = \gamma$.

If $\alpha = 0$, we have

$$\|\psi D^\gamma \varphi_{\varepsilon,k}\|_\infty \leq \big(\sup_{x \in I_{\varepsilon,k}} |\psi(x)|\big)\varepsilon^{-\gamma}\|D^\gamma \varphi\|_\infty \; .$$

By the mean value theorem $|\psi(x)| \leq |\psi(y)| + \|\psi'\|_\infty 4\varepsilon$, whenever $x \in I_{\varepsilon,k}$ and $y \in I_{\varepsilon,k-2}$. Then

$$|I|^{\gamma+1}\|\psi D^\gamma \varphi_{\varepsilon,k}\|_\infty \leq c\varepsilon|\psi(y)| \, \|D^\gamma \varphi\|_\infty + c\varepsilon^2\|\psi'\|_\infty\|D^\gamma \varphi\|_\infty \; .$$

If $\alpha > 0$, we get

$$|I|^{\gamma+1}\|D^\alpha \psi D^\beta \varphi_{\varepsilon,k}\|_\infty \leq c\varepsilon^{\gamma+1-\beta}\|D^\alpha \psi\|_\infty\|D^\beta \varphi\|_\infty \; .$$

Therefore $|I|^{\gamma+1}\|D^\gamma(\psi \varphi_{\varepsilon,k})\|_\infty \leq c\varepsilon|\psi(y)| \, \|D^\gamma \varphi\|_\infty + O(\varepsilon^2)$. Thus, by (1.2)

$$|\langle G, \psi \varphi_{\varepsilon,k} \rangle| \leq c_\gamma G_+^*(y)(\varepsilon|\psi(y)| + O(\varepsilon^2)) \; .$$

Taking the average on $I_{\varepsilon,k-2}$, we have

$$|\langle G, \psi \varphi_{\varepsilon,k} \rangle| \leq c_\gamma \Big(\int_{I_{\varepsilon,k-2}} G_+^*(y)|\psi(y)|dy + \int_{I_{\varepsilon,k-2}} G_+^*(y)dy \cdot O(\varepsilon) \Big) \; .$$

Then, by (3.11)

$$|\langle G, \psi \rangle| \leq c_\gamma \Big(\int G_+^*(y)|\psi(y)|dy + \int_{a-2\varepsilon}^b G_+^*(y)dy \cdot O(\varepsilon) \Big) \; .$$

Letting ε go to zero we obtain

$$|\langle G, \psi \rangle| \leq c_\gamma \int G_+^*(y)|\psi(y)|dy .$$

Thus,

$$|\langle G, \psi \rangle| \leq c_\gamma \|\psi\|_\infty \int_a^b G_+^*(y)dy .$$

We call $\widetilde{G}$ the extension of G to the continuous functions in $[a, b]$ that vanish at a and b. By continuity we get that

$$|\widetilde{G}(g)| \leq c_\gamma \int_a^b G_+^*(y)|g(y)|dy ,$$

holds for every continuous function. Let P and N be the positive and negative parts of $\widetilde{G}$. Then

$$|P(g)| \leq c_\gamma \int_a^b G_+^*(y)|g(y)|dy .$$

Since P is a positive integral it follows that

$$|P(\chi_E)| \leq c_\gamma \int_E G_+^*(y)dy ,$$

holds for any open set $E \subset [a, b]$. Now by the absolute continuity of the integral we get P is absolutely continuous with respect to Lebesgue measure. Similar arguments for the negative part N and the Radon-Nikodym Theorem imply the lemma. ∎

4. Proof of Theorem (2.1). (Approximation of distributions in $H_{+,\gamma}^p(w)$ by functions)

Let

(4.1) $$h(y) = \int_{y/2}^y \rho(t)dt ,$$

where $\rho(t)$ is a non-negative and $C_0^\infty(\mathbb{R})$-function supported on $[1, 2]$. Moreover, we assume that $\int \rho(t)dt = 1$. The function $h(y)$ just defined has the following properties:

(i) $0 \leq h(y) \leq 1$ and $h(y)$ belongs to C^∞,

(ii) the support of $h(y)$ is contained in the interval $[1, 4]$ and

(iii) the function $H(y) = \sum_{j=1}^\infty h(2^j y)$ takes the following values: $H(y) = 0$ if $y \leq 0$ or $y \geq 2$, $H(y) = 1$ if $y \in (0, 1]$ and $H(y) = h(2y)$ when $y \in [1, 2]$.

For $a < b$, and each integer k, let

$$\nu_k(y) = h\Big(\frac{y - b}{2^{-k}(b - a)}\Big) ,$$

where $h(y)$ is the function defined in (4.1). The support of $\nu_k(y)$ is contained in the interval $[b + 2^{-k}(b-a),\ b + 2^{-k+2}(b-a)]$ and by the properties of $h(y)$, we have

$$(4.2) \qquad \sum_{k \in \mathbb{Z}} \nu_k(y) = \chi_{(b,\infty)}(y) \ .$$

LEMMA 4.3. *Let $\gamma \geq 1$ and F belong to $\mathcal{D}'(r,\infty)$, where $-\infty \leq r < a < b$. Then, if $\psi(y)$ belongs to $\mathcal{D}(r,\infty)$, with support contained in $(r,\beta]$, we get*

(i) $\quad |\langle F, \nu_k \psi \rangle| \leq \sum\limits_{s=0}^{\gamma} c_{s,h}\, [2^{-k}(b-a)]^{s+1}\, \|D^s \psi\|_\infty\, F_+^*(b)$.

(ii) $\quad |\sum\limits_{k \in \mathbb{Z}} \langle F, \nu_k \psi \rangle| \leq \sum\limits_{s=0}^{\gamma} c_{s,h}\, (\beta - b)^{s+1}\, \|D^s \psi\|_\infty\, F_+^*(b)$, *holds when $b < \beta$*
$\quad$ *and $\sum_{k \in \mathbb{Z}} \langle F, \nu_k \psi \rangle = 0$, if $\beta \leq b$* .

PROOF. We observe that the support of $\nu_k(y)\psi(y)$ is contained in $J_k = [b, b + 2^{-k+2}(b-a)]$. Therefore

$$|J_k|^{\gamma+1} \|D^\gamma(\nu_k \psi)\|_\infty \leq \sum_{s=0}^{\gamma} c_{s,h} [2^{-k}(b-a)]^{s+1} \|D^s \psi\|_\infty,$$

which, by (1.2), implies (i). Let us prove (ii). If $\beta \leq b$, then $\nu_k(y)\psi(y) \equiv 0$ for all integer k. We assume $b < \beta$ and let k_0 be the unique integer such that

$$b + 2^{-k_0-1}(b-a) \leq \beta < b + 2^{-k_0}(b-a).$$

For $k \leq k_0$, we have $\beta \leq b + 2^{-k}(b-a)$. Then, since the support of $\nu_k(y)$ is contained in $[b + 2^{-k}(b-a),\ b + 2^{-k+2}(b-a)]$, it follows that $\nu_k(y)\psi(y) \equiv 0$. Thus

$$\sum_{k \in \mathbb{Z}} |\langle F, \nu_k \psi \rangle| = \sum_{k > k_0} |\langle F, \nu_k \psi \rangle| \ ,$$

and by part (i) already proved, the last sum is mayorized by

$$\sum_{s=0}^{\gamma} c_{s,h} [2^{-k_0}(b-a)]^{s+1} \|D^s \psi\|_\infty F_+^*(b) \leq \sum_{s=0}^{\gamma} c_{s,h}'(\beta-b)^{s+1} \|D^s \psi\|_\infty F_+^*(b),$$

which proves part (ii) holds. ■

We observe that part (ii) of Lemma (4.3) we just proved allows us to define a distribution F_b in $\mathcal{D}'(r,\infty)$ as

$$(4.4) \qquad \langle F_b, \psi \rangle = \sum_{k \in \mathbb{Z}} \langle F, \nu_k \psi \rangle \ ,$$

provided that $F_+^*(b)$ is finite. This distribution can be thought as the product of the characteristic function of the interval (b,∞) times F.

For $a < b$ and $j \geq 1$, we define

$$\eta_j(y) = h\left(\frac{y-a}{2^{-j}(b-a)} \right) ,$$

where $h(y)$ is the function defined in (4.1). Taking into account the properties of $h(y)$, we get:

(i) $0 \leq \eta_j(y) \leq 1$ and $\eta_j(y)$ belongs to C^∞,

(ii) the support of $\eta_j(y)$ is contained in the interval $N_j = [a + 2^{-j}(b-a), \ a + 2^{-j+2}(b-a)]$ and

(iii) the function $H_{a,b}(y) = \sum_{j=1}^{\infty} \eta_j(y)$ takes the following values: $H_{a,b}(y) = 0$ if $y \leq a$ or $y \geq b + (b-a)$, $H_{a,b}(y) = 1$ if $y \in (a, b]$ and $H_{a,b}(y) = \eta_1(y)$ when $y \in [b, b + (b-a)]$.

If $j > 1$, let P_j be the orthogonal projection of $L^2(\eta_j(y))$ onto the subspace S generated by $1, x, \ldots, x^{\gamma-1}$. Let $\{e_k(y)\}_{k=0}^{\gamma-1}$ be an orthonormal basis of $S \subset L^2(h(y))$. In the sequel, we shall denote for $j > 1$ and $0 \leq k \leq \gamma - 1$,

$$e_k^j(y) = \frac{1}{[2^{-j}(b-a)]^{1/2}} e_k\left(\frac{y-a}{2^{-j}(b-a)}\right).$$

We observe that $\{e_k^j(y)\}_{k=0}^{\gamma-1}$ is an orthonormal basis of $S \subset L^2(\eta_j(y))$.

If $j = 1$, let $\overline{\eta}_1(y) = \eta_1(y)\chi_{(a,b)}(y)$ and let P_1 be the orthogonal projection of $L^2(\overline{\eta}_1(y))$ onto S as a subspace of $L^2(\overline{\eta}_1(y))$. Defining $\overline{h}(y) = h(y)\chi_{(-\infty,2)}(y)$, then $\overline{\eta}_1(y) = \overline{h}\left(\frac{y-a}{2^{-1}(b-a)}\right)$. Given an orthonormal basis $\{\overline{e}_k(y)\}_{k=0}^{\gamma-1}$ of $S \subset L^2(\overline{h}(y))$, we define the polynomials

$$e_k^1(y) = \frac{1}{[2^{-1}(b-a)]^{1/2}} \overline{e}_k\left(\frac{y-a}{2^{-j}(b-a)}\right), \qquad 0 \leq k \leq \gamma - 1.$$

These polynomials form an orthonormal basis of $S \subset L^2(\overline{\eta}_1(y))$.

With these notations, for every $j \geq 1$, we have

(4.5)
$$P_j(\psi)(t) = \sum_{k=0}^{\gamma-1}\left(\int_a^b \psi(s)e_k^j(s)\eta_j(s)ds\right)e_k^j(t), \quad \text{and}$$

$$\sup_{s \in N_j} |e_k^j(s)| \leq c_{h,\gamma}[2^{-j}(b-a)]^{1/2}.$$

We observe that

(4.6)
$$\psi(y)\eta_j(y)\chi_{(a,b)}(y) \equiv 0 \text{ implies } P_j(\psi)(t) \equiv 0.$$

In the following lemma we state some properties of the projection P_j that will be needed.

LEMMA 4.7. Let $0 \leq s \leq \gamma - 1$ and $N_j = [a + 2^{-j}(b-a), \ a + 2^{-j+2}(b-a)]$. We have:

(i) If $\psi(y) \in L^2(\eta_j(y))$ and I_ψ is an interval containing the support if $\psi(y)$, then

$$\sup_{t \in N_j} |D^s[P_j(\psi)](t)| \leq \frac{c_{\gamma,h}}{[2^{-j}(b-a)]^{s+1}}\|\psi\|_\infty|I_\psi|.$$

(ii) *If $\psi(y) \in C^\gamma$, then*

$$\sup_{t \in N_j} |D^s[\psi - P_j(\psi)](t)| \leq c_{\gamma,h} \|D^\gamma \psi\|_\infty [2^{-j}(b-a)]^{\gamma-s}.$$

(iii) *The support of $e_k^j(y)\eta_j(y)$ is contained in the interval $\widetilde{N}_j = [a, a + 2^{-j+2}(b-a)]$ and*

$$|\widetilde{N}_j|^{\gamma+1} \|D^\gamma(e_k^j \eta_j)\|_\infty \leq c_{\gamma,h,k}[2^{-j}(b-a)]^{1/2}.$$

The proof of this lemma is a routine computation and shall be omitted.

Let r, a and b be such that $-\infty \leq r < a < b < \infty$. If $j \geq 1$, let P_j be the projections as in Lemma (4.7) . For any function $\psi(t)$ belonging to $\mathcal{D}(r, \infty)$ we define

$$(4.8) \qquad S_j(\psi)(t) = [\psi - P_j(\psi)](t)\eta_j(t).$$

We observe that $S_j(\psi)(t)$ also belongs to $\mathcal{D}(r, \infty)$ and its support is contained in the interval $N_j = [a + 2^{-j}(b-a), a + 2^{-j+2}(b-a)]$ which contains the support of $\eta_j(t)$.

Let F be a distribution in $\mathcal{D}'(r, \infty)$ such that $F_+^*(b)$ is finite. In the sequel, for $j > 1$, W_j shall be the distribution defined as

$$(4.9) \qquad \langle W_j, \psi \rangle = \langle F, S_j(\psi) \rangle,$$

and for $j = 1$

$$(4.10) \qquad \langle W_1, \psi \rangle = \langle F, S_1(\psi) \rangle - \langle F_b, S_1(\psi) \rangle,$$

where F_b is the distribution defined in (4.4). Since the support of $S_1(\psi)(t)$ is contained in $N_1 = [a + 2^{-1}(b-a), b + (b-a)]$ and the interval $[b + 2^{-k}(b-a), b + 2^{-k+2}(b-a)]$ includes the support of $\nu_k(t)$, if $k \leq 0$ then $\nu_k(t)S_1(\psi)(t)$ is identically zero. Thus, by (4.4) we have

$$(4.11) \qquad \langle F_b, S_1(\psi) \rangle = \sum_{k=1}^{\infty} \langle F, \nu_k S_1(\psi) \rangle.$$

With these notations we have the following proposition.

PROPOSITION 4.12. *Let $\gamma \geq 1$ and F belong to $\mathcal{D}'(r, \infty)$, where $-\infty \leq r < \infty$. Let a and b be such that $r < a < b < \infty$ We shall assume that $F_+^*(b)$ is finite. If $I_j = [a + 2^{-j-1}(b-a), \min\{a + 2^{-j+2}(b-a), b\})$, then*

$$(W_j)_+^*(x) \leq c_{\gamma,h}\{F_+^*(x)\chi_{I_j}(x) + [F_+^*(a) + F_+^*(b)][M^+\chi_{I_j}(x)]^{\gamma+1}\},$$

where the constant $c_{\gamma,h}$ does not depend on a, b and j.

PROOF. Let $x > r$ and let $\psi(y)$ be a function belonging to $\Phi(x)$, with $I_\psi = [x, \beta]$ the interval associated with $\psi(t)$ in this class. First of all, we shall estimate the absolute value of $\langle F, S_j(\psi) \rangle$. This estimation shall be done by considering three cases.

Case 1. $a + 2^{-j+2}(b-a) \leq x$. Since the support of $\eta_j(x)$ is contained in $N_j = [a + 2^{-j}(b-a), a + 2^{-j+2}(b-a)]$ and the interval $[a + 2^{-j+2}(b-a), \infty)$ includes the support of $\psi(y)$, the function $\psi(y)\eta_j(y)$ is identically zero. Then, by (4.6), $P_j(\psi)(t)$ is the null polynomial and by (4.8) we have $S_j(\psi)(t) \equiv 0$. Therefore,

$$(4.13) \qquad \langle F, S_j(\psi) \rangle = 0.$$

Case 2. $a + 2^{-j-1}(b-a) < x < a + 2^{-j+2}(b-a)$. Taking into account that the support of $\eta_j(y)$ is contained in $N_j = [a + 2^{-j}(b-a), a + 2^{-j+2}(b-a)]$ and $I_\psi = [x, \beta]$ contains the support of $\psi(y)$, if $\beta \leq a + 2^{-j}(b-a)$ arguing as in Case 1, we have $S_j(\psi)(t) \equiv 0$. Then, we can assume that

$$(4.14) \qquad a + 2^{-j}(b-a) < \beta.$$

Now, let $\beta_j = \min(\beta, a + 2^{-j+2}(b-a))$ and let $J_j = [x, \beta_j]$. The support of $\psi(y)\eta_j(y)$ is contained in the intersection of I_ψ with N_j and the interval J_j contains this intersection. Besides, the inequality $\beta_j \leq \beta$ implies the inclusion: $J_j \subset I_\psi$. Then, by Lemma (3.6), we get

$$|J_j|^{\gamma+1}\|D^\gamma(\psi\eta_j)\|_\infty \leq \sum_{s=0}^{\gamma} c_s |J_j|^s \|D^s\eta_j\|_\infty.$$

By (4.14), we have that $|J_j| \leq (7/2)2^{-j}(b-a)$. Then,

$$|J_j|^{\gamma+1}\|D^\gamma(\psi\eta_j)\|_\infty \leq \sum_{s=0}^{\gamma} c_s' \|D^s h\|_\infty = c_{\gamma,h}.$$

This shows that $c_{\gamma,h}^{-1}\psi(y)\eta_j(y)$ belongs to the class $\Phi(x)$. Therefore

$$(4.15) \qquad |\langle F, \psi\eta_j \rangle| \leq c_{\gamma,h} F_+^*(x).$$

Applying part (iii) of Lemma (4.7), we get

$$|\langle F, e_k^j \eta_j \rangle| \leq c_{\gamma,h,k}[2^{-j}(b-a)]^{1/2} F_+^*(a),$$

and taking into account that $\|\eta_j\|_\infty \leq 1$ and (4.5), we have

$$|\langle F, P_j(\psi)\eta_j \rangle| \leq \sum_{k=0}^{\gamma-1} \left(\int_a^b |\psi(s)||e_k^j(s)|\eta_j(s)ds \right) |\langle F, e_k^j \eta_j \rangle|$$
$$\leq c_{\gamma,h}\|\psi\|_\infty |I_\psi| F_+^*(a).$$

Now, Lemma (3.6), implies that

$$|\langle F, P_j(\psi)\eta_j \rangle| \leq c_{\gamma,h} F_+^*(a).$$

By this inequality, (4.15) and (4.8), we have

$$(4.16) \qquad |\langle F, S_j(\psi) \rangle| \leq c_{\gamma,h}[F_+^*(x) + F_+^*(a)].$$

Case 3. $r < x \le a + 2^{-j-1}(b-a)$. Arguing as in Case 2, we can assume that (4.14) holds. Consequently,

$$(4.17) \qquad a + 2^{-j}(b-a) - x < |I_\psi|.$$

By (4.8), part (ii) of Lemma (4.7) and Lemma (3.6), we get

$$\|D^\gamma[S_j(\psi)]\|_\infty \le c_{\gamma,h}|I_\psi|^{-\gamma+1}.$$

Then, the inequality (4.17) implies that

$$\|D^\gamma[S_j(\psi)]\|_\infty \le c_{\gamma,h}(a + 2^{-j}(b-a) - x)^{-\gamma+1}.$$

The interval $\widetilde{N}_j = [a, a+2^{-j+2}(b-a)]$ just defined, contains the support of $\eta_j(y)$, therefore it also includes the support of $S_j(\psi)(t)$ and

$$|\widetilde{N}_j|^{\gamma+1}\|D^\gamma[S_j(\psi)]\|_\infty \le c_{\gamma,h}\Big[\frac{2^{-j}(b-a)}{a+2^{-j}(b-a)-x}\Big]^{\gamma+1}.$$

Thus, in Case 3, we get

$$(4.18) \qquad |\langle F, S_j(\psi)\rangle| \le c_{\gamma,h}\Big[\frac{2^{-j}(b-a)}{a+2^{-j}(b-a)-x}\Big]^{\gamma+1} F_+^*(a).$$

For $j > 1$, it is easy to see that:
(i) if $a + 2^{-j-1}(b-a) \le x < a + 2^{-j+2}(b-a)$, then $\frac{6}{7} \le M^+\chi_{I_j}(x) \le 1$ and
(ii) if $x < a + 2^{-j-1}(b-a)$, then

$$\frac{3}{7}\frac{2^{-j}(b-a)}{a+2^{-j}(b-a)-x} \le M^+\chi_{I_j}(x) \le 3\frac{2^{-j}(b-a)}{a+2^{-j}(b-a)-x}.$$

Consequently, considering (4.9), (1.4), (4.13), (4.16) and (4.18), for $j > 1$ the proposition is proved.

Now, let us estimate the maximal function $(W_1)_+^*(x)$. Again we shall consider three cases. First of all, we recall that $\bar{\eta}_1(y) = \eta_1(y)\chi_{(a,b)}(y)$. Besides, P_1 is the orthogonal projection of $L^2(\bar{\eta}_1(y))$ onto its subspace S.

Case 1'. $b \le x$. Since the support of $\nu_k(y)$ is contained in the interval $[b + 2^{-k}(b-a), b + 2^{-k+2}(b-a)]$, we have $\psi(y)\bar{\eta}_1(y) \equiv 0$. Then, by (4.6), $P_1(\psi)(t)$ is the null polynomial and $S_1(\psi)(t) = \psi(t)\eta_1(t)$. Thus, by (4.10) and (4.11)

$$\langle W_1, \psi\rangle = \langle F, \psi\eta_1\rangle - \sum_{k=1}^\infty \langle F, \nu_k\psi\eta_1\rangle.$$

Taking into account that the support of $\eta_1(y)$ is contained in the interval $N_1 = [a+2^{-1}(b-a), b+(b-a)]$, if $b+(b-a) \le x$ the function $\psi(y)\eta_1(y)$ is identically zero and it follows that $\langle W_1, \psi\rangle = 0$. Then, we can assume that $x < b+(b-a)$. Therefore, if $b < x$, there exists a positive integer k_1 such that $b+2^{-k_1+1}(b-a) \le x$. For $k_1 < k$, we have $b+2^{-k+2}(b-a) \le b+2^{-k_1+1}(b-a) \le x$ and recalling that the support of $\nu_k(y)$ is contained in $[b + 2^{-k}(b - a), b + 2^{-k+2}(b - a)]$, the products $\nu_k(y)\psi(y)\eta_1(y)$ are identically equal to zero. Therefore,

$$\langle W_1, \psi \rangle = \langle F, \psi\eta_1 \rangle - \sum_{k=1}^{k_1} \langle F, \nu_k \psi \eta_1 \rangle$$

$$= \langle F, \psi\eta_1 \rangle - \langle F, \sum_{k \in \mathbb{Z}} \nu_k \psi \eta_1 \rangle .$$

Then, by (4.2), we get $\langle W_1, \psi \rangle = 0$. Standard arguments about distributions imply, that for every $x \geq b$,

(4.19) $$(W_1)^*_+(x) = 0.$$

Case 2'. $a + 2^{-2}(b - a) < x < b$. Taking into account (4.8) and (4.11), we have

(4.20) $$\langle F_b, S_1(\psi) \rangle = \sum_{k=1}^{\infty} \langle F, \nu_k \psi \eta_1 \rangle - \sum_{k=1}^{\infty} \langle F, \nu_k P_1(\psi)\eta_1 \rangle = A - B.$$

Since the support of $\nu_k(y)$ is contained in the interval $[b + 2^{-k}(b - a), b + 2^{-k+2}(b - a)]$, if the support of $\psi(y)$ is contained in $I_\psi = [x, \beta]$ with $\beta \leq b$, all the functions $\nu_k(y)\psi(y)\eta_1(y)$ are identically equal to zero and the sum A is equal to zero. Then, we can assume that $b < \beta$. Let $\beta_1 = \min(\beta, b + (b - a))$, and let k_0 be the unique non-negative integer such that

(4.21) $$b + 2^{-k_0-1}(b - a) < \beta_1 \leq b + 2^{-k_0}(b - a).$$

The support of $\psi(y)\eta_1(y)$ is contained in $[x, \beta_1]$. Then, for every $k \leq k_0$, the product $\nu_k(y)\psi(y)\eta_1(y)$ is the null function. By part (i) of Lemma (4.3) and computing the sum, $|A|$ turns out to be bounded by

$$\sum_{k=k_0+1}^{\infty} |\langle F, \nu_k \psi \eta_1 \rangle| \leq \sum_{s=0}^{\gamma} c_{s,h}[2^{-k_0}(b - a)]^{s+1} \|D^s(\psi\eta_1)\|_\infty F^*_+(b).$$

Since $x < b$ and the left sided inequality in (4.21) holds, we get that $2^{-k_0-1}(b - a) < \beta_1 - x \leq |I_\psi|$. Then, by Lemma (3.6) , $\|D^s(\psi\eta_1)\|_\infty \leq \frac{c_{s,h}}{[2^{-k_0}(b-a)]^{s+1}}$. In consequence

$$|A| = \left| \sum_{k=1}^{\infty} \langle F, \nu_k \psi \eta_1 \rangle \right| \leq c_{\gamma,h} F^*_+(b).$$

By part (i) of Lemma (4.3) and computing the sum, $|B|$ is bounded by

$$\sum_{s=0}^{\gamma} c_{s,h}(b - a)^{s+1} \|D^s[P_1(\psi)\eta_1]\|_\infty F^*_+(b).$$

Applying part (i) of Lemma (4.7) , we get that $\|D^s[P_1(\psi)\eta_1]\|_\infty \leq c_{s,h}(b-a)^{-s-1}$ holds. Then, we obtain the estimation

$$|B| = \left| \sum_{k=1}^{\infty} \langle F, \nu_k P_1(\psi)\eta_1 \rangle \right| \leq c_{\gamma,h} F^*_+(b).$$

Now, these estimations and (4.20) show that

$$(4.22) \qquad |\langle F_b, S_1(\psi) \rangle| \le c_{\gamma,h} F_+^*(b) .$$

By (4.10), (4.16) in Case 2 and (4.22), we get

$$(4.23) \qquad (W_1)_+^*(x) \le c_{\gamma,h}[F_+^*(x) + F_+^*(a) + F_+^*(b)] .$$

Case 3'. $r < x \le a + 2^{-2}(b-a)$. Arguing as in Case 3, we suppose that $a + 2^{-1}(b-a) < \beta$, which implies that

$$(4.24) \qquad a + 2^{-1}(b-a) - x < |I_\psi| .$$

By part (i) of Lemma (4.3) and computing the sum

$$\left| \sum_{k=1}^{\infty} \langle F, \nu_k S_1(\psi) \rangle \right| \le \sum_{s=0}^{\gamma} c_{s,h}(b-a)^{s+1} \|D^s[S_1(\psi)]\|_\infty F_+^*(b) .$$

Part (ii) of Lemma (4.7) and Lemma (3.6) imply that $\|D^s[S_1(\psi)]\|_\infty \le c_{s,h}(b-a)^{\gamma-s}|I_\psi|^{-\gamma-1}$. Then, by (4.11) and (4.24)

$$(4.25) \qquad |\langle F_b, S_1(\psi) \rangle| \le c_{\gamma,h} \left[\frac{b-a}{a + 2^{-1}(b-a) - x} \right]^{\gamma+1} F_+^*(b) .$$

This inequality and (4.18) in Case 3, show that

$$(4.26) \qquad (W_1)_+^*(x) \le c_{\gamma,h} \left[\frac{2^{-1}(b-a)}{a + 2^{-1}(b-a) - x} \right]^{\gamma+1} [F_+^*(a) + F_+^*(b)] .$$

Since the maximal function $M^+ \chi_{I_1}(x)$ satisfies
(i) $\frac{2}{3} \le M^+ \chi_{I_1}(x) \le 1$, if $a + 2^{-2}(b-a) < x < a + 2^{-1}(b-a)$ and,
(ii) $\frac{1}{3} \frac{2^{-1}(b-a)}{a+2^{-1}(b-a)-x} \le M^+ \chi_{I_1}(x) \le \frac{2^{-1}(b-a)}{a+2^{-1}(b-a)-x}$, if $x \le a + 2^{-2}(b-a)$,
then, by (4.19), (4.23) and (4.26), we obtain that the proposition holds for the case $j = 1$. ∎

REMARK. Let $w(x) \in A_q^+$, $\gamma \ge 1$ be an integer and $0 < p \le 1$ such that $(\gamma+1)p \ge q > 1$ or $(\gamma+1)p > 1$ if $q = 1$. A point in the interval (a,b) can belong, at most, to three of the intervals I_j defined in Proposition (4.12). Then, by that proposition and parts (i) and (ii) of Lemma (3.1), we get

$$\sum_{j=1}^{\infty} \int_{x-\infty}^{\infty} (W_j)_+^*(x)^p w(x) dx$$

$$\le c_{\gamma,h,w} \left\{ \int_a^b F_+^*(x)^p w(x) dx + [F_+^*(a) + F_+^*(b)]^p w(a,b) \right\} .$$

If the right side of this inequality is finite, by Lemma (3.7) the distribution $W = \sum_{j=1}^{\infty} W_j$ belongs to $H_+^p(w)$. For this distributions W, we have the following proposition.

PROPOSITION 4.27. *Let $w(x) \in A_q^+, \gamma \geq 1$ be an integer and $0 < p \leq 1$ such that $(\gamma+1)p \geq q > 1$ or $(\gamma+1)p > 1$ if $q = 1$. Let a and b satisfy $x_{-\infty} < a < b < \infty$ and let F be a distribution in $\mathcal{D}'(x_{-\infty}, \infty)$. Then,*

(i) $W_+^*(x) \leq \displaystyle\sum_{j=1}^{\infty} (W_j)_+^*(x)$

$$\leq c_{\gamma,h}\Big\{ F_+^*(x)\chi_{(a,b)}(x) + [F_+^*(a) + F_+^*(b)] \sum_{j=1}^{\infty}[M^+\chi_{I_j}(x)]^{\gamma+1}\Big\},$$

where $I_j = [a + 2^{-j-1}(b-a), \ \min\{a + 2^{-j+2}(b-a), b\})$,

(ii) $\displaystyle\int_{x_{-\infty}}^{\infty} W_+^*(x)^p w(x)\,dx \leq c_{\gamma,h,w}\Big\{ \int_a^b F_+^*(x)^p w(x)\,dx + [F_+^*(a) + F_+^*(b)]^p w(a,b)\Big\},$

and

(iii) for every x belonging to (a,b), the inequality

$$(F - W)_+^*(x) \leq c_{\gamma,h}[F_+^*(a) + F_+^*(b)],$$

holds. The constants $c_{\gamma,h}$ and $c_{\gamma,h,w}$ do not depend on neither a nor b.

PROOF. Parts (i) and (ii) are an immediate consequence of Proposition (4.12). Let us prove (iii). Given x in (a,b), there exists $j_0 > 1$ such that $a + 2^{-(j_0+1)+2}(b-a) \leq x$. Let $\psi(y)$ be a function belonging to the class $\Phi(x)$ and let $I_\psi = [x, \beta]$ the interval associated to $\psi(y)$ in this class. If $j_0 + 1 \leq j$ we have, $a + 2^{-j+2}(b-a) \leq a + 2^{-(j_0+1)+2}(b-a) \leq x$. Taking into account that the support of $\eta_j(y)$ is contained in $N_j = [a + 2^{-j}(b-a), a + 2^{-j+2}(b-a)]$, it follows that

(4.28) $\qquad\qquad \psi(y)\eta_j(y) = 0 , \qquad\qquad j \geq j_0 + 1 .$

In consequence, $S_j(\psi)(t) \equiv 0$ and $\langle W_j, \psi \rangle = 0$, whenever $j \geq j_0 + 1$. Then,

$$\Big\langle F - \sum_{j=1}^{\infty} W_j, \psi \Big\rangle = \Big\langle F, \Big(1 - \sum_{j=1}^{j_0} \eta_j\Big)\psi \Big\rangle + \Big\langle F, \sum_{j=1}^{j_0} P_j(\psi)\eta_j \Big\rangle + \langle F_b, S_1(\psi) \rangle$$

(4.29) $\qquad\qquad\qquad = T_1 + T_2 + T_3 .$

In Proposition (4.12) we obtained the estimations (4.22) if $a + 2^{-2}(b-a) < x < b$ and (4.25) for $x_{-\infty} < x \leq a + 2^{-2}(b-a)$. They imply that $|T_3| \leq c_{\gamma,h}F_+^*(b)$. Now, by (4.5)

$$\Big\langle F, \sum_{j=1}^{j_0} P_j(\psi)\eta_j \Big\rangle = \sum_{j=1}^{j_0}\sum_{k=0}^{\gamma-1}\Big(\int_a^b \psi(s)e_k^j(s)\eta_j(s)\,ds\Big)\langle F, e_k^j\eta_j \rangle .$$

Then, applying part (iii) of Lemma (4.7) and, recalling that by (4.5)

$$\sup_{s\in I_j} |e_k^j(s)| \leq c_{\gamma,h}[2^{-j}(b-a)]^{-1/2}$$

and the properties of the functions $n_j(s)$, we have

$$|T_2| \leq c_{\gamma,h} \left(\int_a^b |\psi(s)| \sum_{j=1}^{jo} \eta_j(s)ds \right) F_+^*(a) \leq c_{\gamma,h} F_+^*(a) \ .$$

On the other hand, since $a < x$ and due to the fact that the support of $\psi(y)$ is contained in $I_\psi = [x, \beta]$, by (4.28), we get

$$(4.30) \qquad \psi(y) = \sum_{j=1}^{\infty} \psi(y)\eta_j(y) = \sum_{j=1}^{jo} \psi(y)\eta_j(y) \ ,$$

whenever $y \in I_\psi$ and $y \leq b$. If $\beta \leq b$, then, for every $y \in I_\psi$, (4.30) holds. Thus $<$ $F, \left(1 - \sum_{j=1}^{jo} \eta_j \right)\psi >= 0$. In the case $b < \beta$, the support of $\left[\left(1 - \sum_{j=1}^{jo} \eta_j \right)\psi \right](y)$ is contained in $J = [b, \beta]$ and $\left[\left(1 - \sum_{j=1}^{jo} \eta_j \right)\psi \right](y) = [(1 - \eta_1)\psi](y)$, whenever $b \leq y$. Taking into account that the support of $\eta_1(y)$ is contained in $[a + 2^{-1}(b - a), b + (b - a)]$, we get

$$\left\| D^\gamma \left[\left(1 - \sum_{j=1}^{jo} \eta_j \right)\psi \right] \right\|_\infty$$

$$(4.31) \qquad\qquad \leq \|D^\gamma \psi\|_\infty + \sum_{s=1}^{\gamma} c_s \|D^s \eta_1\|_\infty \sup_{a \leq y \leq b+(b-a)} |D^{\gamma-s}\psi(y)| \ .$$

From the fact that $D^{\gamma-s}\psi(y) = \int_a^y \frac{(y-t)^{s-1}}{(s-1)!} D^\gamma \psi(t)dt$, it follows that (4.31) is bounded by a constant times $\|D^\gamma \psi\|_\infty$. Then by (1.1) we have

$$|J|^{\gamma+1} \left\| D^\gamma \left[\left(1 - \sum_{j=1}^{jo} \eta_j \right)\psi \right] \right\|_\infty \leq c_{\gamma,h} \left(\frac{|J|}{|J_\psi|} \right)^{\gamma+1} \leq c_{\gamma,h} \ ,$$

showing this that $|T_1| \leq c_{\gamma,h} F_+^*(b)$. Thus, by (4.29) and the estimations for $|T_3|$, $|T_2|$ and $|T_1|$, we obtain part (iii) of the proposition. ∎

Now we can prove Theorem (2.1).

PROOF OF THEOREM (2.1). Since

$$\lambda^p w(\{x : x > x_{-\infty}, \ F_+^*(x) > \lambda\}) \leq \|F\|_{H_+^p(w)}^p < \infty,$$

by choosing λ large enough we can make $w(\{x : x > x_{-\infty}, \ F_+^*(x) > \lambda\})$ as small as we please. Thus, in virtue of the absolute continuity of the integral, given $\varepsilon > 0$, there exists $\lambda_1 > 0$ such that

$$\int_{\{x : x > x_{-\infty}, \ F_+^*(x) > \lambda_1\}} F_+^*(x)^p \ w(x)dx < \varepsilon \ .$$

Let $\{\lambda_k\}$ be a strictly increasing sequence, starting at λ_1 and tending to infinity. We denote $\Omega_k = \{x : x > x_{-\infty}, \ F_+^*(x) > \lambda_k\}$. Since $F_+^*(x)$ is lower semicontinuous function, the sets Ω_k are open. Let $I_{k,i} = (a_{k,i}, b_{k,i})$ be the connected components of Ω_k. We observe that $b_{k,i} < \infty$, since $w(I_{k,i}) \leq w(\Omega_k) < \infty$ (see part (iii) of Lemma (3.1)). Given $k \geq 1$, if there is an interval $I_{k,i}$ with $a_{k,i} = x_{-\infty}$, then we shall assume that $i = 1$. If not, we shall assume that $I_{k,1} = \emptyset$. We observe that for $i > 1$, $F_+^*(a_{k,i}) \leq \lambda_k$ and $F_+^*(b_{k,i}) \leq \lambda_k$. Then, taking $a = a_{k,i}$ and $b = b_{k,i}$ in Proposition (4.27), we obtain a distribution $W = \sum_{j \geq 1} W_j$ that we denote by $B_{k,i}$ and intervals $I_{k,i,j}$ that are the intervals I_j of the same Proposition (4.27).

We define the distribution B as

$$B = \sum_{k \geq 1} \sum_{\substack{i > 1 \\ I_{k,i} \subset I_{k-1,1}}} B_{k,i} \ ,$$

where $I_{0,1} = \mathbb{R}$ and we also define a distribution G as

$$G = F - B \ .$$

Let $\mathcal{F} = \{(k, i) : k \geq 1, \ i > 1 \text{ and } I_{k,i} \subset I_{k-1,1}\}$, that is to say, $\mathcal{F}$ is the set of indexes (k, i) involved in the definition of B. The intervals of the family $\{I_{k,i}\}_{(k,i) \in \mathcal{F}}$ are pairwise disjoint and contained in Ω_1. We shall show that the distribution B is well defined and $\|B\|_{H_+^p(w)}^p < c'\varepsilon$. In fact, by part (ii) of Proposition (4.27), we have

$$\int_{x_{-\infty}}^{\infty} B_+^*(x)^p w(x) dx \leq \sum_{(k,i) \in \mathcal{F}} \int_{x_{-\infty}}^{\infty} (B_{k,i})_+^*(x)^p w(x) dx$$

$$\leq c \sum_{(k,i) \in \mathcal{F}} \left\{ \int_{I_{k,i}} F_+^*(x)^p w(x) dx + \lambda_k^p w(I_{k,i}) \right\}$$

$$\leq c' \sum_{(k,i) \in \mathcal{F}} \int_{I_{k,i}} F_+^*(x)^p w(x) dx$$

$$\leq c' \int_{\Omega_1} F_+^*(x)^p w(x) dx < c'\varepsilon \ .$$

Next, we shall estimate $G_+^*(x)$. Let $\Omega = \cup\{I_{k,i} : (k,i) \in \mathcal{F}\}$. If $x \in \Omega$, then $x \in I_{k_0,i_0}$, $(k_0, i_0) \in \mathcal{F}$, and

$$G = F - B_{k_0,i_0} + \sum \{B_{k,i} : (k,i) \in \mathcal{F} \quad \text{and} \quad (k,i) \neq (k_0, i_0)\} \ .$$

By parts (i) and (iii) of Proposition (4.27), we get

$$G_+^*(x) \leq c\lambda_{k_0} + c \sum \left\{ \lambda_k \left[M^+ \chi_{k,i,j}(x) \right]^{\gamma+1} : (k,i) \in \mathcal{F}, \ (k,i) \neq (k_0, i_0), \ j \geq 1 \right\} \ .$$

Since for $x \in I_{k_0,i_0}$, $0 < c" \leq \sum_{j \geq 1} \left[M^+ \chi_{I_{k_0,i_0,j}}(x) \right]^{\gamma+1}$, we obtain

$$G_+^*(x) \leq c \sum_{\substack{(k,i) \in \mathcal{F} \\ j \geq 1}} \lambda_k \left[M^+ \chi_{I_{k_0,i_0,j}}(x) \right]^{\gamma+1} \ .$$

If $x \notin \Omega$, by part (i) of Proposition (4.27), we have

$$G_+^*(x) \leq F_+^*(x)\chi_{c\Omega} + \sum_{\substack{(k,i)\in\mathcal{F}}} (B_{k,i})_+^*(x)$$

$$\leq F_+^*(x)\chi_{c\Omega} + c \sum_{\substack{(k,i)\in\mathcal{F}\\j\geq 1}} \lambda_k \left[M^+ \chi_{I_{k,i,j}}(x) \right]^{\gamma+1} .$$

Therefore, for every $x > x_{-\infty}$

$$G_+^*(x) \leq F_+^*(x)\chi_{c\Omega} + c \sum_{\substack{(k,i)\in\mathcal{F}\\j\geq 1}} \lambda_k \left[M^+ \chi_{I_{k,i,j}}(x) \right]^{\gamma+1} .$$

Now, given $m > x_{-\infty}$, let $h = \min\{k : I_{k,1} \subset (x_{-\infty}, m) \text{ and } k \geq 1\}$. If $x \geq m \geq b_{h,1}$ and $x \notin \Omega$ then $x \notin \Omega_h$, which implies that $F_+^*(x) \leq \lambda_h$. Therefore,

$$(4.32) \qquad \int_m^\infty F_+^*(x)^q \chi_{c\Omega}(x) w(x)dx \leq \lambda_h^{q-p} \int_{x_{-\infty}}^\infty F_+^*(x)^p w(x)dx < \infty .$$

Let $(k,i) \in \mathcal{F}$ with $k > h$, then since $I_{k,i,j} \subset I_{k,i} \subset I_{k-1,1} \subset (x_{-\infty}, m)$, we have $M^+ \chi_{I_{k,i,j}}(x) = 0$, for every $x \geq m$. Thus,

$$\int_m^\infty \left(\sum_{\substack{(k,i)\in\mathcal{F}\\j\geq 1}} \lambda_k \left[M^+ \chi_{I_{k,i,j}}(x) \right]^{\gamma+1} \right)^q w(x)dx$$

$$\leq \lambda_h^q \int \left(\sum_{\substack{(k,i)\in\mathcal{F}\\j\geq 1}} \left[M^+ \chi_{I_{k,i,j}}(x) \right]^{\gamma+1} \right)^q w(x)dx ,$$

which, by Lemma (3.9) is smaller than or equal to $c\lambda_h^q w(\Omega) \leq c\lambda_h^q w(\Omega_1) < \infty$. This result together with (4.32) shows that

$$\int_m^\infty G_+^*(x)^q w(x)dx < \infty .$$

Then, by Hölder's inequality, if $x_{-\infty} < a < b < \infty$, we get

$$\int_a^b G_+^*(x)dx \leq \left(\int_a^b G_+^*(x)^q w(x)dx \right)^{1/q} \left(\int_a^b w(x)^{-\frac{1}{q-1}} dx \right)^{1/q'} < \infty ,$$

showing the local integrability of $G_+^*(x)$ holds. Then, by Lemma (3.10) we have that G coincides with a locally integrable function $g(x)$, ending the proof of the theorem. ∎

5. Proof of the Theorems (2.2) and (2.4) (Decomposition into atoms)

Let us give the proof of Theorem (2.2). By Theorem (2.1), we can assume without loss of generality the distribution F is a locally integrable function $f(x)$ on $(x_{-\infty}, \infty)$.

Let n be an integer and $\Omega_n = \{x : x > x_{-\infty}, f_+^*(x) > 2^n\}$. Since Ω_n is an open set, let $I_{n,i} = (a_{n,i}, b_{n,i})$ be the connected components of Ω_n. As in the proof of Theorem (2.1), if there exists a connected component starting at $x_{-\infty}$, we denote it by $I_{n,1}$, otherwise $I_{n,1} = \emptyset$. Let the functions $\eta_{n,i,j}$ and $e_k^{n,i,j}$, and the intervals $N_{n,i,j}$ and $\widetilde{N}_{n,i,j}$ be equal to the functions η_j and e_k^j, and the intervals N_j and $\widetilde{N}_j$ considered in Lemma (4.7) for $a = a_{n,i}$ and $b = b_{n,i}$, $i > 1$. Moreover, let $B_{n,i,j} = W_j$ the distributions defined in (4.9) and (4.10) for a, b and i as before. Since $f(x)$ is locally integrable on $(x_{-\infty}, \infty)$ we define (as in (4.5)) the polynomial

$$P_{n,i,j}(f)(t) = \sum_{k=0}^{\gamma-1} \left(\int_{a_{n,i}}^{b_{n,i}} f(s) \, e_k^{n,i,j}(s) \, \eta_{n,i,j}(s) ds \right) e_k^{n,i,j}(t) \ .$$

It is simple to show that if

$$b_{n,i,j}(s) = [f(s) - P_{n,i,j}(f)(s)]\eta_{n,i,j}(s)\chi_{I_{n,i}}(s) \ ,$$

then

$$\langle B_{n,i,j}, \psi \rangle = \int b_{n,i,j}(s)\psi(s)ds \ ,$$

for every $\psi(s) \in \mathcal{D}(x_{-\infty}, \infty)$.

Using the notation of Lemma (4.7), we have

$$P_j(f)(t) = \sum_{k=0}^{\gamma-1} \left(\int_a^b f \, e_k^j \, \eta_j ds \right) e_k^j(t) \ .$$

Since $\operatorname{supp}(e_k^j \, \eta_j) \subset \widetilde{N}_j = [a, a + 2^{-j+2}(b-a)]$, it follows that

$$\int_a^b f \, e_k^j \, \eta_j ds = \left(\int_{x_{-\infty}}^{\infty} - \int_b^{\infty} \right) f \, e_k^j \, \eta_j ds = A - B \ .$$

Then, by part (iii) of Lemma (4.7), we obtain

$$|A| \leq c[2^{-j+2}(b-a)]^{1/2} f_+^*(a) \ .$$

Now, taking $\beta = a + 2^{-j+2}(b-a)$ in part (ii) of Lemma (4.3) we get the estimate

$$|B| \leq c \sum_{s=0}^{\gamma} [2^{-j+2}(b-a)]^{s+1} \|D^s(e_k^j \, \eta_j)\|_\infty f_+^*(b) \ .$$

Notice that $|\widetilde{N}_j| = 2^{-j+2}(b-a)$. Then by Lemma (3.6) and part (iii) of Lemma (4.7) we get

$$|B| \leq c[2^{-j}(b-a)]^{1/2} f_+^*(b) \ .$$

Thus, by (4.5), it follows that

$$\sup_{t \in N_j} |P_j(f)(t)| \leq c_\gamma \left(f_+^*(a) + f_+^*(b) \right) \ .$$

If we apply this estimate to $a = a_{n,i}$ and $b = b_{n,i}$, and since $f_+^*(a_{n,i}) \leq 2^n$ and $f_+^*(b_{n,i}) \leq 2^n$, we have that

(5.1) $$\sup_{t \in N_{n,i,j}} |P_{n,i,j}(f)(t)| \leq c \, 2^n \ ,$$

holds for $i > 1$ and $j \geq 1$. It is now easy to check that the decomposition of f,

$$f = f\chi_{c\Omega_n} + \sum_{i>1}\sum_{j\geq 1} P_{n,i,j}(f)\,\eta_{n,i,j}\,\chi_{I_{n,i}} +$$
$$+ \sum_{i>1}\sum_{j\geq 1}(f - P_{n,i,j}(f))\eta_{n,i,j}\,\chi_{I_{n,i}} + f\,\chi_{I_{n,1}}$$

holds pointwisely and in the sense of distribution on $(x_{-\infty}, \infty)$.

Defining $g_n(x)$ as

$$g_n(x) = f(x)\chi_{c\Omega_n}(x) + \sum_{i>1}\sum_{j\geq 1} P_{n,i,j}(f)(x)\,\eta_{n,i,j}(x)\,\chi_{I_{n,i}}(x)\,,$$

by (5.1) and since $|f(x)| \leq c\,f_+^*(x)$ a.e., we obtain that

$$(5.2) \qquad\qquad |g_n(x)| \leq c\,2^n \qquad\qquad \text{a.e. in } (x_{-\infty}, \infty)\,.$$

Next, we shall show that $\lim_{n\to\infty} g_n = f$ in the sense of distributions. In fact, if $W = \sum_{j\geq 1} b_{n,i,j} = b_{n,i}$ then, by part (ii) of Proposition (4.27), we have

$$\left\|\sum_{i>1}\sum_{j\geq 1} b_{n,i,j}\right\|_{H_+^p(w)}^p \leq \sum_{i>1}\|b_{n,i}\|_{H_+^p(w)}^p \leq c\int_{\Omega_n} f_+^*(x)^p w(x)dx < \varepsilon\,,$$

if n is large enough. This implies that for any $\psi \in \mathcal{D}(x_{-\infty}, \infty)$,

$$\lim_{n\to\infty}\int_{x_{-\infty}}^{\infty}(f - g_n)\psi\,dx = \lim_{n\to\infty}\int_{I_{n,1}} f\,\psi\,dx + \lim_{n\to\infty}\sum_{i>1}\int b_{n,i}\psi\,dx$$
$$= \lim_{n\to\infty}\int_{I_{n,1}} f\,\psi\,dx = 0\,,$$

since for n large enough $\mathrm{supp}\psi \cap I_{n,1} = \emptyset$.

On the other hand, by (5.2), $\lim_{n\to\infty}\int g_n\psi dx = 0$. Then,

$$f = \sum_{-\infty}^{\infty}(g_{n+1} - g_n)$$

holds in the sense of distributions on $(x_{-\infty}, \infty)$.

Now, observing that $\mathrm{supp}\,b_{n,i} \subset I_{n,i}$ and since $\Omega_{n+1} \subset \Omega_n$, we obtain

$$g_{n+1} - g_n = \sum_i\left[\sum_{I_{n+1,k}\subset I_{n,i}}(b_{n,i} - b_{n+1,k})\right] +$$
$$+ \left[f\,\chi_{I_{n,1}} - f\,\chi_{I_{n+1,1}} - \sum_{I_{n+1,k}\subset I_{n,1}} b_{n+1,k}\right]$$
$$= \sum_{i>1}\widetilde{A}_{n,i} + \widetilde{A}_{n,1}\,,$$

where the supports of $\widetilde{A}_{n,i}$ are contained in $I_{n,i}$. Since $\|g_{n+1} - g_n\|_\infty \leq c\, 2^{n+1}$, we have $\|\widetilde{A}_{n,i}\|_\infty \leq c\, 2^{n+1}$. Moreover, by the construction of the functions $b_{n,i}$ it turns out that for $i > 1$,

$$\int_{x-\infty}^{\infty} \widetilde{A}_{n,i} x^s dx = 0 \qquad \text{for } 0 \leq s < \gamma .$$

If we define $\lambda_{n,i} = c\, 2^{n+1} w(I_{n,i})^{1/p}$ then the functions $a_{n,i}(x)$ given by $\widetilde{A}_{n,i}(x) = \lambda_{n,i}\, a_{n,i}(x)$ are p-atoms with respect to $w(x)$. Then,

$$f = \sum_{-\infty}^{\infty} (g_{n+1} - g_n) = \sum_n \sum_{i \geq 1} \lambda_{n,i} a_{n,i}$$

and

$$\sum_n \sum_{i \geq 1} \lambda_{n,i}{}^p = c \sum_n \sum_{i \geq 1} 2^{np} w(I_{n,i})$$

$$= c \sum_n 2^{np} w(\Omega_n)$$

$$\leq c' \int_{x-\infty}^{\infty} f_+^*(x)^p w(x) dx ,$$

as we wanted to show. Finally, observe that the first inequality in (2.3) follows from Corollary (3.5). ∎

In order to prove Theorem (2.4), we shall need the following lemma.

LEMMA 5.3. *Let γ be a positive integer, α and M_i, $0 \leq i < \gamma$, real numbers. There exists $r_0 < \alpha$ such that for any $r \leq r_0$, there is a unique polynomial $P_r(x)$ of degree less than γ such that*

(5.4) $$\int_r^\alpha P_r(x)\, x^s dx = -M_s \qquad \text{for } 0 \leq s < \gamma, \qquad \text{and}$$

(5.5) $$\sup_{|x| \leq |r|} |P_r(x)| \leq c_\gamma \max_{0 \leq s < \gamma} |M_s|\, |r|^{-1} .$$

PROOF. Let $P_r(x) = c_0 + \ldots + c_{\gamma-1}\, x^{\gamma-1}$. By (5.4), the coefficients c_i must satisfy

$$\sum_{i=0}^{\gamma-1} c_i (r^{s+i+1} - \alpha^{s+i+1})/(s+i+1) = M_s$$

or, equivalently

$$\sum_{i=0}^{\gamma-1} c_i\, r^i [1 - (\alpha/r)^{s+i+1}]/(s+i+1) = M_s/r^{s+1},\ 0 \leq s < \gamma .$$

Let $\mathcal{M}_\beta$ denote the $\gamma \times \gamma$ matrix $\{m_\beta(i,s)\}$, $0 \leq i, s < \gamma$, where

$$m_\beta(i,s) = (1 - \beta^{s+i+1})/(s+i+1) .$$

Since $\det \mathcal{M}_0 > 0$ (see [7], page 98, Problem 3), then, for $|\beta| < \delta < 1$, $\det \mathcal{M}_\beta > \frac{1}{2} \det \mathcal{M}_0$. On the other hand if $|\beta| \leq 1$ we have $|m_\beta(i,s)| \leq 2$. From these properties of $\mathcal{M}_\beta$ and Cramer's rule we obtain that for $|\alpha/r| < \delta$,

$$(5.6) \qquad |c_i \, r^i| \leq c_\gamma \max_{0 \leq i < \gamma} |M_i|/|r|^{i+1}, \qquad 0 \leq i < \gamma$$

holds. If we define $r_0 = -(|\alpha|\delta^{-1} + 1)$, we see that (5.5) follows from (5.6). ∎

PROOF OF THEOREM (2.4). We are going to show that for any p-atoms $a(x)$ there exists a sequence $\{\mu_k\}$ of real numbers, and a sequence $\{A_k(x)\}$ of p-atoms with bounded associated intervals, such that

$$(5.7) \qquad a(x) = \sum \mu_k A_k(x) \quad \text{and} \quad \sum |\mu_k|^p \leq 2^{p+1}.$$

If we accept this, we can prove the theorem as follows: given $F \in H_+^p(w)$, by Theorem (2.2), we have

$$F = \sum \lambda_i a_i(x) \quad \text{and} \quad \sum |\lambda_i|^p \leq c\|F\|_{H_+^p(w)}^p.$$

By (5.7), for every $a_i(x)$, we have

$$a_i(x) = \sum \mu_{i,k} A_{i,k}(x), \qquad \sum |\mu_{i,k}|^p \leq 2^{p+2}.$$

Then

$$F = \sum_i \sum_k \lambda_i \mu_{i,k} A_{i,k}$$

and $\sum_i \sum_k |\lambda_i \mu_{i,k}|^p \leq 2^{p+1} \sum_i |\lambda_i|^p \leq 2^{p+1} c\|F\|_{H_+^p(w)}^p$, as we wanted to show.

Let us now prove (5.7). If the interval associated to the p-atom $a(x)$ is bounded then (5.7) is trivial (take $\mu_1 = 1$ and $A_1(x) = a(x)$). We may assume that the w-measure of $(-\infty, \beta)$, β real, is finite since otherwise all the atoms have bounded associated intervals. Therefore, let $a(x)$ be a p- atom whose associated interval is $(-\infty, \beta]$, i.e., β real and $w(-\infty, \beta] < \infty$. Applying Lemma (5.3), we define by induction, a sequence $\{P_k(x)\}_1^\infty$ of polynomials of degree less than γ, a sequence $\{\alpha_k\}_0^\infty$ of real numbers with $\alpha_0 = \beta$, $\alpha_1 = -(|\beta| + 1)$ and $\alpha_k \downarrow -\infty$, and a sequence of functions $\{f_k(x)\}_0^\infty$, $f_0(x) = a(x)$, such that

$$(5.8) \qquad \int_{\alpha_{k+2}}^{\alpha_{k+1}} P_{k+1}(x)x^s dx = -\int_{\alpha_{k+1}}^{\alpha_k} f_k(x)x^s dx, \qquad 0 \leq s < \gamma,$$

$$(5.9) \qquad \sup\{|P_{k+1}(x)| : \alpha_{k+2} \leq x \leq \alpha_{k+1}\} \leq 2^{-2^k}\|f_k\|_\infty, \text{ and}$$

$$(5.10) \qquad f_{k+1}(x) = f_k(x)\chi_{(-\infty,\alpha_{k+1}]}(x) - P_{k+1}(x)\chi_{(\alpha_{k+2},\alpha_{k+1}]}(x).$$

Let us define $\widetilde{A}_{k+1}(x)$ as

$$\widetilde{A}_{k+1}(x) = f_k(x)\chi_{(\alpha_{k+1},\alpha_k]}(x) + P_{k+1}(x)\chi_{(\alpha_{k+2},\alpha_{k+1}]}(x).$$

These functions $\widetilde{A}_{k+1}(x)$ have the following properties:
   (i) $\text{supp}\widetilde{A}_{k+1}(x) \subset [\alpha_{k+2}, \alpha_k]$;
by (5.10) and (5.9),

(ii) $\|\widetilde{A}_{k+1}\|_\infty \leq (1 + 2^{-2^k})\|f_k\|_\infty \leq 2\|a\|_\infty$;
by (5.8),

(iii) $\int \widetilde{A}_{k+1}(x)x^s dx = 0$ for $0 \leq s < \gamma$; and
since, by (5.10), $\sup f_k(x) \subset (-\infty, \alpha_k]$, then

(iv) $\widetilde{A}_{k+1}(x) = f_k(x) - f_{k+1}(x)$,
and therefore $a(x) = \sum_{k=0}^{\infty} \widetilde{A}_{k+1}(x)$.

Let $\mu_{k+1} = 2\big(w((\alpha_{k+2}, \alpha_k])/w((-\infty, \beta]))\big)^{1/p}$, define $A_{k+1}(x) = \mu_{k+1}^{-1}\widetilde{A}_{k+1}(x)$.
Then, by (i), (ii) and (iii) we see that $A_{k+1}(x)$ is a p-atom with a bounded
associated interval $[\alpha_{k+2}, \alpha_k]$. Thus, by (iv), $a(x) = \sum_{k=0}^{\infty} \mu_{k+1}A_{k+1}(x)$ and

$$\sum_0^\infty |\mu_{k+1}|^p = 2^p \sum_{k<0}^\infty w((\alpha_{k+2}, \alpha_k])/w((-\infty, \beta]) \leq 2^{p+1} . \blacksquare$$

REFERENCES

1. R. R. Coifman, *A real variable characterization of H^p*, Studia Math. **51** (1974), 267–272.
2. J. García-Cuerva, *Weighted H^p Spaces*, Dissert. Mathematicae 162, Warszana, 1979.
3. J. García-Cuerva and J.L. Rubio de Francia, *Weighted Norm Inequalities and Related Topics*, North-Holland, Amsterdam, 1985.
4. B. Muckenhoupt, *Weighted norm inequalities for the Hardy maximal function*, Trans. Amer. Math. Soc. **165** (1972), 207–226.
5. F.J. Martín-Reyes, *New proofs of weighted inequalities for the one-sided Hardy-Littlewood maximal functions*, Proc. Amer. Math. Soc. **117** (1993), 691–698.
6. R.A. Macías and C. Segovia, *A decomposition into atoms of distributions on spaces of homogeneous type*, Advances in Math. **33** (1979), 271–309.
7. G. Pólya and G. Szegö, *Aufgaben und Lehrsätze aus der Analysis II*, Grund. Math. Wiss. 20, Springer-Verlag, Berlin, 1925.
8. E. Sawyer, *Weighted inequalities for the one-sided Hardy-Littlewood maximal functions*, Trans. Amer. Math. Soc. **297** (1986), 53–61.

DEPARTMENT OF MATHEMATICS, UNIVERSITY OF BUENOS AIRES, BUENOS AIRES, ARGENTINA

Current address: Instituto Argentino de Matemática, Viamonte 1636, Buenos Aires, Argentina 1055

E-mail address: segovia@iamba.edu.ar

Contemporary Mathematics
Volume **189**, 1995

A Direct Proof of a Theorem of Representation of Multilinear Operators on $\prod C_0(T_i)$

IVAN DOBRAKOV AND T. V. PANCHAPAGESAN

Dedicated to Professor Mischa Cotlar on the occasion of his eightieth birthday

ABSTRACT. Let T_i, $i = 1, 2, \ldots, d$, be locally compact Hausdorff spaces and let $C_0(T_i)$ be the Banach space of all scalar valued continuous functions on T_i vanishing at infinity (with the supremum norm). Suppose $U : \prod_1^d C_0(T_i) \to Y$ is a bounded d-linear operator, where either (A) Y is a Banach space such that $c_0 \not\subset Y$, or (B) U is weakly compact. Using the multilinear extension theorem of Pelczyński, Dobrakov obtained in an ealier work a multilinear integral representation of U with respect to a unique Y-valued Baire d-multimeasure on $\prod_1^d \sigma\mathcal{B}_0(T_i)$. The aim of the present note is to provide a direct proof of this representation theorem, without any reference to the said result of Pelczyński. Then the multilinear extension theorem of the latter follows as a corollary.

1. Introduction

In [14] Pelczyński proved the following extension theorem of multilinear operators on $\prod C(S_i)$, where $C(S_i)$ is the Banach space of all scalar valued continuous functions on the compact Hausdorff space S_i.

THEOREM. (Pelczyński). *Suppose $S_1, \ldots, S_d$ are compact Hausdorff spaces and $U : \prod_1^d C(S_i) \to Y$ is a bounded d-linear mapping, where either Y is a Banach space such that $c_0 \not\subset Y$, or U is weakly compact. Then there exists a unique bounded d-linear mapping $U^{**} : \prod_1^d B^\Omega(S_i) \to Y$ such that*

(i) $U^{**} | \prod_1^d C(S_i) = U$, *and*
(ii) *if $(f_{i,n})_{n=1}^\infty \subset \mathcal{B}^\Omega(S_i)$ is such that*

$$\lim_{n \to \infty} f_{i,n}(s_i) = f_i(s_i), \text{ for each } s_i \in S_i,$$

1991 *Mathematics Subject Classification.* Primary 46G99.
The research of the second author was supported by the C.D.C.H.T. project C-586 of ULA and the CONICIT (Venezuela)-CNR (Italy) international cooperation project.
This paper is in final form and no version of it will be submitted for publication elsewhere.

and

$$\sup_{s_i \in S_i, n=1,2,\ldots} |f_{i,n}(s_i)| \le C \ (< \infty)$$

for $i = 1, 2, \ldots, d$, then

$$\lim_{n \to \infty} U^{**}(f_{1,n}, \ldots, f_{d,n}) = U^{**}(f_1, \ldots, f_d).$$

Moreover, in the case (B) the operator U^{**} is also weakly compact.

Using the above theorem, Dobrakov proved (see [8]) that there is a d-multi-measure $\Upsilon : \prod_1^d \sigma\mathcal{B}_0(T_i) \to Y$ such that

$$U^{**}(g_i) = \int_{(T_i)} (g_i)\, d\Upsilon$$

for each $(g_i) \in \prod \mathcal{B}^\Omega(T_i)$, where T_i is a locally compact Hausdorff space, $\sigma\mathcal{B}_0(T_i)$ is the σ-ring of all Baire sets of T_i and $\mathcal{B}^\Omega(T_i)$ is the class of all bounded Baire functions on T_i, for $i = 1, 2, \ldots, d$. Moreover, U^{**} extends U and is a bounded d-linear operator with $||U^{**}|| = ||U|| = ||\Upsilon||(T_i)$, where $||\Upsilon||(T_i)$ is the scalar semivariation of Υ in (T_i). Finally, the range of Υ is relatively weakly compact if and only if U is weakly compact.

The object of the present note is to present a direct proof of the multilinear integral representation theorem of Dobrakov and then to deduce the cited theorem of Pelczyński as a corollary. Then all the results of Dobrakov in [8] remain independent of Pelczyński's multilinear extension theorem.

2. Notation and Terminology

In the sequel, T and T_i, $i = 1, 2, \ldots, d$, are locally compact Hausdorff spaces. $C_0(T)$ is the Banach space of all scalar valued continuous functions on T vanishing at infinity, with the supremum norm $||.||_T$, where $||f||_T = \sup_{t \in T} |f(t)|$. Similarly, we define $C_0(T_i)$, for $i = 1, 2, \ldots, d$.

The family of all compact G_δ's of T is denoted by $\mathcal{K}_0(T)$ and that of T_i by $\mathcal{K}_0(T_i)$, for $i = 1, 2, \ldots, d$. The σ-ring generated by $\mathcal{K}_0(T)$ (resp. $\mathcal{K}_0(T_i)$) is denoted by $\sigma\mathcal{B}_0(T)$ (resp. $\sigma\mathcal{B}_0(T_i)$), whose members are called Baire sets of T (resp. T_i).

The scalar field is denoted by $\mathbb{K}$ ($= \mathbb{R}$ or $\mathbb{C}$). Let Y be a Banach space over $\mathbb{K}$, the scalar field of $C_0(T)$ and $C_0(T_i)$.

DEFINITION 2.1. A mapping $U : \prod_1^d C_0(T_i) \to Y$ is said to be d-linear if it is separately linear on each coordinate. Such a mapping U is said to be bounded if

$$\sup\{|U(f_i, \ldots, f_d)| : ||f_i||_{T_i} \le 1, \ i = 1, 2, \ldots, d\} < \infty$$

where $|.|$ denotes the norm of Y. When U is bounded, the above supremum is denoted by $||U||$. If U is a bounded d-linear mapping and if $\{U(f_i, \ldots, f_d) : ||f_i||_{T_i} \le 1, \ f_i \in C_0(T_i), \ i = 1, 2, \ldots, d\}$ is relatively weakly compact in Y, then U is said to be weakly compact.

We now proceed to state some definitions and results from the theory of multilinear integration of scalar functions. The reader may refer to Dobrakov [5,6,7,8,9].

If $\mathcal{S}_i$, $i = 1, 2, \ldots, d$, are σ-rings of sets in T_i, then let

$$\prod_1^d \mathcal{S}_i = \{(A_1, \ldots, A_d) : A_i \in \mathcal{S}_i, \ i = 1, 2, \ldots, d\}.$$

The rectangle $(A_1, \ldots, A_d)$ is denoted by (A_i).

DEFINITION 2.2. Suppose $\Upsilon : \prod \mathcal{S}_i \to Y$ is a set function such that it is separately σ-additive in norm of Y. Then Υ is called a Y-valued d-multimeasure (or d-polymeasure).

DEFINITION 2.3. Let $f_i = \sum_{j=1}^{r_i} a_{ij} \chi_{A_{ij}}, a_{ij} \in I\!\!K, \ A_{ij} \cap A_{ij'} = \emptyset$ for $j \neq j'$, $A_{ij} \in \mathcal{S}_i$, for $j = 1, 2, \ldots, r_i$ and $i = 1, 2, \ldots, d$. Such functions f_i are called S_i-simple functions and (f_i) is said to be $\prod \mathcal{S}_i$-simple. The set of all $\prod \mathcal{S}_i$-simple functions is denoted by $\prod S(\mathcal{S}_i)$. If $\Upsilon : \prod \mathcal{S}_i \to Y$ is a d-multimeasure, then we define

$$\int_{(A_i)} (f_i) d\Upsilon = \sum_{j_1=1}^{r_1} \cdots \sum_{j_d=1}^{r_d} a_{1j_1} a_{2j_2} \ldots a_{dj_d} \Upsilon(A \cap A_{ij_i})$$

where (f_i) is given as above.

DEFINITION 2.4. For $(A_i) \in \prod \mathcal{S}_i$ and for a d-multimeasure $\Upsilon : \prod \mathcal{S}_i \to Y$, we define the scalar semivariation $||\Upsilon||(A_i)$ by

$$||\Upsilon||(A_i) = \sup\{|\int_{(A_i)} (f_i) d\Upsilon| : (f_i) \in \prod S(\mathcal{S}_i), ||f_i||_{T_i} \leq 1, i = 1, \ldots, d\}$$

and $||\Upsilon||(T_i) = \sup\{||\Upsilon||(A_i) : (A_i) \in \prod \mathcal{S}_i\}$.

THEOREM 2.5. *For a Y-valued d-multimeasure $\Upsilon : \prod \mathcal{S}_i \to Y, ||\Upsilon||(T_i)$ is finite.*

THEOREM 2.6. *Let $\overline{S(\mathcal{S}_i)}$ be the closure of $S(\mathcal{S}_i)$ with respect to the topology of uniform convergence in the space of the bounded scalar functions on T_i. Let $f_i \in \overline{S(\mathcal{S}_i)}$ and let $(f_{i,n_i})_{n_i=1}^{\infty} \subset S(\mathcal{S}_i)$ be such that*

$$\lim_{n_i \to \infty} ||f_i - f_{i,n_i}||_{T_i} = 0$$

for $i = 1, 2, \ldots, d$. If $\Upsilon : \prod \mathcal{S}_i \to Y$ is a d-multimeasure, then

$$\lim_{n_1, n_2, \ldots, n_d \to \infty} \int_{(A_i)} (f_{i,n_i}) d\Upsilon$$

exists in Y, uniformly with respect to $(A_i) \in \prod \mathcal{S}_i$. Moreover, this limit is independent of the converging sequences (f_{i,n_i}).

The above theorem motivates the following

DEFINITION 2.7. Let $f_i \in \overline{S(\mathcal{S}_i)}$ and let $(f_{i,n_i})_{n_i=1}^{\infty}$ and Υ be as in Theorem 2.6. Then we say that (f_i) is Υ-integrable and the Υ-integral of (f_i) over $(A_i) \in \prod \mathcal{S}_i$ is defined as

$$\int_{(A_i)} (f_i) d\Upsilon = \lim_{n_1,\ldots,n_d} \int_{(A_i)} (f_{i,n_i}) d\Upsilon.$$

Moreover, the Υ-integral of (f_i) over (T_i) is defined as that on $\prod N(f_i)$, where $N(f_i) = \{t_i \in T_i : f_i(t_i) \neq 0\}$, $i = 1, 2, \ldots, d$.

We have the following generalized Lebesgue bounded convergence theorem (shortly, LBCT) for the Υ-integrable functions in $\prod \overline{S(\mathcal{S}_i)}$. See Theorem 3 of [8].

THEOREM 2.8. Suppose $f_{i,n_i} \in \overline{S(\mathcal{S}_i)}$ for $n_i = 1, 2, \ldots$ and $f_{i,n_i}(t_i) \to f_i(t_i)$ as $n_i \to \infty$, for each $t_i \in T_i$ and for $i = 1, 2, \ldots, d$. Also suppose

$$\sup_{n_i=1,2,\ldots,\, i=1,2,\ldots,d} \|f_{i,n_i}\|_{T_i} \leq C \,(< \infty).$$

Then

$$\lim_{n_1,\ldots,n_d \to \infty} \int_{(A_i)} (f_{i,n_i}) d\Upsilon = \int_{(A_i)} (f_i) d\Upsilon$$

for all $(A_i) \in \prod \mathcal{S}_i$.

Let $\mathcal{B}^{\Omega}(T_i)$ denote the smallest class of bounded scalar functions on T_i containing $C_0(T_i)$, which is closed under the operation of pointwise limits of uniformly bounded sequences of functions. In other words, if $C_0(T_i) \subset \mathcal{C}$ and if, for $(f_n)_1^{\infty} \subset \mathcal{C}$ with $\sup_n \|f_n\|_{T_i} < \infty$ and with $f_n(t_i) \to f(t_i)$ as $n \to \infty$, for each $t_i \in T_i$, it follows that $f \in \mathcal{C}$, then $\mathcal{B}^{\Omega}(T_i) \subset \mathcal{C}$. Thus $\mathcal{B}^{\Omega}(T_i)$ is the class of all bounded Baire functions on T_i, which also coincides with the family of all bounded Baire measurable scalar functions on T_i.

THEOREM 2.9. Let $\Upsilon : \prod_1^d \sigma \mathcal{B}_0(T_i) \to Y$ be a Baire d-multimeasure. Then:

$$\|\Upsilon\|(T_i) = \sup\{|\int_{(A_i)} (f_i) d\Upsilon| : f_i \in C_0(T_i), \|f_i\|_{A_i} \leq 1, A_i \in \sigma \mathcal{B}_0(T_i)\}.$$

3. A Theorem of Uniqueness

The following theorem states that a Y-valued d-multimeasure Υ on $\prod \sigma \mathcal{B}_0(T_i)$ is determined by the integrals $\int_{(T_i)} (f_i) d\Upsilon$, $(f_i) \in \prod C_0(T_i)$.

THEOREM 3.1.. *Suppose* $\Upsilon_1, \Upsilon_2 : \prod_1^d \sigma \mathcal{B}_0(T_i) \to Y$ *are d-multimeasures such that*

(1)
$$\int_{(T_i)} (f_i) d\Upsilon_1 = \int_{(T_i)} (f_i) d\Upsilon_2$$

for all $(f_i) \in \prod_1^d C_0(T_i)$. *Then* $\Upsilon_1 = \Upsilon_2$.

PROOF. Let $C_i \in \mathcal{K}_0(T_i)$. By Theorem 55.B of Halmos [12] there exists a sequence $\{h_{i,n_i}\}_{n_i=1}^{\infty}$ in $C_0(T_i)$ such that $h_{i,n_i}(t_i) \searrow \chi_{C_i}(t_i)$, for each $t_i \in T_i$ and for $i = 1, 2, \ldots, d$. Then by (1) and by LBCT (Theorem 2.8) we have $\Upsilon_1(C_1) = \Upsilon_2(C_2)$. Thus

$$(2) \qquad \Upsilon_1(C_1, \ldots, C_d) = \Upsilon_2(C_1, \ldots, C_d)$$

for $C_i \in \mathcal{K}_0(T_i)$, $i = 1, 2, \ldots, d$. Let

$$\Sigma_1 = \{E_1 \in \sigma\mathcal{B}_0(T_1) : \Upsilon_1(E_1, C_2, \ldots, C_d) = \Upsilon_2(E_1, C_2, \ldots, C_d)$$
$$\text{for } C_i \in \mathcal{K}_0(T_i), i = 2, 3, \ldots, d\}.$$

By (2) it follows that $\mathcal{K}_0(T_1) \subset \Sigma_1$. Consequently, by the separate finite additivity of Υ_1 and Υ_2 we conclude that $R(\mathcal{K}_0(T_1))$, the ring generated by $\mathcal{K}_0(T_1)$ is contained in Σ_1.

Let $(E_n)_{n=1}^{\infty}$ be a monotone sequence in Σ_1, with $E = \lim_n E_n$. Then by the separate σ-additivity of Υ_1 and Υ_2 we have

$$\begin{aligned}
\Upsilon_1(E, C_2, \ldots, C_d) &= \lim_n \Upsilon_1(E_n, C_2, \ldots, C_d) \\
&= \lim_n \Upsilon_2(E_n, C_2, \ldots, C_d) \\
&= \Upsilon_2(E, C_2, \ldots, C_d)
\end{aligned}$$

for each $C_i \in \mathcal{K}_0(T_i)$, $i = 2, 3, \ldots, d$, since $E_n \in \Sigma_1$ for all n. Thus $E \in \Sigma_1$ and consequently, Σ_1 is a monotone class containing $R(\mathcal{K}_0(T_1))$. Then by Theorem 6.B of Halmos [12], Σ_1 coincides with $\sigma\mathcal{B}_0(T_1)$ and thus

$$(3) \qquad \Upsilon_1(E_1, C_2, \ldots, C_d) = \Upsilon_2(E_1, C_2, \ldots, C_d)$$

for all $E_1 \in \sigma\mathcal{B}_0(T_1)$ and for all $C_i \in \mathcal{K}_0(T_i)$, $i = 2, 3, \ldots, d$.

Now let

$$\Sigma_2 = \{E_2 \in \sigma\mathcal{B}_0(T_2) : \Upsilon_1(E_1, E_2, C_3, \ldots, C_d) = \Upsilon_2(E_1, E_2, C_3, \ldots, C_d)$$
$$\text{for all } E_1 \in \sigma\mathcal{B}_0(T_1) \text{ and for all } C_i \in \mathcal{K}_0(T_i), \quad i = 3, \ldots, d\}.$$

By (3), $\mathcal{K}_0(T_2) \subset \Sigma_2$. By an argument similar to that given above for Σ_1, it is easy to show that $R(\mathcal{K}_0(T_2)) \subset \Sigma_2$ and that Σ_2 is a monotone class. Then by Theorem 6.B of Halmos [12] we conclude that $\Sigma_2 = \sigma\mathcal{B}_0(T_2)$. Continuing this argument step by step, in the dth step we have

$$\Upsilon_1(E_1, E_2, \ldots E_{d-1}, C_d) = \Upsilon_2(E_1, E_2, \ldots, E_{d-1}, C_d)$$

for all $E_i \in \sigma\mathcal{B}_0(T_i)$, $i = 1, 2, \ldots, d-1$ and for $C_d \in \mathcal{K}_0(T_d)$, which shows that $\mathcal{K}_0(T_d) \subset \Sigma_d$, where

$$\Sigma_d = \{E_d \in \sigma\mathcal{B}_0(T_d) : \Upsilon_1(E_1, \ldots, E_{d-1}, E_d) = \Upsilon_2(E_1, \ldots, E_{d-1}, E_d)$$
$$\text{for } E_i \in \sigma\mathcal{B}_0(T_i), i = 1, 2, \ldots, d-1\}.$$

Then, as in the above, Σ_d is a monotone class containing $R(\mathcal{K}_0(T_d))$ and hence $\Sigma_d = \sigma\mathcal{B}_0(T_d)$. This shows that $\Upsilon_1 = \Upsilon_2$.

4. Direct Proof of Theorem 2 of Dobrakov [8]

With the preparation given in the earlier sections, we shall now present a direct proof of the said theorem of Dobrakov (Theorem 2 of [8]) and then deduce the multilinear extension theorem of Pelczyński [14] as a corollary.

THEOREM 4.1. *Let* $U : \prod_1^d C_0(T_i) \to Y$ *be a bounded d-linear operator. Suppose either (A)* $c_0 \not\subset Y$, *or (B)* U *is weakly compact. Then there exists a unique d-multimeasure* Υ *on* $\prod_1^d \sigma\mathcal{B}_0(T_i)$ *with values in* Y *such that*

$$U(f_i) = \int_{(T_i)} (f_i)d\Upsilon, \quad (f_i) \in \prod_1^d C_0(T_i).$$

If $\hat{U}^d : \prod_1^d \mathcal{B}^\Omega(T_i) \to Y$ *is defined by*

$$\hat{U}^d(g_i) = \int_{(T_i)} (g_i)d\Upsilon, \ (g_i) \in \prod_1^d \mathcal{B}^\Omega(T_i)$$

then $\hat{U}^d$ *is well defined, bounded and d-linear. Moreover,* $\hat{U}^d$ *extends* U, $\|\hat{U}^d\| = \|U\| = \|\Upsilon\|(T_i)$ *and* $\hat{U}^d$ *satisfies the following property (P):*
    Let $(f_{i,n_i})_{n_i=1}^\infty \subset \mathcal{B}^\Omega(T_i)$ *with*

$$\sup_{n_i=1,2,\dots} \|f_{i,n_i}\|_{T_i} \le C < \infty$$

and with

$$\lim_{n_i\to\infty} f_{i,n_i}(t_i) = f_i(t_i)$$

for each $t_i \in T_i$ *and for* $i = 1, 2, \dots, d$. *Then*

$$\lim_{n_1,\dots,n_d\to\infty} \hat{U}^d(f_{1,n_1},\dots,f_{d,n_d}) = \hat{U}^d(f_1,\dots,f_d).$$

If U *is weakly compact, then* $\hat{U}^d$ *is also weakly compact.*
    Finally, the bounded d-linear extension $\hat{U}^d$ *is determined uniquely either by property (P) or by the multilinear integral representation; the range of* Υ *is relatively weakly compact if and only if* U *is weakly compact.*

PROOF. Let us prove the theorem by induction on d. Let $d = 1$. Suppose $c_0 \not\subset Y$. Let $\hat{T}_1$ be the Alexandroff compactification of T_1 by adjunction of the point $\{\infty\}$ and let $\hat{U} : C(\hat{T}_1) \to Y$ be defined by $\hat{U}(f) = U(f - f(\infty))$. Then by Theorem VI.2.15 of [2], $\hat{U}$ is weakly compact and hence $U = \hat{U}|C_0(T_1)$ is weakly compact. Thus, in both the cases (A) and (B), by Lemma 2 of Kluvánek [13] there exists a unique Y-valued σ-additive regular Borel measure G on $\mathcal{B}(T_1)$ such that

$$Uf = \int_{T_1} fdG, \qquad f \in C_0(T_1).$$

Let $\Upsilon = G|\sigma\mathcal{B}_0(T_1)$. Then by Theorem 8 of [3], f is Υ- integrable,

$$Uf = \int_{T_1} f d\Upsilon, \qquad f \in C_0(T_1)$$

and by Theorem 1 of [8],$\|\breve{U}\| = \|\Upsilon\|(T_1)$. Clearly, for the second adjoint U^{**} of U we have

$$U^{**}f = \int_{T_1} f d\Upsilon, \qquad f \in \mathcal{B}^\Omega(T_1).$$

Now let us define $\hat{U}^1 = U^{**}|\mathcal{B}^\Omega(T_1)$. Condition (P) holds in virtue of LBCT. As $\mathcal{B}^\Omega(T_1)$ is closed for pointwise limits of bounded sequences, it follows that property (P) implies the uniqueness of the extension $\hat{U}^1$ of U by an argument of transfinite induction. The integral representation of U also determines $\hat{U}^1$ uniquely by Theorem 3.1. Moreover, if U is weakly compact, then U^{**} is weakly compact by the Gantmacher theorem and hence $\hat{U}^1$ is weakly compact.

Up to some stage we closely follow the proof of Pelczyński [14]. Suppose the result holds for $d - 1$. For $(f_i) \in \prod_1^{d-1} C_0(T_i)$, let

$$U_{f_1,\dots,f_{d-1}} : C_0(T_d) \to Y$$

be given by

$$U_{f_1,\dots,f_{d-1}}(f_d) = U(f_1,\dots,f_d), \qquad f_d \in C_0(T_d).$$

Then $U_{f_1,\dots,f_{d-1}}$ is a bounded linear operator on $C_0(T_d)$ with values in Y. Then, by the case $d = 1$ established above, for both the cases (A) and (B) there is a unique σ-additive vector measure

$$\Upsilon_{f_1,\dots,f_{d-1}} : \sigma\mathcal{B}_0(T) \to Y$$

such that

$$U_{f_1,\dots,f_{d-1}}(f_d) = \int_{T_d} f_d \, d\Upsilon_{f_1,\dots,f_{d-1}}, \qquad f_d \in C_0(T_d).$$

Fixing f_d in $C_0(T_d)$, let us define

$$U_{f_d} : \prod_1^{d-1} C_0(T_i) \to Y$$

by

$$U_{f_d}(f_1,\dots,f_{d-1}) = U(f_1,\dots,f_d) \in Y.$$

Clearly, U_{f_d} is a bounded $(d-1)$-linear mapping. When U is weakly compact, clearly U_{f_d} is weakly compact. Thus, when $c_0 \not\subset Y$, or when U is weakly compact, the induction hypothesis implies that there is a $(d-1)$-multimeasure $\Upsilon_{f_d} : \prod_1^{d-1} \sigma\mathcal{B}_0(T_i) \to Y$ such that

$$(1) \qquad U_{f_d}(f_1,\dots,f_{d-1}) = \int_{(T_i)_1^{d-1}} (f_1,\dots,f_{d-1}) d\Upsilon_{f_d}$$

for $(f_i)_1^{d-1} \in \prod_1^{d-1} C_0(T_i)$. On the other hand,

(2) $\quad U_{f_d}(f_1, \ldots, f_{d-1}) = U(f_1, \ldots, f_d) = U_{f_1, \ldots, f_{d-1}}(f_d) = \int_{T_d} f_d \, d\Upsilon_{f_1, \ldots, f_{d-1}}.$

Thus by (1) and (2)

(3) $\qquad \int_{(T_i)_1^{d-1}} (f_1, \ldots, f_{d-1}) \, d\Upsilon_{f_d} = \int_{T_d} f_d \, d\Upsilon_{f_1, \ldots, f_{d-1}}$

for all $f_i \in C_0(T_i)$, $i = 1, 2, \ldots, d$.

For $g_d \in \mathcal{B}^\Omega(T_d)$, let us define

$$U_{g_d} : \prod_1^{d-1} C_0(T_i) \to Y$$

by

(4) $\qquad U_{g_d}(f_1, \ldots, f_{d-1}) = \int_{T_d} g_d \, d\Upsilon_{f_1, \ldots, f_{d-1}}.$

Since g_d is a bounded Baire measurable function, U_{g_d} is well defined. Moreover,

$$\int_{T_d} f_d \, d\Upsilon_{f_1, \ldots, f_{i-1}, \alpha f_i + \beta f_i', f_{i+1}, \ldots, f_{d-1}} = U(f_1, \ldots, f_{i-1}, \alpha f_i + \beta f_i', f_{i+1}, \ldots, f_d)$$

$$= \alpha U(f_1, \ldots, f_d) + \beta U(f_1, \ldots, f_i', \ldots, f_d)$$

$$= \alpha \int_{T_d} f_d \, d\Upsilon_{f_1, \ldots, f_{d-1}}$$

$$+ \qquad + \beta \int_{T_d} f_d \, d\Upsilon_{f_1, \ldots, f_{i-1}, f_i', f_{i+1}, \ldots, f_{d-1}}$$

for $\alpha, \beta \in I\!\!K, f_i \in C_0(T_i)$, $i = 1, 2, \ldots, d$. Then by Theorem 3.1 it follows that

$$\Upsilon_{f_1, \ldots, f_{i-1}, \alpha f_i + \beta f_i', f_{i+1}, \ldots, f_{d-1}} = \alpha \Upsilon_{f_1, \ldots, f_{d-1}} + \beta \Upsilon_{f_1, \ldots, f_{i-1}, f_i', f_{i+1}, \ldots, f_{d-1}}.$$

Using the above equality in (4), we conclude that U_{g_d} is a $(d-1)$-linear operator. We claim that U_{g_d} is bounded. In fact,

$$\|U_{g_d}\| = \sup\{|U_{g_d}(f_1, \ldots, f_{d-1})| : \|f_i\|_{T_i} \leq 1, f_i \in C_0(T_i), 1 \leq i \leq d-1\}$$

$$\leq \|g_d\|_d \sup\{\|\Upsilon_{f_1, \ldots, f_{d-1}}\|(T_d) : \|f_i\|_{T_i} \leq 1, f_i \in C_0(T_i), 1 \leq i \leq d-1\}.$$

Since $U_{f_1, \ldots, f_{d-1}}(f_d) = \int_{T_d} f_d \, d\Upsilon_{f_1, \ldots, f_{d-1}}$ by (2),

$$\|U_{f_1, \ldots, f_{d-1}}\| = \|\Upsilon_{f_1, \ldots, f_{d-1}}\|(T_d).$$

Therefore,

(5) $\qquad \|U_{g_d}\| \leq \|g_d\|_{T_d} \sup\{|U_{f_1, \ldots, f_{d-1}}(f_d)| : \|f_i\|_{T_i} \leq 1, f_i \in C_0(T_i)\}$

$\qquad\qquad = \|g_d\|_{T_d} \|U\| < \infty$

If $c_0 \not\subset Y$, then by induction hypothesis there exists a $(d-1)$-multimeasure

$$\Upsilon_{g_d} : \prod_1^{d-1} \sigma\mathcal{B}_0(T_i) \to Y$$

such that

(6A) $$U_{g_d}(f_1, \ldots, f_{d-1}) = \int_{(T_i)_i^{d-1}} (f_1, \ldots, f_{d-1}) d\Upsilon_{g_d}$$

for each $g_d \in \mathcal{B}^\Omega(T_d)$, $f_i \in C_0(T_i)$, $i = 1, 2, \ldots, d-1$. Moreover, its unique bounded $(d-1)$-linear extension $\tilde{U}_{g_d}$ satisfying property (P) is given by

(7A) $$\tilde{U}_{g_d}(g_1, \ldots, g_{d-1}) = \int_{(T_i)_i^{d-1}} (g_1, \ldots, g_{d-1}) d\Upsilon_{g_d}$$

for $g_i \in \mathcal{B}^\Omega(T_i)$, $i = 1, 2, \ldots, d-1$. Further, by induction hypothesis, $\|U_{g_d}\| = \|\tilde{U}_{g_d}\| = \|\Upsilon_{g_d}\|(T_i)_1^{d-1}$.

Suppose now U is weakly compact. Let $B_i = \{f_i \in C_0(T_i) : \|f_i\|_{T_i} \le 1\}$. Then the range $U(B_1 \times \cdots \times B_d)$ is convex and is relatively weakly compact in Y. Therefore, by Corollary V.3.14 of Dunford and Schwartz [11], the norm closure of $U(B_1 \times \cdots \times B_d)$ is weakly compact. Let $K = $ closure of $U(B_1 \times \cdots \times B_d)$.

We claim that U_{g_d} is weakly compact, whenever U is so. In fact, by (4) it suffices to prove the result for $\|g_d\|_{T_d} \le 1$.

Let $\|g_d\|_{T_d} \le 1$. Let $(f_{d,n})_{n=1}^\infty \subset B_d$ and let $\lim_n f_{d,n}(t) = g_d(t)$, for each $t \in T_d$. Let $f_i \in B_i$, $i = 1, 2, \ldots, d-1$. Then by LBCT and by (3),(4) and (1)

$$\begin{aligned}
U_{g_d}(f_1, \ldots, f_{d-1}) &= \int_{T_d} g_d \, d\Upsilon_{f_1,\ldots,f_{d-1}} \\
&= \lim_n \int_{T_d} f_{d,n} \, d\Upsilon_{f_1,\ldots,f_{d-1}} \\
&= \lim_n \int_{(T_i)_1^{d-1}} (f_1, \ldots, f_{d-1}) d\Upsilon_{f_{d,n}} \\
&= \lim_n U(f_1, \ldots, f_{d-1}, f_{d,n}) \in K.
\end{aligned}$$

Let $\Sigma = \{g_d \in \mathcal{B}^\Omega(T_d) : \|g_d\|_{T_d} \le 1$ and $U_{g_d}(B_1 \times \cdots \times B_{d-1}) \subset K\}$. By the above argument, the closed unit ball of the first Baire class $\mathcal{B}^1(T_d)$ of bounded functions is contained in Σ. By a usual argument of transfinite induction, and by applying LBCT as in the above, it can be shown that Σ coincides with the closed unit ball of $\mathcal{B}^\Omega(T_d)$. Thus U_{g_d} is a weakly compact operator for $g_d \in \mathcal{B}^\Omega(T_d)$ with $\|g_d\|_{T_d} \le 1$ and consequently, for arbitrary $g_d \in \mathcal{B}^\Omega(T_d)$.

Thus in case (B), by induction hypothesis, there is a $(d-1)$-multimeasure $\Upsilon_{g_d} : \prod_1^{d-1} \sigma\mathcal{B}_0(T_i) \to Y$ such that

(6B) $$U_{g_d}(f_1, \ldots, f_{d-1}) = \int_{(T_i)_1^{d-1}} (f_1, \ldots, f_{d-1}) d\Upsilon_{g_d}$$

for each $g_d \in \mathcal{B}^{\Omega}(T_d)$, $f_i \in C_0(T_i)$, $i = 1, 2, \ldots, d-1$. Moreover, its unique bounded $(d-1)$-linear extension $\tilde{U}_{g_d}$ satisfying property (P) is given by

$$(7B) \qquad \tilde{U}_{g_d}(g_1, \ldots, g_{d-1}) = \int_{(T_i)_1^{d-1}} (g_1, \ldots, g_{d-1}) \, d\Upsilon_{g_d}$$

for $g_i \in \mathcal{B}^{\Omega}(T_i)$, $i = 1, 2, \ldots, d-1$, by virtue of LBCT. Further, by induction hypothesis, $\|U_{g_d}\| = \|\tilde{U}_{g_d}\| = \|\Upsilon_{g_d}\|(T_i)_1^{d-1}$.

In the light of (6A) and (7A), or (6B) and (7B), the $(d-1)$-multimeasure Υ_{g_d} is well defined for $g_d \in \mathcal{B}^{\Omega}(T_d)$ and hereafter let us treat cases (A) and (B) simultaneously.

We define

$$U^d : C_0(T_1) \times \cdots \times C_0(T_{d-1}) \times \mathcal{B}^{\Omega}(T_d) \to Y$$

by putting

$$U^d(f_1, \ldots, f_{d-1}, g_d) = U_{g_d}(f_1, \ldots, f_{d-1}), \text{ for } g_d \in \mathcal{B}^{\Omega}(T_d).$$

Since U_d is $(d-1)$-linear, U^d is $(d-1)$-linear in the first $(d-1)$ coordinates. Moreover, by (4) it is clear that U^d is separately linear on the dth coordinate also. Therefore, U^d is d-linear.

$$\begin{aligned}
\|U^d\| &= \sup\{|U_{g_d}(f_1, \ldots, f_{d-1})| : \|f_i\|_{T_i} \leq 1, \\
&\qquad \|g_d\|_{T_d} \leq 1, \, f_i \in C_0(T_i), \, g_d \in \mathcal{B}^{\Omega}(T_d)\} \\
&= \sup\{\|U_{g_d}\| : g_d \in \mathcal{B}^{\Omega}(T_d), \|g_d\|_{T_d} \leq 1\} \\
&\leq \|U\| < \infty
\end{aligned}$$

by (5). Thus U^d is bounded. Moreover, since $U^d | \prod_1^d C_0(T_i) = U$, it follows that $\|U^d\| = \|U\|$.

We define

$$\hat{U}^d : \mathcal{B}^{\Omega}(T_1) \, X \ldots X \, \mathcal{B}^{\Omega}(T_d) \to Y$$

by putting

$$\hat{U}^d(g_1, \ldots, g_d) = \int_{(T_i)_1^{d-1}} (g_1, \ldots, g_{d-1}) \, d\Upsilon_{g_d} = \tilde{U}_{g_d}(g_1, \ldots, g_{d-1}).$$

By 7(A) or 7(B) the operator $\hat{U}^d$ is well defined. By (6A) (resp. (6B)) $\hat{U}^d$ extends U^d. Obviously, $\hat{U}^d$ is separately linear in the first $(d-1)$ coordinates.

Now

$$(8) \qquad \hat{U}^d(g_1, \ldots, g_{d-1}, \alpha g_d + \beta g_d') = \int_{(T_i)_1^{d-1}} (g_1, \ldots, g_{d-1}) d\Upsilon_{\alpha g_d + \beta g_d'}$$

and

$$(9) \quad \alpha \hat{U}^d(g_1, \ldots, g_{d-1}, g_d) + \beta \hat{U}^d(g_1, \ldots, g_{d-1}, g_d')$$
$$= \alpha \int_{(T_i)_1^{d-1}} (g_1, \ldots, g_{d-1}) \, d\Upsilon_{g_d} + \beta \int_{(T_i)_1^{d-1}} (g_1, \ldots, g_{d-1}) d\Upsilon_{g_d'}$$

for $g_i \in \mathcal{B}^\Omega(T_i)$, $i = 1, 2, \ldots, d$, and $g_d' \in \mathcal{B}^\Omega(T_d)$. Let $g_i = f_i \in C_0(T_i)$, for $i = 1, 2, \ldots, d-1$. Then by (6A) (resp.(6B)) and (4)

$$
\begin{aligned}
(10) \qquad \hat{U}^d(f_1, \ldots, f_{d-1}, \alpha g_d + \beta g_d') &= \int_{T_d} (\alpha g_d + \beta g_d') d\Upsilon_{f_1, \ldots, f_{d-1}} \\
&= \alpha \int_{(T_i)_1^{d-1}} (f_1, \ldots, f_{d-1}) d\Upsilon_{g_d} \\
&\quad + \beta \int_{(T_i)_1^{d-1}} (f_1, \ldots, f_{d-1}) d\Upsilon_{g_d'} \\
&= \int_{(T_i)_1^{d-1}} (f_1, \ldots, f_{d-1}) d(\alpha \Upsilon_{g_d} + \beta \Upsilon_{g_d}')
\end{aligned}
$$

Thus by Theorem 3.1, (8) and (10) we have

$$
\alpha \Upsilon_{g_d} + \beta \Upsilon_{g_d'} = \Upsilon_{\alpha g_d + \beta g_d'}
$$

and consequently, by (9) it follows that $\hat{U}^d$ is separately linear on the dth coordinate also. Thus $\hat{U}^d$ is d-linear.

$$
\begin{aligned}
||\hat{U}^d|| &= \sup\{|\hat{U}^d(g_1, \ldots, g_d)| : ||g_i||_{T_i} \le 1, \, g_i \in \mathcal{B}^\Omega(T_i)\} \\
&= \sup\{|\tilde{U}_{g_d}(g_1, \ldots, g_{d-1})| : ||g_i||_{T_i} \le 1, \, g_i \in \mathcal{B}^\Omega(T_i), \, 1 \le i \le d\} \\
&= \sup\{||\tilde{U}_{g_d}|| : ||g_d||_{T_d} \le 1, \, g_d \in \mathcal{B}^\Omega(T_d)\} \\
&= \sup\{||U_{g_d}|| : ||g_d||_{T_d} \le 1, \, g_d \in \mathcal{B}^\Omega(T_d)\} \\
&\le ||U|| < \infty
\end{aligned}
$$

by (5). Thus $\hat{U}^d$ is bounded. Since $\hat{U}^d$ extends U^d and $||U^d|| = ||U||$, we conclude that $||\hat{U}^d|| = ||U||$.

We define $\Upsilon(\cdot) : \prod_1^d \sigma\mathcal{B}_0(T_i) \to Y$ by putting

$$
\Upsilon(A_i) = \Upsilon_{\chi_{A_d}}(A_1, \ldots, A_{d-1}).
$$

Since $\chi_{A_d} \in \mathcal{B}^\Omega(T_d)$,

$$
\Upsilon(A_i) = \int_{(T_i)_1^{d-1}} (\chi_{A_1}, \ldots, \chi_{A_{d-1}}) d\Upsilon_{\chi_{A_d}} = \hat{U}^d(\chi_{A_1}, \ldots, \chi_{A_d}) \in Y
$$

and is well defined. By (4) and (6A) (resp. (6B))

$$
(11) \qquad \int_{T_d} \chi_{A_d} d\Upsilon_{f_1, \ldots, f_{d-1}} = \int_{(T_i)_1^{d-1}} (f_1, \ldots, f_{d-1}) d\Upsilon_{\chi_{A_d}}
$$

for $f_i \in C_0(T_i)$, $i = 1, 2, \ldots, d-1$. By induction hypothesis, $\Upsilon_{\chi_{A_d}}$ is a $(d-1)$-multimeasure on $\prod_1^{d-1} \sigma\mathcal{B}_0(T_i)$ and hence Υ is separately σ-additive in the first $d-1$ coordinates. To show that Υ is separately σ-additive on the dth coordinate also, let us extend (11) to $(g_i)_1^{d-1}$ in $\prod_1^{d-1} \mathcal{B}^\Omega(T_i)$.

Let $(f_{i,n})_{n=1}^\infty \subset C_0(T_i)$ be bounded and let $\lim_n f_{i,n}(t_i) = g_i(t_i)$, for each $t_i \in T_i$, $i = 1, 2, \ldots, d-1$. Then by LBCT and by (11)

$$\lim_{n\to\infty} \int_{T_d} \chi_{A_d} d\Upsilon_{f_{1,n},\ldots,f_{d-1,n}} = \lim_n \int_{(T_1,\ldots,T_{d-1})} (f_{1,n},\ldots,f_{d-1,n}) d\Upsilon_{\chi_{A_d}}$$

$$= \int_{(T_1,\ldots,T_{d-1})} (g_1,\ldots,g_{d-1}) d\Upsilon_{\chi_{A_d}} \in Y.$$

Thus, for the sequence of σ-additive set functions $\{\Upsilon_{f_{1,n},f_{2,n},\ldots,f_{d-1,n}}\}_{n=1}^\infty$,

$$\lim_{n\to\infty} \Upsilon_{f_{1,n},\ldots,f_{d-1,n}}(A_d) \text{ exists in } Y$$

for each $A_d \in \sigma\mathcal{B}_0(T_d)$. Since Theorem I.4.8 of [2] is valid for σ-rings too and since the uniform σ-additivity is the same as uniform strong additivity on σ-rings for σ-additive vector measures, it follows that there exists a σ-additive Y-valued measure $\Upsilon_{g_1,\ldots,g_{d-1}}$ (say) on $\sigma\mathcal{B}_0(T_d)$ such that

$$\lim_{n\to\infty} \Upsilon_{f_{1,n},\ldots,f_{d-1,n}}(A_d) = \Upsilon_{g_1,\ldots,g_{d-1}}(A_d)$$

and the measure $\Upsilon_{g_1,\ldots,g_{d-1}}$ depends solely on $g_1,\ldots,g_{d-1}$ and is independent of the converging sequences $\{f_{i,n}\}_{n=1}^\infty$, $i = 1, 2, \ldots, d-1$. Then

$$\Upsilon_{g_1,\ldots,g_{d-1}}(A_d) = \int_{T_d} \chi_{A_d} d\Upsilon_{g_1,\ldots,g_{d-1}}$$

$$= \lim_n \int_{T_d} \chi_{A_d} d\Upsilon_{f_{1,n},\ldots,f_{d-1,n}}$$

$$= \lim_n \int_{(T_1,\ldots,T_{d-1})} (f_{1,n},\ldots,f_{d-1,n}) d\Upsilon_{A_d}$$

$$= \int_{(T_1,\ldots,T_{d-1})} (g_1,\ldots,g_{d-1}) d\Upsilon_{A_d}$$

by (11) and by LBCT. Thus (11) holds with g_i in place of f_i, for $i = 1, 2, \ldots, d-1$. This shows that (11) is valid for $(g_i)_1^{d-1} \in \prod_1^{d-1} \mathcal{B}^1(T_i)$, where $\mathcal{B}^1(T_i)$ is the first Baire class of bounded functions on T_i. Assuming the validity of (11) for $(g_i)_1^{d-1} \in \prod_1^{d-1} \mathcal{B}^\beta(T_i)$ for all ordinals β strictly less than a countable ordinal α, one can show by the above argument that the equality (11) holds for all $(g_i)_1^{d-1} \in \prod_1^{d-1} \mathcal{B}^\alpha(T_i)$. Now, by transfinite induction we conclude that (11) holds for all $(g_i)_1^{d-1} \in \prod_1^{d-1} \mathcal{B}^\Omega(T_i)$.

Thus $\Upsilon_{g_1,\ldots,g_{d-1}} : \sigma\mathcal{B}_0(T_d) \to Y$ is well defined and is σ-additive for each

$(g_i)_1^{d-1} \in \prod_1^{d-1} \mathcal{B}^{\Omega}(T_i)$. Consequently, for $(A_i) \in \prod_1^d \sigma \mathcal{B}_0(T_i)$,

$$
\begin{aligned}
\Upsilon(A_i) &= \Upsilon_{\chi_{A_d}}(A_1, \ldots, A_{d-1}) \\
&= \int_{(T_i)_1^{d-1}} (\chi_{A_1}, \ldots, \chi_{A_{d-1}}) d\Upsilon_{\chi_{A_d}} \\
&= \int_{T_d} \chi_{A_d} d\Upsilon_{\chi_{A_1}, \ldots, \chi_{A_{d-1}}} \\
&= \Upsilon_{\chi_{A_1}, \ldots, \chi_{A_{d-1}}}(A_d)
\end{aligned}
$$

(12)

because of the validity of (11) for $(\chi_{A_i})_1^{d-1}$. Therefore, Υ is separately σ-additive on the dth coordinate too and thus Υ is a d-multimeasure on $\prod_1^d \sigma \mathcal{B}_0(T_i)$.

By (12) and by the definition of $\hat{U}^d$ we have

$$
\begin{aligned}
\hat{U}^d(\chi_{A_1}, \ldots, \chi_{A_d}) &= \int_{(T_i)_1^{d-1}} (\chi_{A_1}, \ldots, \chi_{A_{d-1}}) d\Upsilon_{\chi_{A_d}} \\
&= \Upsilon(A_1, \ldots, A_d) \\
&= \int_{(T_i)_1^d} (\chi_{A_1}, \ldots, \chi_{A_d}) d\Upsilon
\end{aligned}
$$

(13)

for $A_i \in \sigma \mathcal{B}_0(T_i)$, $i = 1, 2, \ldots, d$. Fixing $A_1, \ldots, A_{d-1}$, and replacing χ_{A_d} by a $\sigma \mathcal{B}_0(T_d)$-simple function s, by the separate linearity of $\hat{U}^d$ and the definition of the integral we deduce from (13) that

$$
\hat{U}^d(\chi_{A_1}, \ldots, \chi_{A_{d-1}}, s) = \int_{(T_i)_1^d} (\chi_{A_1}, \ldots, \chi_{A_{d-1}}, s) d\Upsilon.
$$

Since each $g_d \in \mathcal{B}^{\Omega}(T_d)$ is the uniform limit of a sequence of $\sigma \mathcal{B}_0(T_d)$-simple functions, it then follows by LBCT that

$$
\hat{U}^d(\chi_{A_1}, \ldots, \chi_{A_{d-1}}, g_d) = \int_{(T_i)_1^d} (\chi_{A_1}, \ldots, \chi_{A_{d-1}}, g_d) d\Upsilon.
$$

Similarly, replacing $\chi_{A_{d-1}}$ by $g_{d-1} \in \mathcal{B}^{\Omega}(T_{d-1})$ and keeping $\chi_{A_1}, \ldots, \chi_{A_{d-2}}, g_d$ fixed, it can be shown that

$$
\hat{U}^d(\chi_{A_1}, \ldots, \chi_{A_{d-2}}, g_{d-1}, g_d) = \int_{(T_i)_1^d} (\chi_{A_1}, \ldots, \chi_{A_{d-2}}, g_{d-1}, g_d) d\Upsilon
$$

for $g_{d-1} \in \mathcal{B}^{\Omega}(T_{d-1})$. Proceeding step by step, finally it follows that

$$
\hat{U}^d(g_1, \ldots, g_{d-1}, g_d) = \int_{(T_i)_1^d} (g_1, \ldots, g_{d-1}, g_d) d\Upsilon
$$

for $g_i \in \mathcal{B}^{\Omega}(T_i)$, $i = 1, 2, \ldots, d$.

Since $\hat{U}^d | \prod_1^d C_0(T_i) = U$,

(14)
$$
U(f_i) = \int_{(T_i)} (f_i) d\Upsilon, \qquad (f_i) \in \prod_1^d C_0(T_i).
$$

Property (P) holds for $\hat{U}^d$ by LBCT. Moreover, $\|\hat{U}^d\| = \|U\|$ and $\|U| = \|\Upsilon\|(T_i)$ by Theorem 2.9.

By Theorem 3.1 and by (14), Υ is determined uniquely by U. Consequently, the operator $\hat{U}^d$ is also determined uniquely by U. If $\hat{U}$ is a bounded d-linear extension of U to $\prod_1^d B^\Omega(T_i)$, satisfying property (P), then by an argument of transfinite induction one can show that property (P) implies $\hat{U} = \hat{U}^d$.

Now, let us show that U is weakly compact if and only if the range of Υ is relatively weakly compact. If U is weakly compact, then let $B_1, \ldots, B_d$ and K be as in the above, where we proved the weak compactness of the operator U_{g_d} in case (B). If $(f_{i,n})_{n=1}^\infty \subset C_0(T_i)$ are bounded sequences with $\|f_{i,n}\|_{T_i} \leq 1$, and if $f_{i,n}(t_i) \to g_i(t_i)$ for each $t_i \in T_i$, then by LBCT

$$\hat{U}^d(g_1, \ldots, g_d) = \lim_n \int_{(T_i)_1^d} (f_{i,n}) d\Upsilon \in K.$$

By an argument of transfinite induction it then follows that

$$\{\hat{U}^d(g_1, \ldots, g_d) : g_i \in B^\Omega(T_i), \|g_i\|_{T_i} \leq 1, i = 1, 2, \ldots, d\} \subset K$$

and hence $\hat{U}^d$ is weakly compact. Consequently, the range of Υ, being contained in K, is relatively weakly compact.

If the range of Υ is relatively weakly compact, then by the argument given on p.292 of [8] we conclude that $\hat{U}^d$ and hence U, is weakly compact.

This completes the proof.

COROLLARY 4.2. *Pelczyński's theorem [14] on multilinear extension (see Introduction) holds also for multilinear operators on $\prod_1^d C_0(T_i)$.*

PROOF. If we define $U^{**} = \hat{U}^d$, where $\hat{U}^d$ is as in Theorem 4.1, then U^{**} is the required bounded d-linear extension of U as in Pelczyński's theorem.

COROLLARY 4.3. *Condition (ii) in Pelczyński's theorem (in Introduction) is the same as condition (P) given in Theorem 4.1.*

PROOF. Clearly, condition (P) of Theorem 4.1 implies condition (ii) of Pelczyński's theorem, since $U^{**} = \hat{U}^d$ by Corollary 4.2. Conversely, if condition (ii) of Pelczyński's theorem holds, then as shown by Dobrakov [8],

$$\Upsilon(A_1, \ldots, A_d) = U^{**}(\chi_{A_1}, \ldots, \chi_{A_d})$$

is a separately σ-additive d-multimeasure on $\prod_1^d \mathcal{B}^\Omega(T_i)$ and

$$U^{**}(g_i) = \int_{(T_i)_1^d} (g_1, \ldots, g_d) d\Upsilon.$$

Then U^{**} satisfies property (P) by LBCT of multimeasures.

5. Concluding Remarks

The range of a σ-additive Banach space valued measure defined on a σ-ring of sets is relatively weakly compact. This result is essentially due to Bartle, Dunford and Schwartz [1]. This result implies that, for a σ-additive vector measure $G(\cdot) : \sigma\mathcal{B}_0(T) \to Y$, Y a Banach space, the operator $U : \mathcal{B}^\Omega(T) \to Y$ given by

$$Uf = \int_T f dG$$

is weakly compact, where T is a locally compact Hausdorff space.

Since there are examples of non weakly compact multilinear operators (see p. 385 of [14]) from $C(S)$ into a Hilbert space, S a compact Hausdorff space, the integral representation of a bounded multilinear operator U on $\prod_1^d C_0(T_i)$, with range in a Banach space Y not containing c_0, does not guarantee the weak compactness of the operator U. Consequently, by Theorem 4.1, the range of the associated multimeasure of U is not relatively weakly compact.

Thus the integral representation of a multilinear operator U on $\prod_1^d C_0(T_i)$ with respect to a d-multimeasure Υ on $\prod_1^d \sigma\mathcal{B}_o(T_i)$ with values in Y does not imply that the multilinear operator U is weakly compact. This is contrary to the situation of bounded linear opearators on $C_0(T)$. This observation has motivated our recent note [10].

References

1. Bartle, R.G., Dunford, N., and Schwartz, J.T., *Weak compactness and vector measures*, Canad. J. Math. **7** (1955), 289–305.
2. Diestel, J., and Uhl, J.J., *Vector masures*, Amer. Math Soc. Surveys, No. 15, Providence, 1977.
3. Dobrakov, I., *On integration in Banach spaces, I*, Czech.Math. J. **20** (1970), 511–536.
4. Dobrakov, I., *On representation of linear operators on $C_0(T,X)$*, Czech. Math. J. **21** (1971), 13–30.
5. Dobrakov, I., *On integration in Banach spaces, VIII*, Czech. Math. J. **37** (1987), 487–506.
6. Dobrakov, I., *On integration in Banach spaces, IX*, Czech. Math. J. **38** (1988), 589–601.
7. Dobrakov, I., *On integration in Banach spaces, X*, Czech. Math. J. **38** (1988), 713-725.
8. Dobrakov, I., *Representation of multilinear operators on $XC_0(T_i)$*, Czech. Math. J. **39** (1989), 288-302.
9. Dobrakov, I., *Multilinear integration of sacalar functions*, Lecture Notes (notes written by T. V. Panchapagesan) to be published in Notas de Matemática, Facultad de Ciencias, Universidad de los Andes, Mérida, Venezuela..
10. Dobrakov, I. and Panchapagesan, T. V., *A direct proof of a theorem on representability of operators*, submitted for publication.
11. Dunford, N., and Schwartz, J. T., *Linear Operators, Part I, General Theory*, Interscience, New York, 1958.
12. Halmos, P.R., *Measure Theory*, D. Van Nostrand, New York, 1950.
13. Kluvánek, I., *Characterization of Forier-Stieltjes transforms of vector and operator valued measures*, Czech. Math. J. **17** (1967), 261–277.
14. Pelczyński, A., *A theorem of Dunford-Pettis type for polynomial operator*, Bull. Acad. Polon. Sci. Sér. Sci. Math. Astr. et Phys. **11** (1963), 379–386.

MATHEMATICAL INSTITUTE, SLOVAK ACADEMY OF SCIENCES, 49, ŠTEFANIKOVA, BRATISLAVA, SLOVAKIA.

DEPARTAMENTO DE MATEMÁTICAS, FACULTAD DE CIENCIAS, UNIVERSIDAD DE LOS ANDES, MÉRIDA, VENEZUELA.

E-mail address: panchapa@ciens.ula.ve

Contemporary Mathematics
Volume **189**, 1995

Compact Perturbations of
Operators on Kreĭn Spaces

MICHAEL A. DRITSCHEL

To Mischa Cotlar, in honor of the occasion of his eightieth birthday

ABSTRACT. For a definitizable operator $A \in \mathcal{L}(\mathcal{H})$, $\mathcal{H}$ a Kreĭn space, $K \in \mathfrak{C}_p$, the Schatten p class, $1 \leq p < \infty$, it is shown that the operator $T = A+K$ has nontrivial hyperinvariant subspaces if it is not a constant multiple of the identity. This is used to show that if B and C are bounded operators on a Hilbert space with B selfadjoint, C positive, and $K \in \mathfrak{C}_p$ selfadjoint with $1 \leq p < \infty$, then $B(C + K)$ has a nontrivial hyperinvariant subspace, as does the operator matrix $\left(\begin{smallmatrix} 0 & B \\ C+K & 0 \end{smallmatrix} \right)$.

1. What is to be done

The main goal of this paper is to prove the following theorem:

THEOREM 1.1. *Let $\mathcal{H}$ be a Kreĭn space, A a definitizable selfadjoint or unitary operator in $\mathcal{L}(\mathcal{H})$, and K a compact operator in $\mathfrak{C}_p$, $1 \leq p < \infty$. Then if $T = A + K$ is not a constant multiple of the identity, there is a nontrivial subspace of $\mathcal{H}$ which is invariant for every operator commuting with T.*

A proof of the theorem will only be given when A is selfadjoint, since the argument would be nearly identical in the unitary case. Applications of this result will be given to the problem of the existence of hyperinvariant subspaces for compact perturbations of the product of selfadjoint operators on a Hilbert space and for certain operator matrices related to compact perturbations of (not necessarily definitizable) selfadjoint operators on a Kreĭn space.

A few of the more relevant definitions of mathematical concepts used in the paper, particularly those pertaining to Kreĭn spaces, are given here. For more

1991 *Mathematics Subject Classification.* Primary 47A15, 47A55, 47B50; Secondary 46C20.

The author wishes to specially thank Peter Jonas for insightful discussions related to this work.

details from a source with notation consistent with that of this paper, the reader is referred to [**2**].

A Kreĭn space $\mathcal{H}$ is the direct sum of a Hilbert space and the antispace of a Hilbert space (that is, a Hilbert space where the usual inner product is replaced by its negative). Such a decomposition of a Kreĭn space is called a fundamental decomposition, and though in general such a decomposition will not be unique, the associated Hilbert spaces are equivalent, and consequently a topology can be unambiguously defined on the space making it possible to discuss topologically dependent notions such as convergence in a meaningful way. The transition from Kreĭn space to an associated Hilbert space is effected by the fundamental symmetry J which has the form $\left(\begin{smallmatrix} 1 & 0 \\ 0 & -1 \end{smallmatrix}\right)$ with respect to the chosen fundamental decomposition. The indefinite inner product on $\mathcal{H}$ is denoted by $\langle \cdot, \cdot \rangle$. Throughout, all Hilbert spaces and Kreĭn spaces are assumed to be over the complex field.

The collection of continuous operators from a Kreĭn space $\mathcal{H}$ to a Kreĭn space $\mathcal{K}$ is denoted by $\mathcal{L}(\mathcal{H}, \mathcal{K})$, and from a Kreĭn space $\mathcal{H}$ to itself by $\mathcal{L}(\mathcal{H})$. Every $A \in \mathcal{L}(\mathcal{H}, \mathcal{K})$ has a well-defined adjoint A^* with respect to the indefinite scalar product. The operator $A \in \mathcal{L}(\mathcal{H})$ is said to be selfadjoint if $A^* = A$. It is unitary if both $A^*A = 1$ and $AA^* = 1$. (In general we leave off the descriptor "Kreĭn space" or "with respect to the indefinite scalar product" for the words "selfadjoint" and "unitary," since the exception will be those operators that have such properties in a Hilbert space sense.) A selfadjoint operator A is positive if $\langle Af, f \rangle \geq 0$ for all $f \in \mathcal{H}$.

Recall that a selfadjoint operator $A \in \mathcal{L}(\mathcal{H})$ is definitizable if there is a polynomial p such that $p(A) \geq 0$. It was first shown by Langer [**9**] that these operators have nontrivial invariant subspaces. The reader is referred to such sources as [**11**] and [**1**] for the theory of definitizable operators. It can be shown that on a Kreĭn space where either the positive or negative part of a fundamental decomposition is finite dimensional, all selfadjoint operators are definitizable [**11**]. If, however, we assume we have a Kreĭn space where these dimensions are both countably infinite, the situation is no longer so simple. In this case the positive and negative parts of the given fundamental decomposition are isomorphic, and so we may without loss of generality assume that they are copies of the same space. Then by changing coordinates, it is possible to take the fundamental symmetry to be of the form $J = \left(\begin{smallmatrix} 0 & 1 \\ 1 & 0 \end{smallmatrix}\right)$ on $\mathcal{H}$ viewed as the Hilbert space direct sum $\mathcal{K} \oplus \mathcal{K}$. With respect to the indefinite scalar product, vectors of the form $f \oplus 0$ or $0 \oplus f$ are neutral; that is, they have zero self scalar product. The positive part of the fundamental decomposition is spanned by vectors of the form $f \oplus f$, and the negative part by vectors of the form $f \oplus -f$, $f \in \mathcal{K}$. A straightforward calculation shows that any selfadjoint operator on $\mathcal{H}$ may be written as $\left(\begin{smallmatrix} T & A \\ B & T^\times \end{smallmatrix}\right)$ on $\mathcal{K} \oplus \mathcal{K}$, where T is an arbitrary continuous operator on $\mathcal{K}$, $T^\times$ is the (Hilbert space) adjoint of T, and A and B are (Hilbert space) selfadjoint operators. By considering the case where A and B are zero, it is apparent that any general structure theory for Kreĭn space selfadjoint operators is as difficult as the invariant subspace problem. The case where T is instead zero is also interesting, and we consider it in greater detail later.

Before going any further though, a proof is sketched of the existence of invariant subspaces for definitizable selfadjoint operators, since it contains ideas that will be subsequently useful. The proof makes use of the following two results, proofs of which may be found in [2] and [1], respectively.

LEMMA 1.2 (BOGNÁR AND KRÁMLI). *Let $\mathcal{H}$ be a Kreĭn space, and $A \in \mathcal{L}(\mathcal{H})$ selfadjoint. Then there is a Kreĭn space $\mathcal{D}$, and an operator $D \in \mathcal{L}(\mathcal{D}, \mathcal{H})$ with the property that $\ker A = \{0\}$ and $A = DD^*$.*

Note that in the particular case that $A \geq 0$, the space $\mathcal{D}$ in the last lemma is a Hilbert space.

PROPOSITION 1.3 ([1],Theorem 5). *Let $A \in \mathcal{L}(\mathcal{H})$ be selfadjoint. Then there is a factorization $A = DD^*$, $\mathcal{D}$ a Kreĭn space, $D \in \mathcal{L}(\mathcal{D}, \mathcal{H})$ with zero kernel, such that for any T in the commutant of A, there is an operator $L \in \mathcal{L}(\mathcal{D})$ satisfying*

$$TD = DL,$$
and for the selfadjoint operator $\hat{A} = D^*D$,

$$L\hat{A} = \hat{A}L.$$

The proof that every definitizable selfadjoint operator has a nontrivial invariant (indeed, hyperinvariant) subspace now easily follows. For assume that p is a definitizing polynomial for the operator A; that is, $p(A) \geq 0$. We only consider the case that $p(A) \neq 0$. Using the above lemma and theorem we can factor $p(A) = DD^*$, and for any T commuting with A, T also commutes with $p(A)$, and so there is an operator L such that $AD = DL$. By Proposition 1.3, the operator L commutes with the (Hilbert space) selfadjoint operator D^*D. In the case that $D^*D = \lambda 1_{\mathcal{D}}$, we have $(p(A))^2 = D(D^*D)D^* = \lambda p(A)$; that is, A is algebraic, and we are done. So assume D^*D is not a constant multiple of the identity. Then it has a nontrivial hyperinvariant subspace $\mathcal{M}$. Let $\mathcal{N} = \overline{D\mathcal{M}}$. This is nonzero since $\ker D = \{0\}$, and cannot be all of $\mathcal{H}$, without contradicting the invariance of $\mathcal{M}$ under D^*D. Finally, we see that by the equation $TD = DL$, the space $\mathcal{M}$ is invariant for T.

With somewhat more work, a spectral theory may be developed for definitizable operators. In [11], Proposition 2.1, this spectral theory is used to prove the next proposition (or rather, a special case of it; but the result stated here is a simple observation based on the what is presented there). In what follows, we denote by $k(z)$ the multiplicity of z as a zero of the polynomial p. If z is not a zero of p, then set $k(z) = 0$.

PROPOSITION 1.4. *The nonreal spectrum of a definitizable selfadjoint operator A consists of a finite number of pairs, $\lambda, \overline{\lambda}$, which are eigenvalues of finite*

Riesz index. If $z_0 \in \sigma(A) \cap \mathbb{R}$, and Λ is a sufficiently short open line segment containing z_0 and not contained in $\mathbb{R}$, then there is a constant C and integer k greater than zero (which depend on z_0) such that for all $z \in \Lambda$,

$$\left\| (A-z)^{-1} \right\| \leq C|z - z_0|^{-k}.$$

We will also need one other result about definitizable operators. The author is unfamiliar with a source for it, though it is no doubt known. Our notation is that $\sigma_p(A)$ is the set of point spectra for A.

LEMMA 1.5. *Let A be a definitizable selfadjoint operator. Then $\sigma_p(A)$ is countable.*

PROOF. Assume that $\lambda \in \sigma_p(A)$, and that λ is not a root of the definitizing polynomial or one of the nonreal spectra for A (the totality of such points being a finite set in any case). Then λ is a real number. Let I be a closed interval in the real line containing λ in its interior and containing no root of the definitizing polynomial. Using the spectral theorem for definitizable selfadjoint operators (see, eg. [11] or [1]), there are spectral projections $E(I)$ and $E(\mathbb{R} \setminus I)$ corresponding to this interval and its complement, and the restriction of A to the to the range of one of these projections has spectrum in the corresponding interval. Hence λ is in the point spectrum of $E(I)A|\mathrm{ran}\,(E(I))$. But $\mathrm{ran}\,E(I)$ is either a closed positive definite or closed negative definite subspace (that is, a Hilbert space or the anti-space of a Hilbert space), and so $E(I)A|\mathrm{ran}\,(E(I))$ may be viewed as a Hilbert space selfadjoint operator. For such operators, the point spectra form a countable set. Since the totality of the intervals I with rational endpoints and not including the zeros of the definitizing polynomial covers all but a finite number of points of the spectrum of A, the result follows.

The proof of Theorem 1.1 will mimic the proof of similar results for compact perturbations of normal operators found in §6.2 of *Invariant Subspaces* by Radjavi and Rosenthal [14], with a few twists. We state here several results that may be found in this source, and which will prove useful in the exposition.

WEYL'S THEOREM. *If $A \in \mathcal{L}(\mathcal{H})$, and K compact, then*

$$\sigma(A+K) \subseteq \sigma(A) \cup \sigma_p(A+K).$$

LOMONOSOV'S THEOREM. *Every operator commuting with a nonzero compact operator and which is not a constant multiple of the identity has a nontrivial hyperinvariant subspace.*

In accordance with [14], we define a *smooth Jordan arc* as a one-to-one function $z(t) = x(t) + iy(t)$ mapping the interval $(0,1)$ to the complex numbers which is twice differentiable on $(0,1)$. A bounded operator T is said to have an *exposed arc* Γ in its spectrum if there is an open disk Δ such that $\Delta \cap \sigma(T) = \Gamma$, Γ a smooth Jordan arc.

The next theorem is Theorem 6.3 of [**14**]. The original idea is apparently due to J.T. Schwartz, though a number of other authors, including K. Kitano and J. Stampfli, contributed to the subsequent sequence of theorems presented in [**14**], of which we make free use of below. The reader is referred to the remarks at the end of Chapter 6 of [**14**] for more complete details and citations.

THEOREM 1.6. *Let $T \in \mathcal{L}(\mathcal{H})$ be such that $\sigma(T)$ contains an exposed arc Γ and let k be a positive integer. Suppose that for each point $z_0 \in \Gamma$ and each closed line segment Λ which meets $\sigma(T)$ only at $\{z_0\}$ and which is not tangent to Γ, there exists a constant $\tilde{C}$ such that*

$$\left\|(T - z)^{-1}\right\| \leq exp\left\{\tilde{C}|z - z_0|^{-k}\right\}$$

for all $z \in \Lambda$ not equal to z_0. Then T has a nontrivial hyperinvariant subspace.

We also need the next lemma, which is a variation on Lemma 6.11 of [**14**].

LEMMA 1.7. *Let $T = A + K$, A a continuous definitizable selfadjoint operator, K in $\mathfrak{C}_p$ for some integer $p > 1$. Suppose that $\sigma_p(T) = \emptyset$ and that $\sigma(A)$ contains an open real line segment Γ. Let $z_0 \in \Gamma$ and let Λ be any closed bounded line segment with z_0 as an endpoint which is not tangent to Γ and such that $\Lambda \cap \sigma(A) = z_0$. Then there is a constant C such that*

$$\left\|(T - z)^{-1}\right\| \leq exp\left\{C|z - z_0|^{-p(k_A+1)-1}\right\}$$

for all $z \in \Lambda \setminus \{z_0\}$.

PROOF. The proof is nearly the same as that of Lemma 6.11 of [**14**]. However, instead of using the formula $\left\|(A - z)^{-1}\right\| = 1/d(z, \sigma(A))$ which is valid for normal operators, one uses Proposition 1.4. Other than this, the arguments are identical.

Continuing to follow [**14**], we have the next theorem.

PROPOSITION 1.8. *Let A be a continuous definitizable selfadjoint operator, $K \in \mathfrak{C}_p$ for some $p \geq 1$, and $T = A + K$. If $\sigma(A)$ contains an open segment of the real line, then T has a nontrivial hyperinvariant subspace.*

PROOF. The proof is from Theorem 6.12 of [**14**]. It requires knowing that the point spectrum of A is countable (Lemma 1.5 above). One uses this and Weyl's Theorem, which gives $\sigma(A) \subseteq \sigma(T) \cup \sigma_p(A)$ to conclude that $\sigma(T)$ contains a dense subset of an open segment of the real line, and hence the whole segment. In particular, this means that T is not a constant multiple of the identity. If T has point spectra, the result follows. If it does not, Weyl's Theorem gives $\sigma(T) \subseteq \sigma(A)$, and so $\sigma(T)$ contains an exposed arc. Then the estimate on the norm of the resolvent given in Lemma 1.7 is combined with Theorem 1.6 to yield the result.

COROLLARY 1.9. *Let A be a continuous definitizable selfadjoint operator, $K \in \mathfrak{C}_p$ for some $p \geq 1$, and $T = A + K$. If $\sigma(T)$ contains more than one point, then T has a nontrivial hyperinvariant subspace.*

PROOF. This is similar to Corollary 6.13 of [**14**]. We may assume that $\sigma_p(T) = \emptyset$, since otherwise by the assumption that T is not a constant multiple of the identity, we are done. Then Weyl's theorem tells us that $\sigma(T) \subseteq \sigma(A)$. If $\sigma(T)$ is not a connected subset of the real line, we reach the required conclusion by the Riesz decomposition theorem. Otherwise we may assume that $\sigma(T)$ is a closed real line segment, which is contained in $\sigma(A)$. The result is then a consequence of Proposition 1.8.

Before considering the case where the spectrum of the perturbation is a single point, we digress with the following lemma on positive operators on Kreĭn spaces.

LEMMA 1.10. *Let $A \in \mathcal{L}(\mathcal{H})$ be a positive selfadjoint operator such that the spectrum of A consists of a single point. Then either that point is a nonzero real number, in which case A is a constant multiple of the identity, or it is zero, in which case $A^2 = 0$.*

PROOF. Use Lemma 1.2 to factor $A = DD^*$, where $D \in \mathcal{L}(\mathcal{D}, \mathcal{H})$, $\mathcal{D}$ a Hilbert space, and $\ker D = \{0\}$. We assume that $A \neq 0$ and so $D \neq 0$. Then $AD = DL$, where $L = D^*D$ is Hilbert space selfadjoint. We see that $\sigma(A) \cup \{0\} = \sigma(L) \cup \{0\}$, and since L is Hilbert space selfadjoint, $\sigma(A)$ is contained in the real line. Furthermore, it is not difficult to show that if λ is in the resolvent of A, then it must be in the resolvent of L, and so $\sigma(L) \subseteq \sigma(A)$ (for details see Proposition 7 of [**1**]). Since $D \neq 0$ and $\mathcal{D}$ is a complex Hilbert space, $\sigma(L) \neq \emptyset$, and so $\sigma(L) = \sigma(A)$. But a Hilbert space selfadjoint operator with spectrum a singleton is a constant multiple of the identity.

So suppose the constant, which we call c, is not zero. Then $AD = cD$. Since $c \neq 0$, $(\operatorname{ran} D)^\perp = \ker D^* = \ker A = \{0\}$, and so $A = c1_{\mathcal{H}}$. On the other hand, if $c = 0$, we have

$$\overline{\operatorname{ran} A} = \overline{\operatorname{ran} D} \subseteq \ker D^* = \ker A,$$

and so $A^2 = 0$.

On a Kreĭn space $\mathcal{H}$ we define, as in Hilbert spaces, the ideal $\mathcal{K}(\mathcal{H})$ of compact operators, as well as the the Calkin algebra $\mathcal{C}(\mathcal{H}) = \mathcal{L}(\mathcal{H})/\mathcal{K}(\mathcal{H})$, which is a unital C^* algebra. Denote by π the canonical mapping from $\mathcal{L}(\mathcal{H})$ to $\mathcal{C}(\mathcal{H})$. Then if J is a fundamental symmetry on $\mathcal{H}$, $j = \pi(J)$ is a unitary element of $\mathcal{C}(\mathcal{H})$ with spectrum contained in $\{-1, 1\}$. Furthermore, if $A \in \mathcal{L}(\mathcal{H})$, we can write $A = J\tilde{A}$ ($\tilde{A} = JA$), and the operator A is selfadjoint or positive if and only if $\tilde{A}$ is selfadjoint or positive in the Hilbert space associated to the fundamental symmetry J. If $\tilde{J}$ is the operator which j is mapped to in the GNS representation of $\mathcal{C}(\mathcal{H})$ as a subalgebra of $\mathcal{L}(\tilde{\mathcal{H}})$ for some Hilbert space $\tilde{\mathcal{H}}$, then it may be viewed as a fundamental symmetry making $\tilde{\mathcal{H}}$ into a Kreĭn space. Since Hilbert space selfadjoint and positive elements of $\mathcal{L}(\mathcal{H})$ are sent to Hilbert space selfadjoint and positive elements, respectively, of $\mathcal{L}(\tilde{\mathcal{H}})$, the same is easily seen to be true for Kreĭn space selfadjoint and positive elements.

We are now prepared to look at compact perturbations with a single point in the spectrum.

THEOREM 1.11. *Let $A \in \mathcal{L}(\mathcal{H})$ be a continuous definitizable selfadjoint operator, K compact, and $T = A + K$. If $\sigma(T)$ contains just one point and T is not a constant multiple of the identity, then T has a nontrivial hyperinvariant subspace.*

PROOF. We will assume that $\mathcal{H}$ is not finite dmensional, since the result is then trivial. Now suppose $\sigma(T) = \{\lambda\}$. By Weyl's theorem, $\sigma(A) \subseteq \{\lambda\} \cup \sigma_p(A)$. Let p be any polynomial. Since the compact operators form an ideal, $p(T) = p(A) + S$, where S is compact. Hence $\pi(p(T)) = \pi(p(A))$, where π is the map taking $\mathcal{L}(\mathcal{H})$ into its Calkin algebra. By the spectral mapping theorem, $\sigma(p(T)) = \{p(\lambda)\}$, and since $\sigma(\pi(p(T))) \subseteq \sigma(p(T))$, we conclude that $\sigma(\pi(p(A))) = p(\lambda)$ for any polynomial p. In particular, taking $p(z) = z$ we see by Proposition 1.4 that λ is real.

Let p be the definitizing polynomial for A. Such a polynomial has real coefficients, and so $p(\lambda) \in \mathbb{R}$ as well. We first examine the case where $p(\lambda) \neq 0$. Let $\tilde{A}$ be the representative in the GNS representation of $\pi(A)$ over a Hilbert space $\tilde{\mathcal{H}}$ made into a Kreĭn space in the manner discussed above. Observe that $\sigma(\tilde{A}) = \lambda$. Furthermore, the operator $\tilde{A}$ is selfadjoint, $p(\tilde{A})$ is positive, and $\sigma(p(\tilde{A})) = p(\lambda)$. Hence $p(\tilde{A}) = p(\lambda)1_{\tilde{\mathcal{H}}}$ by Lemma 1.10, an so $\pi(p(A) - p(\lambda)1_{\mathcal{H}}) = 0$. Thus A is polynomially compact and the result follows from Lomonosov's theorem.

Finally consider the case where $p(\lambda) = 0$. Arguing as before, we have $\tilde{A}$ the GNS representative of $\pi(A)$ and $p(\tilde{A})$ a positive operator with spectrum $\{0\}$. We again apply Lemma 1.10 to conclude that $(p(A))^2$ is compact, finishing the proof.

The proof of Theorem 1.1 is now a direct consequence of Corollary 1.9 and Theorem 1.11.

2. Applications

We give next a few applications of these results to other invariant subspace problems. First let us turn our attention to an example which was outlined at the beginning of the paper. Let $\mathcal{H} = \mathcal{K} \oplus \mathcal{K}$, $\mathcal{K}$ a Hilbert space and $T = \begin{pmatrix} 0 & A \\ B & 0 \end{pmatrix}$, with $A, B \in \mathcal{L}(\mathcal{K})$. When does T have a nontrivial invariant subspace? Suppose we know that there is a nontrivial subspace $\mathcal{N} \subset \mathcal{K}$ which is invariant for the product AB. Let $\mathcal{M} = \mathcal{N} \oplus \overline{B\mathcal{N}}$. Then

$$\begin{pmatrix} 0 & A \\ B & 0 \end{pmatrix}\begin{pmatrix} f \\ Bg \end{pmatrix} = \begin{pmatrix} ABg \\ Bf \end{pmatrix} \in \mathcal{M},$$

and so by continuity, $\mathcal{M}$ is a nontrivial invariant subspace for T.

In the special case that one of the two operators A or B is (Hilbert space) positive and the other is selfadjoint, a clever observation made by Radjavi and

Rosenthal [15] shows that AB has hyperinvariant subspaces when it is not a constant multiple of the identity. Take A to be positive. The nontrivial case is when AB and BA have zero kernel. Then A and B have zero kernel, and so

$$(AB)A^{1/2} = A^{1/2}(A^{1/2}BA^{1/2})$$

and

$$A^{1/2}B(AB) = (A^{1/2}BA^{1/2})A^{1/2}B$$

shows that the operators AB and $A^{1/2}BA^{1/2}$ are quasisimilar. By a result Sz.-Nagy and Foiaş [16], if one of the operators has a hyperinvariant subspace, then so does the other. Since $A^{1/2}BA^{1/2}$ is selfadjoint (and a constant multiple of the identity if and only if AB is), the result follows.

We modify the above arguments for the case where $A, B \in \mathcal{L}(\mathcal{K})$, $\mathcal{K}$ a Hilbert space, and A Hilbert space selfadjoint, though not necessarily positive. For the time being we make no assumptions about B other than that the kernels of A and B are zero, both have dense range, and that AB is not a constant multiple of the identity. Use the Bognár and Krámli factorization lemma (Lemma 1.2) to write $A = DD^*$, where $D \in \mathcal{L}(\mathcal{D}, \mathcal{K})$, $\mathcal{D}$ a Kreĭn space, and $\ker D = \{0\}$. Then since AB has dense range and zero kernel,

$$(AB)D = D(D^*BD) \qquad \text{and} \qquad D^*B(AB) = (D^*BD)D^*B,$$

gives that AB is quasisimilar to D^*BD. Likewise

$$(BA)BD = BD(D^*BD) \qquad \text{and} \qquad D^*(BA) = (D^*BD)D^*$$

yields BA also quasisimilar to D^*BD. Observe that our assumptions preclude D^*BD being a constant multiple of the identity since $\operatorname{ran} D$ is dense.

Specializing to the case where B is the sum of a positive Hilbert space selfadjoint operator $\tilde{B}$ and K, a compact operator in $\mathfrak{C}_p$, $1 \leq p < \infty$, we note first of all that $D^*\tilde{B}D$ is definitizable (the definitizing polynomial being $p(z) = z$), and secondly that

$$D^*BD = D^*\tilde{B}D + \tilde{K},$$

where $\tilde{K} = D^*KD$ is in the same Schatten p-class as K. Thus Theorem 1.1 applies, meaning that $D^*\tilde{B}D$ has a nontrivial hyperinvariant subspace. The same will be true for AB and BA if either K is selfadjoint, or if both A and B have dense range and zero kernel.

We summarize the discussion in the following theorem.

THEOREM 2.1. *Suppose A, B in $\mathcal{L}(\mathcal{K})$, $\mathcal{K}$ a Hilbert space, with A selfadjoint, and assume that AB is not a constant multiple of the identity. Factor $A = DD^*$, $D \in \mathcal{L}(\mathcal{D}, \mathcal{K})$, $\mathcal{D}$ a Kreĭn space, and $\ker D = \{0\}$. If either $\ker AB$ or $\ker BA$ is nonzero, then AB has a nontrivial hyperinvariant subspace. If both of these kernels are zero and both A and B have dense range, then AB has a nontrivial hyperinvariant subspace if and only if D^*BD has such a subspace. In particular, if B is the sum of a positive operator C and a compact operator K in $\mathfrak{C}_p$, $1 \leq$*

$k < \infty$, and either K is selfadjoint or $C + K$ has dense range, then AB and BA have nontrivial hyperinvariant subspaces.

It turns out that the operator matrix $\begin{pmatrix} 0 & A \\ B & 0 \end{pmatrix}$ also has nontrivial hyperinvariant subspaces whenever AB does, assuming A to be Hilbert space selfadjoint. In order to show this, we need to first examine operators of the form $T = \begin{pmatrix} 0 & 1 \\ C & 0 \end{pmatrix}$, where C is assumed to have a nontrivial hyperinvariant subspace $\mathcal{N}$. Then $\mathcal{N} \oplus \mathcal{N}$ is a nontrivial hyperinvariant subspace for T. For if $X = \begin{pmatrix} X_1 & X_2 \\ X_3 & X_4 \end{pmatrix}$ commutes with T, then

$$X_1 = X_4, \qquad X_4 C = C X_1, \qquad X_3 = C X_2, \qquad X_2 C = X_3.$$

Hence X_1, X_3, and X_4 all commute with C and so leave $\mathcal{N}$ invariant. Also $X_3 \mathcal{N} = X_2 C \mathcal{N} \subseteq \mathcal{N}$, which gives the result.

Now suppose that A is selfadjoint and that we have factored it as DD^* in the usual way. Then

$$\begin{pmatrix} 0 & A \\ B & 0 \end{pmatrix} \begin{pmatrix} 0 & 1 \\ B & 0 \end{pmatrix} = \begin{pmatrix} 0 & 1 \\ B & 0 \end{pmatrix} \begin{pmatrix} 0 & 1 \\ AB & 0 \end{pmatrix}$$

and

$$\begin{pmatrix} 0 & A \\ AB & 0 \end{pmatrix} \begin{pmatrix} 0 & A \\ B & 0 \end{pmatrix} = \begin{pmatrix} 0 & 1 \\ AB & 0 \end{pmatrix} \begin{pmatrix} 0 & A \\ AB & 0 \end{pmatrix}$$

shows that $T = \begin{pmatrix} 0 & A \\ B & 0 \end{pmatrix}$ and $\begin{pmatrix} 0 & 1 \\ AB & 0 \end{pmatrix}$ are quasisimilar if A and B have zero kernel and dense range. (If this is not the case, then T either has a nontrivial kernel or does not have dense range, and so obviously has a nontrivial hyperinvariant subspace.) By the above discussion then and Theorem 2.1, we have the next theorem.

THEOREM 2.2. *Suppose A, B in $\mathcal{L}(\mathcal{K})$, $\mathcal{K}$ a Hilbert space, with A selfadjoint and A and B not both identically zero. Set $T = \begin{pmatrix} 0 & A \\ B & 0 \end{pmatrix}$. If either $\ker AB$ or $\ker BA$ is nonzero, then T has a nontrivial hyperinvariant subspace. If both of these kernels are zero, then T has a nontrivial hyperinvariant subspace if AB does. In particular, if B is the sum of a positive operator and a compact operator in $\mathfrak{C}_p$, $1 \le p < \infty$, then T has a nontrivial hyperinvariant subspaces.*

In the last theorem it is not necessary to assume that AB is not a constant c times the identity, since in this case $T^2 - c = 0$, and the result is elementary.

We close with a couple of observations. At first glance, it might seem that solving the invariant subspace problem for the product of two selfadjoint operators would be easier than the general invariant subspace problem, but this is probably not so, since any selfadjoint operator on a Kreĭn space is the product of two Hilbert space selfadjoint operators. For example, if $A \in \mathcal{L}(\mathcal{H})$ is Kreĭn space selfadjoint and J is a fundamental symmetry for $\mathcal{H}$, then J is Hilbert space selfadjoint, as is JA, and $A = J(JA)$.

While the main result presented here gives some insight into the structure of compact perturbations of definitizable operators on Kreĭn spaces, for practical applications one would ideally like to know much more. For example, what sort of functional calculus can be developed for such operators? Do there exist maximal positive and negative invariant subspaces? Unfortunately the techniques used here do not appear to give any direct insight into these more difficult problems. There has been some work done in this direction by P. Jonas, H. Langer, and B. Najman. The conditions on the unperturbed operator tend to be more restrictive, as in [10] and [3], where it is assumed that these operators are fundamentally reducible (that is, commuting with a fundamental symmetry). However, the operators are also allowed to be unbounded, and the perturbations, though selfadjoint, may be in the Macaev ideal, which contains all of the compact ideals considered here. The reader is also referred to [4], [7], [8], and [13]. The papers [5] and [12] demonstrate particularly interesting progress in this area. Related to the applications given in this paper, [6] also considers compact perturbations of operator matrices of the form $\left(\begin{smallmatrix} 0 & 1 \\ B & 0 \end{smallmatrix} \right)$, B Hilbert space selfadjoint, in conection with an investigation of the perturbed Klein-Gordon equation.

Finally, it would be interesting to know under what conditions there is some sort of converse to Theorem 2.2; that is, when the existence of hyperinvariant subspaces for matrices of the form $\left(\begin{smallmatrix} 0 & A \\ B & 0 \end{smallmatrix} \right)$ implies the existence of hyperinvariant subspaces for the product AB.

References

1. M. A. Dritschel, *A method for constructing invariant subspaces for some operators on Krein spaces*, Operator Theory, Adv. Appl., Vol. 61, Birkhäuser-Verlag, Basel-Boston, 1993, pp. 85–113.

2. M. A. Dritschel and J. Rovnyak, *Extension theorems for contraction operators on Krein spaces*, Extension and Interpolation of Linear Operators and Matrix Functions, Oper. Theory: Adv. Appl., Vol. 47, Birkhäuser, Basel-Boston, 1990, pp. 221–305, MR 92m:47068.

3. P. Jonas, *Compact perturbations of definitizable operators. II*, J. Operator Theory **8** (1982), 3–18, MR 84d:47046.

4. _____, *On a class of unitary operators in Krein space*, Advances in invariant subspaces and other results of operator theory (Timisoara and Herculane, 1984), Oper. Theory: Adv. Appl., 17, Birkhäuser, Basel-Boston, 1986, pp. 151–172, MR 89a:47056.

5. _____, *On a class of selfadjoint operators in Krein space and their compact perturbations*, Integral Equations Operator Theory **11** (1988), 351–384, MR 89f:47052.

6. _____, *On the spectral theory of operators associated with perturbed Klein-Gordon and wave type equations*, Preprint, P-MATH-29/90 (1990), Akademie der Wissenschaften der DDR, Karl-Weierstrass-Institut fur Mathematik, Berlin.

7. P. Jonas and H. Langer, *Compact perturbations of definitizable operators*, J. Operator Theory **2** (1979), 63–77, MR 81c:47039.

8. _____, *Some questions in the perturbation theory of J-nonnegative operators in Krein spaces*, Math. Nachr. **114** (1983), 205–226, MR 85j:47045.

9. H. Langer, *Spektraltheorie Linearer Operatoren in J-Räumen und Einege Anwendungen auf die Schar $L(\lambda) = \lambda^2 I + \lambda B + C$*, Habilitationsschrift, Fakultät für Mathematik und Naturwissenschaften der Technischen Universität Dresden, Dresden, 1965.

10. _____, *Spektralfunktionen einer Klasse J-selbstadjungierter Operatoren*, Math. Nachr. **33** (1967), 107–120, MR 35#775.

11. _____, *Spectral functions of definitizable operators in Krein spaces*, Functional Analysis (Dubrovnik, 1981) Lecture Notes in Math., Vol. 948, Springer, Berlin-New York, 1982, pp. 1–46, MR 84g:47034.

12. H. Langer and B. Najman, *Perturbation theory for definitizable operators in Krein spaces*, J. Operator Theory **9** (1983), 297–317, MR 85e:47053.
13. B. Najman, *Perturbation theory for selfadjoint operators in Pontrjagin spaces. I*, Glas. Mat. Ser. III **15(35)** (1980), 351–371, MR 82e:47048.
14. H. Radjavi and P. Rosenthal, *Invariant Subspaces*, Springer-Verlag, New York-Heidelberg, 1973, MR 51#3924.
15. _____, *Invariant subspaces for products of Hermitian operators*, Proc. Amer. Math. Soc. **43** (1974), 483–484, MR 58#23663.
16. B. Sz.-Nagy and C. Foiaş, *Harmonic Analysis of Operators on Hilbert Space*, North Holland, New York, 1970, MR 43#947.

DEPARTMENT OF MATHEMATICS, KERCHOF HALL, THE UNIVERSITY OF VIRGINIA CHARLOTTESVILLE, VIRGINIA 22903-3199

E-mail address: mad6n@virginia.edu

Contemporary Mathematics
Volume **189**, 1995

Almost-Orthogonality
and Weighted Inequalities

JAVIER DUOANDIKOETXEA

Dedicated to Mischa Cotlar with affection and respect

ABSTRACT. We study weighted inequalities for a wide class of maximal operators and singular integrals by using almost-orthogonality properties of a suitable decomposition of the operators, uniform boundedness of each term in the decomposition and interpolation with change of measure. The applications include the spherical maximal operator, homogeneous maximal operators and singular integrals and operators along curves.

1. Introduction

The L^2 boundedness of an operator which can be written as a series of mutually orthogonal uniformly bounded L^2 operators is trivially obtained. Out of this ideal situation we can still get L^2 boundedness when the terms in the sum are not orthogonal but almost-orthogonal, which means that the composition of two operators has sufficiently small norm so that a numerical series converges. The concept of almost-orthogonality has a close link with the name of Mischa Cotlar because of his celebrated lemma formulated in 1955 in an article published in the late Revista Matemática Cuyana (see [**S**] for an excellent discussion of the lemma, its variants and applications).

The main drawback of Cotlar's lemma is that its validity is restricted to the Hilbert space setting and some kind of L^p extension was considered necessary in view of several interesting applications. The technique developed in a series of papers starting with [**Ch1**] and [**DR**] can be considered as a possible L^p substitute (see [**CaSe**] for an abstract formulation of the method). It was applied in a great variety of cases, in particular, to prove L^p boundedness of rough operators

1991 *Mathematics Subject Classification.* 42B20, 42B25.
Key words and phrases. Maximal operators, singular integrals, weighted inequalities.
Supported in part by CICYT (Ministerio de Educación, Spain), grant PB 90-187.
This paper is in final form and no version of it will be submitted for publication elsewhere.

like homogeneous singular integrals and Hilbert transforms along curves. Simultaneously, the method could be used to prove L^p boundedness of the associated maximal operators.

The aim of this paper is to apply the same technique in order to get weighted inequalities. For a Calderón-Zygmund singular integral with regular kernel they were obtained by R. Coifman and Ch. Fefferman, early after Muckenhoupt's results for the Hardy-Littlewood maximal operator ([**GR**] will be our main reference for most results concerning weights). Homogeneous kernels of the type p.v. $\Omega(x)|x|^{-n}$ without regularity (Ω is a homogeneous of degree zero function for which only a size condition like $\Omega \in L^r(S^{n-1}), r \geq 1$ is assumed) do not fall under the scope of the Coifman-Fefferman theorem but in [**DR**] we were able to show that if Ω is a bounded function, the singular integral operator is bounded in $L^p(w), \forall w \in A_p$, exactly as if the kernel was regular. After this, [**Wa1**] and [**D2**] extended it to different values of r, showing that the class of weights becomes smaller as r approaches 1. See also [**Ho2**] for a different type of weighted inequalities for rough operators better suited to obtain vector valued inequalities.

Here we state in a general setting some of the above results beginning with maximal functions. This is the general approach mentioned in Section 2 of [**DV**] where it was applied to the particular case of the dyadic spherical maximal operator. Several results in the present paper overlap with those given by D. Watson in [**Wa2**]; although he only considers operators of convolution type, this fact is not essential. Our main result (Theorem 1) is similar to his but the proofs are very different. The factorization theorem (Theorem 3) is essentially the same as in [**Wa2**]. Since only nonextremal weights can be factorized, D. Watson had the clever idea of proving an extrapolation theorem for this class. We incorporate his result into our paper (Theorem 4) in such a way that the proof of the theorem for singular integrals (Theorem 6) becomes simpler than the original one. In Section 4 we consider the problem of finding an operator $\mathcal{N}$ such that the maximal function and the singular integral are bounded from $L^p(\mathcal{N}u)$ into $L^p(u)$, recovering the results of S. Hofmann in [**Ho2**].

In the applications we obtain all the known weighted inequalities for homogeneous singular integrals and maximal functions, spherical dyadic maximal operator, Hardy-Littlewood maximal function and Hilbert transforms along homogeneous curves, etc. and include some new ones. In particular, we can apply Theorem 6 to simplify the proof given by S. Hofmann in [**Ho1**] for commutators of rough singular integrals (which are not of convolution type) and we also give some inequalities for operators along flat convex curves in the plane. In Section 6 we show how to apply some of the preceding results to get boundedness properties in Morrey spaces.

As usual, a weight w is a nonnegative locally integrable function in $\mathbb{R}^n$ and we denote by $L^p(w)$ the space of (classes of) measurable functions such that

$$||f||_{p,w} = \left(\int_{\mathbb{R}^n} |f(x)|^p w(x)\, dx \right)^{1/p} < +\infty .$$

We also use the notation $L^p(l^q, w)$ for the space of sequences of functions $\{f_j\}$ such that $||\{f_j\}||_{l^q}$ is in $L^p(w)$. For an operator S we define

$$W_p(S) = \{w : S \text{ is bounded in } L^p(w)\}, \quad 1 < p < \infty$$

and if S is nonnegative

$$W_1(S) = \{w \geq 0 : Sw(x) \leq Cw(x) \text{ a.e., for some } C\}.$$

Several times we use interpolation with change of measure. This theorem of Stein and Weiss can be found in [**BL**], p. 115.

2. Maximal operators

Let $\{E_j\}_{j \in \mathbb{Z}}$ be a sequence of positive linear operators uniformly bounded in $L^p(\mathbb{R}^n), 1 \leq p \leq \infty$ and $\{E_j^*\}$ their adjoints. We consider the associated maximal operators

$$\mathcal{M}f(x) = \sup_j |E_j f(x)| \quad \text{and} \quad \mathcal{M}^* f(x) = \sup_j |E_j^* f(x)|$$

which are bounded in L^∞. We notice for future use the inequality

$$E_j f \leq C_s (E_j(f^s))^{1/s}, \ s > 1, \ f \geq 0,$$

(C_s depends on s but not on j or f) easily deduced from the uniform boundedness of E_j in L^∞.

Assume we are given another sequence $\{P_j\}$ of positive self-adjoint operators such that $\mathcal{P}f(x) = \sup_j |P_j f(x)|$ is bounded in $L^p, 1 < p \leq \infty$. Denote by $\mathcal{A}_p$ the class of weights $W_p(\mathcal{P})$ and assume that they factorize through $\mathcal{A}_1 = W_1(\mathcal{P})$, i. e., $w \in \mathcal{A}_p$ if and only if there exist $w_0, w_1 \in \mathcal{A}_1$ such that $w = w_0 w_1^{1-p}$ (which in particular implies that $w \in \mathcal{A}_p$ if and only if $w^{-p'/p} \in \mathcal{A}_{p'}$). In general, $\mathcal{P}$ will be equivalent to (a variant of) the Hardy-Littlewood maximal operator and the class $\mathcal{A}_p$ will correspond to (a variant of) the usual A_p class of Muckenhoupt.

Consider also a Littlewood-Paley decomposition $\{Q_j\}$ given by self-adjoint operators such that

$$\sum_{j=-\infty}^{\infty} Q_j^2 = I$$

in the strong operator topology of L^2 and such that the square function

$$g(f) = \left(\sum_{j=-\infty}^{\infty} |Q_j f|^2 \right)^{1/2}$$

defines a bounded operator in $L^p(w), \forall w \in \mathcal{A}_p, 1 < p < \infty$. By duality we also have

$$\int \Big| \sum_{j=-\infty}^{\infty} Q_j f_j \Big|^p w \le C \int \Big(\sum_{j=-\infty}^{\infty} |f_j|^2 \Big)^{p/2} w, \quad \forall w \in \mathcal{A}_p.$$

The assumption of self-adjointness for P_j and Q_j is just for simplicity.

THEOREM 1. *Let* E_j, P_j *and* Q_j *as before and* $w \in \mathcal{A}_p$. *Assume that*

(1) $\|(E_j - P_j)Q_{j+k}f\|_2 \le C \, 2^{-|k|\alpha}\|f\|_2$ *for some* $\alpha > 0$

and

(2) $\|E_j f\|_{p,w} \le C\|f\|_{p,w}.$

Then, $\mathcal{M}$ *is bounded in* $L^p(w^s), 0 \le s < 1$.

PROOF. We have

$$E_j f = P_j f + \sum_k (E_j - P_j)Q_{j+k}^2 f$$

hence

$$\mathcal{M}f(x) \le \mathcal{P}f(x) + \sum_k \sup_j |(E_j - P_j)Q_{j+k}^2 f|.$$

The L^p norm of each term in the sum has exponential decay in k, namely

(3) $\| \sup_j |(E_j - P_j)Q_{j+k}^2 f| \|_p \le C \, 2^{-|k|\alpha(p)}\|f\|_p, \ \alpha(p) > 0, \ 1 < p < \infty.$

For $p = 2$ it is deduced from (1) in the following way

$$\| \sup_j |(E_j - P_j)Q_{j+k}^2 f| \|_2^2 \le \sum_j \|(E_j - P_j)Q_{j+k}^2 f\|_2^2$$

$$\le \sum_j C \, 2^{-2|k|\alpha}\|Q_{j+k}f\|_2^2 = C \, 2^{-2|k|\alpha}\|f\|_2^2.$$

For $p \ne 2$ one can first prove the inequality with an absolute constant and then interpolate with the L^2 norm. This is easy if $p > 2$ because $\mathcal{M}$ is bounded in L^∞ (alternatively, one can use the forthcoming approach with $w \equiv 1$). In the case $p < 2$, it appears in [Ca]. We will not include it here in order to avoid repetitions since we follow an analogous path to prove the weighted case.

Let $p \ge 2$. The following chain of inequalities is deduced from (2)

$$\|(E_j - P_j)Q_{j+k}^2 f\|_{L^p(l^\infty,w)} \le \|(E_j - P_j)Q_{j+k}^2 f\|_{L^p(l^p,w)} \le C\|Q_{j+k}f\|_{L^p(l^p,w)}$$
$$\le C\|Q_{j+k}f\|_{L^p(l^2,w)} \le C\|f\|_{p,w}.$$

By interpolating with change of measure with (3) we get exponential decay for the weights $w^s, 0 \le s < 1$ and summing a geometric series proves the theorem in this case.

Assume now $p < 2$ and consider the finite family $\{E_j\}_{j=-N}^N$. Due to the uniform boundedness of each E_j we have the estimate

$$\text{(4)} \qquad \|\sup_{|j|\leq N} |E_j f|\,\|_{p,w^s} \leq B(N)\|f\|_{p,w^s}$$

where $B(N)$ is the norm of the operator. We want to bound $B(N)$ by an absolute constant independent of N. Consider the mapping $\{g_j\} \longrightarrow \{E_j g_j\}$. It is bounded on $L^p(l^\infty, w^s)$ with constant $B(N)$ due to (4) and the positivity of E_j and in $L^p(l^p, w)$ with constant independent of N due to (2). By interpolation it is bounded in $L^p(l^2, w^r)$ with constant $CB(N)^{1-\epsilon}, \epsilon > 0$ and $s < r < 1$. Taking $g_j = Q_{j+k} f$ we deduce

$$\|\sup_{|j|\leq N} |E_j Q_{j+k} f|\,\|_{p,w^r} \leq CB(N)^{1-\epsilon}\left\|\left(\sum_{j=-N}^N |Q_{j+k}f|^2\right)\right\|_{p,w^r}$$

$$\leq CB(N)^{1-\epsilon}\|f\|_{p,w^r}.$$

Interpolating again with (3) and summing in k we get

$$\|\sup_{|j|\leq N} |E_j f|\,\|_{p,w^s} \leq (C_1 + C_2 B(N)^\theta)\|f\|_{p,w^s} \quad \text{for some } \theta < 1.$$

Comparing with (4) we deduce $B(N) \leq C_1 + C_2 B(N)^\theta$ so that $B(N)$ is bounded independently of N.

COROLLARY 2. *If $\mathcal{M}$ (resp. $\mathcal{M}^*$) is bounded in $L^p(w)$ for some $w \in \mathcal{A}_p$, then $\mathcal{M}^*$ (resp. $\mathcal{M}$) is bounded in $L^{p'}(w^{-sp'/p}), 0 \leq s < 1$.*

PROOF. If $\mathcal{M}$ is bounded in $L^p(w)$, then the operators E_j are uniformly bounded in $L^p(w)$ and by duality the operators E_j^* are uniformly bounded in $L^{p'}(w^{-p'/p})$. The result is now a consequence of Theorem 1.

The weights of a linear positive operator can be completely described by factorization, i.e., E_j is bounded in $L^p(w)$ if and only if there exist w_0, w_1 such that $E_j^*(w_0) \leq Cw_0$ a.e., $E_j(w_1) \leq Cw_1$ a.e. (that is, $w_0 \in W_1(E_j^*)$ and $w_1 \in W_1(E_j)$) and $w = w_0 w_1^{1-p}$ (see for instance [J]). Remark that the condition $E_j^* w \leq Cw$ a.e. (resp. $E_j w \leq Cw$ a.e.) is necessary and sufficient for the boundedness of E_j (resp. E_j^*) in $L^1(w)$. For weights in $W_p(\mathcal{M})$ we obtain the following factorization theorem:

THEOREM 3 (FACTORIZATION). *(i) If $w_0 \in W_1(\mathcal{M}^*)\cap\mathcal{A}_1$ and $w_1 \in W_1(\mathcal{M})\cap \mathcal{A}_1$, then $w = (w_0 w_1^{1-p})^s$ is in $W_p(\mathcal{M})$ for all $0 \leq s < 1$.*

(ii) If $w \in W_p(\mathcal{M})\cap\mathcal{A}_p$, then for all $0 < s < 1$ there exist $w_0 \in W_1(\mathcal{M}^*)\cap\mathcal{A}_1$ and $w_1 \in W_1(\mathcal{M}) \cap \mathcal{A}_1$ such that $w = (w_0 w_1^{1-p})^{1/s}$.

PROOF. Part (i) is immediate from Theorem 1 because under our assumptions $w_0 w_1^{1-p}$ is in $\mathcal{A}_p$ and is an uniform L^p weight for all E_j.

To prove part (ii), apply first Corollary 2 to deduce that $\mathcal{M}^* + \mathcal{P}$ is bounded in $L^{p'}(w^{-sp'/p})$. Then, $w^s \in W_p(\mathcal{M} + \mathcal{P})$ and $(w^s)^{-p'/p} \in W_{p'}(\mathcal{M}^* + \mathcal{P})$ and we can apply the general factorization theorem ([GR] or [J]) to conclude.

The examples below show that in general we cannot take $s = 1$ in none of the above theorems. See nevertheless Theorem 5.

Factorization is an important tool to deduce extrapolation results but Theorem 3 only allows to factorize nonextremal weights, i.e., those $w \in W_p(\mathcal{M} + \mathcal{P})$ such that $w^{1+\epsilon} \in W_p(\mathcal{M} + \mathcal{P})$ for some $\epsilon > 0$. D. Watson proved the extrapolation theorem only for this class of weights, obtaining a result which is very useful for applications. Following his notation we write

$$AW_p(\mathcal{M}) = \{w \in W_p(\mathcal{M}+\mathcal{P}) : w^{1+\epsilon} \in W_p(\mathcal{M}+\mathcal{P}) \text{ for some } \epsilon > 0\}, 1 \le p < \infty$$

and we define $AW_p(\mathcal{M}^*)$ in a similar way. Theorem 3 allows us to write a usual factorization theorem for weights in $AW_p(\mathcal{M})$, namely: $w \in AW_p(\mathcal{M})$ if and only if there exist $w_0 \in AW_1(\mathcal{M}^*)$ and $w_1 \in AW_1(\mathcal{M})$ such that $w = w_0 w_1^{1-p}$. The extrapolation theorem is as follows:

THEOREM 4 (EXTRAPOLATION). *Let $p_0 \ge 1$, and suppose that T is a (sub)-linear operator which is bounded in $L^{p_0}(w)$ for all $w \in AW_{p_0}(\mathcal{M})$ with a constant which depend only on the AW_{p_0} properties of w. Then T is bounded in $L^p(w)$ for all $w \in AW_p(\mathcal{M}), 1 < p < \infty$.*

The proof is exactly the same as in [**Wa2**] and follows the line of the classical extrapolation theorem in [**R**].

In some cases end-point results can be given, that is, we can get $s = 1$ in Theorem 1 for particular weights w. Small modifications in the proof of Proposition 6 in [**DV**] give the following theorem where we use a pseudonorm instead of the euclidean norm because it could be convenient in some examples. A pseudonorm is a norm function ρ for which the triangle inequality is replaced with the more general: $\rho(x + y) \le k(\rho(x) + \rho(y))$ for some constant $k \ge 1$.

THEOREM 5. *Let ρ be a pseudonorm and $E_j f = K_j * f$ where $\operatorname{supp} K_j \subset \{x : \rho(x) \le 2^j\}$. If $\mathcal{M}$ is bounded in L^p and condition (2) in Theorem 1 holds for $w(x) = \rho(x)^{-\beta}$, $\beta > 0$, then $\mathcal{M}$ is bounded in $L^p(w)$.*

The condition we need on w to make the proof work is

$$\sum_{k=l}^{\infty} \sup_{\rho(x)\sim 2^k} w(x) \le C \inf_{\rho(x)\sim 2^l} w(x)$$

but this is essentially equivalent to being a negative power of $\rho(x)$. This result is related to those in [**SoWe**].

3. Singular integral operators

The singular integral operators we will consider can be written as a series each one of whose terms is pointwise bounded by a positive operator which satisfies the hypotheses of Theorem 1. With the appropriate almost-orthogonality conditions on the terms in the series, weighted inequalities with $AW_p(\mathcal{M})$-weights are obtained. The proof becomes very simple due to the extrapolation theorem.

THEOREM 6. *Let E_j be a sequence of operators satisfying the hypotheses in Theorem 1 and $\mathcal{M}$ the associated maximal operator. Let T_j be another sequence of operators such that $|T_j f(x)| \leq E_j(|f|)(x)$ and*

(5) $\qquad ||T_j Q_{j+k} f||_2 + ||Q_{j+k} T_j f||_2 \leq C \, 2^{-|k|\alpha} ||f||_2 \ \text{ for some } \alpha > 0.$

Then $T = \sum_j T_j$ is bounded in $L^p(w), \forall w \in AW_p(\mathcal{M})$.

PROOF. We prove the theorem for $p = 2$ and extrapolate. Set

$$Tf = \sum_{k,k'}\Big(\sum_j Q^2_{j+k} T_j Q^2_{j+k'} f\Big) = \sum_{k,k'} \tilde{T}_{k,k'} f.$$

Condition (5) implies

$$||\tilde{T}_{k,k'} f||^2_2 \leq C \sum_j ||T_j Q^2_{j+k'} f||^2_2 \leq C \, 2^{-2|k'|\alpha} ||f||^2_2$$

and also

$$||\tilde{T}_{k,k'} f||^2_2 \leq C \sum_j ||Q_{j+k} T_j Q^2_{j+k'} f||^2_2$$

$$\leq C \, 2^{-2|k|\alpha} \sum_j ||Q^2_{j+k'} f||^2_2 \leq C \, 2^{-2|k|\alpha} ||f||^2_2$$

so that

$$||\tilde{T}_{k,k'} f||^2_2 \leq C \, 2^{-(|k|+|k'|)\alpha} ||f||^2_2.$$

Let $w \in AW_2(\mathcal{M})$ and $\epsilon > 0$ such that $w^{1+\epsilon} \in W_2(\mathcal{M}) \cap \mathcal{A}_2$. Then

$$||\tilde{T}_{k,k'} f||^2_{2,w^{1+\epsilon}} \leq C \sum_j ||T_j Q^2_{j+k'} f||^2_{2,w^{1+\epsilon}} \leq C \sum_j ||E_j(Q^2_{j+k'} f)||^2_{2,w^{1+\epsilon}}$$

$$\leq C \sum_j ||Q^2_{j+k'} f||^2_{2,w^{1+\epsilon}} \leq C \, ||f||^2_{2,w^{1+\epsilon}}.$$

Interpolating with change of measure and summing in k and k' the theorem is proved.

Almost everywhere convergence of truncated integrals in the case of operators defined as a principal value integral is related to the boundedness of the maximal operator given by

$$T_* f(x) = \sup_{N,M}\Big|\sum_{j=N}^M T_j f(x)\Big|.$$

We are not able to prove the $L^p(w)$ boundedness of T_* (nor even for $w \equiv 1$) under the general conditions of Theorem 6. The best we can do is to consider convolution operators $T_j f = K_j * f$ with $\operatorname{supp} K_j \subset \{x : \rho(x) \leq 2^j\}$ (ρ is a pseudonorm as above) but then the proof is essentially the same as in [**Ch1**] and [**DR**]. It consists in a modification of the usual Cotlar's inequality for singular integrals with regular kernel

$$T_* f(x) \leq C(M(Tf)(x) + Mf(x))$$

(again a result published in one of his papers in Revista Matemática Cuyana in 1955). In our present situation we need to add to the right-hand side one more term which is a series whose terms have small norm in L^2 (or L^p) and uniform norm in the weighted space.

4. Two-weighted and vector valued inequalities

In many cases it is useful to determine an operator $\mathcal{N}$ such that

$$\int |\mathcal{M}f|^p u \leq C \int |f|^p \mathcal{N}u$$

and the same for T ($\mathcal{N}$ depends on $\mathcal{M}$ and T, of course). Inequalities of this type are useful to derive vector-valued inequalities and they will also be of interest in Section 6. For some of the examples below inequalities of this type were obtained by S. Hofmann [**Ho2**] and more generally by Watson in [**Wa2**]. With some of the applications in mind we will make the following assumptions for simplicity:

(6) $\mathcal{P}(\mathcal{P}u^s)^{1/s} \leq C_s (\mathcal{P}u^s)^{1/s}$ $(s > 1)$ a.e.   and $u \leq C'\mathcal{P}u$ a.e.,

with constants independent of u. The first inequality implies that $(\mathcal{P}u^s)^{1/s}$ is an $\mathcal{A}_1$-weight and combined with the second one provides

$$||\mathcal{P}f||_{p,u} + ||g(f)||_{p,u} \leq C||f||_{p,(\mathcal{P}u^s)^{1/s}}$$

$$\int \Big|\sum_j Q_j g_j\Big|^p u \leq C \int \Big(\sum_j |g_j|^2\Big)^{p/2} (\mathcal{P}u^s)^{1/s}.$$

THEOREM 7. *If E_j and T_j are two sequences of operators satisfying (1) and (5) and $\mathcal{P}$ satisfies (6), given $s > 1$ the following inequalities hold*

(7) $$\int |\mathcal{M}f|^p u \leq C \int |f|^p (\mathcal{P}\mathcal{M}^* u^s)^{1/s}$$

(8) $$\int |Tf|^p u \leq C \int |f|^p (\mathcal{P}\mathcal{M}^* \mathcal{P}u^s)^{1/s}.$$

In the right-hand side we can replace $\mathcal{M}^*$ with any other operator $\mathcal{L}$ satisfying $||E_j f||_{p,u} \leq C||f||_{p,\mathcal{L}u}$.

The proofs are adapted from those of Theorems 1 and 6. In the case of T we cannot use extrapolation and the proof given for Theorem 6 will only provide the result for $p = 2$. Essentially the same proof works for $p > 2$ and for $p < 2$ we need to use the following vector-valued inequality

$$\int \Big(\sum_j |\mathcal{M}g_j|^2\Big)^{p/2} u \leq C \int \Big(\sum_j |g_j|^2\Big)^{p/2} (\mathcal{P}\mathcal{M}^* u^s)^{1/s}.$$

Since this inequality with l^p and l^∞ instead of l^2 is easily deduced from (7) and the positivity of $\mathcal{M}$ we can get l^2 by interpolation.

Different kinds of two-weighted inequalities for $\mathcal{P}$ and g would provide more general statements as can be seen in the application of Section 5.5.

Vector-valued inequalities are obtained from Theorem 7 in the standard way (see [**GR**]):

COROLLARY 8. *Under the hypotheses of the preceding theorem the following vector-valued (weighted) inequalities hold*

$$||\{\mathcal{M}f_j + |Tf_j|\}||_{L^p(l^q,w)} \leq C||\{f_j\}||_{L^p(l^q,w)}$$

for all $w \in AW_p(\mathcal{M})$ *and* $1 < p, q < \infty$.

5. Applications

5.1. Spherical dyadic maximal operator. The spherical dyadic maximal operator fall under the scope of Theorem 1. $E_j f(x)$ is defined as the mean value of f on the sphere of radius 2^j centered at x. Condition (1) is obtained from the size properties of the Fourier transform of the normalized Lebesgue measure on the sphere. All these results appear in [**DV**]. In this example $\mathcal{A}_p$ (which is exactly the Muckenhoupt class A_p) contains $W_p(\mathcal{M})$ making the statement of the theorems simpler.

This example shows that the results in Theorem 1 and Corollary 2 are sharp in the sense that $s = 1$ is forbidden. In fact, E_j satisfies (2) for $w(x) = |x|^{(n-1)(p-1)}$ but $\mathcal{M}$ is not bounded in $L^p(w)$; nevertheless it is bounded in $L^p(|x|^{1-n})$.

5.2. Homogeneous singular integrals and associated maximal operators. Let Ω be a homogeneous function of degree zero whose restriction to the unit sphere is in $L^q(S^{n-1})$ for some $q > 1$. We consider the maximal operator

$$M_\Omega f(x) = \sup_{r>0} r^{-n} \int_{|y|<r} |\Omega(y)f(x-y)|dy$$

and the singular integral (with the additional assumption that Ω has integral zero over S^{n-1})

$$T_\Omega f(x) = \text{p.v.} \int_{\mathbb{R}^n} \frac{\Omega(y)}{|y|^n} f(x-y)dy.$$

Both are classical operators whose L^p estimates can be obtained by the method of rotations. E_j is defined by

$$E_j f(x) = 2^{-jn} \int_{2^j \leq |y| < 2^{j+1}} |\Omega(y)| f(x-y)dy$$

and it is clear that M_Ω and the maximal operator $\mathcal{M}$ associated to these E_j are equivalent $(2^{-n}\mathcal{M}(|f|)(x) \leq M_\Omega f(x) \leq 2^n \mathcal{M}(|f|)(x))$. E_j^* is similar to E_j with $\tilde{\Omega}(y) = \overline{\Omega(-y)}$ and $\mathcal{M}^*$ is equivalent to $M_{\tilde{\Omega}}$. T_j is defined likewise as the restriction to $2^j \leq |y| < 2^{j+1}$ of the integral for T_Ω. We recover from the above theorems all the weighted inequalities in [**D2**] and [**Wa1**] and extend them to include all weights in $AW_p(M_\Omega)$ (as in [**Wa2**]). P_j can be taken as the dyadic dilation of the Poisson kernel and $\mathcal{P}$ is equivalent to the usual Hardy-Littlewood maximal function (denoted subsequently by M). Q_j is also a usual Littlewood-Paley decomposition given by multiplier operators with regular multipliers supported in $2^{j-1} \leq |\xi| \leq 2^{j+1}$. $\mathcal{A}_p$ is the usual A_p class and under some conditions on Ω (for instance, $|\Omega(u)| \geq c > 0$), $W_p(M_\Omega)$ is contained in A_p. Estimates (1) and (5) are easily obtained by using Fourier transform and can be seen in [**DR**].

The two weighted inequality given by Theorem 7 is now

$$(9) \qquad \int_{\mathbb{R}^n} |M_\Omega f(x)|^p u(x)\, dx \leq C \int_{\mathbb{R}^n} |f(x)|^p (MM_{\tilde{\Omega}} u^s(x))^{1/s}\, dx \quad (s > 1)$$

and similarly

$$(10) \qquad \int_{\mathbb{R}^n} |T_\Omega f(x)|^p u(x)\, dx \leq C \int_{\mathbb{R}^n} |f(x)|^p (MM_{\tilde{\Omega}} Mu^s(x))^{1/s}\, dx \quad (s > 1).$$

Nevertheless we can improve this inequality by choosing on the right hand side a different operator for each p, more precisely

$$(11) \qquad \int_{\mathbb{R}^n} |M_\Omega f(x)|^p u(x)\, dx \leq C \int_{\mathbb{R}^n} |f(x)|^p (MM_{\tilde{\Omega}^{r(p,q)}} u^s(x))^{1/s}\, dx \quad (s > 1)$$

where $r(p,q) = p(1-q)+q$ if $p < q'$ since these operators satisfy the requirements of the operator $\mathcal{L}$ in Theorem 7 as Hölder's inequality shows. In the case $p > q'$, $M_\Omega f(x) \leq C(Mf^{q'}(x))^{1/q'}$ and we obtain

$$(12) \qquad \int_{\mathbb{R}^n} |M_\Omega f(x)|^p u(x)\, dx \leq C \int_{\mathbb{R}^n} |f(x)|^p Mu(x)\, dx$$

from the well-known inequality of Fefferman-Stein for the Hardy-Littlewood maximal function (see [**GR**]). Similar modifications apply to T_Ω.

Checking that $M_{\tilde{\Omega}^{r(p,q)}}(|x|^\alpha) \leq C|x|^\alpha$ for $-1-(n-1)p/q' < \alpha \leq 0$ (as in [**D2**]), the preceding inequalities together with Corollary 2 provide sharp power weights: M_Ω and T_Ω are bounded in $L^p(w)$ for $w(x) = |x|^\alpha$ if $\max\{-n, -1-(n-1)p/q'\} < \alpha < \min\{n(p-1), p-1+(n-1)p/q'\}$. This result appears first in [**MWh**] and different approaches appear in [**D2**] and [**SoWe**]. Given q it is possible to find $\Omega \in L^q \setminus \bigcup_{r>q} L^r$ such that M_Ω is bounded in $L^p(|x|^\alpha)$ for $\alpha = -1 - (n-1)p/q'$ (using Theorem 5). This shows that $AW_p(M_\Omega)$ can be strictly contained in $W_p(M_\Omega)$.

5.3. Operators along homogeneous curves. Let A be a real $n \times n$ matrix whose eigenvalues have positive real part. We can associate to A a group of dilations $\delta_t x = t^A x$ and a pseudonorm ρ such that $\rho(\delta_t x) = t\rho(x)$ (see [**SW**]). A two-sided homogeneous curve in $\mathbb{R}^n$ is given by

$$\Gamma(t) = \delta_t \omega_1 \ (t > 0), \quad \Gamma(t) = \delta_{-t} \omega_2 \ (t < 0), \quad \Gamma(0) = 0$$

where $\omega_1, \omega_2 \in \mathbb{R}^n \setminus \{0\}$ and both $\{\Gamma(t), t > 0\}$ and $\{\Gamma(t), t < 0\}$ span $\mathbb{R}^n$. Associated to a curve Γ we define the maximal function and Hilbert transform along it as

$$M_\Gamma f(x) = \sup_{h>0} \frac{1}{h} \left| \int_0^h f(x - \Gamma(t)) \, dt \right| \quad \text{and} \quad H_\Gamma f(x) = \text{p.v.} \int_{-\infty}^\infty f(x - \Gamma(t)) \frac{dt}{t} .$$

E_j is defined by integrating over $2^j \le |t| < 2^{j+1}$ and E_j^* is similar to E_j with $\tilde{\Gamma}(t) = -\Gamma(t)$ instead of Γ. To apply the above theorems we need to adapt P_j and Q_j to the dilation structure of $\mathbb{R}^n$ associated to the matrix A. The Hardy-Littlewood maximal operator (denoted again by M) and Littlewood-Paley theory which correspond to the operators $\mathcal{P}$ and g in the present situation have similar properties to the isotropic case (see [**Ho2**]). To check conditions (1) and (5) we use Fourier transform (see [**SW**], where also the original proof for the unweighted case can be found). Then we can describe weights for M_Γ and H_Γ by factorization and we also obtain the results of [**Ho2**]

$$\int_{\mathbb{R}^n} |M_\Gamma f(x)|^p u(x) \, dx \le C \int_{\mathbb{R}^n} |f(x)|^p (MM_{\tilde{\Gamma}} u^s(x))^{1/s} \, dx \quad (s > 1)$$

$$\int_{\mathbb{R}^n} |H_\Gamma f(x)|^p u(x) \, dx \le C \int_{\mathbb{R}^n} |f(x)|^p (MM_{\tilde{\Gamma}} M u^s(x))^{1/s} \, dx \quad (s > 1).$$

5.4. Commutators of rough singular integrals. In [**Ho1**], S. Hofmann studied weighted inequalities for the operator

$$\mathcal{C}_{k,\Omega} f(x) = \text{p.v.} \int \frac{\Omega(x-y)}{|x-y|^{n+k}} [a(x) - a(y)]^k f(y) \, dy$$

where a is Lipschitz and Ω homogeneous of degree zero. If we define T_j as the part of the integral in $2^j \le |x - y| < 2^{j+1}$, we clearly have $|T_j f(x)| \le C\|\nabla a\|_\infty^k E_j(|f|)(x)$ (E_j as in 5.2). To apply Theorem 6 we only need to check that condition (5) holds. This is the same as inequality (2.18) in page 1285 of [**Ho1**]. As a consequence the proof of Theorems 1.3 and 1.3a in Hofmann's paper can be considerably simplified and the result extended as to include all the nonextremal weights of M_Ω for the same Ω appearing in the definition of $\mathcal{C}_{k,\Omega}$.

5.5. Operators along flat convex curves in the plane. We show next how to write weighted inequalities for some other operators. To simplify matters we will consider convex curves in the plane satisfying the conditions of Corollary 5.3 in [**DR**]. Let $\Gamma(t) = (t, \varphi(t))$ be a $\mathcal{C}^1$ curve in the plane such that $\varphi(0) = \varphi'(0) = 0$. Assume that φ is either even or odd and that $\varphi'(t)$ is a convex increasing function for $t > 0$.

E_j is given by convolution with a singular measure $E_j f = \mu_j * f$ whose Fourier transform only decays in the first variable (see [**DR**]). We choose a Littlewood-Paley decomposition acting only on the first variable and P_{j_s} will the composition of two operators: one of them acts on the first variable and its multiplier is given by a (dilation of a) Schwartz function, the other one acts on the second variable and its multiplier is $\hat{\mu}_j(0, \xi_2)$. The corresponding maximal operator is bounded by the composition of M_1 and M_2, the Hardy-Littlewood maximal operators along the directions of the coordinate axes. The weighted inequalities for $\mathcal{P}$ and g are not the same but the reader will easily modify the above theorems to adapt them to the present case. In particular, we can get the following two-weighted inequalities:

$$\int_{\mathbb{R}^2} |M_\Gamma f(x)|^p u(x)\, dx \leq C \int_{\mathbb{R}^2} |f(x)|^p (M_1(M_{\tilde{\Gamma}} + M_2 M_1) u^s(x))^{1/s}\, dx \quad (s > 1)$$

$$\int_{\mathbb{R}^2} |H_\Gamma f(x)|^p u(x)\, dx \leq C \int_{\mathbb{R}^2} |f(x)|^p (M_1(M_{\tilde{\Gamma}} + M_2 M_1) M_1 u^s(x))^{1/s}\, dx \quad (s > 1).$$

The weight $w(x) = |x|^\alpha$ belongs to $AW_1(\mathcal{M})$ for $-1 < \alpha \leq 0$ so that M_Γ and T_Γ are bounded in $L^p(|x|^\alpha)$ for $-1 < \alpha < p - 1$.

6. Boundedness in Morrey spaces

Given $0 < \lambda < n$ and $p \geq 1$ we define the Morrey space $L^{p,\lambda}(\mathbb{R}^n)$ as the set of functions $f \in L^p_{\text{loc}}$ such that

$$\|f\|_{p,\lambda} = \sup_{r > 0, x \in \mathbb{R}^n} \left(r^{-\lambda} \int_{|y| < r} |f(y)|^p\, dy \right)^{1/p} < +\infty.$$

In [**ChiF**], F. Chiarenza and M. Frasca proved that an operator which is bounded in $L^p(w)$ for all $w \in A_1$, is bounded in $L^{p,\lambda}$ for $0 < \lambda < n$. Their proof can be generalized to the following situation

LEMMA   9. *Let $x_0 \in \mathbb{R}^n$, $R > 0$ and $B(x_0, R)$ the ball centered at x_0 with radius R. If an operator S satisfies the inequality*

$$(13) \qquad \int_{B(x_0,R)} |Sf(x)|^p\, dx \leq C \int_{\mathbb{R}^n} |f(x)|^p \left(\frac{R}{R + |x - x_0|} \right)^\beta dx$$

with C independent of x_0 and R, it is bounded in $L^{p,\lambda}$ for $0 \leq \lambda < \beta$.

To prove the lemma just decompose the right-hand side as a sum where each term is given by the integral over $\{x : 2^j R \leq |x - x_0| < 2^{j+1} R\}$. An operator which is bounded in $L^p(w)$ for $w \in A_1$ satisfies (13) with $\beta = n/s, s > 1$ because $M(\chi_{B(x_0,R)})^{1/s} \sim (R/(R + |x - x_0|))^{-n/s}$ is in A_1. We can use two-weighted inequalities with $u = \chi_{B(x_0,R)}$ to deduce inequalities like (13) and obtain boundedness in Morrey spaces. We have, for instance

THEOREM 10. *(i) The spherical dyadic maximal operator is bounded in $L^{p,\lambda}(\mathbb{R}^n)$ for $0 \leq \lambda < n - 1$.*

(ii) The operators M_Ω, T_Ω and the commutator $\mathcal{C}_{k,\Omega}$ of Section 5.4 with $\Omega \in L^q(S^{n-1})$ are bounded in $L^{p,\lambda}$ for $0 \leq \lambda < \min(n, 1 + (n-1)p/q')$.

The extension to the generalized Morrey spaces of [**M**] is straightforward.

7. Further results and comments

Apart from the applications in Section 5, we can mention the following:

1. More general surfaces can be considered in 5.1, the only requisite being the decay of the Fourier transform of the Lebesgue measure on the surface.

2. In the operators of Section 5.2 we can include a radial function into the kernel, that is, $h(|x|)\Omega(x)$ for the maximal operator and $h(|x|)\Omega(x)|x|^{-n}$ for the singular integral, where h is in $L^s(0,\infty;dr/r)$. The results depend also on s. See [**Wa1**].

3. Non-isotropic analogues of the operators in 5.2 can also be considered. This means kernels of the form $\Omega(x)\rho(x)^{-d}$ where ρ is the pseudonorm defined through a matrix A as in 5.3 and d the associated dimension ($=$ trace of A). Ω has to be homogeneous with respect to the generalized dilations, that is, $\Omega(\delta_t x) = \Omega(x)$. A function $h(\rho(x))$ can also appear in the kernel.

4. The above homogeneous operators and operators along curves can be considered in the context of nilpotent Lie groups as in [**Ch1**].

5. For flat convex curves in the plane, we can generalize the results of Section 5.5 and consider the curves of [**CCC**] or [**CoR**]. P_j and Q_j are again as in 5.5 but we do not have (1); instead we need to introduce a lacunary (smooth) Littlewood-Paley decomposition in the plane given by operators S_j and substitute

$$||(E_j - P_j)(I - S_j)Q_{j+k}f||_2 \le C\,2^{-|k|\alpha}||f||_2$$

for (1). E_j is decomposed now as

$$E_j f = P_j f + (E_j - P_j)S_j f + \sum_k (E_j - P_j)(I - S_j)Q_{j+k}^2 f.$$

We leave the modifications to the reader.

6. The results in Sections 2 and 3 can be extended to the mutiparametric setting. See in [**D1**] the unweighted case.

7. Most examples satisfy the requirements at the end of Section 3, allowing us to treat maximal operators associated to singular integrals.

8. Weighted weak-type (1,1) inequalities are known only for the operators in Section 5.2. In general, even the unweighted inequalities are unknown (see [**Ch2**] for some partial results). The weighted weak type inequalities for M_Ω and T_Ω are similar to (9) and (10) and they imply the one-weight results with $AW_1(M_\Omega)$ weights ([**Va**] and [**Wa3**]). It is possible to obtain (11) from these results and Hölder's inequality.

REFERENCES

[BL]   J. Bergh and J. Löfström, *Interpolation spaces. An introduction*, Springer-Verlag, Berlin, 1976.

[Ca]   A. Carbery, *An almost–orthogonality principle with applications to maximal functions associated to convex bodies*, Bull. Amer. Math. Soc. **14** (1986), 269–273.

[CaSe] A. Carbery and A. Seeger, *Conditionally convergent series of linear operators on L^p-spaces and L^p-estimates for pseudodifferential operators*, Proc. London Math. Soc. **57** (1988), 481–510.

[CCC]   H. Carlsson, M. Christ, A. Córdoba, J. Duoandikoetxea, J.L. Rubio de Francia, J. Vance, S. Wainger and D. Weinberg, *L^p estimates for maximal functions and Hilbert transforms along flat convex curves in $\mathbb{R}^2$*, Bull. Amer. Math. Soc. **14** (1986), 263–267.

[ChiF] F. Chiarenza and M. Frasca, *Morrey spaces and Hardy-Littlewood maximal function*, Rend. Mat. Appl. (7) **7** (1987), 273-279.

[Ch1]   M. Christ, *Hilbert transforms along curves, I: Nilpotent groups*, Annals Math. **122** (1985), 575–596.

[Ch2]   ———, *Weak type (1,1) bounds for rough operators*, Annals of Math. **128** (1988), 19–42.

[ChR]   M. Christ and J.L. Rubio de Francia, *Weak type (1,1) bounds for rough operators II*, Invent. Math. **93** (1988), 225–237.

[CoR]   A. Córdoba and J.L. Rubio de Francia, *Estimates for Wainger's singular integrals along curves*, Rev. Mat. Iberoamericana (1986), 105–117.

[D1]   J. Duoandikoetxea, *Multiple singular integrals and maximal functions along hypersurfaces*, Ann. Inst. Fourier **36**, **4** (1986), 185–206.

[D2]   ———, *Weighted norm inequalities for homogeneous singular integrals*, Trans. Amer. Math. Soc. **336** (1993), 869–880.

[DR]   J. Duoandikoetxea and J.L. Rubio de Francia, *Maximal and singular integral operators via Fourier transform estimates*, Invent. Math. **84** (1986), 541–561.

[DV]   J. Duoandikoetxea and L. Vega, *Spherical means and weighted inequalities*, to appear, J. London Math. Soc..

[GR]   J. García–Cuerva and J. L. Rubio de Francia, *Weighted Norm Inequalities and related Topics*, North–Holland, Amsterdam, 1985.

[Ho1]   S. Hofmann, *Weighted inequalities for commutators of rough singular integrals*, Indiana Univ. Math. J. **39** (1990), 1275–1304.

[Ho2]   ———, *Weighted norm inequalities and vector-valued inequalities for certain rough operators*, Indiana Univ. Math. J. **42** (1993), 1-14.

[J]   B. Jawerth, *Weighted inequalities for maximal operators: linearization, localization and factorization*, Amer. J. Math. **108** (1986), 361-414.

[M]   T. Mizuhara, *Boundedness of some classical operators on generalized Morrey spaces*, Harmonic Analysis, ICM-90 Satellite Conference Proceedings (S. Igari, ed.), Springer-Verlag, Tokio, 1991, pp. 183-189.

[MWh]   B. Muckenhoupt and R.L. Wheeden, *Weighted norm inequalities for singular and fractional integrals*, Trans. Amer. Math. Soc. **161** (1971), 249–258.

[R]   J.L. Rubio de Francia, *Factorization theory and A_p weights*, Amer. J. Math. **106** (1984), 533–547.

[SoWe] F. Soria and G. Weiss, *A remark on singular integral and power weights*, Indiana Univ. Math. J. **43** (1994), 187-204.

[S]   E.M. Stein, *Harmonic Analysis: Real-variable methods, orthogonality and oscillatory integrals*, Princeton Univ. Press, Princeton, N.J., 1993.

[SW]   E.M. Stein and S. Wainger, *Problems in Harmonic Analysis related to curvature*, Bull. Amer. Math. Soc. **84** (1978), 1239–1295.

[Va]   A. Vargas, *Weighted weak type (1,1) bounds for rough operators*, to appear, J. London Math. Soc..

[Wa1]   D. Watson, *Weighted estimates for singular integrals via Fourier transform estimates*, Duke Math. J. **60** (1990), 389-399.

[Wa2]   ———, *Vector-valued inequalities, factorization and extrapolation for a family of rough operators*, J. Funct. Anal. **121** (1994), 389-415.

[Wa3]   ———, *A_1 weights and weak type (1,1) estimates for rough operators*, Preprint.

DEPARTAMENTO DE MATEMÁTICAS, UNIVERSIDAD DEL PAÍS VASCO-EUSKAL HERRIKO UNIBERTSITATEA, APARTADO 644, 48080 BILBAO (SPAIN).

E-mail address: mtpduzuj@lg.ehu.es

Contemporary Mathematics
Volume **189**, 1995

On The Extension of Intertwining Operators

CIPRIAN FOIAS

February, 1994

To Professor Micha Cotlar with best wishes for many active years and many more seminal papers alone or with his students.

ABSTRACT. This is a presentation of the present state of an important and interesting problem in both pure and applied operator theory, namely that of the existence of intertwining extensions of intertwining operators.

Introduction

Extensions of operators which intertwine restrictions of operators on Hilbert spaces have recently occurred as essential tools in nonlinear analytic robust control [FGT1]. The aim of this note is to present and discuss the central extension problem for intertwining operators, with the aim of encouraging the search for a satisfactory solution to it. The problem is the following: Let T' on $\mathcal{H}'$ be an operator with an invariant subspace $\mathcal{H}'_0 \subset \mathcal{H}'$ and let $T'_0 = T'/\mathcal{H}'_0$ be the restriction of T' to $\mathcal{H}'_0$. Let T be another operator on $\mathcal{H}$ and let A be an operator from $\mathcal{H}'_0$ into $\mathcal{H}$ intertwining T and T', that is $AT'_0 = TA$. (Here, as throughout, all spaces are Hilbert spaces.) Find an operator B from $\mathcal{H}'$ into $\mathcal{H}$ intertwining T and T', i.e. $BT' = TB$, extending A. In fact this problem is not always solvable; so one has to find necessary and sufficient conditions for the existence of such an intertwining extension B and then the minimal norm of all such intertwining extensions. This extension problem has been solved in some particular interesting cases. The most useful case is that when T' and T are coisometries (that is when the adjoints T'^* and T^* are isometries) for which the theory of the commutant

---

1991 *Mathematics Subject Classification.* Primary 47A20; Secondary 47A99.
Key words and phrases. Intertwining extension, minimal isometric dilation, commutant lifting theorem, causal commutant lifting theorem.
This research was partially supported by the National Science Foundation grant no. NSF DMS-9024769, the Office of Naval Research grant no. NAVY N00014-91-J-1140, and the Research Fund of Indiana University.

lifting theorem provides a complete answer (see [FF]). In fact, in each new case for which a positive answer to the extension problem exists, it is worth trying to develop for that case the full analogue of the theory of the commutant lifting theorem (uniqueness, parametrization and Schur type parameters, connections to systems theory and to interpolation theory, etc.). Finally, to conclude this short introduction, we note that we can always assume throughout, that the operators T' and T are contractions (that is their norm $\|T'\|$, $\|T\|$ are less or equal to 1).

1. The Problem

Let $T' : \mathcal{H}' \mapsto \mathcal{H}'$ and $T : \mathcal{H} \mapsto \mathcal{H}$ be two contractions and let $\mathcal{H}'_0$ be a subspace of $\mathcal{H}'$, invariant to T'; denote by $T'_0 = T'|\mathcal{H}'_0$ the restriction of T' to $\mathcal{H}'_0$. The orthogonal projection of $\mathcal{H}'$ onto $\mathcal{H}'_0$ will be denoted by $P_{\mathcal{H}'_0}$.

Extension formulation.

Let $A : \mathcal{H}'_0 \mapsto \mathcal{H}$ be an operator intertwining T and T'_0, i.e. $TA = AT'_0$. Find

$$\mu(A; T, T'; \mathcal{H}'_0) = \mu(A) = \min\{\|B\|; B : \mathcal{H}' \mapsto \mathcal{H}, BT' = TB, B|\mathcal{H}'_0 = A\} .$$

(If no such B exists, set $\mu(A) = \infty$.)

Lifting (dilation) formulation.

Let T, T' be as above *but* let $\mathcal{H}'_0$ be invariant to T'^* (instead of being invariant to T'). Let $A : \mathcal{H} \mapsto \mathcal{H}'_0$ satisfy $AT = T'_0 A$ where $T'_0 = P_{\mathcal{H}'_0} T'|\mathcal{H}'_0$; so $T'^*_0 = T'^*|\mathcal{H}'_0$. Find

$$\nu(A; T, T'; \mathcal{H}'_0) = \nu(A) = \min\{\|B\|; B : \mathcal{H} \mapsto \mathcal{H}', BT = T'B, P_{\mathcal{H}'_0} B = A\} .$$

(Again, if no such B exists, set $\nu(A) = \infty$.)

The two formulations are equivalent since it is obvious that

$$\mu(A; T, T'; \mathcal{H}'_0) = \nu(A^*, T^*, T'^*; \mathcal{H}'_0) .$$

Where no confusion can occur, we shall drop in the notation some of the arguments of μ or ν; for instance we will write $\mu(A)$, $\mu(A; T)$, etc.

2. First Example

The Commutant Lifting Theorem (CLThm) ([Sz.-NF1], [Sz.-NF2]; see also [S], [DMP], [P]). *If T, T' are coisometries (i.e. T^*, T'^* are isometries), then $\mu(A) = \|A\|$. If T, T' are isometries, then $\nu(A) = \|A\|$.*

This theorem has many connections with classical and modern analysis and with H^∞–Control Theory (e.g. see [FF]). For further discussion, we need only a couple of comments. First for T and T' isometries we have

$$\nu(A; T, T'; \mathcal{H}'_0) = \nu(A; T, T'|\mathcal{H}'_1; \mathcal{H}'_0)$$

where

$$\mathcal{H}_1' = \bigvee_{n=0}^{\infty} T'^n \mathcal{H}_0' .$$

The reason is that $\mathcal{H}_1'$ reduces T' ([P], [Sz.-NF2]) and thus if $BT = T'B$ then $P_{\mathcal{H}_1'} BT = (T'|\mathcal{H}_1')P_{\mathcal{H}_1} B$. We also note that

$$\mathcal{H}_1' = \mathcal{H}_0' \oplus \underbrace{((T' - T_0')\mathcal{H}_0')^-}_{} \oplus T'((T' - T_0')\mathcal{H}_0')^- \oplus \cdots$$

where each $T'^n((T' - T_0')\mathcal{H}_0')^-$ can be identified with

$$\mathcal{D}_{T_0'} = (D_{T_0'}\mathcal{H}_0')^- \text{ where } D_{T_0'} = (I - T_0'^* T_0')^{1/2} .$$

After this identification $T'|\mathcal{H}_1'$ takes the matrix form

$$\begin{bmatrix} T_0' & & & 0 \\ D_{T_0'} & 0 & & \\ & I & 0 & \\ 0 & & \ddots & \ddots \end{bmatrix} \text{ on } \mathcal{H}_0' \oplus \mathcal{D}_{T_0'} \oplus \mathcal{D}_{T_0'} \oplus \cdots$$

In this sense $T'|\mathcal{H}_1'$ is unique and is called the minimal isometric dilation of T_0' [Sz.-N].

3. Second Example

Generalization (corollary) of the CLThm.

$$T = \text{coisometry} \iff \mu(A;T) = \|A\| \ \forall A \text{ (see [G])}$$
$$T = \text{isometry} \iff \nu(A;T) = \|A\| \ \forall A$$

PROOF. It suffices to prove the second statement. First assume that T is isometric. Let U' on $\mathcal{K}'$ be the minimal isometric dilation of T'. Then $U'^*|\mathcal{H}' = T'^*$, hence $U'^*|\mathcal{H}_0' = T_0'^*$. Apply the CLThm to T, U', A obtaining

$$B' : \mathcal{H} \mapsto \mathcal{K}' , \ U'B' = B'T ,$$
$$\|B'\| = \|A\| , \ A = P_{\mathcal{H}_0'} B' .$$

Set $B = P_{\mathcal{H}'} B'$. Then $BT = T'B$, $P_{\mathcal{H}_0'} B = A$, $\|B\| = \|A\|$, thus $\nu(A,T) \leq \|A\|$.

For the converse implication take $T_0' = T, T'$ on $\mathcal{H}'$ the minimal isometric dilation of T and $A = I$. Then since $\nu(A;T,T';\mathcal{H}')(= \nu(A;T)) = \|A\|$ there exists an operator $B : \mathcal{H} \mapsto \mathcal{H}'$ satisfying the relations $BT = T'B$, $\|B\| = 1$ and $P_{\mathcal{H}} B = I$.

The last two relations imply $B = I$ and consequently the first relation above implies $T'|\mathcal{H} = T$. So we conclude that $T = T'$ is isometric.

4. Reduction

For computing μ we can always assume that T' is an isometry [FGT1]. Precisely:

Let U' on $\mathcal{K}'$ be the minimal isometric dilation of T' and set $\mathcal{K}'_0 = (\mathcal{K}' \ominus \mathcal{H}') + \mathcal{H}'_0$, $U'_0 = U'|\mathcal{K}'_0$, $A' = A P_{\mathcal{H}'_0}|\mathcal{K}'_0$. Then

$$TA' = A'U'_0$$

and

$$\mu(A;T,T';\mathcal{H}'_0) = \mu(A';T,U';\mathcal{K}'_0)$$

for $\forall T$ and $\forall A$ such that

$$TA = AT'_0 \, .$$

By duality it is now clear that for the computation of ν one can always assume that T' is a coisometry.

5. Proof of the CLThm

Probably the most direct proof of the CLThm comes out from Arocena's coupling approach [A] as independently outlined by Cotlar and Sadosky in [CSa], §2. Therefore it is appropriate to give it here. As we already noticed, we can assume that T'_0 is a coisometry. We consider the case when $\|A\| < 1$. Then $\mathcal{H} \times \mathcal{H}'$ endowed with the scalar product

$$(\{h,h'\}, \{k,k'\}) = (h,k) + (h',k') + (Ah,k') + (h',Ak)$$

is a Hilbert space. By identifying $\mathcal{H}$ and $\mathcal{H}'$ with $\mathcal{H} \times \{0\}$ and $\{0\} \times \mathcal{H}'$ respectively, we can assume $\mathcal{H}, \mathcal{H}' \subset \mathcal{H} \times \mathcal{H}'$. Consider the subspaces

$$\mathcal{M} = \mathcal{H} \times (T'^*_0 \mathcal{H}'_0) \, , \; \mathcal{N} = T\mathcal{H} \times \mathcal{H}'_0$$

and for

$$m = \{h,h'\} \in \mathcal{M} \text{ define } Vm = \{Th, T'_0 h'\} \, .$$

Then V is unitary operator from $\mathcal{M}$ onto $\mathcal{N}$ and moreover its adjoint is given by

$$V^*\{h,h'\} = \{T^*h, T'^*_0 h'\}, \{h,h'\} \in \mathcal{N} \, .$$

Note also that $\mathcal{H} \subset \mathcal{M}, V|\mathcal{H} = T$ and $\mathcal{H}'_0 \subset \mathcal{N}, V^*|\mathcal{H}'_0 = T'^*_0$.

Let U on $\mathcal{K}$ be any isometric extension of V with $\mathcal{K} \supset \mathcal{M}, \mathcal{N}$; e.g. take

$$\mathcal{K} = \cdots \oplus \mathcal{N}^\perp \oplus \mathcal{N}^\perp \oplus (\mathcal{H} \times \mathcal{H}'_0) \oplus \mathcal{M}^\perp \oplus \mathcal{M}^\perp \oplus \cdots$$
$$= \cdots \oplus \mathcal{N}^\perp \oplus \mathcal{N}^\perp \oplus (\mathcal{M} \oplus \mathcal{M}^\perp) \oplus \mathcal{M}^\perp \oplus \mathcal{M}^\perp \oplus \cdots$$
$$= \cdots \oplus \mathcal{N}^\perp \oplus \mathcal{N}^\perp \oplus (\mathcal{N} \oplus \mathcal{N}^\perp) \oplus \mathcal{M}^\perp \oplus \mathcal{M}^\perp \oplus \cdots$$

(where $\mathcal{M}^\perp = (\mathcal{H} \times \mathcal{H}'_0) \ominus \mathcal{M}$, $\mathcal{N}^\perp = (\mathcal{H} \times \mathcal{H}'_0) \ominus \mathcal{N}$) and define U by

$$U(\cdots \oplus n^\perp_2 \oplus n^\perp_1 \oplus (m \oplus m^\perp_0) \oplus m^\perp_1 \oplus m^\perp_2 \oplus \cdots) =$$
$$= (\cdots \oplus n^\perp_3 \oplus n^\perp_2 \oplus (n^\perp_1 \oplus Vm) \oplus m^\perp_0 \oplus m^\perp_1 \oplus \ldots) \, .$$

So U is an extension of T and U^* is an extension of $T_0'^*$. Therefore if we set $\mathcal{H}_1' = \bigvee_{n=0}^{\infty} U^n \mathcal{H}_0'$, then $U|\mathcal{H}_1'$ is the minimal isometric dilation of T_0'.

Set $B = P_{\mathcal{H}_1'}|\mathcal{H}$. It is clear that $BT = UB$, $P_{\mathcal{H}_0'}B = A$, $\|B\| \leq 1$. Thus $\nu(A) \leq 1$. Obviously we can now conclude that $\nu(A) = \|A\|$.

6. Third Example

The Commutant Extension Theorem ([CS], [C]). *If T, T' are isometries, then*

$$\mu(A) = \min M$$

where M runs over all positive numbers satisfying

$$\begin{cases} \|A\| \leq M \\ \|(I - T^n T^{*n})Ah_0'\| \leq M\|(I - T'^n T'^*)h_0'\| \ \forall h' \in \mathcal{H}_0', \forall n = 1, 2, \dots \end{cases}$$

COROLLARY. *If $T = $ unitary, then $\mu(A) = \|A\|$.*

Remark [FGT1]. $\mu(A, T) = \|A\|$, $\forall T$ *isometric and* $\forall A : \mathcal{H}_0' \mapsto \mathcal{H}$ *such that $TA = AT_0'$ iff $\mathcal{H}_0'$ reduces T' (i.e. $T'^*\mathcal{H}_0' \subset \mathcal{H}_0'$ too).*

In other words, the analogue of the CLThm is valid in the extension theorem above only in a trivial case.

7. Fourth Example

A trivial generalization of the second example is the following fact: *Let T be similar to a coisometry (i.e. $T = XT_1X^{-1}$ where T_1 is a coisometry). Then*

$$\mu(A; T) \leq \|X\| \|X^{-1}\| \|A\| .$$

PROOF. Indeed if $TA = AT_0'$ then $T_1X^{-1}A = X^{-1}AT_0'$; by §3, there exists an operator B_1 such that $T_1B_1 = B_1T_0'$, $\|B_1\| = \|X^{-1}A\|$, $B_1|\mathcal{H}_0' = X^{-1}A$. Let $B = XB_1$. Then $\|B\| \leq \|X\| \|X^{-1}\| \|A\|$ and $TB = BT'$, $B|\mathcal{H}_0' = A$. So $\mu(A) \leq \|X\| \|X^{-1}\| \|A\|$.

If we do an analagous modification in the third example, a similar conclusion holds; precisely if T is similar to an isometry, $(T = XT_1X^{-1}$ where T_1 is an isometry), then

$$\mu(A; T) \leq \|X\| \ \mu(X^{-1}A; T_1) .$$

(Recall that by virtue of §4 we can always consider only the case when T' is an isometry.) However, unlike the previous case there is no simple estimate of $\mu(X^{-1}A; T_1)$ in terms of A.

8. Fifth Example

Introduce
$$\mathcal{F} = \mathcal{H}' \ominus \mathcal{H}_0' \, , \; C = P_{\mathcal{F}} T' | \mathcal{F}$$
where $P_{\mathcal{F}}$ denotes the orthogonal projection of $\mathcal{H}'$ onto $\mathcal{F}$. Obviously
$$C = \text{contraction} \, , \; P_{\mathcal{F}} T' = C P_{\mathcal{F}} \, .$$

Recall that [Sz.-NF3] C is similar to an isometry iff $\exists \alpha > 0$ such that
$$\|C^n f\| \geq \alpha \|f\| \quad \forall f \in \mathcal{F}, \forall n = 1, 2, \dots \, .$$
In this case $\mu(A) \leq (1/\beta)\|A\|$, *where* $\beta = \alpha(1 + \alpha^2)^{-1/2}$ [FGT1].
Is there any connection between this example and the previous example?

9. Sixth Example

A useful particular case [FGT1], relevant to nonlinear control theory and to the causal version of the CLThm, is the following:
$$\mathcal{F} = \left(\bigcup_{n=1}^{\infty} \ker C^n \right)^- \, .$$

Notice
$$\mathcal{H}_0' + \ker C^n = \{h' \in \mathcal{H}' : T'^n \mathcal{H}' \in \mathcal{H}_0'\} \quad \forall n = 1, 2, \dots$$
so
$$\mathcal{H}' = \left(\bigcup_1^{\infty} \mathcal{H}_{-n}' \right)^- \, ,$$
where
$$\mathcal{H}_{-n}' = \{h' \in \mathcal{H}' : T'^n h' \in \mathcal{H}_0'\} \quad \forall n = 0, 1, 2, \dots \, .$$
If $\ker T = \{0\}$ *then*
$$\mu(A) = \min M$$
where $\mathcal{M}$ *runs over all positive number satisfying*
$$\begin{cases} \|A\| \leq M \\ \|P_{\mathcal{H}_{-n}'} T'^{*n} A^* h\| \leq M \|T^{*n} h\| \quad \forall h \in \mathcal{H}, \forall n = 1, 2, \dots \, . \end{cases}$$

The condition $\ker T = \{0\}$ is generic in the applications to nonlinear H^∞-control theory. In these applications, the general case is $\ker T^n = \ker T^{n+1}$ (for some n). (See [FT], [FGT2], [G].) There is an obvious superficial resemblance between the above result and the commutant extension theorem in Section 6. Is there any deeper reason for that resemblance?

10. Connection with the Causal Commutant Lifting Theorem

Let

$$\mathcal{G}_n = T'^n \mathcal{H}'_{-n} \quad (n = 0, 1, \dots).$$

Then

$$\mathcal{G}_0 = \mathcal{H}'_0 \supset \mathcal{G}_1 \supset \mathcal{G}_2 \supset \cdots \text{ and } T'\mathcal{G}_n \subset \mathcal{G}_{n+1}, \quad \forall n = 0, 1, 2, \dots$$

(hence $T'^n \mathcal{G}_0 \subset \mathcal{G}_n \ \forall n = 0, 1, \dots$).

Let U on $\mathcal{K}$ be the minimal isometric dilation of T and

$$B : \mathcal{H}'_0 \mapsto \mathcal{K}, \ BT'_0 = UB.$$

Such an intertwining operator B is called causal (with respect to $\{\mathcal{G}_n\}_{n=0}^\infty$) if

$$B\mathcal{G}_n \subset U^n \mathcal{K} \quad \forall n = 0, 1, 2, \dots.$$

The Causal CLThm is concerned with the computation of the following value

$$\nu_c(A; T'_0, U; \mathcal{H}) = \min\{\|B\|; B : \mathcal{H}'_0 \mapsto \mathcal{K}, \ BT'_0 = UB, \ B \text{ causal}, \ P_{\mathcal{H}}B = A\}.$$

Note that the difference between ν_c and ν resides in the request of causality for the intertwining dilations B.

Causal CLThm [FGT1]. *With the above notation, we have*

$$\nu_c(A; T'_0, U; \mathcal{H}) = \mu(A; T, T'; \mathcal{H}'_0).$$

This theorem shows that the causal version of the commutant lifting theorem reduces to the extension problem for the particular case considered in §9.

11. Seventh Example (Foias and Gu '94)

Connection to Commutators. *If the spectra of C and T are disjoint, then*

$$\mu(A) < \infty \quad \forall A : \mathcal{H}'_0 \mapsto \mathcal{H}, \ AT'_0 = TA.$$

PROOF. The operator equation

$$(*) \qquad XC - TX = AP_{\mathcal{H}'_0}T'|\mathcal{F}$$

is solvable [R]. To obtain the desired intertwining extension of A set $B = SP_{\mathcal{F}} + AP_{\mathcal{H}'_0}$. Thus $\mu(A) < \infty$.

Remark. $\mu(A) < \infty$ iff $(*) =$ solvable.

Recall the following easy consequence of the Hahn–Banach theorem: If $\Xi : \mathcal{X} \mapsto \mathcal{X}$ is a linear continuous operator in a Banach space $\mathcal{X}$, then

$$\xi^* \in \text{Ran } \Xi^* \iff \exists M : |\xi^*(\xi)| \leq M\|\Xi\xi\| \quad \forall \xi \in \mathcal{X}.$$

Apply this to the case $\mathcal{X} = \sigma_1(\mathcal{H}, \mathcal{F})$ the space of all finite trace operators from $\mathcal{H}$ into $\mathcal{F}$ and $\Xi(Y) = CY - YT, Y \in \mathcal{X}$. It follows that $\mu(A) < \infty$ iff $\exists M < \infty$ such that

$$(**) \qquad |\text{Trace}\,(AP_{\mathcal{H}_0'}T'Y)| \leq M\|CY - YT\|_1 \quad \forall Y \in \sigma_1(\mathcal{H}, \mathcal{F})$$

Although this is a definite characterization of the solvability of the extension problem it is also a bad answer to that problem. Indeed one cannot use (**) in a simple way to deduce the theorems which were presented in the six preceeding examples. Therefore, the extension (or dilation) problem stated in §1 must be considered still open.

REFERENCES

[A]   R. Arocena, *Unitary extensions of isometries and contractive intertwining dilations*, Oper. Theory: Adv. and Appl, Vol. 41, Birkhauser (Boston) (1989), 13-23.

[CS]   J.G.W. Carswell and C.F. Schubert, *Lifting of operators that commute with shifts*, Mich. Math. J. **22** (1975), 65-69.

[C]   Z. Ceausescu, *Lifting of a contraction intertwining two isometries*, Mich. Math. J. **26** (1979), 231-241.

[CSa]   M. Cotlar and C. Sadosky, *Transference of metrics induced by unitary couplings, a Sarason theorem for the bidimensional torus and a Sz.-Nagy-Foias theorem for two pairs of dilations*, J. Funct. Anal. **111** (1993), 473-488.

[DMP]   R.G. Douglas, P.S. Muhly and C. Pearcy, *Lifting commuting operators*, Mich. Math. J. **15** (1968), 385-395.

[FF]   C. Foias and A.E. Frazho, *The Commutant Lifting Approach to Interpolation Problems*, Oper. Theory: Adv. and Appl., Vol. 44, Birkhauser (Boston), 1990.

[FG]   C. Foias and C. Gu, *Intertwining extensions and a two-sided corona theorem*, (in preparation).

[FGT1] C. Foias, C. Gu and A. Tannenbaum, *Intertwining dilations, intertwining extension and causality*, Acta Sci. Math. **57** (1993), 101-123.

[FGT2] _____, *On a causal linear optimization theorem*, J. of Math. Anal. and Appl. (to appear).

[FT]   C. Foias and A. Tannenbaum, *Causality in commutant lifting theory*, J. Funct. Anal. **2** (1993), 407-441.

[G]   C. Gu, *A generalization of Cowen's characterization of hyponormal Toeplitz operators*, J. Funct. Anal. (to appear).

[P]   S. Parrott, *Unitary dilations for commuting contractions*, Pacific J. Math. **34** (1970), 481-490.

[R]   M. Rosenblum, *On the operator equation $BX - XA = Q$*, Duke Math. J. **23** (1956), 263-270.

[S]   D. Sarason, *Generalized interpolation in H^∞*, Trans. AMS **127** (1967), 179-203.

[Sz.-N] B. Sz.-Nagy, *On Schäffer's construction of unitary dilations*, Ann. Univ. Budapest (sect. math.) **3-4** (1960/61), 343-346; (announced in), Deuxième congres math. hongrois, Budapest, (août 1960), 24-31.

[Sz.-NF1] C. Foias and B. Sz.-Nagy, *Dilatation des commutants d'opérateurs*, C.R. Acad Sci. Paris, Serie A **266** (1968), 493-495.

[Sz.-NF2] _____, *The "lifting theorem" for intertwining operators and some new applications*, I.U. Math. J. **20** (1970), 901-904.

[Sz.-NF3] _____, *On contractions similar to isometries and Toeplitz operators*, Ann. Acad. Sci. Finn. Math. **2** (1976), 553-564.

DEPARTMENT OF MATHEMATICS, INDIANA UNIVERSITY, BLOOMINGTON, IN 47408

E-mail address: foias@ucs.indiana.edu

Contemporary Mathematics
Volume **189**, 1995

A New Atomic Decomposition
For The Triebel-Lizorkin Spaces

Y.-S. HAN, M. PALUSZYŃSKI AND G. WEISS

ABSTRACT. The Triebel-Lizorkin spaces $\dot{F}_p^{\alpha,q}$, $\alpha \in \mathbf{R}$, $0 < p \leq q < \infty$, $p \leq 1$ can be characterized by means of an "atomic" decomposition that is completely analogous to the "classical" one for the Hardy spaces $H^p(\mathbf{R}^n)$, $0 < p \leq 1$. More specifically, the coefficients of this decomposition belong to the space ℓ^p, which is a much simpler condition than the one obtained by using "smooth" atoms.

1. Introduction

The purpose of this paper is to study the class of Triebel-Lizorkin spaces $\dot{F}_p^{\alpha,q}$, $\alpha \in \mathbf{R}$, $0 < p \leq q < \infty$, $p \leq 1$ by means of an "atomic" characterization that is completely analogous to the "classical" one for the Hardy spaces $H^p(\mathbf{R}^n)$, $0 < p \leq 1$.

We begin by describing the principal features of this last decomposition. An *atom* for $H^p(\mathbf{R}^n)$ is a function a satisfying:

$$\text{supp } a \subset Q \text{ (a cube with sides parallel to the axes)},$$
$$\|a\|_2 \leq |Q|^{1/2-1/p},$$
$$\int a(x)x^\beta dx = 0 \text{ for } 0 \leq |\beta| \leq [n(1/p-1)].$$

The basic characterization theorem is ([CW], [L]):

1991 *Mathematics Subject Classification.* Primary 42B30; Secondary 43A15.

The second author was supported by a grant fund from the Southwestern Bell Company; the third author was supported by NSF grant DMS-9007491.

THEOREM (1.1). *A tempered distribution f belongs to $H^p(\mathbf{R}^n)$, $0 < p \le 1$ if and only if $f = \sum_k \lambda_k a_k$, where each a_k is an atom for $H^p(\mathbf{R}^n)$, and $\sum_k |\lambda_k|^p < \infty$. Moreover, the classical norm $\|f\|_{H^p}$ is equivalent to the norm*

$$\inf \{ (\sum_k |\lambda_k|^p)^{1/p} : \text{all possible representations } f = \sum_k \lambda_k a_k \}.$$

REMARKS. (1) The atoms considered here are called $(p, 2)$ atoms in [CW]. In that paper (p, q)-atoms were considered for all q such that $0 < p \le 1 \le q$ and $p < q$ (and they, also, characterize H^p).
(2) The Triebel-Lizorkin spaces with $\alpha = 0$, $q = 2$ coincide with the Hardy spaces $H^p(\mathbf{R}^n)$, $0 < p \le 1$: $\dot{F}_p^{0,2} = H^p(\mathbf{R}^n)$ ([FJ], [FJW]).

QUESTION. Do there exist "atoms" that represent the elements of the spaces $\dot{F}_p^{\alpha,q}$ in a way completely analogous to the one just described for the spaces $H^p(\mathbf{R}^n)$?

Frazier and Jawerth obtain a "smooth" atomic decomposition for the Triebel-Lizorkin spaces based on their "ϕ- & ψ- transform" techniques. More specifically, they start with the following (Peetre) definition of the spaces $\dot{F}_p^{\alpha,q}$.

DEFINITION (1.2). Suppose that $\phi \in \mathcal{S}$ (the space of tempered test functions) satisfies supp $\hat{\phi} \subset \{\xi \in \mathbf{R}^n : 1/2 \le |\xi| \le 2\}$ and $|\hat{\phi}(\xi)| \ge c > 0$ for $3/5 \le |\xi| \le 5/3$. The *homogeneous Triebel-Lizorkin space* $\dot{F}_p^{\alpha,q}$ is defined to be the collection of all $f \in \mathcal{S}'/\mathcal{P}$ (the space of tempered distributions modulo polynomials), such, that

$$(*) \qquad \|f\|_{\dot{F}_p^{\alpha,q}} = \left\| \left\{ \sum_{k \in \mathbf{Z}} \left(2^{k\alpha} |\phi_k * f| \right)^q \right\}^{1/q} \right\|_p < \infty,$$

where $\phi_k(x) = 2^{kn}\phi(2^k x)$ for $k \in \mathbf{Z}$ and $x \in \mathbf{R}^n$.

It is well known that the defining equality $(*)$ can also involve a compactly supported ϕ, rather than the band-limited ϕ we just used.

Frazier and Jawerth, in [FJ] introduced the following *smooth atoms*. Let $\mathcal{Q}$ denote the set of all dyadic cubes in $\mathbf{R}^n$, that is, all cubes $Q = Q_{k,j}$ of the form

$$Q_{k,j} = \left\{ x \in \mathbf{R}^n : 2^{-k} j_l \le x_l \le 2^{-k}(j_l + 1), k \in \mathbf{Z}, j = (j_1, \dots, j_n) \in \mathbf{Z}^n \right\}.$$

DEFINITION (1.3). A function $a_Q \in \mathcal{D}$ (the elements of $\mathcal{S}$ having compact support) is said to ba a *smooth N-atom* ($N = N(\alpha, q, p)$) for $Q \in \mathcal{Q}$, if

(i) supp $a_Q \subset 3Q$,
(ii) $|D_Q^\gamma(x)| \le |Q|^{-|\gamma|/n - 1/2}$ for all n-tuples of non negative integers $\gamma = (\gamma_1, \dots, \gamma_n)$ satisfying $0 \le \gamma_1 + \gamma_2 + \cdots + \gamma_n = |\gamma|$ with $|\gamma| \le [\alpha + 1]_+$,
(iii) $\int a_Q(x) x^\gamma dx = 0$ for $0 \le |\gamma| \le N = [J - n - \alpha]$, $J = n/\min(1, p, q)$.

They also introduced the following sequence spaces $\dot{f}_p^{\alpha,q}$:

DEFINITION (1.4). *A sequence* $s = \{s_Q\}$, $Q \in \mathcal{Q}$, *belongs to* $\dot{f}_p^{\alpha,q}$, $\alpha \in$ **R**, $0 < p, q \leq \infty$, $p < \infty$, *if and only if*

$$\|\{s_Q\}\|_{\dot{f}_p^{\alpha,q}} = \left\| \left\{ \sum_Q \left[|Q|^{-\alpha/n-1/2} |s_Q| \chi_Q \right]^q \right\}^{1/q} \right\|_p < \infty.$$

From the Calderòn reproducing formula $f = \sum_\nu \psi_\nu * \phi_\nu * f$, for an appropriate ψ, they then obtain an "atomic" decomposition having the form $f = \sum_{Q \in \mathcal{Q}} s_Q a_Q$ ([FJ] or [FJW]) for the space $\dot{F}_p^{\alpha,q}$:

THEOREM (1.5). *Suppose that* $\alpha \in$ **R**, $0 < p < \infty$, $0 < q \leq \infty$ *and* $N = [J - n - \alpha]$. *Then* $f \in \dot{F}_p^{\alpha,q}$ *if and only if there exists a sequence* $s = \{s_Q\}$ *and smooth* N-*atoms* $\{a_Q\}$, $Q \in \mathcal{Q}$, *such that* $f = \sum_Q s_Q a_Q$ *in* $\mathcal{S}'/\mathcal{P}$ *and*

$$\|f\|_{\dot{F}_p^{\alpha,q}} \sim \left\| \left\{ \sum_Q \left[|Q|^{-\alpha/n-1/2} |s_Q| \chi_Q \right]^q \right\}^{1/q} \right\|_p,$$

where χ_Q *is the characteristic function of* Q.

Let us observe that the sequence $\{s_Q\}$ is not in l^p; in fact, the formula defining its norm in (1.4) is rather complicated. Our principal goal, in this paper, is to obtain a different atomic decomposition where the coefficients, as in the Hardy spaces case, belong to l^p, $0 < p \leq 1$. This requires a different notion of an "atom" that includes the $(p, 2)$-atoms used in the Hardy spaces case. We do want to point out two things, however: our decomposition will be seen to be valid only when $0 < p \leq 1$. The one just described is valid for indices $p > 1$ as well. Secondly, in [FJ] a decomposition involving l^p sequences for coefficients is also presented. The atoms we shall introduce here, however, seem to be more natural since they have a more direct definition (it is easily checked that each of their atoms is one of ours).

DEFINITION (1.6). Suppose $\alpha \in$ **R**, $0 < p \leq q < \infty$, $p \leq 1$. A distribution a is said to be a (p, q, α) atom if
(i) supp $a \subset Q$, Q is a cube in **R**^{n},
(ii) $\|a\|_{\dot{F}_q^{\alpha,q}} \leq |Q|^{1/q-1/p}$,
(iii) For every $g \in \mathcal{S}$, a polynomial P of degree at most $N = [n(1/p - 1) - \alpha]$, and a smooth cutoff function $\eta_Q \in \mathcal{S}$ such, that $\eta_Q \equiv 1$ on Q and $\eta_Q \equiv 0$ outside $2Q$, we have
$$\langle a, g \rangle = \langle a, (g - P)\eta_Q \rangle.$$
In the above formulas, $\langle f, \phi \rangle$ denotes the usual pairing of $f \in \mathcal{S}'$, $\phi \in \mathcal{S}$ ($\langle f, \phi \rangle = \int f\phi$ when f is a function).

We remark, that this definition was first introduced in [H] for the case when $\alpha = 0$ and $0 < p \leq 1 \leq q \leq 2$. In that case, the $(p, q, 0)$ atoms are L^q functions; hence, condition (iii) of (1.6) is precisely the usual cancellation condition.

If (p, q, α) atoms are locally integrable, then one can see that condition (iii) of (1.6) is, again, the usual cancellation condition. This observatiom applies to the case $\alpha > 0$, since then (p, q, α) atoms are locally integrable functions. But, in general, (p, q, α) atoms, $\alpha \leq 0$, are distributions and, hence, the usual cancellation condition does not make sense. It is easy to see that the $(p, 2, 0)$ atoms are the $(p, 2)$ atoms discussed in the first "Remark" above.

One of the main results in this paper is the following atomic decomposition of $\dot{F}_p^{\alpha,q}$ for $\alpha \in \mathbf{R}$ and $0 < p \leq q < \infty$, $p \leq 1$ that uses these atoms:

THEOREM 1. *If $\alpha \in \mathbf{R}$, $0 < p \leq q < \infty$, $p \leq 1$, then $f \in \dot{F}_p^{\alpha,q}$ if and only if there exist a sequence of numbers $\{\lambda_k\}$ in l^p and (p, q, α) atoms $\{a_k\}$ such, that $f = \sum_k \lambda_k a_k$ in $\mathcal{S}'/\mathcal{P}$. Moreover,*
(1.7)
$$\|f\|_{\dot{F}_p^{\alpha,q}} \sim \inf \left\{ \left(\sum_{k=1}^{\infty} |\lambda_k|^p \right)^{1/p} : \text{ all possible representations } f = \sum_{k=1}^{\infty} \lambda_k a_k \right\}.$$

This result was first proved in [H] for the case where $\alpha = 0$ and $0 < p \leq 1 \leq q \leq 2$. In particular, when $\alpha = 0$, $0 < p \leq 1$, and $q = 2$ this atomic decomposition provides the usual atomic decomposition of the Hardy spaces H^p for $0 < p \leq 1$.

This atomic decomposition, together with the fact that $(\dot{F}_p^{0,q})^* = \dot{F}_\infty^{n(1/p-1),\infty}$ for $0 < p < 1 \leq q < \infty$, provides a new characterization of the Lipschitz spaces:

THEOREM 2. *Suppose that $0 < p < 1 \leq q < \infty$, $s = [n(1/p - 1)]$. Then $f \in \dot{F}_\infty^{n(1/p-1),\infty}$ if and only if*
$$\sup_Q |Q|^{1-1/p} \left(\frac{1}{|Q|} \left\| (f - P_Q^s(f)) \eta_Q \right\|_{\dot{F}_q^{0,q}}^q \right)^{1/q} \leq c,$$

where $P_Q^s(f)$, the Gramm-Schmidt polynomial, is is the unique polynomial of degree at most s such that
$$\int_Q \left(f(x) - P_Q^s(f)(x) \right) x^\gamma dx = 0$$

for all γ, $0 \leq |\gamma| \leq s$, and η_Q is the smooth cutoff function we used before.

We remark, that in [TW] the Lipschitz space $L(\beta, q, s)$, s a positive integer, $0 < [n\beta] < s$, $1 \leq q \leq \infty$, was defined as follows:

If g is locally integrable on $\mathbf{R}^n$ and Q is a cube, let $P_Q^s(g)$ be the corresponding Gramm-Schmidt polynomial; then, the space $L(\beta, q, s)$ is defined by the norm
$$\|g\|_{L(\beta,q,s)} = \sup_Q |Q|^{-\beta} \left\{ \frac{1}{|Q|} \int_Q |g - P_Q^s(g)|^q \right\}^{1/q}.$$

It is easy to see that if $q = 2$, $\beta = 1/p - 1$ and $0 < n\beta < 1$ the above is equivalent to the condition in Theorem 2. We will make use of the fact that these spaces are independent (up to equivalence of norms) of q, $1 \leq q \leq \infty$.

2. Proof of Theorem 1.

("If" part). Let a be a (p, q, α) atom, where $0 < p \le q < \infty$, $p \le 1$, and let $N = [n(1/p - 1) - \alpha]$. We are going to show that

$$\|a\|_{\dot{F}_p^{\alpha,q}} \le c,$$

where c is a constant depending only on p, q, α and n. Given this, if

$$f = \sum_{k=1}^{\infty} \lambda_k a_k,$$

we have

$$\|f\|_{\dot{F}_p^{\alpha,q}}^p \le \sum_{k=1}^{\infty} \lambda_k^p \|a_k\|_{\dot{F}_p^{\alpha,q}}^p \le c^p \sum_{k=1}^{\infty} \lambda_k^p,$$

and the theorem is proved. The next observation is that it is enough to consider an atom a associated with the unit cube $Q = Q(0, 1)$, a cube centered at the origin, with side length 1. To see this, let a be an atom associated with a cube $Q(x_0, t)$, centered at x_0 and of side length t. Define

$$a'(x) = t^{n/p - \alpha} a(t(x + x_0/t)),$$

where we used notation valid if a were a function; however, it is easy to interpret all this appropriately in case a is only a tempered distribution. It is well known, that there exists a constant $c > 0$, independent of a, such that

$$c^{-1} \|a\|_{\dot{F}_p^{\alpha,q}} \le \|a'\|_{\dot{F}_p^{\alpha,q}} \le c \|a\|_{\dot{F}_p^{\alpha,q}}.$$

Thus, if $\|a\|_{\dot{F}_q^{\alpha,q}} \le |Q|^{1/q - 1/p}$, then $\|a'\|_{\dot{F}_q^{\alpha,q}} \le c t^{n/p - n/q} |Q|^{1/q - 1/p} = c$, and a' is a c multiple of an atom associated with the unit cube. If we have our result for this case, we conclude that $\|a'\|_{\dot{F}_p^{\alpha,q}} \le c$; thus $\|a\|_{\dot{F}_p^{\alpha,q}} \le c$. Thus, we may restrict ourselves to an atom a associated with the unit cube. Let us choose ϕ with supp $\phi \subset Q(0, 1)$ so that

$$\|a\|_{\dot{F}_p^{\alpha,q}} = \left\| \left(\sum_{l \in \mathbf{Z}} \left(2^{l\alpha} |\phi_l * a| \right)^q \right)^{1/q} \right\|_p.$$

(See comment following $(1.2)(*)$). Hence,

$$\|a\|_{\dot{F}_p^{\alpha,q}}^p = \int_{2\sqrt{n}Q} \left(\sum_{l \in \mathbf{Z}} \left(2^{l\alpha} |\phi_l * a(x)| \right)^q \right)^{p/q} dx +$$

$$+ \int_{(2\sqrt{n}Q)^c} \left(\sum_{l \in \mathbf{Z}} \left(2^{l\alpha} |\phi_l * a(x)| \right)^q \right)^{p/q} dx$$

$$= I + II.$$

Observe, that if $p = q$, there is nothing to prove. Thus, assume $p < q$. By Hölder's inequality, with q/p,

$$I \leq \left(\int_{2\sqrt{n}Q} \sum_{l \in \mathbf{Z}} \left(2^{l\alpha} |\phi_l * a(x)| \right)^q dx \right)^{p/q} \left(\int_{2\sqrt{n}Q} dx \right)^{1-p/q}$$

$$\leq \|a\|_{\dot{F}_q^{\alpha,q}}^p |2\sqrt{n}Q|^{1-p/q}$$

$$\leq c.$$

To estimate II we make use of the following

CLAIM. *Suppose* $\alpha < 0$, $N_0 = [n(1/p - 1) - \alpha] + 1$. *We then have*

$$|\phi_l * a(x)| \leq c 2^{l(N_0 + n)}.$$

We shall prove this claim later; we assume it for the moment. Let us first assume $\alpha < 0$. If $x \in (2\sqrt{n}Q)^c$, and $2^l > \sqrt{n}/|x|$, then $\phi_l * a(x) = 0$. To see this, let $y \in Q$, and $l \leq 0$.

$$2^l |x - y| \geq 2^l |x| - 2^l |y| > \sqrt{n} - 2^l |y| \geq \sqrt{n} - 2^l \sqrt{n}/2 \geq \sqrt{n}/2,$$

where we used $2^l > \sqrt{n}/|x|$. Now let $l > 0$. Using $x \in (2\sqrt{n}Q)^c$, and $y \in Q$,

$$|x - y| \geq |x| - |y| \geq \sqrt{n} - \sqrt{n}/2 = \sqrt{n}/2 > 2^{-l}\sqrt{n}/2.$$

So,

$$|\phi_l * a(x)| = \left| \int_{Q(0,1)} \phi_l(x - y) a(y) dy \right| = 0,$$

since supp $\phi_l \subset \{x : |x| \leq 2^{-l}\sqrt{n}/2\}$. Thus,

$$II = \int_{(2\sqrt{n}Q)^c} \left(\sum_{l \in \mathbf{Z}} \left(2^{l\alpha} |\phi_l * a(x)| \right)^q \right)^{p/q} dx$$

$$\leq \int_{(2\sqrt{n}Q)^c} \left(\sum_{2^l \leq \sqrt{n}/|x|} 2^{l(\alpha + N_0 + n)q} \right)^{p/q} dx$$

$$= \int_{(2\sqrt{n}Q)^c} \left(\sum_{l=-\infty}^{[\log_2(\sqrt{n}/|x|)]} 2^{l(\alpha + N_0 + n)q} \right)^{p/q} dx$$

$$= \int_{(2\sqrt{n}Q)^c} 2^{(\alpha + N_0 + n)[\log_2(\sqrt{n}/|x|)]p} \left(\sum_{l=-\infty}^{0} 2^{l(\alpha + N_0 + n)q} \right)^{p/q} dx$$

$$\leq c \int_{(2\sqrt{n}Q)^c} \left(\frac{\sqrt{n}}{|x|} \right)^{(\alpha + N_0 + n)p} dx$$

$$= c,$$

since $(\alpha + N_0 + n)p > (\alpha + n(1/p - 1) - \alpha + n)p = n$.

We now prove the claim:

Using the cancellation condition for a, we obtain

$$|\phi_l * a(x)| = |\langle a, \phi_l(x - \cdot)\rangle| = |\langle a, (\phi_l(x - \cdot) - P_{N_0-1}(\phi_l)(x, \cdot))\eta(\cdot)\rangle|,$$

where $P_{N_0-1}(x, \cdot)$ is the Taylor polynomial of $\phi_l(x - \cdot)$ of degree $N_0 - 1$. By duality,

$$|\phi_l * a(x)| \leq \|a\|_{\dot{F}_q^{\alpha,q}} \|(\phi_l(x - \cdot) - P_{N_0-1}(\phi_l)(x, \cdot))\eta(\cdot)\|_{\dot{F}_{q'}^{\beta,q'}}.$$

If $q \geq 1$ then q' is the dual index, and $\beta = -\alpha$. If $q < 1$, then $q' = \infty$, $\beta = -\alpha + n(1/q - 1)$ (see [FJ], p.61). We claim that b_l, the expression inside the above norm is a smooth atom for $\dot{F}_{q'}^{\beta,q'}$, multiplied by $2^{l(N_0+n)}$. Observe, that the moment condition for the smooth atom is empty, since $\alpha < 0$. Also, η is supported on $Q(0, 2)$, so the only condition to check is:

$$(*) \qquad |D^\gamma b_l(x)| \leq c 2^{l(N_0+n)}, \quad 0 \leq |\gamma| \leq [\beta + 1]_+ \leq N_0.$$

First of all, using the integral form of the remainder of the Taylor series, we obtain for $\psi \in \mathcal{S}$, $M = 0, 1, \ldots$

$$\sup_{h \in Q(0,2)} |\psi(x - h) - P_M(\psi)(x, h)|$$

$$= \sup_{h \in Q(0,2)} \left| (M + 1) \sum_{|\delta|=M+1} \frac{(-h)^\delta}{\delta!} \int_0^1 (1 - t)^M \psi^{(\delta)}(x - ht) dt \right|$$

$$(**)$$

$$\leq c \sum_{|\delta|=M+1} \left\| \psi^{(\delta)} \right\|_\infty, \qquad c = c(M, n).$$

Consider $(*)$, with $|\gamma| \leq N_0$:

$$|D^\gamma b_l(x)| = \left| \sum_{\delta \leq \gamma} \binom{\gamma}{\delta} D_h^\delta \left(\phi_l(x - h) - P_{N_0-1}(\phi_l)(x, h) \right) D_h^{\gamma-\delta} \eta(h) \right|$$

$$= \left| \sum_{\delta \leq \gamma} \binom{\gamma}{\delta} (-1)^{|\delta|} \left((\phi_l)^{(\delta)}(x - h) - P_{N_0-1-|\delta|}((\phi_l)^{(\delta)})(x, h) \right) D_h^{\gamma-\delta} \eta(h) \right|.$$

We have used the convention

$$P_{-1}(\psi)(x, h) \equiv 0.$$

Observe, that the estimate $(**)$ still holds, if $M = -1$, and we use the above convention. Thus,

$$
\begin{aligned}
|D^\gamma b_l(x)| &\leq c \sum_{\delta \leq \gamma} \binom{\gamma}{\delta} \sup_{h \in Q(0,2)} \left| (\phi_l)^{(\delta)}(x - h) - P_{N_0 - 1 - |\delta|}((\phi_l)^{(\delta)})(x, h) \right) \\
&\leq c \sum_{\delta \leq \gamma} \binom{\gamma}{\delta} \sum_{|\omega| = N_0 - |\delta|} \left\| (\phi_l)^{(\delta + \omega)} \right\|_\infty \\
&= c \sum_{\delta \leq \gamma} \binom{\gamma}{\delta} \sum_{|\omega| = N_0 - |\delta|} 2^{l(N_0 + n)} \left\| \phi^{(\delta + \omega)} \right\|_\infty \\
&\leq c 2^{l(N_0 + n)}, \quad c = c(\phi, N_0, \gamma, n).
\end{aligned}
$$

This proves the claim; hence the "if" part of Theorem 1 is proved in the case $\alpha < 0$.

We now reduce the general case to this one. Suppose $\alpha < 1$, and a is a (p, q, α) atom. Then it is easy to see, that $D^i a$ is a constant multiple of a $(p, q, \alpha - 1)$ atom, where D^i is the partial derivative, $i = 1, \ldots, n$. To see this, we use

$$
(***) \qquad \frac{1}{c} \|f\|_{\dot{F}_p^{\alpha,q}} \leq \sum_{i=1}^n \|D^i f\|_{\dot{F}_p^{\alpha-1,q}} \leq c \|f\|_{\dot{F}_p^{\alpha,q}}, \quad \forall \alpha \in \mathbf{R}, \; 0 < p, q \leq \infty.
$$

(See [FJW]). From the part of the theorem already proved, we obtain

$$
\|D^i a\|_{\dot{F}_p^{\alpha-1,q}} \leq c, \quad i = 1, \ldots, n.
$$

Applying $(***)$ again, we obtain

$$
\|a\|_{\dot{F}_p^{\alpha,q}} \leq c.
$$

We proceed inductively to obtain the "if" part of the theorem for every $\alpha \in \mathbf{R}$.

To prove the "only if" part of Theorem 1, we use the Littlewood-Paley S-function characterization of the Triebel-Lizorkin spaces, namely, let

$$
S_q^\alpha(f)(x) = \left\{ \sum_{k \in \mathbf{Z}} \int_{|x-y| \leq c 2^{-k}} 2^{nk} \left(2^{k\alpha} |(\phi_k * f)(y)| \right)^q dy \right\}^{1/q},
$$

where c is some fixed constant, and ϕ_k are the same as above (see [T]). Then,

$$
\begin{aligned}
\|f\|_{\dot{F}_p^{\alpha,q}} &\sim \left\| \left\{ \sum_{k \in \mathbf{Z}} \int_{|x-y| \leq c 2^{-k}} 2^{nk} \left(2^{k\alpha} |(\phi_k * f)(y)| \right)^q dy \right\}^{1/q} \right\|_p \\
&= \left\| S_q^\alpha(f) \right\|_p.
\end{aligned}
$$

(2.1)

Suppose that $\alpha \in \mathbf{R}$, $0 < p \leq q < \infty$, $p \leq 1$, and $f \in \dot{F}_p^{\alpha,q}$. Then $S_q^\alpha(f) \in L^p$. We follow an idea in [CF]. Let

$$
\Omega_k = \{x \in \mathbf{R}^n : S_q^\alpha(f)(x) > 2^k\}
$$

and set $\mathcal{B}_k = \{Q \in \mathcal{Q} : |Q \cap \Omega_k| \geq 1/2|Q| \text{ and } |Q \cap \Omega_{k+1}| < 1/2|Q|\}$. It is easy to see that for each dyadic cube in $\mathbf{R}^n$ there is a unique $k \in \mathbf{Z}$ such that $Q \in \mathcal{B}_k$ and, for each $Q \in \mathcal{B}_k$, there is a unique maximal dyadic cube $Q' \in \mathcal{B}_k$ such that $Q \subset Q'$. Denote the collection of all such dyadic cubes in $\mathcal{B}_k$ by $\{Q_k^i\}_{i \in I_k}$, the index set I_k depends on k (and is possibly finite). We then have

$$(2.2) \qquad \{Q\}_{Q \text{ dyadic}} = \bigcup_k \bigcup_{i \in I_k} \{Q\}_{Q \subset Q_k^i, Q \in \mathcal{B}_k}.$$

Let $\phi_Q = \phi_l$ whenever $l(Q) = 2^{-l}$. (The symbol "ϕ_Q" should not be confused with the same symbol used in the ϕ- and ψ- transform literature. Our function ϕ_Q only takes into account the side length of Q and not its location. The latter is accounted for by the location of the integration.) Applying the Calderòn reproducing formula (see [FJW]), we obtain

$$(2.3) \quad f(x) = \sum_{k \in \mathbf{Z}} \psi_k * \phi_k * f(x) = \sum_k \sum_i \left(\sum_{\substack{Q \subset Q_k^i \\ Q \in \mathcal{B}_k}} \int_Q \psi_Q(x - y)(\phi_Q * f)(y) dy \right),$$

where the series converges in $\mathcal{S}'/\mathcal{P}$. To obtain a regular atomic decomposition of f, we need the following lemma:

LEMMA (2.4). *Suppose that the set $\mathcal{B}_k$ is as above. Then there exists a constant c such, that*

$$(2.5) \qquad \sum_{Q \in \mathcal{B}_k} \int_Q \left(l(Q)^{-\alpha} |(\phi_Q * f)(y)| \right)^q dy \leq c 2^{kq} |\Omega_k|.$$

PROOF. Let $\bar{\Omega}_k = \{x \in \mathbf{R}^n; M(\chi_{\Omega_k})(x) > 1/2\}$, where M is the Hardy-Littlewood maximal function. By the Hardy-Littlewood maximal function theorem, $|\bar{\Omega}_k| \leq c|\Omega_k|$. Thus,

$$(2.6) \qquad \int_{\bar{\Omega}_k \setminus \Omega_{k+1}} (S_k^\alpha(f)(x))^q dx \leq 2^{(k+1)q} |\bar{\Omega}_k| \leq c 2^{kq} |\Omega_k|.$$

On the other hand,

$$\int_{\bar{\Omega}_k \setminus \Omega_{k+1}} (S_k^\alpha(f)(x))^q dx =$$

$$(2.7)$$

$$= \sum_{l \in \mathbf{Z}} \int \left(2^{l\alpha} |(\phi_l * f)(y)| \right)^q 2^{ln} |\{x \in \bar{\Omega}_k \setminus \Omega_{k+1} : |x - y| \leq c 2^{-l}\}| dy$$

$$\geq \sum_{Q \in \mathcal{B}_k} \int_Q \left(l(Q)^{-\alpha} |(\phi_Q * f)(y)| \right)^q |Q|^{-1} |\{x \in \bar{\Omega}_k \setminus \Omega_{k+1} : |x - y| \leq cl(Q)\}| dy.$$

If $y \in \mathcal{B}_k$ then $Q \subset \bar{\Omega}_k$; thus, there exists a constant c such that if $Q \in \mathcal{B}_k$

$$(2.8) \qquad |\{x \in \bar{\Omega}_k \setminus \Omega_{k+1} : |x - y| \le cl(Q)\}| \ge |\{(Q \cap \bar{\Omega}_k) \setminus \Omega_{k+1}\}|$$
$$= |\{Q \cap \bar{\Omega}_k\}| - |\{Q \cap \Omega_{k+1}\}| \ge |Q| - 1/2|Q| = 1/2|Q|.$$

Substituting (2.8) into (2.7) and combining with (2.6) gives us (2.5); hence, Lemma (2.4) is proved.

Now set

$$\lambda_{k,i} = |Q|^{1/p-1/q} \left(\sum_{\substack{Q \subset Q_k^i \\ Q \in \mathcal{B}_k}} \int_Q \left(l(Q)^{-\alpha} |(\phi_Q * f)(y)| \right)^q dy \right)^{1/q}$$

and

$$a_{k,i} = \frac{1}{\lambda_{k,i}} \left(\sum_{\substack{Q \subset Q_k^i \\ Q \in \mathcal{B}_k}} \int_Q \psi_Q(x - y)(\phi_Q * f)(y) dy \right).$$

This will be the decomposition if $q \ge 1$. If $q < 1$, we will define slightly different $\lambda_{k,i}$'s. By Lemma (2.4),

$$\sum_{\substack{Q \subset Q_k^i \\ Q \in \mathcal{B}_k}} \int_Q \left(l(Q)^{-\alpha} |(\phi_Q * f)(y)| \right)^q dy \le \sum_{Q \in \mathcal{B}_k} \int_Q \left(l(Q)^{-\alpha} |(\phi_Q * f)(y)| \right)^q dy$$

$$\le c2^{kq}|\Omega_k| < \infty,$$

which shows that $\lambda_{k,i}$ and $a_{k,i}$ are well defined. Thus, by (2.3), $f = \sum \lambda_{k,i} a_{k,i}$. It is easy to check that $a_{k,i}$ satisfy the condition (i) of (1.6). To see that the $a_{k,i}$ satisfy condition (ii), if $1 \le q < \infty$, we use the duality argument:

$$\|a_{k,i}\|_{\dot{F}_q^{\alpha,q}} = \sup_{\|g\|_{\dot{F}_{q'}^{-\alpha,q'}} \le 1} |\langle a_{k,i}, g \rangle|$$

$$= \sup_{\|g\|_{\dot{F}_{q'}^{-\alpha,q'}} \le 1} \frac{1}{\lambda_{k,i}} \left(\sum_{\substack{Q \subset Q_k^i \\ Q \in \mathcal{B}_k}} \int_Q |(\psi_Q * g)(y)| |(\phi_Q * f)(y) dy \right)$$

$$\le \sup_{\|g\|_{\dot{F}_{q'}^{-\alpha,q'}} \le 1} |Q_k^i|^{1/q-1/p} \left(\sum_{\substack{Q \subset Q_k^i \\ Q \in \mathcal{B}_k}} \int_Q \left(l(Q)^{\alpha} |\psi_Q * g(y)| \right)^{q'} dy \right)^{1/q'}$$

$$\le |Q_k^i|^{1/q-1/p}.$$

The above proof also shows that the $a_{k,i}$ satisfy the cancellation condition (iii) of (1.6) since the ψ_Q's are functions in $\mathcal{S}(\mathbf{R}^n)$ satisfying appropriate moment cancellation properties. Finally, we check that $\sum |\lambda_{k,i}|^p < \infty$.

$$\sum_{k,i} |\lambda_{k,i}|^p = \sum_{k,i} |Q|^{1-p/q} \left(\sum_{\substack{Q \subset Q_k^i \\ Q \in \mathcal{B}_k}} \int_Q \left(l(Q)^{-\alpha} |(\phi_Q * f)(y)| \right)^q dy \right)^{p/q}$$

$$\leq \sum_k \left(\sum_i |Q_k^i| \right)^{(1-p/q)} \left(\sum_i \sum_{\substack{Q \subset Q_k^i \\ Q \in \mathcal{B}_k}} \int_Q \left(l(Q)^{-\alpha} |(\phi_Q * f)(y)| \right)^q dy \right)^{p/q}$$

$$\leq \sum_k |\Omega_k|^{(1-p/q)} \left(\sum_{Q \in \mathcal{B}_k} \int_Q \left(l(Q)^{-\alpha} |(\phi_Q * f)(y)| \right)^q dy \right)^{p/q},$$

since $|\Omega_k \cap Q_k^i| \geq 1/2 |Q_k^i|$ and Q_k^i, $i \in I_k$ are disjoint. Thus,

$$\sum_{k,i} |\lambda_{k,i}|^p \leq c \sum_k |\Omega_k|^{(1-p/q)} 2^{kp} |\Omega_k|^{p/q} \leq c \sum_k 2^{kp} |\Omega_k| \leq c \|S_q^\alpha(f)\|_p^p,$$

which shows that f has an atomic decomposition.

Now we consider the case of $q < 1$. To do so, we consider the variant $\bar{S}$ of the S function:

$$\bar{S}_q^\alpha(f)(x) = \left(\sum_{l \in \mathbf{Z}} 2^{l\alpha q} \sup_{|x-y| \leq \sqrt{n} 2^{-l}} |(\phi_l * f)(y)|^q \right)^{1/q}.$$

It is easy to observe, that $\bar{S}_q^\alpha$ also characterizes $\dot{F}_p^{\alpha,q}$. In fact, for every $\lambda > 0$

$$\sup_{|x-y| \leq \sqrt{n} 2^{-l}} |(\phi_l * f)(y)| \leq (1 + \sqrt{n})^\lambda \sup_{y \in \mathbf{R}^n} \frac{|(\phi_l * f)(x - y)|}{(1 + 2^l |y|)^\lambda}.$$

The right-hand side is the Peetre's maximal function, and thus we obtain the characterization (see [FJW], p. 44). We proceed to construct Ω_k, $\bar{\Omega}_k$, $\mathcal{B}_k$ and Q_k^i as before, with $\bar{S}$ in the place of S. We define

$$\lambda_{k,i} = |Q_k^i|^{1/p-1/q} \left(\sum_{\substack{Q \subset Q_k^i \\ Q \in \mathcal{B}_k}} |Q| \sup_{y \in Q} \left(l(Q)^{-\alpha} |(\phi_Q * f)(y)| \right)^q \right)^{1/q},$$

and

$$a_{k,i}(x) = \frac{1}{\lambda_{k,i}} \left(\sum_{\substack{Q \subset Q_k^i \\ Q \in \mathcal{B}_k}} \int_Q \psi_Q(x - y)(\phi_Q * f)(y) dy \right).$$

We have

$$
\left\| |Q_k^i|^{1/p-1/q} a_{k,i} \right\|_{\dot{F}_q^{\alpha,q}}^q = \frac{\left\| \sum\limits_{\substack{Q \subset Q_k^i \\ Q \in \mathcal{B}_k}} \int_Q \psi_Q(x-y)(\phi_Q * f)(y)dy \right\|_{\dot{F}_q^{\alpha,q}}^q}{\sum\limits_{\substack{Q \subset Q_k^i \\ Q \in \mathcal{B}_k}} |Q| \sup_{y \in Q} (l(Q)^{-\alpha}|(\phi_Q * f)(y)|)^q}
$$

$$
\leq \frac{\sum\limits_{\substack{Q \subset Q_k^i \\ Q \in \mathcal{B}_k}} \left\| \int_Q \psi_Q(x-y)(\phi_Q * f)(y)dy \right\|_{\dot{F}_q^{\alpha,q}}^q}{\sum\limits_{\substack{Q \subset Q_k^i \\ Q \in \mathcal{B}_k}} |Q| \sup_{y \in Q} (l(Q)^{-\alpha}|(\phi_Q * f)(y)|)^q},
$$

where we have used the q-triangle inequality. The summand in the numerator is

$$
\int_{\mathbf{R}^n} \sum_{l \in \mathbf{Z}} 2^{l\alpha q} \left| \left(\phi_l * \int_Q \psi_Q(\cdot - y)(\phi_Q * f)(y)dy \right)(x) \right|^q dx
$$

$$
\leq \sup_{y \in Q} |(\phi_Q * f)(y)|^q \int_{\mathbf{R}^n} \sum_{l \in \mathbf{Z}} 2^{l\alpha q} \left(\int_Q |(\phi_l * \psi_Q)(x-y)|dy \right)^q dx
$$

$$
= \sup_{y \in Q} |(\phi_Q * f)(y)|^q \cdot I.
$$

A change of variables gives us

$$
I = l(Q)^{n-\alpha q} \int_{\mathbf{R}^n} \sum_{l \in \mathbf{Z}} 2^{l\alpha q} \left(\int_{Q(0,1)} |(\phi_l * \psi)(x-y)|dy \right)^q dx
$$

$$
= c l(Q)^{n-\alpha q}.
$$

It is not difficult to see, again using the $\bar{S}$-function, that

$$
c \leq \|\psi\|_{\dot{F}_q^{\alpha,q}}^q + \|\psi\|_{\dot{F}_q^{\alpha+n/q,q}}^q < \infty.
$$

Returning to our estimate for $a_{k,i}$, we obtain

$$
\left\| |Q_k^i|^{1/p-1/q} a_{k,i} \right\|_{\dot{F}_q^{\alpha,q}}^q \leq c.
$$

Now, we will show that the l^p norm of the coefficients $\lambda_{k,i}$ can be controlled by the appropriate norm of f. The proof is identical to that in the case $q \geq 1$, once we establish the following version of Lemma (2.4):

LEMMA (2.4'). *We have, using the above notation,*

$$
\sum_{Q \in \mathcal{B}_k} |Q|^{1-\alpha q/n} \sup_{y \in Q} |(\phi_Q * f)(y)|^q \leq c 2^{kq} |\Omega_k|.
$$

PROOF. For each cube $Q \in \mathcal{B}_k$ pick a point $x_Q \in (\bar{\Omega}_k \setminus \Omega_{k+1}) \cap Q$. Let this x_Q be chosen such, that

$$
\sup_{|x_Q - y| \leq \sqrt{n} l(Q)} |(\phi_Q * f)(y)|
$$

is minimal. Then,

$$\sup_{y\in Q}|(\phi_Q * f)(y)|^q \le \sup_{|x_Q-y|\le\sqrt{n}l(Q)}|(\phi_Q * f)(y)|^q,$$

since the supremum on the right is taken over a larger set. Then,

$$\sum_{Q\in\mathcal{B}_k}|Q|^{1-\alpha q/n}\sup_{y\in Q}|(\phi_Q * f)(y)|^q$$

$$\le \sum_{l\in\mathbf{Z}}\sum_{\substack{Q\in\mathcal{B}_k\\l(Q)=2^{-l}}}2^{-ln+\alpha ql}\sup_{|x_Q-y|\le\sqrt{n}l(Q)}|(\phi_l * f)(y)|^q$$

$$\le \sum_{l\in\mathbf{Z}}\sum_{\substack{Q\in\mathcal{B}_k\\l(Q)=2^{-l}}}2^{-ln+\alpha ql}\frac{1}{|(\bar\Omega_k\setminus\Omega_{k+1})\cap Q|}\times$$

$$\times\int_{(\bar\Omega_k\setminus\Omega_{k+1})\cap Q}\sup_{|x-y|\le\sqrt{n}l(Q)}|(\phi_l * f)(y)|^q dx$$

$$\le 2\sum_{l\in\mathbf{Z}}\sum_{\substack{Q\in\mathcal{B}_k\\l(Q)=2^{-l}}}2^{\alpha ql}\int_{(\bar\Omega_k\setminus\Omega_{k+1})\cap Q}\sup_{|x-y|\le\sqrt{n}l(Q)}|(\phi_l * f)(y)|^q dx$$

$$\le 2\sum_{l\in\mathbf{Z}}2^{\alpha ql}\int_{\bar\Omega_k\setminus\Omega_{k+1}}\sup_{|x-y|\le\sqrt{n}l(Q)}|(\phi_l * f)(y)|^q dx$$

$$= 2\int_{\bar\Omega_k\setminus\Omega_{k+1}}\left(\sum_{l\in\mathbf{Z}}2^{\alpha ql}\sup_{|x-y|\le\sqrt{n}l(Q)}|(\phi_l * f)(y)|^q\right)dx$$

$$= 2\int_{\bar\Omega_k\setminus\Omega_{k+1}}|\bar S_q^\alpha(f)(x)|^q dx$$

$$\le 22^{(k+1)q}|\bar\Omega_k\setminus\Omega_{k+1}| = c2^{kq}|\Omega_k|.$$

Lemma (2.4') is proved, and hence, Theorem 1 is proved. $\square$

3. Proof of Theorem 2.

First we prove the "if" part. By the duality

$$\left(\dot F_p^{0,q'}\right)^* = \dot F_\infty^{n(1/p-1),\infty},\text{ for }0<p<1\le q'\le\infty,$$

it suffices to show that there exists a constant c such that for each $g\in\dot F_p^{0,q'}$

$$(3.1)\qquad |\langle g,f\rangle|\le c\|g\|_{\dot F_p^{0,q'}}.$$

To see this, by our atomic decomposition of $\dot F_p^{0,q'}$ we have $g=\sum_k\sum_i\lambda_{k,i}a_{k,i}$ where, since $\alpha=0$,

$$\lambda_{k,i}=|Q_k^i|^{1/p-1/q'}\left(\sum_{\substack{Q\subset Q_k^i\\Q\in\mathcal{B}_k}}\int_Q(|\phi_Q * g(y)|)^{q'}dy\right)^{1/q'},$$

and

$$a_{k,i}(x) = \frac{1}{\lambda_{k,i}} \left(\sum_{\substack{Q \subset Q_k^i \\ Q \in \mathcal{B}_k}} \int_Q \psi_Q(x-y) \phi_Q * g(y) dy \right).$$

Thus,

$$|\langle g, f \rangle| \le \sum_k \sum_i |\lambda_{k,i}| |\langle a_{k,i}, f \rangle|$$

$$= \sum_k \sum_i |\lambda_{k,i}| |Q_k^i|^{1/q'-1/p} \frac{\left(\sum_{\substack{Q \subset Q_k^i \\ Q \in \mathcal{B}_k}} \int_Q |\langle \psi_Q(\cdot - y), f \rangle| |\phi_Q * g(y)| dy \right)}{\left(\sum_{\substack{Q \subset Q_k^i \\ Q \in \mathcal{B}_k}} \int_Q (|\phi_Q * g(y)|)^{q'} dy \right)^{1/q'}}$$

$$= \sum_k \sum_i |\lambda_{k,i}| \frac{\left(\sum_{\substack{Q \subset Q_k^i \\ Q \in \mathcal{B}_k}} \int_Q |\langle \psi_Q(\cdot - y), (f - P_{3Q_k^i}^s)\eta_{3Q_k^i} \rangle| |\phi_Q * g(y)| dy \right)}{|Q_k^i|^{1/p-1/q'} \left(\sum_{\substack{Q \subset Q_k^i \\ Q \in \mathcal{B}_k}} \int_Q (|\phi_Q * g(y)|)^{q'} dy \right)^{1/q'}}$$

$$\le \sum_k \sum_i |\lambda_{k,i}| |Q_k^i|^{1/q'-1/p} \left\| (f - P_{3Q_k^i}^s)\eta_{3Q_k^i} \right\|_{\dot{F}_q^{0,q}} \le c \sum_k \sum_i |\lambda_{k,i}|$$

$$\le c \left(\sum_k \sum_i |\lambda_{k,i}|^p \right)^{1/p} \le c \|g\|_{\dot{F}_p^{0,q'}},$$

which shows that $f \in (\dot{F}_p^{0,q})^* = \dot{F}_\infty^{n(1/p-1),\infty}$.

To prove the "only if" part of the theorem, we use the fact that $\dot{F}_\infty^{n(1/p-1),\infty} = \text{Lip}(n(1/p-1))$ with equivalent norms. We will show, that $(f - P_Q^s(f))\eta_Q \in \dot{F}_q^{0,q}$, and

$$\left\| (f - P_Q^s(f))\eta_Q \right\|_{\dot{F}_q^{0,q}} \le c|Q|^{1/q+1/p-1}.$$

Using the homogeneity of $\dot{F}_q^{0,q}$ and $\text{Lip}(1/p-1)$, we can reduce the problem to showing

$$\left\| (f - P_Q^s(f))\eta_Q \right\|_{\dot{F}_q^{0,q}} \le c \|f\|_{\text{Lip}(n(1/p-1))},$$

where Q is the unit cube $Q(0,1)$. Pick r, $1 < r < q$, and appropriate ϵ, $0 < \epsilon < 1$, so that $\epsilon = n(1/r - 1/q)$. By the Sobolev-Besov embedding, for any F,

$$\|F\|_{\dot{F}_q^{0,q}} \le c \|F\|_{\dot{F}_r^{\epsilon,r}}.$$

It suffices to show, that $(f - P_Q^s(f))\eta_Q$ is a smooth molecule for $\dot{F}_r^{\epsilon,r}$, for the unit cube $Q(0,1)$, with molecular norm bounded by $c\|f\|_{\text{Lip}(n(1/p-1))}$. There are two conditions to verify (see [FJW], p. 47):

$$|(f(x) - P_Q^s(f)(x))\eta_Q(x)| \leq c \|f\|_{\mathrm{Lip}(n(1/p-1))} (1 + x)^{-M},$$
$$|(f(x) - P_Q^s(f)(x))\eta_Q(x) - (f(y) - P_Q^s(f)(y))\eta_Q(y)|$$
$$\leq c \|f\|_{\mathrm{Lip}(n(1/p-1))} |x - y|^\delta \sup_{|z| \leq |x-y|} (1 + |x - z|)^{-M},$$

for some $M > n$ and δ, $\epsilon < \delta < 1$. Both inequalities follow immediately from the work of Taibleson and Weiss ([TW]). This finishes the proof of Theorem 2. $\square$

References

[CF]   S.-Y. A. Chang and R. Fefferman, *A continuous version of duality of H^1 and* **BMO** *on the bidisc*, Ann. of Math. **112** (1980), 179–201.

[CW]   R. Coifman and G. Weiss, *Extensions of Hardy Spaces and their use in analysis*, Bull. Amer. Math. Soc. **83** (1977), 569–645.

[FJ]   M. Frazier and B. Jawerth, *A discrete transform and decompositions of distribution spaces*, J. Func. Anal. **93** (1990), 34–170.

[FJW]   M. Frazier, B. Jawerth and G. Weiss, *Littlewood-Paley Theory and the Study of Function Spaces*, CBMS Regional Conference Series in Math. **79**.

[H]   Y.-S. Han, *Certain Hardy-type spaces that can be characterized by maximal functions and variations of the square functions*, Ph. Thesis, Washington University, St. Louis, 1984.

[L]   R. H. Latter, *A characterization of $H^p(\mathbf{R}^n)$ in terms of atoms*, Studia Math. **62** (1978), 93–101.

[T]   H. Triebel, *Theory of Function Spaces*, Birkhäuser, Basel-Boston-Stuttgart, 1983.

[TW]   M. Taibleson and G. Weiss, *The molecular characterization of certain Hardy spaces*, Asterisque **77** (1980), 67–149.

AUBURN UNIVERSITY, AUBURN, ALABAMA

E-mail address: hanyong@mail.auburn.edu

UNIVERSITY OF WROCŁAW, WROCŁAW, POLAND AND WASHINGTON UNIVERSITY, ST. LOUIS, MISSOURI

E-mail address: mpal@math.uni.wroc.pl

WASHINGTON UNIVERSITY, ST. LOUIS, MISSOURI

E-mail address: guido@artsci.wustl.edu

Contemporary Mathematics
Volume 189, 1995

Commutators of Parabolic Singular Integrals

STEVE HOFMANN

ABSTRACT. We obtain L^p bounds, $1 < p < \infty$, for the parabolic analogues of the Calderón Commutators, and also show that the commutator $[K^{(j)}, A]$ (where $K^{(j)}$ denotes a "generic" parabolic Calderón commutator of order $j \geq 0$) is "parabolically smoothing", i.e. that it maps L^p into a homogeneous Sobolev space having full spatial derivative and a half-order time derivative in L^p. The condition which we impose on $A(x,t)$ is sharp in the sense that it is necessary and sufficient for L^2 boundedness of $[\sqrt{\triangle - \frac{\partial}{\partial t}}, A]$. The theory of these operators is distinguished from the analogous elliptic theory, and also the parabolic theory in cylinder domains, by the fact that one is forced to consider "rough" singular integrals, for which $T1 \notin BMO$.

INTRODUCTION

The theory of those multilinear singular integral operators which arise in elliptic partial differential equations is now quite well understood; see, for example, Calderón [Ca 1], [Ca 2], David and Journé [DJ], Coifman, MacIntosh and Meyer [CMM], Coifman and Meyer [CM], and Christ and Journé [CJ sections 1-4]. The fundamental modern tool for handling such operators is the "T1" Theorem [DJ], which permits one to control a k-linear operator by means of a $(k-1)$-linear operator, thus setting up a natural induction scheme.

In contrast, the multilinear and non-linear singular integrals which arise in parabolic PDE have not been nearly so well understood. These "parabolic" singular integrals are invariant with respect to a non-isotropic group of dilations, but it is not this "mixed homogeneity", per se, which causes problems: a well-developed theory of convolution operators of parabolic type has existed for some time; see, e.g., B.F. Jones [Jns B], Fabes and Riviere [FR 1], [FR 2], and Riviere

1991 *Mathematics Subject Classification.* 42B20.

The author was supported by NSF grant DMS-9203930 at Wright State University, Dayton, Ohio.

[R]. Furthermore, a T1 Theorem does exist in this setting, and even in the more general context of spaces of homogeneous type [R. Coifman, unpublished], but see also, e.g., Lemarie, [L] Lewis and Murray [LM 1] and David, Journé, Semmes [DJS]. The real difficulty turns out to be that the parabolic multilinear singular integrals are "rough " in a sense which will be made more precise below. In particular, one is forced to treat operators for which the fundamental premise of the T1 theorem, namely that T1 $\in BMO$, fails. Thus, a direct attack via the non-isotropic T1 theorem does not work. However, we shall see that a local substitute for the condition that T1 $\in BMO$ does hold. This condition is a sort of quasi-Carleson measure condition, and is similar to a condition used in previous work of the author [H1] to give a general boundedness criteron for singular integral operators for which T1 $\notin BMO$. The general criterion of [H1] is not directly applicable to the operators of the present paper, but the use of Littlewood-Paley Theory to reduce matters to an appropriate local estimate is in the same spirit (in this vein, see also [CJ, section 7]).

The paper is organized as follows. In section 1, we set notation, and introduce the multilinear parabolic singular integrals that are our principal concern here. We also discuss known results for such operators, in particular some recent work of J.L. Lewis and M.A.M. Murray [LM1], [LM 2], and also the author [H5], [H2]. The latter reference establishes the sharpness of the results of the present paper. Section 2 describes the basic strategy of the proof, and reduces matters to one main estimate, plus the "quasi-Carleson measure" condition. In section 3, the operator under consideration in this main estimate is split into two parts, one "standard" and one "rough", and the standard part is treated. In section 4, we treat the rough part. This reduces matters to the quasi-Carleson measure estimate, which we prove in section 5. Section 6 is an appendix, where we give the proof of a technical lemma. We note that the strategy of our proof follows in part along the same lines as the arguments in recently communicated work of the author [H5] treating parabolic singular integrals and smoothing operators of "Calderon-type" (such as, e.g., the double and single layer potentials for the heat equation). Indeed, we had found the results of the present paper first, and then had pushed those methods further to include the "non-linear" operators of [H5]. The estimates for the multilinear operators considered here may be of some independent interest, given that we are able to treat "generic" parabolic commutators (the results of [H5] yield estimates only for a smaller class of commutators, namely those satisfying a cancellation condition which in the elliptic case amounts to antisymmetry of the kernel). In addition, the present paper may be a bit more "accessible" than [H5], in that certain technical difficulties arising in the non-linear case do not occur in the multilinear case.

Acknowledgements

I am very grateful to John L. Lewis and Margaret Murray, for introducing me to the problem, for generously sharing their ideas with me, and for several helpful and encouraging conversations. I also thank Jill Pipher for a helpful conversation.

1. Notation, History, and Statement of Main Theorems

Our results will be obtained in $R^n = R^{n-1} \times R$, endowed with the group of parabolic dilations $(x,t) \to (\lambda x, \lambda^2 t)$. We shall denote points in R^n by $z \equiv (x,t), w \equiv (y,s)$ (and also sometimes by u and v), and we indicate our dilations by the multi-index notation

$$\text{(1)} \qquad \lambda^\alpha z \equiv (\lambda x, \lambda^2 t)$$

(i.e., in multi-index notation $\alpha \equiv (1, ..., 1, 2)$). Points on the Fourier transform side will be denoted $\zeta \equiv (\xi, \tau) \in R^{n-1} \times R$: Points on the sphere S^{n-1} shall be denoted by $\sigma \equiv (\sigma_1, \sigma_2, ... \sigma_n)$ and for such σ, we define $\sigma' \equiv (\sigma, \sigma_2, ... \sigma_{n-1})$. Similarly, for $z = (x,t), w = (y,s)$ we shall also sometimes find it convenient to write $z' = x, w' \equiv y$.

The spatial gradient will be denoted by $\nabla_x \equiv (\frac{\partial}{\partial x_1}, \frac{\partial}{\partial x_2}, \cdots \frac{\partial}{\partial x_{n-1}})$, and the full n-dimensional gradient by $\nabla_n \equiv (\nabla_x, \frac{\partial}{\partial t})$. Following [FR1], the 1/2 order time derivative D_n is defined by

$$\text{(2)} \qquad (\widehat{D_n f}(\xi, \tau) \equiv \frac{\tau}{\| (\xi, \tau) \|} \hat{f}(\xi, \tau),$$

where, associated to the group of dilations (1) is a non-isotropic "norm" which we denote $\| z \|$, defined as the unique positive solution of

$$\text{(3)} \qquad 1 \equiv \sum_{j=1}^{n-1} \frac{x_j^2}{\| (x,t) \|^2} + \frac{t^2}{\| (x,t) \|^4},$$

and having the dilation invariance property $\| \lambda^\alpha z \| \equiv \lambda \| z \|$. Endowed with the metric induced by $\| \cdot \|$, R^n is a space of homogeneous type in the sense of Coifman and Weiss [CW], with homogeneous dimension $d \equiv | \alpha | \equiv n + 1$. We have the polar decomposition

$$z = \rho^\alpha \sigma, \ dz \equiv \rho^{d-1} d\rho \, J(\sigma) \, d\sigma,$$

where $J(\sigma)$ is a smooth, positive function, even in each of $\sigma_1, \sigma_2, ... \sigma_n$ separately (see, e.g. [FR1,2] or [R]), and $\rho \equiv \| z \|$. (In fact, $J(\sigma) \equiv 1 + \sigma_n^2$).

We indicate the order 1 parabolic fractional integration and differentiation operators by I_{par} and D_{par} respectively, with Fourier multipliers

$$\text{(4)} \qquad (I_{par})^\wedge(\zeta) \equiv \frac{1}{\| \zeta \|}, \quad (D_{par})^\wedge(\zeta) \equiv \| \zeta \|.$$

(We remark that D_{par} is essentially comparable to the square root of $\triangle - \frac{\partial}{\partial t}$).

Throughout the paper, we shall, without further comment, write $T(z,w)$ for the kernel of an operator T (or simply $T(z)$ in the convolution case), whenever this is not likely to cause confusion. We shall also use the convention that the generic constant C may vary at different occurences, but will depend only on dimension and possibly other harmless parameters (such as, e.g., our particular choice of Littlewood-Paley functions).

J.L. Lewis and M.A.M. Muray [LM1] have recently considered the following caloric double layer potential on the boundary of a time dependent domain D, with ∂D given by the graph $(x, t, A(x, t))$. Let D_t denote the cross section of D with t fixed, and let

$$W(X, t) \equiv C_n t^{-n/2} \exp(-\mid X \mid^2 /4t) \chi_{[0, \infty)}(t)$$

denote the fundamental solution of the heat equation in $\mathbb{R}^n \times \mathbb{R}$. The boundary double layer potential is given by

$$Kf(P, t) \equiv p.v. \int_{-\infty}^{t} \int_{\partial D_s} \frac{\partial}{\partial N_Q} W(P - Q, t - s) \, f(Q, s) d\sigma_s \, ds,$$

where $P \in \partial D_t, Q \in \partial D_s, d\sigma_s$ is surface measure on ∂D_s, and $\partial/\partial N_Q$ denotes differentiation in the direction of the outer unit normal to D_s at the point Q. With a slight abuse of notation, we re-write Kf in graph co-ordinates as $K_A f(x, t) \equiv$

$$(5) \quad C_n \; p.v. \int_{-\infty}^{t} \int_{\mathbb{R}^{n-1}} \frac{A(x, t) - A(y, s) - (x - y) \cdot \nabla_y A(y, s)}{(t - s)^{1+n/2}} \times$$

$$\exp\left[\frac{-\mid x - y \mid^2 - [A(x, t) - A(y, s)]^2}{4(t - s)}\right] f(y, s) \, dyds.$$

Similarly, the boundary single layer potential may be written as $S_A f(x, t) \equiv$

$$(6) C_n \int_{-\infty}^{t} \int_{\mathbb{R}^{n-1}} \frac{\exp\left[\frac{-|x-y|^2 - [A(x,t) - A(y,s)]^2}{4(t-s)}\right]}{(t - s)^{n/2}} f(y, s) \sqrt{1 + \mid \nabla_y A(y, s) \mid^2} \, dyds.$$

In a remarkable technical achievement, Lewis and Murray [LM1] obtained L^2 (and hence L^p) boundedness of K_A for functions $A(x, t)$ which were Lipschitz in the space variable (uniformly in time), and whose 1-dimensional 1/2 order derivative

$$\mid D_t \mid^{1/2} A(x, \cdot)(t) \equiv (\mid \tau \mid^{1/2} \hat{A}(x, \tau))^{\vee}(t)$$

belonged to one-dimensional BMO, uniformly in x (here only, but *not* in the sequel, $\wedge$ and $\vee$ denote, respectively, the 1-dimensional Fourier and inverse Fourier transforms in the time variable). Their method of proof was to expand (5) into a series of commutators $K_A f(x, t) \equiv$

$$(7) \quad C_n \sum_{m=0}^{\infty} \frac{(-1)^m}{4^m m!} \int_{-\infty}^{t} \int_{\mathbb{R}^{n-1}} \frac{[A(x, t) - A(y, s)]^{2m}}{(t - s)^{1+m+n/2}} \exp\left(-\frac{\mid x - y \mid^2}{4(t - s)}\right) \cdot$$

$$\nabla_y A(y, s) \cdot (x - y) \, f(y_1 s) dyds$$

$$+$$

$$C_n \sum_{m=0}^{\infty} \frac{(-1)^m}{4^m m!} \int_{-\infty}^{t} \int_{\mathbb{R}^{n-1}} \frac{[A(x,t) - A(y,s)]^{2m+1}}{(t-s)^{1+m+n/2}} \exp\left(-\frac{\mid x-y\mid^2}{4(t-s)}\right) f(y,s) dy ds$$

$$\equiv \sum_{m=0}^{\infty} \frac{(-1)^m}{4^m m!} \left\{ \vec{K}_0^{(2m)}(\nabla_y Af)(x,t) + K_0^{(2m+1)}(f)(x,t) \right\}.$$

(These integrals are interpreted as principal values). Since $\nabla_y A(y,s) \in L^{\infty}$, by assumption, Lewis and Murray were able to obtain a real-analyticity result for K in a small neighborhood of zero, and, in particular, L^2 boundedness in the small constant case, by showing that the operator norm of $K_0^{(2m+1)}$ grew on the order of

$$C^m [m! \parallel \nabla_x A \parallel_{\infty}^{2m+1} + C^m \sup_x \parallel \mid D_t \mid^{1/2} A(x,\cdot) \parallel_{BMO(\mathbb{R},dt)}^{2m+1}],$$

and that a similar estimate held also for $\vec{K}_0^{(2m)}$. In [LM2], they were then able to remove the "small constant" restriction by a bootstrapping argument à la G. David (but with considerably more difficulty than in the elliptic case). Further perturbation arguments permitted them to consider more general domains.

In [H2], the author found a necessary and sufficient condition for the L^2 boundedness of the first commutator $K_0^{(1)}$ (we remark that

$$K_0^{(1)} \equiv C[\sqrt{H}, A],$$

where $H \equiv \triangle - \frac{\partial}{\partial t}$, as may be verified by a Fourier transform argument; see [H2]). Furthermore, the sufficiency result still held if the kernel of $\sqrt{H}$ was replaced by a "generic" parabolic kernel with the same homogeneity, smooth away from the origin, and having vanishing first spatial moments on the sphere. This necessary and sufficient condition for L^2 boundedness is that, again $A(x,t)$ is Lipschitz in x, uniformly in time, but now the uniform 1-dimensional time condition is replaced by the assumption that the usual "n-dimensional" $1/2$ order time derivative (2) belongs to n-dimensional parabolic BMO. The latter is the space of functions modulo constants with norm

$$\parallel b \parallel_* \equiv \sup \frac{1}{\mid \triangle_r(z) \mid} \int_{\triangle_r(z)} \mid b(w) - m_{\triangle_r(z)} b \mid dw,$$

where the supremum runs over all positive r and all z in R^n,

(8) $$\triangle_r(z) \equiv \{ w \epsilon \mathbb{R}^n : \parallel z - w \parallel < r \}$$

and for a ball $\triangle$,

$$m_{\triangle} b \equiv \frac{1}{\mid \triangle \mid} \int_{\triangle} b.$$

This still left open the question of whether or not the single layer potential (6) was "parabolically smoothing" on L^p of the boundary. In [H5], the author has resolved this question in the affirmative by showing that $D_{par} S_A$ is bounded on L^p.

It has been subsequently realized that estimates for the commutators in the expansion of (7), obtained via perturbation in [LM2], were obtained for A satisfying a condition equivalent to that of [H2], (and therefore weaker than that of [LM1]) in slightly disguised form. In the present paper, we give a direct proof (without perturbation) of the boundedness of "generic" parabolic commutators including, in particular, $\vec{K}_0^{(2m)}$ and $K_0^{(2m+1)}$, for A satisfying the sharp condition of [H2], with operator norms on the order of $C^m m! \parallel A \parallel_{comm}^{2m}$ or $C^m m! \parallel A \parallel_{comm}^{2m+1}$, respectively, where $\parallel A \parallel_{comm} \equiv \parallel \nabla_x A \parallel_{L^\infty(\mathbb{R}^n)} + \parallel D_n A \parallel_*$ (by [H2], $\parallel A \parallel_{comm} \approx \parallel [\sqrt{H}, A] \parallel_{op}$). These estimates yield real analyticity of the mapping $A \to K_A$, taking values in the space of bounded operators on L^2, under sharp conditions on A, in a small neighborhood of 0.

Let us now discuss the particulars. Suppose

$$(9) \qquad\qquad H^{(j)}(\lambda^\alpha z) \equiv \lambda^{-d-j} H^{(j)}(z)$$

(we recall that $d \equiv n + 1$), and that $H^{(j)}$ satisfies the cancellation property

$$(10) \qquad\qquad \int_{S^{n-1}} H^{(j)}(\sigma)(\sigma')^\gamma \sigma_n^\kappa \, J(\sigma) d\sigma \equiv 0,$$

for any integer $\kappa \geq 0$, and any non-negative, n-1 dimensional multi-index γ, such that $|\gamma| + 2\kappa \equiv j$. In particular, this holds in the case that $H^{(j)}(x,t)$ is even (resp., odd) in x, for each fixed t, when j is odd (resp. even), which case is included in the results of [H5]. Furthermore, we shall assume that $H^{(j)} \in C^2(\mathbb{R}^n \backslash \{0\})$. We now give a quantitative measurement of the size and smoothness of $H^{(j)}$. Let κ denote a non-negative integer, and let γ_1, γ_2 denote non-negative n-1 dimensional multi-indices. We define

$$(11) \qquad\qquad | H^{(j)} |_{C-Z,j} \equiv$$

$$\sup_{\substack{|\gamma_2| + \kappa \leq 2 \\ |\gamma_1| \leq j \\ (x,t) \neq 0}} \{ \parallel (x,t) \parallel^{d+j+|\gamma_2|+2\kappa-|\gamma_1|} | \frac{\partial}{\partial x}^{\gamma_2} \frac{\partial}{\partial t}^\kappa [x^{\gamma_1} H^{(j)}(x,t)] | \}.$$

The meaning of the quantity (11) is that it measures size and smoothness, not only of $H^{(j)}$, but also of lower order $H^{(k)}, k < j$ which will arise in our induction scheme. We shall assume that there exists a fixed integer k_0, and a fixed constant β_0, such that for all integers $N \in [0, m]$, we have

$$(12) \qquad | t^N H^{(2m)} |_{C-Z,2(m-N)} + | t^N H^{(2m+1)} |_{C-Z,2(m-N)+1}$$

$$\leq \beta_0^{m-N}(m - N + k_0)!$$

(again we are taking into account kernels that arise in the induction process). We note that, in particular, the kernels

$$\vec{H}_0^{(2m)}(x,t) \equiv \frac{x}{t^{1+m+n/2}} \exp \frac{-|x|^2}{4t} \chi\{t > 0\}$$
$$H_0^{(2m+1)}(x,t) \equiv \frac{1}{t^{1+m+n/2}} \exp \frac{-|x|^2}{4t} \chi\{t > 0\}$$

satisfy (9) (obviously), (10) (because the resulting integrand is odd in x, for each fixed t), and, up to a purely dimensional multiplicative constant, (12). The latter bound follows from Stirling's formula, the fact that $\| (x,t) \| \approx | x | + | t |^{1/2}$, and the elementary observation that the maximum value of the function $r^a e^{-r/4}$, with $r \geq 0$, is $(4a)^a e^{-a}$. We leave the details to the reader.

In order to similify the notation, we shall find it convenient to normalize our kernels. This amounts to absorbing into the kernel the factor of $(4^m m!)^{-1}$ in the series expansion (7). Actually, it is even more convenient to divide by $(m + k_0)!$, for the same k_0 as in (12); to compensate, we need only introduce a harmless polynomial growth factor into the series expansion. In general, we divide by $\beta_0^m (m + k_0)!$, for β_0 as in (12). It is easily verified that the resulting kernels satisfy

(13)
$$| \beta_0^N \frac{(m + k_0)!}{(m - N + k_0)!} t^N H^{(2m)} |_{C-z,2(m-N)} +$$

$$| \beta_0^N \frac{(m + k_0)!}{(m - N + k_0)!} t^N H^{(2m+1)} |_{C-Z,2(m-N)+1}$$

$$\leq 1,$$

for all integers $N \in [0, m]$. We observe that, in particular, we then have $| H^{(j)} |_{C-Z,j} \leq 1$. It is not hard to check that (13) for the normalized kernels is equivalent to (12) for the original kernels.

We now define, for such $H^{(j)}$,

(14)
$$K^{(j)}(A, B) f(z) \equiv$$

$$p.v. \int [B(z) - B(w)][A(z) - A(w)]^{j-1} H^j(z - w) f(w) dw,$$

or if $j = 1$, we simply write $K^{(1)}(B)$ or just $K^{(1)}$; $K^{(0)}$ is the usual convolution operator. To simplify the notation, in the case $A \equiv B$, we shall frequently write $K^{(j)}(A, A) \equiv K^{(j)}$. We recall that $\| A \|_{comm} \equiv \| \nabla_x A \|_\infty + \| D_n A \|_*$. Our first result is the following generalization to generic parabolic commutators of a result of [LM2] (we repeat the cautionary remark that, in [LM2], the condition on A appears in a somewhat different, albeit equivalent, form).

THEOREM 1. *Suppose that* $H^{(2m)}, H^{(2m+1)} \in C^2(\mathbb{R}^n \backslash \{0\})$ *satisfy (9), (10) and (13), for every non-negative integer* m. *Then for all weights* $\omega \in A_2$, *there exists a constant* $C_0(A_2)$ *(depending only on dimension and the* A_2 *constant of of* ω*) such that*

$$\| K^{(2m)}(A, B) f \|_{2,\omega} \leq (C_0(A_2))^{m+1/2} \| B \|_{comm} \| A \|_{comm}^{2m-1} \| f \|_{2,\omega},$$

$$\| K^{(2m+1)}(A,B) f \|_{2,\omega} \le (C_0(A_2))^{m+1} \| B \|_{comm} \| A \|_{comm}^{2m-1} \| f \|_{2,\omega} .$$

We remark that by the usual Calderón-Zygmund theory, it would suffice to prove the unweighted case $\omega \equiv 1$. For the rough operators of the next theorem, however, the standard Calderón-Zygmund theory does not apply, and it will be convenient to obtain weighted estimates directly. We shall return to this point momentarily.

The proof of Theorem 1 will turn out to be a fairly routine consequence of the (parabolic) T1 Theorem, plus the following, which is really the main result of this paper. We set $[T, B] \equiv TB - BT$, where, with a slight abuse of notation, we indicate by B the operation of multiplication by the function B. We have

THEOREM 2. *Under the same hypothesis as above, there exists* $C_1(A_2)$, *depending only on dimension and the* A_2 *constant of* ω, *such that, for the same* C_0 *as in Theorem 1, we have*

$$\| D_{par}[K^{(2m)}, B] f \|_{2,\omega}$$

$$\le C_1(A_2)(C_0(A_2))^{m+1/2}(m+1) \| B \|_{comm} \| A \|_{comm}^{2m} \| f \|_{2,\omega},$$

$$\| D_{par}[K^{(2m+1)}, B] f \|_{2,\omega}$$

$$\le C_1(A_2)(C_0(A_2))^{m+1}(m+1) \| B \|_{comm} \| A \|_{comm}^{2m+1} \| f \|_{2,\omega} .$$

Since $K^{(j)}$ *is a generic parabolic commutator, similar estimates for* $[K^{(j)}, B]D_{par}$ *follow by duality.*

Remarks: By extrapolating (see [GR]), we may extend Theorem 2 immediately to L^p for all $p, 1 < p < \infty$. Indeed, the weighted estimates are a "rough operator" substitute for the usual Calderon-Zygmund Theory. The case $j = 1, A \equiv B$ is an n-dimensional version of a 1-dimensional operator considered in [LM3]. These operators are rough enough that for $j \ge 1$, $D_{par}[K^{(j)}, A]1$ need not in general, belong to BMO (see, e.g., [LM3, appendix]). Thus the T1 theorem does not apply, nor can one follow the usual Calderón-Zygmund program to obtain L^p estimates by interpolating with an endpoint result, even supposing that one had obtained (unweighted) L^2 bounds. The roughness of these operators reflects the roughness of the operator $D_{par}A \to K^{(j)}f$; in fact, we shall see that $K^{(j)}1 = [K^{(j-2)}, A]D_n a +$ other terms , where $a = D_{par}A$. (We remark that the first order case of Theorem 2, i.e. $D_{par}[K^{(0)}, A]$, is somewhat easier, being rather directly related to $K^{(1)}$. We shall not pursue this point further, since the desired bound will also follow from our general approach).

We have already remarked that the (normalized) kernels arising in the series expansion (7) satisfy the hypotheses of Theorem 1, so that we have real-analyticity of the L^2 operator valued mapping $A \to K_A$, if $\| \nabla_x A \|_\infty + \| D_n A \|_*$ is sufficiently small. More generally, we have

COROLLARY 3. *There exists $\epsilon_0 > 0$ such that if $\| \nabla_x A \|_\infty + \| D_n A \|_* < \epsilon_0$, and if $K^{(j)}$ satisfies the hypotheses of Theorem 1, then for any fixed non-negative integer k the operator defined by*

$$K(A,B)\,f(x,t) \equiv$$
$$\sum_{m=0}^{\infty} (1+m^k)\{K^{(2m)}\,(\nabla_y Bf)(x,t) + K^{(2m+1)}\,(A,B)\,f\,(x,t)\}$$

satisfies

$$\| K(A,B) \|_{op} \leq C(\| \nabla_x B \|_\infty + \| D_n B \|_*).$$

Moreover, for each fixed B, the operator valued mapping $A \rightarrow K(A,B)$ is real analytic in a neighborhood of zero.

Also, it is routine to check that

$$(15) \qquad S_A f \equiv \sum_{m=1}^{\infty} \frac{(-1)^m}{4^m m!}\,[K_0^{(2m-1)}, A]\,g + I_H g,$$

where $g(y,s) \equiv \sqrt{1 + |\nabla_y A_{(y,s)}|^2}\; f(y,s)$, and I_H is the parabolic smoothing operator

$$g \rightarrow \int W(x - y, t - s)\,g(y,s)dyds.$$

We remark that $I_H \equiv (\triangle - \partial/\partial t)^{-1/2}$ (see, e.g., [H2]). Note that $[(\triangle - \partial/\partial t)^{1/2}]^\wedge(\xi, \tau) \equiv (|\xi|^2 - i\tau)^{1/2}$, so that D_{par} and $\sqrt{H} \equiv (\triangle - \partial/\partial t)^{1/2}$ are, by symbolic calculus (see Fabes and Riviere [FR1]), equivalent up to a "nice" bounded, invertible parabolic singular integral. More generally, we can consider

$$(16) \qquad S(A,B) \equiv \sum_{j=1}^{\infty} j^k\,[K^{(j)}, B],$$

where $K^{(j)}$ satisfies the hypotheses of Theorems 1 and 2.

We have the following immediate corollary of Theorem 2:

COROLLARY 4. *There exists ϵ_0 such that if $\| \nabla A_x \|_\infty + \| D_n A \|_* < \epsilon_0$, then, for all $\omega \in A$*

$$\| D_{par} S(A,B)f \|_{2,\omega} + \| S(A,B)D_{par}f \|_{2,\omega} \leq C(A_2)\,\| B \|_{comm}\| f \|_{2,\omega}\,.$$

Moreover, the operator valued mappings $A \rightarrow S_A, A \rightarrow S(A,B)$ are real-analytic in a neighborhood of 0.

At this point, before going on to the proofs of the Theorems, we shall find it convenient to assemble some size and smoothness estimates for our kernels, which shall be used repeatedly in the sequel. We begin by noting that if, say $\| v - z \| \leq \| z \| /10$, then $\| v \| \approx \| z \|$, and in particular $\| u \|^{-k} \leq \beta_1^k \| z \|^{-k}$, for some absolute constant β_1, whenever u lies on the line segment joining v and z. Thus, by Taylor's Theorem, our assumptions (9) and (13), and the fact that

$|t| \leq \| (x,t) \|^2$ (so that one time derivative is worth two space derivatives - see (iii) below), we have, whenever $\| v - z \| \leq \| z \| /10$,

(17)
$$\begin{aligned}
&(i) \quad | H^{(k)}(z) | \leq \| z \|^{-d-k} \\
&(ii) \quad | H^{(k)}(v) - H^{(k)}(z) | \leq C\beta_1^k \| z - v \| \| z \|^{-d-k-1} \\
&(iii) \quad | H^{(k)}(v) - H^{(k)}(z) - (v' - z') \cdot \nabla_x H^{(k)}(z) \leq \\
&\qquad\qquad\qquad\qquad\qquad C\beta_1^k \| z - v \|^2 \| z \|^{-d-k-2}
\end{aligned}$$

As a consequence of (17) (iii), we also have that if $\| z - v \|, \| u - v \| \leq$ $\| z - u \| /100, \| z - w \| /100$, then

(18)
$$\begin{aligned}
&| [H^k(v - u) - H^{(k)}(z - u)] - [H^{(k)}(v - w) - H^{(k)}(z - w)] | \leq \\
&\qquad C\beta_1^k [\| z - v \|^2 + \| z - v \| \| u - w \|] \| z - w \|^{-d-k-2} .
\end{aligned}$$

We also remark that the same estimates hold whenever $H^{(k)}(v)$ is multiplied by a smooth cut-off factor $[\eta(\frac{\|v\|}{R}) - \eta(\frac{\|v\|}{\epsilon})]$, where $\eta \equiv 1$ on $\{\| v \| < 100\}$, and vanishes on $\{\| v \| > 200\}$. The bounds obtained are still on the order of $C\beta_1^k$, where C may depend on η and dimension, but is of course independent of $0 \leq \epsilon < R \leq \infty$. We shall use this well known fact about smoothly truncated kernels repeatedly in the sequel, without further comment.

2. Proofs of the Theorems: Preliminary Arguments

We start with Theorem 1, and reduce its proof to that of Theorem 2. It is enough to show that for all $f, g, \in C_0^\infty$, and all A, B, with $\| A \|_{comm} \leq 1, \| B \|_{comm} \leq 1$, then

(1)
$$\begin{aligned}
&|< K^{(2m)}(A,B)f, g >| \leq C_0^{m+1/2} \| f \|_2 \| g \|_2, \\
&|< K^{(2m+1)}(A,B)f, g >| \leq C_0^{m+1} \| f \|_2 \| g \|_2 .
\end{aligned}$$

A (parabolic) regularization of the identity (which may be removed by standard limiting arguments once the a priori bounds (1) have been established) permits us to assume that $A, B \in C^\infty$.

We note that we have

(2)
$$\| A \|_{comm} \approx \| \nabla_x A \|_\infty + \| D_{par} A \|_* ,$$

since one may pass back and forth between the two quantities by means of the parabolic Riesz Transforms, which are defined by the Fourier multipliers

(3)
$$\begin{aligned}
&(R_j)^\wedge(\xi, \tau) \equiv \frac{\xi_j}{\|(\xi,\tau)\|}, 1 \leq j \leq n - 1 \\
&(R_n)^\wedge(\xi, \tau) \equiv \frac{\tau}{\|(\xi,\tau)\|^2} .
\end{aligned}$$

In fact, by the identity (3, Section 1) applied on the Fourier transform side, we have

(4)
$$D_{par} \equiv \sum_{j=1}^{n-1} \frac{\partial}{\partial x_j} R_j + D_n R_n .$$

Also, it is clear that, by definition (see (2, section 1))

$$(5) \qquad D_n \equiv R_n D_{par} \equiv \frac{\partial}{\partial t} I_{par}$$

The reader may now easily verify the equivalence (2).

Furthermore, by [H2, Lemma 3.1, and inequality (15)], and (2), we have, for some dimensional constant β_2,

$$(6) \qquad \mid A(z) - A(w) \mid \le \beta_2 \parallel A \parallel_{comm} \parallel z - w \parallel .$$

In particular, (6) implies that the kernel $K^{(j)}(z,w)$ is a standard Calderón-Zygmund kernel, and thus by the parabolic T1 Theorem (see, e.g. Lemarie [L]; alternatively, the reader could easily adapt the argument in Coifman and Meyer [CM] to the parabolic setting), it suffices to show that $K^{(j)}1 \in BMO$, with appropriate bounds (the transpose $K^{(j)*}$ is of exactly the same form as $K^{(j)}$) and that $K^{(j)}$ satisfies the Weak Boundedness Property (WBP), also with appropriate bounds. We define WBP as follows. Let $\Lambda(z,r)$ denote the space of all $\varphi \in C_0^\infty$, supported in $\triangle_r(z)$, and satisfying

$$(7) \qquad \begin{array}{l} (i) \parallel \varphi \parallel_\infty \le 1 \\ (ii) \mid \varphi(u) - \varphi(w) \mid \le \parallel u - w \parallel /r \\ (iii) \sup_{|\nu|+\kappa \le 2} \{ r^{2\kappa+|\nu|} \parallel \left(\frac{\partial}{\partial t}\right)^\nu \left(\frac{\partial}{\partial t}\right)^\kappa \varphi \parallel_\infty \le 1, \end{array}$$

where $\kappa \in \{0,1,2\}$, and ν is an n-1 dimensional multi-index.

Then we say that T satisfies WBP if for all $z \in \mathbb{R}^n$, $r > 0$, and $\psi, \varphi \in \Lambda(z,r)$, we have

$$(8) \qquad |< \psi, T\varphi >| \le C_T r^d.$$

The smallest C_T for which (8) holds will be denoted $\mid T \mid_{WBP}$.

To prove estimate (1), we may assume that $\parallel B \parallel_{comm} \equiv 1 \equiv \parallel A \parallel_{comm}$. The case $j = 0$ is well known [FR 2], [Jns B], and our proof will give $j = 1$ directly. Our strategy is then to prove that for $j \ge 2$, and for a purely dimensional constant C_2,

$$(9)$$
$$\parallel K^{(j)}(A,B)1 \parallel_* \le C[C_2^j + \parallel \vec{K}^{(j-1)}(A,\cdot) \parallel_{bi-op} + \parallel [\tilde{K}^{(j-2)},\cdot]D_{par} \parallel_{bi-op}]$$

and that the WBP constant in (8) satisfies

$$(10)$$

$$\mid K^{(j)}(A,B) \mid_{WBP} \le C[C_2^j + \parallel \vec{K}^{(j-1)}(A,\cdot) \parallel_{bi-op} + \parallel [\tilde{K}^{(j-2)},\cdot]D_{par} \parallel_{bi-op}],$$

where $\vec{H}^{(j-1)}(x,t) \equiv xH^{(j)}$, and $\tilde{H}^{(j-2)}(x,t) \equiv tH^{(j)}$, and where the bi-linear operator (bi-op) norm and weighted $(bi-op, \omega)$ norm are defined respectively, by

$$
\begin{array}{ll}
(11) & \| T(\cdot)\|_{bi-op} \equiv \sup_{\|f\|_2=1, \|B\|_{comm}=1} \| T(B)f \|_2 \\
& \| T(\cdot)\|_{bi-op,\omega} \equiv \sup_{\|f\|_{2,\omega}=1, \|B\|_{comm}=1} \| T(B)f \|_{2,\omega} \, .
\end{array}
$$

We remark that the T1 Theorem immediately implies that $\| K^{(j)}(A,\cdot)\|_{bi-op}$ satisfies a bound on the order of that in (9).

We now introduce the quasi-Carleson measure condition which substitutes for $T1 \in BMO$. We choose a smooth cut-off function $\eta \in C_0^\infty([-200,200])$ such that $0 \le \eta \le 1$, and $\eta \equiv 1$ on $[-100,100]$ (the choice of η is immaterial, so long as it is fixed). For any given parabolic ball $\triangle_r(z_0)$, we set

$$
(12) \qquad
\begin{array}{l}
\eta_1(z) \equiv \eta_{1,\triangle_r(z_0)}(z) \equiv \eta \frac{\|z-z_0\|}{r} \\
\eta_2(z) \equiv \eta_{2,\triangle_r(z_0)}(z) \equiv 1 - \eta_1(z).
\end{array}
$$

We also select a (non-trivial) parabolic "approximation to the zero operator" Q_δ^* (again, this choice will be fixed) such that $Q_\delta^*(z) \equiv \delta^{-d}Q^*(\delta^{-\alpha}z)$, where

$$
(13) \qquad
\begin{array}{l}
(i) \quad Q_\delta^*1 = 0 \\
(ii) \, | Q^*(z) | \le (1+ \| z \|)^{-d-1} \\
(iii) \, | Q^*(z) - Q^*(w) | \le \frac{\|z-w\|}{(1+\|z\|)^{d+1}}, \; if \; \| z \| > 2 \| z - w \| \, .
\end{array}
$$

We now define the quasi-Carleson (q-C) and weighted (q-C, ω) norm (actually semi-norm) of T by

$$
(14) \; \| T \|_{q-C} \equiv \sup_{z_0 \in R^n, r>0} \left(\frac{1}{|\triangle_r(z_0)|} \int_0^r \int_{\triangle_r(z_0)} | Q_\delta^* T\eta_1(z) |^2 \, (z) \, dz \frac{d\delta}{\delta} \right)^{\frac{1}{2}} \, .
$$

$$
\| T \|_{q-C,\omega} \equiv \sup_{z_0 \in R^n, r>0} \left(\frac{1}{\omega(\triangle_r(z_0))} \int_0^r \int_{\triangle_r(z_0)} | Q_\delta^* T\eta_1(z) |^2 \, \omega(z) \, dz \frac{d\delta}{\delta} \right)^{\frac{1}{2}} \, .
$$

Our strategy in the proof of Theorem 2 is then to establish, for all integers $k > 0$, our "main" estimate

$$
(15) \quad \| D_{par}[K^{(k)}, \cdot] \|_{bi-op, \omega}
$$

$$
\le C(A_2)[C_2^k + \| D_{par}[K^{(k)}, \cdot] \|_{q-C, \omega} + k \| K^{(k)}(A,\cdot) \|_{bi-op, \omega} \, .
$$

(If $k = 0$ or 1, then $K^{(k)}(A,\cdot)$ is replaced by $K^{(0)}$ or $K^{(1)}(\cdot)$, respectively). The latter estimate, which will reduce matters to controlling the $q-C$ norm of $D_{par}[K^{(k)}, \cdot]$, will be proved in sections 3 and 4. Finally, in section 5 we shall prove the following bound for the $q-C$ norms, with $k \ge 1$,

$$(16) \qquad \| D_{par}[K^{(k)}, \cdot] \|_{q-C,\omega} \le C(A_2)\{C_2^k + \| K^{(k)} \|_{op}$$

$$+ k \| K^{(k)}(A, \cdot) \|_{bi-op,\omega} + k \| \vec{K}^{(k-1)} \|_{op,\omega} + \| D_{par}[\vec{K}^{(k-1)}, \cdot] \|_{bi-op,\omega}\},$$

where $\vec{H}^{(k-1)} \equiv x\, H^{(k)}$. The case $k = 0$ of (16) shall be treated directly.

Theorems 1 and 2 then follow by a rather straightforward, albeit slightly complicated, induction argument. We shall not bore the reader with the routine details.

In the rest of this section, we shall prove (9) and (10). This will not be difficult, and the proofs will proceed simultaneously. To prove (9), we let ψ be a parabolic H^1 atom, supported in a parabolic ball $\triangle(z_0)$ which, by dilation invariance, we may assume to have radius $r = 1$. Then $\| \psi \|_\infty \le 1$, and $\int \psi = 0$. We select a smooth cut-off function η as above, and define η_1 and η_2 as in (12). Then, since ψ has mean value zero, a very standard argument from the Calderon-Zygmund theory shows that

$$|< \psi, K^{(j)}(A, B)\eta_2 >| \le C_2^j,$$

for C_2 large enough, where we have used ((17, Section 1) (ii), (6)) and the normalization $\| A \|_{comm} \equiv 1 \equiv \| B \|_{comm}$. Next we consider $< \psi, K^{(j)}(A, B)\eta_1 >$, and in this case we shall no longer require that ψ have mean value zero, but only that ψ and η_1 be bounded by 1 and have support in a fixed (parabolic) multiple of a unit ball. Thus, the treatment of this term will simultaneously establish WPB (10). We write $K^{(j)}(A, B)\eta_1(z)$ in polar co-ordinates $z - w = \rho^\alpha \sigma$ (we recall that in multi-index notation $\rho^\alpha(x, t) \equiv (\rho x, \rho^2 t)$), and then integrate by parts in the ρ variable to obtain

$$(17) \qquad K^{(j)}(A, B)\eta_1(z) \equiv$$

$$p.v. \int_{S^{n-1}} H^{(j)}(\sigma) \int_0^\infty [B(z)-B(z-\rho^\alpha\sigma)][A(z)-A(z-\rho^\alpha\sigma)]^{j-1}\eta_1(z-\rho^\alpha\sigma)\frac{d\rho}{\rho^{j+1}}J(\sigma)d\sigma$$

$$\equiv p.v. \frac{1}{j} \int_{S^{n-1}} H^{(j)}(\sigma)\sigma' \cdot \int_0^\infty \{[A(z) - A(z - \rho^\alpha\sigma)]^{j-1}\nabla_x B(z - \rho^\alpha\sigma)$$

$$+(j-1)[B(z)-B(z-\rho^\alpha\sigma)][A(z)-A(z-\rho^\alpha\sigma)]^{j-2}\nabla_x A(z-\rho^\alpha\sigma)\}\eta_1(z-\rho^\alpha\sigma)\frac{d\rho}{\rho^j}J(\sigma)d\sigma$$

$$+$$

$$\frac{2}{j} \int_{S^{n-1}} H^{(j)}(\sigma)\sigma_n \int_0^\infty \{[A(z) - A(z - \rho^\alpha\sigma)]^{j-1}\frac{\partial}{\partial t}B(z - \rho^\alpha\sigma)$$

$$+(j-1)[B(z)-B(z-\rho^\alpha\sigma)][A(z)-A(z-\rho^\alpha\sigma)]^{j-2}\frac{\partial}{\partial t}A(z-\rho^\alpha\sigma)\}\eta_1(z-\rho^\alpha\sigma)\frac{d\rho}{\rho^{j-1}}J(\sigma)d\sigma$$

$$+$$

$$\frac{1}{j}\int\limits_{S^{n-1}} H^{(j)}(\sigma)\sigma' \cdot \int\limits_0^\infty [B(z)-B(z-\rho^\alpha\sigma)][A(z)-A(z-\rho^\alpha\sigma)]^{j-1}\nabla_x\eta_1(z-\rho^\alpha\sigma)\frac{d\rho}{\rho^j}J(\sigma)d\sigma$$

$$+$$

$$\frac{2}{j}\int\limits_{S^{n-1}} H^{(j)}(\sigma)\sigma_n \int\limits_0^\infty [B(z)-B(z-\rho^\alpha\sigma)][A(z)-A(z-\rho^\alpha\sigma)]^{j-1}\frac{\partial}{\partial t}\eta_1(z-\rho^\alpha\sigma)\frac{d\rho}{\rho^{j-1}}J(\sigma)d\sigma$$

$$\equiv I + II + III + IV.$$

Setting $\vec{H}^{(j-1)} \equiv xH^{(j)}$, we see that I is a polar co-ordinate representation of

$$\frac{1}{j}\{\vec{K}^{(j-1)}(A,A)(\nabla_x B\eta_1) + (j-1)\vec{K}^{(j-1)}(A,B)(\nabla_x A\eta_1)\},$$

whose L^2 norm is no larger than $\|\vec{K}^{(j-1)}(A,\cdot)\|_{bi-op}$. In the case $j=1$, I reduces to the convolution operator $\vec{K}^{(0)}(\nabla_x B\eta_1)$, whose L^2 norm can be directly controlled by a constant. In either case, we can handle $|<\psi, I>|$ by Schwarz. We can control III and IV rather trivially by the support condition on η_1 and the fact that the singularities in III and IV are weak. In fact

$$\|III\|_{L^\infty(\triangle)} + \|IV\|_{L^\infty(\triangle)} \le C\beta_2^j,$$

as the reader may readily verify.

The only term which presents any difficulty is II. Setting $\tilde{H}^{(j-2)} \equiv t H^{(j)}$, we see that II is a polar co-ordinate representation of

$$(18) \qquad \frac{2}{j}\{[\tilde{K}^{(j-2)}, A](\frac{\partial}{\partial t}B)\eta_1 + (j-1)[\tilde{K}^{(j-2)}, B](\frac{\partial}{\partial t}A)\eta_1\},$$

at least if $j \ge 2$. If $j = 1$, then $II \equiv \tilde{I}_{par}(\frac{\partial}{\partial t}B\eta_1)$, where $\tilde{I}_{par}$ denotes convolution with $\tilde{I}_{par}(x,t) \equiv tH^{(1)}(x,t)$ (thus, $\tilde{I}_{par}$ is on the order of a parabolic fractional integral). Let us set $a \equiv D_{par}A, b \equiv D_{par}B$. Then by (5),

$$(19) \qquad \frac{\partial}{\partial t}B \equiv D_n b, \quad \frac{\partial}{\partial t}A \equiv D_n a.$$

By [FR1, pp. 111-113], the kernel of D_n is odd, has vanishing first moments on the sphere, belongs to $C^\infty(\mathbb{R}^n\backslash\{0\})$, and satisfies the homogeneity property $D_n(\lambda^\alpha w) \equiv \lambda^{-d-1}D_n(w)$. We note that by our a priori regularization, $a, b \in BMO \cap C^\infty$, so that (19) is well defined in the principal value sense. To see this, we begin by observing that since D_n is odd, we have

$$D_n 1(w) \equiv \lim_{\epsilon\to 0, R\to\infty} \int\limits_{\epsilon<\|w-v\|<R} D_n(w-v)dv = 0.$$

Thus,

$$(20) \qquad D_n a(w) \equiv p.v. \int D_n(w-v)[a(v) - m_{\triangle(z_0)}a]\varphi(v)dv$$

$$+ \int D_n(w-v)[a(v) - m_{\triangle(z_0)}a][1 - \varphi(v)]dv$$

$$\equiv D_n a_1(w) + D_n a_2(w),$$

where $\varphi \in C_0^\infty(\triangle_{1000}(z_0)), \varphi \equiv 1$ on $\triangle_{500}(z_0)$. Then $D_n a_1$ is well defined by [FR1], and for $w \in \triangle_{200}(z_0)$, we have

(21)
$$| D_n a_2(w) | \le C \int \frac{1}{1 + \|z_0 - v\|^{d+1}} | a(v) - m_{\triangle(z_0)}a | dv$$
$$\le C \| a \|_* .$$

The last inequality is the parabolic version of a well known result of Fefferman and Stein [FS], and the proof is the same as in the elliptic case. We set $J^{(j-2)}(z,w) \equiv [\tilde{K}^{(j-2)}, A](z,w)(\text{or } \tilde{I}_{par}(z-w), \text{ if } j = 1)$, so that the first term in (18) equals

(22)
$$\int J^{(j-2)}(z,w) \int D_n(w-v)a_1(v)[\eta_1(w) - \eta_1(v)]dvdw$$

$$+$$

$$\int J^{(j-2)}(z,w)(p.v.) \int D_n(w-v)a_1(v)\eta_1(v)dvdw$$

$$+$$

$$\int J^{(j-2)}(z,w)D_n a_2(w)\eta_1(w)dw$$

$$\equiv II_1 + II_2 + II_3.$$

(The second term in (18) can be handled in exactly the same fashion). Since η_1 is supported in $\triangle_{200}(z_0)$, by (6), (17, Section 1)(i) and (21), and the normalization $\| a \|_* \le C \| A \|_{comm} \equiv C$, we have

(23)
$$| II_3 | \le C\beta_2^{j-1} \| a \|_* \int_{\|z-w\| \le C} \| z - w \|^{1-d} dw \le C_2^j.$$

Thus,

$$|< \psi, II_3 >| \le CC_2^j$$

also.

Next, by definition,

$$II_2 \equiv [\tilde{K}^{(j-2)}, A]D_n(a_1\eta_1), \text{ if } j \ge 2; \text{ or } \tilde{I}_p D_n(a_1\eta_1), \text{ if } j = 1.$$

But by (5), $D_n \equiv D_{par} R_n$, where R_n is bounded on L^2. Furthermore

$$\| a_1\eta_1 \|_2 \le C \| a \|_* \le C.$$

Thus, by Schwarz, if $j \ge 2$,

$$|< \psi, II_2 >| \le \| [\tilde{K}^{(j-2)}, \cdot]D_{par\|_{bi-op}}.$$

The case $j = 1$ of this last estimate will be treated in the appendix; we remark that in this case we are considering $\tilde{I}_{par} D_{par}$, which is known to by bounded by symbolic calculus if $\tilde{I}_{par}(z) \in C^\infty(\mathbb{R}^n \backslash \{0\})$.

Finally, $II_1(z) = \int J^{(j-2)}(z,w)[D_n, \eta_1]a_1(w)dw$. Now $J^{(j-z)}(z,w) \le C\beta_2^{j-1}$
$\| z - w \|^{1-d}$. Thus, in particular, $J^{(j-2)} \in L^1 + L^2$. Also, by [FR1], the
commutator $[D_n, \eta_1]$ is bounded on L^2 since η_1 is smooth. Alternatively, it is
not hard to prove the latter fact directly, as we shall do in a moment. Thus,
$II_1 \in L^2 + L^\infty$, with the same bound as that satisfied by II_3 in (23). Estimates
(9) and (10) now follow. To establish the L^2 boundedness of $[D_n, \eta_1]$, we write,
for $f \in C_0^\infty$, $[D_n, \eta_1]f(z) \equiv$

$$\int_{\|w-v\|\le 1} D_n(w-v)[\eta_1(w) - \eta_1(v) - (w'-v') \cdot \nabla_x\eta_1(v)]f(v)dv$$

$$+ p.v. \int_{\|w-v\|\le 1} D_n(w-v)(w'-v') \cdot \nabla_x\eta_1(v)f(v)dv$$

$$+ \int_{\|w-v\|>1} D_n(w-v)[\eta_1(w) - \eta_1(v)]f(v)dv$$

$$\equiv \text{"1"} + \text{"2"} + \text{"3"}.$$

But "2" is a truncation of a "nice" parabolic singular integral acting on the L^2
function $(\nabla_x\eta_1)f$. Since $\| \eta_1 \|_\infty \le 1$, "3" is dominated in absolute value by

$$C \int \frac{1}{(1+ \| w - v \|)^{d+1}} \mid f(v) \mid dv$$

and the L^2 bound follows by Young's convolution inequality. Finally, in "1", we
use the higher order smoothness of η_1, and the fact that $\mid t - s \mid \le \| (x,t) - (y,s) \|^2$, to weaken the singularity of the integral so that

$$\mid \text{"1"} \mid \le C \int_{\|w-v\|\le 1} \| w - v \|^{1-d} \mid f(v) \mid dv.$$

The L^2 boundedness of $[D_n, \eta_1]$ follows.

3. Proof of the "Main" Estimate: The "Standard" Part

We start by describing an appropriate parabolic Littlewood-Paley decomposition. For the most part, this is quite standard, except for one particular feature
which we shall require. Following [CM], we choose $P(z) \in C_0^\infty$, even, non-
negative, supported in the ball $\mid z \mid \le 1$, and normalized so that $\int P(z)dz \equiv 1$.
We define $P_\delta(z) \equiv \delta^{-d}P(\delta^{-\alpha}z)$ and denote also by P_δ the operator given by
convolution with $P_\delta(z)$. We let T denote $D_{par}[K^{(k)}, B]$, and, as usual, we then
have

(1) $$T \equiv - \int_0^\infty \frac{\partial}{\partial\delta} \left(P_\delta^2 T P_\delta^2 \right) d\delta,$$

in an appropriate sense (we shall be more precise about this point momentarily). Let us consider, for the moment, the operator $\delta P_\delta \frac{\partial}{\partial \delta} P_\delta$. A simple computation shows that, on the Fourier Transform side, the symbol of this operator is

$$(2) \qquad \hat{P}(\delta\xi, \delta^2\tau)(\delta\xi, 2\delta^2\tau) \cdot ((x,t)P(x,t))^\wedge (\delta\xi, \delta^2\tau).$$

Since P is even, its first moments are zero. Thus, as usual, (2) is the Fourier Transform of

$$(3) \qquad \delta P_\delta \frac{\partial}{\partial \delta} P_\delta \equiv \vec{Q}_\delta^{(1)} \cdot \vec{Q}_\delta^{(2)},$$

where each of the operators $\vec{Q}_\delta^{(i)}, i = 1, 2$, denotes convolution with a vector valued kernel $\delta^{-d} \vec{Q}^{(i)}(\delta^{-\alpha}z) \in C_0^\infty$, which has mean value zero. It will also be useful for us, however, to write

$$(4) \qquad \delta P_\delta \frac{\partial}{\partial \delta} P_\delta \equiv P_\delta Q_\delta^{(0)},$$

where the kernel $Q_\delta^{(0)}(z) \in C_0^\infty$ not only has mean value zero, but also zero first moments. The fact that $Q_\delta^{(0)} \equiv \delta \frac{\partial}{\partial \delta} P_\delta$ has mean value zero is obvious. Furthermore, since P_δ is even, it has vanishing first moments, so that also

$$(5) \qquad \int z_j\, Q_\delta^{(0)}(z)\, dz \equiv \delta \frac{\partial}{\partial \delta} \int z_j\, P_\delta(z)dz \equiv 0.$$

The first moment condition satisfied by $Q^{(0)}$ will be important, as one would expect: we are really showing that $[K^{(k)}, A]$ maps $L^p \to I_{par}(L^p)$, and one requires vanishing first moments to give a Littlewood-Paley characterization of Sobolev spaces having a full order of smoothness (as one would expect, it would really suffice to have vanishing first moments in the space variable only, but the construction gives them in the time variable as well; in the sequel we shall only use the vanishing spatial moments). Thus, ignoring multiplication by absolute constants, we see that it is enough to consider

$$(6) \qquad \int_0^\infty < P_\delta\, Q_\delta^{(0)} T\, P_\delta^2 f, g > \frac{d\delta}{\delta}$$

and

$$(7) \qquad \int_0^\infty < P_\delta^2\, T\, \vec{Q}_\delta^{(1)} \cdot \vec{Q}_\delta^{(2)} f, g > \frac{d\delta}{\delta},$$

where $f, g \in C_0^\infty$. To be more precise, one actually considers $\int_\epsilon^R \dots \frac{d\delta}{\delta}$, and obtains bounds independent of the truncation. One can then pass to the limit by standard arguments. We shall suppress the truncation in the sequel, in order not to tire the reader with routine details.

In this section, we treat (7). This will be the easier of the two, and will be handled by "standard" estimates. In the sequel, we shall use the notational convention that Q_δ denotes convolution with a generic, compactly supported, parabolic approximation to the zero operator, not necessarily the same at each occurrence, but chosen from a finite stock of such operators depending only on our choice of P_δ. Special Q_δ, such as $Q_\delta^{(0)}$ of (5), or those whose kernels may not have compact support, will be designated individually. With this convention, and by duality, (7) is a sum of terms of the form

$$(8) \qquad \int_0^\infty < Q_\delta f, Q_\delta [K^{(k)}, B] D_{par} P_\delta^2 g > \frac{d\delta}{\delta}.$$

(Actually, we would have the transpose of $[K^{(k)}, B]$, but since the two are of essentially the same form we write $[K^{(k)}, B]$ for the sake of notational convenience). By (4, Section 2), and the boundedness of the parabolic Riesz Transforms R_j, it is enough to replace $D_{par} P_\delta$ by

$$(9) \qquad \frac{1}{\delta} Q_\delta \equiv \frac{\partial}{\partial x_j} P_\delta, \ 1 \le j \le n - 1$$

$$\frac{1}{\delta} \tilde{Q}_\delta \equiv D_n P_\delta.$$

We define

$$(10) \qquad S_\delta^{(k)} \equiv \frac{1}{\delta} Q_\delta [K^{(k)}, B] Q_\delta P_\delta$$

$$\tilde{S}_\delta^{(k)} \equiv \frac{1}{\delta} Q_\delta [K^{(k)}, B] \tilde{Q}_\delta P_\delta.$$

We note that $S_\delta^{(k)} 1 \equiv 0 \equiv \tilde{S}_\delta^{(k)} 1$. The square function $f \to \left(\int_0^\infty | Q_\delta f |^2 \frac{d\delta}{\delta} \right)^{1/2}$ is bounded on $L^p (1 < p < \infty)$ (as is well known), and also on L_ω^2, $\omega \in A_2$ (a proof of the latter fact in the non-isotropic case is given, for example, in [H3]). Thus, by the usual "almost orthogonality" arguments (e.g., by expanding $g \equiv \int_0^\infty \vec{Q}_\theta^{(1)} \cdot \vec{Q}_\theta^{(2)} g \frac{d\theta}{\theta}$ and obtaining $| S_\delta^{(k)} Q_\theta h | \le C \min \left(\frac{\theta}{\delta}, \frac{\delta}{\theta} \right)^\epsilon Mh$, etc.), it suffices to prove the following

LEMMA 1. *We assume that* $\| A \|_{comm} \equiv 1 \equiv \| B \|_{comm}$. *The kernels of* $S_\delta^{(k)}$ *and* $\tilde{S}_\delta^{(k)}$ *satisfy*

$$(i) \qquad | S_\delta^{(k)}(z, w) | \le C C_2^k \frac{\delta}{(\delta + \| z - w \|)^{d+1}}$$

$$| \tilde{S}_\delta^{(k)}(z, w) | \le C C_2^k \frac{\delta^{1/2}}{(\delta + \| z - w \|)^{d+1/2}}$$

$$(ii) \qquad | S_\delta^{(k)}(z, w) - S_\delta^{(k)}(z, v) | \le C C_2^k \frac{\| v - w \|}{(\delta + \| z - w \|)^{d+1}}$$

$$| \tilde{S}_\delta^{(k)}(z,w) - \tilde{S}_\delta^{(k)}(z,v) | \le CC_2^k \frac{\| v - w \|^{1/2}}{(\delta + \| z - w \|)^{d+1/2}},$$

where the estimate (ii) holds whenever $\| v - w \| \le 1/2 \ \min(\delta, \| z - w \|)$.

Proof of the Lemma

By replacing P_δ by its n-dimensional gradient, we see that it is enough to prove (i). We set

$$(11) \qquad\qquad T_\delta^{(k)} \equiv \frac{1}{\delta} Q_\delta [K^{(k)}, B],$$

and to prove (i) it is clearly enough to prove that $T_\delta^{(k)} Q_\delta$ and $T_\delta^{(k)} \tilde{Q}_\delta$ satisfy (i) in place of $S_\delta^{(k)}$ and $\tilde{S}_\delta^{(k)}$ respectively. This last estimate, in turn, will be an easy consequence of the following

LEMMA 2. *The kernel of* $T_\delta^{(k)}$ *satisfies*

$$(i) \qquad | T_\delta^{(k)}(z,u) | \le C_2^k \ (\delta + \| z - u \|)^{-d}$$

$$(ii) \qquad | T_\delta^{(k)}(z,u) - T_\delta^{(k)}(z,w) | \le C_2^k \frac{\| u - w \| + \delta}{(\delta + \| z - w \|^{d+1})}$$

where (ii) holds whenever $(\delta + \| u - w \|) \le \frac{1}{500} \| z - u \|$.

Let us assume, for the moment, the truth of Lemma 2, and deduce Lemma 1. There are two cases.

Case 1: $\| z - w \| \le 10,000 \ \delta$

In this case, we apply crudely the size estimate (i) of Lemma 2 to obtain

$$| T_\delta^{(k)} Q_\delta(z,w) | + | T_\delta^{(k)} \tilde{Q}_\delta(z,w) | \le C_2^k \ \delta^{-d} (\| Q \|_{L^1} + \| \tilde{Q} \|_{L^1}),$$

which is the desired estimate in the present case.

Case 2: $\| z - w \| > 10,000 \ \delta$

Since $\int Q = 0 = \int \tilde{Q}$, we have

$$(12) \qquad T_\delta^{(k)} Q_\delta(z,w) \equiv \int [T_\delta^{(k)}(z,u) - T_\delta^{(k)}(z,w)] Q_\delta(u-w) du$$

$$(13) \qquad T_\delta^{(k)} \tilde{Q}_\delta(z,w) \equiv \int [T_\delta^{(k)}(z,u) - T_\delta^{(k)}(z,w)] \tilde{Q}_\delta(u-w) du.$$

Since Q_δ has compact support, the desired bound for (12) is an immediate consequence of Lemma 2 (ii). We have to work a little harder to handle (13). We split the integral into two parts: (13) $\equiv$

$$\int_{\|u-w\| \le \|z-u\|/600} + \int_{\|u-w\| > \|z-u\|/600} \equiv I + II.$$

In I, $\| z-u \| \approx \| z-w \| >> \delta$, so we can apply Lemma 2 (ii) to obtain $| I | \leq CC_2^k$ times

$$\frac{\delta}{\delta + \| z-w \|^{d+1}} \| \tilde{Q} \|_{L^1} + \frac{\delta^{1/2}}{(\delta + \| z-w \|)^{d+1/2}} \int \frac{\| u-w \|^{1/2}}{\delta^{1/2}} \tilde{Q}_\delta(u-w) du,$$

where we have used that $\| u-w \|^{1/2} / (\delta + \| z-w \|)^{1/2} \leq 1$. The desired bound for I then follows from dilation invariance and the fact that

$$(14) \qquad | \tilde{Q}(u-w) | \equiv | \text{p.v.} \int D_n(u-v) P(v-w) dv |$$

$$\leq \int_{\| u-v \| > 10} \frac{C}{\| u-v \|^{d+1}} P(v-w) dv$$

$$+ \int_{\| u-v \| \leq 10} | D_n(u-v)[P(v-w) - P(u-w) - (v'-u') \cdot \nabla_x P(u-w)] | \, dw$$

$$\leq C \left[\frac{1}{\| u-w \|^{d+1}} \chi(\| u-w \| \geq 9) + \chi(\| u-w \| \leq 11) \right]$$

$$\leq C/(1 + \| u-w \|)^{d+1},$$

where in the last series of estimates we have used properties of D_n (see our remarks following (19, Section 2) above), the fact that $P(v)$ is supported in the unit ball, and the Taylor's Theorem estimate (with $\| u-v \| \leq 10$)

$$(15) \qquad | P(v) - P(u) - (v'-u') \cdot \nabla_x P(u) | \leq | v_n - u_n | \| \frac{\partial}{\partial t} P \|_\infty$$

$$+ | v-u |^2 \sum_{1 \leq i,j \leq n} \| \frac{\partial^2}{\partial u_i \partial u_j} P(u) \|_\infty \leq C \| v-u \|^2 .$$

Next, we consider II, whose absolute value is dominated by

$$\int_{\| u-w \| > \| z-u \|/600} | T_\delta^{(k)}(z,u) | \, | \tilde{Q}_\delta(u,w) | \, du \qquad +$$

$$\int_{\| u-w \| > \| z-u \|/600} | \tilde{Q}_\delta(u,w) | \, du \, | T_\delta^{(k)}(z,w) |$$

$$\equiv II_1 + II_2.$$

Since $\| z-w \| \leq \| z-u \| + \| u-w \| \leq 601 \| u-w \|$, by Lemma 2 (i), and (14) plus dilation invariance, II_2 is no larger than CC_2^k times

$$\| z-w \|^{-d} \int_{\| u-w \| > C \| z-w \|} \frac{\delta}{\| u-w \|^{d+1}} du = \frac{C\delta}{\| z-w \|^{d+1}},$$

which is the desired bound for case 2. Finally, we treat II_1, which since

$$\| u-w \| \geq C \| z-w \|,$$

is by Lemma 2 (i) no larger than CC_2^k times

$$\frac{\delta}{(\delta+\|z-w\|)^{d+1}}\left\{\delta^{-d}\int_{\|z-u\|\leq\delta}1\,du+\int_{\delta\leq\|z-u\|\leq1000\|z-w\|}\|z-u\|^{-d}\,du\right\}$$

$$+\|z-w\|^{-d}\int_{\|u-w\|>C\|z-w\|}\frac{\delta}{\|u-w\|^{d+1}}du.$$

The last summand can be treated exactly like II_2, and the expression in curly brackets is dominated by $C(1+\log\frac{\|z-w\|}{\delta})$. We have therefore reduced the proof of Lemma 1 to that of Lemma 2.

Proof of Lemma 2

Part (i) is quite easy. There are two cases.

Case 1: $\|z-u\|>1000\delta$.

By definition (11), and the fact that $Q_\delta 1=0$, we have

$$T_\delta^{(k)}(z,u)\equiv\frac{1}{\delta}\int Q_\delta(z-v)[J^{(k)}(v,u)-J^{(k)}(z,u)]dv,$$

where $J^{(k)}$ is the kernel of $[K^{(k)},B]$. By (17, Section 1) (ii) and (2.6), the last expression is bounded in absolute value by $CC_2^k\|z-u\|^{-d}$, for C_2 chosen large enough.

Case 2: $\|z-u\|\leq1000\delta$.

In this case, crude size estimates for $Q_\delta(z)$ and $J^{(k)}$ yield the bound

$$|T_\delta^{(k)}(z,u)|\leq CC_2^k\,\delta^{-d-1}\int_{\|v-u\|\leq C\delta}\|v-u\|^{1-d}\,dv$$

$$=CC_2^k\,\delta^{-d}.$$

Next, we proceed to the proof of Lemma 2 (ii). Again, by definition, and the fact that $Q_\delta 1=0$, we have

$$(16)\qquad T_\delta^{(k)}(z,u)\equiv\frac{1}{\delta}\int Q_\delta(z-v)\left\{J^{(k)}(v,u)-J^{(k)}(z,u)\right\}dv,$$

where $J^{(k)}$ is the kernel of $[K^{(k)},B]$. The expression in curly brackets equals

$$[H^{(k)}(v-u)-H^{(k)}(z-u)][B(v)-B(u)][A(v)-A(u)]^k$$

$$+$$

$$H^{(k)}(z-u)\{[B(v)-B(u)][A(v)-A(u)]^k-[B(z)-B(u)][A(z)-A(u)]^k\}$$

$$\equiv\Psi_1(v,z,u)+\Psi_2(v,z,u).$$

We have, since $\|v-z\|\leq\delta,|\Psi_1(v,z,u)-\Psi_1(v,z,w)|\leq$

$$(17)\qquad\begin{array}{c}|[H^{(k)}(v-u)-H^{(k)}(z-u)]-[H^{(k)}(v-w)-H^{(k)}(z-w)]|\\\cdot|B(v)-B(u)||A(v)-A(u)|^k\end{array}$$

$$+$$

$$| H^{(k)}(v-w) - H^{(k)}(z-w) | \cdot | [B(v) - B(u)][A(v) - A(u)]^k$$

$$-[B(v) - B(w)][A(v) - A(w)]^k \le C\beta_1^k \beta_2^{k+1} k \frac{\delta^2 + \delta \parallel u - w \parallel}{\parallel z - w \parallel^{d+1}}.$$

where the inequality follows from (17 and 18, Section 1), and (6). If we plug (17) into (16) in place of the expression in curly brackets, we obtain Lemma 2 (ii) for this part of $T_\delta^{(k)}$, for a sufficiently large C_2.

To handle $\Psi_2(v, z, u)$, we use the identity

$$(18) \qquad [B(v) - B(u)][A(v) - A(u)]^k - [B(z) - B(u)][A(z) - A(u)]^k \equiv$$

$$[B(v) - B(z)]\{[A(v) - A(u)]^k - [A(z) - A(u)]^k\}$$

$$+$$

$$[B(v) - B(z)][A(z) - A(u)]^k$$

$$+$$

$$[B(z) - B(u)]\{[A(v) - A(u)]^k - [A(z) - A(u)]^k\}.$$

$$\equiv I(v, z, u) + II(v, z, u) + III(v, z, u).$$

(If $k = 0$, we have only the term $II \equiv B(v) - B(z)$). By (6, Section 2), and the fact that $\parallel v - z \parallel \le \delta << \parallel z - u \parallel$,

$$| I(v, z, u) | \le Ck\delta^2 \beta_2^{k+1} \max(\parallel v - u \parallel, \parallel z - u \parallel)^{k-1}.$$

For the part of (16) corresponding to $H^{(k)}(z - u) I(v, z, u)$, the desired bound Lemma 2 (ii) follows by crude size estimates and the fact that $\parallel v - u \parallel \approx \parallel z - u \parallel \approx \parallel z - w \parallel$. By (6) and (17, Section 1)(ii),

$$| H^{(k)}(z - u) II(v, z, u) - H^{(k)}(z - w) II(v, z, w) |$$
$$\le C(k + 1)\beta_1^k \beta_2^{k+1} \delta \parallel u - w \parallel \parallel z - w \parallel^{-d-1},$$

and Lemma 2 (ii) follows for this term also. Finally, by Taylor's Theorem applied to the function $x \to x^k$, $III(v, z, u) \equiv$

$$(19) \qquad [B(z) - B(u)][A(v) - A(z)]k[A(z) - A(u)]^{k-1}$$

$$+$$

$$[B(z) - B(u)][A(v) - A(z)]^2 k(k - 1)E(v, z, u) \equiv III_1 + III_2,$$

(The term III_2 is vacuous unless $k \ge 2$), where

$$| E(v, z, u) | \le C^{k-2} \parallel z - u \parallel^{k-2}.$$

The term III_1 can be handled exactly like II, and the term III_2 can be handled exactly like I. This concludes the proof of Lemma 2, and thus also our treatment of the "standard" part of $D_{par}[K^{(k)}, B]$, namely (7).

4. Proof of the "Main" Estimate: The "Rough" Part

We begin with some elementary observations. Recall that (6, Section 3) equals

$$(1) \qquad \int_0^\infty < \frac{1}{\delta} \, Q_\delta^{(0)} \, [K^{(k)}, B] P_\delta^2 \, f, \delta \, D_{par} \, P_\delta \, g > \frac{d\delta}{\delta},$$

where we have used that D_{par} and $Q_\delta^{(0)}$ commute. As above, (see (9, Section 3)), we can use (4, Section 2) and the boundedness of the parabolic Riesz Transforms R_j to replace $\delta D_{par} P_\delta$ by $Q_\delta \equiv \delta \frac{\partial}{\partial x_j} P_\delta, 1 \leq j \leq n-1$, or $\tilde{Q}_\delta \equiv \delta D_n P_\delta$. By dilation invariance and estimate (14, Section 3) (which can be applied also with P replaced by its n-dimensional gradient to given smoothness estimates for $\tilde{Q}_\delta$) and the fact that $\tilde{Q}_\delta 1 \equiv 0$, we have L_ω^2 boundedness of the square function $g \to \left(\int_0^\infty | \tilde{Q}_\delta g |^2 \frac{d\delta}{\delta} \right)^{1/2}$. Of course, a similar estimate also holds with $\tilde{Q}_\delta$ replaced by Q_δ.

Our initial strategy, then, is to split $\frac{1}{\delta} Q_\delta^{(0)} [K^{(k)}, B] P_\delta^2$ into two parts, one satisfying standard estimates on the order of the bounds in Lemma 1 of Section 3, and the other rough, but treatable by well known Carleson measure techniques. We now proceed to give the details. To obtain standard estimates, on the order of those proved for $S_\delta^{(k)}$ in Section 3, Lemma 1 (i) and (ii), for

$$(standard \ part \ of \ \frac{1}{\delta} \, Q_\delta^{(0)} \, [K^{(k)}, B]) \, P_\delta^2 \equiv R_\delta^{(k)} P_\delta^2,$$

it is clearly enough to show that $R_\delta^{(k)}$ satisfies the same size estimate as $S_\delta^{(k)}$ in Lemma 1 (i); i.e. $R_\delta^{(k)}(z, u)$ will be defined as a sum of terms, each satisfying the bound

$$(2) \qquad | \, R_\delta^{(k)}(z,u) \, | \leq CC_2^k \, \frac{\delta}{(\delta + \| \, z - u \, \|^{d+1})}, \quad \| \, A \, \|_{comm} \leq 1.$$

We recall that $J^{(k)}$ denotes the kernel of $[K^{(k)}, B]$. As above, we choose a smooth cut-off function $\eta \in C_0^\infty[-200, 200], \eta \equiv 1$ on $[-100, 100]$, and write $H^{(k)}(v-u) \equiv$

$$(3) \qquad H^{(k)}(v-u) \eta \left(\frac{\| \, v-u \, \|}{\delta} \right) + H^{(k)}(v-u)[1 - \eta \left(\frac{\| \, v-u \, \|}{\delta} \right)]$$

$$\equiv h_\delta^{(k)}(v-u) + H_\delta^{(k)}(v-u),$$

and this induces a corresponding splitting of $J^{(k)}(v,u)$ into $j_\delta^{(k)}(v,u) + J_\delta^{(k)}(v,u)$. By crude size estimates, (since the singularity of $J^{(k)}$ is weak), we have

$$(4) \qquad | \, \frac{1}{\delta} \int Q_\delta^{(0)}(z-v) j_\delta^{(k)}(v,u) dv \, | \leq C_2^k \, \delta^{-d} \chi \{ \| \, z - u \, \| < C\delta \}$$

which is clearly controlled by the right side of (2). As usual, we use that $Q_\delta^{(0)} 1 = 0$ to write $\frac{1}{\delta} Q_\delta^{(0)} J_\delta^{(k)}(z, u) \equiv$

(5) $$\frac{1}{\delta} \int Q_\delta^{(0)}(z - v)\{J_\delta^{(k)}(v, u) - J_\delta^{(k)}(z, u)\}dv,$$

We split the expression in curly brackets into

(6) $$[B(v) - B(u)][A(v) - A(u)]^k [H_\delta^{(k)}(v - u) - H_\delta^{(k)}(z - u)]$$

$$+$$

(7)
$$H_\delta^{(k)}(z - u)\{[B(v) - B(u)][A(v) - A(u)]^k - [B(z) - B(u)][A(z) - A(u)]^k\}.$$

(A similar splitting of the analogous 1-dimensional kernel appears in [LM3]). If we plug (6) into (5) in place of the expression in curly brackets, we can use the fact that $Q_k^{(0)}$ has zero first moments (see (5, Section 3)) to obtain

$$\frac{1}{\delta} \int Q_\delta^{(0)}(z - v)\{[B(v) - B(u)][A(v) - A(u)]^k - [B(z) - B(u)][A(z) - A(u)]^k\}$$

$$\times \quad [H_\delta^{(k)}(v - u) - H_\delta^{(k)}(z - u)]dv$$

(8) $$+$$

$$[B(z) - B(u)][A(z) - A(u)]^k \frac{1}{\delta} \int Q_\delta^{(0)}(z - v)\{H_\delta^{(k)}(v - u) - H_\delta^{(k)}(z - u)$$

$$-(v' - z') \cdot \nabla_x H_\delta^{(k)}(z - u)\}dv.$$

These two terms are dominated in absolute value by the right side of (2), by virtue of (6, Section 2) and the fact that $H_\delta^{(k)}$ also satisfies (17, Section 1) with bounds independent of δ.

Next, we again use the identity (18, Section 3), so that (7) equals

(9) $H_\delta^{(k)}(z - u) I(v, z, u) + H_\delta^{(k)}(z - u) II(v, z, u) + H_\delta^{(k)}(z - u) III(v, z, u),$

where I, II, III are defined as in (18, Section 3). As before, there is a gain of $\| v - z \|^2 \leq \delta^2$ in I, so the first term in (9) can be handled like (8). The second term, when plugged into (5) in place of curly brackets, yields precisely

(10) $$\frac{1}{\delta} Q_\delta^{(0)} B(z) K_\delta^{(k)}(z, u) \equiv Q_\delta^* b(z) K_\delta^{(k)}(z, u),$$

where $b \equiv D_{par} B, Q_\delta^* \equiv \frac{1}{\delta} I_{par} Q_\delta^{(0)}$, and where $K_\delta^{(k)}(z, u)$ is the kernel of $K^{(k)}$ multiplied by the truncation factor $1 - \eta\left[\frac{\|z - u\|}{\delta}\right]$. Let us leave aside this term, for the moment, and return to the third term in (9). We use (19, Section 3) to write $H_\delta^{(k)}(z - u) III(v, z, u) \equiv$

(11) $$H_\delta^{(k)}(z - u) III_1(v, z, u) + H_\delta^{(k)}(z - u) III_2(v, z, u).$$

Since III_2 also has a gain of $\| v - z \|^2 \leq \delta^2$, the second term in (11) can also be handled like (8) and the first term in (9). On the other hand, the first term in (11), when plugged into (5), yields, just as in (10), the term

$$(12) \qquad k \frac{1}{\delta} \, Q_\delta^{(0)} A(z) \, K_\delta^{(k)}(A, B)(z, u) \equiv k \, Q_\delta^* \, a(z) \, K_\delta^{(k)}(A, B)(z, u),$$

where $a \equiv D_{par} \, A$, and $K_\delta^{(k)}(A, B)$ is the smooth truncation of the kernel of $K^{(k)}(A, B)$.

We now define $R_\delta^{(k)}$ by

$$(13) \qquad \frac{1}{\delta} \, Q_\delta^{(0)} \, [K^{(k)}, B](z, u) \equiv R_\delta^{(k)} \, (z, u) + (10) \, + (12).$$

We have just shown that $R_\delta^{(k)}$ satisfies (2).

We dispose of (10) and (12) first. We begin by claiming that $Q_\delta^* \equiv \frac{1}{\delta} I_{par} \, Q_\delta^{(0)}$ satisfies

$$(14) \qquad (i) \quad | \, Q_\delta^*(z) \, | \leq \frac{\delta}{(\delta + \| \, z \, \|)^{d+1}}$$

$$(ii) \, | \, Q_\delta^*(z) - Q_\delta^*(w) \, | \leq \frac{\| \, z - w \, \|}{(\delta + \| \, z \, \|)^{d+1}}, \; if \; \| \, z - w \, \| \leq \delta.$$

It is enough to prove (i), since we could replace $Q_\delta^{(0)}$ by its n-dimensional gradient to obtain (ii). By dilation invariance, we may assume $\delta \equiv 1$. For our usual smooth cut-off function η, we set $k_1(z) \equiv I_{par} \, (z) \eta(\| \, z \, \|)$, and $k_2 \equiv I_{par} - k_1$. Then, since $Q^{(0)}$ has both mean value zero and first moments equal to zero, we have

$$Q^*(z) \equiv \int k_1(w) Q^{(0)} \, (z - w) \, dw$$

$$+ \int [k_2(w) - k_2(z) - (w' - z') \cdot \nabla_x \, k_2(z)] Q^{(0)} \, (z - w) \, dw.$$

Since $I_{par} \, (\lambda^\alpha z) \equiv \lambda^{1-d} I_{par} \, (z)$, the claim follows by straight forward computation. We note at this point that it is this particular choice of Q_δ^* which will appear in the $q - C$ condition (14, Section 2).

If we replace $\frac{1}{\delta} Q_\delta^{(0)} [K^{(k)}, B]$ in (1) by (10) + (12), and apply Schwarz, we obtain a bound of

$$(15) \qquad \left(\int\limits_{R^n} \int\limits_0^\infty | \, \delta D_{par} \, P_\delta g \, |^2 \, \frac{d\delta}{\delta} \, \omega^{-1}(z) \, dz \right)^{1/2} \qquad \times$$

$$\{ k (\int\limits_0^\infty \int\limits_{R^n} | \, Q_\delta^* \, a(z) \, |^2 | \, K_\delta^{(k)}(A, B) \, P_\delta^2 \, f(z) \, |^2 \, \frac{d\delta}{\delta} \, \omega(z) \, dz)^{1/2}$$

$$+$$

$$\left(\int\limits_0^\infty \int\limits_{\mathbb{R}^n} \mid Q_\delta^* \, b(z) \mid^2 \mid K_\delta^{(k)} P_\delta^2 \, f(z) \mid^2 \, \frac{d\delta}{\delta} \, \omega(z) \, dz \right)^{1/2} \}.$$

We have already observed that the first factor is no larger than $C \parallel g \parallel_{2,1/\omega}$. Also, the non-tangential maximal function

$$\sup_{\|z-z_0\|<\delta} \mid K_\delta^{(k)}(A, B) P_\delta^2 f(z) \mid \leq$$

$$C(C^k \, Mf(z_0) + \sup_\delta \mid K_\delta^{(k)}(A, B) \, f(z_0) \mid \leq C(C^k \, Mf(z_0) + M(K^{(k)}(A, B) \, f)(z_0),$$

where the first inequality follows by writing $P_\delta^2 = P_\delta^2 - I + I$, subtracting off and adding back in $K_\delta^{(k)}(A, B)f(z_0)$, and using the "standard" kernel conditions; the second inequality uses the parabolic version of Cotlar's inequality whose proof (see, e.g. Journé [J., p. 56]) is the same as in the elliptic case. An analogous bound holds for $K^{(k)} \equiv K^{(k)}(A, A)$. Thus, by the usual Carleson measure theory, the second factor in (15) is no larger than $\parallel f \parallel_{2,\omega}$ times

(16) $$C(A_2)(C_2^k + (k+1) \parallel K^{(k)}(A, \cdot) \parallel_{bi-op,\omega}).$$

Next, we turn to $R_\delta^{(k)}$. Following the strategy of [CM] (see also [CJ]), we write

$$R_\delta^{(k)} P_\delta^2 \equiv (R_\delta^{(k)} - R_\delta^{(k)}1)P_\delta^2 + R_\delta^{(k)}1P_\delta^2$$

$$\Sigma_\delta^{(k)} \equiv + R_\delta^{(k)}1P_\delta^2.$$

Now, $\Sigma_\delta^{(k)} 1 \equiv 0$, and by (2), $\Sigma_\delta^{(k)}(z, w)$ satisfies the same standard bounds as $S_\delta^{(k)}$ in Section 3 Lemma 1 (i) and (ii). Thus, the part of (1) corresponding to $\Sigma_\delta^{(k)}$ can be treated by well known almost orthogonality arguments, just like those used to handle $S_\delta^{(k)}$, to give a norm estimate of $C(A_2)C_2^k$.

Next, by the usual arguments involving the non-tangential maximal function, the part of (1) containing $R_\delta^{(k)}1P_\delta^2$ is controlled by $C(A_2) \parallel f \parallel_{2,\omega}$ $\parallel g \parallel_{2,1/\omega}$ times

(17) $$\sup_{r>0, \, z_0 \in \mathbb{R}^n} \left\{ \frac{1}{\omega(\triangle_r(z_0))} \int_0^r \int_{\triangle_r(z_0)} \mid R_\delta^{(k)}1(z) \mid^2 \, \omega(z) \, dz \, \frac{d\delta}{\delta} \right\}^{1/2} .$$

For $\triangle_r(z_0)$ fixed, we write $1 \equiv \eta_1 + \eta_2$ as in (12, Section 2). By a routine computation involving (2), and the fact that η_2 is supported far from $\triangle_r(z_0)$, the square root of the part of the expression in curly brackets in (17), with η_2 in place of 1 is no larger than $C(A_2)C_2^k$. To treat the part of (17) with η_1 in place of 1, we use the identity (13) to obtain the bound

(18)

$$
\sup_{z_0 \in \mathbb{R}^n, r>0} \left\{ \frac{1}{\omega(\triangle_r(z_0))} \int_0^r \int_{\triangle_r(z_0)} \left| \frac{1}{\delta} Q_\delta^{(0)} [K^{(k)}, B] \eta_1(z) \right|^2 \omega(z) \, dz \, \frac{d\delta}{\delta} \right\}^{1/2}
$$

$$
+
$$

$$
\sup_{z_0 \in \mathbb{R}^n, r>0} \left\{ \frac{1}{\omega(\triangle_r(z_0))} \int_0^r \int_{\triangle_r(z_0)} \left| [(10) + (12)] \eta_1 \right|^2 \omega(z) \, dz \, \frac{d\delta}{\delta} \right\}^{1/2} .
$$

The second term in (18) can be handled exactly as before (see the second factor in (15)), if we extend the domain of integration to be $(0, \infty) \times \mathbb{R}^n$. Furthermore, since $\| \eta_1 \|_{2,\omega} \le C(A_2) (\omega(\triangle_r(z_0)))^{1/2}$ (by the doubling property of A_p weights), we obtain the same bound (16).

Finally, since $\frac{1}{\delta} Q_\delta^{(0)} \equiv Q_\delta^* D_{par}$, the first term in (19) is precisely

$$
\| D_{par} [K^{(k)}, B] \|_{q-C, \omega}
$$

(see (14, Section 2)) and this will be treated in the next section. Adding up the various bounds, we complete the proof of estimate (15, Section 2).

5. Proof of the "q-C" Estimate.

In the spirit of the approach of P. Jones [Jns P] to the Cauchy integral (see also Christ [Ch]) we shall ultimately reduce the proof of (16, Section 2) to a parabolic version of certain expressions arising in the work of Dorronsoro [Do] on potential spaces. We fix a ball $\triangle_r(z_0)$. The only characteristics of η_1 which will affect our estimates are its L^∞ norm (which is one), and that it is supported on a fixed multiple of $\triangle_r(z_0)$. By dilation invariance, then, we may take $r \equiv 1$. We need to consider

$$
\frac{1}{\omega(\triangle(z_0))} \int_0^1 \int_{\triangle(z_0)} \left| \delta^{-1} Q_\delta^{(0)} [K^{(k)}, B] \eta_1(z) \right|^2 \omega(z) \, dz \, \frac{d\delta}{\delta} .
$$

By the following localization lemma, we may replace B by $\tilde{B}$, with

$$
\| \tilde{B} \|_{comm} \le C, \| D_{par} \tilde{B} \|_q \le C_q, 1 < q < \infty.
$$

LEMMA 1. *Suppose* $\| B \|_{comm} < \infty$. *Then, given a ball* $\triangle_r (z_0)$, *there exists a function* $\tilde{B}$, *supported on* $\triangle_{600r} (z_0)$, *such that* $B(v) - B(u) \equiv \tilde{B}(v) - \tilde{B}(u)$ *on* $\triangle_{300r} (z_0) \times \triangle_{300r} (z_0)$, *and such that*

(i) $\| \tilde{B} \|_{comm} \le C \| B \|_{comm}$

(ii) $(r^{-d} \int_{\mathbb{R}^n} | D_{par} \tilde{B} |^q)^{1/q} \le C_q \| B \|_{comm} .$

The proof of his Lemma may be found in [H5, Section 6].

We recall that the kernel

$$\delta^{-1} Q_\delta^{(0)}[K^{(k)}, \tilde{B}](z, u) \equiv \int \delta^{-1} Q_\delta(z - v) \, K^{(k)}(v, u)[\tilde{B}(v) - \tilde{B}(u)]dv.$$

We now replace $\tilde{B}(v) - \tilde{B}(u)$ by

$$(1) \qquad\qquad \tilde{B}(v) - \tilde{B}(u) - (v' - u') \cdot P_\delta \nabla_x \, \tilde{B}(z).$$

The error committed by doing so (at least for $k \geq 1$; we discuss the case $k = 0$ momentarily) is controlled by

$$(2) \quad \frac{1}{\omega(\triangle(z_0))} \int_{\triangle(z_0)} \int_0^1 \mid \delta^{-1} Q_\delta^{(0)}[\vec{K}^{(k-1)}, A]\eta_1(z) \mid^2 \mid P_\delta \nabla_x \tilde{B}(z) \mid^2 \omega(z) \frac{d\delta}{\delta} \, dz,$$

where $\vec{H}^{(k-1)}(x, t) \equiv x H^{(k)}$. Since $\parallel \nabla_x \tilde{B} \parallel_\infty \leq C$, and since $\delta^{-1} Q_\delta^{(0)} \equiv Q_\delta^* D_{par}$, Littlewood-Paley theory implies that the square root of (2) is dominated by

$$C(A_2) \parallel D_{par}[\vec{K}^{(k-1)}, \cdot] \parallel_{bi-op, \omega},$$

which is allowable in (16, Section 2). We have used here that $\int \mid \eta_1 \mid^2 \omega \leq C\omega(\triangle(z_0))$, by the doubling property of A_p weights. If $k = 0$, instead of $D_{par}[\vec{K}^{(k-1)}, A]$ we have simply $D_{par}\vec{I}_{par}$, where $\vec{I}_{par}$ denotes convolution with the kernel $xK^{(0)}(x, t)$. The boundedness of $D_{par}\vec{I}_{par}$ (which is routine) will be treated in the appendix.

Next, we write (1) as

$$(3) \qquad\qquad \tilde{B}(v) - \tilde{B}(z) - (v' - z') \cdot P_\delta \nabla_x \tilde{B}(z)$$

$$+$$

$$(4) \qquad\qquad \tilde{B}(z) - \tilde{B}(u) - (z' - u') \cdot P_\delta \nabla_x \tilde{B}(z).$$

The contribution of (3) is no larger than

$$(5) \qquad\qquad \frac{1}{\omega(\triangle(z_0))} \int_{\triangle(z_0)} \int_0^1 \mid T_\delta \, K^{(k)} \, \eta_1(z) \mid^2 \omega(z)\frac{d\delta}{\delta} \, dz$$

$$+$$

$$\frac{1}{\omega(\triangle(z_0))} \int_{\triangle(z_0)} \int_0^1 \mid \vec{Q}_\delta \, K^{(k)} \, \eta_1(z) \mid^2 \mid P_\delta \nabla_x \tilde{B}(z) \mid^2 \omega(z) \frac{d\delta}{\delta} \, dz,$$

where

$$(6) \qquad\qquad T_\delta(z, v) \equiv \delta^{-1} Q_\delta^{(0)}(z - v)[\tilde{B}(z) - \bar{B}(v)]$$

and

$$\vec{Q}_\delta(z - v) \equiv \delta^{-1} Q_\delta^{(0)}(z - v)[z' - v'].$$

The moment condition of $Q_\delta^{(0)}$, Littlewood-Paley Theory and the uniform bound-edness of $P_\delta \nabla_x \tilde{B}$ immediately imply that the second expression in (5) is no larger than $C(A_2) \parallel K^{(k)} \parallel_{op,\omega}$. Also,

$$| T_\delta(z,v) | \leq C\delta^{-d} \chi\{\parallel z - v \parallel \leq \delta\}$$

$$| T_\delta(z,v) - T_\delta(z,w) | \leq C\frac{\parallel v - w \parallel}{\delta} \delta^{-d} \chi\{\parallel z - v \parallel \leq 2\delta\},$$

with the second bound holding whenever $\parallel v - w \parallel < \delta$. Thus, by the same orthogonality arguments used to treat S_δ in section 3, the square function

$$f \to \left(\int_0^\infty | (T_\delta - T_\delta 1 P_\delta)f |^2 \frac{d\delta}{\delta} \right)^{1/2}$$

is bounded on L^2_ω. The square root of the first term in (5) therefore yields a bound of $C(A_2) \parallel K^{(k)} \parallel_{op,\omega}$, if we replace T_δ by $(T_\delta - T_\delta 1 P_\delta)$. The error committed is dominated by,

$$\frac{1}{\omega(\triangle(z_0))} \int_0^1 \int_{\triangle(z_0)} | T_\delta 1(z) |^2 | P_\delta K^{(k)} \eta_1(z) |^2 \omega(z) \, dz \, \frac{d\delta}{\delta}.$$

By extending the domain of integration to all of $[0,\infty] \times \mathbf{R}^n$, we see that the usual non-tangential maximal function arguments reduce matters to showing that $| T_\delta 1(z) |^2 \omega(z)dz\frac{d\delta}{\delta}$ is a weighted Carleson measure, but this is immediate by definition (6) since $T_\delta 1(z) \equiv \delta^{-1} Q_\delta^{(0)} \tilde{B}(z) \equiv Q_\delta^* \tilde{b}(z)$, where $\tilde{b} \equiv D_{par}\tilde{B}$. For the square root of this term also, then, we obtain a bound of $C(A_2) \parallel K^{(k)} \parallel_{op,\omega}$.

We have therefore reduced matters to treating the contribution of (4). It is enough, then, to consider

$$(7) \qquad \frac{1}{\omega(\triangle(z_0))} \int_0^1 \int_{\triangle(z_0)} | \delta^{-1} Q_\delta^{(0)} K^{(k)}(F(z,\cdot) \eta_1(\cdot))(z) |^2 \omega(z) \, dz \, \frac{d\delta}{\delta},$$

where $F(z,u)$ is precisely (4). We perform the usual splitting of $K^{(k)}$ into

$$K^{(k)}(v,u) \eta \left(\frac{\parallel v - u \parallel}{\delta} \right) + K^{(k)}(v,u)[1 - \eta \left(\frac{\parallel v - u \parallel}{\delta} \right)]$$

$$\equiv k_\delta^{(k)}(v,u) + K_\delta^{(k)}(v,u).$$

For the contribution of $k_\delta^{(k)}(v,u)$, we may multiply $F(z,u)$ by $\eta\left(\frac{\parallel z-u\parallel}{5\delta}\right)$. By Schwarz,

$$| \delta^{-1} \int Q_\delta^{(0)}(z-v) \int k_\delta^{(k)}(v,u) F(z,u) \eta \left(\frac{\parallel z - u \parallel}{5\delta} \right) \eta_1(u)du \, dv |^2$$

$$\leq C\delta^{-d-2} \int | \int k_\delta^{(k)}(v,u) F(z,u) \eta \left(\frac{\parallel z - u \parallel}{5\delta} \right) \eta_1(u)du |^2 \, dv,$$

and since Cotlar's inequality implies in particular that all truncations of an L^2 bounded Calderón-Zygmund operator are also bounded, the last expression is no larger than

(8)

$$C(C^k+\parallel K^{(k)}\parallel_{op})^2\,\delta^{-d-2}\int_{\parallel z-v\parallel\leq 1000\delta}\mid\tilde{B}(z)-\tilde{B}(v)-(z'-v')\cdot P_\delta\nabla_x\tilde{B}(z)\mid^2 dv.$$

We remark that variants of this sort of expression have appeared previously in connection with Carleson measures and the Cauchy Integral on Lipschitz graphs, in work of P. Jones [Jns P] and, before that, in connection with characterizations of potential spaces in work of Dorronsoro [Do]. See also Fang [Fng], and the expository monograph of M. Christ [Ch, pp. 32-33]. Let us leave this term aside for the moment, and discuss the contribution of $K_\delta^{(k)}$. We note that

$$\int\delta^{-1}Q_\delta^{(0)}(z-v)K_\delta^{(k)}(v,u)dv$$

can be analyzed exactly like $\delta^{-1}Q_\delta^{(0)}[K_\delta^{(k)},B](z,u)$, which was considered in Section 4. The only difference is that, in the present case, the order of decay is larger by 1. Thus, we have a decomposition similar to that of (13, Section 4), but now the term analogous to $R_\delta^{(k)}$, call it $Z_\delta^{(k)}$, satisfies

(9)
$$\mid Z_\delta^{(k)}(z,u)\mid\leq C^k\frac{\delta}{(\delta+\parallel z-u\parallel)^{d+2}}$$

(compare with (2, Section 4)). Also, the term analogous to (10 and 12, Section 4) is

(10)
$$(k+1)Q_\delta^*\,a(z)\,H_\delta^{(k)}(z-u)[A(z)-A(u)]^{k-1}$$

(if $k=0$, this last term is vacuous). If (10) is integrated against

$$[\tilde{B}(z)-\tilde{B}(u)-(z'-u')\cdot P_\delta\nabla_x\tilde{B}(z)]\,\eta_1(u)\,du,$$

the result is precisely

$$(k+1)Q_\delta^*\,a(z)\{K_\delta^{(k)}(A,\tilde{B})\,\eta_1(z)-[\vec{K}_\delta^{(k-1)}\eta_1(z)]\cdot P_\delta\nabla_x\tilde{B}(z)\}.$$

By extending the domain of integration to all of $(0,\infty)\times\mathbf{R}^n$, we see that the contribution of this last expression to the square root of (7) is dominated by

$$(k+1)\{C^k+\parallel K^{(k)}(A,\cdot)\parallel_{bi-op,\omega}+\parallel\vec{K}^{(k-1)}\parallel_{op,\omega}\}$$

(this follows by the usual Carleson measure - non-tangential maximal function arguments). As to (9), we have that

$$\int\mid Z_\delta^{(k)}(z,u)\mid[\tilde{B}(z)-\tilde{B}(u)-(z'-u')\cdot P_\delta\nabla_x\,\tilde{B}(z)\mid du\,\leq$$

$$C^k \{\delta^{-d-1} \int_{\|z-u\|\leq\delta} | \tilde{B}(z) - \tilde{B}(u) - (z'-u') \cdot P_\delta \nabla_x \tilde{B}(z) | \, du$$

$$+ \sum_{\ell=1}^{\infty} 2^{-\ell} (2^\ell \delta)^{-d-1} \int_{2^{\ell-1}\delta\leq\|z-u\|\leq 2^\ell\delta} | \tilde{B}(z) - \tilde{B}(u) - (z'-u') \cdot P_\delta \nabla_x \tilde{B}(z) | \, du\}.$$

By Schwarz, the square of this last expression is no larger than C^k (with a different C) times

$$(11) \qquad \delta^{-d-2} \int_{\|z-u\|\leq\delta} | \tilde{B}(z) - \tilde{B}(u) - (z'-u') \cdot P_\delta \nabla_x \tilde{B}(z) |^2 \, du$$

$$+$$

$$(12)$$

$$\sum_{\ell=1}^{\infty} 2^{-\ell} (2^\ell\delta)^{-d-2} \int_{2^{\ell-1}\delta\leq\|z-u\|\leq 2^\ell\delta} | \tilde{B}(z) - \tilde{B}(u) - (z'-u') \cdot P_\delta \nabla_x \tilde{B}(z) |^2 \, du.$$

We then plug (9), (11) and (12) into (7), in place of

$$| \delta^{-1} Q_\delta^{(0)} K^{(k)} (F(z,\cdot) \, \eta_1(\cdot)) (z) |^2,$$

and extend the domain of integration in δ to be all of $(0,\infty)$. By appropriate changes of variable in the $d\delta/\delta$ integral, this reduces matters to considering

$$(13)$$

$$\sum_{\ell=0}^{\infty} \frac{2^{-\ell}}{\omega(\Delta(z_0))} \int_{\Delta(z_0)} \int_0^\infty \delta^{-d-2} \int_{\|z-u\|\leq\delta} | \tilde{B}(z)-\tilde{B}(u)-(z'-u')\cdot P_{2^{-\ell}\delta}\nabla_x \tilde{B}(z) |^2 \, du \, \frac{d\delta}{\delta} \, \omega(z) dz$$

(multiplied, of course, by $C(C^k+ \| K^{(k)} \|_{op})^2)$. Thus, it is enough to show that (13) is dominated by $C(A_2)$. In the unweighted case $\omega \equiv 1$, this will be an easy consequence of Plancherel's Theorem. We omit the proof in the (more complicated) weighted case - the details of the proof of the desired bound for (13) in that case may be found in [H5, Section 6]. We remark that were the $d\delta/\delta$ integral restricted to the interval $(0,1)$, we would have (with $\omega \equiv 1$) precisely the parabolic version of the functional described in [Ch, pp. 32-33].

To treat the case $\omega \equiv 1$, we make the change of variable $u \equiv z + h$, then extend the dz integral to all of $\mathbb{R}^n$, and apply Plancherel. Since $\tilde{B} \equiv I_{par} \tilde{b}$, the unweighted version of (13) is then equivalent to

$$(14)$$

$$\sum_{\ell=0}^{\infty} 2^{-\ell} \int_{\mathbb{R}^n} \int_0^\infty \delta^{-d-2} \int_{\|h\|\leq\delta} \frac{| e^{i\varsigma\cdot h} - 1 - ih' \cdot \varsigma' \hat{P}((2^{-\ell}\delta)^\alpha\varsigma) |^2}{\| \varsigma \|^2} | \tilde{b}(\varsigma) |^2 \, dh \, \frac{d\delta}{\delta} d\varsigma.$$

We make the (self-adjoint) transformation $h \to \delta^\alpha h$, so that (14) equals

(15)
$$\sum_{\ell=0}^{\infty} 2^{-\ell} \int\limits_{\|h\| \leq 1} \int\limits_{R^n} |\tilde{b}(\varsigma)|^2 \int\limits_0^\infty \frac{\left| e^{i\delta^\alpha \varsigma \cdot h} - 1 - (ih' \cdot \delta\varsigma')\hat{P}((2^{-\ell}\delta)^\alpha \varsigma) \right|^2}{(\delta \|\varsigma\|)^2} \frac{d\delta}{\delta} \, d\varsigma \, dh.$$

Now, $P \in C_0^\infty$ so in particular, $\hat{P} \in C^\infty$ and $\hat{P}(\varsigma) \leq C_\epsilon (1+ \|\varsigma\|)^{-\epsilon}$. Also $\hat{P}(0) = 1$. Thus, by Taylor's Theorem, and the fact that

$$|x|^2 + |t| \approx \|(x,t)\|^2,$$

we have

$$(16) \frac{\left| e^{i\delta^\alpha \varsigma \cdot h} - 1 - (ih' \cdot \delta\varsigma')\hat{P}((2^{-\ell}\delta)^\alpha \varsigma) \right|}{\delta \|\varsigma\|} \leq C \min\left(\delta \|\varsigma\|, \frac{1}{(2^{-\ell}\delta \|\varsigma\|)^\epsilon} \right).$$

By an elementary computation, (15) is therefore dominated by $C \|\tilde{b}\|_2^2 \leq C$. This concludes the proof.

6. Appendix

Boundedness of $D_{par}\tilde{I}_{par}, D_{par}\vec{I}_{par}$

We recall that $\tilde{I}_{par}(x,t) \equiv tH^{(1)}(x,t)$, $\vec{I}_{par}(x,t) \equiv xH^{(0)}(x,t)$, ; and the operators $\tilde{I}_{par}, \vec{I}_{par}$ denote convolution with these kernels. The L^2 boundedness of $D_{par}\tilde{I}_{par}$ is needed to finish the proof of the case $j = 1$ of estimates (9 and 10, Section 2) ($D_{par}\tilde{I}_{par}$ arises in the case $j = 1$ in place of $[\tilde{K}^{j-2}, \cdot]$. The boundedness on L^2_ω, $\omega \in A_2$ of $D_{par}\tilde{I}_{par}$ is needed to complete the proof of estimate (16, Section 2) in the case $k = 0$ ($\vec{I}_{par}$ appears in place of $[\vec{K}^{(k-1)}, \cdot]$. More generally, we have

LEMMA 1. *Suppose* $I(z) \in C^2(R^n \backslash \{0\})$, $I(\lambda^\alpha z) \equiv \lambda^{1-d}I(z)$. *Let* I *also denote convolution with the kernel* $I(z)$. *Then*

$$\|D_{par} I f\|_{2,\omega} \leq C(A_2) \|I\|_{C^2} \|f\|_{2,\omega},$$

where $\|I\|_{C^2} \equiv \sup\limits_{|\nu| \leq 2} \sup\limits_{z \in S^{n-1}} \left| \left(\frac{\partial}{\partial z}\right)^\gamma I(z) \right|.$

Proof

We may assume $\|I\|_{C^2} = 1$. We use our usual Littlewood-Paley decomposition. We have

$$\frac{1}{2} < D_{par} If, g > \equiv \int\limits_0^\infty < \delta^{-1} Q_\delta^{(0)} If, \delta D_{par} P_\delta \, g > \frac{d\delta}{\delta},$$

where, as before

$$Q_\delta^{(0)} \equiv \delta \frac{\partial}{\partial \delta} P_\delta$$

satisfies $\int Q_\delta^{(0)}(z)\,dz = 0$, $\int z_j\,Q_\delta^{(0)}(z)\,dz = 0$, $1 \le j \le n-1$. We split I into

$$I_1(z) + I_2(z) \equiv I(z)\eta\left(\frac{\|z\|}{\delta}\right) + I(z)\left[1 - \eta\left(\frac{\|z\|}{\delta}\right)\right]$$

By a crude size estimate,

$$\delta^{-1}Q_\delta^{(0)}\,I_1(z) \le C\delta^{-d}\chi\{\|z\| \le C\delta\}.$$

By the cancellation conditions of $Q_\delta^{(0)}$, we have

$$\delta^{-1}Q_\delta^{(0)}\,I_2(z) \equiv \delta^{-1}\int Q_\delta^{(0)}(z-v)[I_2(v) - I_2(z) - (v'-z')\cdot\nabla_x I_2(z)]dv.$$

By our usual Taylor's Theorem estimate, the last expression is bounded in absolute value by

$$C\frac{\delta}{\|z\|^{d+1}}\chi\{\|z\| > C\delta\}.$$

Altogether, then,

$$|\,\delta^{-1}Q_\delta^{(0)}\,I(z)\,| \le C\frac{\delta}{(\delta+\|z\|)^{d+1}},$$

and

$$|\,\delta^{-1}Q_\delta^{(0)}\,I(z) - \delta^{-1}Q_\delta^{(0)}\,I(w)\,| \le C\frac{\|z-w\|}{(\delta+\|z\|)^{d+1}},$$

The latter estimate, which may be verified by replacing $Q_\delta^{(0)}$ by its full n-dimensional gradient, holds whenever $\|z-w\| \le \delta$. Note also that $\delta^{-1}Q_\delta^{(0)}\,I1 \equiv \delta^{-1}I Q_\delta^{(0)}\,1 = 0$. We expand $f \equiv \int_0^\infty Q_\theta^{(1)}\cdot Q_\theta^{(2)}f\frac{d\theta}{\theta}$, and the Lemma then follows by Schwarz, the boundedness on $L_{1/w}^2$ of the square function

$$g \to \left(\int_0^\infty |\,\delta D_{par}P_\delta g\,|^2\,\frac{d\delta}{\delta}\right)^{1/2},$$

and the fact that

$$|\,\delta^{-1}Q_\delta^{(0)}I\,Q_\theta^{(1)}\,Q_\theta^{(2)}f\,| \le C\min\left(\frac{\theta}{\delta},\frac{\delta}{\theta}\right)M\,(Q_\theta^{(2)}f).$$

We leave the routine details to the reader.

7. Bibliography

1 [Ca 1] A.P. Calderon. Commutators of Singular Integral Operators, Proc. Nat. Acad. Sci., USA, 53 (1965), 1092-1099.

2 [Ca 2] Cauchy integrals on Lipschitz curves and related operators, Proc. Nat. Acad. Sci., USA, 74 (1977), 1324-1327.

3 [CF] R. Coifman and C. Fefferman, *Weighted norm inequalities for maximal functions and singular integrals*, Studia Math 51 (1974), 241-250.

4 [Ch] M. Christ, Lectures on Singular Integral Operators, CBMS regional conference series, #77, Amer. Math. Soc., Providence, 1990.

5 [CJ] ___, and J.L. Journe, Polynomial growth estimates for multilinear singular integral operators. Acta Math. 159 (1987), 51-80.

6 [CM] R. Coifman and Y. Meyer, *Non-linear harmonic analysis, operator theory, and P.D.E.*, Beijing Lectures in Harmonic Analysis, E.M. Stein, Ed., Princeton Univ. Press, Princeton, N.J., 1986, pp. 3-45.

7 [CMM] R. Coifman, A. MacIntosh and Y. Meyer, L'integrale de Cauchy definit un opērateur borné sur L^2.

8 [Co] J. Cohen, A sharp estimate for a multilinear singular integral in $\mathbb{R}^n$, Indiana Univ. Math. J. Vol. 30, No. 5 (1981), 693-702.

9 [CR] R. Coifman and R. Rochberg, *Another characterization of BMO*, Proc. Amer. Math. Soc. 79 (1980), 249-254.

10 [CW] R. Coifman and G. Weiss, *Analyse harmonique non-commutatiave sur certains espaces homogènes*, Lecture Notes in Math., Vol. 242, Springer-Verlag, Berlin, 1971.

11 [DJ] G. David and J.L. Journé, *A boundedness criterion for generalized Calderón-Zygmund operators*, Ann. of Math. 120 (1984), 371-397.

12 [DJS] G. David, J.L. Journé, and S. Semmes, *Opérateurs de Calderón-Zygmund, fonctions para-accrétives et interpolation*, Rev. Mat. Iberoamericana 1 (1985), 1-56.

13 [Do] J. Dorronsoro, *A characterization of potential spaces,* Proc. Amer. Math. Soc. 95 (1985), 21-31.

14 [DR] J. Duoandikoetxea and J. L. Rubio de Francia, *Maaximal and Singular integrals via Fourier transform estimates*, Invent. Math. 84 (1986), 541-561.

15 [FR1] E.B. Fabes and N.M. Riviere, *Symbolic Calculus of Kernels with Mixed Homogeneity*, Singular Integrals, A.P. Calderón, Ed., Proc. Symp. Pure Math., Vol. 10, Amer. Math. Soc., Providence, 1967, pp. 106-127.

16 [FR2]—, *Singular Integrals with mixed homogeneity*, Studia Math. 27 (1966), 19-38.

17 [Fng] X.Fang, Ph.D Thesis, Yale University, 1990

18 [FS] C. Fefferman and E.M. Stein, *H^p spaces of several variables*, Acta. Math. 129 (1972) 137-193.

19 [GR] J. Garcia-Cuerva and J.L. Rubio de Francia, *Weighted Norm In-*

equalities and Related Topics, math. Studies, Vol. 116, North-Holland, Amsterdam, 1985.

20 [H1] S. Hofmann, *Boundedness criteria for rough singular integrals*, to appear, Proc. Lond. Math. Soc..

21 [H2] ___, *A characterization of commutators of parabolic singular integrals*, to appear, Proceedings of the Conference on Harmonic Analysis and PDE held at Miraflores de la Sierra, Spain, 1992.

22 [H3] ___, *Weighted norm inequalities and vector-valued inequalities for certain rough operators*, Indiana U. Math. J..

23 [H4] ___, *Singular integrals of Calderon-type in $\mathbb{R}^n$, and BMO*, to appear, Rev. Mat. Iberoamericana.

24 [H5] ___, *Parabolic singular integrals of Calderon-type, rough operators, and caloric layer potentials*, preprint.

25 [J] J.L. Journe, *Calderón-Zygmund Operators, Pseudo-differential Operators, and the Cauchy Integral of Calderón*, Lecture Notes in Math., vol. 994, Springer-Verlag, Berlin 1983.

26 [Jns Bf] B.F. Jones, *A Class of Singular Integrals*, Amer. J. Math. 86 (1964), 441-462.

27 [Jns P] P. Jones, Square functions, *Cauchy integrals, analytic capacity, and harmonic measure*, Harmonic Analysis and Partial Differential Equations (J. Garcia-Cuerva, ed.) Lecture Notes in Math, Vol. 1384, Springer-Verlag, Berlin, 1989, pp. 24-68.

28 [L] P.G. Lemarie, *Algebras d'opérateurs et semi-groupes de Poisson sur un espace de nature homogène*, Publications Mathématiques d'Orsay, université de Paris-Sud. Department de Mathématique, Orsay (1984), #84-03.

29 [LM1] J.L. Lewis and M.A.M. Murray, T*he Method of Layer Potentials for the Heat Equation in Time-Varying Domains I: Singular Integrals*, preprint.

30 [LM2] ___, *The Method of layer potentials for the heat equation in time-varying domains II: the David buildup scheme*, preprint.

31 [LM3] ___, *Regularity properties of commutators and layer potentials associated to the heat equation*, Trans. Amer. Math. Soc.

32 [R] N.M. Riviere, *Singular Integrals and Multiplier Operators*, Ark. Math. 9 (1971), pp. 243-278.

33 [S] E.M. Stein, *Singular Integrals and Differentiability Properties of Functions*, Princeton, 1970.

DEPARTMENT OF MATHEMATICS AND STATISTICS, WRIGHT STATE UNIVERSITY, DAYTON, OHIO 45435

Current address: Department of Mathematics, University of Missouri, Columbia, Missouri 65211

E-mail address: hofmann@msindy2.cs.missouri.edu

Contemporary Mathematics
Volume **189**, 1995

Local behavior of Riemann's function

Stéphane Jaffard

ABSTRACT: We investigate how local Hölder regularity and local scalings of a function can be determined from its wavelet transform; we apply these results to Riemann's function.

1. Introduction

Let $x_0 \in \mathbb{R}$; by definition, a function f is $C^\alpha(x_0)$ if there exists a polynomial P of order at most α such that

$$(1) \qquad |f(x_0 + h) - P(h)| \leq C|h|^\alpha.$$

Even if this upper bound is optimal, it does not describe the behavior of f in the neighbourhood of x_0. For instance *trigonometric chirps* studied in [10] describe a very oscillatory behavior, of the form

$$f(x_0 + h) - P(h) \sim |h|^\alpha \sin \frac{1}{|h|^\beta}.$$

A less oscillatory behavior appears when f is "approximately selfsimilar",i.e. satisfies (for $0 < \alpha < 1$ and $\lambda < 1$)

$$f(x_0 + h) - P(h) = \lambda^{-\alpha} \left(f(x_0 + \lambda h) - P(\lambda h)\right) + o(|h|^\alpha).$$

This equality is clearly equivalent to

$$(2) \qquad f(x_0 + h) - P(h) = h^\alpha G_\pm(\log(\pm h)) + o(|h|^\alpha)$$

where $\pm$ is the sign of h, and G_+ and G_- are $\log \lambda$ periodic. Thus, if any of these conditions is satisfied, we will say that f has a *logarithmic chirp* of order (α, λ) at x_0; partial necessary conditions on the wavelet transform of f can be found in [6].

1991 *Mathematics Subject Classification.* Primary 26A15; Secondary 11J70, 11F03.

This paper is in final form and no version of it will be submitted for publication elsewhere.

If the functions G_+ and G_- are C^γ, the chirp is said to have the regularity γ. Our purpose in the first part is to find simple necessary or sufficient conditions of selfsimilarity on the wavelet transform of f. We will then apply these criteria to several examples. Let us mention two of them; the first one is the "selfsimilar functions" introduced in [8], for which we will determine the regularity of the chirps and an explicit formula for the Fourier coefficients of the functions G_+ and G_-; for instance, Weierstrass functions

$$\sum_1^\infty 2^{-\alpha j} \sin 2\pi 2^j x$$

have logarithmic chirps at the rational points where the Fourier coefficients of G_+ and G_- are values of the Gamma function (given by (44)). The second example is "Riemann's function"

$$\varphi(x) = \sum_1^\infty \frac{1}{n^2} \sin \pi n^2 x.$$

In this case, J.J.Duistermaat proved in [1] the existence of logarithmic chirps at the *quadratic irrationals* (the irrational numbers roots of a polynomial of degree two with integer coefficients). We will complete this study by showing that the chirps have a regularity $1/2$ at each of these points. We will also show how to determine the Hölder regularity of Riemann's function at every point. It may be interesting to recall briefly the history of this function.

Riemann proposed it as an example of continuous nowhere differentiable function. Actually, Hardy and Littlewood proved that φ is nowhere $C^{3/4}$ except perhaps at the rational points of the form $(2p+1)/(2q+1)$, $p, q \in \mathbb{Z}$ ([4]). Their proof anticipates wavelet methods: they remark that the function

$$C(a,b) = \frac{a}{2}(\theta(b+ia) - 1)$$

(where θ is the Jacobi function $\theta(z) = \sum_{n\in\mathbb{Z}} e^{i\pi n^2 z}$) is the convolution of φ with contractions (by a factor a) of

$$\psi(x) = \frac{1}{\pi(x-i)^2}.$$

Since ψ has a vanishing integral, an Abel type theorem (which we will state precisely in Proposition 1) shows that, if φ is smooth at x_0, $C(a,b)$ must have a certain decay when $a \to 0$ and $b \to x_0$. This method will thus yield upper bounds for the pointwise Hölder exponent

$$\alpha(x_0) = \sup\{\beta : \varphi \in C^\beta(x_0)\}.$$

This method actually yields the following more precise result which relates the pointwise behavior of φ at x_0 to the Diophantine approximation properties of x_0. Let $x_0 \notin \mathbb{Q}$, let p_n/q_n be its continued fraction expansion and define

$$\tau(x_0) = \sup\{\tau : \mid x_0 - \frac{p_m}{q_m} \mid \leq \frac{1}{q_m^\tau}\}$$

for infinitely many m's such that p_m and q_m are not both odd. Then

$$\alpha(x_0) \leq \frac{1}{2} + \frac{1}{2\tau(x_0)}.$$

a result which is actually stated by J.J.Duistermaat [1] where a more direct proof is given.

Converse results, which would yield an information about the pointwise behavior of φ from estimates on its convolutions are more difficult to obtain since they are of tauberian type. We will state such results in Proposition 1.

Finally Gerver proved the differentiability at the rational points of the form $(2p + 1)/(2q + 1)$ [2] (where we now know that φ is exactly $C^{3/2}$, see [10]). The analysis of the behavior of φ near such a rational point x_0 has been considerably sharpened since; in [10], a complete "chirp" asymptotic expansion which describes the oscillations of φ is exhibited:

$$\varphi(x) = u(x) + \sum_{n \geq 0} (x - x_0)^{\frac{3}{2}+n} v_+^n(\frac{1}{x - x_0}) \quad \text{if } x \geq x_0$$

$$\varphi(x) = u(x) + \sum_{n \geq 0} |x - x_0|^{\frac{3}{2}+n} v_-^n(\frac{1}{|x - x_0|}) \quad \text{if } x \leq x_0$$

where u is C^∞, the $v_\pm^n$ are 2π periodic, with a vanishing integral, and are $C^{\frac{1}{2}+n}$ (we will actually show that these points are the only ones where a chirp expansion exists).

The results of Hardy and Gerver left open the problem of the determination of the exact regularity of φ at irrational points; one of our purposes is to do this determination using the wavelet method we sketched.

Our main results concerning Riemann's function are stated in the following theorem.

Theorem 1 *Let $x \notin \mathbb{Q}$ and let p_n/q_n be the sequence of its approximations by continued fractions. Let*

$$\tau(x) = \sup\{\tau : \mid x - \frac{p_m}{q_m} \mid \leq \frac{1}{q_m^\tau}\}$$

for infinitely many m's such that p_m and q_m are not both odd.
Then

$$\alpha(x) = \frac{1}{2} + \frac{1}{2\tau(x)}.$$

If x is a quadratic irrational, φ has a logarithmic chirp at x of regularity $1/2$.

Thus Riemann's function is interesting not only for historical reasons but also because it displays many different local behaviors: a whole range of different Hölder exponents, chirps at some rationals and logarithmic chirps at quadratic irrationals. Thus it is a perfect example on which one can test the efficiency of local analysis methods; and, in our opinion, it clearly shows the efficiency of wavelet methods.

This paper is partly a rewiew paper and partly contains new results: the analysis of the local regularity by wavelet methods can be found in [7] and the pointwise Hölder exponent of Riemann's function is determined in [9]. The results concerning the analysis of local scalings is new, and its application for Riemann's function sharpens some results of J.J.Duistermaat in [1].

2. The tools: Wavelet transform and two-microlocalization.

Let $\alpha > 0$; it will be the largest order of Hölder regularity we will be interested in. Let $k = [\alpha]$. Suppose that a function ψ is nonvanishing and satisfies the following asumptions

$$(3) \qquad |\psi(x)| + |\psi'(x)| + \cdots + |\psi^{(k+1)}(x)| \leq C(1 + |x|)^{-k-2},$$

$$(4) \qquad \int \psi(x)dx = \int x\psi(x)dx = \cdots = \int x^k \psi(x)dx = 0$$

and either

$$(5) \qquad \int_0^\infty |\hat{\psi}(\xi)|^2 \frac{d\xi}{\xi} = \int_0^\infty |\hat{\psi}(-\xi)|^2 \frac{d\xi}{\xi} = 1$$

or

$$(6) \qquad \hat{\psi}(\xi) = 0 \text{ if } \xi < 0 \text{ and } \int_0^\infty |\hat{\psi}(\xi)|^2 \frac{d\xi}{\xi} = 1$$

In the last case the wavelet is said to be analytic. The wavelet transform of an L^∞ function f is defined by

$$C(a,b)(f) = \frac{1}{a} \int f(t)\bar{\psi}(\frac{t-b}{a})dt.$$

We will consider the three following settings: In the first one the analyzed function f is real valued, and the wavelet satisfies (6); in the second one, f is complex valued and the wavelet satisfies (5); in the third one, $\hat{f}(\xi) = 0$ if $\xi < 0$ and the wavelet satisfies (6).

In each case, the following results concerning the relationships between the size of the wavelet transform and the regularity of the function hold.

Proposition 1 *Under the previous hypotheses if a function f is $C^\alpha(x_0)$,*

$$(7) \qquad |C(a,b)(f)| \leq Ca^\alpha(1 + \frac{|b - x_0|}{a})^\alpha$$

Conversely, if

$$(8) \qquad |C(a,b)(f)| \leq Ca^\alpha(1 + \frac{|b - x_0|}{a})^{\alpha'} \text{ for an } \alpha' < \alpha$$

then

$$(9) \qquad |f(x) - f(x_0)| \leq C|x - x_0|^\alpha$$

for all x such that $|x - x_0| \leq 1/2$. Furthermore, large O's can be replaced by little o's in these results, i.e. if

$$f(x_0 + h) - P(h) = o\left(|h|^\alpha\right)$$

then

(10)
$$C(a, b)(f) = o\left(a^\alpha(1 + \frac{|b - x_0|}{a})^\alpha\right);$$

and conversely, if

(11)
$$C(a, b)(f) = o\left(a^\alpha(1 + \frac{|b - x_0|}{a})^{\alpha'}\right) \quad \text{for an} \ \alpha' < \alpha.$$

then

$$f(x_0 + h) - P(h) = o\left(|h|^\alpha\right).$$

Let us now recall the definition of the two-microlocal spaces $C^{\alpha,\alpha'}(x_0)$ (see [7]). The function f belongs to $C^{\alpha,\alpha'}(x_0)$ if its wavelet transform satisfies

$$|C(a, b)(f)| \leq Ca^\alpha(1 + \frac{|b - x_0|}{a})^{-\alpha'}.$$

The conditions above are clearly expressed in terms of two-microlocal spaces; the connexion will even be more obvious for the conditions that imply the existence of a logarithmic chirp of a given regularity.

Proof of Proposition 1: We only consider the case $0 < \alpha < 1$. The reader will easily extend it to any $\alpha > 0$. If $f \in C^\alpha(x_0)$,

$$
\begin{aligned}
|C(a, b)(f)| &= \frac{1}{a}\left| \int f(x)\psi(\frac{x - b}{a})dx \right| \\
&= \frac{1}{a}\left| \int (f(x) - f(x_0))\psi(\frac{x - b}{a})dx \right| \\
&\leq \frac{C}{a} \int |x - x_0|^\alpha \left(\frac{1}{1 + |\frac{x-b}{a}|}\right)^2 dx \\
&\leq \frac{C}{a} \int \frac{|x - b|^\alpha}{(1 + |\frac{x-b}{a}|)^2}dx + |b - x_0|^\alpha \frac{C}{a} \int \frac{dx}{(1 + |\frac{x-b}{a\cdot}|)^2} \\
&\leq Ca^\alpha \left(1 + |\frac{b - x_0}{a}|\right)^\alpha
\end{aligned}
$$

Suppose now that we are in the first or the third case. In that case, f is reconstructed from its wavelet transform by

$$f(x) = \int \int C(a, b)(f)\psi(\frac{x - b}{a})\frac{da\,db}{a^2}.$$

Let

$$\omega(a,x) = \int C(a,b)(f)\psi(\frac{x-b}{a})\frac{db}{a};$$

if (8) holds,

$$|\omega(a,x)| \leq Ca^\alpha(1 + \frac{|x-x_0|}{a})^{\alpha'}$$

and

$$|\frac{\partial\omega(a,x)}{\partial x}| \leq Ca^{\alpha-1}(1 + \frac{|x-x_0|}{a})^{\alpha'}.$$

Using the second estimate (and the mean value theorem) for $a \geq |x - x_0|$ and the first estimate for $a \leq |x - x_0|$, we obtain

$$|f(x) - f(x_0)| \leq \int_{a \geq |x-x_0|} Ca^{\alpha-1}\left(1 + |\frac{x-x_0}{a}|\right)^{\alpha'}|x-x_0|\frac{da}{a}\dots$$

$$+ \int_{a \leq |x-x_0|} Ca^\alpha\left(1 + |\frac{x-x_0}{a}|\right)^{\alpha'}\frac{da}{a} \leq C|x-x_0|^\alpha.$$

This also implies the result in the second case by superposing reconstruction formulas for $\xi \geq 0$ and $\xi \leq 0$. The proof for little o's instead of big o's is exactly the same, and we leave it to the reader.

There exists some similar results for negative values of α. The reader is reffered to [8] for a discussion on negative Hölder exponents, and the corresponding conditions on the wavelet transform.

3. Logarithmic chirps and wavelets

We will prove the following result.

Theorem 2 *If f has a logarithmic chirp of order (α, λ) and of regularity $\gamma \geq 0$ at x_0, then its wavelet transform satisfies*

$$(12) \qquad C(a,b) = a^\alpha H(\log a, \frac{b-x_0}{a}) + o(a^\alpha + (\frac{|b-x_0|}{a})^\alpha).$$

Where H is $Log\lambda$ periodic in the first variable and satisfies

$$(13) \qquad \| H(.,y) \|_{L^\infty} \leq C(1 + |y|)^{\alpha-\gamma}$$

$$(14) \qquad \| H(.,y) \|_{C^\gamma} \leq C(1 + |y|)^\alpha$$

Conversely, if

$$(15) \qquad C(\lambda a, x_0 + \lambda b) = \lambda^{-\alpha}C(a, x_0 + b) + o(a^\alpha(1 + \frac{|b-x_0|}{a})^{\alpha'})$$

where $\lambda < 1$, and $\alpha' < \alpha$; and if furthermore $f \in C^{\alpha,-\alpha+\gamma}(x_0)$, then f has a logarithmic chirp of order (α, λ) and of regularity $\gamma' \geq 0$ at x_0, for any $\gamma' < \gamma$.

Let us make a few comments. First, we can formulate the "algebraic" selfsimilarity condition satisfied by the wavelet transform either by a selfsimilarity condition (as in (15)) or through the existence of a periodic function (as in (12)). This equivalence is the purpose of the following lemma. The interesting point about using the periodic function formulation is that we can write explicit regularity asumptions about this function.

Lemma 1 *Suppose that there exists $\lambda < 0$ and $\alpha, \alpha' > 0$ such that*

(16) $$C(a, x_0 + b) = \lambda^{-\alpha} C(\lambda a, x_0 + \lambda b) + O(a^{\alpha'} + b^{\alpha'})$$

then

(17) $$C(a, b) = a^{\alpha} H(\log a, \frac{b - x_0}{a}) + O(a^{\alpha'} + b^{\alpha'})$$

where H is periodic of period $\log \lambda$ in the first variable.

Proof: In order to simplify the notations, suppose that $x_0 = 0$. The sequence $\lambda^{-n\alpha} C(\lambda^n a, \lambda^n b)$ converges uniformy on any compact that does not contain a point of the real axis $a = 0$ because (16) implies that

$$\lambda^{-n\alpha} C(\lambda^n a, \lambda^n b) - \lambda^{-(n+1)\alpha} C(\lambda^{n+1} a, \lambda^{n+1} b) = O((\lambda^n a)^{\alpha'} + (\lambda^n b)^{\alpha'})$$

and the limit function $d(a, b)$ satisfies

(18) $$d(a, b) = \lambda^{-\alpha} d(\lambda a, \lambda b)$$

so that the lemma reduces to the change of notations

$$H(x, y) = e^{-\alpha x} d(e^x, y e^x).$$

We remark that this proof shows that we will be able to obtain size and regularity estimates on H from similar uniform estimates on the sequence $\lambda^{-n\alpha} C(\lambda^n a, \lambda^n b)$.

Let us now prove the direct part in Theorem 2.
Suppose that (2) holds. Because ψ has vanishing moments,

$$C(a, b) = \frac{1}{a} \int (f(x) - P(x - x_0)) \psi(\frac{x - b}{a}) dx.$$

Consider the integral for $x \geq x_0$ (the other part is similar); it is written

$$\frac{1}{a} \int |x - x_0|^{\alpha} G_+(\log |x - x_0|) \psi(\frac{x - b}{a}) dx + o\left(\frac{1}{a} \int |x - x_0|^{\alpha} |\psi(\frac{x - b}{a})| dx\right).$$

The second term is clearly bounded by $o(a^{\alpha} + |b - x_0|^{\alpha})$ and the first one is

(19) $$a^{\alpha} \int |u + \frac{b - x_0}{a}|^{\alpha} G_+(\log a + \log |u + \frac{b - x_0}{a}|) \psi(u) du$$

which is of the form $a^{\alpha} H(\log a, \frac{b - x_0}{a})$ where H is $\log \lambda$ periodic in the first variable. Furthermore, this term is bounded by

$$C a^{\alpha} \int (|u|^{\alpha} + |\frac{b - x_0}{a}|^{\alpha}) \frac{du}{1 + u^{k+2}} \leq C a^{\alpha} (1 + |\frac{b - x_0}{a}|)^{\alpha}$$

In order to prove (13), we can thus suppose that $|y| \geq 1$.

Once (12) is proved, it is clear that (13) is just a two-microlocal estimate for the term $a^\alpha H(\log a, \frac{b-x_0}{a})$. Since such estimates do not depend on the wavelet (see [7]), we can suppose that $supp\psi \subset [-1,1]$. From (19), we get

$$H(x,y) = \int |u+y|^\alpha G_+(x+\log|u+y|)\psi(u)du.$$

Integrating $[\gamma]$ times by parts, we obtain

$$H(x,y) = \int |u+y|^{\alpha'} \tilde{G}(x+\log|u+y|)\tilde\psi(u)du$$

where $\alpha' = \alpha - [\gamma]$, $\tilde\psi$ is the ψ integrated $[\gamma]$ times, and

$$\tilde{G}(x+\log|u+y|) = \sum_0^{[\gamma]} c_l G_+^{(l)}(x+\log|u+y|),$$

so that $\tilde{G}$ is $C^{\gamma-[\gamma]}$. We have

$$|u+y|^{\alpha'} = |y|^{\alpha'} + ug(u,y)$$

where

$$|g(u,y)| \leq Cy^{\alpha'-1} \text{ if } u \leq y$$

(which is always the case because $|u| \leq 1 \leq |y|$. Thus H is the sum of two terms, the second one being

$$\int ug(u,y)\tilde{G}(x+\log|u+y|)\tilde\psi(u)du$$

which is bounded if $u \leq y$ by

$$\int_{u\leq y} |y|^{\alpha'-1}u|\tilde\psi(u)|du \leq C|y|^{\alpha'-1}(1+|\log y|).$$

The first term of H is

(20) $$|y|^{\alpha'} \int \tilde{G}(x+\log|u+y|)\tilde\psi(u)du.$$

If $|y| \leq 1$, it is bounded by a constant, and if $|y| \geq 1$, we write

$$\tilde{G}(x+\log|u+y|) = \tilde{G}(x+\log|y| + \log|1+\frac{u}{y}|);$$

after substracting $\tilde{G}(x+\log|y|)$; we bound (20) by

$$C|y|^{\alpha'} \int |\log(1+\frac{u}{y})|^{\gamma-[\gamma]}|\tilde\psi(u)|du \leq C|y|^{\alpha-\gamma}.$$

Concerning the second estimate of Theorem 2, we have

$$H(z,y) - \sum_{|k|\leq[\alpha]} \frac{h^k}{k!}\partial_x^k H(x,y) =$$

$$\int |u+y|^\alpha \left(G_+(z+\log|u+y|) - \sum_{|k|\leq[\alpha]} \frac{h^k}{k!}\partial_x^k G_+(x+\log|u+y|) \right) \psi(u)du$$

which is bounded by

$$C\int |u+y|^\alpha |x-z|^\gamma \frac{du}{1+u^{k+2}} \leq C|x-z|^\gamma |1+y|^\alpha.$$

In order to prove the converse part of Theorem 2, we will need the following proposition, which is interesting in its own right.

Proposition 2 *Suppose there exists $\alpha' < \alpha$ such that*

$$(21) \qquad C(a,b) = a^\alpha H(\log a, \frac{b-x_0}{a}) + o\left(a^\alpha (1 + \frac{b-x_0}{a})^{\alpha'} \right)$$

where H is $\log \lambda$ periodic in the first variable and satisfies (13) and (14), then f has a logarithmic chirp of order (α, λ) and of regularity γ' at x_0 for any $\gamma' < \gamma$.

Proof: f is reconstructed using the formula

$$f(x) = \int C(a,b)\psi(\frac{x-b}{a})\frac{dadb}{a^2}$$

By a classical argument (see [7]) the second term in (21) brings a contribution to $f(x) - P(x - x_0)$ which is $o(|x - x_0|^\alpha)$; it can thus be neglected. We choose for P:

$$P(x-x_0) = \sum_{l=0}^{[\alpha]} \frac{(x-x_0)^l}{l!} \int C(a,b)\psi^{(l)}(\frac{x_0-b}{a})\frac{dadb}{a^{l+2}}$$

(one easily checks that these coefficients are well defined). Thus

$$f(x) - P(x-x_0)$$

$$= \int\int a^\alpha H(\log a, \frac{b-x_0}{a}) \left(\psi(\frac{x-b}{a}) - \sum \frac{1}{l!}(\frac{x-x_0}{a})^l \psi^{(l)}(\frac{x_0-b}{a}) \right) \frac{dadb}{a^2}$$

$$= \int\int (\frac{x-x_0}{u})^\alpha H(\log|x-x_0| - \log u, b) \left(\psi(u-b) - \sum \frac{u^l}{l!}\psi^{(l)}(-b) \right) \frac{dudb}{u}$$

which is of the form $|x-x_0|^\alpha G(\log|x-x_0|)$ with g periodic of period $\log \lambda$. Let us prove the estimates on G. We have

$$G(x) = \int\int H(\log x - \log u, b) \left(\psi(u-b) - \sum \frac{u^l}{l!}\psi^{(l)}(-b) \right) \frac{dudb}{u^{\alpha+1}}$$

First, let us prove that G is bounded. In the integral, if $u \leq 1$, we bound $\psi(u - b) - \sum \frac{u^l}{l!}\psi^{(l)}(-b)$ by $\frac{Cu^{k+1}}{(1+|b|)^{k+2}}$ (here $k = [\alpha]$); and the bound follows from (13). If $u \geq 1$, it is bounded by $\frac{C}{(1+|u-b|)^{l+2}} + C\sum \frac{u^l}{(1+|b|)^{k+2}}$ and the bound still follows from (13). In order to estimate $\| G(.,y) \|_{C^{\gamma'}}$, one cannot use directly (14) because the factor $(1+|y|)^\alpha$ yields divergent integrals; one remarks that (13) and (14) imply that $\forall \gamma' < \gamma$, there exists $\alpha' < \alpha$ such that

$$\| H(.,y) \|_{C^{\gamma'}} \leq C(1 + |y|)^{\alpha'}$$

and the bound for $\| G(.,y) \|_{C^{\gamma'}}$ is obtained as above.

Let us now check that the converse part of Theorem 2 is an easy consequence of Proposition 2. Suppose that (15) holds. Lemma 1 shows that the limit

$$\lim_{n\to\infty} \lambda^{-n\alpha} C(\lambda^n a, \lambda^n b) = d(a,b)$$

exists and that

$$H(x,y) = e^{-\alpha x} d(e^x, ye^x)$$

is $\log \lambda$ periodic in the first variable. We still have to check (13) and (14). The two-microlocal hypothesis implies that

$$|\lambda^{-n\alpha} C(\lambda^n a, \lambda^n b)| \leq C\lambda^{-n\alpha}(\lambda^n a)^\alpha (1 + |\frac{b}{a}|)^{\alpha'} = Ca^\alpha(1 + |\frac{b}{a}|)^{\alpha'},$$

hence the same estimate for $d(a,b)$, so that

(22) $$|H(x,y)| \leq Ce^{-\alpha x}e^{\alpha x}(1 + |y|)^{\alpha'} \leq C(1 + |y|)^{\alpha'}.$$

Let us now estimate the regularity of H in the first variable. Here too we write the proof only for $\alpha < 1$, the general case follows easily. Remark that

$$\frac{\partial}{\partial b}C(a,b) = \frac{1}{a}C_1(a,b),$$

where C_1 is the wavelet transform of f using the admissible wavelet $-\psi'$. Similarly

$$\frac{\partial}{\partial a}C(a,b) = \frac{1}{a}C_2(a,b),$$

where C_2 is the wavelet transform of f using the admissible wavelet $-\psi - x\psi'$. Since two-microlocal estimates are independant of the wavelet that is chosen (see [7]),

$$\nabla d(a,b) = \lim_{n\to\infty} \lambda^{-n\alpha}\nabla C(\lambda^n a, \lambda^n b)$$

so that

$$|\nabla d(a,b)| \leq \lambda^{-n\alpha}\lambda^n \frac{1}{\lambda^n a}(\lambda^n a)^\alpha (1 + |\frac{b}{a}|)^{\alpha'} = a^{\alpha-1}(1 + |\frac{b}{a}|)^{\alpha'}$$

But

$$\partial_x H(x,y) = -\alpha H(x,y) + e^{(1-\alpha)x}\left(\partial_1 d(e^x, ye^x) + y\partial_2 d(e^x, ye^x)\right).$$

So that

$$|\partial_x H(x,y)| \leq C(1+|y|)^{\alpha'} + e^{(1-\alpha)x}\left(e^{(\alpha-1)x}(1+|y|)^{\alpha'} + ye^{(\alpha-1)x}(1+|y|)^{\alpha'}\right).$$

Thus

$$|\partial_x H(x,y)| \leq C(1+|y|)^{\alpha'+1}$$

Thus (14) follows from this estimate and (22). (The proof for higher order derivatives is exactly similar). We are now in the position to apply Proposition 2, hence Theorem 2.

4. Theta Jacobi function and continued fractions

In this section we establish or recall some prerequisites about the *Jacobi theta function* which will be necessary for the determination of the Hölder regularity of φ and of its local scalings.

Using Cauchy's formula, we obtain that (using the wavelet $\psi(x) = (x-i)^{-2}$) the wavelet transform of $\varphi(x)$ is $2ia(\theta(b+ia)-1)/2$. Since we want to determine Hölder exponents between $1/2$ and $3/4$, because of (8) we can add a term ia and the study of the pointwise regularity of φ reduces to obtaining estimates similar to (7) for the function

(23) $$C(a,b) = a\theta(b+ia).$$

The *theta modular group* is obtained by composing the two transforms

$$x \to x+2 \quad \text{and} \quad x \to -1/x$$

It is composed of the fractional linear transformations

$$\gamma(x) = \frac{rx+s}{qx-p}$$

where $rp + sq = -1$, r, s, p, q are integers and the matrix

(24) $$\begin{pmatrix} r & s \\ q & p \end{pmatrix} \text{ is of the form } \begin{pmatrix} even & odd \\ odd & even \end{pmatrix} \text{ or } \begin{pmatrix} odd & even \\ even & odd \end{pmatrix}$$

When γ belongs to the theta modular group, θ is transformed following the formula (cf [1])

(25) $$\theta(z) = \theta(\gamma(z))e^{im\pi/4}q^{-1/2}(z - \frac{p}{q})^{-1/2}$$

where m is an integer which depends on r, s, p, q.

Let $\rho \notin \mathbb{Q}$ and p_n/q_n the sequence of its approximations by continued fractions. The idea of the proof of Theorem 1 is to use (25), which will allow us to deduce the behavior of $\theta(z)$ near p_n/q_n (hence near ρ) from its behavior near 0 or 1. Because of (24), we will have to separate two cases depending on whether p_n and q_n are both odd or not; but let us first derive some straightforward estimates for θ near 0 and 1.

First remark that

(26) $$|\theta(z) - 1| \leq \frac{1}{2} \quad \text{if } Imz \geq 1$$

because in this case,

$$|\theta(z) - 1| \le 2 \sum_{n \ge 1} e^{-\pi n^2 Imz} \le \frac{2e^{-\pi Imz}}{1 - e^{-\pi Imz}} \le \frac{1}{2}$$

We also have
(27) $$|\theta(z)| \le C|Imz|^{-1/2} \quad \text{if} \quad Im(z) \le 1$$

because $|\theta(z)| \le \sum e^{-n^2 Im(z)}$; the sum for $n \le Im(z)^{-1/2}$ is bounded trivially by $Im(z)^{-1/2} + 1$ and the same bound holds for $n > Im(z)^{-1/2}$ (by comparison with an integral).

Let us now obtain the behavior of θ near the point 1. Recall that θ satisfies

(28) $$\theta(1 + z) = \sqrt{\frac{i}{z}} \left(\theta(-\frac{1}{4z}) - \theta(-\frac{1}{z}) \right)$$

so that

$$\theta(1 + z) = 2\sqrt{\frac{i}{z}} \left(A(4z) - A(z) \right)$$

where $A(z) = \sum_1^\infty e^{-i\pi n^2/z}$. If $Im(\frac{-1}{z}) \ge 1$, then

$$|A(z)| \le 2 \exp(-\pi Im(\frac{-1}{z})),$$

so that in that case

(29) $$|\theta(1 + z)| \le C|z|^{-1/2} \exp(-\pi Im(\frac{-1}{z})).$$

Proposition 3 *Let* $\dfrac{p_n}{q_n}$ *be the sequence of approximations of* ρ *by continued fractions; let* τ_n *be defined by*
(30) $$|\rho - \frac{p_n}{q_n}| = (\frac{1}{q_n})^{\tau_n}.$$

If a, b, *and* n *are such that*

(31) $$3|\rho - \frac{p_n}{q_n}| \le |b - \rho + ia| \le 3|\rho - \frac{p_{n-1}}{q_{n-1}}|,$$

the following estimates hold:
    If p_n *and* q_n *are not both odd but* p_{n-1} *and* q_{n-1} *are both odd*

(32) $$|C(a,b)| \le Ca^{\frac{1}{2} + \frac{1}{2\tau_n}} (1 + \frac{|b - \rho|}{a})^{\frac{1}{2\tau_n}}.$$

    If p_n *and* q_n *are not both odd and* p_{n-1} *and* q_{n-1} *are not both odd*

(33) $$|C(a,b)| \le Ca^{\frac{1}{2} + \frac{1}{2\tau_n}} (1 + \frac{|b - \rho|}{a})^{\frac{1}{2\tau_n}}$$

or
(34) $$|C(a,b)| \le Ca^{\frac{1}{2} + \frac{1}{2\tau_{n-1}}} (1 + \frac{|b - \rho|}{a})^{\frac{1}{2\tau_{n-1}}}.$$

If p_n and q_n are both odd

$$(35) \qquad |C(a,b)| \leq Ca^{\frac{1}{2}+\frac{1}{2\tau_{n-1}}}(1+\frac{|b-\rho|}{a})^{\frac{1}{2\tau_{n-1}}}.$$

Furthermore, if p_n and q_n are not both odd, these estimates are optimal, which means that there exists a point in the domain (31) where (32) or (33) are equalities.

We remark that, since $\tau_n \geq 2$, this result together with Proposition 1 implies Hardy's result that $\varphi(x) - \varphi(x_0)$ is nowhere $o(|x-x_0|)^{3/4}$ except perhaps at the rational points that are quotients of two odd numbers. More precisely, we have

Corollary 1 *Let $\rho \notin \mathbb{Q}$; if there exists an infinity of integers n such that p_n and q_n are not both odd and such that $\tau_n \geq \tau$, then*

$$\varphi(x) - \varphi(x_0) \quad \text{is not} \quad o(|x-x_0|)^{\frac{1}{2}+\frac{1}{2\tau}},$$

but if there exists N such that $\tau_n \leq \tau$ for any $n \geq N$ (for which p_n and q_n are not both odd), then

$$\varphi(x) - \varphi(x_0) = O(|x-x_0|)^{\frac{1}{2}+\frac{1}{2\tau}}.$$

Define $\Gamma^\alpha(x_0)$ as the set of functions f such that

$$\begin{cases} \forall \beta > \alpha \quad f \notin C^\beta(x_0) \\ \forall \beta < \alpha \quad f \in C^\beta(x_0). \end{cases}$$

If $\eta(\rho) = \limsup \tau_n(\rho)$, where the lim sup bears only on the n's such that p_n and q_n are not both odd, this result implies that $\varphi \in \Gamma^{\frac{1}{2}+\frac{1}{2\eta(\rho)}}(\rho)$. We will prove this proposition in the two next sections, and in the following one, we will show how to derive the local scalings at quadratic irrationals.

5. The case when p_n and q_n are not both odd

Let us now recall a few properties of approximations by continued fractions. Since

$$(36) \qquad p_n q_{n-1} - p_{n-1} q_n = (-1)^{n-1},$$

we have

$$\frac{1}{q_n q_{n+1}} = |\frac{p_n}{q_n} - \frac{p_{n+1}}{q_{n+1}}| \geq |\rho - \frac{p_n}{q_n}|$$

because (see [10]) $\frac{p_n}{q_n}$ and $\frac{p_{n+1}}{q_{n+1}}$ are not on the same side of ρ; on the other hand,

$$\frac{1}{q_n q_{n+1}} = |\frac{p_n}{q_n} - \frac{p_{n+1}}{q_{n+1}}| \leq 2|\rho - \frac{p_n}{q_n}|$$

so that

$$(37) \qquad (\frac{1}{q_n})^{\tau_n-1} \leq \frac{1}{q_{n+1}} \leq 2(\frac{1}{q_n})^{\tau_n-1}.$$

We first determine $\gamma_n = \dfrac{r_n x + s_n}{q_n x - p_n}$ in the theta modular group such that the pole of γ_n is p_n/q_n. Because of (36) if p_{n-1} and q_{n-1} are not both odd, we can choose

$$r_n = (-1)^n q_{n-1}, \; s_n = (-1)^{n+1} p_{n-1};$$

the corresponding transform satisfies (24) and thus belongs to the theta modular group, and if p_{n-1} and q_{n-1} are both odd, we can choose

$$r_n = (-1)^n q_{n-1} + q_n, \; s_n = (-1)^{n+1} p_{n-1} - p_n.$$

Since

$$\gamma_n\left(\frac{p_n}{q_n} + z\right) = \frac{r_n}{q_n} - \frac{1}{q_n^2 z},$$

applying (25) to $\dfrac{p_n}{q_n} + z$ and γ_n leads to

(38)
$$\left|\theta\left(\frac{p_n}{q_n} + z\right)\right| = \left|\theta\left(\frac{r_n}{q_n} - \frac{1}{q_n^2 z}\right)\right| \frac{1}{\sqrt{q_n|z|}}$$

Since $Im\left(\dfrac{-1}{q_n^2 z}\right) = \dfrac{Im(z)}{q_n^2 |z|^2}$, we consider the two following cases.

FIRST CASE: $\dfrac{Im(z)}{q_n^2 |z|^2} \geq 1$; then (38) and (26) imply that

$$\left|\theta\left(\frac{p_n}{q_n} + z\right)\right| \sim \frac{1}{\sqrt{q_n|z|}}$$

so that

$$|C(a,b)| \sim \frac{Ca}{\sqrt{q_n(a + |b - \rho|)}}$$

(note that here and hereafter, $\sim$ means that the two quantities are equivalent, the constants in the equivalence being independent of n). Because of (31),

$$(a + |b - \rho|) \geq \frac{1}{q_n^{\tau_n}},$$

so that

$$|C(a,b)| \leq Ca^{\frac{1}{2\tau_n} + \frac{1}{2}}\left(1 + \frac{|b - \rho|}{a}\right)^{\frac{1}{2\tau_n} - \frac{1}{2}};$$

and because of (26) this upper bound becomes an equality if we choose $a = \frac{1}{q_n^{\tau_n}}$, $b = 0$. This proves (32) and (33) in this case, as well as their optimality.

SECOND CASE: $\dfrac{Im(z)}{q_n^2 |z|^2} \leq 1$; we separate this case into two subcases:

First Subcase: p_{n-1} and q_{n-1} are not both odd; then

(39)
$$\left|\theta\left(\frac{p_n}{q_n} + z\right)\right| \leq \frac{1}{\sqrt{q_n|z|}}\left(\frac{Im(z)}{q_n^2 |z|^2}\right)^{-1/2} = \sqrt{\frac{q_n|z|}{Im(z)}}$$

so that, since $|z| \geq 2|\rho - \frac{p_n}{q_n}|$,

$$(40) \qquad |C(a,b)| \leq 2\sqrt{aq_n(a + |b - \rho|)}.$$

Because of (31),

$$a + |b - \rho| \leq 6|\rho - \frac{p_{n-1}}{q_{n-1}}| \leq 6\left(\frac{1}{q_{n-1}}\right)^{\tau_{n-1}} \leq 6\left(\frac{1}{q_n}\right)^{\frac{\tau_{n-1}}{\tau_{n-1} - 1}};$$

thus

$$(41) \qquad \begin{aligned} |C(a,b)| &\leq Caq_n^{1/2}(1 + \frac{|b - \rho|}{a})^{1/2} \\ &\leq Ca\left(\frac{1}{a + |b - \rho|}\right)^{\frac{\tau_{n-1} - 1}{2\tau_{n-1}}}(1 + \frac{|b - \rho|}{a})^{1/2} \\ &\leq Ca^{\frac{1}{2} + \frac{1}{2\tau_{n-1}}}(1 + \frac{|b - \rho|}{a})^{\frac{1}{2\tau_{n-1}}}; \end{aligned}$$

which proves (34).

Second Subcase: p_{n-1} and q_{n-1} are both odd; then

$$r_n = (-1)^n q_{n-1} + q_n, \quad s_n = (-1)^{n+1} p_{n-1} - p_n.$$

We now want to estimate θ near the points p_n/q_n where p_n and q_n are both odd; we will deduce this estimate from (29).
We see that (38) becomes

$$|\theta(\frac{p_n}{q_n} + z)| = |\theta(\frac{(-1)^n q_{n-1}}{q_n} - \frac{1}{q_n^2 z} + 1)|\frac{1}{\sqrt{q_n|z|}}$$

let

$$g_n(z) = \frac{(-1)^n q_n q_{n-1} z - 1}{q_n^2 z}$$

From (28), we get

$$|\theta(\frac{p_n}{q_n} + z)| = \frac{1}{\sqrt{q_n|z||g_n(z)|}}|\theta(\frac{-1}{4g_n(z)}) - \theta(\frac{-1}{g_n(z)})|.$$

Remark that

$$Im(\frac{-1}{g_n(z)}) = \frac{Im(g_n(z))}{|g_n(z)|^2} = \frac{Im(z)}{q_n^2|z|^2|g_n(z)|^2}.$$

If $Im(\dfrac{-1}{g_n(z)}) \geq 1$, using (29),

$$|\theta(\dfrac{p_n}{q_n} + z)| \leq \dfrac{1}{\sqrt{q_n|z||g_n(z)|}} \exp(-\pi \dfrac{Im(z)}{q_n^2|z|^2|g_n(z)|^2})$$

$$\leq \dfrac{1}{\sqrt{q_n|z||g_n(z)|}} \left(\dfrac{q_n^2|z|^2|g_n(z)|^2}{Im(z)} \right)^{1/4}$$

$$\leq \left(\dfrac{1}{Im(z)} \right)^{1/4}$$

so that $|C(a,b)| \leq a^{3/4}$; hence (32) in that case.

Suppose now that $Im(\dfrac{-1}{g_n(z)}) \leq 1$. Then

$$|\theta(\dfrac{p_n}{q_n} + z)| \leq \dfrac{1}{\sqrt{q_n|z||g_n(z)|}} \left(Im(\dfrac{-1}{g_n(z)}) \right)^{-1/2}$$

$$\leq \dfrac{1}{\sqrt{q_n|z||g_n(z)|}} \dfrac{q_n|z||g_n(z)|}{\sqrt{Im(z)}}$$

$$\leq \sqrt{\dfrac{q_n|z||g_n(z)|}{Im(z)}}.$$

Because of (31), $|z| \leq \dfrac{6}{q_n q_{n-1}}$, so that $|g_n(z)| \leq \dfrac{7}{q_n^2|z|}$ and

$$|\theta(\dfrac{p_n}{q_n} + z)| \leq \sqrt{\dfrac{q_n|z|}{q_n^2|z|Im(z)}} \leq \dfrac{1}{\sqrt{q_n Im(z)}}.$$

Thus

$$|C(a,b)| \leq \dfrac{\sqrt{a}}{\sqrt{q_n}}$$

From (31), we have $(1/q_n)^{\tau_n} \leq (a + |b - \rho|)$ so that

$$|C(a,b)| \leq a^{\frac{1}{2} + \frac{1}{2\tau_n}} (1 + \dfrac{|b - \rho|}{a})^{1\frac{1}{2\tau_n}},$$

which proves (32) in this case.

6. The case when p_n and q_n are not both odd

Following the same procedure as in the previous section, we first determine $\gamma_n = \dfrac{ax + b}{cx + d}$ such that $\gamma_n(p_n/q_n) = 1$. We choose either $r_n = q_{n+1}$, $s_n = p_{n+1}$ or $r_n = -q_{n+1}$, $s_n = -p_{n+1}$ such that

$$p_n r_n - s_n q_n = 1$$

(which is possible because of (36)). Now γ_n is defined by the coefficients

$$\begin{cases} a = & q_n + r_n & b = & -p_n - s_n \\ c = & r_n & d = & -s_n. \end{cases}$$

One easily checks that (24) holds and thus γ_n belongs to the theta modular group;

$$\gamma_n(\frac{p_n}{q_n} + z) = \frac{1 + q_n(r_n + q_n)z}{1 + r_n q_n z} = 1 + f_n(z)$$

with

$$f_n(z) = \frac{q_n^2 z}{1 + r_n q_n z}.$$

Because of (25), it follows that

$$|\theta(\frac{p_n}{q_n} + z)| = |\theta(1 + f_n(z))| \frac{1}{r_n^{1/2}(z + \frac{p_n}{q_n} - \frac{p_{n+1}}{q_{n+1}})^{1/2}};$$

on the other hand (31) implies

$$|z| \geq 3|\rho - \frac{p_n}{q_n}| \geq \frac{3}{2}|\frac{p_n}{q_n} - \frac{p_{n+1}}{q_{n+1}}|,$$

so that

(42)
$$|\theta(\frac{p_n}{q_n} + z)| \leq \frac{|\theta(1 + f_n(z))|}{r_n^{1/2}|z|^{1/2}}$$

Remark that the condition $Im(\frac{-1}{f_n(z)}) \geq 1$ is equivalent to $Im(z) \geq q_n^2|z|^2$. Thus we now consider the two following cases.

First Case: $Im(z) \geq q_n^2|z|^2$. In this case, because of (29),

$$|\theta(\frac{p_n}{q_n} + z)| \leq C \exp(\frac{-\pi Im(z)}{q_n^2|z|^2}) \left(\frac{1}{|f_n(z)r_n z|}\right)^{1/2}$$

$$\leq C \exp(\frac{-\pi Im(z)}{q_n^2|z|^2}) \frac{|1 + r_n q_n z|^{1/2}}{(|r_n z|)^{1/2}|q_n^2 z|^{1/2}}.$$

Because of (30), $\frac{1}{|r_n q_n|} \leq (\frac{2}{q_n^{\tau_n}})$ and thus $|r_n q_n z| \geq 3/2$. Thus

$$|\theta(\frac{p_n}{q_n} + z)| \leq C \exp(\frac{-\pi Im(z)}{q_n^2|z|^2}) \frac{1}{|z|^{1/2}q_n^{1/2}}$$

$$\leq C \left(\frac{q_n^2|z|^2}{Im(z)}\right)^{\frac{1}{4}} \frac{1}{|z|^{1/2}q_n^{1/2}} = \frac{C}{Im(z)^{1/4}}$$

so that $|C(a,b)| \leq a^{3/4}$ which proves (35) in this case

Second Case: $Im(z) \leq q_n^2 |z|^2$. In this case, from (27), (28) and (42), we obtain

$$|\theta(\frac{p_n}{q_n} + z)| \leq \frac{1}{\sqrt{|f_n(z)r_n z|}} |\theta(\frac{-1}{4f_n(z)}) - \theta(\frac{-1}{f_n(z)})|$$

$$\leq C \frac{q_n |z|^{1/2}}{\sqrt{r_n |f_n(z)| Im(z)}}$$

$$\leq C \frac{|1 + r_n q_n z|^{1/2}}{\sqrt{r_n Im(z)}} \leq C \frac{|q_n z|^{1/2}}{\sqrt{Im(z)}}.$$

As above $|r_n q_n z| \geq 3/2$, so that

$$|C(a,b)| \leq C a^{1/2} \sqrt{q_n(a + |b - \rho|)}.$$

Because of (31), $a + |b - \rho| \leq 3(\frac{1}{q_n})^{\frac{\tau_{n-1}}{\tau_{n-1}-1}}$ so that

$$|C(a,b)| \leq C a^{\frac{1}{2} + \frac{1}{2\tau_{n-1}}} (1 + \frac{|b - \rho|}{a})^{\frac{1}{2\tau_{n-1}}}.$$

Thus (35) holds in this case and Proposition 3 is proved.

The fact that (32) and (35) cannot be improved in a cone

$$Im(z - \rho) \geq C Re(z - \rho)$$

yields a slightly more precise information than the fact that φ is not smoother than $\frac{1}{2} + \frac{1}{2\eta(\rho)}$ at ρ because it shows that φ has no chirp expansion at an irrational point ρ (see [9]); thus the only points where φ has a chirp expansion are the rationals of the form odd / odd.

The existence of these sets of smooth points is not only a consequence of the kind of lacunarity introduced by the frequencies n^2, but also of the very special coefficents that are chosen, which creates an exceptional behavior, as shown by the following remark: if the coefficients were multiplied by Independant Identically Distributed Gaussians or Rademacher series (± 1), Theorem 4 (chap 8) of [11] shows that the corresponding random function would be almost surely nowhere $C^{1/2}$. It is hard to have an idea about how specific coefficients modify the spectrum or even the regularity at a given point. Under this respect, the following example is interesting. Consider the function

$$\varphi_\tau(x) = \sum e^{i\pi n\tau} \frac{1}{n^2} e^{i\pi n^2 x}$$

where τ is a parameter. For $\tau = 0$ we recover φ (or rather its analytic part, but this does not change its regularity at any point). For a generic τ one might think that we introduce phases which behave quite randomly. This is not the case and actually φ_τ has the same regularity as φ at any point. Let us sketch the proof of this result.

Using the same wavelet as above, we obtain that the wavelet transform of φ_τ is (with the same notations)

$$C(a,b) = a \sum e^{i\pi n^2 z} e^{i\pi n \tau}.$$

Let $\theta(z,\tau) = \sum e^{i\pi n^2 z} e^{i\pi n \tau}$; the changes $n \to n + l$ and $\tau \to \tau + 2m$ show that

$$|\theta(z,\tau)| = |\theta(z, \tau + 2lz + 2m)|$$

If z is irrational, the values of $lz + 2m$ are dense; hence by continuity,

$$|\theta(z,\tau)| = |\theta(z,0)|$$

which, again by continuity also holds if z is rational. Thus the wavelet transforms of φ_τ and φ have the same modulus everywhere, so that these functions have everywhere the same regularity.

7. Logarithmic chirps of Riemann's function and selfsimilar functions

We will prove first the following theorem.

Theorem 3 *Riemann's function has at every quadratic irrational a logarithmic chirp of regularity 1/2.*

Proof: If ω is a quadratic irrational, there exists $\gamma(z)$ in the theta modular group,

$$\gamma(z) = \frac{rz + s}{pz + q},$$

such that ω is invariant by γ (see [1]). Using (25), we see that

$$\theta(\omega + z) = \theta(\omega + \frac{r - p\omega}{q + p\omega} z + O(z^2)) \frac{e^{im\pi/4}}{\sqrt{q\omega + p}} (1 + O(z))$$

since

$$|\theta'(z)| \le \frac{C}{(Im\ z)^{3/2}},$$

we have

$$\theta(\omega + z) = A\theta(\omega + Bz) + O(\frac{z^2}{(Im\ z)^{3/2}})$$

where B is a real number different from 1 or -1 (because $B = \pm 1$ would imply that ω is a rational). Iterating this relation, we can suppose that A is real. Thus, if $z = b - \omega + ia$,

$$\begin{aligned} C(a, \omega + b) &= a\theta(\omega + z) \\[2ex] &= Aa\theta(\omega + Bz) + O(a\frac{z^2}{a^{3/2}}) \\[2ex] &= \frac{A}{B}C(BA, \omega + Bb) + a^{-1/2}O(a^2 + b^2) \end{aligned}$$

These last two terms are $O(a^{1/2})$ because φ is $C^{1/2}$ on $\mathbf{R}$, so that the rest is simultaneously $O(a^{1/2})$ and $O(a^{-1/2}(a^2 + b^2))$. Hence by a weighted geometric average, it is $O(a^\alpha(1 + \frac{|b|}{a})^{\alpha'}$, where we can choose $\alpha' < \alpha$ and α arbitrarily close to 1. Since Proposition 3 implies that φ is $C^{3/4,-1/4}$ at quadratic irrationals, Theorem 2 applies, and the conclusion of Theorem 3 follows.

Let us now consider the following functions

$$(43) \qquad f(x) = \sum_1^\infty \lambda^{-\alpha j} g(\lambda^j x)$$

where g has some Hölder continuity, vanishes at the origin and has at most polynomial growth. This situation includes of course Weierstrass functions (where $g(x) = \sin 2\pi x$), but also many examples which appear in the literature about "fractal functions". They are particular examples of the "selfsimilar functions introduced and studied in [8], and the reader will easily check that the following study immediately extends to this setting.

Proposition 4 *The function (43) has a logarithmic chirp at the origin of order* (α, λ) *with the same regularity as g. The Fourier coefficients of the associated function G are given by the Mellin transform Mg of g as follows:*

$$c_n = Mg(-\alpha - \frac{2i\pi n}{\log \lambda})$$

Proof: Note that f is almost written as a logarithmic chirp, so that by inspection, one gets

$$G(x) = \sum_{j=-\infty}^{+\infty} \lambda^{-\alpha j} e^{-\alpha x} g(\lambda^{\alpha j} e^x)$$

Since G is obtained by a periodization of $e^{-\alpha x} g(e^x)$, its Fourier coefficients are

$$c_n = \int_R e^{-\alpha x} g(e^x) e^{-2i\pi n x \log \lambda} dx = \int_0^\infty u^{-\alpha} g(u) u^{-2i\pi n \log \lambda} \frac{du}{u}$$

Thus for Weierstrass functions

$$w_\alpha(x) = \sum_1^\infty 2^{-\alpha j} \sin 2\pi 2^j x$$

and we obtain in this case

$$(44) \qquad c_n = \Gamma(-\alpha - \frac{2i\pi n}{\log 2}) \sin \frac{\pi}{2}(-\alpha - \frac{2i\pi n}{\log 2})$$

One easily checks that, if g is periodic of period 1 (which is the case for Weierstrass functions), f has logarithmic chirps at the rational points whose Fourier coefficients are given by the Mellin transform of a finite linear combination of g and some of its dilations and translations.

References

1. J.J.Duistermaat, *Selfsimilarity of Riemann's non-differentiable function.* Nieuw Archief voor Wiskunde 9 (1991) 303-337.

2. J. Gerver, *The differentiability of the Riemann function at certain rational multiples of π.* Am. J. Math. 92 (1970), 33-55.

3. G.H.Hardy, *Weierstrass's non-differentiable function* . Trans. AMS 17 (1916), 301-325 .

4. G.H.Hardy and J.E.Littlewood, *Some problems of Diophantine approximation.* Acta Mathematica 37 (1914), 193-239.

5. M. Holschneider & P. Tchamitchian, *Pointwise analysis of Riemann's "non differentiable" function.* Inventiones Mathematicae 105 (1991), 157-176.

6. M. Holschneider, *On the wavelet transformation of fractal objects* Journal of Statistical Physics 5/6 (1988), 963-993.

7. S.Jaffard, *Pointwise smoothness, two-microlocalization and wavelet coefficients* Publicacions Matematiques 35 (1991), p.155-168.

8. S.Jaffard, *Multifractal Formalism for functions Part I: Results valid for all functions and Part II: Selfsimilar functions* Preprint (1993).

9. S.Jaffard S, *The spectrum of singularities of Riemann's function* Preprint (1994).

10. S.Jaffard and Y.Meyer Y, *Pointwise behavior of functions* Preprint (1993).

11. J.P.Kahane, *Some Random Series of Functions* Cambridge Studies in Advanced Mathematics 5 (1993).

12. G.Valiron, *Théorie des fonctions* Masson (1942).

C.E.R.M.A., ECOLE NATIONALE DES PONTS ET CHAUSSÉES, LA COURTINE, 93167 NOISY-LE-GRAND, FRANCE
AND C.M.L.A., E.N.S. CACHAN, 61 AV. DU PRÉSIDENT WILSON, 94235 CACHAN CEDEX, FRANCE
E-mail address: sj@cerma.enpc.fr

Contemporary Mathematics
Volume **189**, 1995

On the spectra of the higher dimensional Maxwell operators on nonsmooth domains

BJÖRN JAWERTH AND MARIUS MITREA

Dedicated to Mischa Cotlar, on his 80th birthday

ABSTRACT. In this paper we discuss the spectral properties of the boundary singular integral operators arising in connection with the basic boundary value problems of the electromagnetic scattering theory by Lipschitz obstacles in $\mathbf{R}^m$.

1. Introduction and statement of results.

The time-independent form of the (reduced) Maxwell equations describe the wave propagation of the scattered electromagnetic fields E, H in a homogeneous, isotropic, and perfectly conducting medium, occupying a domain Ω in $\mathbf{R}^3$. The boundary value problem naturally associated with these equations reads

$$(\mathcal{M}_3) \begin{cases} curl\, E - ikH = 0 \ on\ \Omega, \\ curl\, H + ikE = 0 \ on\ \Omega, \\ n \times E|_{\partial\Omega} = A. \end{cases}$$

Here $k \in \mathbf{C}$ depends only on the electric and magnetic characteristics of the medium, n is the outward unit normal to $\partial\Omega$ and the Cauchy datum A is a given tangential vector field on $\partial\Omega$ (see [**3**]). For simplicity, throughout this paper we shall assume that k satisfies $Im\ k > |Re\ k|$.

1991 *Mathematics Subject Classification.* 42B20, 47A10; Secondary 45F15, 35J55.

Key words and phrases. Maxwell equations, Lipschitz domains, spectral theory.

The first author was supported in part by ONR Grant #N0001-90-J-1343 and ARPA Grant #749620-93-1-0083

The second author was supported in part by ONR Grant #N0001-90-J-1343

For a bounded domain $\Omega \subset \mathrm{R}^3$ with smooth boundary, i. e., $\partial\Omega \in C^2$, the Maxwell boundary value problem $(\mathcal{M}_3)$ was solved in the early 1950's by using layer potential techniques and relying on compactness arguments ([**15**], [**12**], [**13**], [**1**]). Also, Weyl [**16**] indicated how to extend this work to higher dimensions.

More recently, the smoothness assumptions have been relaxed to include C^1 domains in the three dimensional Euclidean space ([**11**]) and even domains with Lipschitz boundaries in R^3 ([**10**]). Furthermore, a comprehensive approach to the C^1 and Lipschitz case in the higher dimensional setting has been developed in [**8**].

For general Lipschitz boundaries, the main difficulty is that compactness type arguments cannot be used directly. This difficulty was overcome in [**10**] by establishing the relevant so-called Rellich identity for vector fields satisfying the Maxwell system in R^3. Subsequently, these have been generalized to include differential forms of arbitrary degree in [**8**]. We also refer to this latter paper for more on the history of the subject.

The research reported here is a natural continuation of the work in [**11**], [**10**] and [**8**]. To explain our main results, we first need to introduce some notation.

Let us note that $(\mathcal{M}_3)$ can be regarded as a special case $(m = 3)$ of the following general Maxwell boundary value problem in R^m which we shall consider. Given a Lipschitz domain Ω in R^m, $m \geq 3$, $0 \leq l \leq m$, determine a $(l + 1)$-form E and a l-form H, smooth in Ω and such that

$$(\mathcal{M}_m) \begin{cases} \delta E - ikH = 0 \text{ on } \Omega, \\ dH + ikE = 0 \text{ on } \Omega, \\ E^*, \, H^* \in L^p(\partial\Omega), \\ n \vee E = A. \end{cases}$$

Here d and δ are the exterior derivative operator and its formal transpose, respectively, and $(\cdot)^*$ denotes the usual nontangential maximal operator.

As explained in [**8**], the solvability of $(\mathcal{M}_m)$ and other related problems rest on the invertibility of $\pm\frac{1}{2} + M_l$ on $\Lambda^l L_t^{p,\delta}(\partial\Omega)$ and $\Lambda^l L_t^p(\partial\Omega)$, respectively. Here $\Lambda^l L_t^p(\partial\Omega)$ is the linear space of tangential l-forms with coefficients in $L^p(\partial\Omega)$, $\Lambda^l L_t^{p,\delta}(\partial\Omega)$ is the Sobolev space of l-forms A in $\Lambda^l L_t^p(\partial\Omega)$ for which $\delta_\partial A$ belongs to $\Lambda^{l-1} L_t^p(\partial\Omega)$ (δ_∂ is essentially the transpose of the restriction of d to $\partial\Omega$; precise definitions will be given later). Also, the Maxwell boundary singular integral operator M_l is given by the principal-value integral operator acting on a l-form A on $\partial\Omega$ by

$$M_l A(X) := \lim_{\epsilon \to 0} \left(n(X) \vee d \int_{\substack{Y \in \partial\Omega \\ |X-Y| \geq \epsilon}} \Phi(X - Y)A(Y) \right), \quad X \in \partial\Omega.$$

Here Φ is a certain radial fundamental solution for the Helmholtz operator $\triangle + k^2$ in R^m (see §3).

For X a Banach space and $T : X \to X$ a bounded, linear operator, let $\sigma(T; X)$, $\sigma_p(T; X) \subset \mathrm{C}$ stand for the spectrum and the point spectrum of T, respectively.

Also, let $\sigma_\mu(T; X)$ stand for the collection of all complex λ so that $\lambda - T$ does not have a closed range or a finite dimensional kernel. We also set $\rho(T; X) := \mathbf{C} \setminus \sigma(T; X)$, $\rho_\mu(T; X) := \mathbf{C} \setminus \sigma_\mu(T; X)$, $r(T; X) := max\{|\lambda|; \lambda \in \sigma(T; X)\}$. For an arbitrary subset G of $\mathbf{C}$, we denote by $[G]$ the complement of the unbounded component of $\mathbf{C} \setminus G$. Finally, let $D(0; \frac{1}{2}) \subset \mathbf{C}$ denote the *open* disk of radius $\frac{1}{2}$ centered at the origin of $\mathbf{C}$. Note that for each $\lambda \in \sigma_\mu$, the operator $\lambda - T$ is semi-Fredholm. Consequently, using the invariance of the index to homotopic transformations within the class of all semi-Fredholm operators, we have

$$(1.1) \qquad \sigma(T; X) \subseteq [\sigma_p(T; X) \cup \sigma_\mu(T; X)].$$

Also, $\lambda \in \sigma_\mu(T; X)$ if and only if $\lambda - T$ is bounded from below on X modulo compact operators.

As we shall mostly deal in the sequel with the spectra of the operators M_l on the spaces $\Lambda^l L_t^{p,\delta}(\partial\Omega)$ and $\Lambda^l L_t^p(\partial\Omega)$, in order to simplify notation we shall occasionally write $\sigma(M_l; L_t^{p,\delta}(\partial\Omega))$ and $\sigma(M_l; L_t^p(\partial\Omega))$ in place of $\sigma(M_l; \Lambda^l L_t^{p,\delta}(\partial\Omega))$ and $\sigma(M_l; \Lambda^l L_t^p(\partial\Omega))$, respectively. Also, we shall employ similar conventions for σ_p, σ_μ, ρ, and ρ_μ.

With the above notation and conventions, the main result of [**8**] reads as follows.

THEOREM 1.1. *Let Ω be a bounded Lipschitz domain in R^m. For $0 \le l \le m$, we have that $\sigma_p(M_l; L_t^{2,\delta}(\partial\Omega)) \subset D(0; 1/2)$ and*

$$\sigma(M_l; L_t^{2,\delta}(\partial\Omega)) \cap R \subset \left(-\frac{1}{2}, \frac{1}{2}\right).$$

In this paper we shall be concerned with the analysis of the fine properties of the spectra of the operators M_l, $l = 0, 1, \cdots, m$, on Lipschitz and C^1 domains. Specifically, we will prove the following.

THEOREM 1.2. *Let Ω be a bounded Lipschitz domain in R^m.*
(1) For each $0 \le l \le m-1$, we have

$$\sigma_\mu(M_l; L_t^{2,\delta}(\partial\Omega)) = -\sigma_\mu(M_{m-l-1}; L_t^{2,\delta}(\partial\Omega)).$$

(2) For each $1 \le l \le m-1$, we have

$$\sigma_\mu(M_l; L_t^{2,\delta}(\partial\Omega)) \subseteq \sigma_\mu(M_{l-1}; L_t^{2,\delta}(\partial\Omega)) \cup \sigma_\mu(M_{l+1}; L_t^{2,\delta}(\partial\Omega)).$$

(3) For each $0 \le l \le m-1$, and each $1 < p, q < \infty$ with $\frac{1}{p} + \frac{1}{q} = 1$, we have

$$\sigma(M_l; L_t^p(\partial\Omega)) = -\sigma(M_{m-l-1}; L_t^q(\partial\Omega)).$$

Similar results with σ_p or σ_μ in place of σ are also valid.
(4) For each $0 \le l \le m$, $1 < p < \infty$, we have that $\sigma_p(M_l; L_t^p(\partial\Omega))$ is included in the set of all complex λ's for which the operator $\lambda - M_l$ does not have a dense range on

$\Lambda^l L_t^{p,\delta}(\partial\Omega)$. *In particular,*

$$\sigma_p(M_l; L_t^p(\partial\Omega)) \subseteq \sigma(M_l; L_t^{p,\delta}(\partial\Omega)).$$

(5) For each $0 \le l \le m$ and each $1 < p < \infty$ we have that

$$\sigma_p(M_l; L_t^p(\partial\Omega)) \subseteq \sigma_p(M_l; L_t^{p,\delta}(\partial\Omega)) \cup (-\sigma(M_{m-l}; L_t^{p,\delta}(\partial\Omega))).$$

(6) For each $0 \le l \le m-1$ and each $1 < p < \infty$ we have that

$$\sigma_p(M_l; L_t^{p,\delta}(\partial\Omega)) = -\sigma_p(M_{m-l-1}; L_t^{p,\delta}(\partial\Omega)).$$

This theorem has several consequences of independent interest, among which we single out the following corollaries.

COROLLARY 1.3. *Let Ω be a bounded Lipschitz domain in R^m, and assume that $A \in \Lambda^l L_t^2(\partial\Omega)$ has the property that $(\frac{1}{2} + M_l)A \in \Lambda^l L_t^{2,\delta}(\partial\Omega)$. Then actually A belongs to $\Lambda^l L_t^{2,\delta}(\partial\Omega)$. In particular, the operator $\frac{1}{2} + M_l$ is injective and has dense range on $\Lambda^l L_t^2(\partial\Omega)$.*

Analogous results hold for the operator $-\frac{1}{2} + M_l$.

COROLLARY 1.4. *Let Ω be a bounded Lipschitz domain in R^m, $0 \le l \le m-1$. Then*

$$r(M_l; L_t^{2,\delta}(\partial\Omega)) = r(M_{m-l-1}; L_t^{2,\delta}(\partial\Omega)).$$

COROLLARY 1.5. *Assume that Ω is a bounded domain with C^1 boundary in R^m and that $1 < p < \infty$. Then, for each $0 \le l \le m$, we have that*

$$\sigma(M_l; L_t^{p,\delta}(\partial\Omega)) = \sigma(M_l; L_t^p(\partial\Omega)) \subset D(0; 1/2).$$

In particular, the spectral radius of $M_l : \Lambda^l L_t^p(\partial\Omega) \to \Lambda^l L_t^p(\partial\Omega)$ and the spectral radius of $M_l : \Lambda^l L_t^{p,\delta}(\partial\Omega) \to \Lambda^l L_t^{p,\delta}(\partial\Omega)$ are $< \frac{1}{2}$. Consequently,

$$\left(\pm\frac{1}{2} + M_l\right)^{-1} = \pm 2 \sum_{j=0}^{\infty} (\mp 2M_l)^j$$

with convergence in the strong operator norm.

COROLLARY 1.6. *Assume that Ω is a bounded C^1 domain in R^m and that $1 < p < \infty$. Then, for each $A \in \Lambda^l L_t^p(\partial\Omega)$, the boundary value problem*

$$\begin{cases} E \in \Lambda^{l+1} C^\infty(\Omega), \\ (\triangle + k^2)E = 0 \text{ on } \Omega, \\ dE = 0 \text{ on } \Omega, \\ E^* \in L^p(\partial\Omega), \\ n \vee E|_{\partial\Omega} = A, \end{cases}$$

has a unique solution. Moreover, $(\delta E)^* \in L^p(\partial\Omega)$ if and only if $A \in \Lambda^l L_t^{p,\delta}(\partial\Omega)$.

Similar statements for the exterior or the dual problem are valid as well.

COROLLARY 1.7. *(a) Let Ω be a bounded Lipschitz domain in R^3. Then $\partial\sigma(M_1; L_t^{2,\delta}(\partial\Omega))$ is symmetric with respect to the origin of C.*

(b) Assume that Ω is either a bounded convex domain in R^3 or a polyhedral domain in R^3. Then

$$\sigma(M_1; L_t^{2,\delta}(\partial\Omega)) \subset D(0; 1/2) \quad and \quad \sigma_p(M_1; L_t^2(\partial\Omega)) \subset D(0; 1/2).$$

In particular, the spectral radius of $M_1 : \Lambda^1 L_t^{2,\delta}(\partial\Omega) \to \Lambda^1 L_t^{2,\delta}(\partial\Omega)$ is $< \frac{1}{2}$, so that

$$\left(\pm\frac{1}{2} + M_1\right)^{-1} = \pm 2 \sum_{j=0}^{\infty} (\mp 2M_1)^j$$

with convergence in the strong operator norm.

Before we start the major part of this work, we introduce some notation. Recall that $\Omega \subset R^m$ is called a Lipschitz (or C^1) domain if its boundary is given locally by the graph of a Lipschitz function (or C^1 function, respectively). We set $\Omega_+ := \Omega$ and $\Omega_- := R^m \setminus \bar{\Omega}$. For a differential form E defined in $\Omega_\pm$, *the nontangential maximal function E^* is given by $E^*(X) := \sup_{Y \in \Gamma_\pm(X)} |E(Y)|$,* where $|\cdot|$ refers to the Euclidean norm. Here $\Gamma_\pm(X)$ denote the interiors of the two components (in Ω_+ and in Ω_-) of a regular family of circular, doubly truncated cones $\{\Gamma(X); X \in \partial\Omega\}$, with vertex at X, as defined in e.g. [**14**]. Also, the boundary traces of differential forms defined in $\Omega_\pm$ are assumed to be taken as nontangential limits almost everywhere with respect to surface measure on the boundary.

In the sequel, it will be important to approximate the Lipschitz domain Ω, from the interior or from the exterior, with sequences of smooth domains $\{\Omega_j\}_j$ as in [**14**]. We shall denote such approximating sequences by $\Omega_j \uparrow \Omega$ and $\Omega_j \downarrow \Omega$, respectively.

2. The algebra of differential forms.

Let $\mathcal{A}$ be an algebra of complex valued functions defined in a given subset of R^m. We shall let $\Lambda^l \mathcal{A}$ denote the vector space of all differential l-forms, $0 \le l \le m$, with coefficients in $\mathcal{A}$, i.e. $E \in \Lambda^l \mathcal{A}$ if and only if

$$E = \sum_{|I|=l}' E_I \, dx^I,$$

with $E_I \in \mathcal{A}$ for all multi-indices $I \in \{1, 2, ..., m\}^l$. We shall adopt the convention that $\sum'$ indicates that the sum is performed only over strictly increasing multi-indices I. Furthermore, dx^I stands for $dx_{i_1} \wedge dx_{i_2} \wedge \ldots \wedge dx_{i_l}$ if $I = (i_1, i_2, ..., i_l)$. We shall also allow multi-indices I which are not necessarily increasing in which case we assume that the coefficients $\{E_I\}_I$ are skew-symmetric in I. More specifically, if ϵ_J^I is the sign of the permutation taking I into J, we shall assume that $E_I = \epsilon_J^I E_J$, for any I, J.

For $E, F \in \Lambda^l \mathcal{A}$ we introduce $\langle E, F \rangle := \sum_{|I|=l}' E_I F_I$, the conjugation $\bar{E} :=$ $\sum_I' \bar{E}_I dx^I$, and $|E| := \langle E, \bar{E} \rangle^{1/2}$. We recall that the Hodge $*$-operator, $* :$ $\Lambda^l \mathcal{A} \to \Lambda^{m-l} \mathcal{A}$, is defined by

$$*E := \sum_{|I|=l}' \epsilon_{\{1,...,m\}}^{II^c} E_I dx^{I^c},$$

where $I^c := \{1, 2, ..., m\} \setminus I$. The Hodge $*$-operator is an isomorphism since $** = (-1)^{l(m+1)}$ on $\Lambda^l \mathcal{A}$.

In the sequel, we shall identify $\mathcal{A}^m$ with $\Lambda^1 \mathcal{A}$, i.e. we shall identify the vector field $\alpha = (\alpha_1, \alpha_2, ..., \alpha_m)$ with the 1-form $\alpha = \alpha_1 dx_1 + ... + \alpha_m dx_m$. In this fashion, the exterior product $\alpha \wedge E$ is well-defined whenever $\alpha \in \mathcal{A}^m$ and $E \in \Lambda^l \mathcal{A}$. Also, $\alpha \vee E$ will denote the interior product of α and E.

The basic elementary properties of these operators are collected in the following lemma.

LEMMA 2.1. *For* $\alpha, \beta \in \mathcal{A}^m$ *and* $E \in \Lambda^l \mathcal{A}$, $F \in \Lambda^{l-1} \mathcal{A}$, *the following are true:*
  i) $\alpha \wedge (\alpha \wedge E) = 0$, $\alpha \vee (\alpha \vee E) = 0$, $*(\alpha \wedge E) = (-1)^l \alpha \vee *E$;
  ii) $\alpha \wedge (\beta \vee E) + \beta \vee (\alpha \wedge E) = \langle \alpha, \beta \rangle E$;
  iii) $\langle \alpha \vee E, F \rangle = \langle E, \alpha \wedge F \rangle$.

We recall that the classical exterior derivative operator $d := \sum_{i=1}^m \partial_i dx_i$, acts on forms $E \in \Lambda^l \mathcal{A}$, $E = \sum_I' E_I dx^I$, by

$$dE = \sum_I' \sum_i \partial_i E_I dx_i \wedge dx^I.$$

Its formal transpose δ is given by

$$\delta E = - \sum_{|J|=l-1}' \sum_i \partial_i E_{iJ} dx^J.$$

Recall that $d^2 = 0$, $\delta^2 = 0$ and $-(d\delta + \delta d) = \triangle$, the Laplacian in R^m. Also, $\delta = (-1)^{m(l+1)+1} * d*$ and $*\delta = (-1)^l d*$, $\delta* = (-1)^{l+1} * d$ on $\Lambda^l \mathcal{A}$.

3. The boundary behavior of the single layer acoustic potential operator.

For $k \in \mathrm{C}$ as in §1 and $m \geq 3$, let us consider the function Φ defined for $X \in \mathrm{R}^m \setminus \{0\}$ by

$$\Phi(X) := -\frac{1}{\omega_m} \frac{(-ik)^{m-2}}{(m-2)!} \int_1^\infty (t^2 - 1)^{\frac{m-3}{2}} e^{ik|X|t} dt,$$

where ω_m denotes the area of the unit sphere in R^m. It is not difficult to check that Φ is an elementary solution of the Helmholtz operator $\triangle + k^2$ in R^m.

Let Ω be a bounded Lipschitz domain in R^m. We recall that the single layer acoustic potential operator is defined for a l-form $A \in \Lambda^l L^p(\partial\Omega)$ by

$$\mathcal{S}A(X) := \int_{\partial\Omega} \Phi(X - Y)A(Y)\,d\sigma(Y), \quad X \in \mathrm{R}^m \setminus \partial\Omega.$$

From [2] and classical arguments, we obtain the following.

LEMMA 3.1. *Let Ω be a bounded Lipschitz domain in R^m, and let A be a l-form on $\partial\Omega$, $0 \le l \le m$, with coefficients in $L^p(\partial\Omega)$, for some fixed $1 < p < \infty$. Then, at almost every point on the boundary,*

$$\lim_{\substack{X \to P \\ X \in \Gamma_+(P)}} \mathcal{S}A(X) = \lim_{\substack{X \to P \\ X \in \Gamma_-(P)}} \mathcal{S}A(X) =: SA(P), \ \ P \in \partial\Omega,$$

$$\lim_{\substack{X \to P \\ X \in \Gamma_\pm(P)}} d\mathcal{S}A(X) = \mp\frac{1}{2}(n \wedge A)(P) + p.v. \left(d\int_{\partial\Omega} \Phi(P - Y)A(Y) \right), \ \ P \in \partial\Omega,$$

$$\lim_{\substack{X \to P \\ X \in \Gamma_\pm(P)}} \delta\mathcal{S}A(X) = \pm\frac{1}{2}(n \vee A)(P) + p.v. \left(\delta\int_{\partial\Omega} \Phi(P - Y)A(Y) \right), \ \ P \in \partial\Omega.$$

Moreover,

$$\|(\mathcal{S}A)^*\|_{L^p(\partial\Omega)} + \|(d\mathcal{S}A)^*\|_{L^p(\partial\Omega)} + \|(\delta\mathcal{S}A)^*\|_{L^p(\partial\Omega)} \le C\|A\|_{L^p(\partial\Omega)}.$$

Next, we describe a Green type integral representation formula for differential forms with metaharmonic coefficients.

LEMMA 3.2. *Let $E \in \Lambda^l C^\infty(\Omega_\pm)$ (in the case in which E is defined in Ω_-, E is also assumed to decay at infinity) be such that $(\triangle + k^2)E = 0$ in $\Omega_\pm$ and, for some $1 < p < \infty$, $E^*, (dE)^*, (\delta E)^* \in L^p(\partial\Omega)$. Then $E, dE, \delta E$ have nontangential boundary traces in $L^p(\partial\Omega)$ and*

$$(3.2) \qquad \pm E = -dS(n \vee E) + \delta S(n \wedge E) + S(n \wedge \delta E) - S(n \vee dE)$$

in $\Omega_\pm$.

This follows from the fact that the identity is valid in the smooth case plus a familiar limiting argument, cf. [8].

4. Sobolev spaces of differential forms on the boundary.

Let Ω be a bounded Lipschitz domain in $\mathbf{R}^m$. A differential form E defined a.e. on $\partial\Omega$ is called *tangential* if $n \vee E = 0$ a.e. on $\partial\Omega$, and *normal* if $n \wedge E = 0$ a.e. on $\partial\Omega$. *The tangential component* of E, E_t, is given by $E_t := n \vee (n \wedge E)$, whereas *the normal component* is given by $E_n := n \wedge (n \vee E)$. Note that $E = E_t + E_n$ and $\langle E_t, E_n \rangle = 0$, so that $|E|^2 = |E_t|^2 + |E_n|^2$.

For a tangential differential form $A \in \Lambda^l L^1(\partial\Omega)$ we shall write

$$\delta_\partial A \in \Lambda^{l-1} L^1(\partial\Omega)$$

if there exists a $(l-1)$-form $\delta_\partial A$ with coefficients in $L^1(\partial\Omega)$ such that

$$\int_{\partial\Omega} \langle d\psi, A \rangle = \int_{\partial\Omega} \langle \psi, \delta_\partial A \rangle,$$

for all $\psi \in \Lambda^{l-1} C^\infty(\mathbf{R}^m)$. Similarly, for a normal differential form $A \in \Lambda^l L^1(\partial\Omega)$ we shall write that $d_\partial A \in \Lambda^{l+1} L^1(\partial\Omega)$ if there exists a $(l+1)$-form $d_\partial A$ with coefficients in $L^1(\partial\Omega)$ such that

$$\int_{\partial\Omega} \langle \delta\psi, A \rangle = \int_{\partial\Omega} \langle \psi, d_\partial A \rangle,$$

for all $\psi \in \Lambda^{l+1} C^\infty(\mathbf{R}^m)$.

We record the following lemmas for later use.

LEMMA 4.1. *Let Ω be a bounded Lipschitz domain in $\mathbf{R}^m$. If $E \in \Lambda^l C^\infty(\Omega)$ is such that $E^*, (\delta E)^* \in L^p(\partial\Omega)$, for some fixed $1 < p < \infty$, and such that $E, \delta E$ have nontangential boundary traces on $\partial\Omega$, then*

$$(4.3) \qquad\qquad \delta_\partial (n \vee E) = -n \vee \delta E.$$

Analogously, if $E^*, (dE)^* \in L^p(\partial\Omega)$ and E, dE have nontangential boundary traces on $\partial\Omega$, then

$$(4.4) \qquad\qquad d_\partial (n \wedge E) = -n \wedge dE.$$

Similar results are valid in $\mathbf{R}^m \setminus \bar{\Omega}$, too.

LEMMA 4.2. *For any $A \in \Lambda^l L_t^{p,\delta}(\partial\Omega)$ and any $B \in \Lambda^l L_n^{p,d}(\partial\Omega)$, we have*

$$(4.5) \qquad\qquad \delta SA = S(\delta_\partial A), \quad dSB = S(d_\partial B).$$

Let $\Lambda^l L_t^p(\partial\Omega)$ denote the vector space of tangential l-forms with coefficients from $L^p(\partial\Omega)$, and let $\Lambda^l L_n^p(\partial\Omega)$ stand for the vector space of normal l-forms with coefficients from $L^p(\partial\Omega)$. We set

$$\Lambda^l L_t^{p,\delta}(\partial\Omega) := \{A \in \Lambda^l L_t^p(\partial\Omega) \, ; \, \delta_\partial A \in \Lambda^{l-1} L_t^p(\partial\Omega)\},$$

and

$$\Lambda^l L_n^{p,d}(\partial\Omega) := \{A \in \Lambda^l L_n^p(\partial\Omega) \, ; \, d_\partial A \in \Lambda^{l+1} L_n^p(\partial\Omega)\}.$$

For each $1 < p < \infty$ these become Banach spaces when endowed with the norms

$$\|A\|_{\Lambda^l L_t^{p,\delta}(\partial\Omega)} := \|A\|_{\Lambda^l L^p(\partial\Omega)} + \|\delta_\partial A\|_{\Lambda^{l-1} L^p(\partial\Omega)},$$

and

$$\|A\|_{\Lambda^l L_n^{p,d}(\partial\Omega)} := \|A\|_{\Lambda^l L^p(\partial\Omega)} + \|d_\partial A\|_{\Lambda^{l+1} L^p(\partial\Omega)},$$

respectively. Clearly, the Hodge $*$-operator is an isomorphism between $\Lambda^l L_t^p(\partial\Omega)$ and $\Lambda^{m-l} L_n^p(\partial\Omega)$, and between $\Lambda^l L_t^{p,\delta}(\partial\Omega)$ and $\Lambda^{m-l} L_n^{p,d}(\partial\Omega)$.

Next, we define two principal-value integral operators acting on a l−form A on $\partial\Omega$ by

$$M_l A(X) := \lim_{\epsilon \to 0} \left(n(X) \vee d \int_{\substack{Y \in \partial\Omega \\ |X-Y| \geq \epsilon}} \Phi(X-Y) A(Y) \right), \quad X \in \partial\Omega,$$

$$N_l A(X) := \lim_{\epsilon \to 0} \left(n(X) \wedge \delta \int_{\substack{Y \in \partial\Omega \\ |X-Y| \geq \epsilon}} \Phi(X-Y) A(Y) \right), \quad X \in \partial\Omega.$$

LEMMA 4.3. *For each $1 < p < \infty$ and each $0 \leq l \leq m$, the operator M_l is well-defined and bounded on $\Lambda^l L_t^p(\partial\Omega)$, and on $\Lambda^l L_t^{p,\delta}(\partial\Omega)$. Similarly, the operator N_l is well-defined and bounded on $\Lambda^l L_n^p(\partial\Omega)$, and on $\Lambda^l L_n^{p,d}(\partial\Omega)$.*

Proof.
Let us first note that it actually suffices to consider only the case of the operator M_l. This is a consequence of the easily verified identity

$$(4.6) \qquad\qquad M_{m-l} * = - * N_l,$$

The boundedness on the space $\Lambda^l L_t^p(\partial\Omega)$ follows by well-known techniques from the fundamental results of [**2**]. As for the space $\Lambda^l L_t^{p,\delta}(\partial\Omega)$, we observe that for $A \in \Lambda^l L_t^{p,\delta}(\partial\Omega)$, a combination of Lemma 4.2 and (4.1) yields

$$(4.7) \quad \delta_\partial M_l A = -k^2 n \vee SA + n \vee dS(\delta_\partial A) = -k^2 n \vee SA + M_{l-1}(\delta_\partial A).$$

With this at hand, the conclusion follows once again from [**2**]. $\qquad\square$

Later on, we shall also find it useful to have the following density lemma (cf. [**10**] for a proof in the case $m = 3$, $l = 1$).

LEMMA 4.4. *For any $0 \leq l \leq m$ and $1 < p < \infty$, the space $\Lambda^l L_t^{p,\delta}(\partial\Omega)$ is densely included into $\Lambda^l L_t^p(\partial\Omega)$.*

5. Rellich identities and boundary estimates for differential forms.

A crucial ingredient for obtaining estimates concerning the operators M_l is the following family of Rellich type identities from [**8**].

LEMMA 5.1. *For each $0 \leq l \leq m$, each differential form $E \in \Lambda^l C^\infty(\bar{\Omega})$ and each vector field $\Theta = \{\Theta_j\}_j$ in R^m, with smooth, real-valued components, we have*

$$Re \iint_\Omega \langle \delta E, \Theta \vee \bar{E} \rangle + \langle dE, \Theta \wedge \bar{E} \rangle + \mathcal{O}(|E|^2)$$

$$= \int_{\partial\Omega} \frac{1}{2}|E|^2 \langle \Theta, n \rangle - Re \int_{\partial\Omega} \langle \Theta \vee \bar{E}, n \vee E \rangle.$$

$$= \int_{\partial\Omega} \frac{1}{2}|E|^2 \langle \Theta, n \rangle + Re \int_{\partial\Omega} \langle \Theta \wedge \bar{E}, n \wedge E \rangle.$$

Proof.
Let $\{e_j\}_{j=1}^m$ be the standard basis of R^m. If $E \in \Lambda^l C^\infty(\Omega)$ and $\Theta = \{\Theta_j\}_j$ is a vector field in R^m (also identified with the $1-$form $\Theta = \sum_j \Theta_j dx_j$) with smooth, real valued-components, then by direct calculation

$$div \left\{ \frac{1}{2}|E|^2 \Theta - \left(Re \sideset{}{'}\sum_I \sum_i \bar{E}_{iI} E_{jI} \Theta_i \right)_{1 \leq j \leq m} \right\}$$

$$= Re \left\{ \frac{1}{2}|E|^2 div\, \Theta + \langle \delta E, \Theta \vee \bar{E} \rangle + \langle dE, \Theta \wedge \bar{E} \rangle - \sum_i \langle \nabla \Theta_i \vee E, e_i \vee \bar{E} \rangle \right\}.$$

Consequently, the lemma follows from this and the divergence theorem. □

LEMMA 5.2. *For each form E defined in Ω such that $(\triangle + k^2)E = 0$ on Ω, and $E^*, (dE)^*, (\delta E)^* \in L^2(\partial\Omega)$, we have*

$$\|E\|_{L^2(\partial\Omega)} \leq C\,min\left\{\|n \wedge E\|_{L^2(\partial\Omega)}, \|n \vee E\|_{L^2(\partial\Omega)}\right\}$$

$$(5.8) \qquad\qquad +C\left|\int_{\partial\Omega} \langle \delta E, n \vee \bar{E} \rangle\right|^{1/2} + C\left|\int_{\partial\Omega} \langle dE, n \wedge \bar{E} \rangle\right|^{1/2}.$$

Proof.
Analyzing the real and the imaginary parts in the identity

$$\iint_\Omega |dE|^2 + |\delta E|^2 - k^2|E|^2 = \int_{\partial\Omega} \langle dE, n \wedge \bar{E} \rangle - \langle \delta E, n \vee \bar{E} \rangle$$

we obtain the energy estimate

$$\iint_\Omega |dE|^2 + |\delta E|^2 + |E|^2 \leq C\left|\int_{\partial\Omega} \langle dE, n \wedge \bar{E} \rangle\right| + C\left|\int_{\partial\Omega} \langle \delta E, n \vee \bar{E} \rangle\right|.$$

Let Θ be a real-valued, compactly supported, smooth vector field in R^m such that $\langle \Theta, n \rangle \geq c > 0$ almost everywhere on $\partial\Omega$. Now, if E is smooth up to the boundary of Ω, our lemma follows by combining this estimate with those derived from Lemma 5.1. The general case follows by working on a sequence of approximating domains $\Omega_j \uparrow \Omega$ and by using a standard limiting argument. $\square$

6. The proof of Theorem 1.2.

Before we proceed with the proof of the theorem, let us explain some notational conventions which we shall use several times. Below, $A \ll B$ will stand for $A(s) \leq CB(s)$ for some positive constant C, uniformly in the parameter s. Also, (small) and (large) will denote positive constants which we may take, respectively, as small and as large as we want. Finally, Comp will stand for generic compact operators on $L^2(\partial\Omega)$.

Proof of (1).

Let $\mathcal{X}$ be a certain subset of $\mathbb{C}$ (to be specified shortly). The estimate we shall eventually prove is that for each fixed $\lambda \in \mathcal{X}$, we have

$$
\begin{aligned}
\|A\|_{\Lambda^l L_t^{2,\delta}(\partial\Omega)} \ll \|(\lambda - M_l)A\|_{\Lambda^l L_t^{2,\delta}(\partial\Omega)} \quad &+ \quad \|Comp\,(A)\|_{L^2(\partial\Omega)} \\
&+ \quad \|Comp\,(\delta_\partial A)\|_{L^2(\partial\Omega)},
\end{aligned}
$$
(6.9)

uniformly for $A \in \Lambda^l L_t^{2,\delta}(\partial\Omega)$. Assuming this, it follows that $\lambda - M_l$ is bounded from below on $\Lambda^l L_t^{2,\delta}(\partial\Omega)$, modulo compact operators and, in particular, $\lambda - M_l$ has closed range and finite dimensional kernel on $\Lambda^l L_t^{2,\delta}(\partial\Omega)$. Therefore, the above analysis shows that $\sigma_\mu(M_l; L_t^{2,\delta}(\partial\Omega)) \subseteq \mathbb{C} \setminus \mathcal{X}$.

Consider now $A \in \Lambda^l L_t^{2,\delta}(\partial\Omega)$. The estimate (6.9) will be an obvious consequence of

$$
\begin{aligned}
\|A\|_{L^2} \ll \|(\lambda - M_l)A\|_{L^2} + \|Comp\,(A)\|_{L^2} \quad &+ \quad \|Comp\,(\delta_\partial A)\|_{L^2} \\
&+ \quad (small)\,\|\delta_\partial A\|_{L^2}
\end{aligned}
$$
(6.10)

and

$$
\begin{aligned}
\|\delta_\partial A\|_{L^2} \ll \|\delta_\partial[(\lambda - M_l)A]\|_{L^2} + \|Comp\,(A)\|_{L^2} \quad &+ \quad \|Comp\,(\delta_\partial A)\|_{L^2} \\
&+ \quad (small)\,\|A\|_{L^2}.
\end{aligned}
$$
(6.11)

Let us first consider (6.10). For $A \in \Lambda^l L_t^{2,\delta}(\partial\Omega)$, set $E := dSA$ and let I, II denote the right-most two integrals in the right hand side of (5.8). Obviously, $II = 0$. As for I, by Lemma 4.2 we have that $\delta E = k^2 SA - dS(\delta_\partial A)$. Hence, by applying (4.3)-(4.5) repeatedly, we obtain

$$
\begin{aligned}
|I| \ll (small)\,\|\delta_\partial A\|_{L^2} \quad &+ \quad (large)\,\|Comp\,(A)\|_{L^2} \\
&+ \quad (large)\,\|Comp\,(\delta_\partial A)\|_{L^2}.
\end{aligned}
$$
(6.12)

Using the Green representation formula (3.2) for E, going to the boundary and applying $n \wedge \cdot$ on both sides, we finally arrive at

$$
(\lambda - N_{l+2})(n \wedge E) = n \wedge dS[(\lambda - M_l)A] + Comp\,(A) + Comp\,(\delta_\partial A).
$$

Since $d_\partial N_{l+2} = N_{l+3} d_\partial + Comp$, and since $d_\partial (n \wedge E) = 0$, we have that

$$\|n \wedge E\|_{L^2} \ll \|(\lambda - N_{l+2})(n \wedge E)\|_{L^2} + \|Comp\,(A)\|_{L^2}$$
$$\ll \|(\lambda - M_l)A\|_{L^2} + \|Comp\,(A)\|_{L^2} + \|Comp\,(\delta_\partial A)\|_{L^2}$$

for any $\lambda \in \rho_\mu(N_{l+2}; \Lambda^{l+2} L_n^{2,d}(\partial\Omega))$. Moreover, as $n \vee E = (\lambda - \frac{1}{2})A - (\lambda - M_l)A$, the above estimate together with (6.12) and (5.8) yield (6.10). Consequently, (6.10) holds for each $\lambda \in \rho_\mu(N_{l+2}; \Lambda^{l+2} L_n^{2,d}(\partial\Omega))$.

On the other hand, noting that

$$(\lambda - N_{l+2})(n \wedge E) = -d_\partial[(\lambda - N_{l+1})(n \wedge SA)] + Comp\,(A),$$

and that $(\lambda - N_{l+1})(n \wedge SA) = Comp\,(A)$, we can also estimate $n \wedge E$ as

$$\|n \wedge E\|_{L^2} \ll \|(\lambda - N_{l+1})(n \wedge SA)\|_{\Lambda^{l+1} L_n^{2,d}(\partial\Omega)} + \|Comp\,(A)\|_{L^2}$$
$$\ll \|(\lambda - N_{l+2})(n \wedge E)\|_{L^2} + \|Comp\,(A)\|_{L^2} + \|Comp\,(\delta_\partial A)\|_{L^2}$$

for $\lambda \in \rho_\mu(N_{l+1}; \Lambda^{l+1} L_n^{2,d}(\partial\Omega))$. Thus, arguing as before, we conclude that (6.10) holds true for $\lambda \in \rho_\mu(N_{l+1}; \Lambda^{l+1} L_n^{2,d}(\partial\Omega))$ as well.

Next, we turn our attention to (6.11). We have that $\delta_\partial A \in \Lambda^{l-1} L_t^{2,\delta}(\partial\Omega)$ and, by (4.7), $\delta_\partial M_l A = M_{l-1}(\delta_\partial A) + Comp\,(A)$. Therefore, working with $\delta_\partial A$ in place of A it follows that the estimate (6.11) holds as long as the $l-1$ version of (6.10) does so.

Alternatively, assuming that $\lambda \in \rho_\mu(N_{l+1}; \Lambda^{l+1} L_n^{2,d}(\partial\Omega))$, we may use (4.7) and (6.10) to obtain (6.11).

Summarizing, at this point we have shown that (6.10) is valid for

$$\lambda \in \rho_\mu(N_{l+1}; \Lambda^{l+1} L_n^{2,d}(\partial\Omega)) \cup \rho_\mu(N_{l+2}; \Lambda^{l+2} L_n^{2,d}(\partial\Omega)),$$

whereas (6.11) is valid for

$$\lambda \in \rho_\mu(N_l; \Lambda^l L_n^{2,d}(\partial\Omega)) \cup \rho_\mu(N_{l+1}; \Lambda^{l+1} L_n^{2,d}(\partial\Omega)) \cup \rho_\mu(M_{l-1}; L_t^{2,\delta}(\partial\Omega)).$$

In particular, both (6.10) and (6.11) are valid for $\lambda \in \rho_\mu(N_{l+1}; \Lambda^{l+1} L_n^{2,d}(\partial\Omega))$. If we now observe that, by Hodge duality,

$$\sigma_\mu(M_l; \Lambda^l L_t^{2,\delta}(\partial\Omega)) = -\sigma_\mu(N_{m-l}; \Lambda^{m-l} L_n^{2,d}(\partial\Omega)),$$

we obtain that $\sigma_\mu(M_l; L_t^{2,\delta}(\partial\Omega)) \subseteq -\sigma_\mu(M_{m-l-1}; L_t^{2,\delta}(\partial\Omega))$. The point (1) in the Theorem 1.2 then easily follows from this. $\qquad \square$

Proof of (2).
The fact that both (6.10) and (6.11) are valid for $\lambda \in \rho_\mu(N_{l+2}; \Lambda^{l+2} L_n^{2,d}(\partial\Omega)) \cap \rho_\mu(M_{l-1}; L_t^{2,\delta}(\partial\Omega))$ implies that

$$\rho_\mu(N_{l+2}; \Lambda^{l+2} L_n^{2,d}(\partial\Omega)) \cap \rho_\mu(M_{l-1}; L_t^{2,\delta}(\partial\Omega)) \subseteq \rho_\mu(M_l; L_t^{2,\delta}(\partial\Omega)).$$

By using Hodge duality and the first part of (1) we see that

$$\rho_\mu(N_{l+2}; \Lambda^{l+2} L_n^{2,d}(\partial\Omega)) = \rho_\mu(M_{l+1}; L_t^{2,\delta}(\partial\Omega)).$$

With this at hand, (2) follows.   □

Proof of (3).

The key observation is that the formal transpose of M_l, acting on $\Lambda^l L_t^p(\partial\Omega)$, is $n \vee N_{l+1}(n \wedge \cdot)$. This is easily seen from

$$\int_{\partial\Omega} \langle n \vee dSA, B \rangle = \int_{\partial\Omega} \langle dSA, n \wedge B \rangle$$

$$= \int_{\partial\Omega} \left\langle \int_{\partial\Omega} \nabla_X \Phi(X-Y) \wedge A(Y), n(X) \wedge B(X) \right\rangle$$

$$= -\int_{\partial\Omega} \int_{\partial\Omega} \langle A(Y), \nabla_Y \Phi(X-Y) \vee (n(X) \wedge B(X)) \rangle$$

$$= \int_{\partial\Omega} \langle A, n \vee (n \wedge \delta S)(n \wedge B) \rangle .$$

Using this, the fact that the operators $n \wedge \cdot : \Lambda^l L_t^p(\partial\Omega) \to \Lambda^{l+1} L_n^p(\partial\Omega)$ and $n \vee \cdot : \Lambda^{l+1} L_n^p(\partial\Omega) \to \Lambda^l L_t^p(\partial\Omega)$ are inverse to each other, and (4.6), we get

$$\sigma(M_l; L_t^p(\partial\Omega)) = \sigma(N_{l+1}; \Lambda^{l+1} L_n^q(\partial\Omega)) = -\sigma(M_{m-l-1}; \Lambda^{m-l-1} L_t^q \partial\Omega)),$$

hence we are done.   □

Proof of (4).

Consider $1 < q < \infty$ with $\frac{1}{p} + \frac{1}{q} = 1$, and let us assume that $\lambda \in \mathbb{C}$ is such that $\lambda - M_l$ has a dense range as an operator on $\Lambda^l L_t^{p,\delta}(\partial\Omega)$. From the density lemma 4.4 we infer that $\lambda - M_l$ as an operator on $\Lambda^l L_t^p(\partial\Omega)$ has a dense range too. Passing to the dual spaces and then using Hodge duality, we obtain, as in the proof of (3), that $\lambda + M_{m-l-1}$ is injective on the space $\Lambda^{m-l-1} L_t^q(\partial\Omega)$. This together with (3) show that

$$\lambda \in -\sigma_p(M_{m-l-1}; L_t^q(\partial\Omega)) = \sigma_p(M_l; L_t^p(\partial\Omega)),$$

which concludes the proof of (4).   □

Proof of (5).

Assume that

$$\lambda \notin \sigma_p(M_l; L_t^{p,\delta}(\partial\Omega)) \cup \sigma(N_l; L_n^{p,d}(\partial\Omega)).$$

By (4), we infer that $\lambda \notin \sigma_p(N_l; L_n^p(\partial\Omega))$ as well. Let now $A \in \Lambda^l L_t^p(\partial\Omega)$ so that $(\lambda - M_l)A = 0$, and set $U := SA$ in $\mathbb{R}^m \setminus \partial\Omega$. Note that

(6.13) $$n \vee dU|_{\partial\Omega_\pm} = \left(\mp\frac{1}{2} + \lambda\right) A.$$

Next, using the representation (3.2) for U, applying δ to both sides, going to the boundary, and finally taking $n \wedge \cdot$ yields

$$(\lambda - N_l)(n \wedge \delta U) = k^2 n \wedge S(n \vee U) - n \wedge dS(n \vee \delta U) \in \Lambda^l L_n^{p,d}(\partial\Omega).$$

Consequently, $n \wedge \delta U \in \Lambda^l L_n^{p,d}(\partial\Omega)$. Using this, (3.2) and (4.4), we get

$$(\pm 1/2 - \lambda)\delta dU = k^2 \delta S(n \wedge U) + k^2 S(n \wedge \delta U) + \delta S(d(n \wedge \delta U)).$$

If now $\lambda \neq \pm\frac{1}{2}$, then this implies that $(\delta dU)^* \in L^p(\partial\Omega)$. In particular, by (4.3) and Lemma 3.1,

$$A = (n \vee dU)|_{\partial\Omega_-} - (n \vee dU)|_{\partial\Omega_+} \in \Lambda^l L_t^{p,\delta}(\partial\Omega).$$

Using (6.13) we may show that this remains valid if $\lambda = \pm\frac{1}{2}$ as well. Since $\lambda \notin \sigma_p(M_l; L_t^{p,\delta}(\partial\Omega))$, this implies $A = 0$. Hence, at this point, we have shown that

$$\sigma(M_l; L_t^p(\partial\Omega)) \subseteq \sigma_p(M_l; L_t^{p,\delta}(\partial\Omega)) \cup \sigma(N_l; L_n^{p,d}(\partial\Omega)).$$

Using Hodge duality, the proof of (5) is complete. $\qquad\square$

Proof of (6).

Let $\lambda \in \sigma_p(M_l; L_t^{p,\delta}(\partial\Omega))$, and let $0 \neq A \in \Lambda^l L_t^{p,\delta}(\partial\Omega)$ so that $(\lambda - M_l)A = 0$. Set $E := dSA$, $H := \delta E$. Then, Green's formula for E yields

$$(\lambda + 1/2)E = \delta S(n \wedge E) + S(n \wedge \delta E).$$

Applying δ to both sides and going to the boundary, we get

$$(\lambda - N_{l+1})(n \wedge H) = 0.$$

Thus, as $n \wedge H \neq 0$, $\lambda \in \sigma_p(N_{l+1}; L_n^{p,d}(\partial\Omega)) = -\sigma_p(M_{m-l-1}; L_t^{p,\delta}(\partial\Omega))$. The proof of (6) and, hence, the proof of the Theorem 1.2 is therefore finished. $\qquad\square$

7. The proofs of the corollaries.

This section is devoted to presenting the proofs of the corollaries of the main theorem stated in the introduction. However, before doing so we first make a couple of remarks.

First, we note that if Ω is a bounded C^1 domain in $\mathbf{R}^m$, then the operator M_l is compact on $\Lambda^l L_t^p(\partial\Omega)$ and on $\Lambda^l L_t^{p,\delta}(\partial\Omega)$, for $1 < p < \infty$, $0 \leq l \leq m$.

To see this, we note that for an arbitrary tangential form A the integrand $n(X) \vee (\nabla_X \Phi(X - Y) \wedge A(Y))$ in $M_l A(X)$ can be decomposed as

$$\frac{\partial \Phi(X - Y)}{\partial n(X)} A(Y) - \nabla_X \Phi(X - Y) \wedge ((n(X) - n(Y)) \vee A(Y))$$

(here we have used the fact that $n \vee A = 0$ a. e. on $\partial\Omega$). It is well-known that the first term corresponds to a compact operator on L^p ([**6**]). As for the second term, standard arguments show that this kernel also yields a compact operator on L^p ([**6**]) since $n(X) - n(Y) = o(|X - Y|)$ as $|X - Y| \to 0$. Hence, M_l is compact on $\Lambda^l L_t^p(\partial\Omega)$, $1 < p < \infty$. Using this, (4.7), and the fact that $\delta_\partial A \in \Lambda^{l-1} L_t^{p,\delta}(\partial\Omega)$, we can proceed inductively on l to see that M_l is compact on $\Lambda^l L_t^{p,\delta}(\partial\Omega)$, $1 < p < \infty$, too.

Our second remark concerns the spectrum of K^*, the transposed of the double layer acoustic potential operator on $\partial\Omega$ (cf. e.g. [3]). We shall need that

$$\sigma_\mu(K^*; L^p(\partial\Omega)) \subset D(0; 1/2)$$

if $\partial\Omega \in C^1$, $1 < p < \infty$, or if $p = 2$ and Ω is either a convex or a polyhedral domain in R^3. This can be seen by utilizing the corresponding results for K_0^*, the transposed of the double layer harmonic potential operator (cf. [5] and [4]), and the fact that $K^* - K_0^*$ is a compact operator. The important thing for us is that M_0 can be canonically identified with K^*.

Proof of Corollary 1.3.
The first part is a direct consequence of Theorem 1.1 and the point (4) in Theorem 1.2. The second part follows from Theorem 1.1 and the density Lemma 4.4.
□

Proof of Corollary 1.4.
This follows from the points (1) and (6) in Theorem 1.2 with the help of (1.1).
□

Proof of Corollary 1.5.
Since M_l is compact on $\Lambda^l L_t^p(\partial\Omega)$ and on $\Lambda^l L_t^{p,\delta}(\partial\Omega)$, the corollary follows from the point (4) of Theorem 1.2. Note that the spectrum of M_l on $\Lambda^l L_t^p(\partial\Omega)$ is actually independent of $p \in (1, \infty)$, as one can check by using the techniques of [6]. □

Proof of Corollary 1.6.
Taking $E := dSB$ with $B \in \Lambda^l L_t^p(\partial\Omega)$ and using Lemma 3.1, the existence part is readily seen to reduce to the invertibility of the operator $-\frac{1}{2} + M_l$ on $\Lambda^l L_t^p(\partial\Omega)$. We have already established this in Corollary 1.3.

Let us deal now with the uniqueness part. Since $\partial\Omega \in C^1$, the results in [2] can be used to show that the operator norm of $-\frac{1}{2} + M_l$ on $\Lambda^l L_t^p(\partial\Omega)$ is bounded by a constant which depends exclusively on the C^1 constant of Ω. Using this and the fact that for smooth domains our problem has a unique solution, a well-known argument then yields the conclusion (see [11] for more details in a similar situation). Finally, the last part of the statement is easily seen from the Corollary 1.5. □

Proof of Corollary 1.7.
(a) This follows from the points (1) and (6) in Theorem 1.2 if we keep in mind that, in general, $\partial\sigma(T; X) \subseteq \sigma_\mu(T; X) \cup \sigma_p(T; X)$.
(b) From the results of [7], [4] and the second remark made at the beginning of this section we have that $\sigma_\mu(M_0; L_t^{2,\delta}(\partial\Omega)) \subseteq D(0; \frac{1}{2})$. Consequently, using (1) and (2) in Theorem 1.2 we obtain that $\sigma_\mu(M_1; L_t^{2,\delta}(\partial\Omega)) \subseteq D(0; \frac{1}{2})$. Finally, by (1.1), we get

$$\sigma(M_1; L_t^{2,\delta}(\partial\Omega)) \subseteq [\sigma_p(M_1; L_t^{2,\delta}(\partial\Omega)) \cup \sigma_\mu(M_1; L_t^{2,\delta}(\partial\Omega))] \subseteq D(0; 1/2)$$

and the first inclusion follows. The second one will follow from this and the point (4) in Theorem 1.2.   □

References

1. A. P. Calderón, *The multipole expansion of radiation fields*, J. Rat. Mech. Anal., 3 (1954), 523–537.
2. R. Coifman, A. McIntosh, and Y. Meyer, *L'intégrale de Cauchy définit un opérateur borné sur L^2 pour les courbes Lipschitziennes*, Ann. of Math., 116 (1982), 361–387.
3. D. Colton and R. Kress, *Integral equation methods in scattering theory*, Wiley, New York, 1983.
4. J. Elschner, *The double layer potential operator over polyhedral domains I: Solvability in weighted Sobolev spaces*, preprint (1992).
5. L. Escauriaza, E. Fabes and G. Verchota, *On a regularity theorem for weak solutions to transmission problems with internal Lipschitz boundaries*, Proc. Amer. Math. Soc., 115 (1992), 1069–1076.
6. E. Fabes, M. Jodeit, and N. Rivière, *Potential techniques for boundary value problems on C^1 domains*, Acta Math., 141 (1978), 165–186.
7. E. Fabes, M. Sand and J. Seo, *The spectral radius of the classical layer potentials on convex domains*, in: *Partial differential equations with minimal smoothness and applications*, 129–137, Springer, New York, 1992.
8. B. Jawerth and M. Mitrea, *Higher dimensional scattering theory on C^1 and Lipschitz domains*, submitted.
9. M. Mitrea, *Clifford Wavelets, Singular Integrals, and Hardy Spaces*, Lecture Notes in Mathematics, No. 1575, Springer-Verlag, 1994.
10. _____, *The method of layer potentials in electro-magnetic scattering theory on non-smooth domains*, to appear in Duke Math. J., 77 (1994).
11. M. Mitrea, R. Torres, and G. Welland, *Layer potential techniques in electromagnetism*, preprint.
12. C. Müller, *Über die Beugung elektromagnetischer Schwingungen an endlichen homogenen Körpern*, Math. Ann., 123 (1951) 345–378.
13. W. K. Saunders, *On solutions of Maxwell equations in an exterior region*, Proc. Nat. Acad. Sci. U.S.A., 38 (1952), 342–348.
14. G. Verchota, *Layer potentials and boundary value problems for Laplace's equation in Lipschitz domains*, J. Funct. Anal., 59 (1984), 572–611.
15. H. Weyl, *Kapazität von Strahlungsfeldern*, Math. Zeit., 55 (1952), 187–198.
16. _____, *Die natürlichen Randwertaufgaben im Aussenraum für Strahlungsfeldern beliebiger Dimensionen und beliebiger Ranges*, Math. Zeit., 56 (1952), 105–119.

DEPARTMENT OF MATHEMATICS, UNIVERSITY OF SOUTH CAROLINA, CO-
LUMBIA, SC 29208, USA
E-mail address: bj@loki.scarolina.edu

INSTITUTE OF MATHEMATICS OF THE ROMANIAN ACADEMY, P.O. BOX
1-764 RO-70700 BUCHAREST, ROMANIA
Current address: School of Mathematics, University of Minnesota, 127 VinH,
206 Church St. SE, Minneapolis, MN 55455, USA
E-mail address: mitrea@math.umn.edu

Contemporary Mathematics
Volume **189**, 1995

Almost Periodic Factorization:
An Analogue of Chebotarev's Algorithm

YURI KARLOVICH AND ILYA SPITKOVSKY

Dedicated to Professor Mischa Cotlar
on the occasion of his 80th birthday

ABSTRACT. Explicit formulas are found for almost periodic factorization of certain triangular matrices arising in the theory of convolution type equations on finite intervals. Results are based on continued fraction expansions of almost periodic polynomials.

1. Introduction

1.1. Definitions and notation. Let AP be the Banach algebra of all Bohr almost periodic functions, that is, a C^*-subalgebra of $L^\infty(\mathbb{R})$ spanned by the functions $e_\mu(x) = e^{i\mu x}$, $\mu \in \mathbb{R}$. We will use the following properties and notations belonging to the theory of AP functions; see [**12, 13**] for their detailed consideration.

Each $f \in AP$ has the so called *mean value*

$$\mathbf{M}(f) = \lim_{T \to \infty} \frac{1}{2T} \int_{-T}^{T} f(x)\, dx.$$

In particular, for all $\mu \in \mathbb{R}$ there exist $\hat{f}(\mu) = \mathbf{M}(e_{-\mu}f)$. The set

$$\Omega(f) = \{\mu \in \mathbb{R} \colon \hat{f}(\mu) \neq 0\}$$

is not more than countable. It is called the *Fourier spectrum* of f, and

(1.1) $$\sum_{\mu \in \Omega(f)} \hat{f}(\mu)e^{i\mu x}$$

1991 *Mathematics Subject Classification.* Primary 47A68; Secondary 11A55, 40A15, 42A75.
The second author was supported in part by NSF Grant #9401848.

is, respectively, the *Fourier series* of f.

We say that f is an *almost periodic (AP) polynomial* if its Fourier spectrum is finite, and $f \in AP_W$ if its Fourier series (1.1) converges absolutely, that is,

$$\sum_{\mu \in \Omega(f)} |\hat{f}(\mu)| < \infty.$$

Of course, all AP polynomials are in AP_W. Finally, let

$$AP^{\pm} = \{f \in AP \colon \Omega(f) \in \mathbb{R}_{\pm}\},$$

where, as usual, $\mathbb{R}_{\pm} = \{x \in \mathbb{R} \colon \pm x \geq 0\}$. For all function classes X mentioned above, inclusions $f \in X$, when f is a matrix function, are understood elementwise.

The main subject of our considerations is an *AP-factorization* of $n \times n$ matrix functions G, that is, their representation in the form

$$(1.2) \qquad\qquad G(x) = G_+(x)\Lambda(x)G_-(x),$$

where

$$(1.3) \qquad\qquad G_+^{\pm 1} \in AP^+, \quad G_-^{\pm 1} \in AP^-,$$

$\Lambda(x) = \text{diag}[e^{i\lambda_1 x}, \dots e^{i\lambda_n x}]$, and the *partial AP-indices* $\lambda_1, \dots, \lambda_n$ of G are real numbers.

We call (1.2) an *AP_W-factorization* of G if in (1.3) the classes $AP^{\pm}$ are substituted by $AP_W^{\pm} = AP^{\pm} \cap AP_W$. Of course, for AP- (AP_W-) factorization to exist, G must be an invertible element of AP (respectively, AP_W).

1.2. Known results. *AP*-factorization of matrix functions arises as a natural (and inevitable) step in the study of Toeplitz operators with almost periodic and so called semi-almost periodic matrix symbols (see [**7, 1**]). More surprisingly, it also arises in the theory of convolution type equations on finite intervals, even when equations themselves are scalar, and Fourier transforms $\mathcal{F}k$ of their kernels k are continuous or piecewise continuous functions [**15, 6, 7, 8**]. In particular, consideration of scalar convolution type equations on a finite interval of length λ leads to *AP*-factorization of matrices

$$(1.4) \qquad\qquad G(x) = \begin{bmatrix} e^{i\lambda x} & 0 \\ f(x) & e^{-i\lambda x} \end{bmatrix}.$$

For actual use of *AP*-factorization in the Fredholm theory of operators and equations mentioned, not only is the existence of this factorization important, but also conditions under which the partial *AP*-indices equal zero and, when these conditions are satisfied, the explicit formulas for

$$\mathbf{d}(G) = \mathbf{M}(G_+)\mathbf{M}(G_-)$$

(according to [**7**], partial AP indices of G are defined up to their order and, when they all equal zero, $\mathbf{d}(G)$ is defined uniquely in spite of the fact that $G_\pm$ themselves are not unique).

If G is periodic with a period equal, say, μ, then the change of variable $t = e^{2\pi i x/\mu}$ reduces (see [**7**]) the AP-factorization problem for G to the *Wiener-Hopf factorization*

$$(1.5) \qquad F(t) = F_+(t)M(t)F_-(t), \quad t \in \mathbb{T},$$

of the continuous on the unit circle $\mathbb{T}$ matrix function $F(t) = G(\frac{\mu \log t}{2\pi i})$. In (1.5), $M(t) = \mathrm{diag}[t^{\kappa_1}, \ldots, t^{\kappa_n}]$, $\kappa_1, \ldots, \kappa_n \in \mathbb{Z}$, and continuous on $\mathbb{T}$ matrix functions $F_\pm$ admit analytic and nonsingular continuation to the interior/exterior of $\mathbb{T}$. From classical results of I. Gohberg and M. Krein [**3**] it follows that factorization (1.5) exists for every nonsingular on $\mathbb{T}$ matrix function F with absolutely convergent there Laurent series $\sum_{j=-\infty}^{+\infty} F_j t^j$; the multiples $F_\pm$ automatically have the same property. Hence, for periodic nonsingular on $\mathbb{R}$ $n \times n$ matrix functions $G \in AP_W$ there always exists an AP_W-factorization. On the other hand, explicit formulas for multiples $F_\pm$ and partial indices $\kappa_1, \ldots, \kappa_n$ are not known for general $n \times n$ matrix functions F, no matter what their smoothness properties are. So, there is no hope to obtain such formulas applicable to all (even periodic) AP_W matrix functions G.

The situation changes, however, if additional algebraic structure is imposed. Let, in particular, G be periodic of the form (1.4), that is,

$$(1.6) \qquad G(x) = \begin{bmatrix} e^{iN\mu x} & 0 \\ \sum_{n=-\infty}^{+\infty} a_n e^{in\mu x} & e^{-iN\mu x} \end{bmatrix}.$$

The corresponding matrix function F is

$$(1.7) \qquad F(x) = \begin{bmatrix} t^N & 0 \\ \sum_{n=-\infty}^{+\infty} a_n t^n & t^{-N} \end{bmatrix},$$

and an application of G. Chebotarev's constructive procedure [**2**] (see also [**14**, Section 4.3]) leads to algorithms allowing to find Λ, $G_\pm$, and (if $\Lambda = I$) $\mathbf{d}(G)$. This procedure is based on the continued fraction expansion

$$(1.8)$$

$$\Phi = \cfrac{1}{Q_0 + \cfrac{1}{Q_1 + \cfrac{1}{\cdots +}}} \qquad (Q_0, Q_1, \ldots \text{ are polynomials}, \deg Q_0 < \deg Q_1 < \cdots)$$

of the rational function $\Phi(z) = \sum_{n=-N+1}^{N-1} a_n z^{n-N}$.

More recently [11], an explicit necessary and sufficient condition was found for matrices (1.7) to have zero partial indices $\kappa_1, \ldots, \kappa_n$. After transformation from (1.7) to (1.6), this result reads as follows.

LEMMA 1.1. *For partial AP-indices of the matrix function (1.6) to be equal zero, it is necessary and sufficient that the finite Toeplitz matrix*

$$A_N(G) = (a_{i-j})_{i,j=1}^N$$

be invertible.

1.3. About this paper. In our paper, we consider AP_W matrices of the form (1.4) which we do not assume to be periodic. Even existence of AP-factorization in this case becomes a non-trivial problem, see [7, 9] for examples of non-factorizable AP-polynomial matrices of the form (1.4). Nevertheless, we will modify here an expansion (1.8) in such a way that it becomes applicable to almost periodic polynomials Φ. The Chebotarev's procedure, when modified respectively, becomes an effective device for AP-factorization of matrices (1.4) at least for some classes of AP polynomials f. The procedure itself is explained in Section 2. Its applicability to functions f with commensurable distances between all the points of $\Omega(f)$, as well as criteria for AP-indices to be equal zero and formulas for $\mathbf{d}(G)$ in that case, are given in Section 3. They cover, in particular, the case of binomials f, a different approach to which has been developed earlier in [7]. Finally, Section 4 contains the corresponding results for trinomials f under certain additional restrictions on the structure of $\Omega(f)$.

2. General scheme

For any $f \in AP$ put $\operatorname{ord} f = \sup\{\lambda : \lambda \in \Omega(f)\}$. We say that $\operatorname{ord} f$ is an *order* of the function f. According to this definition, $\operatorname{ord} f = -\infty$ if and only if $f = 0$. On the other hand, there are many AP functions f with $\operatorname{ord} f = +\infty$. Of course, $\operatorname{ord} f$ is finite for all nonzero AP polynomials.

Considering AP-factorization of matrices (1.4), we can always multiply them by $\begin{bmatrix} 1 & 0 \\ g^+(x) & 1 \end{bmatrix}$ from the left and by $\begin{bmatrix} 1 & 0 \\ g^-(x) & 1 \end{bmatrix}$ from the right. For any $g^\pm \in AP^\pm$, such a multiplication does not change neither the existence of AP-factorization, nor the values of partial AP-indices and the matrix $\mathbf{d}(G)$. If $f \in AP_W$, let us choose

$$g^\pm = -e^{\mp i\lambda x} \sum_{\mu \in \Omega(f), \pm\mu \geq \lambda} \hat{f}(\mu) e^{i\mu x}.$$

Then the multiplication described above changes (1.4) into

$$(2.1) \qquad G(x) = \begin{bmatrix} e^{i\lambda x} & 0 \\ r_0(x) & e^{-i\lambda x} \end{bmatrix},$$

where

$$r_0(x) = \sum_{\mu \in \Omega(f), \, |\mu| < \lambda} \hat{f}(\mu) e^{i\mu x}.$$

Therefore, without loss of generality we may consider, instead of (1.4), matrices (2.1) in which $\Omega(r_0) \subset (-\lambda, \lambda)$.

Put $r_{-1} = e^{i\lambda x}$, and define a recursive sequence of functions $r_j \in AP$, $\operatorname{ord} r_j = \mu_j$, such that

$$
(2.2) \qquad
\begin{aligned}
r_{j-2} &= q_j^+ r_{j-1} + r_j, \\
q_j^+ &\in AP^+, \ \mu_j < \mu_{j-1} \quad (j = 1, 2, \dots).
\end{aligned}
$$

The functions q_j^+ and r_j are defined by (2.2) uniquely. Indeed, if $r_{j-2} = q_j' r_{j-1} + r_j'$, where $q_j' \in AP^+$, $q_j' \neq q_j^+$, and $\operatorname{ord} r_j' < \mu_{j-1}$, then

$$
\operatorname{ord}((q_j^+ - q_j') r_{j-1}) \geq \operatorname{ord} r_{j-1} > \operatorname{ord}(r_j - r_j'),
$$

which is a contradiction with $(q_j^+ - q_j') r_{j-1} = r_j' - r_j$.

Let us now show that such functions q_j^+ and r_j exist, and, moreover, are AP polynomials. To that end it suffices to construct such AP polynomials q_j^+ and r_j when it is given that r_{j-2} and r_{j-1} are AP polynomials, and $\mu_{j-2} > \mu_{j-1}$. Thus, suppose that

$$
(2.3) \qquad r_{j-2} = \sum_{s=1}^{t} a_s e^{i\nu_s x}, \quad r_{j-1} = b_0 e^{i\mu_{j-1} x} \left(1 - \sum_{l=1}^{m} b_l e^{i\gamma_l x} \right),
$$

in which $\mu_{j-2} = \nu_1 > \dots > \nu_{s_0} \geq \mu_{j-1} > \nu_{s_0+1} > \dots > \nu_t$ for a certain $s_0 \in \{2, \dots, t\}$; $0 > \gamma_1 > \dots > \gamma_m$ and $a_1 a_2 \cdots a_t b_0 \neq 0$.

Let us denote $\gamma = (\gamma_1, \dots, \gamma_m)$, $n = (n_1, \dots, n_m)$, $\langle n, \gamma \rangle = \sum_{l=1}^{m} n_l \gamma_l$,

$$
N(\nu, r_{j-1}) = \{ n = (n_1, \dots, n_m) \colon n_1, \dots, n_m \in \mathbb{Z}_+; \nu + \langle n, \gamma \rangle - \mu_{j-1} \geq 0 \},
$$
$$
N_l(\nu, r_{j-1}) = \{ n \in N(\nu, r_{j-1}) \colon n_l = f(n_1, \dots, n_{l-1}, n_{l+1}, \dots, n_m) \},
$$

$l = 1, \dots, m$, where

$$
f(n_1, \dots, n_{l-1}, n_{l+1}, \dots, n_m) = \sup\{ n_l \colon \nu + \langle n, \gamma \rangle - \mu_{j-1} \geq 0 \}
$$

when all components of the vector n, except n_l, are fixed. Consider AP polynomials

$$
(2.4) \qquad
\begin{aligned}
q_j^+ &= \sum_{s=1}^{s_0} a_s b_0^{-1} e^{i(\nu_s - \mu_{j-1})x} \sum_{n \in N(\nu_s, r_{j-1})} c_n e^{i\langle n, \gamma \rangle x}, \\
r_j &= \sum_{s=s_0+1}^{t} a_s e^{i\nu_s x} + \sum_{s=1}^{s_0} a_s e^{i\nu_s x} \sum_{l=1}^{m} \sum_{n \in N_l(\nu_s, r_{j-1})} c_n b_l e^{i(\langle n, \gamma \rangle + \gamma_l)x}
\end{aligned}
$$

(as usual, a sum with an empty set of summation indices is assumed to be equal zero), in which

$$
(2.5) \qquad c_n = \frac{(n_1 + n_2 + \dots + n_m)!}{n_1! n_2! \cdots n_m!} b_1^{n_1} b_2^{n_2} \cdots b_m^{n_m}.
$$

Obviously, $q_j^+ \in AP^+$ and $\mu_j < \mu_{j-1}$. Further, for $n \in N(\nu, r_{j-1}) \setminus \{0\}$, the following recursive relation holds:

$$(2.6) \qquad c_n = \sum_{l=1}^{m} c_{n^{(l)}} b_l,$$

where $n^{(l)} = (n_1, \ldots, n_{l-1}, n_l - 1, n_{l+1}, \ldots, n_m)$ for $n_l > 0$, and $c_{n^{(l)}} = 0$ for $n_l = 0$. Let $\tilde{N}_l(\nu) = \{n \in N(\nu, r_{j-1}) : n_l \geq 1\}$, $N_l'(\nu) = N(\nu, r_{j-1}) \setminus N_l(\nu, r_{j-1})$. According to (2.6),

$$\sum_{l=1}^{m} \sum_{n \in N_l'(\nu)} c_n b_l e^{i(\langle n, \gamma \rangle + \gamma_l)x} = \sum_{l=1}^{m} \sum_{n \in \tilde{N}_l(\nu)} c_{n^{(l)}} b_l e^{i\langle n, \gamma \rangle x} =$$

$$\sum_{n \in N(\nu, r_{j-1}) \setminus \{0\}} \sum_{l=1}^{m} c_{n^{(l)}} b_l e^{i\langle n, \gamma \rangle x} = \sum_{n \in N(\nu, r_{j-1}) \setminus \{0\}} c_n e^{i\langle n, \gamma \rangle x}.$$

Since $c_0 = 1$, it follows from here that

$$r_{j-2} - q_j^+ r_{j-1} - r_j =$$

$$\sum_{s=1}^{s_0} a_s e^{i\nu_s x} \left(1 - \sum_{n \in N(\nu_s, r_{j-1})} c_n e^{i\langle n, \gamma \rangle x} + \sum_{l=1}^{m} \sum_{n \in N_l'(\nu_s)} c_n b_l e^{i(\langle n, \gamma \rangle + \gamma_l)x} \right) = 0.$$

Therefore, the recursive relations (2.2) define the AP polynomials q_j^+ and r_j uniquely, and these polynomials are delivered by formulas (2.4), if r_{j-2} and r_{j-1} are of the form (2.3).

If at a certain step of this recursion we arrive at $r_k = 0$, then the corresponding value of μ_k is $-\infty$, and the process terminates. Otherwise we can continue infinitely, and the resulting sequence $\{\mu_j\}_{j=0}^{\infty}$, being decreasing, has a limit $\mu_\infty \geq -\infty$. Let us denote by K the set of indices of the sequence $\{\mu_j\}$, that is, the segment $\{0, \ldots, k\}$ in the first case, and $\mathbb{Z}_+$ in the second.

Relations (2.2) are equivalent to the representation of a function r_{-1}/r_0 in a form of a continued fraction

$$\frac{r_{-1}}{r_0} = q_1^+ + \cfrac{1}{q_2^+ + \cfrac{1}{\ddots \, q_k^+ + \cfrac{r_k}{r_{k-1}}}}.$$

If $\mu_k = -\infty$, the latter equality means that

$$\frac{r_{-1}}{r_0} = q_1^+ + \cfrac{1}{q_2^+ + \cfrac{1}{\ddots \, q_k^+}} \qquad (\overset{\text{def}}{=} [q_1^+, \ldots, q_k^+]).$$

If $K = \mathbb{Z}_+$, we will associate with the function $\dfrac{r_{-1}}{r_0}$ a formal infinite continued fraction

$$\frac{r_{-1}}{r_0} \sim q_1^+ + \cfrac{1}{q_2^+ + \cfrac{1}{\ddots \; q_k^+ + \cdots}} \qquad (\overset{\text{def}}{=} [q_1^+, \dots, q_k^+, \dots]),$$

irrespective to the convergence of the latter. Denote by p_j^+ the numerator of j-th convergent $[q_1^+, \dots, q_j^+]$ of the continued fraction $[q_1^+, \dots, q_j^+, \dots]$. These numerators satisfy the following recursive relations:

$$(2.7) \qquad p_j^+ = q_j^+ p_{j-1}^+ + p_{j-2}^+ \ (j = 1, 2, \dots), \quad p_{-1}^+ = 0, p_0^+ = 1,$$

and are therefore AP polynomials lying in AP^+ together with q_j^+. In addition,

$$\operatorname{ord} p_j^+ = \operatorname{ord}(q_j^+ \cdots q_1^+) = \lambda - \mu_{j-1}.$$

Put also

$$Q_j^+ = \left[\begin{array}{cc} q_j^+ & 1 \\ 1 & 0 \end{array} \right].$$

LEMMA 2.1. *For all* $k \in K$ *the matrix function* (2.1) *admits a representation*

$$(2.8) \qquad\qquad G = G_k^+ \Lambda_k G_k^-,$$

in which

$$(2.9) \quad \begin{aligned} & G_k^+ = Q_1^+ \cdots Q_k^+, \quad \Lambda_k = \operatorname{diag}[e^{i\mu_{k-1}x}, e^{-i\mu_{k-1}x}], \\ & G_k^- = \left[\begin{array}{cc} r_{k-1}(x)e^{-i\mu_{k-1}x} & (-1)^{k-1}p_{k-1}^+(x)e^{-i(\mu_{k-1}+\lambda)x} \\ r_k(x)e^{i\mu_{k-1}x} & (-1)^k p_k^+(x)e^{i(\mu_{k-1}-\lambda)x} \end{array} \right]. \end{aligned}$$

PROOF. For $k = 0$, the equality (2.8) takes the form

$$G(x) = \left[\begin{array}{cc} e^{i\lambda x} & 0 \\ 0 & e^{-i\lambda x} \end{array} \right] \left[\begin{array}{cc} 1 & 0 \\ r_e(x)e^{i\lambda x} & 1 \end{array} \right],$$

and is therefore valid. With the use of (2.2) and (2.7) it can be verified that $G_j^- = (Q_j^+ \Lambda_j)^{-1} \Lambda_{j-1} G_{j-1}^-$. Hence, if (2.8) holds for $k = j - 1$, then

$$G_j^+ \Lambda_j G_j^- = G_{j-1}^+ Q_j^+ \Lambda_j (Q_j^+ \Lambda_j)^{-1} \Lambda_{j-1} G_{j-1}^- = G_{j-1}^+ \Lambda_{j-1} G_{j-1}^- = G,$$

that is, (2.8) is also valid for $k = j$. $\square$

In (2.8), a factor G_k^+ is an AP polynomial with $\Omega(G_k^+) \subset [0, \lambda - \mu_{k-1}]$ and $\det G_k^+ = (-1)^k$. Hence, $(G_k^+)^{\pm 1} \in AP^+$. The multiple G_k^- has the same determinant $(-1)^k$ as G_k^+ since $\det G_k^+ G_k^- = \det G / \det \Lambda_k = 1$. Further, elements of G_k^- are AP polynomials, orders of which equal exactly $\mu_k + \mu_{k-1}$, 0, and $-(\mu_{k-1} + \mu_{k-2})$. Therefore, (2.8) delivers an AP-factorization of the matrix (2.1) if and only if

$$(2.10) \qquad\qquad \mu_{k-2} + \mu_{k-1} \geq 0 \geq \mu_{k-1} + \mu_k.$$

From here follows

THEOREM 2.1. *Let in* (2.1) r_0 *be an AP polynomial with its spectrum* $\Omega(r_0)$ *located in* $(-\lambda, \lambda)$; *and let* $\{r_j\}$ *be a sequence of AP polynomials of orders* $\operatorname{ord} r_j = \mu_j$ *constructed according to* (2.3)–(2.4). *Suppose that for a certain* k (2.10) *is satisfied. Then the matrix* (2.1) *is AP-factorizable, its partial AP-indices equal* $\pm \mu_{k-1}$, *and the factors* $G_\pm (= G_k^\pm)$ *can be constructed by formulas* (2.9).

Obviously, the desired value of k exists if and only if the sequence $\{\mu_j\}$ terminates, or its limit μ_∞ is nonpositive. If, in addition, the sequence $\mu_{j-1} + \mu_j$ skips 0, then such a value of k is unique, and both inequalities in (2.10) are strict. On the other hand, if there exists (unique, due to strict monotonicity of μ_j) j_0 such that $\mu_{j_0-1} + \mu_{j_0} = 0$, then both $k = j_0$ and $k = j_0 + 1$ are acceptable for our purposes, and respectively the right or the left inequality in (2.10) turns into equality. From computational point of view, it is more convenient to choose $k = j_0$, that is, the minimal possible k such that $\mu_{k-1} + \mu_k \leq 0$.

Using explicit formulas for the factors of the AP-factorization (2.8), it is possible now to find a necessary and sufficient condition for partial indices of G to be equal 0, and to derive also a formula for $\mathbf{d}(G)$. Denote by a_j the leading coefficient of q_j^+: $a_j = \mathbf{M}(q_j^+ e^{i(\mu_{j-1} - \mu_{j-2})x})$; put also $c_j = \mathbf{M}(q_j^+)$.

COROLLARY 2.1. *Under the conditions of Theorem 2.1, the partial AP-indices of the matrix* (2.1) *equal zero if and only if a sequence* μ_j *assumes value* 0, *that is, there exists* $k \in \mathbb{Z}_+$ *such that* $\mu_{k-1} = 0$. *If this condition is satisfied, then*

$$(2.11) \quad \mathbf{d}(G) = \begin{bmatrix} c_1 & 1 \\ 1 & 0 \end{bmatrix} \cdots \begin{bmatrix} c_k & 1 \\ 1 & 0 \end{bmatrix} \begin{bmatrix} a_k^{-1} \cdots a_1^{-1} & 0 \\ 0 & (-1)^k a_k \cdots a_1 \end{bmatrix}.$$

Observe that $a_k^{-1} \cdots a_1^{-1}$ is nothing but the leading coefficient of r_{k-1}.

Put now $J = \begin{bmatrix} 0 & 1 \\ 1 & 0 \end{bmatrix}$. From factorization $G = G_+ \Lambda G_-$ of the matrix function G it immediately follows the factorization $J(G_-)^* \Lambda^* (G_+)^* J$ of JG^*J, and vice versa. At the same time, for matrices G of the form (2.1), the mapping

$$(2.12) \qquad\qquad\qquad G \mapsto JG^*J$$

is simply a substitution of r_0 by its complex conjugate $\overline{r_0}$. Applying Theorem 2.1 and Corollary 2.1 to JG^*J, we get the following results in terms of the original matrix G:

THEOREM 2.2. *Let in the matrix* (2.1) r_0 *be an AP polynomial with its spectrum located in* $(-\lambda, \lambda)$. *Define AP polynomials* r_j' *by the recursive formulas*

$$(2.13) \qquad r_{j-2}' = q_j^- r_{j-1}' + r_j', \quad q_j^- \in AP^-, \quad r_0' = r_0, \ r_{-1}' = e^{-i\lambda x},$$

where the sequence $\mu_j' = -\operatorname{ord} r_j'$ *strictly increases. If there exists* k *such that* $\mu_{k-2}' + \mu_{k-1}' \leq 0 \leq \mu_{k-1}' + \mu_k'$, *then the matrix* (2.1) *is AP-factorizable, its partial AP-indices equal* $\pm\mu'_{k-1}$, *and its factorization multiples may be chosen in the form*

$$G_+ = \begin{bmatrix} (-1)^{k-1}p^-_{k-1}e^{i(\lambda-\mu'_{k-1})x} & (-1)^k p^-_k e^{i(\lambda+\mu'_{k-1})x} \\ r'_{k-1}e^{-i\mu'_{k-1}x} & r'_k e^{i\mu'_{k-1}x} \end{bmatrix},$$

$$\Lambda = \begin{bmatrix} e^{i\mu'_{k-1}x} & 0 \\ 0 & e^{-i\mu'_{k-1}x} \end{bmatrix}, G_- = Q^-_k \cdots Q^-_1 J.$$

Here Q^-_j and p^-_j are obtained from Q^+_j and p^+_j respectively by the substitution of q^+_l to q^-_l.

Of course, q^-_j and r'_j are defined by (2.13) uniquely, and coincide respectively with $\overline{q^+_j}$ and $\overline{r_j}$, where q^+_j and r_j are given by the formulas (2.4).

COROLLARY 2.2. *Under the conditions of Theorem 2.2, the partial AP-indices of the matrix (2.1) equal zero if and only if there exists* $k \in \mathbb{Z}_+$ *such that* $\mu'_{k-1} = 0$. *If this condition is satisfied, then*

(2.14)

$$\mathbf{d}(G) = \begin{bmatrix} (-1)^k a'_1 \cdots a'_k & 0 \\ 0 & (a'_1)^{-1} \cdots (a'_k)^{-1} \end{bmatrix} \begin{bmatrix} 0 & 1 \\ 1 & c'_k \end{bmatrix} \cdots \begin{bmatrix} 0 & 1 \\ 1 & c'_1 \end{bmatrix},$$

where $c'_j = \mathbf{M}(q^-_j)$ *are constant terms, and* $a'_j = \mathbf{M}(q^-_j e^{i(\mu'_{j-1}-\mu'_{j-2})x})$ — *the lowest coefficients of the AP polynomials* q^-_j.

3. Commensurable distances between points of $\Omega(r_0)$

3.1. Applicability of Chebotarev's algorithm. In [7, Theorem 2.4], we have considered AP-factorization of matrices (2.1) with AP polynomials r_0 such that $\Omega(r_0) \subset (-\lambda, 0]$ or $\Omega(r_0) \subset [0, \lambda)$. Though we did not use Chebotarev's algorithm there explicitly, the results obtained actually mean that Theorem 2.1 is applicable in the first case, Theorem 2.2 — in the second and, moreover, the procedure terminates in one step.

In this section, we consider more involved case where no restriction is imposed on signs of elements of $\Omega(r_0)$, but the distances between them are supposed to be commensurable. Then r_0 can be represented as

(3.1) $$r_0(x) = e^{i\mu x} \sum_{n=0}^{m} b_n e^{in\alpha x},$$

where $b_0 \neq 0$, $\lambda > \mu > \mu + m\alpha > -\lambda$. We will find a similar representation for all the functions r_j. More precisely, we will show that

(3.2) $$r_j(x) = e^{i\mu_j x} \sum_{n=0}^{m-\eta_j} b^{(j)}_n e^{in\alpha x},$$

where $b_0^{(j)} \neq 0$, $\eta_j \leq m$, and η_j are related with $\psi_j = (\mu_{j-1} - \mu_j)/|\alpha|$ according
to the formula

$$(3.3) \qquad \eta_j - \eta_{j-1} = \begin{cases} s_j + 1 & \text{if } \psi_j \text{ is not integer and } j \text{ is odd,} \\ s_j & \text{otherwise.} \end{cases}$$

Here $s_j = E(\psi_j)$, and $E(x)$ stands for an *integer part* of x. Representation (3.2)
holds for $j = -1$ and $j = 0$ with $\mu_{-1} = \lambda$, $\mu_0 = \mu$, $\eta_{-1} = -s_0$, $\eta_0 = 0$, $b_0^{(-1)} = 1$,
$b_n^{(-1)} = 0$ for $n = 1, \ldots, m + s_0$, $b_n^{(0)} = b_n$ for $n = 0, \ldots, m$. In addition,
$\eta_0 - \eta_{-1} = s_0$, whence (3.3) is fulfilled for $j = 0$.

Suppose now that (3.2) holds for $j = k - 2$ and $j = k - 1$, and for the latter
value of j (3.3) is also true. Then, in particular,

$$s_{k-1} \leq s_{k-1} + (m - \eta_{k-1}) \leq (\eta_{k-1} - \eta_{k-2}) + (m - \eta_{k-1}) = m - \eta_{k-2},$$

so that $t_{k-1} = m - \eta_{k-2} - s_{k-1} - 1 \geq -1$. According to the recursive relations
(2.4), the AP polynomials $q_j^+ \in AP^+$ and $r_j \in AP$ for $j = k$ should be searched
for in the form

$$(3.4) \qquad \begin{aligned} q_j^+(x) &= e^{i(\mu_{j-2} - \mu_{j-1})x} \sum_{n=0}^{s_{j-1}} a_n^{(j)} e^{in\alpha x}, \\ r_j(x) &= e^{i(\mu_{j-2} + (s_{j-1}+1)\alpha)x} \sum_{n=0}^{t_{j-1}} d_n^{(j)} e^{in\alpha x}, \end{aligned}$$

where r_j is regarded to be 0 for $t_{j-1} = -1$. According to (3.4), the recursive
relation (2.2) can be rewritten as

$$\sum_{n=0}^{m-\eta_{j-2}} b_n^{(j-2)} e^{in\alpha x} = \sum_{n=0}^{s_{j-1}} a_n^{(j)} e^{in\alpha x} \sum_{\nu=0}^{m-\eta_{j-1}} b_\nu^{(j-1)} e^{i\nu\alpha x} + e^{i(s_{j-1}+1)\alpha x} \sum_{n=0}^{t_{j-1}} d_n^{(j)} e^{in\alpha x},$$

whence

$$(3.5) \qquad \sum_{\nu=0}^{n} a_\nu^{(j)} b_{n-\nu}^{(j-1)} = b_n^{(j-2)} \qquad (n = 0, \ldots, s_{j-1}),$$

and for $s_{j-1} < m - \eta_{j-2}$, in addition,

$$(3.6) \qquad \sum_{\nu=0}^{s_{j-1}} a_\nu^{(j)} b_{s_{j-1}+1+n-\nu}^{(j-1)} + d_n^{(j)} = b_{s_{j-1}+1+n}^{(j-2)} \qquad (n = 0, \ldots, t_{j-1}),$$

where it is assumed that $b_\nu^{(j-1)} = 0$ for $\nu > m - \eta_{j-1}$. Solving the system of linear equations (3.5)–(3.6), we find that

(3.7)

$$a_0^{(j)} = \left(b_0^{(j-1)}\right)^{-1} b_0^{(j-2)},$$

$$a_n^{(j)} = \left(b_0^{(j-1)}\right)^{-n-1} \det \left[\begin{array}{cccc|c} b_0^{(j-1)} & & & & b_0^{(j-2)} \\ b_1^{(j-1)} & b_0^{(j-1)} & & & b_1^{(j-2)} \\ \vdots & \vdots & \ddots & & \vdots \\ b_{n-1}^{(j-1)} & b_{n-2}^{(j-1)} & \cdots & b_0^{(j-1)} & b_{n-1}^{(j-2)} \\ b_n^{(j-1)} & b_{n-1}^{(j-1)} & \cdots & b_1^{(j-1)} & b_n^{(j-2)} \end{array}\right]$$

$$(n = 1, \dots, s_{j-1}),$$

$$d_n^{(j)} = \left(b_0^{(j-1)}\right)^{-s_{j-1}-1} \det \left[\begin{array}{cccc|c} b_0^{(j-1)} & & & & b_0^{(j-2)} \\ b_1^{(j-1)} & b_0^{(j-1)} & & & b_1^{(j-2)} \\ \vdots & \vdots & \ddots & & \vdots \\ b_{s_{j-1}}^{(j-1)} & b_{s_{j-1}-1}^{(j-1)} & \cdots & b_0^{(j-1)} & b_{s_{j-1}}^{(j-2)} \\ \hline b_{s_{j-1}+n+1}^{(j-1)} & b_{s_{j-1}+n}^{(j-1)} & \cdots & b_{1+n}^{(j-1)} & b_{s_{j-1}+n+1}^{(j-2)} \end{array}\right].$$

$$(n = 0, \dots, t_{j-1}).$$

If $r_j \not\equiv 0$, that is, $s_{j-1} < m - \eta_{j-2}$ and not all $d_n^{(j)}$ are equal zero, let us put

(3.8)

$$\eta_j = \eta_{j-2} + s_j' + s_{j-1} + 1, \ \mu_j = \mu_{j-2} + (s_j' + s_{j-1} + 1)\alpha, \ b_n^{(j)} = d_{s_j'+n}^{(j)},$$

where s_j' is the smallest value of ν for which $d_\nu^{(j)} \neq 0$. Then r_j is of the form (3.2), where $b_0^{(j)} \neq 0$, $\eta_j \leq m$, and, due to (3.8) and the inequality $\mu_{j-2} - \mu_{j-1} + (s_{j-1} + 1)\alpha < 0$,

$$\mu_j - \mu_{j-1} = (\mu_{j-2} - \mu_{j-1} + (s_{j-1} + 1)\alpha) + s_j'\alpha < 0.$$

Using (3.8) again:

(3.9)

$$s_j = E(\psi_j) = E(-\psi_{j-1}) + s_j' + s_{j-1} + 1 = \begin{cases} s_j' + 1 & \text{if } \psi_{j-1} \text{ is integer,} \\ s_j' & \text{otherwise.} \end{cases}$$

Further, it follows from (3.8), (3.3) and (3.9) that

$$\eta_j - \eta_{j-1} = 1 + s_j' + s_{j-1} + \eta_{j-2} - \eta_{j-1} =$$

$$\begin{cases} 1 + s_j' = s_j & \text{if } \psi_{j-1} \text{ is integer} \\ 1 + s_j' - \frac{1}{2}\left(1 + (-1)^j\right) = s_j + \frac{1+(-1)^{j+1}}{2} & \text{otherwise} \end{cases}$$

Since numbers ψ_{j-1} and ψ_j are integers only simultaneously, it follows from here that (3.3) holds when $j = k - 1$ is substituted by $j = k$.

Therefore, if $r_k \not\equiv 0$, then the procedure described above can be applied to r_{k-1} and r_k. It allows to find almost periodic polynomials r_j of the form (3.2) and $q_j^+ \in AP^+$ successively for $j = 1, 2, \ldots$.

From (3.3) and the equality $\mu_j - \mu_{j-2} = (\eta_j - \eta_{j-2})\alpha$ (see (3.8)), it follows that

$$
\eta_j = \begin{cases} \displaystyle\sum_{l=1}^{j} s_l, & \text{if } \dfrac{\lambda - \mu}{|\alpha|} \text{ is integer} \\[4mm] \displaystyle\sum_{l=1}^{j} s_l + E\left(\dfrac{j+1}{2}\right) & \text{otherwise,} \end{cases}
$$

$$
(3.10) \qquad\qquad \mu_j = \begin{cases} \lambda + (s_0 + \eta_j)\alpha & \text{if } j \text{ is odd} \\ \mu + \eta_j \alpha & \text{otherwise.} \end{cases}
$$

Let us find a lower bound for $\Delta_j = (\mu_j + \mu_{j-1}) - (\mu_{j-1} + \mu_{j-2})$. According to (3.8) and (3.9),

$$
\Delta_j = \mu_j - \mu_{j-2} = (s'_j + s_{j-1} + 1)\alpha = \begin{cases} (s'_j + s'_{j-1} + 2)\alpha & \text{if } \dfrac{\lambda - \mu}{|\alpha|} \text{ is integer,} \\[3mm] (s'_j + s'_{j-1} + 1)\alpha & \text{otherwise} \end{cases}
$$

with the upper equality valid for $j > 1$. Hence, at any step of our procedure, beginning from the second one, the value of $\mu_j + \mu_{j-1}$ decreases at least by $|\alpha|$ (by $2|\alpha|$, if $(\lambda - \mu)/|\alpha|$ is integer). As the final result, in finite number $k \ (\leq E\left(\dfrac{\mu + \lambda}{|\alpha|}\right) + 1)$ of steps, the quantity $\mu_k + \mu_{k-1}$ will become nonpositive. Hence, the following theorem holds:

THEOREM 3.1. *If r_0 is an almost periodic polynomial with commensurable distances between its exponents, then the matrix (2.1) is AP-factorizable. Its AP-factorization is delivered by formulas (2.9), in which r_j and q_j^+ are defined by (3.2), (3.4), (3.7) and (3.8).*

Of course, Theorem 2.2 also can be applied in the case under consideration. The upper bound for the number of steps required by this approach is $E\left(\dfrac{\lambda - \mu'}{|\alpha|}\right) + 1$ where μ' is the very left point of $\Omega(r_0)$.

3.2. Explicit formulas. It follows from (3.1) and formulas (3.10) for μ_j that for partial AP-indices of the matrix function (2.1) to be equal zero, it is necessary that at least one of the numbers $\frac{\mu}{\alpha}$ or $\frac{\lambda}{\alpha}$ is integer. Suppose that this condition is satisfied.

If $\frac{\mu}{\alpha} \in \mathbb{Z}$, then the function r_0 can be represented as

$$
(3.11) \qquad\qquad r_0 = \sum_{n=-N}^{N} c_n e^{in\beta x},
$$

where $\beta = |\alpha|$,

$$(3.12) \qquad N = \begin{cases} E(\frac{\lambda}{\beta}) & \text{if } \frac{\lambda}{\beta} \notin \mathbb{N} \\ \frac{\lambda}{\beta} - 1 & \text{if } \frac{\lambda}{\beta} \in \mathbb{N}. \end{cases}$$

If $\frac{\mu}{\alpha} \notin \mathbb{Z}$, then, under condition $\frac{\lambda}{\alpha} \in \mathbb{Z}$, the function r_0 can be represented as

$$(3.13) \qquad r_0(x) = \sum_{n=-N}^{N-1} c_n e^{i(\nu + n\beta)x},$$

where $0 < \nu < \beta < \nu + \lambda$ and $N = E\left(\frac{\nu+\lambda}{\beta}\right)$.

The criterion for partial AP-indices of the matrix (2.1) with r_0 given by (3.11) or (3.13) to be equal 0 rests on the following approximating technique. This technique, as well as Theorem 3.2 of this subsection, was developed by the authors jointly with P. Tishin (see [16, 17, 5]).

Let A be an $n \times n$ AP matrix function with the finite spectrum $\Omega(A)$. Consider

$$\text{span } \Omega(A) = \{\sum m_k \omega_k : m_k \in \mathbb{Z}, \omega_k \in \Omega(A)\}.$$

Choose a basis $H = \{h_1, \dots, h_s\}$ in $\text{span } \Omega(A)$ linearly independent over $\mathbb{Z}$. Then a lattice

$$\text{span } H = \{\sum_{k=1}^{s} m_k h_k : m_k \in \mathbb{Z}, \ k = 1, \dots, s\}$$

coincides with $\text{span } \Omega(A)$, and A can be represented as an AP polynomial

$$A(x) = \sum_{m_1, \dots, m_s} a_{m_1, \dots, m_s} e^{i(m_1 h_1 + \dots + m_s h_s)x}$$

with the coefficients $a_{m_1, \dots, m_s} \in \mathbb{C}^{n \times n}$.

Along with A, for any set $H' = \{h'_1, \dots, h'_s\}$ consisting of s real numbers, let us consider a matrix function

$$A'(x) = \sum_{m_1, \dots, m_s} a_{m_1, \dots, m_s} e^{i(m_1 h'_1 + \dots + m_s h'_s)x}$$

with the finite spectrum $\Omega(A') \subset \text{span } H'$. The mapping $A \mapsto A'$ can be extended to a homomorphism $\varphi_{H'} : AP_W(H) \to AP_W(H')$, where

$$AP_W(X) = \{B \in AP_W : \Omega(B) \subset \text{span } X\}.$$

LEMMA 3.1. *Let A admit an AP-factorization $A = A_+ \Lambda A_-$, where $A_\pm$, $A_\pm^{-1}$ are AP polynomials in $AP_W^\pm$; $\Omega(A_\pm), \Omega(\Lambda) \subset \text{span}\{h_1, \dots, h_s\}$, and $h_1, \dots, h_s$ are linearly independent over $\mathbb{Z}$. Then there exists $\epsilon > 0$ such that for all $H' = \{h'_1, \dots, h'_s\}$ satisfying the condition*

$$(3.14) \qquad |h_j - h'_j| < \epsilon \qquad (j = 1, \dots, s),$$

AP-factorizations of matrix functions $A' = \varphi_{H'}(A)$ *exist and are delivered by*

$$(3.15) \qquad\qquad A' = \varphi_{H'}(A_+)\varphi_{H'}(\Lambda)\varphi_{H'}(A_-).$$

PROOF. The equality (3.15) follows from the AP-factorization of A under the action of the homomorphism $\varphi_{H'}$, and therefore holds for any set H'. It remains to show that

$$(3.16) \qquad\qquad \Omega(\varphi_{H'}(A_\pm)) \subset \mathbb{R}_\pm, \quad \Omega(\varphi_{H'}(A_\pm^{-1})) \subset \mathbb{R}_\pm$$

if (3.14) is satisfied and ϵ is small enough.

To that end, notice that $\Omega(\varphi_{H'}(A_\pm))$ is an image of $\Omega(A_\pm)$ under the mapping

$$(3.17) \qquad\qquad \chi: m_1 h_1 + \cdots + m_s h_s \mapsto m_1 h_1' + \cdots + m_s h_s'.$$

If $\lambda (\in \Omega(A_\pm)) = 0$, then in its representation $\lambda = m_1 h_1 + \cdots + m_s h_s$, all the coefficients m_j equal 0 (due to linear independence of h_j over $\mathbb{Z}$), and hence $\chi(\lambda) = 0$. For any nonzero point $\lambda \in \Omega(A_\pm)$, there exists such an ϵ_λ (due to continuity of χ) that signs of λ and $\chi(\lambda)$ coincide. Since the sets $\Omega(A_\pm)$ are finite, the first of conditions (3.16) will be ensured if we set $\epsilon = \min\{\epsilon_\lambda \colon \lambda \in \Omega(A_\pm)\}$. The same reasoning applies to the second of conditions (3.16), because $\Omega(A_\pm^{-1}) \subset \mathrm{span}\{h_1, \ldots, h_n\}$ together with $\Omega(A_\pm)$. $\square$

According to (3.15), the partial AP-indices of matrix functions A' are obtained from the partial AP-indices of A by applying the mapping (3.17). Therefore, for ϵ small enough, the number of positive, negative, and zero partial AP-indices is the same for A' and A. From here follows:

COROLLARY 3.1. *In the setting of Lemma 3.1, the following statements are equivalent:*

  (i) *the matrix A admits AP-factorization with zero partial AP-indices;*
  (ii) *there exist sequences $h_{j,n} \to h_j$ $(j = 1, \ldots, s)$ such that all the matrices $A_n = \varphi_{H_n}(A)$, where $H_n = \{h_{1,n}, \ldots h_{s,n}\}$, are AP-factorizable with zero AP-indices;*
  (iii) *for all sequences $h_{j,n} \to h_j$ $(j = 1, \ldots, s)$ and n large enough, matrices $A_n = \varphi_{H_n}(A)$ are AP-factorizable with zero AP-indices.*

In the case (3.11), consider Toeplitz matrices $T_n = (c_{i-j})_{i,j=1}^n$ with $n = N$, $N + 1$, $N + 2$ (for the latter one we agree to set $c_{N+1} = c_{-N-1} = 0$). Denote by $A_{N+2,1}$ and $A_{1,N+2}$ the cofactors of the corner elements on the secondary diagonal of T_{N+2}:

$$A_{N+2,1} = (-1)^{N+1} \begin{vmatrix} c_{-1} & \cdots & c_{-N} & 0 \\ c_0 & \ddots & c_{-N+1} & c_{-N} \\ \vdots & \ddots & \ddots & \vdots \\ c_{N-1} & \cdots & c_0 & c_{-1} \end{vmatrix},$$

$$A_{1,N+2} = (-1)^{N+1} \begin{vmatrix} c_1 & c_0 & \cdots & c_{-N+1} \\ \vdots & \ddots & \ddots & \vdots \\ c_N & c_{N-1} & \ddots & c_0 \\ 0 & c_N & \cdots & c_1 \end{vmatrix}.$$

THEOREM 3.2. *Let*

$$(3.18) \qquad G(x) = \begin{pmatrix} e^{i\lambda x} & 0 \\ \sum_{n=-N}^{N} c_n e^{in\beta x} & e^{-i\lambda x} \end{pmatrix},$$

where $0 < \beta < \lambda$ and N is given by (3.12). Then

(i) *for all partial AP-indices of G to be equal zero, it is necessary and sufficient that*

(3.19) $\det T_{N+1} \neq 0$ *if* $\dfrac{\lambda}{\beta} = N+1$, *and* $\det T_N \det T_{N+1} \neq 0$ *if* $\dfrac{\lambda}{\beta} \notin \mathbb{Z}$.

(ii) *If condition (3.19) is satisfied, then*

(3.20)

$$\mathbf{d}(G) = \begin{bmatrix} (\det T_{N+1})^{-1} A_{N+2,1} & -(\det T_{N+1})^{-1} \det T_N \\ (\det T_{N+1})^{-1} \det T_{N+2} & (\det T_{N+1})^{-1} A_{1,N+2} \end{bmatrix} \text{ if } \frac{\lambda}{\beta} = N+1,$$

and

(3.21)

$$\mathbf{d}(G) = \begin{bmatrix} 0 & -(\det T_{N+1})^{-1} \det T_N \\ (\det T_N)^{-1} \det T_{N+1} & 0 \end{bmatrix} \text{ if } \frac{\lambda}{\beta} \notin \mathbb{Z}.$$

PROOF. Consider a rational number $\frac{p}{q} \in (N, N+1]$ and a matrix function $G_{p,q}$ obtained from G by a substitution $\lambda \mapsto \beta \frac{p}{q}$. Such $G_{p,q}$ is a periodic matrix function, and, according to Lemma 1.1, its partial AP-indices equal zero if and only if the Toeplitz matrix

$$(3.22) \qquad A_p(G_{p,q}) = \left[\begin{array}{c|c} T_N \otimes I_q & \begin{bmatrix} c_{-N} \\ \vdots \\ c_{-1} \end{bmatrix} \otimes \begin{bmatrix} I_{p-Nq} \\ 0 \end{bmatrix} \\ \hline [c_N \ldots c_1] \otimes [I_{p-Nq} \ 0] & c_0 \otimes I_{p-Nq} \end{array} \right]$$

is invertible. Notice that $A_p(G_{p,q})$ is similar to the direct sum of $(N+1)q - p$ copies of T_N with $p - Nq$ copies of T_{N+1}.

Let us discuss now the cases of integer, fractional rational, and irrational $\frac{\lambda}{\beta}$ separately. If $\frac{\lambda}{\beta}$ is integer, then, according to (3.12), it equals $N+1$. Setting $\frac{p}{q} = N+1$, we see then that $G_{p,q}$ coincides with G, $\det A_p(G_{p,q})$ equals $(\det T_{N+1})^q$, and therefore partial AP-indices of G equal zero if and only if T_{N+1} is invertible.

In the case $\frac{\lambda}{\beta} \notin \mathbb{Z}$ consider $\frac{p}{q}$ strictly between $N(=E(\frac{\lambda}{\beta}))$ and $N+1$. Both integers $(N+1)q - p$ and $p - Nq$ are then positive, and therefore partial AP-indices of $G_{p,q}$ equal zero if and only if

$$(3.23) \qquad \det T_N \det T_{N+1} \neq 0.$$

If $\frac{\lambda}{\beta}$ is rational, then we may again set $\frac{p}{q} = \frac{\lambda}{\beta}$, so that G coincides with $G_{p,q}$. Hence, for rational, but not integer $\frac{\lambda}{\beta}$, the partial AP-indices of G equal zero if and only if (3.23) is satisfied.

Finally, let $\frac{\lambda}{\beta}$ be irrational. The matrix G is AP-factorizable due to Theorem 2.1, with multiples $G_\pm$ in its AP-factorization (1.2) and their inverses $G_\pm^{-1}$ being AP-polynomials, and $\Omega(G_\pm), \Omega(\Lambda) \subset \mathrm{span}\{\lambda, \beta\}$. In notations of Lemma 3.1, $G_{p,q} = \varphi_{H'}(G)$, with $H = \{\lambda, \beta\}$ and $H' = \{\frac{p}{q}\beta, \beta\}$ (remind that λ and β are linearly independent over $\mathbb{Z}$ due to irrationality of $\frac{\lambda}{\beta}$). Choosing now a sequence $\frac{p_k}{q_k}$ convergent to $\frac{\lambda}{\beta}$ and applying Corollary 3.1, we find that condition (3.23) is also necessary and sufficient for partial indices of G to be equal 0.

This concludes the proof of statement (i). To prove (ii), notice first of all that in case of zero partial AP-indices (1.2) implies

$$(3.24) \qquad G_+^{-1}G = G_-.$$

Denoting

$$(3.25) \qquad G_+^{-1} = \left(g_{kj}^+\right)_{k,j=1}^2, \quad G_- = \left(g_{kj}^-\right)_{k,j=1}^2,$$

rewrite (3.24) as a system

$$(3.26) \qquad \begin{cases} g_{k1}^+(x)e^{i\lambda x} + g_{k2}^+(x)\sum_{n=-N}^N c_n e^{in\beta x} = g_{k1}^-(x) \\ g_{k2}^+(x)e^{-i\lambda x} = g_{k2}^-(x) \end{cases}, \quad k=1,2.$$

Without loss of generality we may suppose that $\mathbf{M}(G_-) = I$, that is,

$$(3.27) \qquad \mathbf{M}(g_{11}^-) = \mathbf{M}(g_{22}^-) = 1, \ \mathbf{M}(g_{12}^-) = \mathbf{M}(g_{21}^-) = 0.$$

Then $\mathbf{d}(G) = \mathbf{M}(G_+)$, and, since $\det \mathbf{d}(G) = \mathbf{d}(\det(G)) = 1$,

$$(3.28) \qquad \mathbf{d}(G) = \begin{bmatrix} \mathbf{M}(g_{22}^+) & -\mathbf{M}(g_{12}^+) \\ -\mathbf{M}(g_{21}^+) & \mathbf{M}(g_{11}^+) \end{bmatrix}.$$

Let at first $\frac{\lambda}{\beta} \notin \mathbb{N}$. Represent g_{k2}^+ in the form

$$g_{k2}^+(x) = \sum_{m=0}^{N} a_{km} e^{im\beta x} + \tilde{g}_{k2}^+(x), \quad k = 1, 2,$$

where $\Omega(\tilde{g}_{k2}^+) = \Omega(g_{k2}^+) \setminus \{m\beta : m = 0, 1, \ldots, N\}$. Equating coefficients at $e^{im\beta x}$, $m = 0, 1, \ldots, N$, in the first of equations (3.26), we arrive to the following systems of algebraic linear equations:

$$(3.29) \qquad T_{N+1} \begin{bmatrix} a_{k0} \\ a_{k1} \\ \vdots \\ a_{kN} \end{bmatrix} = \begin{bmatrix} \mathbf{M}(g_{k1}^-) \\ 0 \\ \vdots \\ 0 \end{bmatrix}, \qquad k = 1, 2.$$

Since $\det T_{N+1} \neq 0$, (3.29) yields

$$\mathbf{M}(g_{k2}^+) = a_{k0} = \mathbf{M}(g_{k1}^-) \det T_N / \det T_{N+1} \quad (k = 1, 2).$$

From here and (3.27):

$$\mathbf{M}(g_{12}^+) = \det T_N / \det T_{N+1}, \quad \mathbf{M}(g_{22}^+) = 0,$$

and so (3.28) takes form:

$$(3.30) \qquad \mathbf{d}(G) = \begin{bmatrix} 0 & -\det T_N / \det T_{N+1} \\ * & * \end{bmatrix}.$$

On the other hand, from (3.4), (3.10) and $\frac{\lambda}{\beta} \notin \mathbb{Z}$, it follows that $\mathbf{M}(q_j^+) = 0$. Corollary 2.1 implies then that for a certain k,

$$(3.31) \qquad \mathbf{d}(G) = \begin{pmatrix} 0 & 1 \\ 1 & 0 \end{pmatrix}^k \operatorname{diag}[a_1^{-1} \cdots a_k^{-1}, (-1)^k a_1 \cdots a_k].$$

Comparing the latter formula with (3.30), we find that this is possible only if k is odd and $a_1 \cdots a_k = \det T_N / \det T_{N+1}$. Plugging these values back into (3.31), we arrive to the desired formula (3.21).

Let now $\frac{\lambda}{\beta} \in \mathbb{N}$. Then, according to (3.12), $\frac{\lambda}{\beta} = N+1$, and the first equations (3.26) can be rewritten as

$$(3.32) \qquad g_{k1}^+(x) e^{i(N+1)\beta x} + g_{k2}^+(x) \sum_{n=-N}^{N} c_n e^{in\beta x} = g_{k1}^-(x), \quad k = 1, 2.$$

But from (3.27) and the second equations (3.26), it follows that

$$g_{12}^+(x) = \mathbf{M}(g_{12}^+) + \sum_{m=1}^{N} a_{1m} e^{im\beta x}, \quad g_{22}^+(x) = \mathbf{M}(g_{22}^+) + \sum_{m=1}^{N} a_{2m} e^{im\beta x} + e^{i(N+1)\beta x}.$$

Plugging these expansions into (3.32) and equating coefficients at $e^{im\beta x}$ ($m = 0, 1, \ldots, N+1$), we arrive to linear systems

$$
\begin{bmatrix}
c_0 & c_{-1} & \cdots & c_{-N} & 0 \\
c_1 & c_0 & \cdots & c_{-N+1} & 0 \\
\vdots & \vdots & \ddots & \vdots & \vdots \\
c_N & c_{N-1} & \cdots & c_0 & 0 \\
0 & c_N & \cdots & c_1 & 1
\end{bmatrix}
\begin{bmatrix}
\mathbf{M}(g_{k2}^+) \\
a_{k1} \\
\vdots \\
a_{kN} \\
\mathbf{M}(g_{k1}^+)
\end{bmatrix}
= B_k \quad (k = 1, 2),
$$

where $B_1 = \begin{bmatrix} 1 \\ 0 \\ \vdots \\ 0 \\ 0 \end{bmatrix}$, $B_2 = \begin{bmatrix} 0 \\ -c_{-N} \\ \vdots \\ -c_{-1} \\ -c_0 \end{bmatrix}$. Solving these systems, we find that

$$
\mathbf{M}(g_{11}^+) = A_{1,N+2} / \det T_{N+1}, \quad \mathbf{M}(g_{12}^+) = \det T_N / \det T_{N+1},
$$
$$
\mathbf{M}(g_{21}^+) = -\det T_{N+2} / \det T_{N+1}, \quad \mathbf{M}(g_{22}^+) = A_{N+2,1} / \det T_{N+1}.
$$

From here and (3.28) follows (3.20). $\square$

Combining Theorem 3.2 with [11, Theorem 2 and Remark 2], we obtain

COROLLARY 3.2. *Let in* (3.18) *the number* $\frac{\lambda}{\beta}$ *be not integer,* $c_{-N} = \cdots = c_{-2} = 0$, $c_{-1} \neq 0$, *and*

$$
\text{(3.33)} \qquad\qquad \frac{1}{r_0(x)} = \sum_{k=0}^{\infty} \gamma_k e^{i(k+1)\beta x}.
$$

Then for partial AP-indices of the matrix function G to be equal zero, it is necessary and sufficient that $\gamma_N \gamma_{N+1} \neq 0$.

In case of integer $\frac{\lambda}{\beta}$ and r_0 satisfying (3.33) it follows from [11, Theorem 2] directly that partial AP-indices of (3.18) equal zero if and only if $\gamma_{N+1} \neq 0$.

Consider now the case (3.13). Introduce Toeplitz matrices

$$
\Delta_1 =
\begin{bmatrix}
c_0 & c_{-1} & \cdots & c_{-N+1} \\
c_1 & c_0 & \cdots & c_{-N+2} \\
\vdots & \vdots & \ddots & \vdots \\
c_{N-1} & c_{N-2} & \cdots & c_0
\end{bmatrix},
\quad
\Delta_2 =
\begin{bmatrix}
c_{-1} & c_{-2} & \cdots & c_{-N} \\
c_0 & c_{-1} & \cdots & c_{-N+1} \\
\vdots & \vdots & \ddots & \vdots \\
c_{N-2} & c_{N-3} & \cdots & c_{-1}
\end{bmatrix},
$$

obtained from T_{N+1} by deleting its $(N+1)$-st row and $(N+1)$-st (respectively, 1-st) column.

THEOREM 3.3. *Let* $0 < \nu < \beta < \lambda + \nu$ *and* $N = E\left(\frac{\lambda+\nu}{\beta}\right)$. *Then the matrix function*

$$(3.34) \qquad G(x) = \begin{pmatrix} e^{i\lambda x} & 0 \\ \sum_{n=-N}^{N-1} c_n e^{i(\nu+n\beta)x} & e^{-i\lambda x} \end{pmatrix},$$

is AP-factorizable with zero partial AP-indices if and only if

$$(3.35) \qquad \lambda = N\beta \text{ and } \det \Delta_1 \det \Delta_2 \neq 0.$$

If (3.35) *is satisfied, then*

$$(3.36) \qquad \mathbf{d}(G) = (-1)^N \operatorname{diag}[\det \Delta_2 / \det \Delta_1, \det \Delta_1 / \det \Delta_2].$$

PROOF. According to (3.10), for partial AP-indices of (3.34) to be equal zero, it is necessary that $\frac{\lambda}{\beta} \in \mathbb{Z}$. This, along with inequalities $0 < \nu < \beta < \lambda + \nu$, implies

$$\frac{\lambda}{\beta} = E\left(\frac{\lambda + \nu}{\beta}\right) = N \geq 1.$$

Suppose therefore that condition $\lambda = N\beta$ is satisfied. For any rational number $\frac{p}{q} \in (0,1)$, let

$$G_{p,q}(x) = \begin{pmatrix} e^{iN\beta x} & 0 \\ \sum_{n=-N}^{N-1} c_n e^{i(\frac{p}{q}+n)\beta x} & e^{-iN\beta x} \end{pmatrix}.$$

According to Lemma 1.1, the partial AP-indices of this matrix function equal zero if and only if

$$(3.37) \qquad \det A_{Nq}(G_{p,q}) \neq 0,$$

where

$$A_{Nq}(G_{p,q}) = \left(\sum_{-N}^{N-1} c_n \delta_{i,j+p+nq}\right)_{i,j=1}^{Nq}.$$

Multiplying $A_{Nq}(G_{p,q})$ by nonsingular matrices $\begin{bmatrix} I_N \otimes \begin{bmatrix} 0_{(q-p)\times p} & I_{q-p} \end{bmatrix} \\ I_N \otimes \begin{bmatrix} I_p & 0_{p\times(q-p)} \end{bmatrix} \end{bmatrix}$ from

the left, and $\begin{bmatrix} I_N \otimes \begin{bmatrix} I_{q-p} \\ 0_{p\times(q-p)} \end{bmatrix} & I_N \otimes \begin{bmatrix} 0_{(q-p)\times p} \\ I_p \end{bmatrix} \end{bmatrix}$ from the right, we obtain a matrix

$$\operatorname{diag}[\Delta_1 \otimes I_{q-p}, \Delta_2 \otimes I_p].$$

Therefore, condition (3.37) is equivalent to

$$(3.38) \qquad \det \Delta_1 \det \Delta_2 \neq 0$$

and does not depend on the choice of $p, q \in \mathbb{N}$ as long as $p < q$. Setting $\frac{p}{q} = \frac{\nu}{\beta}$ in case of rational $\frac{\nu}{\beta}$, and applying Corollary 3.1 otherwise, we obtain from here that (3.38) is a necessary and sufficient condition not only for AP-indices of $G_{p,q}$, but also for partial AP-indices of G itself to be equal zero (compare with the corresponding part of the proof of Theorem 3.2).

It remains to obtain a formula for $\mathbf{d}(G)$ when (3.35) is satisfied. As in Theorem 3.2, plug (3.25) in (3.24):

$$(3.39) \quad \begin{cases} g_{k1}^+(x)e^{iN\beta x} + g_{k2}^+(x)\sum_{n=-N}^{N-1} c_n e^{i(\nu+n\beta)x} = g_{k1}^-(x) \\ g_{k2}^+(x)e^{-iN\beta x} = g_{k2}^-(x) \end{cases}, \quad k = 1,2.$$

Without loss of generality we may suppose that (3.27) is satisfied. Then, as in the proof of Theorem 3.2, formula (3.28) applies. Representing g_{k2}^+ in the form

$$g_{12}^+(x) = \sum_{m=0}^{N-1} a_{1m}e^{im\beta x} + \tilde{g}_{12}^+(x), \; g_{22}^+(x) = \sum_{m=0}^{N} a_{2m}e^{im\beta x} + e^{iN\beta x} + \tilde{g}_{22}^+(x),$$

where $\Omega(\tilde{g}_{12}^+) = \Omega(g_{12}^+)\backslash\{m\beta\colon m = 0,1,\dots,N-1\}$, $\Omega(\tilde{g}_{22}^+) = \Omega(g_{22}^+)\backslash\{m\beta\colon m = 0,1,\dots,N\}$, and equating the coefficients of $e^{i(\nu+m\beta)x}$ ($m = 0,1,\dots,N-1$) in the left and right sides of the first equations in (3.39), we arrive to linear algebraic systems:

$$\Delta_1 \begin{bmatrix} a_{10} \\ \vdots \\ a_{1,N-1} \end{bmatrix} = \begin{bmatrix} 0 \\ \vdots \\ 0 \end{bmatrix}, \quad \Delta_1 \begin{bmatrix} a_{20} \\ \vdots \\ a_{2,N-1} \end{bmatrix} = - \begin{bmatrix} c_{-N} \\ \vdots \\ c_{-1} \end{bmatrix}.$$

From here and (3.38):

$$(3.40) \qquad \mathbf{M}(g_{12}^+) = a_{10} = 0, \quad \mathbf{M}(g_{22}^+) = a_{20} = (-1)^N \det\Delta_2/\det\Delta_1,$$

so that (3.28) implies

$$(3.41) \qquad\qquad \mathbf{d}(G) = \begin{bmatrix} (-1)^N \det\Delta_2/\det\Delta_1 & 0 \\ * & * \end{bmatrix}.$$

On the other hand, from $0 < \nu < \beta$, $\lambda = N\beta$, (3.4) and (3.10), it follows that $\mathbf{M}(q_j^+) = 0$, and, according to Corollary 2.1, the matrix $\mathbf{d}(G)$ has structure (3.31). Comparing that with (3.41), we find that it is possible only if k is even, and $a_1^{-1}\cdots a_k^{-1} = (-1)^N \det\Delta_2/\det\Delta_1$. Finally, from here and (3.31) follows (3.36). $\square$

4. Trinomials r_0

Theory developed in the previous section applies, in particular, to matrices (2.1) with binomial term r_0, since in this case the set of all distances between points of $\Omega(r_0)$ consists of just one element. The AP-factorization (1.2) can be obtained in $k \leq 3$ steps by applying either Theorem 2.1 or Theorem 2.2. By a clever choice between these theorems, it is always possible to reduce the number k of steps required to 2. The direct computation of $G_\pm$, partial AP-indices, and $\mathbf{d}(G)$ in this situation has been carried out in [7, Theorem 2.3].

In this section we consider the case of trinomial r_0 with non-commensurable distances between points of $\Omega(r_0)$ (otherwise results of Section 3 would still be applicable). In other words,

$$(4.1) \qquad r_0 = b_0 e^{i\mu x}(1 - b_1 e^{i\gamma_1 x} - b_2 e^{i\gamma_2 x}), \quad b_0 b_1 b_2 \neq 0,$$

where

$$\lambda > \mu > \mu + \gamma_1 > \mu + \gamma_2 > -\lambda \text{ and } \gamma_2/\gamma_1 \text{ is irrational.}$$

Suppose also that

$$(4.2) \qquad \mu + \gamma_1 < 0, \quad \gamma_2 = \mu - \lambda.$$

Expanding γ_2/γ_1 into a continued fraction $[k_1, k_2, \ldots]$, $k_i \in \mathbb{N}$ (see [10, 4]), put

$$(4.3) \qquad \beta_{-1} = \gamma_2, \ \beta_0 = \gamma_1, \ \beta_j = \beta_{j-2} - k_j \beta_{j-1} \quad (j \in \mathbb{N}).$$

The sequence $\{\beta_j\}$ is increasing, and assumes negative values only. Hence, it converges to the limit $\beta_\infty = \lim_{j \to \infty} \beta_j \leq 0$. If we suppose that $\beta_\infty < 0$, then, due to (4.3), $\beta_j - \beta_{j-2} = k_j|\beta_{j-1}| > |\beta_\infty|$, which is a contradiction with negativity of all β_j. Therefore, $\beta_j \uparrow 0$.

According to (2.4),

$$r_1 = \sum_{n \in N_1(\lambda, r_0)} c_n b_1 e^{i(\gamma_1 + \langle n, \gamma \rangle)x} + \sum_{n \in N_2(\lambda, r_0)} c_n b_2 e^{i(\gamma_2 + \langle n, \gamma \rangle)x},$$

where $\gamma = (\gamma_1, \gamma_2)$, c_n are determined by (2.5), and, due to (4.2)–(4.3),

$$N_1(\lambda, r_0) = \{(k_1, 0), (0, 1)\}, N_2(\lambda, r_0) = \{(1, 0), \ldots, (k_1, 0), (0, 1)\}.$$

Therefore,

$$r_1 = b_1^{k_1+1} e^{i(\lambda + (k_1+1)\gamma_1)x} + 2b_1 b_2 e^{i(\lambda + \gamma_1 + \gamma_2)x} +$$

$$\sum_{k=2}^{k_1} b_1^k b_2 e^{i(\lambda + k\gamma_1 + \gamma_2)x} + b_2^2 e^{i(\lambda + 2\gamma_2)x}, \quad \mu_1 = \lambda + (k_1 + 1)\gamma_1 = \mu + \beta_0 - \beta_1.$$

Consider the case $k_1 > 1$ at first. Using (4.3), one gets

$$(4.4) \qquad \begin{aligned} r_0 &= b_0 e^{i\mu x}(1 - b_1 e^{i\beta_0 x}) + \varphi_0, \\ r_1 &= b_{01} e^{i(\mu + \beta_0 - \beta_1)x}(1 - b_{11} e^{i\beta_1 x} - b_{21} e^{i(\beta_1 + \beta_0)x}) + \psi_1, \end{aligned}$$

where $b_{01} = b_1^{k_1+1}$, $b_{11} = -2b_1^{-k_1} b_2$, $b_{21} = -b_1^{-k_1+1} b_2$, and φ_0, ψ_1 are AP polynomials of orders $\mathrm{ord}\,\varphi_0 < \mu + \beta_0 + \beta_1$ and $\mathrm{ord}\,\psi_1 < \mu + 2\beta_0$ respectively.

Let us now show that if

$$(4.5)$$
$$r_{j-1} = b_{0,j-1} e^{i(\mu + \beta_0 - \beta_{j-1})x}(1 - b_{1,j-1} e^{i\beta_{j-1}x} - d_{j-1} e^{i(\beta_{j-1} + \beta_j)x}) + \varphi_{j-1},$$

$$(4.6) \qquad r_j = b_{0,j} e^{i(\mu + \beta_0 - \beta_j)x}(1 - b_{1,j} e^{i\beta_j x} - b_{2,j} e^{i(\beta_{j-1} + \beta_j)x}) + \psi_j,$$

where $b_{0,j-1}b_{1,j-1} \neq 0$, $b_{0,j}b_{1,j}b_{2,j} \neq 0$, φ_{j-1} and $\dot{\psi}_j$ are AP polynomials of orders $\operatorname{ord} \varphi_{j-1} < \mu + \beta_0 + \beta_j$ and $\operatorname{ord} \psi_j < \mu + \beta_0 + \beta_{j-1}$ respectively, then under the condition $d_{j-1} \neq b_{2,j}$, functions r_s $(s = j+1, j+2, \dots)$ are of the form (4.6). Indeed, $\beta_j + \beta_{j-1} < \beta_{j-1} - \beta_j = \mu_j - \mu_{j-1}$, and so $N(\mu_{j-1}, r_j) = \{(k, 0, \dots, 0): k = 0, \dots k_{j+1} - 1\}$. From here according to (2.4) follows:

$$q_{j+1}^+ = b_{0,j-1}b_{0,j}^{-1} e^{i(\beta_j - \beta_{j-1})x} \sum_{k=0}^{k_{j+1}-1} b_{1,j}^k e^{ik\beta_j x},$$

and therefore

$$r_{j+1} = r_{j-1} - q_{j+1}^+ r_j = b_{0,j-1} b_{1,j}^{k_{j+1}} e^{i(\mu + \beta_0 - \beta_{j+1})x} - $$
$$b_{0,j-1} b_{1,j-1} e^{i(\mu+\beta_0)x} + b_{0,j-1}(b_{2,j} - d_{j-1}) e^{i(\mu+\beta_0+\beta_j)x} + \psi_{j+1},$$

where

$$\psi_{j+1} = \varphi_{j-1} + b_{0,j-1} b_{2,j} e^{i(\mu+\beta_0+\beta_j)x} \sum_{k=1}^{k_{j+1}-1} b_{1,j}^k e^{ik\beta_j x} + q_{j+1}^+ \psi_j$$

and $\operatorname{ord} \psi_{j+1} < \mu + \beta_0 + \beta_j$. Since r_{j+1} has the form (4.6) and r_j can be represented as (4.5) with $d_j = 0$ and $\varphi_j = -b_{0,j}b_{2,j} e^{i(\mu+\beta_0+\beta_{j-1})x} + \psi_j$, then r_{j+2} also has the form (4.6). According to the induction principle, it means that the representation (4.6) is valid for all r_s $(s \geq j)$.

Equalities (4.4) mean that the representation (4.6) holds for $j = 1$. Therefore, (4.6) is true for all $j \in \mathbb{N}$, whence $\mu_j = \mu + \beta_0 - \beta_j$. But then $\lim_{j\to\infty} \mu_j = \mu + \beta_0 = \mu + \gamma_1 < 0$.

Let now $k_1 = 1$. Then

$$r_1 = e^{i(\mu + \beta_0 - \beta_1)x} (b_1^2 + 2b_1 b_2 e^{i\beta_1 x} + b_2^2 e^{2i\beta_1 x}).$$

The direct substitution into (2.2) yields

$$q_2^+ = e^{i(\beta_1 - \beta_0)x} \sum_{k=0}^{k_2 - 1} (-1)^k (k+1) b_1^{-k-2} b_2^k b_0 e^{ik\beta_1 x}, \quad r_2 = -b_0 b_1 e^{i(\mu+\gamma_1)x} - $$
$$b_0 b_2 e^{i(\mu+\gamma_2)x} + e^{i(\mu+k_2\beta_1)x} (-1)^{k_2} b_1^{-k_2} b_2^{k_2} b_0 (k_2 + 1 + k_2 b_1^{-1} b_2 e^{i\beta_1 x}).$$

Representing r_2 in the form

$$r_2 = b_{02} e^{i(\mu+\beta_0-\beta_2)x} (1 - b_{12} e^{i\beta_2 x} - b_{22} e^{i\beta_1 x} - b_{32} e^{i(\beta_2+\beta_1)x}),$$

where

(4.7)
$$b_{02} = (-b_1^{-1} b_2)^{k_2} b_0 (k_2 + 1), \quad b_{12} = b_{02}^{-1} b_0 b_1,$$
$$b_{22} = -b_1^{-1} b_2 (k_2 + 1)^{-1} k_2, \quad b_{32} = b_{02}^{-1} b_0 b_2,$$

we obtain

$$q_3^+ = b_1^2 b_{02}^{-1} e^{i(\beta_2 - \beta_1)x} \sum_{k=0}^{k_3-1} b_{12}^k e^{ki\beta_2 x}, \ r_3 = b_1^2 b_{12}^{k_3} e^{i(\mu+\beta_0-\beta_3)x} + b_1^2 (b_{22} e^{i(\mu+\beta_0)x} +$$

$$b_{32} e^{i(\mu+\beta_0+\beta_2)x}) \sum_{k=0}^{k_3-1} b_{12}^k e^{ik\beta_2 x} + 2b_1 b_2 e^{i(\mu+\beta_0)x} + b_2^2 e^{i(\mu+\beta_0+\beta_1)x}.$$

Note that r_2 and r_3 are represented in the form (4.5) and (4.6), respectively, with

$$b_{03} = b_1^2 b_{12}^{k_3}, \ b_{13} = -b_{12}^{-k_3}(2b_1^{-1}b_2 + b_{22}),$$

$$d_2 = \begin{cases} b_{22}, \\ 0 \end{cases} \quad b_{23} = \begin{cases} -b_{12}^{-1} b_{32} \\ -b_{12}^{-k_3}(b_{32} + b_{22}b_{12}) \end{cases} \quad \text{if} \quad \begin{cases} k_3 = 1 \\ k_3 > 1, \end{cases}$$

and, according to (4.7),

$$b_{13} = -b_{12}^{-k_3}(2b_1^{-1}b_2 - b_1^{-1}b_2(k_2+1)^{-1}k_2) \neq 0,$$

$$d_2 - b_{23} = b_1^{-1}b_2(1 - (k_2+1)^{-1}k_2) \neq 0, \text{ if } k_3 = 1,$$

$$d_2 - b_{23} = -b_{23} = b_{12}^{-k_3}b_{02}^{-1}b_0 b_2(1 - (k_2+1)^{-1}k_2) \neq 0, \text{ if } k_3 > 1.$$

Therefore, r_j has a form (4.6) for all $j = 3, 4, \ldots$, and the value of $\mu_\infty = \lim_{j \to \infty} \mu_j$ is again negative.

The same way it is possible to prove negativity of μ_∞ in the case $\mu + \gamma_1 < 0$, $\gamma_2 = \mu - \lambda + \gamma_1$. If $\mu + \gamma_1 < 0$ and $\mu - \lambda = \gamma_1 E(\gamma_2/\gamma_1)$, then the functions r_j (at least, for $j \geq 4$) can be represented by formulas (4.6), in which parameters β_j are defined by (4.3) with β_{-1} substituted by $(2E(\gamma_2/\gamma_1) + 1)\gamma_1 - \gamma_2$. Hence, $\mu_\infty < 0$ in this case also.

According to the reasoning above, the following result holds:

THEOREM 4.1. *Let r_0 in the matrix (2.1) be given by the formula (4.1), in which $\mu + \gamma_1 < 0$ and $\mu - \lambda$ coincides with one of the numbers γ_2, $\gamma_2 - \gamma_1$, $\gamma_1 E(\gamma_2/\gamma_1)$. Then the matrix function (2.1) is AP-factorizable, and its AP-factorization can be constructed explicitly according to Theorem 2.1:*

Using the transformation (2.12) (compare with the reasoning preceding Theorem 2.2), we obtain the following analogue of Theorem 4.1.

THEOREM 4.2. *Let r_0 in the matrix (2.1) be given by the formula (4.1), in which $\mu + \gamma_1 > 0$ and $\mu + \lambda + \gamma_2$ coincides with one of the numbers $-\gamma_2$, $-\gamma_1$, $(\gamma_1 - \gamma_2)E(\gamma_2/(\gamma_2 - \gamma_1))$. Then the matrix function (2.1) is AP-factorizable, and its AP-factorization can be constructed explicitly according to Theorem 2.2.*

Based on Corollaries 2.1 and 2.2, it is possible to write down the necessary and sufficient conditions for partial AP-indices to be equal zero, and give the explicit formulas for $\mathbf{d}(G)$ when these conditions are satisfied, in the setting of Theorems 4.1, 4.2. Let us do it for the case (4.2) only.

COROLLARY 4.1. *Suppose that conditions (4.1)–(4.2) are satisfied, and γ_2/γ_1 is irrational. Then for a matrix (2.1) to be AP-factorizable with zero partial AP-indices, it is necessary and sufficient that the sequence (4.3) assume value $\mu + \gamma_1$: $\mu + \gamma_1 = \beta_k$ for some $k \in \mathbb{N}$* [1]. *In this case,*

$$(4.8) \qquad \mathbf{d}(G) = \begin{bmatrix} b_0^{-1}b_2 & 1 \\ 1 & 0 \end{bmatrix} \begin{bmatrix} 0 & 1 \\ 1 & 0 \end{bmatrix}^k \begin{bmatrix} b_{0,k} & 0 \\ 0 & (-1)^{k+1}b_{0,k}^{-1} \end{bmatrix},$$

where $b_{0,0} = b_0$, $b_{1,0} = b_1$, $b_{0,1} = b_1^{k_1+1}$, $b_{1,1} = -2b_1^{-k_1}b_2$; for $k_1 = 1$ $b_{0,2} = (-b_1^{-1}b_2)^{k_2}b_0(k_2+1)$, $b_{1,2} = b_{0,2}^{-1}b_0b_1$, $b_{0,3} = b_1^2 b_{1,2}^{k_3}$, $b_{1,3} = -b_{1,2}^{-k_3}(2b_1^{-1}b_2 - b_1^{-1}b_2(k_2+1)^{-1}k_2)$, and, finally, $b_{0,j} = b_{0,j-2}b_{1,j-1}^{k_j}$, $b_{1,j} = b_{1,j-1}^{-k_j}b_{1,j-2}$ for $j \geq 2$ when $k_1 > 1$, and for $j \geq 4$ when $k_1 = 1$.

The proof of (4.8) is based, in particular, on the fact that $\mathbf{M}(q_1^+) = b_0^{-1}b_2$, since

$$q_1^+ = b_0^{-1}e^{i(\lambda-\mu)x}(b_2 e^{i\gamma_2 x} + \sum_{n=0}^{k_1} b_1^n e^{in\gamma_1 x}) \text{ and } \lambda - \mu = \gamma_2,$$

and $\mathbf{M}(q_j^+) = 0$ for $j > 1$.

Let us conclude this section with an example of a matrix (2.1) with the element r_0 of the type (4.1) the AP-factorization of which cannot be constructed applying Theorems 2.1 and 2.2. Namely, put

$$(4.9) \qquad G(x) = \begin{bmatrix} e^{i\frac{1+\sqrt{5}}{2}x} & 0 \\ e^{-i\frac{\sqrt{5}-1}{2}x} - 1 + ce^{ix} & e^{-i\frac{1+\sqrt{5}}{2}x} \end{bmatrix},$$

where $c \neq 0$. In the notations of (4.1), for this matrix $\mu = 1$, $\gamma_1 = -1$, $\gamma_2 = -\frac{1+\sqrt{5}}{2}$, $\lambda = \frac{1+\sqrt{5}}{2}$, and therefore

$$(4.10) \qquad \mu + \gamma_1 = 0 \text{ and } \gamma_2 = -\lambda.$$

Such a relation between exponents is a limiting case for the situations $\pm(\mu+\gamma_1) = \gamma_2 + \lambda < 0$ considered in Theorems 4.1, 4.2.

It is well known (see [10, 4]) that a continued fraction for $\frac{1+\sqrt{5}}{2}$ is $[1, 1, \ldots]$. According to that, $\mu_{-1} = \lambda$, $\mu_0 = 1$ and $\mu_j = \mu_{j-2} - \mu_{j-1}$ $(j = 1, 2, \ldots)$. Relations (2.2) in our situation yield

$$q_j^+ = a_{j-2}a_{j-1}^{-1}e^{i\mu_j x}, \quad r_j = a_j e^{i\mu_j x} - d_j + \psi_j(x) \quad (j \geq 1)$$

where $a_{-1} = 1$, $a_0 = c$,

$$a_j = \begin{cases} a_{j-2}a_{j-1}^{-1} \\ a_{j-2}a_{j-1}^{-1}c^{-1}, \end{cases} \quad d_j = \begin{cases} c^{-1} \\ 1 \end{cases} \text{ if } \begin{cases} j \text{ is odd} \\ j \text{ is even,} \end{cases} \quad j \geq 1,$$

---

[1]It is impossible for $\mu + \gamma_1$ to coincide with β_0 or β_{-1} due to (4.1)–(4.2).

$\psi_1 = 0$, $\psi_j \in AP$ and ord $\psi_j = -\mu_{j-1}$ for $j > 1$. Since $\mu_j + \mu_{j-1} > 0$ for all $j \in \mathbb{N}$, Theorem 2.1 is not applicable for the investigation of AP-factorizability of the matrix (4.9). Analogously, based on the decomposition $\lambda/\alpha = (3 + \sqrt{5})/2 = [2, 1, 1, \ldots]$, it is possible to demonstrate the non-applicability of Theorem 2.2 to the construction of AP-factorization for the matrix (4.9).

It can be shown [9] that when $|c| = 1$, the matrix (4.9) is indeed not AP-factorizable. For $|c| \neq 1$, the AP-factorization of this matrix exists, though with factors $G_{\pm}$ not being AP polynomials.

REFERENCES

1. A. Böttcher, Yu. Karlovich, and I. Spitkovsky, *Toeplitz operators with semi-almost periodic symbols on spaces with Muckenhoupt weight*, Integr. Equat. and Oper. Theory **18** (1994), 261–276.
2. G. N. Chebotarev, *Partial indices of the Riemann boundary value problem with a second order triangular matrix coefficient*, Uspehi Mat. Nauk **11** (1956), no. 3, 192–202, in Russsian.
3. I. Gohberg and M. G. Krein, *Systems of integral equations on a half-line with kernel depending upon the difference of the arguments*, Uspehi Mat. Nauk **13** (1958), no. 2, 3–72 (in Russian), English translation in *Amer. Math. Soc. Transl.* **14** (1960), no. 2, 217–287.
4. W. B. Jones and W. J. Thron, *Continued fractions. Analytic theory and applications*, Addison-Wesley, 1980.
5. Yu. I. Karlovich, *The continuous invertibility of functional operators in Banach spaces*, Poloniya Matematica (1994), to appear.
6. Yu. I. Karlovich and I. M. Spitkovsky, *On the theory of systems of equations of convolution type with semi-almost-periodic symbols in spaces of Bessel potentials*, Soviet Math. Dokl. **33** (1986), 145–149.
7. _____, *Factorization of almost periodic matrix-valued functions and the Noether theory for certain classes of equations of convolution type*, Izv. Akad. Nauk SSSR, Ser. Mat **53** (1989), no. 2, 276–308 (in Russian), English translation in Mathematics of the USSR, Izvestiya **34** (1990), 281–316.
8. _____, *(Semi)-Fredholmness of convolution operators on the spaces of Bessel potentials*, Operator Theory: Advances and Applications, Birkhäuser-Verlag, 1993.
9. _____, *Factorization of almost periodic matrix functions*, J. Math. Anal. Appl., to appear.
10. A. Ya. Khintchine, *Continued fractions*, GIFML, Moscow, 1961 (in Russian), English translation: P. Noordhoff, Groningen, 1963.
11. N. Krupnik and I. Feldman, *Relations between factorization and invertibility of finite Toeplitz matrices*, Izvestiya Akademii Nauk Moldavskoi SSR. Serya fiziko-tehnicheskih i matematicheskih nauk (1985), no. 3, 20–26 (in Russian).
12. B. M. Levitan, *Almost periodic functions*, GITTL, Moscow, 1953 (in Russian).
13. B. M. Levitan and V. V. Zhikov, *Almost periodic functions and differential equations*, Cambridge University Press, 1982.
14. G. S. Litvinchuk and I. M. Spitkovsky, *Factorization of measurable matrix functions*, Birkhäuser Verlag, Basel and Boston, 1987.
15. I. Spitkovsky, *Factorization of several classes of semi-almost periodic matrix functions and applications to systems of convolution equations*, Izvestiya VUZ., Mat. (1983), no. 4, 88–94 (in Russian), English translation in *Soviet Math. – Iz. VUZ* **27** (1983), 383–388.
16. I. Spitkovsky and P. M. Tishin, *Factorization of new classes of almost-periodic matrix functions*, Reports of the extended sessions of a seminar of the I. N. Vekua Institute for Applied Mathematics **3** (1989), no. 1, 170–173 (in Russian).
17. P. M. Tishin, *Factorization and formulas for partial indices of certain classes of matrix functions*, Ph.D. thesis, Odessa State University, 1990.

Hydroacoustic Department, Marine Hydrophysical Institute, Ukrainian Academy of Sciences, Soviet Army Str. 3, 270100 Odessa, Ukraine
Current address: Fachbereich Mathematik, TU Chemnitz–Zwickau, PSF 964, D-09009, Chemnitz, Germany
E-mail address: karlovic@mathematik.tu-chemnitz.de

Department of Mathematics, The College of William and Mary, Williamsburg, VA 23187-8795, USA
E-mail address: ilya@cs.wm.edu

Contemporary Mathematics
Volume **189**, 1995

On the IVP for the Nonlinear Schrödinger Equations

CARLOS E. KENIG, GUSTAVO PONCE AND LUIS VEGA

1. Introduction

In this paper we shall study the initial value problem (IVP) for nonlinear Schrödinger equations of the form

$$(1.1) \qquad \begin{cases} \partial_t u = i\Delta u + F(u, \nabla_x u, \overline{u}, \nabla_x \overline{u}), & t \in \mathbb{R},\ x \in \mathbb{R}^n \\ u(x,0) = u_0(x) \end{cases}$$

where $u = u(x,t)$ is a complex valued function and

$$(1.2) \qquad F : \mathbb{C}^{2n+2} \longrightarrow \mathbb{C}$$

represents the nonlinearity.

Our main purpose here is to review the local and global theory of (1.1), with emphasis on some recent results, prove some new ones as well as announce others. Also we shall present some open questions.

Let $\{e^{it\Delta}\}_{-\infty}^{\infty}$ denote the unitary group describing the solution of the associated linear IVP to (1.1)

$$(1.3) \qquad \begin{cases} \partial_t u = i\Delta u, & t \in \mathbb{R},\ x \in \mathbb{R}^n \\ u(x,0) = u_0(x) \end{cases}$$

where

$$(1.4) \qquad u(x,t) = e^{it\Delta} u_0(x) = S_t * u_0(x)$$

with $S_t(\cdot)$ defined by the oscillatory integral

$$(1.5) \qquad S_t(x) = c \int_{\mathbb{R}^n} e^{i\langle x\cdot\xi\rangle} e^{it|\xi|^2}\, d\xi.$$

This paper is organized in the following manner. In section 2 we study the semi-linear case, i.e. $F = F(u, \overline{u})$ in (1.1). Section 3 is concerned with problem (1.1). Finally, section 4 contains a short discussion of some physical models.

1991 *Mathematics Subject Classification.* Primary 35K22; Secondary 35P05.

Key words and phrases. Schrödinger equation, initial value problem, well-posedness.

C. E. Kenig and G. Ponce were supported by NSF grants. L. Vega was supported by a DGICYT grant.

2. The semi-linear case

In the particular semi-linear case

(2.1) $F(u, \overline{u}) = f(|u|)u$ (with $f(\cdot)$ a real valued function)

the IVP (1.1) has been extensively studied. In particular local and global theories and blow up results have been established. These depend upon the regularity and size of the data, the degree and sign of $f(\cdot)$ and the dimension n (see [**T1**], [**T2**], [**CzW**], and [**Cz**] for a complete list of references).

We will review these results in the case

(2.2) $F(u, \overline{u}) = -\lambda|u|^{\alpha}u$,   with $\lambda \in \mathbb{R} - \{0\}$ and $\alpha > 0$.

Thus we rewrite (1.1) as

(2.3) $\begin{cases} i\partial_t u + \Delta u + \lambda|u|^{\alpha}u = 0, & t \in \mathbb{R}, \ x \in \mathbb{R}^n, \ \lambda \in \mathbb{R} - \{0\}, \ \alpha > 0 \\ u(x, 0) = u_0(x). \end{cases}$

Formally, solutions of (2.3) satisfy, at least, two conservation laws

(2.4) $\|u(\cdot, t)\|_2 = \|u_0\|_2,$

and

(2.5) $\int_{\mathbb{R}^n} \left(|\nabla_x u(x, t)|^2 - \frac{2\lambda}{\alpha + 2}|u(x, t)|^{\alpha+2} \right) dx = \|\nabla u_0\|_2^2 - \frac{2\lambda}{\alpha + 1}\|u_0\|_{\alpha+2}^{\alpha+2}.$

Hence, if one can show that the IVP (2.3) is locally well-posed in $L^2(\mathbb{R}^n)$ or in $H^1(\mathbb{R}^n)$, with the time of existence depending only on the size of the data, then (2.4)-(2.5) guarantee that the local solutions can be extended to a global one.

The best local well-posedness results known for the IVP (2.3), see [**T2**], [**CzW2**], is contained in the following theorem.

THEOREM 2.1.
   Let $0 < \alpha < \infty$, $s > s_\alpha = n/2 - 2/\alpha$ and $s \geq 0$, with $[s] < \alpha$ if α is not an even integer. Given $u_0 \in H^s(\mathbb{R}^n)$ there exists $T = T(\|u_0\|_{s,2}, s) > 0$ (with $T(\cdot, s)$ a nondecreasing continuous function such that $T(\rho, s) \to \infty$ as $\rho \to 0$) and a unique strong solution $u(\cdot)$ of the IVP (2.3) satisfying

(2.6) $u \in C([-T, T] : H^s(\mathbb{R}^n)) \cap W_{s,n}^T = \mathcal{W}_{s,n}^T.$

Moreover, given $T' \in (0, T)$ there exists a constant $r = r(\|D^s u_0\|_{s,2}, s, T') > 0$ such that the map

(2.7) $\{\widetilde{u}_0 \in H^s(\mathbb{R}^n) \mid \|u_0 - \widetilde{u}_0\|_{s,2} < r\} \to \mathcal{W}_{s,n}^T$

is Lipschitz.

The proof of this theorem, as well as the precise definition of the space $W_{s,n}^T$ in (2.6) and those below $Y_{s_k}^T$, Z_α in (2.13), (2.16) respectively, are based on the so called Strichartz [**Sr**] estimates. They describe a regularization effect present in solutions of the homogeneous linear problem (1.2) and in its inhomogeneous version and guarantee that for these low regularity solutions of (2.3) the nonlinear term makes sense.

THEOREM 2.2.

The group $\{e^{it\Delta}\}_{-\infty}^{\infty}$ satisfies the following inequalities:

$$(2.8) \qquad \left(\int_{-\infty}^{\infty} \|e^{it\Delta}f\|_p^q dt \right)^{1/q} \leq c\|f\|_2,$$

and

$$(2.9) \qquad \left(\int_{-\infty}^{\infty} \| \int_{-\infty}^{\infty} e^{i(t-t')\Delta} g(\cdot,t')dt' \|_p^q dt \right)^{1/q} \leq c \left(\int_{-\infty}^{\infty} \|g(\cdot,t)\|_{p'}^{q'} dt \right)^{1/q'}.$$

where

$$(2.10) \qquad \begin{cases} 2 \leq p < \dfrac{2n}{n-2}, & if \quad n \geq 3, \\[2mm] 2 \leq p < \infty, & if \quad n = 2, \\[2mm] 2 \leq p \leq \infty, & if \quad n = 1, \\[2mm] and \quad \dfrac{2}{q} = \dfrac{n}{2} - \dfrac{n}{p} \end{cases}$$

where the constant $c = c(p,n)$ depends only on p and n.

For the proof of this theorem we refer to [**GiVe2**]. The value $s_\alpha = n/2 - 2/\alpha$ in the statement of Theorem 2.1, which we shall call the "critical case", can be explained by the following scaling argument. If $u(x,t)$ denotes a solution of the IVP in (2.1), then

$$(2.11) \qquad u_\beta(x,t) = \beta^{2/\alpha} u(\beta x, \beta^2 t), \quad \beta > 0.$$

solves the same equation with data $u_{0,\beta}(x) = \beta^{2/\alpha} u_0(\beta x)$, for which one has that

$$(2.12) \qquad \|D^s u_{0,\beta}\|_2 \sim \beta^{s-s_\alpha}.$$

Therefore, this is independent of β only when $s = s_\alpha$.

In the critical case one has the following local and global results found in [**CzW**].

THEOREM 2.3.

Let $4/n \leq \alpha < \infty$ and $s_\alpha = n/2 - 2/\alpha \geq 0$, with $[s] < \alpha$ if α is not an even integer. Given $u_0 \in \dot{H}^{s_\alpha}(\mathbb{R}^n)$ there exist $T = T(D^{s_\alpha} u_0) > 0$ and a unique strong solution $u(\cdot)$ of the IVP (2.3) satisfying

$$(2.13) \qquad u \in C([-T,T] : \dot{H}^{s_\alpha}(\mathbb{R}^n)) \cap Y_{s_k}^T = \mathcal{Y}_{s_k}^T.$$

Moreover, given $T' \in (0,T)$ there exists a constant $\widetilde{r}_\alpha = \widetilde{r}_\alpha(D^{s_\alpha} u_0) > 0$ such that the map

$$(2.14) \qquad \{\widetilde{u}_0 \in \dot{H}^{s_\alpha}(\mathbb{R}^n) \mid \|D^{s_\alpha}(u_0 - \widetilde{u}_0)\|_2 < r_\alpha\} \to \mathcal{Y}_{s_k}^{T'}$$

is Lipschitz.

THEOREM 2.4.

Let $4/n \leq \alpha < \infty$ and $s_\alpha = n/2 - 2/\alpha \geq 0$. There exists $\delta_\alpha > 0$ such that for any $u_0 \in \dot{H}^{s_\alpha}(\mathbb{R}^n)$ with

$$\|D^{s_\alpha} u_0\|_2 < \delta_\alpha \tag{2.15}$$

there exists a unique global strong solution $u(\cdot)$ of the IVP (2.3) such that

$$u \in C(\mathbb{R} : \dot{H}^{s_\alpha}(\mathbb{R}^n)) \cap Z_\alpha = \mathcal{Z}_\alpha. \tag{2.16}$$

Moreover, the map

$$\{\widetilde{u}_0 \in \dot{H}^{s_\alpha}(\mathbb{R}^n) \mid \|D^{s_\alpha} \widetilde{u}_0\|_2 < \delta_\alpha\} \to \mathcal{Z}_\alpha \tag{2.17}$$

is Lipschitz.

We observe that the above local well-posedness results extend to any semi-linear non-linearity F such that

$$F(\beta u, \beta \bar{u}) = \beta^{\alpha+1} F(u, \bar{u}). \tag{2.18}$$

Also notice that above it was always assumed that the Sobolev exponent s was non-negative. In this regard we remark that local well posedness results for Sobolev spaces of negative order have been established in other dispersive models. For example, it was shown in [**KPV5**] that the IVP for the Korteweg-de Vries (KdV) equation

$$\begin{cases} \partial_t u + \partial_x^3 u + \partial_x(u^2) = 0, & t, x \in \mathbb{R}, \\ u(x, 0) = u_0(x) \end{cases} \tag{2.19}$$

is locally well posed in $H^s(\mathbb{R})$ for $s > -3/4$.

Therefore the following question presents itself.

Question 1: Are there values α, n for which the IVP (2.3) is locally well posed in negative Sobolev spaces ?

In a forthcoming paper [**KPV6**], we shall prove that in the one dimensional case, $n = 1$, with nonlinearities F of the following type

$$F = u^2, \ u\bar{u}, \ (\bar{u})^2, \ (\bar{u})^3, \ldots\ldots \tag{2.20}$$

question 1 has a positive answer. For example, in the case $F = u^2$ and $n = 1$ one has that the IVP (1.1) is locally well-posed in $H^s(\mathbb{R})$ for $s > -3/4$.

In the opposite direction, it was shown in [**BPS**] that the IVP (2.3) with $\alpha = 2$ and $n = 1$ is ill-posed in $H^s(\mathbb{R})$ for $s < -1/2$.

On the other hand, if we restrict ourselves to the case $s \geq 0$ one has the following result found in [**BKPSV**], which shows that the results in Theorem 2.1 are optimal.

THEOREM 2.5.
- *Theorem 2.1 fails for $s = s_\alpha \geq 0$ and $\lambda > 0$.*
- *Theorem 2.3 fails for $\lambda > 0$, $s = s_\alpha \geq 0$ and $\delta_\alpha \geq C(\lambda, \alpha, n)$.*

The proof of Theorem 2.5 uses special properties of some solutions of the IVP (2.3) with $\lambda > 0$. For $n = 1$ one has the solitary waves described by the formula (see [St2], [ZS])

$$(2.21) \qquad v_{\lambda,\alpha,c,b}(x,t) = \phi_{\lambda,\alpha,\beta}(x - ct)e^{ic(x-bt)/2}, \quad c > 0, \ b \in \mathbb{R}$$

where

$$(2.22) \qquad \phi_{\lambda,\alpha,\beta}(x) = \left\{ \frac{(\alpha+2)\beta}{2\lambda} \, sech^2 \left(\frac{\alpha}{2}\sqrt{\beta}\, x \right) \right\}^{1/\alpha}.$$

and

$$(2.23) \qquad \beta = \frac{c}{2}\left(\frac{c}{2} - b \right) > 0.$$

For the case $n \geq 1$ and $s \leq 1$ fixing the value $\lambda = 1$ and using the results in [BeL], [St1] one has that the IVP (2.3) has solutions in the form of "ground state", i.e.

$$(2.24) \qquad u_\mu(x,t) = e^{i\mu t}\Phi_\mu(x),$$

where the function $\Phi_\mu(\cdot)$ solves the nonlinear elliptic eigenvalue problem

$$(2.25) \qquad \begin{cases} \Delta\Phi - \mu\Phi + |\Phi|^\alpha\Phi = 0, & x \in \mathbb{R}^n \\ \Phi \in H^1(\mathbb{R}^n) \end{cases}$$

with $0 < \alpha < 4/(n-2)$, if $n > 2$. In fact, it was established in [BeL], [St1] that (2.25) has a solution $\Phi_\mu(\cdot) \in C^2(\mathbb{R}^n)$ which decays, as well as its derivatives up to order two, exponentially at infinity.

Finally for the values $s \geq 1$ one combines the scaling argument in (2.11)-(2.12) with the blow-up result found in [G], (see Theorem 2.7 below).

Question 2: Do the results in Theorem 2.6 extend to the case $\lambda < 0$?

Now we turn our attention to the global well posedness results to the IVP (2.3). Due to the form of the conservation laws (2.4)-(2.5) it is convenient to consider L^2 and H^1 solutions.

THEOREM 2.6.
(a) Let $0 < \alpha < 4/n$. Then for any $u_0 \in L^2(\mathbb{R}^n)$ the corresponding local solution $u(x,t)$ of the IVP (2.3) provided by Theorem 2.1 extends to global one in the same class (2.13).

(b) Let $0 < \alpha < \alpha_n$ where $\alpha_n = 4/(n-2)$ if $n \geq 3$. If α satisfies one of the following conditions

  i) $\lambda < 0$,
  ii) $\lambda > 0$ and $\alpha < 4/n$,

or

iii) $\lambda > 0$, $\alpha \geq 4/n$ *and* $\|u_0\|_{1,2} \leq K(\lambda, \alpha)$,

then for any $u_0 \in H^1(\mathbb{R}^n)$ *the corresponding local solution* $u(x,t)$ *of the IVP (2.3) provided by Theorem 2.1 extends to a global one in the same class (2.13).*

The proof of Theorem 2.6 follows directly by combining Theorem 2.3, the identities (2.4)-(2.5) and the Gagliardo-Nirenberg inequality. We observe that $\alpha = 4/n$ and $\alpha = 4/(n-2)$, if $n \geq 3$, are the critical case for $L^2(\mathbb{R}^n)$ and $H^1(\mathbb{R}^n)$ respectively (see (2.11)-(2.12)).

To complement the long time behavior of H^1-solutions of (2.3) we have the following blow-up result found in [**G**].

THEOREM 2.7.

Let $1 + 4/n \leq \alpha < \alpha_n$, *where* $\alpha_n < (n+2)/(n-2)$ *if* $n \geq 3$. *If none of the hypothesis i)-iii) of Theorem 2.6 hold then there exists* $u_0 \in H^1(\mathbb{R}^n)$ *such that the corresponding local solution* $u(x,t)$ *of the IVP (2.3) provided by Theorem 2.3 blows-up in finite time, i.e. there exists* $T_0 > 0$ *such that*

$$(2.26) \qquad \lim_{t \uparrow T_0} \|\nabla u(t)\|_2 = \infty.$$

The description of the behavior of the solution near the blow-up time has motivated several works, both theoretical and numerical. Numerical computations, see [**RRy**],[**MPaSuSl**] and references therein, indicate that the solution blows up at an earlier time than T_0 given in the proof of Theorem 2.7. On the other hand, under appropriate assumptions on the form of the data and on the order of the nonlinearity α, specially for the value $\alpha = 1 + 4/\alpha$, more precise description of this blow up result has been obtained in [**Me**], [**MeT**].

Collecting the information in Theorems 2.2-2.7 the following questions appear naturally.

Question 3: Assume $\lambda < 0$, $s_\alpha = n/2 - 2/\alpha \geq 0$ and $u_0 \in H^{s_\alpha}(\mathbb{R}^n)$ is arbitrary. Does the solution $u(x,t)$ of the IVP (2.3) provided by Theorem 2.3 exist for all time ?

Question 4: Assume $\lambda < 0$, $s_\alpha = n/2 - 2/\alpha = 1$, α an even integer and $u_0 \in H^2(\mathbb{R}^n)$ is arbitrary. Does the solution $u(x,t)$ of the IVP (2.3) provided by Theorem 2.1 satisfy that $u \in L^\infty(\mathbb{R} : H^2(\mathbb{R}^n))$?

We observe that question 3 has an affirmative answer if, in addition, one assumes that $\|D^{s_\alpha}u_0\|_2$ is small enough or if $u_0 \in H^1(\mathbb{R}^n)$ (Theorem 2.6).

3. The general case

The general IVP (1.1) has been mainly treated when one of the following hypothesis hold:

i) energy estimates are available

or

ii) analytic data (analytic solutions).

Energy estimates for the equation in (1.1) can be established when using integration by parts one can show that

$$(3.1) \qquad \left| \sum_{|\alpha| \leq s} \int_{\mathbb{R}^n} \partial_x^\alpha F(u, \nabla_x u, \overline{u}, \nabla_x \overline{u}) \partial_x^\alpha u \, dx \right| \leq c_s (1 + \|u\|_{s,2}^\rho) \|u\|_{s,2}^2$$

for any $u \in H^s(\mathbb{R}^n)$ with $s > n/2 + 1$ and $\rho = \rho(P) \in \mathbb{Z}^+$.

It is clear that the estimate (3.1) requires some some symmetry of the nonlinearity $F(\cdot)$, i.e. the derivatives of order $s + 1$ disappear after integration by parts. In the following three examples condition (3.1) holds:

$$(3.2) \qquad n = 1 \quad \text{and} \quad F = \partial_x(|u|^k u), \qquad k \in \mathbb{Z}^+ \quad (\text{see } [\mathbf{TsF1}],[\mathbf{TsF2}]),$$

$$(3.3) \qquad n \geq 1 \quad \text{and} \quad F = F(u, \overline{u}, \nabla \overline{u}),$$

and for $n \geq 1$ and $F(u, \overline{u}, \nabla u, \nabla \overline{u})$ where

$$(3.4) \quad D_{\partial_{x_j} u} F, \quad D_{\partial_{x_j} \overline{u}} F \quad \text{j=1, \ldots , n} \quad \text{are real-valued} \quad (\text{see } [\mathbf{Kl}],[\mathbf{KlP}], [\mathbf{Sh}]).$$

When $F(\cdot)$ satisfies (3.1) the proof of the local existence theory in $H^s(\mathbb{R}^n)$ with $s > n/2 + 1$ follows the argument used for quasi-linear symmetric hyperbolic systems (see [$\mathbf{Ka1}$]). Indeed, this local result does not use the dispersive structure of the equations.

The other approach uses analytic function techniques to overcome the loss of derivatives introduced by the analytic nonlinearity $F(\cdot)$. Thus under appropriate assumptions on the analytic data u_0 and the nonlinearity $F(\cdot)$ local and global analytic results have been obtained (see [$\mathbf{H1}$], [$\mathbf{SiTa}$]).

In [$\mathbf{KPV2}$] a different approach was given. This was based on the dispersive character of the associated linear problem (1.3). More precisely, it relies in a crucial manner on estimates describing the smoothing effects in the group $\{e^{it\Delta}\}_{-\infty}^\infty$.

These smoothing effects were first established by T. Kato in [$\mathbf{Ka2}$] and used there to prove global existence of L^2 solutions for the KdV equation (see (2.19)). The corresponding homogeneous version for the free Schrödinger group $\{e^{it\Delta}\}_{-\infty}^\infty$ was simultaneously obtained in [$\mathbf{CoSa}$],[$\mathbf{Sj}$],[$\mathbf{V}$]. This asserts that

$$(3.5) \qquad \sup_{\alpha \in \mathbb{Z}^n} \left(\int_{Q_\alpha} \int_{-\infty}^\infty |D_x^{1/2} e^{it\Delta} u_0(x)|^2 dt \, dx \right)^{1/2} \leq cR\|u_0\|_2$$

where $D_x = (-\Delta)^{1/2}$. In [$\mathbf{KPV1}$] we show that in the one dimensional case (3.5) can be improved, i.e.

$$(3.6) \qquad \sup_x \int_{-\infty}^\infty |D_x^{1/2} e^{it\partial_x^2} u_0|^2 dt \leq c\|u_0\|_2^2$$

and that this estimate is sharp in the sense that equality holds for a class of data in $L^2(\mathbb{R})$.

However, the gain of half of a derivative presents in (3.5)-(3.6) does not allow control of the loss of derivatives, equal to one, introduced by the nonlinear term F. More precisely, for the integral equation associated to the IVP (1.1)

$$(3.7) \qquad u(t) = e^{it\Delta}u_0 + \int_0^t e^{i(t-t')\Delta} F(u, \nabla_x u, \overline{u}, \nabla_x \overline{u})(t')dt'.$$

with general F the estimates (3.5)-(3.6) do not provide any estimate for the solution $u(\cdot)$ of (3.7). Observe that the Strichartz estimates for the free Schrödinger group, see Theorem 2.2, do not involve any gain of derivatives.

For this non-linear problem one needs to study the smoothing effect in the inhomogeneous problem

$$(3.8) \qquad \begin{cases} \partial_t u = i\Delta u + g(x,t), & t \in \mathbb{R}, \ x \in \mathbb{R}^n \\ u(x,0) = 0. \end{cases}$$

In this direction one has the following result found in [**KPV2**].

THEOREM 3.1.

(a) *When $n = 1$ the solution $u(x,t)$ of the IVP (3.8) satisfies*

$$(3.9) \qquad \sup_x \left(\int_{-\infty}^{\infty} |\partial_x u(x,t)|^2 dt \right)^{1/2} \leq c \int_{-\infty}^{\infty} \left(\int_{-\infty}^{\infty} |g(x,t)|^2 dt \right)^{1/2} dx.$$

(b) *When $n \geq 2$ the solution $u(x,t)$ of the IVP (3.8) satisfies*
(3.10)

$$\sup_{\alpha \in \mathbb{Z}^n} \left(\int_{Q_\alpha} \int_{-\infty}^{\infty} |\nabla_x u(x,t)|^2 dt\, dx \right)^{1/2} \leq c\, R \sum_{\alpha \in \mathbb{Z}^n} \left(\int_{Q_\alpha} \int_{-\infty}^{\infty} |g(x,t)|^2 dt\, dx \right)^{1/2}$$

where $\{Q_\alpha\}_{\alpha \in \mathbb{Z}^n}$ denotes a family of cubes of size R with disjoint interior such that $\mathbb{R}^n = \bigcup_{\alpha \in \mathbb{Z}^n} Q_\alpha$.

Heuristically (3.9)-(3.10) tell us that the gain of derivatives in the inhomogeneous case is twice that obtained for the homogeneous problem (3.5)-(3.6). To complement these inequalities one needs to use estimates related with the maximal function:

$$(3.11) \qquad \sup_{[0,T]} |e^{it\Delta}u_0|.$$

In this regards one has the following global estimate established in [**C**], [**KR**]

$$(3.12) \qquad \left(\int_{-\infty}^{\infty} \sup_{(-\infty,\infty)} |e^{it\partial_x^2}u_0(x)|^4 dx \right)^{1/4} \leq c\|D_x^{1/4}u_0\|_2.$$

It is interesting to remark that the estimate in (3.12) cannot hold in L^p with $p < 4$, independent of the number of derivatives considered in its right hand side. The n dimensional version

(3.13)
$$\left(\int_{\mathbb{R}^n} \sup_{(-\infty,\infty)} |e^{it\Delta}u_0(x)|^4 dx \right)^{1/4} \leq c\|D_x^{n/4}u_0\|_2$$

as well as its inhomogeneous version
(3.14)
$$\left(\int_{\mathbb{R}^n} \sup_{(-\infty,\infty)} \left| \int_0^t e^{i(t-t')\Delta}g(\cdot,t')dt' \right|^4 dx \right)^{1/4} \leq c \left(\int_{-\infty}^{\infty} \|D_x^{n/2}g(\cdot,t)\|_1^{4/3} \right)^{3/4}$$

can be obtained by combining the arguments in [C], [KR], [KPV2].

THEOREM 3.2.
Given a smooth nonlinear function F with $F(0) = \nabla F(0) = 0$ there exist $\delta = \delta(F,n) > 0$ and $s,m \in \mathbb{Z}^+$ such that for any

(3.15)
$$u_0 \in H^s(\mathbb{R}^n) \cap L^2(\mathbb{R}^n : |x|^{2m}dx) = X_{s,m}$$

with

(3.16)
$$\|u_0\|_{X_{s,m}} = \|u_0\|_{s,2} + \||x|^m u_0\|_2 < \delta$$

the IVP (1.1) has a unique solution $u(\cdot)$ defined in the time interval $[0,T]$, ($T = T(\|u_0\|_{s,2} + \||x|^m u_0\|_2) > 0$ with $T(\theta) \to \infty$ as $\theta \to 0$) satisfying

(3.17)
$$u \in C([0,T] : X_{s,m}),$$

and

(3.18)
$$\sup_{\alpha \in \mathbb{Z}^+} \int_0^T \int_{Q_\alpha} |(-\Delta)^{s/2+1/2}u(x,t)|^2 dxdt < \infty,$$

for any $\alpha \in \mathbb{Z}^+$.
Moreover, for any $T' \in (0,T)$ there exists $\epsilon > 0$ such that the map $\widetilde{u}_0 \to \widetilde{u}(t)$ from $\{\widetilde{u}_0 \in X_{s,m} \mid \|\widetilde{u}_0 - u_0\|_{X_{s,m}} < \epsilon\}$ into the class defined in (3.14)-(3.15), with T' instead of T is Lipschitz.
If in addition $\partial^2 F(0) = 0$ one can take $m = 0$.

For the proof of Theorem 3.2 we refer to [KPV2].
In [HO1], for the one dimensional case $n = 1$ the smallness assumption on the data in Theorem 3.2 was removed. The proof was based on a "gauge transformation" which allows one to write the 1-D equation in (1.1) into an equivalent system for which energy estimates work.
Very recently, in [Ch], a combination of the idea of a "gauge transformation" with some arguments involving pseudo-differential operators were used to remove the smallness assumption in Theorem 3.2 for any dimension n.

Question 5: Does the local solution of the IVP (1.1) found in [Ch] have locally the same regularity that the solution of the associated linear problem $e^{it\Delta}u_0$, i.e. does it satisfy the local version of the smoothing effect described in (3.18)?

We may remark that in the case of the local solutions discussed in section 2 the answer is positive.

The problem of the global existence of "small" solutions of (1.1) has also motivated several recent works. For example, in [KtT] for the one dimensional case $n = 1$ and under the additional hypothesis on the nonlinearity F,

$$(a)\ \ F(e^{i\theta}u, e^{i\theta}\partial_x u, e^{-i\theta}\bar{u}, e^{-i\theta}\partial_x \bar{u}) = e^{i\theta}F(u, \partial_x u, \bar{u}, \partial_x \bar{u}),$$

and

$$(b)F = \partial_x(|u|^2)(\nu u + \eta \partial_x u) + \ \text{higher order terms},$$

the solution of the IVP (1.1) provided by Theorem 3.2 extends to a global one.

For similar results in higher dimensions, i.e. global "small" solutions of (1.1) under some restriction on the form and degree of the nonlinearity F, see for example [GiH], [H2], [HO2].

We shall prove here that without any further restrictions, for higher nonlinearity the "small" local solution obtained in Theorem 3.2 extends to a global one.

THEOREM 3.3.
   Given a smooth nonlinear function F with $\partial^\alpha F(0) = 0$ for $|\alpha| \leq 4$ there exist $\delta_0 \in (0, \delta)$ and $s \in \mathbb{Z}^+$ such that for any $u_0 \in H^s(\mathbb{R}^n)$ with

$$(3.19) \qquad\qquad \|u_0\|_{s,2} < \delta_0$$

the solution $u(x,t)$ of the IVP (1.1) provided by Theorem 3.2 extends to any time interval $[-T, T]$ with

$$(3.20) \qquad \sup_{\alpha \in \mathbb{Z}^+} \int_{-\infty}^{\infty} \int_{Q_\alpha} |(-\Delta)^{s/2+1/2}u(x,t)|^2 dxdt < \infty.$$

Moreover, the map $\tilde{u}_0 \to \tilde{u}(t)$ from $\{\tilde{u}_0 \in H^s(\mathbb{R}^n) \mid \|\tilde{u}_0\|_{s,2} < \delta_0\}$ into the class $C(\mathbb{R} : H^s(\mathbb{R}^n))$ intersected with (3.18) is Lipschitz.

PROOF (SKETCH).
   For $u \in H^s(\mathbb{R}^n)$ with $s_0 = s + 1/2$ an even integer and $\|u_0\|_{s_0,2} = \delta_0$ (to be determined below) we consider the linear IVP

$$(3.21) \qquad\qquad \begin{cases} \partial_t u = i\Delta u + F(v, \nabla_x v, \bar{v}, \nabla_x \bar{v}) \\ u(x,0) = u_0(x) \end{cases}$$

for

$$(3.22) \qquad\qquad v \in \mathcal{E}_a = \{v : \mathbb{R}^n \times \mathbb{R} \to C \mid \Lambda(v) \leq a\ \}$$

where the λ_j's are defined as

$$(3.23) \qquad \lambda_1(v) = \sup_{(-\infty,\infty)} \|v(t)\|_{s_0,2},$$

$$(3.24) \qquad \lambda_2(v) = \sum_{1 \le |\beta| \le s_0+1/2} \sup_{\alpha \in \mathbb{Z}^+} \left(\int_{-\infty}^{\infty} \int_{Q_\alpha} |\partial_x^\beta v(x,t)|^2 dx\, dt \right)^{1/2},$$

$$(3.25) \qquad \lambda_3(v) = \left(\sum_{|\beta| \le s_0/2} \sum_{\alpha \in \mathbb{Z}^+} \sup_{(-\infty,\infty)} \sup_{Q_\alpha} |\partial_x^\beta v(x,t)|^4 \right)^{1/4}.$$

and

$$(3.26) \qquad \Lambda(v) = \sup\{\lambda_j(v) \mid j = 1,2,3\}.$$

It will be shown that there exist constants δ_0 and $a = a(\delta_0) > 0$ such that if $v \in \mathcal{E}^a$ so does the solution $u(\cdot)$ of (4.20), and that the map $\Phi(v) = u$ is a contraction.

We shall rely on the integral equation

$$(3.27) \qquad u(t) = \Phi(v)(t) = e^{it\Delta}u_o + \int_0^t e^{i(t-\tau)\Delta} F(v, \nabla_x v, \bar{v}, \nabla_x \bar{v})(\tau) d\tau.$$

Combining estimates (3.9)-(3.10) and some calculus of inequalities involving fractional derivatives (see [**KPV3**]) we find that

$$(3.28) \qquad \begin{aligned} \lambda_2(\Phi(v)) &\le c\delta + c\lambda_2(v) \left(\lambda_3(v)\right)^4 (1 + (\lambda_1(v))^r) \\ &\le c\delta + c\Lambda(v) \left(\Lambda(v)\right)^4 (1 + (\Lambda(v))^r) \end{aligned}$$

where $r \in \mathbb{Z}^+$ depends on F.

Next, using the dual version of (3.10) and the group properties it follows that

$$(3.29) \qquad \lambda_1(\Phi(v)) \le c\delta + c\lambda_2(v) \left(\lambda_3(v)\right)^4 (1 + (\lambda_1(v))^r).$$

Finally, from (3.13)-(3.14) we obtain that

$$(3.30) \qquad \lambda_3(\Phi(v)) \le c\delta + c\Lambda(v) \left(\Lambda(v)\right)^4 (1 + (\Lambda(v))^r).$$

Collecting (3.27)-(3.29) we write

$$(3.31) \qquad \Lambda(\Phi(v)) \le c\delta + c\Lambda(v) \left(\Lambda(v)\right)^4 (1 + (\Lambda(v))^r).$$

Hence, for a sufficiently small

$$(3.32) \qquad \Phi(\mathcal{E}_a) \subseteq \mathcal{E}_a.$$

Similarly, one has that

$$(3.33) \qquad \Lambda(\Phi(v) - \Phi(\widetilde{v})) \le \frac{1}{2}\Lambda(v - \widetilde{v})$$

which basically completes the proof of the Theorem.

The main difference between Theorem 3.3 and other results concerning the asymptotic behavior of small solutions, which are perturbations of classical linear ones, is that in the later ones the dimension always plays an essential role.

Question 6: Without further restrictions, is Theorem 3.3 the best possible result?

4. Some examples

Consider the initial value problem IVP for the Davey-Stewartson (D-S) system

$$
(4.1) \quad
\begin{cases}
i\partial_t u + c_0 \partial_x^2 u + \partial_y^2 u = c_1 |u|^2 u + c_2 u \partial_x \varphi, & x,\, y \in \mathbb{R},\ t > 0 \\
\partial_x^2 \varphi + c_3 \partial_y^2 \varphi = \partial_x |u|^2 \\
u(x,y,0) = u_0(x,y)
\end{cases}
$$

where $u = u(x,y,t)$ is a complex-valued function, $\varphi = \varphi(x,y,t)$ is a real-valued function $\partial_t = \partial/\partial t$ and $c_0, .., c_3$ are real parameters.

A system of this kind was first derived by Davey and Stewartson [DSw] in their work on two-dimensional long waves over finite depth liquids. Independently Ablowitz and Haberman [AHa] obtained a particular form of (4.1) as an example of a completely integrable two-dimensional model which generalizes the cubic Schrödinger equation

$$
(4.2) \quad i\partial_t u + \partial^2 u + |u|^2 u, \qquad x, t \in \mathbb{R}.
$$

which was proved in [ZSb] to be completely integrable. Since then several works have been devoted to studying special forms of system (4.1) using the inverse scattering approach. In fact when $(c_0, c_1, c_2, c_3) = (-1, 1, -2, 1)$ or $(1, -1, 2, -1)$ system (4.1) is known in inverse scattering theory as the DSI and DSII respectively.

In the case $c_3 > 0$ the evolution equation in (4.1) has a nonlinear term of order zero, therefore most of the comments given in section 2 applies, for details see [GhSa1].

If $c_3 < 0$ the situation is quite different. In [LiP] the arguments in [KPV2] were used to establish a result similar to that described in Theorem 3.2 for the IVP (1.1).

The above 2-D system appears is as special case of the IVP for the Zakharov-Schulman system

$$
(4.3) \quad
\begin{cases}
i\partial_t u + \mathcal{L}_1 u = \varphi u, & x \in \mathbb{R}^n,\ t > 0, \\
\mathcal{L}_2 \varphi = \mathcal{L}_3 |u|^2, \\
u(x,0) = u_0(x),
\end{cases}
$$

where $u : \mathbb{R}^n \times [0,\infty) \to \mathbb{C}$, $\varphi : \mathbb{R}^d \times [0,\infty) \to \mathbb{R}$ and

$$
(4.4) \quad \mathcal{L}_k = \sum_{i,j=1}^{n} a_{i,j}^k \partial_{x_i x_j}^2, \quad k = 1, 2, 3
$$

with $a_{i,j}^k = a_{j,i}^k$ real constants. These systems describe the interactions of small amplitude, high frequency waves with acoustic type waves. It is also known that in the 3-D case none of them are completely integrable, see [ZSc].

In [KPV4] the argument in [KPV2] were extended to the IVP for the Zakharov-Schulman system. More precisely, we showed that the local "small" data well posedness result described in Theorem 3.2 applies for the IVP (4.3) in its general form.

Question 7: Is the IVP (4.3) locally well posed for "large" data?

Finally we consider the Ishimori system, for more physical examples of systems of type described in (1.1) we refer to [**GhSa2**]. This provides a generalization of the Heisenberg equation in ferromagnetism and can be written in stereographic variables $u : \mathbb{R}^2 \to \mathbb{C}$ in the following form

(4.5)

$$
\begin{cases}
i\partial_t u + \partial_x^2 u \mp \partial_y^2 u = \dfrac{2\bar{u}}{1 + |u|^2}((\partial_x u)^2 - (\partial_y u)^2) + ib(\partial_x \phi \partial_y u + \partial_y \phi \partial_x u), \\[2mm]
\partial_x^2 \phi \pm \partial_y^2 \phi = 4i\, \dfrac{\partial_x u \partial_y \bar{u} - \partial_x \bar{u} \partial_y u}{(1 + |u|^2)^2}, \\[2mm]
u(x, y, 0) = u_0(x, y),
\end{cases}
$$

The case $b = 0$ was studied in [**B-Su-Sl**]. In [**KoMa**] it was shown that for $b = 1$ the equations in (4.5) are completely integrable. In the same case $b = 1$ with the signs $\mp$, i.e. $-$ and $+$ in the first and second equation in (4.5) respectively, was considered in [**So**]. There the existence of global solution corresponding to "small" data as well as its asymptotic behavior were obtained.

Question 8: Under which conditions is the IVP (4.5) with $+$ and $-$ in the first and second equation respectively locally well-posed ?

REFERENCES

[AHa]   Ablowitz, M. J. and Haberman, R., *Nonlinear evolution equations in two and three dimensions*, Phys. Rev. Lett. **35** (1975), 1185–1188.

[BaSuSl]   Bardos, C., Sulem, C. and Sulem, P. L., *On a continuous limit for a system of classical spin*, Comm. Math. Phys. **107** (1986), 431–450.

[BeL]   Berestycki, H. and Lions, P. L., *Nonlinear scalar field equations*, Arch. Rational Mech. Anal. **82** (1983), 313–376.

[BPS]   Birnir, B., Ponce, G. and Svanstedt, N., *to appear*.

[BKPSV]   Birnir, B., Kenig, C. E, Ponce, G., Svanstedt, N. and Vega, L., *On the ill-posedness of the IVP for the generalized Korteweg-de Vries and Schrödinger equations*, pre-print.

[C]   Carleson, L., *Some analytical problems related to statistical mechanics,*, Lecture Notes in Math **779 Springer-Verlag** (1979), 9–45.

[Cz]   Cazenave, T., *An introduction to nonlinear Schrödinger equations*, Textos de Métodos Matemáticos **22**, Universidade Federal do Rio de Janeiro.

[CzW1]   Cazenave, T. and Weissler, F. B., *Some remarks on the nonlinear Schrödinger equation in the critical case*, Lecture Notes in Math **1392 Springer-Verlag** (1989), 18–29.

[CzW2]   Cazenave, T. and Weissler, F. B., *The Cauchy problem for the critical nonlinear Schrödinger equation in H^s.*, Nonlinear Anal. TMA **14** (1990), 807–836.

[Ch]   Chihara, H., *Local existence for the semi-linear Schrödinger equations*, pre-print.

[CoSa]   Constantin, P. and Saut, J. C., *Local smoothing properties of dispersive equations*, J. Amer. Math. Soc. **1** (1989), 413–446.

[DSw]   Davey, A. and Stewartson, K., *On three-dimensional packets of surface waves*, Proc. Royal London Soc. A **338** (1974), 101-110.

[GiVe1]   Ginibre, J. and Velo, G., *On a class of Schrödinger equations*, J. Funct. Anal. **32** (1979), 1–71.

[GiVe2]   Ginibre, J. and Velo, G., *Scattering theory in the energy space for a class of non-linear Schrödinger equations*, J. Math. Pure Appl. **64** (1985), 363–401.

[GiH]   Ginibre, J. and Hayashi, N., *Almost global existence of small solutions to quadratic nonlinear Schrödinger equations in three space dimensions*, to appear in Math. Z..

[GhSa1]   Ghidaglia, J. M., and Saut, J. C., *On the initial value problem for the Davey-Stewartson systems*, Nonlinearity **3** (1990), 475–506.

[GhSa2]   Ghidaglia, J. M. and Saut, J. C., *On the Zakharov-Schulman equations*, to appear in Nonlinear Dispersive Waves, L. Debnath ed. World Scientific (1992).

[G]   Glassey, R. T., *On the blowing up solutions to the Cauchy problem for nonlinear Schrödinger equations*, J. Math. Phys. **18** (1979), 1794–1797.

[H1]   Hayashi, N., *Global existence of small analytic solutions to nonlinear Schrödinger equations*, Duke Math. J **62** (1991), 575–592.

[H2]   Hayashi, N., *Global and almost global solutions to quadratic nonlinear Schrödinger equations with small initial data*, pre-print.

[HO1]   Hayashi, N.and Ozawa, T., *Remarks on nonlinear Schrödinger equations in one space dimension*, Diff. Integral Eqs **2** (1994), 453–461.

[HO2]   Hayashi, N.and Ozawa, T., *Global, small radially symmetric solutions to nonlinear Schrödinger equations and a gauge transformation*, pre-print.

[I]   Ishimori, Y., *Multi vortex solutions of a two dimensional nonlinear wave equation*, Progr. Theor. Phys **72** (1984), 33–37.

[KtT]   Katayama, S. and Tsutsumi, Y., *Global existence of solutions for nonlinear Schrödinger equation in one space dimension*, pre-print.

[Ka1]   Kato, T., *Quasi-linear evolution equation, with applications to partial differential equations*, Lecture Notes in Math. **448 Springer-Verlag** (1975), 27–50.

[Ka2]   Kato, T., *On the Cauchy problem for the (generalized) Korteweg-de Vries equation*, Advances in Math. Supp. Studies, Studies in Applied Math. **8** (1983), 93–128.

[Ka3]   Kato, T., *Nonlinear Schrödinger equation*, Schrödinger operators, H. Holden and A. Jensen (Eds), Lecture Notes in Physics **345 Springer-Verlag** (1989), 218–263.

[KuN]   Kaup, D. J. and Newell, A. C., *An exact solution for a derivative nonlinear Schrödinger equation*, J. Math. Phys. **19** (1978), 798–801.

[KPV1]   Kenig, C. E., Ponce, G. and Vega, L., *Oscillatory integrals and regularity of dispersive equations*, Indiana University Math. J. **40** (1991), 33–69.

[KPV2]   Kenig, C. E., Ponce, G. and Vega, L., *Small solutions to nonlinear Schrödinger equations*, Annales de l'I.H.P. **10** (1993), 255–288.

[KPV3]   Kenig, C. E., Ponce, G. and Vega, L., *Well-posedness and scattering results for the generalized Korteweg-de Vries equation via the contraction principle*, Comm. Pure Appl. Math. **46** (1993), 527–620.

[KPV4]   Kenig, C. E., Ponce, G. and Vega, L., *On the Zakharov and Zakharov-Schulman systems*, to appear in J. Funct. Anal..

[KPV5]   Kenig, C. E., Ponce, G. and Vega, L., *Bilinear estimates with applications to the KdV equation*, pre-print.

[KPV6]   Kenig, C. E., Ponce, G. and Vega, L., to appear.

[KR]   Kenig, C. E. and Ruiz, A., *A strong type (2,2) estimate for the maximal function associated to the Schrödinger equation*, Trans. Amer. Math. Soc. **280** (1983), 239–246.

[Kl]   Klainerman, S., *Long time behavior of solutions to nonlinear evolutions equations*, Arch. Ration. Mech. and Analysis **78** (1981), 73–98.

[KlP]   Klainerman, S. and Ponce, G., *Global small amplitude solutions to nonlinear evolution equations*, Comm. Pure Appl. Math. **36** (1983), 133–141.

[KoMa]   Konopelchenko, B. G. and Matkarimov, B. T., *On the inverse scattering transform of the Ishimori equations*, Phys. Lett. A **135** (1989), 183–189.

[LiP]   F. Linares, and G. Ponce, *On the Davey-Stewartson systems*, Annales de l'I.H.P. Analyse non linéaire **10** (1993), 523–548.

[MPSuSl]   D. MacLaughlin, G. Papanicolaou, C. Sulem and P. L. Sulem, *The focusing singularity of the cubic Schrödinger equation*, Phys. Rev. A **34** (1986), 1200–1210.

[Me]   F. Merle, *Construction of solutions with exactly k blow-up points for the Schrödinger equation with critical nonlinearity*, Comm. Math. Phys **129** (1990), 223–240.

[MeT]   F. Merle and Y. Tsutsumi, L^2 *-concentration of blow-up solutions for the nonlinear Schrödinger equation with critical power nonlinearity*, J. Diff. Eqs **84** (1990), 205–214.

[RRy]   Rasmussen, J. and Rypdal, K., *Blow-up of nonlinear Schrödinger equations*, Phys. Scripta **33** (1986), 481–504.

[Sh]   Shatah, J., *Global existence of small solutions to nonlinear evolution equations*, J. Diff. Eqs. **46** (1982), 409–423.

[SiTa] Simon, J. and Taflin, E., *Wave operators and analytic solutions for systems of nonlinear Klein-Gordon equations and of non-linear Schrödinger equations*, Comm. Math. Phys. **99** (1985), 541–562.

[Sj]   Sjölin, P., *Regularity of solutions to the Schrödinger equations*, Duke Math. J. **55** (1987), 699–715.

[So]   Soyeur, A., *The Cauchy problem for the Ishimori equations*, J. Funct. Anal. **105** (1992), 233–255.

[St1]   Strauss, W. A., *Existence of solitary waves in higher dimensions*, Comm. Math. Phys. **55** (1977), 149–162.

[St2]   Strauss, W. A., *Nonlinear scattering theory at low energy*, J. Funct. Anal. **41** (1981), 110–133.

[St3]   Strauss, W. A., *Nonlinear wave equations*, CBMS Regional Conference **73** (1989).

[Sr]   Strichartz, R. S., *Restriction of Fourier transform to quadratic surfaces and decay of solutions of wave equations*, Duke Math. J. **44** (1977), 705–714.

[T1]   Tsutsumi, Y., *Global strong solutions for nonlinear Schrödinger equation*, Nonlinear Anal. **11** (1987), 1143–1154.

[T2]   Tsutsumi, Y., L^2-*solutions for nonlinear Schrödinger equations and nonlinear groups* **30** (1987), 115-125.

[TsF1] Tsutsumi, M. and Fukuda, I., *On solutions of the derivative nonlinear Schrödinger equation. Existence and Uniqueness Theorem*, Funkcialaj Ekvacioj **23** (1980), 259-277.

[TsF2] Tsutsumi, M. and Fukuda, I., *On solutions of the derivative nonlinear Schrödinger equation. II*, Funkcialaj Ekvacioj **24** (1981), 85–94.

[V]   Vega, L., *The Schrödinger equation: pointwise convergence to the initial data*, Proc. Amer. Math. Soc. **102** (1988), 874–878.

[Wi]   Weinstein, M. I., *On the solitary traveling wave of the generalized Korteweg-de Vries equation*, Lectures Appl. Math. **23** (1986), 23–29.

[Z]   V. E. Zakharov, *Collapse of Langmuir waves*, Sov. Phys. JETP **35** (1972), 908-914.

[ZSb]   Zakharov, V. E. and Shabat, A. B., *Exact theory of two dimensional self-focusing and one-dimensional self-modulation of waves in non-linear media*, Soviet Physics JETP **34** (1972), 62–69.

[ZKu]   V. E. Zakharov, and E. A. Kuznetson, *Multi-scale expansions in the theory of systems integrable by the inverse scattering method*, Physica D **18** (1986), 455–463.

[ZSc]   V. E. Zakharov, and E. I. Schulman, *Degenerated dispersion laws, motion invariant and kinetic equations*, Physica **1D** (1980), 185-250.

DEPARTMENT OF MATHEMATICS, UNIVERSITY OF CHICAGO, CHICAGO, IL 60637, USA

CEK@MATH.UCHICAGO.EDU

DEPARTMENT OF MATHEMATICS, UNIVERSITY OF CALIFORNIA, SANTA BARBARA, CA 93106, USA

PONCE@MATH.UCSB.EDU

DEPARTAMENTO DE MATEMATICAS, UNIVERSIDAD DEL PAIS VASCO, APARTADO 644, 48080 BILBAO, SPAIN

MTPVEGOL@LG.EHU.ES

Contemporary Mathematics
Volume **189**, 1995

Unitary Dilations of Contractive Representations of $H^\infty(\mathbb{D}^2)$

MAREK KOSIEK

Dedicated to Professor Mischa Cotlar

Weak*-continuous contractive representations of dual algebras have recently been studied in connection with invariant subspaces and reflexivity problems. H. Bercovici and W.S. Li constructed in [**BL**] a special unitary dilation for a contractive weak*-continuous representation. We give below a more general result. It is a consequence of a Lebesgue-type decomposition of pairs of commuting operators [**K1**] and of the functional calculus given in [**KP**]. Our method simplifies the argument of [**BL**].

By $H^\infty(\mathbb{D}^2)$ we denote the algebra of all bounded analytic functions on the open unit bidisc $\mathbb{D}^2$ and by $A(\mathbb{D}^2)$ its subalgebra consisting of holomorphic functions on $\mathbb{D}^2$ that extend continuously to the closed bidisc, $\overline{\mathbb{D}^2}$. The algebra $H^\infty(\mathbb{D}^2)$ can be considered as a dual algebra (see [**KP**], Sec.3). Recall that a positive measure ν_z on $\overline{\mathbb{D}^2}$ is a *representing measure* for $z \in \mathbb{D}^2$ (with respect to the algebra $A(\mathbb{D}^2)$) if $\int u\, d\nu_z = u(z)$, for $u \in A(\mathbb{D}^2)$. The algebra $L(\mathcal{H})$ of all linear bounded operators on a Hilbert space $\mathcal{H}$ is the dual of the space of all trace class operators on $\mathcal{H}$. Hence we can consider the weak* topology in $L(\mathcal{H})$.

Recall that by a *contractive representation* of a Banach algebra A we mean any homomorphism $\Phi : A \to L(\mathcal{H})$ such that $\|\Phi(h)\| \leq \|h\|$ for $h \in A$. Now, if $A = A(\mathbb{D}^2)$, then standard considerations yield a collection $\{\mu_{x,y}\}$ of complex Borel measures on $\overline{\mathbb{D}^2}$, called *elementary measures* for Φ, such that

$$(1) \qquad (\Phi(u)x, y) = \int u\, d\mu_{x,y} , \qquad \text{for } x, y \in \mathcal{H} .$$

The measure $\mu_{x,x}$ will be denoted by μ_x.

1991 *Mathematics Subject Classification.* Primary 47A20, 47A60; Secondary 46J25.
Key words and phrases. Representation, unitary dilation, absolute continuity.
The author was supported in part by KBN Grant # 2 1251 9101.
This paper is in final form and no version of it will be submitted for publication elsewhere.

We say that a representation of $A(\mathbb{D}^2)$ is *absolutely continuous* if it has a system of elementary measures such that each element of the system is absolutely continuous (notation: $\ll$) with respect to some measure ν_z representing for a point $z \in \mathbb{D}^2$ (cf.[**KP**]). For $i = 1, 2$ denote $T_i = \Phi(\chi_i)$, where χ_1, χ_2 are the coordinate functions on $\mathbb{D}^2$ (i.e. $\chi_i(z_1, z_2) = z_i$ for $(z_1, z_2) \in \mathbb{D}^2$, $i = 1, 2$).

Using the Proposition of [**K1**] we get the following decomposition of $\mathcal{H}$ into an orthogonal sum of subspaces reducing Φ :

$$(2) \qquad\qquad \mathcal{H} = \mathcal{H}_0 \oplus \mathcal{H}_1 \oplus \mathcal{H}_2 \oplus \mathcal{H}_3 ,$$

where

$\Phi|_{\mathcal{H}_0}$ is absolutely continuous,

$\Phi(\chi_i)|_{\mathcal{H}_i}$ $(i = 1, 2)$ is unitary singular,

$\Phi(\chi_j)|_{\mathcal{H}_i}$ $(i, j = 1, 2, \ i \neq j)$ is an absolutely continuous contraction i.e. the spectral measure of its minimal unitary dilation is absolutely continuous with respect to the Lebesgue measure on the unit circle,

$\Phi|_{\mathcal{H}_3}$ is completely singular i.e. has a spectral measure which is singular to all measures orthogonal to $A(\mathbb{D}^2)$.

By the Polarization Formula the space $\mathcal{H}_0$ is precisely the set

$$(3) \quad \mathcal{H}_0 = \{x \in \mathcal{H} : \exists \mu_x \text{ elementary}, \ \exists \nu_z \text{ representing for } z \in \mathbb{D}^2, \ \mu_x \ll \nu_z\}$$

By Proposition 4.1 of [**KP**], we can extend Φ to a contractive representation $\Phi : H^\infty(\mathbb{D}^2) \to L(\mathcal{H})$ continuous, when each space is endowed with its respective weak*-topology (see also Remark 3.7 of [**KP**]). On the other hand the restriction to $A(\mathbb{D}^2)$ of every weak*-continuous representation of $H^\infty(\mathbb{D}^2)$ is absolutely continuous (weak*-star continuous functionals can be identified with complex mesures absolutely continuous with respect to representing measures of some points in $\mathbb{D}^2$).

Let us start with a representation $\Phi : H^\infty(\mathbb{D}^2) \to L(\mathcal{H})$.

DEFINITION. *A weak*-continuous (contractive) representation $\Psi : H^\infty(\mathbb{D}^2) \to L(\mathcal{K})$ is called a unitary dilation of Φ if*
(*) $\mathcal{K}$ *is a Hilbert space containing $\mathcal{H}$,*
(**) $\Phi(u) = P_{\mathcal{H}} \Psi(u)|\mathcal{H}$ *for all $u \in H^\infty(\mathbb{D}^2)$,*
(***) $\Psi(\chi_1), \Psi(\chi_2)$ *are unitary operators.*

THEOREM 1. *Every minimal unitary dilation (U_1, U_2) of (T_1, T_2) generates a contractive weak*-continuous representation of $H^\infty(\mathbb{D}^2)$ which is a unitary dilation of Φ.*

PROOF. Since Φ is a weak*-continuous representation of $H^\infty(\mathbb{D}^2)$, its restriction to $A(\mathbb{D}^2)$ is absolutely continuous. Let now $\Psi : A(\mathbb{D}^2) \to L(\mathcal{K})$ be the representation generated by (U_1, U_2). Let $\mathcal{K} = \mathcal{K}_0 \oplus \mathcal{K}_1 \oplus \mathcal{K}_2 \oplus \mathcal{K}_3$ be the decomposition of type (2) for Ψ. Since (U_1, U_2) is a dilation of (T_1, T_2), we get, first for polynomials, and, hence, for $u \in A(\mathbb{D}^2)$ the following

$$(4) \qquad (\Psi(u)x, y) = (\Phi(u)x, y) = \int u\, d\mu_{x,y} \,, \qquad \text{for } x, y \in \mathcal{H}.$$

By the absolute continuity of Φ, for each $x \in \mathcal{H}$, μ_x is absolutely continuous with respect to some representing measure of a point in $\mathbb{D}^2$. So, by (4) and the characterization (3) for $\mathcal{K}_0$, we get $\mathcal{H} \subset \mathcal{K}_0$. Hence, by the minimality of (U_1, U_2), we have $\mathcal{K} = \mathcal{K}_0$, which means that Ψ is absolutely continuous. By Proposition 4.1 and Remark 3.7 of [**KP**], Ψ can be extended to a contractive weak*-continuous representation of $H^\infty(\mathbb{D}^2)$. So the conditions (*) and (***) are satisfied. The condition (**) we get from the first equality in (4), the weak*-density of $A(\mathbb{D}^2)$ in $H^\infty(\mathbb{D}^2)$ and the weak*-continuity of both representations. This completes the proof.

By Ando's theorem we get

COROLLARY. *Every weak*-continuous Hilbert space representation of $H^\infty(\mathbb{D}^2)$ has a unitary dilation.*

By results of [**K2**], we can establish an N-dimensional version of Theorem 1. But for $N > 2$ the Corollary will be no longer valid.

Let $\Phi : H^\infty(\mathbb{D}^N) \to L(\mathcal{H})$ be a contractive weak*-continuous representation. Denote as before $T_i = \Phi(\chi_i)$ $(i = 1, ..., N)$ where χ_i are the coordinate functions on $\mathbb{D}^N$. Then we have

THEOREM 2. *Every minimal unitary dilation $(U_1, ..., U_N)$ of $(T_1, ..., T_2)$ generates a contractive weak*-continuous representation of $H^\infty(\mathbb{D}^N)$ which is a unitary dilation of Φ.*

REFERENCES

[BL] H.Becovici, and W.S.Li, *Isometric Functional Calculus on the Bidisk*, Bull. London Math. Soc. **25** (1993), 582–590.

[K1] M.Kosiek, *Lebesgue-Type Decomposition of a Pair of Commuting Hilbert Space Operators*, Bull. Acad. Polon. Sci., Ser. Sci. Math. **27** (1979), 583–589.

[K2] M.Kosiek, *Representation generated by a finite number of Hilbert space operators*, Ann. Polon. Math. **44** (1984), 309–315.

[KP] M.Kosiek, and M.Ptak, *Reflexivity of N- tuples of Contractions with Rich Joint Left Essential Spectrum*, Integral Eqs.Op. Theory **13** (1990), 395–420.

INSTITUTE OF MATHEMATICS, JAGIELLONIAN UNIVERSITY, REYMONTA 4, 30-059 KRAKÓW, POLAND

E-mail address: mko@im.uj.edu.pl

Contemporary Mathematics
Volume **189**, 1995

Noether and normalization theories for a class of singular integral operators with Carleman shift and unbounded coefficients.

V. G. Kravchenko, A. B. Lebre, G. S. Litvinchuk and F. S. Teixeira

Abstract. A criterion for the Noetherity of singular integral operators with Carleman shift in $L_p(\Gamma)$ is obtained, where Γ is either the unit circle or the real line. The approach allows to consider unbounded coefficients in a class related to that of quasi-continuous functions and reduces the given operator to a singular integral operator with bounded coefficients but, in general, with degenerate symbol. In connection with this the normalization problem for the above mentioned operators is considered. Applications to Wiener-Hopf-Hankel type operators and operators with linear fractional Carleman shift on $\mathbb{R}$ are included.

1. Introduction.

Let Γ be the unit circle $\mathbb{T}$ or the real line $\mathbb{R}$ and let $S : L_p(\Gamma) \to L_p(\Gamma), 1 < p < \infty$, denote the operator of singular integration

$$(S\varphi)(t) = \frac{1}{\pi i} \int_\Gamma \frac{1}{\tau - t} \, \varphi(\tau) \, d\tau, \qquad (1.1)$$

to which we associate the Cauchy projection operators

$$P_\pm = \frac{1}{2}(I \pm S). \qquad (1.2)$$

Further, let $\alpha : \Gamma \to \Gamma$ be a Carleman shift, which changes the orientation on Γ, whose derivative satisfies the usual Holder condition $\alpha' \in H_\mu(\Gamma)$ with $\alpha'(t) \neq 0$ for all $t \in \Gamma$. Consider also the weighted Carleman shift operator $U : L_p(\Gamma) \to L_p(\Gamma)$ given by

$$(U\varphi)(t) = k(t)(V\varphi)(t), \qquad (1.3)$$

AMS Subject Classification: Primary 45E05, Secondary 47A53.

This research was supported by JNICT under the grant PBIC/C/CEN/1040/92. The main results of this paper were presented in the International Conference "Harmonic Analysis and Operator Theory", Caracas, Venezuela, in January of 1994.

where $(V\varphi)(t) = \varphi(\alpha(t))$ is the usual Carleman shift operator and the weighted factor $k(t)$ is such that it guarantees its boundeness, the Carleman condition $U^2 = I$ and the anticommutating property

$$US = -SU. \tag{1.4}$$

For the sake of simplicity, and without loss of generality, we will take as a model the case $\Gamma = \mathbb{T}$ and the flip operator

$$(U\varphi)(t) = \frac{1}{t}\varphi(\frac{1}{t}) \ , \ t \in \mathbb{T}. \tag{1.5}$$

Let $H_\infty = H_\infty(\mathbb{T})$ denote the space of boundary values of bounded analytic functions on the unit disc $\mathbb{T}^+$ and let $C = C(\mathbb{T})$ be the class of continuous functions on $\mathbb{T}$. Consider the Douglas algebra $H_\infty + C$ and denote as usual by $QC = QC(\mathbb{T})$ the class of quasicontinuous functions on $\mathbb{T}$:

$$QC = H_\infty + C \cap \overline{H_\infty + C}$$

With each measurable function f on $\mathbb{T}$ we associate a function $\check{f}$, defined by

$$\check{f}(t) = \begin{cases} 1 & \text{if } |f(t)| \le 1 \\ f(t) & \text{if } |f(t)| > 1 \end{cases}, \tag{1.6}$$

which will be called the non–bounded part of f. Assuming that $\ln|\check{f}| \in L_1(= L_1(\mathbb{T}))$, we further define a function $e_{|\check{f}|}$ on the unit disc by

$$e_{|\check{f}|}(z) = \exp\left\{\frac{1}{2\pi}\int_{\mathbb{T}} \frac{\tau+z}{\tau-z} \ln|\check{f}(\tau)| \, |d\tau|\right\},$$

which will be called the outer non–bounded part of f, since it is an outer function with modulus $|\check{f}|$ on $\mathbb{T}$ (see [4]). Now we introduce two classes of measurable functions defined on $\mathbb{T}$, $\check{Q}C^+$ and $\check{Q}C^-$, which will be frequently used in the sequel:

$$\check{Q}C^+ = \left\{f : \ln|f| \in L_1 \text{ and } f e_{|\check{f}|}^{-1} \in QC\right\},$$
$$\check{Q}C^- = \left\{f : \ln|f| \in L_1 \text{ and } f \overline{e_{|\check{f}|}^{-1}} \in QC\right\},$$

where $e_{|\check{f}|}^{-1}(t), t \in \mathbb{T}$, are the boundary values in H_∞ of the $H_\infty(\mathbb{T}^+)$ function $e_{|\check{f}|}^{-1}(z) = e_{|\check{f}|^{-1}}(z), z \in \mathbb{T}^+$, and, as it is usual, $\overline{e_{|\check{f}|}^{-1}(t)} = \overline{e_{|\check{f}|}^{-1}}(t), t \in \mathbb{T}$, are the boundary values in $\overline{H_\infty}$ of the $H_\infty(\mathbb{T}^-)$ function $e_{|\check{f}|}^{-1}(\frac{1}{z})$. Here $\mathbb{T}^-$ is the exterior of the unit circle and the bar denotes complex conjugation.

We shall use the following notation: $L_p = L_p(\mathbb{T})$. Given a closed operator $T : Dom(T) \subset L_p \to L_p$ we denote by $D(T)$ the Banach space which is obtained by introducing in $Dom(T)$ the graph–norm

$$\|x\|_{D(T)} = \|x\|_{L_p} + \|Tx\|_{L_p}.$$

We are going to study singular integral operators with shift (S.I.O.S.) of the form

$$R(A) = P_+ + AP_- : Dom(R(A)) \subset L_p \to L_p, \tag{1.7}$$

where

$$A = a\,I + b\,U \tag{1.8}$$

with $a \in L_\infty(\Gamma)$ and $b \in \check{Q}\check{C}^\pm$, for what it is convenient to consider the singular integral operator (S. I. O.) without shift

$$R(a) = P_+ + a\,P_- : L_p \to L_p. \tag{1.9}$$

Since $a \in L_\infty$ the operator $R(a)$ is bounded in L_p. Also UP_- is a bounded operator and the multiplication operator by a function $b \in \check{Q}\check{C}^+$ is closed. Thus, $R(A)$ is closed in L_p.

In section 2 we obtain a criterion for the Noetherity of the operator (1.7). It turns out that the reasoning used in section 2 enables to reduce the operator (1.7) to an operator, say $R(\tilde{a})$, where $\tilde{a}$ can be degenerated on $\mathbb{T}$. Therefore in section 3 we formulate the abstract normalization problem and in section 4 we solve the corresponding normalization problem for S. I. O. S.. If one consider the linear fractional Carleman shift operator

$$(V\varphi)(x) = \frac{\mu x - \nu}{x - \mu}\ ,\ x \in \mathbb{R}, \tag{1.10}$$

we see that is not a bounded operator in $L_p(\mathbb{R})$. This particular case is studied in section 5. We transfer the non–boundedness of the operator V to the the non-boundedness of the coefficient b and consider afterwards the Noether theory and the normalization problem for closed (in general non bounded) S. I. O. S. on $\mathbb{R}$. We also show that some operators of Wiener-Hopf-Hankel type arising in applications, namely in diffraction theory (see [14] and the references therein), are included in this theory.

The results presented herein are based on the papers [9] and [10].

2. Noether theory for a class of unbounded S.I.O.S. on the unit circle.

It is well known (see, e.g. [15]) that $P_- aP_+$ is a compact operator on L_p if $a \in C(\mathbb{T})$. There is a generalization of this result for the case where $a \in QC$ (Hartman's theorem). It turns out that functions in the classes $\check{Q}\check{C}^\pm$ possess the following important property.

LEMMA 2.1   1. If $f \in \check{Q}\check{C}^+$ then the operator $P_- fP_+$ is compact relative to fP_+.

2. If $f \in \check{Q}\check{C}^-$ then the operator $P_+ fP_-$ is compact relative to fP_-.

The definitions of relative boundedness and relative compactness can be seen in [6], for instance, and the proof in [9].

LEMMA 2.2 $D(R(A)) = U\,D(bP_+)$.

The proof consists in showing the equivalence of the two norms $\|\varphi\|_{D(R(A))}$ and $\|\psi\|_{D(bP_+)}$, where $\varphi = U\psi$ (see [9]).

LEMMA 2.3 Let $a \in L_\infty$ and $b \in \check{Q}\check{C}^+$. Then $R(A) : D(R(A)) \to L_p$ is a Noether operator if and only if $R(a) : D(R(A)) \to L_p$ is a Noether operator. For the Noether operator $R(A) : Dom(R(A)) \to L_p$ there holds

$$\text{ind}\,(R(A) : Dom(R(A)) \to L_p) = \text{ind}\,(R(a) : D(R(A)) \to L_p). \qquad (2.1)$$

PROOF. According to Lemma 2.1 the operator $P_- b U P_- = P_- b P_+ U :$ $Dom(R(A)) \to L_p$ is compact relative to the operator $b P_+ U$ (note that $U P_- = P_+ U$, because the shift changes the orientation on $\mathbb{T}$) and, hence, it is compact relative to $R(A)$. Also, $P_- b U P_-$ is bounded relative to $R(A)$ (see [6]). Thus, after introducing in $Dom(R(A))$ the graph-norm, the operator

$$P_- b U P_- : D(R(A)) \to L_p$$

becomes a bounded and compact operator (see [6]). Consequently, the Noetherity of the operator $R(A)$ is equivalent to the Noetherity of the operator $R(A) - P_- b U P_-$ (see, for instance, [3] or [6]).

Now, for any $\varphi \in D(R(A))$, we have

$$(R(A) - P_- b U P_-)(I - P_+ b U P_-)\varphi = R(a)\varphi. \qquad (2.2)$$

Hence, to show that the Noetherity of the operator $R(A)$ is equivalent to the Noetherity of the operator $R(a)$ it is sufficient to prove that the operator $I - P_+ b U P_-$ is bounded and continuously invertible in $D(R(A))$. But the latter can be proved easily if we note that, for any $\varphi \in D(R(A))$, we have $\psi = (I - P_+ b U P_-)\varphi \in D(R(A))$ and $\varphi = (I + P_+ b U P_-)\psi$, which can be seen by direct verification.

Since $D(R(A))$ coincides with the domain of definition of the operator $R(A)$, we have $\text{ind}\,(R(A) : Dom(R(A)) \to L_p) = \text{ind}\,(R(A) : D(R(A)) \to L_p)$. Taking into account (2.2) and the fact that $(I + P_+ b U P_-)$ is invertible in $D(R(A))$, we obtain (2.1). $\qquad \square$

Given a measurable function b such that $\ln |\check{b}| \in L_1$, where $\check{b}$ be the non-bounded part of b, we define

$$u_-(t) \stackrel{\text{def}}{=} e_{|\check{b}|}^{-1}(\frac{1}{t}) = \lim_{\substack{z \to 1/t \\ z \in \mathbb{T}^+}} e_{|\check{b}|}^{-1}(z)\,, t \in \mathbb{T}. \qquad (2.3)$$

where $\mathbb{T}^+$ is the open unit disc and the limit is consider for directions non-tangential to $\mathbb{T}$. Further, let

$$\tilde{u}_-(t) = u_-(1/t)\,, t \in \mathbb{T} \qquad (2.4)$$

Then $\tilde{u}_-$ is an outer function in H_∞, due to $|\tilde{u}_-(t)| = |\check{b}(t)|^{-1} \leq 1\,, t \in \mathbb{T}$ (recall the definition of $\check{b}$) and, thus, $u_- \in \overline{H}_\infty$.

LEMMA 2.4 $R(u_-) = P_+ + u_- P_- : L_p \to D(R(A))$ is a bounded and continuously invertible operator and $R^{-1}(u_-) = P_+ + u_-^{-1} P_- : D(R(A)) \to L_p..$

PROOF. It is easy to show that $R(u_-)$ is well defined on the whole L_p and $R(A)R(u_-)\varphi \in L_p$. Consequently, $im\,R(u_-) \subset D(R(A))$. The reverse inclusion $im\,R(u_-) \supset D(R(A))$ can be proved with the help of Lemmas 2.1 and 2.2.

To conclude the proof it remains to note that, since $u_- \in \overline{H}_\infty$, $R(u_-)$ is an invertible operator and, since $im\, R(u_-) = D(R(A))$, by the closed graph theorem the inverse operator is continuous. $\quad\square$

LEMMA 2.5 $R(a) : D(R(A)) \to L_p$ is a Noether operator if and only if $R(au_-) :$ $L_p \to L_p$ is a Noether operator. If one of these operators is a Noether operator then

$$\mathrm{ind}\,(R(a) : D(R(A)) \to L_p) = \mathrm{ind}\,(R(au_-) : L_p \to L_p) \qquad (2.5)$$

PROOF. This lemma is a consequence of the following identity

$$R(a)\varphi = R(au_-)R^{-1}(u_-)\varphi, \qquad (2.6)$$

valid for any $\varphi \in D(R(A))$, which can be shown by a direct and straightforward computation. $\quad\square$

THEOREM 2.6 Let $a \in L_\infty$ and $b \in \check{Q}C^+$. Then

$$R(A) : Dom(R(A)) \subset L_p \to L_p \qquad (2.7)$$

is a Noether operator if and only if $b \in L_\infty$ and $R(a) : L_p \to L_p$ is a Noether operator. If $R(A)$ is a Noether operator then

$$\mathrm{ind}\, R(A) = \mathrm{ind}\, R(a). \qquad (2.8)$$

PROOF. Suppose that $R(A) : Dom(R(A)) \to L_p$ is a Noether operator. From Lemma 2.3 this implies that $R(a) : D(R(A)) \to L_p$ is a Noether operator. But, from Lemma 2.5, this means that $R(au_-) : L_p \to L_p$ is a Noether operator. Therefore (see [8]) ess inf $|a(t)u_-(t)| > 0$ and, consequently, ess inf $|u_-(t)| > 0$. But according to (2.3) the latter means that $b \in L_\infty$.

Conversely, assume that $b \in L_\infty$ and $R(a) : L_p \to L_p$ is a Noether operator. Then the first condition guarantees that $D(R(A)) = L_p$ and the second one, together with Lemma 2.3, implies that $R(A) : L_p \to L_p$ is a Noether operator.

Finally, (2.8) follows from (2.5) and Lemma 2.3. $\quad\square$

Due to Lemma 2.4 we can consider the operator $R(A)R(u_-) : L_p \to L_p$ instead of the operator $R(A)$. Indeed, these two operators satisfy or not, simultaneously, the Noether property and their indices coincide. Moreover, their defect numbers also coincide and the following equality holds:

$$im\, \dot{R}(A) = im\, R(A)R(u_-).$$

We have yet

$$R(Au_-) = P_+ + (au_-\, I + b\tilde{u}_-\, U)P_-. \qquad (2.9)$$

Hence the problem due to the non–boundedness of the coefficient b in (1.8) is already overcome, because in the above operator $au_-, b\tilde{u}_- \in L_\infty$. Now, from Theorem 2.6

we know that the Noether properties of $R(Au_-)$ are those of the operator $R(au_-)$: $L_p \xrightarrow{\cdot} L_p$, given by

$$R(au_-) = P_+ + au_- P_-.$$

Therefore, it is clear that $R(Au_-)$ is not Noether in L_p whenever au_- has zeros on $\mathbb{T}$. So being interested precisely in that case, it is natural to use a normalization theory for the operator $R(Au_-)$.

For the normalization theory of singular integral operators without shift and convolution type operators we refer to the pioneer work of S. Prössdorf and the works of V. Dybin, M. Khaikin and B. Silbermann, whose references can be found in [15] (see also [16]). The approach that we are going to use in some sense generalizes the basic ideas of these authors and constitutes an application of the main abstract results of the normalization theory developed recently by V. G. Kravchenko et al. (see [7][8] and also [1][11][12][13]).

REMARK 2.7 The results in this section also hold if instead of the space L_p we consider weighted spaces $L_p(\mathbb{T}, r)$, with weights r such that the operator S remains bounded.

The same can be said if the shift $\frac{1}{t}$ is replaced by any Carleman shift α, which changes the orientation of $\mathbb{T}$, and such that the corresponding operator U, defined by $U\varphi(t) = (\alpha'(t))^{\frac{1}{p}}\varphi(\alpha(t))$, is anticommutating with S (see (1.4)). In particular, if we now define the function u_- by

$$u_-(t) = e_{|\check{b}|}^{-1}(\alpha(t)) \tag{2.10}$$

instead of (2.3), the proofs can be carried out using the same kind of arguments.

REMARK 2.8 The class of singular integral operators with shift studied, with the coefficient b in $\check{Q}C^+$, can be easily enlarged. Indeed, all the results obtained above are kept if, instead of $\check{Q}C^+$ ($\check{Q}C^-$), we take the class of all measurable functions f, defined on $\mathbb{T}$, such that, for each f, there exists an outer function e_f with $f e_f^{-1} \in QC$ ($f \overline{e_f^{-1}} \in QC$). Obviously, the classes $\check{Q}C^{\pm}$, introduced in the begining of this section, are just concrete realizations of the latter ones.

3. The normalization problem.

In this section we just summarize some relevant results which were obtained by V. Kravchenko in [8].

Let X_0, Y_0 be Banach spaces, $\mathcal{C}(X_0, Y_0)$ be the set of closed linear operators from X_0 into Y_0 and M be some subset of $\mathcal{C}(X_0, Y_0)$. The abstract normalization problem we are interested in can be stated as follows:

Normalization problem $\mathcal{N}(M)$: Find a Banach space Y_1 such that Y_1 is continuously embeded in Y_0 and such that for any $A \in M$ it holds

$$imA \subset Y_1 \ , \ imA = \overline{imA}^{1},$$

where $\overline{imA}^{1}$ is the closure of the image of A in the topology of the space Y_1.

If the space Y_1 satisfies these condition we shall say that the pair (X_0, Y_1) solves the normalization problem $\mathcal{N}(M)$ and we will write $(X_0, Y_1) \in \mathcal{N}(M)$.

Let us particularize this normalization problem for a special class of operators from $\mathcal{C}(X)$, the space of closed linear operators from X into X. Here X is a Banach space in which we assume to be defined two complementary bounded linear projectors, P_+ and $P_- = I - P_+$, such that $X = X_+ \oplus X_-$ with $X_{\pm} = P_{\pm}X$.

For each $A \in \mathcal{C}(X)$ we define closed operators $\widetilde{A}$ and $R(A)$ by the formulas

$$\widetilde{A} = A\, P_+ + A\, P_-, \tag{3.1}$$

$$R(A) = P_+ + A\, P_-. \tag{3.2}$$

It is clear that $\widetilde{R(A)} = R(A)$ and that $\widetilde{A} \subset A$, i.e., $Dom(\widetilde{A}) \subset Dom(A)$ and $\widetilde{A}x = A\,x$ for $x \in Dom(\widetilde{A})$. We will denote by $\mathcal{C}_+(X)$ the set of operators $A \in \mathcal{C}(X)$ such that $Dom(\widetilde{A})$ is dense in X and $P_-AP_+ = 0$. Let $V_+ \in \mathcal{C}(X)$ be an operator such that $\widetilde{V}_+$ is invertible and $\widetilde{V}_+^{\pm 1} \in \mathcal{C}_+(X)$. We shall say that an operator $A \in \mathcal{C}(X)$ satisfies the condition (V_+) and we write $A \in (V_+)$ if $im A \subset im \widetilde{V}_+$. Let $X(A)$ denote the Banach space which is obtained by introducing in $Dom(A)$ the graph–norm, $\|x\|_{X(A)} = \|x\|_X + \|A\,x\|_X$, and put $X_-(A) = X(A|_{X_-})$.

Suppose that $A \in (V_+)$. As $im \widetilde{V}_+ = Dom(\widetilde{V}_+^{-1})$ it follows that $im\widetilde{A} \subset X(\widetilde{V}_+^{-1})$ and, consequently,

$$im R(A) \subset X_+ + X(\widetilde{V}_+^{-1}).$$

Hence $(X, X_+ + X(\widetilde{V}_+^{-1})) \in \mathcal{N}(R(A))$ if and only if the set $im R(A)$ is closed in $X_+ + X(\widetilde{V}_+^{-1})$. The main result, which constitutes an useful necessary and sufficient condition, is stated in the next theorem, where $\hat{R}(A) : Dom(R(A)) \to L_p$ is the operator given by

$$\hat{R}(A)\,x = R(A)\,x \ , \ x \in Dom(R(A)).$$

Its proof, rather lengthy, can be found in [8] or [1].

THEOREM 3.1 *If $A \in (V_+)$ then*

$$(X, X_+ + X(\widetilde{V}_+^{-1})) \in \mathcal{N}(R(A))$$

if and only if the operator

$$R \overset{\text{def}}{=} R(\widetilde{V}_+^{-1}A) : X_+ + X_-(A) \to X_+ + X(\widetilde{V}_+)$$

is normally solvable. Moreover, if R is normally solvable then

$$\dim \ker \hat{R}(A) = \dim \ker R \ , \ \dim \operatorname{coker} \hat{R}(A) = \dim \operatorname{coker} R.$$

4. Normalization of the operator $R(A)$ on the unit circle.

Let $L_p^+ = P_+L_p(\mathbb{T})$ be the usual Hardy space on the disc and let $L_p(\rho) = L_p(\mathbb{T}, \rho)$ denotes the weighted L_p-space with weight function ρ. Further, given a measurable function a such that $\ln|a| \in L_1$, let $QC(e_{|a|}^{-1})$ be the class of functions that after multiplication by the outer function $e_{|a|}^{-1}$ belong to QC. We start with the following lemma.

LEMMA 4.1 *Let the coefficients a and b of the operator A, given by (1.8), be such that* *(i)* $a \in L_\infty$, $\ln |a| \in L_1$,

(ii) $b \in H_\infty + QC(e_{|a|}^{-1})$,

and, consequently, $Dom(A) = Dom(R(A)) = L_p$.

Then

$$(L_p, L_p^+ + L_p(|a|^{-1})) \in \mathcal{N}(R(A)) \tag{4.1}$$

if and only if

$$(L_p, L_p) \in \mathcal{N}(R(e_{|a|}^{-1} a)). \tag{4.2}$$

Moreover, if $R(A)$ *is Noetherian then*

$$\text{ind } R(A) = \text{ind } R(e_{|a|}^{-1} a). \tag{4.3}$$

PROOF. According to the hypothesis there exists $w \in H_\infty$ such that $b - w \in QC(e_{|a|}^{-1})$. Moreover,

$$
\begin{aligned}
R(A)(I - wP_+U) &= P_+ - wP_+U + (a\,I + b\,U)P_- \\
&= P_+ + (a\,I + (b - w)\,U)P_- \\
&\stackrel{\text{def}}{=} R(A_w),
\end{aligned}
$$

where A_w is defined by

$$A_w = a\,I + (b - w)\,U.$$

Obviously the operator $I - wP_+U$ is bounded and continuously invertible in L_p, with inverse given by $I + wP_+U$ (cf. (1.4)). Therefore, $R(A)$ is a Noether (respectively normally solvable) operator if and only if $R(A_w)$ is a Noether (respectively normally solvable) operator. Furthermore,

$$imR(A) = imR(A_w).$$

Let $V_+ = e_{|a|} I$. Note that, since $a \in L_\infty$ and $\ln |a| \in L_1$, $e_{|a|}$ is an outer function in H_∞ and so $Dom(V_+) = L_p$. Using the notation of the previous section, in this case we have $V_+ = \widetilde{V_+}$. Since $e_{|a|} \in H_\infty$, we have $P_- V_+ P_+ = 0$ and since $e_{|a|}^{-1} = e_{|a|^{-1}}$ is an outer function (although not necessarily in H_∞) it follows that $P_- V_+^{-1} P_+ = 0$ (see, for instance, [2], Theorem 2.11). Next we prove that A_w satisfies condition (V_+), i.e., $A_w \in (V_+)$. Let $f \in imA_w$ be arbitrary. Then $f = a\varphi + (b - w)U\varphi$, for some $\varphi \in L_p$. Therefore, puting $f = e_{|a|}g$, we have

$$g = e_{|a|}^{-1}a\varphi + e_{|a|}^{-1}(b - w)U\varphi = V_+^{-1}A_w\varphi \in L_p,$$

since $e_{|a|}^{-1}a \in L_\infty$ and $(b - w) \in QC(e_{|a|}^{-1})$. Thus $f \in imV_+$ and, since $f \in imA_w$ is arbitrary, it follows that $imA_w \subset imV_+$. Hence, we have $A_w \in (V_+)$.

Taking $X = L_p$ we have, in the notation of the preceding section, $X_+ = L_p^+$, $X(V_+^{-1}) = L_p(|a|^{-1})$ (note that $|e_{|a|}^{-1}(t)| = |a(t)|^{-1}, t \in \mathbb{T}$) and, since $A_w \in (V_+)$, we can use Theorem 3.1. From it we conclude that

$$(L_p, L_p^+ + L_p(|a|^{-1})) \in \mathcal{N}(R(A_w))$$

if and only if

$$(L_p, L_p) \in \mathcal{N}(R(V_+^{-1}A_w)).$$

Noting that

$$R(V_+^{-1} A_w) = P_+ + V_+^{-1} A_w P_- = P_+ + (e_{|a|}^{-1} a\, I + e_{|a|}^{-1}(b - w)\, U) P_-$$

and, since the coefficients of this operator are essentially bounded functions, we have

$$(L_p, L_p) \in \mathcal{N}(R(V_+^{-1} A_w))$$

if and only if

$$(L_p, L_p) \in \mathcal{N}(R(e_{|a|}^{-1} a)),$$

that is, (4.1) is equivalent to (4.2). The equality (4.3) is also a consequence of $\square$

THEOREM 4.2 *Let the coefficients of the operator A, given by (1.8), be such that*

(i) $a \in L_\infty$, $\ln |a| \in L_1$,

(ii) $\ln |\check{b}| \in L_1$, $b\,\tilde{u}_- \in H_\infty + QC(e_{|au_-|}^{-1})$,

where u_- *and* $\tilde{u}_-$ *are the functions defined in (2.3) and (2.4).*
    Then

$$(L_p, L_p^+ + L_p(|au_-|^{-1})) \in \mathcal{N}(R(A)) \tag{4.4}$$

if and only if

$$(L_p, L_p) \in \mathcal{N}(R(e_{|au_-|}^{-1} au_-)). \tag{4.5}$$

Moreover, if $R(A)$ is Noetherian then

$$\operatorname{ind} R(A) = \operatorname{ind} R(e_{|au_-|}^{-1} au_-). \tag{4.6}$$

PROOF. We first note that the hypothesis of this theorem imply that $b \in \check{Q}C^+$. As mentioned before, the operator $R(A) : Dom(R(A)) \to L_p$ and $R(Au_-) = R(A)R(u_-) : L_p \to L_p$, are simultaneously Noether or not Noether operators and, in the case of Noetherity, their indices as well as defect numbers coincide. Therefore, instead of $R(A)$, we consider the operator

$$R(Au_-) = P_+ + (au_-\, I + b\tilde{u}_- U) P_-.$$

Now, the coefficients of this operator satisfy the conditions of the preceding lemma, if there we replace a and b by au_- and $b\tilde{u}_-$, respectively. In fact, as $u_- \in L_\infty$ and $\ln |u_-| \in L_1$ it follows that $au_- \in L_\infty$ and $\ln |au_-| \in L_1$. Also, by hypothesis $b\tilde{u}_- \in H_\infty + QC(e_{|au_-|}^{-1})$. Then, from Lemma 4.1, we conclude that

$$(L_p, L_p^+ + L_p(|au_-|^{-1}) \in \mathcal{N}(R(Au_-))$$

if and only if

$$(L_p, L_p) \in \mathcal{N}(R(e_{|au_-|}^{-1} au_-))$$

and, if $R(A)$ is Noetherian, relation (4.6) holds. $\square$

REMARK 4.3 In the last two propositions it was assumed that the shift operator is just the flip operator (1.5), to which it corresponds the auxiliary function u_-, given by (2.3). However, it is clear that all the results are kept if one takes any Carleman shift operator satisfying (1.3) and (1.4), making simultaneously $u_-(t) = e_{|\check{b}|}^{-1}(\alpha(t))$.

5. The real line case. Applications.

In this section we present the Noether and normalization theories for singular integral operators with Carleman shift on the real line.

Let $QC(\mathbb{R})$ be the class of quasicontinuous functions on $\mathbb{R}$ and for each measurable function g on $\mathbb{R}$ let $\breve{g}$ denote, as before, its non–bounded part (cf. (1.6)). For the classes of measurable functions on $\mathbb{R}$

$$\breve{Q}C^{\pm}(\mathbb{R}) = \left\{ g : g \circ \theta^{-1} \in \breve{Q}C^{\mp} \right\}$$

we get

$$\breve{Q}C^{-}(\mathbb{R}) = \left\{ g : \ln|\breve{g}| \in L_1(\mathbb{R}, \rho) \text{ and } g\, E_{|\breve{g}|}^{-1} \in QC(\mathbb{R}) \right\},$$

$$\breve{Q}C^{+}(\mathbb{R}) = \left\{ g : \ln|\breve{g}| \in L_1(\mathbb{R}, \rho) \text{ and } g\, \overline{E_{|\breve{g}|}^{-1}} \in QC(\mathbb{R}) \right\},$$

where $\rho(x) = (x^2 + 1)^{-1}, x \in \mathbb{R}$, and $E_{|\breve{g}|}$ is the boundary value of the outer non–bounded part of g, defined on $\mathbb{C}^-$ by

$$E_{|\breve{g}|}(w) = \exp\left\{ \frac{1}{\pi i} \int_{\mathbb{R}} \frac{1 + wy}{w - y} \ln|\breve{g}(y)|\, \frac{dy}{y^2 + 1} \right\}.$$

REMARK 5.1 We note that $E_{|\breve{g}|}^{-1}(x), x \in \mathbb{R}$, is the boundary value of the function $E_{|\breve{g}|}^{-1}(w) = E_{|\breve{g}|^{-1}}(w), w \in \mathbb{C}^-$, which belongs to $H_\infty(\mathbb{C}^-)$. Furthermore, $\overline{E_{|\breve{g}|}^{-1}}(x) = \overline{E_{|\breve{g}|}^{-1}}(x), x \in \mathbb{R}$, are the boundary values of the function $\overline{E_{|\breve{g}|}^{-1}(\overline{w})}$, which (as a function of w) belongs to $H_\infty(\mathbb{C}^+)$.

THEOREM 5.2 Let $a \in L_\infty(\mathbb{R}), b \in \breve{Q}C^{-}(\mathbb{R})$ and let $A : Dom(A) \to L_p(\mathbb{R}), p > 1$, be the functional operator

$$A = a\, I + b\, V,$$

where $V : L_p(\mathbb{R}) \to L_p(\mathbb{R})$ is any bounded Carleman shift operator which anticommutes with $S_\mathbb{R}$ (see (1.4)). Then

$$\mathcal{R}(A) = P_{\mathbb{R}}^- + AP_{\mathbb{R}}^+ : Dom(\mathcal{R}(A)) \to L_p(\mathbb{R}) \tag{5.1}$$

is a Noether operator if and only if $b \in L_\infty(\mathbb{R})$ and

$$\mathcal{R}(a) = P_{\mathbb{R}}^- + aP_{\mathbb{R}}^+ : L_p(\mathbb{R}) \to L_p(\mathbb{R})$$

is a Noether operator. If $\mathcal{R}(A)$ is a Noether operator then

$$\operatorname{ind} \mathcal{R}(A) = \operatorname{ind} \mathcal{R}(a). \tag{5.2}$$

PROOF. The result is a direct consequence of Theorem 2.6, Remark 4.3 and the preceding considerations. □

If one takes the flip operator on $\mathbb{T}$ defined in the previous section and which was used as a model for the development of the theory, the correspondent operator V on $\mathbb{R}$ is $V = -J$, where $J : L_p(\mathbb{R}) \to L_p(\mathbb{R})$ is the reflection operator, given by

$$J\psi(x) = \psi(-x), \ x \in \mathbb{R}.$$

In this case, the operator $\mathcal{R}(A)$ given by (5.1) is said to be of Wiener–Hopf–Hankel type. Indeed, according to the above theorem, for coefficients $a \in L_\infty(\mathbb{R})$ and $b \in \check{Q}C^-(\mathbb{R})$, $\mathcal{R}(A)$ is a operator Noether only if b is an essentially bounded function, and in this case the operator has an equivalent representation in terms of a Wiener-Hopf-Hankel operator on $L_p^+(\mathbb{R}) = \{\psi \in L_p : \operatorname{supp} \psi \subset \overline{\mathbb{R}^+}\}$ defined by

$$\mathcal{W}(a) + \mathcal{H}(b) : L_p^+(\mathbb{R}) \to L_p^+(\mathbb{R}), \tag{5.3}$$

where $\mathcal{W}(a)$ is the Wiener-Hopf operator and $\mathcal{H}(b)$ is the Hankel operator, given by

$$\mathcal{W}(a) = \mathcal{P}^+ \mathcal{F}^{-1} a \mathcal{F} : L_p^+(\mathbb{R}) \to L_p^+(\mathbb{R}), \mathcal{H}(b) = \mathcal{P}^+ \mathcal{F}^{-1} b \mathcal{F} J : L_p^+(\mathbb{R}) \to L_p^+(\mathbb{R}) \tag{5.4}$$

for $\mathcal{F}$ denoting the Fourier transformation in the Schwartz space $\mathcal{S}'(\mathbb{R})$ and $\mathcal{P}^+ = \mathcal{F}^{-1} P_+ \mathcal{F}$ (see [14]).

As a remarkable consequence of the present theory, it turns out that for symbols $a \in L_\infty(\mathbb{R})$ and $b \in L_\infty(\mathbb{R}) \cap \check{Q}C^-(\mathbb{R})$ the Noether property of the Wiener–Hopf–Hankel operator $\mathcal{W}(a) + \mathcal{H}(b)$ is equivalent to the Noether property of the Wiener–Hopf operator $\mathcal{W}(a)$.

Another application that we are interested in refers to singular integral operators with non–bounded shift acting in $L_p(\mathbb{R})$ of the form

$$T = P_{\mathbb{R}}^+ + (a\, I + b\, W_\alpha) P_{\mathbb{R}}^-. \tag{5.5}$$

where W_α is the linear fractional Carleman shift

$$W_\alpha \varphi(x) = \varphi(\alpha(x))$$

with $\alpha(x) = \frac{\mu x - \nu}{x - \mu}$ for $\mu, \nu \in \mathbb{R}$ such that $\mu^2 - \nu > 0$. These type of operators where studied before by N. K. Karapetyants and S. G. Samko [5] in the case $a, b \in C(\dot{\mathbb{R}})$.

Let us modify slightly the way of writing the operator T, in order to have a bounded Carleman shift operator in $L_p(\mathbb{R})$, which anticommutes with $S_{\mathbb{R}}$, so that Theorem 5.2 can be applied. Define $U_\alpha : L_p(\mathbb{R}) \to L_p(\mathbb{R})$ by

$$U_\alpha \varphi(x) = v(x) W_\alpha \varphi(x),$$

where

$$v(x) = (\frac{k}{x - \mu})^{2/p}, \tag{5.6}$$

with $k = i\sqrt{\mu^2 - \nu}$. Then U_α is a bounded operator in $L_p(\mathbb{R}), U_\alpha^2 = I$ and $U_\alpha S_{\mathbb{R}} = -S_{\mathbb{R}} U_\alpha$, the latter being consequence of the assumption $\mu^2 - \nu > 0$. Now, rewrite the operator T as

$$T = P_{\mathbb{R}}^+ + (a\, I + b_0\, U_\alpha) P_{\mathbb{R}}^- \tag{5.7}$$

with $b_0 = b/v$. We have, as a direct consequence of Theorem 5.2:

PROPOSITION 5.3 *Let $a \in L_\infty(\mathbb{R}), b_0 \in \check{Q}C^-(\mathbb{R})$ and $T : Dom(T) \to L_p(\mathbb{R})$ be the operator defined by (5.7). Then T is a Noether operator if and only if $\mathcal{R}(a)$ is a Noether operator and $b_0 \in L_\infty(\mathbb{R})$. If T is a Noether operator then*

$$\operatorname{ind} T = \operatorname{ind} \mathcal{R}(a).$$

In particular, the following proposition holds for the initial operator T.

COROLLARY 5.4 *Let* $a \in C(\dot{\mathbb{R}})$ *and* $b \in C(\dot{\mathbb{R}})$ *be such that* $b(\infty) = \lim_{\xi \to \pm\infty} b(\xi) \neq 0$. *Then* T, *defined by (5.5), is not a Noether operator.*

The above Corollary can also be obtained directly from the results of N. K. Karapetyants and S. G. Samko (see [5], Theorems 14.2 and 24.1′).

Let $\widehat{L_p^-}(\mathbb{R}) = P_{\mathbb{R}}^- L_p(\mathbb{R})$, $H_\infty^-(\mathbb{R})$ is the space of the boundary values of $H_\infty(\mathbb{C}^-)$ functions and, for $a \in L_\infty(\mathbb{R})$ such that $\ln|a| \in L_1(\mathbb{R}, \rho)$ for $\rho(x) = (x^2+1)^{-1}$, $x \in \mathbb{R}$, $QC(E_{|a|}^{-1})$ denotes the class of all measurable functions that after multiplication by $E_{|a|}^{-1}$ belong to $QC(\mathbb{R})$. The following result holds (cf. [10]).

THEOREM 5.5 *Let the coefficients of the operator A, given by (1.8), be such that*

  (i) $a \in L_\infty(\mathbb{R})$, $\ln|a| \in L_1(\mathbb{R}, \rho)$,

  (ii) $\ln|\breve{b}| \in L_1(\mathbb{R}, \rho)$, $b\,\tilde{u}_+ \in H_\infty^-(\mathbb{R}) + QC(E_{|au_+|}^{-1})$,

where u_+ and $\tilde{u}_+$ are the functions defined by $u_+(x) = E_{|\breve{b}|}^{-1}(\alpha(x))$, $\tilde{u}_+(x) = u_+(\alpha(x))$, $x \in \mathbb{R}$.
Then

$$(L_p(\mathbb{R}), \widehat{L_p^-}(\mathbb{R}) + L_p(\mathbb{R}, |au_+|^{-1})) \in \mathcal{N}(\mathcal{R}(A)) \tag{5.8}$$

if and only if

$$(L_p(\mathbb{R}), L_p(\mathbb{R})) \in \mathcal{N}(\mathcal{R}(E_{|au_+|}^{-1}\, au_+)). \tag{5.9}$$

Moreover, if $\mathcal{R}(A)$ is Noetherian then

$$\operatorname{ind} \mathcal{R}(A) = \operatorname{ind} \mathcal{R}(E_{|au_+|}^{-1} au_+). \tag{5.10}$$

The above result can be applied to the normalization of some important classes of S.I.O.S.. In particular, when we take for U the reflection operator $U\varphi(x) = \varphi(-x)$, $x \in \mathbb{R}$, we get the so-called Wiener-Hopf-Hankel operators [14]). Another relevant case is that of singular integral operators with linear fractional shift on $\mathbb{R}$ (see, for instance, [13]).

References

[1] Drekova, G. N. and Kravchenko, V. G.: To the theory of normalization of the Riemann problem. Izv. Vyssh. Uchebn. Zaved. Mat. n⁰ 9 (1991) 20-28 (in Russian).

[2] Duren, P. L.: Theory of H^p Spaces, Academic Press, 1970.

[3] Gohberg, I. and Krein, M. G.: The basic propositions on defect numbers, root numbers and indices of linear operators. Uspekhi Mat. Nauk.12, n⁰ 2 (1957) 44-118 (in Russian); Translation: Amer. Math. Soc. Trans.13, n⁰ 2 (1960) 185-264.

[4] Hoffman, K.: Banach Spaces of Analytic Functions, Dover Publications, 1988.

[5] Karapetyants, N. K. and Samko, S. G.: Equations with Involutory Operators and their Applications, Rostov Don University, 1988 (in Russian).

[6] Kato, T.: Perturbation Theory for Linear Operators, Springer–Verlag, 1984.

[7] Kravchenko, V. G.: On the normalization of singular integral operators. Dokl. Akad. Nauk SSSR 285, n° 6 (1985) 1314–1317; Translation: Soviet Math. Dokl. 32, n° 3 (1985) 880-883.

[8] Kravchenko, V. G.: The Noether theory of systems of singular integral equations with shift. Dr. Sc. Dissertation, Kiev, 1987 (in Russian).

[9] Kravchenko, V. G., Lebre, A. B., Litvinchuk, G. S. and Teixeira, F. S.: Fredholm theory for a class of singular integral operators with Carleman shift and unbounded coefficients. To appear in Math. Nachr. 170 (1994).

[10] Kravchenko, V. G., Lebre, A. B., Litvinchuk, G. S. and Teixeira, F. S.: A normalization problem for a class of singular integral operators with Carleman shift and unbounded coefficient. To appear in Integral Equations and Operator Theory.

[11] Kravchenko, V. G. and Litvinchuk, G. S.: On the problem of normalization of singular integral operators with shift. Uspekhi Mat. Nauk 39, n° 4 (1984) 126–126 (in Russian).

[12] Kravchenko, V. G. and Litvinchuk, G. S.: On the normalization of some equations of Wiener–Hopf type. Differentsial'nye Uravneniya 26, n° 10 (1990) 1824–1826 (in Russian).

[13] Kravchenko, V. G. and Shaev, A. K.: On the normalization of singular integral equations with Carleman shift. In: Modern Analysis and its Apllications, Kiev, 1989 , 74–82 (In Russian).

[14] Lebre, A.B., Meister,E. and Teixeira, F.S.: Some results on the invertibility of Wiener-Hopf-Hankel operators. Z. Anal. Angew. 11 (1992) 57-76.

[15] Prössdorf, S.: Some Classes of Singular Equations. Akademie–Verlag, 1974 (In German). Translation: North–Holland, 1978.

[16] Teixeira, F. S.: Wiener–Hopf Operators in Sobolev Spaces and Applications to Diffraction Theory. Ph.D. Thesis, Instituto Superior Técnico U.T.L., Lisbon 1989 (in Portuguese).

V. G. Kravchenko
Unidade de Ciências Exactas e Humanas da Universidade do Algarve, Campus de Gambelas 8000 Faro, Portugal.

A. B. Lebre and F. S. Teixeira
Departamento de Matemática do Instituto Superior Técnico, Av. Rovisco Pais, 1096 Lisboa Codex, Portugal.

G. S. Litvinchuk
Unidade de Matemática da Universidade da Madeira, Largo do Colégio, 9000 Funchal, Portugal.

Contemporary Mathematics
Volume **189**, 1995

Estimates for generalized Poisson integrals in a half space

TRINIDAD MENÁRGUEZ AND FERNANDO SORIA

ABSTRACT. In this work we study conditions on Poisson type kernels which are sufficient to ensure weak L^1 estimates on the corresponding Poisson transforms. These transforms map functions defined on $\mathbf{R}^n$ into functions on the half space $\mathbf{R}_+^{n+1}$. Our results extend previous work by P. Sjögren.

1. Introduction

Given a positive measurable function K defined on $\mathbf{R}^n$, we consider, for a fixed real number $\gamma > 0$, the dilations

$$K_t^\gamma(x) = K_t(x) = t^{-n\gamma} K(\frac{x}{t^\gamma}), \ x \in \mathbf{R}^n, \ t > 0.$$

In the sequel, we will assume that K is always integrable and, therefore, we can define for any finite measure μ on $\mathbf{R}^n$

$$K\mu(x,t) = K_t * \mu(x).$$

When

$$K(x) = \frac{c_n}{(1+|x|^2)^{\frac{n+1}{2}}}, \ \gamma = 1$$

$K\mu(x,t)$ represents the usual Poisson integral of the measure μ. Another well known example is given by $K(x) = e^{-\frac{x^2}{4}}$, $\gamma = \frac{1}{2}$. In that case, the uniparametric family of functions $\{K_t\}_{t>0}$ represents the heat semigroup, and the functions $u(x,t) = K\mu(x,t)$ all satisfy the heat equation on the half space $\mathbf{R}_+^{n+1}$.

With more generality, if K is radial and decreasing, we can look at $u(x,t) = K\mu(x,t)$ as the generalized Poisson integral of μ, a special case of the one considered by Calderón and Torchinsky, [C-T], in conection with parabolic Hardy spaces.

1991 *Mathematics Subject Classification.* 42B20.
Key words and phrases. Maximal Functions, Multiplier Operators, Weighted Inequalities.
Both authors were supported by DGICYT

Following Sjögren [Sj2], an associated notion of K-harmonicity can be defined and then, as in classical potential theory, one is lead to the problem of determining when a given function $u(x,t)$ is K-harmonic. This problem is always intimately connected to the condition $t^{-1}u(x,t) \in L^{1,\infty}(\mathbf{R}_+^{n+1})$ (see [Sj1]). This, in turn, motivates the study of the boundedness of the operator

$$f \longrightarrow t^{-1}Kf(x,t)$$

from $L^1(\mathbf{R}^n)$ into $L^{1,\infty}(\mathbf{R}_+^{n+1})$ under certain conditions on the kernel K.

Our main result here is presented in the following

THEOREM 1. *Let K be positive, radial and decreasing. Let us assume also that K satisfies the condition*

$$(1.1) \qquad \int_{\mathbf{R}^n} K(x)|\log|\log|x|||dx < \infty.$$

Then, for every $\gamma > 0$, the transformation $f \to t^{-1}Kf(x,t) = t^{-1}K_t^\gamma * f(x)$ with K_t^γ defined as above maps $L^1(\mathbf{R}^n)$ into $L^{1,\infty}(\mathbf{R}_+^{n+1})$. That is, there exists a constant C, which only depends on K, γ and n, such that

$$(1.2) \qquad \int_0^\infty |\{x : |t^{-1}Kf(x,t)| > \lambda\}|dt \le \frac{C}{\lambda} \int_{\mathbf{R}^n} |f(x)|dx, \quad \forall \lambda > 0.$$

Here $|\cdot|$ denotes Lebesgue measure on $\mathbf{R}^n$.

Before going into the proof of Theorem 1, let us make several remarks. First, we observe that, due to the monotonicity of K, the integrability condition (1.1) is only to be considered into account either for $|x| \to 0$ (small values of x) or for $|x| \to \infty$ (large values of x). In other words, the monotonicity condition tells us that (1.1) is equivalent to the following two conditions together:

$$(1.3) \qquad \int_{|x|<\frac{1}{100}} K(x) \log\log\frac{1}{|x|}dx < \infty$$

$$(1.4) \qquad \int_{|x|>100} K(x) \log\log|x|dx < \infty.$$

As in [Sj2], our result extends easily to the following situation: Let F be a compact null set and let ϕ be a positive decreasing function defined on $[0,\infty)$. Then, the conclusion of Theorem 1 remains true for the kernel

$$K(x) = \phi(\text{dist}(x,F))$$

provided (1.1) is replaced by

$$\int_{\mathbf{R}^n} K(x)|\log|\log \text{ dist}(x,F)||dx < \infty.$$

For simplicity, we will only consider the case $F = \{0\}$ stated in the theorem.

Theorem 1 improves the results obtained by P. Sjögren in [Sj2] where the integrability condition is given in the stronger form

$$\int_{\mathbf{R}^n} K(x)|\log|x||dx < \infty.$$

We must add, nevertheless, that we were highly motivated by the work of Sjögren and that in fact what we have done is to study possible solutions to a question raised by that author regarding the weakening of the hypothesis on K so that the boundedness still holds, (see [Sj 2], page 86). Our proof of the main result, however, is of a complete different nature and relies on some ideas developed by the two authors about discretization. This amounts to the study of the behavior of the K-transform on linear combinations of Dirac deltas (see [M-S]).

Finally, it is with great pleasure that we dedicate this work to Mischa Cotlar on the occasion of his 80th birthday.

2. Proof of Theorem 1.

Let us start by observing that by making the change of variables $t^\gamma = s$, estimate (1.2) is equivalent to showing that the transformation

$$L_K\mu(x,s) = s^\alpha \frac{1}{s^n} K(\frac{\cdot}{s}) * \mu(x)$$

takes every finite measure μ, of total variation $\|\mu\| = 1$, into the space

$$L^{1,\infty}(\mathbf{R}_+^{n+1}, dx\frac{ds}{s^{\alpha+1}}),$$

where α (which takes the role of $-\frac{1}{\gamma}$) is a fixed negative real number.

Let us prove first the following particular example of Theorem 1:

LEMMA 2. *Let $\bar{B}$ be any ball in $\mathbf{R}^n$ centered at the origin and let $K = \chi_{\bar{B}}$. Then, with the notation above and setting $d\sigma_\alpha = dx\frac{ds}{s^{\alpha+1}}$, $\alpha < 0$, one has*

(2.1) $$\sigma_\alpha(\{(x,s) : L_K\mu(x,s) > \lambda\}) \le \frac{C}{\lambda}\|\mu\||\bar{B}|, \quad \forall \lambda > 0.$$

PROOF. We may assume, without loss of generality, that μ is positive. Also, by dilation invariance, we can assume $\bar{B} = B_1 = \{x : |x| \le 1\}$. Now, we observe

$$L_K\mu(x,s) = s^\alpha \frac{1}{s^n} K(\frac{\cdot}{s}) * \mu(x) = c_n \frac{1}{|B_s|^{1-\frac{\alpha}{n}}}\mu(x + B_s),$$

where $B_s = \{x : |x| \le s\}$.

Therefore, given $\lambda > 0$, if $L_K\mu(x,s) > \lambda$ then there exists a ball B containing the point x, of radius s and so that

(2.2) $$\frac{1}{|B|^{1-\frac{\alpha}{n}}}\mu(B) > \frac{\lambda}{c_n}.$$

Since $|B| = \omega_n s^n$, where ω_n is the Lebesgue measure of the unit ball in $\mathbf{R}^n$, we note that the "cylinder" in $\mathbf{R}^{n+1}_+$

$$\hat{B} = B \times [0, (\frac{|B|}{\omega_n})^{\frac{1}{n}}]$$

contains in particular the point (x, s).

This says that the level set

$$\{(x, s) : L_K \mu(x, s) > \lambda\}$$

is contained in the union of cylinders

$$\bigcup_j \{\hat{B}_j : B_j \text{ is a ball satisfying (2.2)}\}.$$

By a simple covering lemma, applied on $\mathbf{R}^n$ to the family of balls $\{B_j\}_j$, there exists a subfamily of mutually disjoint balls, $\{B'_k\}_k$, so that

$$\bigcup_j \hat{B}_j \subset \bigcup_k \widehat{3B'_k}.$$

Thus,

$$\sigma_\alpha(\{(x, s) : L_K \mu(x, s) > \lambda\}) \le C \sum_k \sigma_\alpha(\hat{B}'_k) = C \sum_k \int_0^{(\frac{|B'_k|}{\omega_n})^{\frac{1}{n}}} \int_{B'_k} dx \frac{ds}{s^{\alpha+1}}.$$

Using that α is negative, this sum can be estimated as

$$c_{\alpha,n} \sum_k |B'_k|^{1-\frac{\alpha}{n}}.$$

Now, using (2.2) and the fact that the B'_k are disjoint, we get

$$\sigma_\alpha(\{(x, s) : L_K \mu(x, s) > \lambda\}) \le \frac{c_{\alpha,n}}{\lambda} \sum_k \mu(B'_k) \le \frac{c_{\alpha,n}}{\lambda} \|\mu\|.$$

q.e.d.

By a slight abuse of notation, let us denote by L_B the operator associated to the kernel $K(x) = \frac{1}{|B|}\chi_B(x)$, where B is again a given ball of $\mathbf{R}^n$ centered at the origin. With this definition, (2.1) becomes

(2.3) $$\sigma_\alpha(\{(x, s) : L_B \mu(x, s) > \lambda\}) \le \frac{C}{\lambda}\|\mu\|,$$

with C a constant depending on α and n, but independent of λ, $\|\mu\|$ and the ball B.

Before finishing with the proof of Theorem 1, we state the following well-known result due to E. Stein and N. Weiss [S-W], which describes the adding up process in the quasi-Banach space $L^{1,\infty}$.

LEMMA 3. (Stein-N. Weiss) *Let $\{g_k\}$ be a sequence of functions in the unit ball of $L^{1,\infty}(d\sigma)$. Then, given a numerical sequence $\{c_k\}$ with*

$$(2.4) \qquad N(\{c_k\}) = \sum_k |c_k|(1 + \log \frac{\sum_j |c_j|}{|c_j|}) < \infty$$

we have $\sum c_k g_k \in L^{1,\infty}(d\sigma)$ and in fact

$$(2.5) \qquad \sigma(\{x : |\sum_k c_k g_k(x)| > \lambda\}) \leq \frac{6}{\lambda} N(\{c_k\}).$$

The last part of Theorem 1 is based on the following

LEMMA 4. *If K is positive, radial and decreasing, and satisfies (1.1), then there is a sequence of numbers $\{c_k\}$ satisfying (2.4) and there exists a sequence of balls $\{B_k\}$ centered at the origin, so that*

$$K(x) \leq \sum_k c_k \frac{1}{|B_k|} \chi_{B_k}(x), \quad \forall x \in \mathbf{R}^n.$$

Before proving Lemma 4, let us also show how this finishes with the proof of Theorem 1. To see it, we note that if μ is positive, then with the notation of Lemma 4

$$L_K \mu(x,s) \leq \sum_k c_k L_{B_k} \mu(x,s).$$

Since the $\{c_k\}$ satisfy (2.4), and (2.3) tells us that $g_k = \frac{1}{C\|\mu\|} L_{B_k} \mu$ belongs to the unit ball of $L^{1,\infty}(d\sigma_\alpha)$, then we can apply (2.5) and Theorem 1 follows.

It only remains to give a

PROOF OF LEMMA 4. Since K is monotonic, (1.1) is equivalent to the condition

$$(2.6) \qquad \sum_{k \in \mathbf{Z}^*} K_0(2^k) 2^{kn} \log |k| < \infty,$$

where K_0 is defined by $K(x) = K_0(|x|)$, and where $\mathbf{Z}^* = \mathbf{Z}\backslash\{0\}$. Now, we have the estimate

$$K(x) \leq \sum_{k \in \mathbf{Z}} K_0(2^k) \chi_{\{2^k \leq |x| \leq 2^{k+1}\}} \leq C \sum_{k \in \mathbf{Z}} K_0(2^k) 2^{kn} \frac{1}{|B_k|} \chi_{B_k},$$

where B_k denotes the ball centered at the origin and radius 2^{k+1}. Therefore, we only need to show, with $c_k = K_0(2^k) 2^{kn}$, that

$$N = N(\{c_k\}) = \sum_k K_0(2^k) 2^{kn}(1 + \log \frac{\sum_j K_0(2^j) 2^{jn}}{K_0(2^k) 2^{kn}}) < \infty.$$

We may assume that $\sum_j K_0(2^j) 2^{jn} = 1$. This amounts to consider

$$\int_{\mathbf{R}^n} K(x)dx \sim 1.$$

Define $I = \{k \in \mathbf{Z} : K(2^k)2^{kn} \leq \frac{1}{1+k^2}\}$, and split

$$N = \sum_{k \in I} + \sum_{k \notin I}.$$

Using that $x(1 + \log \frac{1}{x})$ is increasing on $(0,1]$, we get

$$\sum_{k \in I} \leq \sum_{k \in \mathbf{Z}} \frac{1}{1 + k^2}(1 + \log(1 + k^2)) < \infty.$$

On the other hand, (2.6) gives us directly

$$\sum_{k \notin I} \leq \sum_{k \in \mathbf{Z}} K_0(2^k)2^{kn}(1 + \log(1 + k^2)) < \infty.$$

q.e.d.

REFERENCES

[C-T] A.P. Calderón and A. Torchinski, *Parabolic maximal functions associated with a distribution*, Adv. in Math. **16** (1975), 1–64.

[M-S] T. Menárguez and F. Soria, *Weak type inequalities for maximal convolution operators*, Rend. Circ. Mat. Palermo **41** (1992), 324–352.

[Sj1] P. Sjögren, *Weak L^1 characterizations of Poisson integrals, Green potentials and H^p spaces*, Trans. Amer. Math. Soc. **233** (1977), 179–196.

[Sj2] ———, *Generalized Poisson integrals in a half-space and weak L^1*, J. London Math. Soc. **27** (1983), 85–96.

[S-W] E. Stein and N. Weiss, *On the convergence of Poisson integrals*, Trans. Amer. Math. Soc. **140** (1969), 35–54.

DEPTO. DE MATEMÁTICAS, UNIVERSIDAD AUTÓNOMA DE MADRID, 28049 - MADRID, SPAIN

Contemporary Mathematics
Volume **189**, 1995

Wavelet analysis, local Fourier analysis, and 2-microlocalization

Y. MEYER

en hommage à Misha COTLAR

ABSTRACT. When a function satisfies a Hölder condition at a given point, its wavelet coefficients satisfy a simple size estimate around this point. The converse statement is a Tauberian theorem which was proved by S. Jaffard. We give a new approach to Jaffard's theorem.

I. Introduction.

Many tools have been constructed for analyzing the singularities of a function or a distribution. The main motivation was to investigate the propagation of singularities for evolution equations in a PDE context. This line of research led to the definition of the wave-front set of a distribution. J. M. Bony developed the so-called para-differential operators in order to investigate the creation of singularities in non-linear evolution equations. J. M. Bony introduced a remarkable family of functional spaces which permit a refined study of these singularities. These spaces $C_{x_0}^{s,s'}$ will be defined in this paper and can be used to detect the few points where a function (or a distribution) is regular in a non-smooth environment. Let us provide the reader with an example.

The trigonometric series $\sigma(x) = \sum_1^\infty n^{-2} \sin\left(n^2 x\right)$ was proposed by Riemann as an example of a continuous function which is nowhere differentiable. G. H. Hardy could prove the following fact : if x_0 is not of the form $\frac{2p+1}{2q+1}\pi$ ($p \in \mathbb{N}$, $q \in \mathbb{N}$) then the Hölder exponent $\alpha(x_0)$ cannot exceed $3/4$. We define this pointwise Hölder exponent as the supremum of all β's such that $|f(x) - f(x_0)| = 0(|x - x_0|)^\beta$ as x tends to x_0.

J. Gerver proved in 1970 that $\sigma(x)$ is differentiable in x_0 whenever $x_0 = \frac{2p+1}{2q+1}\pi$ and Stéphane Jaffard found that $\sigma(x)$ is a multi-fractal function.

The Riemann function σ belongs to $C^{1/2}(\mathbb{R})$ and $\alpha(x_0) = 1/2$ when $x_0 = p/q\,\pi$ and p, q are not both odd integers. It means that a point (like $x_0 = \pi$) where σ

1991 *Mathematics Subject Classification*. Primary 35A27, 35S05, 44A15.

is differentiable is surrounded by points where σ is much worse. J. Gerver was looking for points of regularity hidden in a bunch of cusps and other singularities It happens that the J. M. Bony 2-microlocal analysis is a remarkable tool for this type of search.

Let us recall the definition of the two-microlocal spaces $C_{x_0}^{s,s'}$.

Let us fix a function φ in the Schwartz class $\mathcal{S}(\mathbb{R}^n)$ with the following properties: the Fourier transform $\hat{\varphi}(\xi)$ of φ vanishes outside the unit ball and $\hat{\varphi}(\xi) = 1$ when $|\xi| \leq 1/2$. For each $j \in \mathbb{Z}$, let us denote by $\varphi_j(x)$ the function $2^{nj}\varphi(2^j x)$ and by S_j the convolution operator with this φ_j : $S_j(f) = f * \varphi_j$. Next $\Delta_j = S_{j+1} - S_j$ and the so-called Littlewood-Paley decomposition is the following decomposition of the identity

$$(1.1) \qquad\qquad I = S_0 + \sum_0^\infty \Delta_j \ .$$

Some authors also consider the decomposition

$$(1.2) \qquad\qquad I = \sum_{-\infty}^\infty \Delta_j$$

but the latter is not valid in all circumstances. The problem is to give a meaning to the limit of $S_j(f)$ when $j \to -\infty$ and when f is a tempered distribution. One can prove that there exists an integer $N = N(f)$ and a "floating polynomial" $P_j(f)(x)$ whose degree does not exceed N such that $S_j(f) - P_j(f)$ tends to 0 uniformly on compact sets as j tends to $-\infty$. People say that (1.2) is valid modulo (floating) polynomials.

We now define the Bony spaces $C_{x_0}^{s,s'}$.

Let s and s' be two real numbers and x_0 a point in $\mathbb{R}^n$. Then $C_{x_0}^{s,s'}$ is a Banach space consisting of all the tempered distributions $u \in \mathcal{S}'(\mathbb{R}^n)$ such that

$$(1.3) \qquad\qquad |S_0(u)(x)| \leq C(1 + |x - x_0|)^{-s'}$$

together with

$$(1.4) \qquad\qquad |\Delta_j(u)(x)| \leq C 2^{-js}(1 + 2^j|x - x_0|)^{-s'}$$

for each $x \in \mathbb{R}^n$, $j \in \mathbb{N}$.

The norm of u in $C_{x_0}^{s,s'}$ is defined as the lower bound of all the constants C for which (1.3) and (1.4) are valid. Different choices of the Littlewood-Paley analysis yield equivalent norms.

These spaces have been tailored for improving the analysis of the behavior of f at x_0. An ordinary analysis is provided by the space $C^s(x_0)$ which is defined by

$$(1.5) \qquad\qquad |f(x) - P_m(x - x_0)| \leq C|x - x_0|^s$$

when $m \in \mathbb{N}$, $m < s < m + 1$ and where the degree of P_m does not exceed m.

The main goal of this paper is to relate the sophisticated 2-microlocal analysis using the $C_{x_0}^{s,s'}$ spaces to the crude analysis which uses the $C^s(x_0)$ space. But in both cases we are only interested in knowing what is happening near x_0. It means that it is quite natural to replace the Banach spaces $C_{x_0}^{s,s'}$ by the corresponding local spaces, denoted by $C^{s,s'}(x_0)$ which will now be defined.

Let us first observe that if x_0 does not belong to the singular support of a distribution u, then (1.4) is satisfied for all $j \in \mathbb{N}$ and all x in some neighborhood of x_0.

A distribution u, defined in some neighborhood of x_0, belongs to the local version $C^{s,s'}(x_0)$ of $C_{x_0}^{s,s'}$ if one of the two following conditions is satisfied

(1.6)   there exists a neighborhood V of x_0 such that (1.4) holds for
$$j \in \mathbb{N} \text{ and } x \in V$$

(1.7)   there exists a distribution $f \in \mathcal{S}'(\mathbb{R}^n)$ such that $f \in C_{x_0}^{s,s'}$
(the global space) and $f = u$ in some neighborhood of x_0.

A few other definitions are needed before stating our main results.

Let B be the open ball, centered at x_0, with radius ρ. We want to define, for any real number s, the Banach space $C^s(B)$.

Let us start with the homogeneous Hölder space $\dot{C}^s(\mathbb{R}^n)$ whose definition will be reminded.

Whenever $s \geq 0$ we denote by m the integral part of s ($m \in \mathbb{N}$ and $m \leq s < m+1$) and $\dot{C}^s(\mathbb{R}^n)$ is a space of classes of functions, modulo the polynomials of degree not exceeding m.

If $0 < s < 1$, $f \in \dot{C}^s(\mathbb{R}^n)$ simply means that $|f(y) - f(x)| \leq C|y-x|^s$, $x \in \mathbb{R}^n$, $y \in \mathbb{R}^n$.

If $s = 1$, $f \in \dot{C}^1(\mathbb{R}^n)$ means $|f(x+y) + f(x-y) - 2f(x)| \leq C|y|$ for $x \in \mathbb{R}^n$, $y \in \mathbb{R}^n$. This space is the celebrated Zygmund class.

If $m < s \leq m+1$, one writes $s = m + \sigma$ where $0 < \sigma \leq 1$. Then $f \in \dot{C}^s(\mathbb{R}^n)$ means $\partial^\alpha f \in \dot{C}^\sigma(\mathbb{R}^n)$ for all multi-indices α such that $|\alpha| = \alpha_1 + \cdots + \alpha_n = m$.

When $s = 0$, $\dot{C}^0(\mathbb{R}^n)$ is a space of distributions modulo constant functions. Indeed $f \in \dot{C}^0(\mathbb{R}^n)$ if and only if $f = \partial_1 f_1 + \cdots + \partial_n f_n$ where $f_1, \ldots, f_n$ belong to the Zygmund class and $\partial_j = \partial/\partial x_j$.

An equivalent definition is given by $\|\Delta_j(f)\|_\infty \leq C$, $j \in \mathbb{Z}$.

Finally when $s < 0$, one has $f \in \dot{C}^s(\mathbb{R}^n)$ if and only if $f = \partial_1 f_1 + \cdots + \partial_n f_n$ where $f_j \in \dot{C}^{s+1}(\mathbb{R}^n)$, $1 \leq j \leq n$. Starting with the case when $0 < s \leq 1$, one obtains the general case $s \leq 0$. An equivalent definition is given by $\|\Delta_j(f)\|_\infty \leq C2^{-js}$, $j \in \mathbb{Z}$.

We now return to the definition of $C^s(B)$ where B is some open ball. The space $C^s(B)$ consists in the restrictions to B of the functions (or distributions) in $\dot{C}^s(\mathbb{R}^n)$. If $s < 0$, there is nothing else to say and the norm of f in $C^s(B)$ is the infimum of $\|\tilde{f}\|_{\dot{C}^s(\mathbb{R}^n)}$ where $\tilde{f} = f$ on B.

If $s \geq 0$, we first define $\|f\|_{\Delta^s(B)}$ as the infimum of $\|\tilde{f}\|_{\dot{C}^s(\mathbb{R}^n)}$ where $\tilde{f} = f$ on B. Then, for $s > 0$

(1.8) $$\|f\|_{C^s(B)} = \|f\|_{\Delta^s(B)} + \rho^{-s}\|f\|_{L^\infty(B)}$$

and if $s = 0$

(1.9) $$\|f\|_{C^s(B)} = \|f\|_{\Delta^s(B)} + |<f, \omega_B>|$$

where $\omega_B(x) = \rho^{-n}\,\omega(\frac{x}{\rho})$, $\omega(x)$ being a test function, supported by the unit ball and such that $\int \omega(x)\,dx = 1$.

This definition has three interesting properties. When $s > 0$ and $m < s < m + 1$, there exists a unique polynomial $P_m(x - x_0)$ of degree not exceeding m such that $\|f(x) - P_m(x - x_0)\|_{L^\infty(B)} \leq C|f|_{\Delta^s(B)}$. It means that within the equivalence class of f in the (homogeneous) Hölder space $\dot{C}^s(\mathbb{R}^n)$, there is a function for which the second term in the right-hand side is not needed.

A second observation in the invariance through dilations. If $\gamma(x) = \alpha x + b$, $\alpha > 0$, $b \in \mathbb{R}^n$ and if $f(x) \in C^s(B)$, then $f \circ \gamma$ will belong to $C^s(\gamma^{-1}(B))$ with a norm given by $\alpha^s\|f\|_{C^s(B)}$. This remarkable property permits to reduce any computation to the case when B is the unit ball.

A third observation is the following fact : if f belongs to $C^s(B)$, then f is the restriction to B of a function (or a distribution) u such that

(1.10) $$u(x) \quad \text{is supported by} \quad |x - x_0| \leq 2\rho$$

(1.11) $$\|u\|_{\dot{C}^s(\mathbb{R}^n)} \leq C\|u\|_{C^s(B)} .$$

With these notations and definitions in mind, one has the following statement

Theorem 1. *Let $s > 0$, $s \notin \mathbb{N}$ and $\sigma < s$ be two real numbers (negative σ's are allowed). Then for each $x_0 \in \mathbb{R}^n$ and any distribution f defined in some neighborhood of x_0, the two following properties (A) and (B) are equivalent*

(A) *there exists a polynomial P with degree not exceeding s, a real number ρ_0 and a constant C such that*

(1.12) $$\|f(x) - P(x - x_0)\|_{C^\sigma(B_\rho)} \leq C\rho^{s-\sigma} \qquad (0 < \rho < \rho_0)$$

(B) *$f(x)$ belongs to $C^{s,\sigma-s}(x_0)$.*

Before proving this theorem, a few remarks might be welcome.

When $0 < \sigma < s$, this theorem describes the search for a few regular points inside a non smooth global behavior. Indeed this global behavior is expressed by a C^σ norm. This $C^\sigma(B_\rho)$ norm includes the term $\rho^{-\sigma}\|f(x) - P(x - x_0)\|_{L^\infty(B_\rho)}$ and (1.12) implies

(1.13) $$|f(x) - P(x - x_0)| \leq C|x - x_0|^s$$

of $f \in C^s(x_0)$.

The normalization which has been adopted for the $C^\sigma(B_\rho)$ norm implies that the norm of $(x - x_0)^\alpha$ is $C_\alpha \rho^{|\alpha|-\sigma}$ where $c_\alpha \neq 0$.

When σ is negative, $C^\sigma(B)$ contains quite a few distributions f and the $C^\sigma(B)$ norm is giving some information about the oscillations of these distributions. An example will clarify this matter. If $n = 1$, $x_0 = 0$ and $f(x) = x^{-2} \sin(1/x)$, then $f(x)$ is interpreted as the distributional derivative of $\cos(1/x)$ and, if $\sigma \leq -1$, one obtains

$$\|f\|_{C^\sigma(I_\rho)} = 0(\rho^{-2(1+\sigma)}) \, .$$

For proving theorem 1, we will use the characterization obtained by S. Jaffard of the space $C^{s,s'}_{x_0}$ by a wavelet analysis.

The definition of orthonormal wavelets will be given for the reader's convenience.

We consider for each integer $N \geq 1$ the pair (φ, ψ) of the scaling function (or the father wavelet) φ and the mother wavelet ψ. These two functions $\varphi = \varphi_N$ and $\psi = \psi_N$ are supported by the interval $[0, 2N - 1]$ and belong to the Hölder space $C^r(\mathbb{R})$ where $r = r_N$ tends to infinity with N.

Let F be the finite set $\{0, 1\}^n \setminus (0, 0, \ldots, 0)$. If $\varepsilon = (\varepsilon_1, \ldots, \varepsilon_n) \in F$, we define

(1.14)
$$\begin{cases} \psi^{(\varepsilon)}(x) = \varphi^{(\varepsilon_1)}(x_1) \ldots \varphi^{(\varepsilon_n)}(x_n) \\ \varphi^{(0)}(t) = \varphi(t) \quad , \quad \varphi^{(1)} = \psi(t) \\ \psi^{(\varepsilon)}_{j,k}(x) = 2^{nj/2} \psi^{(\varepsilon)}(2^j x - k) \, , \quad j \in \mathbb{Z} \, , \ k \in \mathbb{Z}^n \\ \text{and} \\ \varphi(x) = \varphi(x_1) \ldots \varphi(x_n) \end{cases}$$

Then for each fixed j, the functions $\psi^{(\varepsilon)}_{j,k}$, $\varepsilon \in F$, $k \in \mathbb{Z}^n$, are an orthonormal sequence and the closed linear span of this sequence is denoted by W_j. These "frequency channels" are pairwise orthogonal and $L^2(\mathbb{R}^n)$ is the (hilbertian) direct sum of these frequency channels W_j, $j \in \mathbb{Z}$.

Similarly $\varphi(x - k)$, $k \in \mathbb{Z}^n$, is an orthonormal sequence spanning the closed space V_0. Then V_j is defined by $f(x) \in V_0 \Leftrightarrow f(2^j x) \in V_j$.

The main statement in this construction is the following fact : $V_{j+1} = V_j \oplus W_j$. An equivalent statement is

(1.15) $$L^2(\mathbb{R}^n) = V_j \oplus W_j \oplus W_{j+1} \oplus \cdots \oplus W_m \oplus \cdots \, .$$

With these notations, we have

Theorem 2. *Keeping theorem 1's notations, the two following properties of a distribution f are equivalent*

(B) *f belongs to the 2-microlocal space $C^{s,s'}(x_0)$*

(C) *the wavelet coefficients $\alpha(j, k, \varepsilon) = <f, \psi^{(\varepsilon)}_{j,k}>$ satisfy the estimates*

$$|\alpha(j, k, \varepsilon)| \leq C 2^{-(\sigma+n/2)j} \big(2^{-j} + |k2^{-j} - x_0|\big)^{s-\sigma}$$

whenever $2^{-j} + |k2^{-j} - x_0| < \eta$ for some $\eta > 0$.

This theorem is proved in [1].

2. The implication (A) $\Rightarrow$ (C).

This implication is obvious. On the open ball, B_ρ, one has $f(x) = g(x) + P(x)$ where $g(x)$ is the restriction to B_ρ of a function belonging to the (homogeneous) Hölder space $\dot{C}^\sigma(\mathbb{R}^n)$. The support of $\psi_{j,k}^{(\varepsilon)}$ is denoted by $S(\varepsilon, j, k)$. There exists a constant M such that $S(\varepsilon, j, k) \subset B_\rho$ whenever $|k2^{-j} - x_0| + M2^{-j} \leq \rho$.

Since the degree of P does not exceed $s < r = r_N$, one has $\int P(x) \overline{\psi}_{j,k}^{(\varepsilon)}(x) \, dx = 0$. Therefore $\alpha(j, k, \varepsilon) = \int f(x) \overline{\psi}_{j,k}^{(\varepsilon)}(x) \, dx = \int g(x) \overline{\psi}_{j,k}^{(\varepsilon)}(x) \, dx$. Then the characterization of the (global) Hölder spaces

$$\dot{C}^\sigma(\mathbb{R}^n)$$

yields $|\alpha(j, k, \varepsilon)| \leq C 2^{-j(\sigma + n/2)} \rho^{s-\sigma}$ whenever $\rho \geq |k2^{-j} - x_0| + M2^{-j}$. In order to obtain (C) it suffices to take $\rho = |k2^{-j} - x_0| + M2^{-j}$.

Since theorem 2 has been proved by S. Jaffard, theorem 1 will be proved if one can show that (C) $\Rightarrow$ (A). This will be achieved in the next section.

3. The implication (C) $\Rightarrow$ (A).

Let E_j denote the orthonormal projection on V_j and $D_j = E_{j+1} - E_j$ the orthogonal projection on W_j. Then one uses (1.15) which amounts the same to writing

$$(3.1) \qquad\qquad I = E_j + D_j + D_{j+1} + \cdots .$$

In order to restrict this decomposition to B_ρ, one defines j_0 by $2^{-(j_0+2)} \leq \rho < 2^{-(j_0+1)}$.

Let Ω be the closed cube $[0, 2]^n$. We define $k = k_j \in \mathbb{Z}^n$ by the following condition: $\Omega_j = 2^{-j}(k + \Omega)$ should contain the ball $|x - x_0| \leq 2^{-j-1}$. When several choices exist, we systematically select the least k (in the lexicographical ordering). This yields $\Omega_{j+1} \subset \Omega_j$ for each j (this is easily checked but not trivial) and $B_\rho \subset \Omega_{j_0}$.

There exists an integer M such that the restrictions to Ω_{j_0} of $\psi_{j,k}^{(\varepsilon)}(x)$ vanish identically whenever $j \geq j_0$ and $|k2^{-j} - x_0| \geq M \cdot 2^{-j_0}$. Let us denote by $F(j_0)$ the collection of all pairs (j, k) such that $j \geq j_0$ and $|k2^{-j} - x_0| < M \cdot 2^{-j_0}$ and observe that, if $x \in \Omega_{j_0}$ one has

$$\sum_{j \geq j_0} D_j(f)(x) = \sum_{\varepsilon \in F} \sum \sum_{(j,k) \in F_{j_0}} \alpha(\varepsilon, j, k) \psi_{(j,k)}^{(\varepsilon)}(x) = h_{j_0}(x) .$$

Then (C) yields

$$(3.2) \qquad\qquad \|h_{j_0}\|_{L^\infty(\mathbb{R}^n)} \leq C 2^{-j_0 s}$$

and

(3.3)
$$\|h_{j_0}\|_{\dot{C}^\sigma(\mathbb{R}^n)} \leq C\, 2^{-j_0(s-\sigma)}\,.$$

Finally on Ω_{j_0} one obtains from (3.1)

(3.4)
$$f(x) \;=\; g_{j_0}(x) + h_{j_0}(x)$$

where $g_{j_0}(x) \in V_{j_0}$.

We can forget the precise definition of g_{j_0} and h_{j_0} and only keep in mind (3.2), (3.3) and $g_{j_0} \in V_{j_0}$. Then one can replace j_0 by j since the index j will not play any further role.

Since $\Omega_{j+1} \subset \Omega_j$, one has on Ω_{j+1}

(3.5)
$$f(x) \;=\; g_j(x) + h_j(x) \;=\; g_{j+1}(x) + h_{j+1}(x)$$

which implies

(3.6)
$$r_{j+1}(x) \;=\; g_{j+1}(x) - g_j(x) \;\in\; V_{j+1}$$

together with

(3.7)
$$\|r_{j+1}\|_{L^\infty(\Omega_{j+1})} \leq C\, 2^{-js}\,.$$

For estimating $\|\partial^\alpha r_j\|_{L^\infty(\Omega_j)}$ one needs a local version of the celebrated S. Bernstein's inequalities. This version is given by the two following lemmata.

Lemma 1. *There exists a constant C such that, for any $f \in V_0$ and every multi-index α such that $|\alpha| < r$, one has*

(3.8)
$$\sup_{x \in \Omega} |\partial^\alpha f(x)| \;\leq\; C \sup_{x \in \Omega} |f(x)|\,.$$

Indeed, once restricted to Ω, V_0 has a finite dimension. A basis of this finite dimensional space is the collection of the $(2N)^n$ functions $\varphi(x - k)$, $k \in \mathbb{Z}^n$, whose support has a non empty intersection with the open cube $(0,2)^n$.

On such a finite dimensional space, all norms are equivalent which yields (3.8).

Since V_0 is invariant by integral translations, (3.8) is still valid, with the same constant C, if Ω is replaced by $\Omega + k$, $k \in \mathbb{Z}^n$.

Then an obvious rescaling yields the following estimate

Lemma 2. *There exists a constant C (the same as in (3.8)) such that for each $j \in \mathbb{N}$ and each $f \in V_j$, one has*

(3.9)
$$\sup_{x \in \Omega_j} |\partial^\alpha f(x)| \;\leq\; C 2^{j|\alpha|} \sup_{x \in \Omega_j} |f(x)|$$

whenever $|\alpha| < r$.

Since $r_j(x)$ belongs to V_j and since $x_0 \in \Omega_j$, one has

(3.10)
$$|\partial^\alpha r_j(x_0)| \;\leq\; C\, 2^{j(|\alpha|-s)}\,, \qquad |\alpha| < r\,.$$

Therefore $\partial^\alpha g_j(x_0)$ tends to a limit λ_α as $j \to +\infty$ and

$$(3.11) \qquad\qquad |\partial^\alpha g_j(x_0) - \lambda_\alpha| \leq C' 2^{j(|\alpha|-s)} \;.$$

We define $P(x) = \sum_{|\alpha|<s} \lambda_\alpha x^\alpha / \alpha!$ and, similarly,

$$P_j(x) = \sum_{|\alpha|<s} \partial^\alpha g_j(x_0) x^\alpha / \alpha! \;.$$

We want to prove the following estimate

$$(3.12) \qquad\qquad \|f(x) - P(x-x_0)\|_{C^\sigma(B_\rho)} \leq C\rho^{s-\sigma} \;.$$

As it was mentioned, j and ρ satisfy $2^{-j-2} \leq \rho < 2^{-j-1}$ and it suffices to replace B_ρ by Ω_j.

One writes $f(x)-P(x-x_0) = [g_j(x)-P_j(x-x_0)]+[P_j(x-x_0)-P(x-x_0)]+h_j(x) = A_j(x) + B_j(x) + h_j(x)$ on Ω_j.

For estimating A_j, the following lemma will be our main tool.

Lemma 3. *Let $m \in \mathbb{N}$ and $t \in (m, m+1)$. If the function $u(x)$ belongs to the homogeneous Hölder space $\dot{C}^t(\mathbb{R}^n)$, one has for each $\tau \in (0,t)$ and each ball B centered at x_0 with radius ρ*

$$(3.13) \qquad\qquad \|u(x) - P_m(x-x_0)\|_{C^\tau(B)} \leq C\rho^{t-\tau}\|u\|_{\dot{C}^t(\mathbb{R}^n)}$$

where

$$P_m(x) = \sum_{|\alpha| \leq m} \partial^\alpha u(x_0) x^\alpha / \alpha! \;.$$

This lemma will be applied to $u(x) = g_j(x)$, $B = \Omega_j$, $t \in (s, m+1)$ and $t \in (s, r)$ and finally $\tau = \sigma$. The optimal right-hand side is $\|g_j\|_{\Delta^t(\Omega_j)}$ which will now be estimated.

We first use the obvious generalization of lemma 2. Keeping lemma 2's notations, we have

$$(3.14) \qquad\qquad \|f\|_{C^t(\Omega_j)} \leq C\, 2^{jt} \|f\|_{L^\infty(\Omega_j)} \;.$$

This estimate is applied to $r_j(x)$ and one obtains $\|r_j\|_{C^t(\Omega_j)} \leq C\, 2^{j(t-s)}$.

A drawback of the norm $C^t(B)$ is the fact that $B' \subset B$ does not imply $\|f\|_{C^t(B')} \leq \|f\|_{C^t(B)}$. However this holds when $C^t(B')$ is replaced by $\Delta^t(B')$.

Therefore,

$$\|g_j\|_{\Delta^t(\Omega_j)} \leq \|r_j\|_{\Delta^t(\Omega_j)} + \|g_{j-1}\|_{\Delta^t(\Omega_j)} \leq \|r_j\|_{\Delta^t(\Omega_j)} + \|g_{J-1}\|_{\Delta^t(\Omega_{j-1})}$$

and we can proceed. We end with

$$\|g_j\|_{\Delta^t(\Omega_j)} \leq C' 2^{j(t-s)} + \|g_0\|_{\Delta^t(\Omega_0)} = C' 2^{j(t-s)} + C_0 \;.$$

Returning to (3.13), one obtains

$$(3.15) \qquad \|g_j(x) - P_j(x-x_0)\|_{C^\sigma(\Omega_j)} \leq C\,2^{-j(t-\sigma)}\,2^{j(t-s)} = C\,2^{-j(s-\sigma)}\,.$$

For estimating $\|P_j(x-x_0) - P(x-x_0)\|_{C^\sigma(\Omega_j)}$ we observe that the norm of $(x-x_0)^\alpha$ in $C^\sigma(\Omega_j)$ is $C\,2^{-j(|\alpha|-\sigma)}$. Then (3.11) is used and provides

$$\|P_j(x-x_0) - P(x-x_0)\|_{C^\sigma(\Omega_j)} \leq C\,2^{-j(s-\sigma)}\,.$$

Finally (3.3) gives

$$\|h_j\|_{C^\sigma(\Omega_j)} = \|h_j\|_{\dot{C}^\sigma(\Omega_j)} + C2^{j\sigma}\|h_j\|_{L^\infty(\Omega_j)} \leq C'2^{-j(s-\sigma)}\,.$$

This ends the proof.

[1] S. Jaffard. *Pointwise smoothness, two-microlocalization and wavelet coefficients*. Publicacions Matemàtiques, Vol. **35** (1991), 155–168.

CEREMADE UNIVERSITÉ PARIS-DAUPHINE, 75775 PARIS CEDEX 16

Contemporary Mathematics
Volume **189**, 1995

The Role of Cancellation in Interpolation Theory

Mario Milman   Richard Rochberg

November 7, 1994

This paper is dedicated to our teacher and friend Mischa Cotlar with affection and admiration. His point of view of looking at the unity of mathematics has been most influential in our work.

1   Background

In recent years there has been a great deal of interest in estimating the size of certain non-linear multi-term expressions. Among many examples are ([26], [8], [6], and [13]). Although the details and methods vary there are certain unifying themes. A central issue is that the multi-term expression has better integrability than the individual terms due to (sometimes hidden) internal cancellation.

One of the tools which has been used to obtain such results are commutator estimates from interpolation theory. Our goal here is to present those commutator results, and other new commutator results, in a way which we hope exhibits clearly the role of cancellation.

1991 *Mathematics Subject Classification*: Primary 46M35; Secondary 42B30

The first author was partially supported by an International NSF grant; the second author was supported by NSF grant DMS-9302828

2   Introduction

There is a natural dichotomy between the abstract methods of interpolation theory and its possible applications in analysis. In interpolation theory one tries to obtain estimates for operators using a priori estimates that must be obtained in each specific instance before the machinery of interpolation theory can be applied. Thus, given operators with the same a priori estimates, interpolation theory will produce the same conclusions. On the other hand, in classical analysis we study specific operators and many times the boundedness of these operators in suitable function spaces is due to subtle cancellations.

A theory of commutator estimates in interpolation theory has been emerging for the last decade or so (cf. [28], [15], [10], [11], [16], [17], [25], [14], [20], [21], [22], [4], [5]). Its purpose is to study the role of cancellation in interpolation theory and its applications. The model for these developments are "commutator theorems": the issue here is that a commutator between a bounded operator on a scale of spaces and certain unbounded operators, which are naturally associated with a given scale, are bounded because of cancellation.

Let us discuss some examples in the familiar setting of L^p spaces. Here the main result of [28] states that if $T : L^{p_i} \to L^{p_i}$, $\|T\|_{L^{p_i}} \le 1, i = 1, 2$, then for each p, $p_1 < p < p_2$, there exists a constant $c_p > 0$ such that,

$$\|[T, \Omega]f\|_p \le c_p \|f\|_p \tag{1}$$

where

$$\Omega f = f \log |f|$$

Here is an application of (a variation on) (1). Let K be a Calderón-Zygmund operator, and for $b \in BMO$ consider the multiplication operator $M_b f = bf$. A result of Coifman, Rochberg and Weiss (cf. [9]) gives that the commutator

$$[K, M_b]f = K(M_b f) - M_b(Kf) : L^p \to L^p \tag{2}$$

is bounded, where , $1 < p < \infty$. This estimate is important in the theory of compensated compactness (cf. [7]). In [28] the authors combine the ideas which lead to (1) and the theory of A_p weights and its relationship with the space BMO to give a new proof of (2), showing that it is a variant of (1).

In the setting of L^p spaces, or some more general lattices, Kalton [16], has shown that the conclusion of (1) persists for more general operators Ω. Iwaniec and Sbordone [14] have obtained a variant of (1) with interesting applications to nonlinear PDE's. Let T be an operator $T : L^p \to L^p, p \in [r_1, r_2]$, where $1 \le r_1 < r_2 < \infty$, and let $\frac{p}{r_2} - 1 \le \varepsilon \le \frac{p}{r_1} - 1$, then

$$\|[T, \Omega_\varepsilon]f\|_{\frac{p}{1+\varepsilon}} \le c_p |\varepsilon| \|f\|_p^{1+\varepsilon} \tag{3}$$

where

$$\Omega_\varepsilon(f) = |f|^\varepsilon f.$$

and

$$c_p = \frac{2p(r_2 - r_1)}{(p - r_1)(r_2 - p)} \sup_{r_1 \le s \le r_2} \|T\|_s$$

Since

$$\frac{[T, \Omega_\varepsilon]f}{\varepsilon} = T(\frac{|f|^\varepsilon f - f}{\varepsilon}) - \frac{|Tf|^\varepsilon Tf - Tf}{\varepsilon}$$

letting $\varepsilon \to 0$ in (3) we obtain (1). The crucial fact is the factor $|\varepsilon|$ in (3) which, in some cases, allows the left hand side of the inequality to be treated as a harmless error term.

The abstract theory of commutator estimates captures these results and provides new estimates for other scales of spaces. The theory provides a general set up were the arguments and the role of the cancellations of commutators become apparent and are not obscured by the richness of the specific environments to which it can be applied. It has found applications to logarithmic Sobolev estimates (cf. [11]), the study of H^p spaces (cf. [17]), operator ideals (cf. [18]), nonlinear PDE's and compensated compactness (cf. [14], [19], [23], [24]), the functional calculus of a positive operator [22].

Recent work of the authors (cf. [27] and [21]) has shown that suitable extensions can be obtained for higher order commutators. In these results the individual terms of the expressions that can be controlled are progressively more unbounded but their combination exhibits enough cancellation to effect the necessary control. For example, in the setting of L^p spaces, the results of [27] (and also [21]) show that if T is a bounded operator in L^1 and L^∞ then, for $1 < p < \infty$, there exist absolute constants c_p such that

$$\|[T, \Omega_2]f - \Omega_1[T, \Omega_1]f\|_{L^p} \le c_p \|f\|_{L^p}$$

where

$$\Omega_1 f = f \log \frac{f}{\|f\|_{L^p}}$$

$$\Omega_2 f = \frac{1}{2}f \left(\log \frac{f}{\|f\|_{L^p}} \right)^2$$

More generally, if we let

$$\Omega_n f = \frac{1}{n!}f \left(\log \frac{f}{\|f\|_{L^p}} \right)^n$$

and define inductively

$$C_0(Tf) = Tf$$
$$C_1(Tf) = [T, \Omega_1]f$$
$$C_n(Tf) = [T, \Omega_n]f - \sum_{k=1}^{n-1} \Omega_k C_{n-k}(Tf)$$

then

$$\|C_n(Tf)\|_{L^p} \le c_p \|f\|_{L^p}$$

The theory developed in [21] also gives results for *individual* higher order $\Omega'_n s$. For example, and continuing our discussion of commutator theorems for operators acting on L^p spaces, we have, under the current assumptions, that

$$\|[T, \Omega_n]f\|_{L^p} \le c_p \left(\|f\|_{L^p} + \|\Omega_{n-1}f\|_{L^p} \right)$$

In some sense one should view these developments as a first step towards understanding the rôle of cancellation in interpolation theory. One could argue that a second step would be to understand why operators that exhibit certain type of cancellations, whether in commutator form or not, are bounded in suitable interpolation spaces. But in interpolation theory the central objects of study are the spaces and one should therefore seek to formulate those conditions in terms of the constructions of the interpolation spaces themselves. The first results in this direction were obtained in [4] in the context of the complex method and in this paper we continue this study and consider as well the real method of interpolation. Our point of view can be summarized as follows. Given a pair of spaces we can construct parameterized families of interpolation spaces, the classical constructions being just examples. The elements in these interpolation scales can be represented in specific ways as transforms of simpler elements, or values of analytic functions with certain properties, and we measure size in the interpolation spaces in terms of the corresponding representations. It turns out that when the representation of an element exhibits extra cancellations then we can ascertain that the element belongs to a better space (i.e. another scale) than the one we could have predicted a priori only through size considerations. We show that this qualitative improvement is what is behind the theory of commutator estimates for the real and complex methods of interpolation.

This paper is organized as follows. In Section **3**, which is largely expository, we review and extend slightly the results for the complex method obtained in [4]. It turns out that in formulating commutator results it is important to consider an extension of the complex method due to Schechter [29]. In the section after that we give a new approach to the study of commutators and other cancellation phenomena for the real method of interpolation. It is not our intention to present here the most general results but rather to emphasize our new point of view. In a forthcoming paper we shall study a number of related issues and applications. In particular we shall show how to construct "conditional" interpolation spaces and discuss weak compactness in this general setting.

3   Complex Method

We briefly review the basic constructions of the complex method of interpolation. In fact, following [12] it will be convenient for our purposes here to formulate these constructions in a somewhat unfamiliar but equivalent form.

Let $\bar{A}$ be a Banach pair, let $P(D, \Delta(\bar{A}))$ be the set of all polynomials on the unit disk D, with coefficients in $\Delta(\bar{A})$, and let $A(D, \bar{A})$ be the closure of $P(D, \Delta(\bar{A}))$ with the norm

$$\|h\|_{A(D,\bar{A})} = \max\{\sup_{\Gamma_0} \|h(\gamma)\|_{A_0}, \sup_{\Gamma_1} \|h(\gamma)\|_{A_1}\}$$

where $\{\Gamma_0, \Gamma_1\}$ is a partition of $\Gamma = \{z : |z| = 1\}$.

Given $\theta \in D$, $n = 0, 1...$, the interpolation spaces $[\bar{A}]_{\theta,n}$ are defined by

$$[\bar{A}]_{\theta,n} = \{x : \exists f \in A(D, \bar{A}) \text{ such that } \frac{1}{n!} f^{(n)}(\theta) = x\}$$

and we let

$$\|a\|_{[\bar{A}]_{\theta,n}} = \inf\{\|f\|_{A(D,\bar{A})} : f \in A(D,\bar{A}), \frac{1}{n!}f^{(n)}(\theta) = x\}$$

These spaces can be identified with the usual interpolation spaces of Calderón, Lions and Schechter. For example, using the y axis divide Γ into two equal parts, $\Gamma = \Gamma_0 \cup \Gamma_1$(left side union right side), then for $\theta = 0$, we have

$$[\bar{A}]_{0,n} = \begin{cases} [\bar{A}]_{\frac{1}{2}} & \text{if } n = 0 \\ [\bar{A}]_{\delta^{(n)}(\frac{1}{2})} & \text{if } n > 0 \end{cases}$$

where the spaces to the left hand side are the classical constructions of Calderón and Schechter. Likewise it is possible to show that for each $\theta \in (0,1)$, we can associate a point $\hat{\theta} \in D$, such that

$$[\bar{A}]_{\hat{\theta},n} = \begin{cases} [\bar{A}]_{\theta} & \text{if } n = 0 \\ [\bar{A}]_{\delta^{(n)}(\theta)} & \text{if } n > 0 \end{cases}$$

(cf. [12] for the details).

It is also easy to see that if $n < m$

$$[\bar{A}]_{\theta,n} \subset [\bar{A}]_{\theta,m}$$

Conversely, if an element $a \in [\bar{A}]_{\theta,m}$ can be represented by

$$a = \frac{1}{(n+1)!}f^{(m)}(\theta), f \in A(D,\bar{A}), f^{(k)}(\theta) = 0, k = 0...n-1$$

then $(z-\theta)^{-n}f$ is also in $A(D,\bar{A})$, and hence we see that

$$a \in [\bar{A}]_{\theta,n}$$

In other words we can *detect* the elements of $[\bar{A}]_{\theta,n}$ in $[\bar{A}]_{\theta,m}$ because they can be represented by representations exhibiting special cancellations.

Following [4] let us apply these considerations to the study of commutator estimates. Let $c > 1$ be a fixed constant and for each $a \in [\bar{A}]_{\theta}$ select an analytic function $h_a \in A(D,\bar{A})$ such that

$$h_a(\theta) = a, \text{and } \|h_a\|_{A(D,\bar{A})} \le c\|a\|_{[\bar{A}]_{\theta}}$$

Define an operator by

$$\Omega_{\bar{A}}a = \Omega a = h_a'(\theta) \tag{4}$$

It was shown in [28] that if $T : \bar{A} \to \bar{B}$ is a bounded operator between two Banach pairs, $\theta \in D$, then the commutator defined by

$$[T,\Omega]x = T\Omega_{\bar{A}}x - \Omega_{\bar{B}}Tx$$

is bounded, $[T, \Omega] : [\bar{A}]_\theta \to [\bar{B}]_\theta$. That is, there exists $c > 0$ such that

$$\|[T, \Omega]\, a\|_{[\bar{A}]_\theta} \le c \,\|a\|_{[\bar{B}]_\theta}\,.$$

We illustrate our ideas providing a short proof of this result. Let $a \in [\bar{A}]_\theta$, pick $h_a \in A(D, \bar{A})$, $h_{Ta} \in A(D, \bar{B})$, such that $h_a(\theta) = a, h_{Ta}(\theta) = Ta$, and

$$\|h_a\|_{A(D, \bar{A})} \approx \|a\|_{[\bar{A}]_\theta}\,, \quad \|h_{Ta}\|_{A(D, \bar{B})} \approx \|Ta\|_{[\bar{B}]_\theta}$$

Then $\gamma(z) = T h_a(z) - h_{Ta}(z)$ is in $A(D, \bar{B})$, and since $\gamma(\theta) = 0$, the discussion above shows that

$$\gamma'(\theta) = [T, \Omega]\, a \in [\bar{B}]_\theta$$

and moreover,

$$\|[T, \Omega]\, a\|_{[\bar{A}]_\theta} \le c \,\|a\|_{[\bar{B}]_\theta}\,.$$

From our present viewpoint the important point is that the cancellation $\gamma(\theta) = 0$ allows us to recognize that the element $\gamma'(\theta)$ of $[\bar{B}]_{\theta,1}$ is actually in the better space, $[\bar{B}]_\theta$.

Let us proceed further and show how these ideas can be used to study higher order commutators, i.e. operations defined in a similar fashion using higher order derivatives. The point is that to control these operations we require higher order cancellations. Let us start by defining for $n = 1$.....

$$\Omega_n a = \frac{1}{n!} h_a^{(n)}(\theta)$$

Then, the following result was obtained in [27] using a different approach.

Theorem 1 *Let $\bar{A}$, $\bar{B}$, be Banach pairs, $\theta \in D$, and let $T : \bar{A} \to \bar{B}$, be a bounded linear operator. For $n \in N$, and $a \in \left[\bar{A}\right]_\theta$, let $\Omega_n a = \frac{1}{n!} h_a^{(n)}(\theta)$, and let C_n be defined by*

$$C_n a = \begin{cases} [T, \Omega_1] a & \dots \quad n = 1 \\ [T, \Omega_2]\, a - \Omega_1 [T, \Omega_1] a & \dots \quad n = 2 \\ \dots & \dots \quad \dots \\ [T, \Omega_n]\, a - \sum_{k=1}^{n-1} \Omega_k C_{n-k} a & \dots \quad n > 2 \end{cases}$$

Then, there exists $c > 0$ such that

$$\|C_n a\|_{[\bar{B}]_\theta} \le c \,\|a\|_{[\bar{A}]_\theta}\,.$$

A proof, using our present point of view, can be achieved by exhibiting $C_n a$ as an element of the space $[\bar{B}]_{\theta,n}$ admitting a representation with sufficient cancellation to guarantee that it is actually in the better space $[\bar{B}]_\theta$. Let us consider the case $n = 2$ in detail. Suppose that $T : \bar{A} \to \bar{B}$ is a bounded linear operator, $\theta \in D$, we need to show that the there exists $c > 0$, such that

$$\|[T, \Omega_2]\, a - \Omega_1 [T, \Omega_1] a\|_{[\bar{B}]_\theta} \le c \,\|a\|_{[\bar{A}]_\theta}$$

There is no loss of generality assuming that $\theta = 0$. Let $b_1 = [T, \Omega_1] a$, and define

$$\gamma_1(z) = \frac{1}{2}(T h_a(z) - h_{Ta}(z))$$

$$\gamma_2(z) = \gamma_1(z) - \frac{z}{2}h_{b_1}(z)$$

Then, $\gamma_2(0) = 0$, $\gamma_2'(z) = \gamma_1'(z) - \frac{1}{2}h_{b_1}(z) - \frac{z}{2}h_{b_1}(z)$, and therefore $\gamma_2'(0) = \frac{1}{2}b_1 - \frac{1}{2}b_1 - 0 = 0$. Thus, although we would normally only be able to claim that

$$\gamma_2''(0) = [T, \Omega_2]\,a - \frac{1}{2}\Omega_1 b_1 - \frac{1}{2}\Omega_1 b_1 = [T, \Omega_2]\,a - \Omega_1[T, \Omega_1]a \in [\bar{A}]_{\theta,2}$$

the cancellations give us

$$[T, \Omega_2]\,a - \Omega_1[T, \Omega_1]a \in [\bar{A}]_\theta$$

with appropriate norm control. The general n^{th} case can be obtained in a similar fashion.

Let K be a Calderón Zygmund operator, and for $b \in BMO(R^n)$ denote by M_b the operator defined by $M_b f = fb$. Then, a result of [9] states that $[K, M_b]$ is a bounded operator,

$$[K, M_b] : L^p(R^n) \to L^p(R^n). \tag{5}$$

In [28], Rochberg and Weiss give an interesting application of the theory of commutators to prove this result as follows. By homogeneity we can assume that $\|b\|_{BMO}$ is sufficiently small so that e^b and e^{-b} are A_p weights. Therefore,

$$K : (L^p(e^b), L^p(e^{-b})) \to (L^p(e^b), L^p(e^{-b})) \tag{6}$$

Now, since we can choose $\Omega_{(L^p(e^b), L^p(e^{-b}))}f = M_b f$, (5) follows. In fact (5) corresponds to the case $n = 1$ of Theorem 1. The case $n = 2$ gives the boundedness of $[[K, M_b], M_b]$, and the obvious pattern continues.

4   The Real Method

Recall that given a Banach pair $\bar{A} = (A_0, A_1)$, and an element $a \in \Sigma(\bar{A}) = A_0 + A_1$, the K functional of a is defined, for $t > 0$, by

$$K(t, a; \bar{A}) = \inf\{\|a_0\|_{A_0} + t\|a_1\|_{A_1} : a = a_0 + a_1\}$$

For a fixed constant $c > 0$, let us say that a decomposition $a = a_0(t) + a_1(t)$ is *almost optimal* for the K method if

$$\|a_0(t)\|_{A_0} + t\|a_1(t)\|_{A_1} \leq cK(t, a; \bar{A}) \tag{7}$$

We shall then write $D_K(t; \bar{A})a = D_K(t)a = a_0(t)$.

One possible approach to real interpolation is through the study of spaces defined in terms of the K functional as follows. For $\theta \in (0, 1)$, $0 < q \leq \infty$, $\beta \in R$, and for measurable functions on the half line, let us define

$$\Phi_{\theta, q, \log^\beta}(f) = \left\{ \int_0^\infty |f(t)t^{-\theta}(1 + |\log t|)^\beta|^q \frac{dt}{t} \right\}^{\frac{1}{q}}$$

We also let $\Phi_{\theta,q} = \Phi_{\theta,q,\log^0}$. Then, the interpolation spaces $\bar{A}_{\theta,q;\log^\beta;K}$, $0 < \theta < 1$, $0 < q \leq \infty$, $\beta \in R$, are defined by

$$\bar{A}_{\theta,q;\log^\beta;K} = \{a : \|a\|_{\bar{A}_{\theta,q;\log^\beta;K}} = \Phi_{\theta,q,\log^\beta}\left(K(.,a,\bar{A})\right) < \infty\} \qquad (8)$$

For $n \in N$, we define operators $\Omega_{\bar{A},n;K} = \Omega_{n;K}$ associated with almost optimal decompositions as follows,

$$\Omega_{n;K}a = \frac{1}{(n-1)!}\left(\int_0^1 (\log t)^{n-1} D_K(t)a\frac{dt}{t} - \int_1^\infty (\log t)^{n-1}(I - D_K(t))a\frac{dt}{t}\right) \qquad (9)$$

We now define the corresponding operators $\Omega_{n;J}$ associated with the J method. For a Banach pair $\bar{A}$, the spaces $\bar{A}_{\theta,q;\log^\beta;J}$, $0 < \theta < 1$, $0 < q \leq \infty$, $\beta \in R$, are defined using the quasi- norms

$$\|a\|_{\bar{A}_{\theta,q;\log^\beta;J}} = \inf\{\Phi_{\theta,q,\log^\beta}\left(J(.,u(.),\bar{A})\right) : a = \int_0^\infty u(s)\frac{ds}{s}\}$$

where $u : (0,\infty) \to \Delta(\bar{A})$, and the J functional is defined, for $h \in \Delta(\bar{A}), t > 0$, by

$$J(t,h;\bar{A}) = \max\{\|h\|_{A_0}, t\|h\|_{A_1}\}$$

Let us say that $u(t)$ is an almost optimal decomposition of a for the J method, if

$$a = \int_0^\infty u(s)\frac{ds}{s}, \quad \|a\|_{\bar{A}_{\theta,q;\log^\beta;J}} \approx \Phi_{\theta,q,\log^\beta}\left(J(.,u(.),\bar{A})\right) \qquad (10)$$

In which case we write $D_J(t)a = u(t)$, and the corresponding $\Omega_{\bar{A},n;J} = \Omega_{n;J}$ operators are defined by

$$\Omega_{n;J}a = \frac{1}{n!}\int_0^\infty D_J(t)a\,(\log t)^n\,\frac{dt}{t} \qquad (11)$$

(There are analogous operators associated with the E and F methods but we won't discuss them here.)

Whenever no confusion arises we shall drop the subindex that indicates which method of interpolation we are using in the definition of the $\Omega's$.

Let $\bar{A}, \bar{B}$, be Banach pairs, and let $T : \bar{A} \to \bar{B}$ be a bounded linear operator. We let $[T, \Omega_1] = T\Omega_{\bar{A},1} - \Omega_{\bar{B},1}T$. The analog of the complex commutator theorem is given by the following result of [15]:

$$\|[T, \Omega_1]f\|_{\bar{B}_{\theta,q}} \leq c \|f\|_{\bar{A}_{\theta,q}}. \qquad (12)$$

More generally we also have a higher order version of this result obtained in [22].
For $n = 0, 1, 2....$define

$$
\begin{aligned}
C_n f \;&=\; C_{n,K} f \\
&= \begin{cases}
Tf & ... \quad n = 0 \\
[T, \Omega_{1,F}]f & ... \quad n = 1 \\
... & ... \quad .... \\
[T, \Omega_{n,K}]f + \Omega_{1,K}C_{n-1,K}f + ... + \Omega_{n-1,K}C_{1,K}f & ... \quad n \geq 2
\end{cases}
\end{aligned}
$$

$$
C_n f = C_{n,J} f = \begin{cases}
Tf & ... \quad n = 0 \\
[T, \Omega_{1,J}]f & ... \quad n = 1 \\
... & ... \quad .... \\
[T, \Omega_{n,J}]f - \Omega_{1,F}C_{n-1,J}f - ... - \Omega_{n-1,J}C_{1,J}f & ... \quad n \geq 2
\end{cases}
$$

Then, for $0 < \theta < 1, 1 \leq q \leq \infty$, there exist absolute constants $c = c(\theta, q, n, T) > 0$ such that

$$\|C_n f\|_{\bar{B}_{\theta,q}} \leq c \|f\|_{\bar{A}_{\theta,q}}.$$

We shall now indicate a proof that parallels the one we gave in the previous section for the complex method.

Let us start by noting that from the definitions it follows readily that

$$\Omega_{1,J} : \bar{A}_{\theta,q;J} \to \bar{A}_{\theta,q,\log^{-1};J}$$

(in fact $\bar{A}_{\theta,q,\log^{-1};J}$ is *the* range space in a suitable technical sense, (cf [11]).) If we combine this with the fact that, by interpolation, $T : \bar{A}_{\theta,q,\log^{-1};J} \to \bar{B}_{\theta,q,\log^{-1};J}$, we see that

$$[T, \Omega] : \bar{A}_{\theta,q;J} \to \bar{B}_{\theta,q,\log^{-1};J}.$$

In fact, we can see this directly from the definitions, since we have

$$[T, \Omega]f = \int_0^\infty (D_J(t)(Tf) - T(D_J(t)(f))) \log t \frac{dt}{t} \tag{13}$$

with

$$\Phi_{\theta,q}(J(t, D_J(t)(Tf) - T(D_J(t)(f)); \bar{B})) \leq c\Phi_{\theta,q}(J(t, D_J(t)f; \bar{A}))$$
$$\leq c \|f\|_{\bar{A}_{\theta,q;J}}$$

However, since the representation (13) exhibits the cancellation

$$\int_0^\infty (D_J(t)(Tf) - T(D_J(t)(f))) \frac{dt}{t} = 0$$

we deduce that in fact

$$[T, \Omega] : \bar{A}_{\theta,q;J} \to \bar{B}_{\theta,q;J}$$

This is the content of the next

Theorem 2 *Let $\bar{H}$ be a Banach pair, and suppose that*

$$a = \int_0^\infty u(s) \log s \frac{ds}{s}$$

with

$$\int_0^\infty u(s) \frac{ds}{s} = 0, \quad \Phi_{\theta,q}(J(t, u(t); \bar{H})) < \infty$$

then,

$$a \in \bar{H}_{\theta,q;J},$$

and moreover,

$$\|a\|_{\bar{H}_{\theta,q;J}} \leq c\Phi_{\theta,q}(J(t, u(t); \bar{H}))$$

Remark. We would normally be able to conclude that $a \in \bar{H}_{\theta,q;\log^{-1};J}$, which is a somewhat larger space than $\bar{H}_{\theta,q;J}$. The condition $\int \ldots = 0$ is the crucial cancellation that lets us improve that. At the conceptual level this is the analog of the fact that in the complex method we had a function $\gamma(z)$ with $\gamma(0) = 0$. (Note that this is also a mean value zero condition; $\int_{|z|=1} \gamma(e^{i\theta}) d\theta = 0$.) Many of the results on compensated compactness which make no reference to interpolation also make crucial use of such similar cancellation conditions.

Proof.   Integrating by parts let us write

$$a = \int_0^\infty \left(\int_0^t u(s) \frac{ds}{s} \right) \frac{dt}{t}$$

Now,

$$J\left(t, \int_0^t u(s) \frac{ds}{s}; \bar{A}\right) \leq \left\| \int_0^t u(s) \frac{ds}{s} \right\|_{H_0} + t \left\| \int_0^t u(s) \frac{ds}{s} \right\|_{H_1}$$

The first term can be estimated directly as follows

$$\left\| \int_0^t u(s) \frac{ds}{s} \right\|_{H_0} \leq \int_0^t \|u(s)\|_{H_0} \frac{ds}{s}$$

$$\leq \int_0^t J(s, u(s); \bar{H}) \frac{ds}{s}$$

Now, because of the cancellation

$$\int_0^t u(s) \frac{ds}{s} = - \int_t^\infty u(s) \frac{ds}{s}$$

we can estimate

$$t \left\| \int_0^t u(s) \frac{ds}{s} \right\|_{H_1} = t \left\| \int_t^\infty u(s) \frac{ds}{s} \right\|_{H_1}$$

$$\leq t \int_t^\infty \|u(s)\|_{H_1} \frac{ds}{s}$$

$$\leq t \int_t^\infty J(s, u(s); \bar{H}) \frac{ds}{s^2}$$

So that all in all

$$J(t, \int_0^t u(s) \frac{ds}{s}; \bar{H}) \leq \int_0^t J(s, u(s); \bar{H}) \frac{ds}{s} + t \int_t^\infty J(s, u(s); \bar{H}) \frac{ds}{s^2}$$

Hardy's inequality applied to the last estimate shows that shows that

$$\Phi_{\theta,q}(J(t, \int_0^t u(s) \frac{ds}{s}; \bar{H})) \leq c\Phi_{\theta,q}(J(s, u(s); \bar{H}))$$

and therefore

$$\|a\|_{\bar{H}_{\theta,q;J}} \leq c\Phi_{\theta,q}(J(t, u(t); \bar{H}))$$

□

The commutator theorem for the real method is now a consequence of Theorem 2 and our previous discussion.

To deal with higher order cancellation let us state and prove the following

Theorem 3 *If*

$$a = \int_0^\infty u(s)(\log s)^n \frac{ds}{s}$$

with

$$\int_0^\infty u(s)(\log s)^k \frac{ds}{s} = 0, \quad k = 0, ...n-1; \quad \Phi_{\theta,q}(J(t, u(t); \bar{H})) < \infty$$

then,

$$a \in \bar{H}_{\theta,q;J},$$

and moreover,

$$\|a\|_{\bar{H}_{\theta,q;J}} \leq c\Phi_{\theta,q}(J(t, u(t); \bar{H}))$$

Proof. Lets us consider in detail the case $n = 2$. Suppose that

$$a = \int_0^\infty u(s)(\log s)^2 \frac{ds}{s}$$

with

$$\int_0^\infty u(s) \frac{ds}{s} = 0, \quad \int_0^\infty u(s)(\log s) \frac{ds}{s} = 0, \quad \Phi_{\theta,q}(J(t, u(t); \bar{H})) < \infty$$

Given this we can write

$$a = - \int_0^\infty (\int_0^t u(s) \log s \frac{ds}{s}) \frac{dt}{t}$$

further integration by parts in the inner integral, using the cancellation conditions, gives

$$a = 2 \int_0^\infty \left(\int_0^t u(r) \log \frac{t}{r} \frac{dr}{r} \right) \frac{dt}{t}$$

Thus, we estimate

$$J\left(t, \int_0^t u(r) \log \frac{t}{r} \frac{dr}{r}; \bar{H}\right) \leq \left\| \int_0^t u(r) \log \frac{t}{r} \frac{dr}{r} \right\|_{H_0} + t \left\| \int_0^t u(r) \log \frac{t}{r} \frac{dr}{r} \right\|_{H_1}$$

To estimate the first term we apply Minkowski's inequality to get

$$\left\| \int_0^t u(r) \log \frac{t}{r} \frac{dr}{r} \right\|_{H_0} \leq \int_0^t \|u(r)\|_{H_0} \log \frac{t}{r} \frac{dr}{r}$$

To estimate the second term we write

$$\int_0^t u(r) \log \frac{t}{r} \frac{dr}{r} = \int_0^t u(r) \log t \frac{dr}{r} - \int_0^t u(r) \log r \frac{dr}{r}$$

Now, because of the cancellation $\int_0^\infty u(s) \frac{ds}{s} = 0$,

$$\int_0^t u(r) \log t \frac{dr}{r} = - \int_t^\infty u(r) \log t \frac{dr}{r}$$

while the cancellation $\int_0^\infty u(s) \log s \frac{ds}{s} = 0$, gives

$$- \int_0^t u(r) \log r \frac{dr}{r} = \int_t^\infty u(r) \log r \frac{dr}{r}$$

Therefore,

$$t \left\| \int_0^t u(r) \log \frac{t}{r} \frac{dr}{r} \right\|_{H_1} = t \left\| \int_t^\infty u(r) \log \frac{r}{t} \frac{dr}{r} \right\|_{H_1}$$

$$\leq t \int_t^\infty \|u(r)\|_{H_1} \log \frac{r}{t} \frac{dr}{r}$$

and

$$J\left(t, \int_0^t u(r) \log \frac{t}{r} \frac{dr}{r}; \bar{H}\right) \leq \int_0^t J(r, u(r); \bar{H}) \log \frac{t}{r} \frac{dr}{r} + t \int_t^\infty J(r, u(r); \bar{H}) \log \frac{r}{t} \frac{dr}{r^2}$$

Consequently, by an iterate of Hardy's inequality, we get

$$\Phi_{\theta,q}\left(J\left(t, \int_0^t u(r) \log \frac{t}{r} \frac{dr}{r}; \bar{H}\right)\right) \leq c \Phi_{\theta,q}(J(t, u(t); \bar{H}))$$

and we are done.

By induction, the previous argument actually shows that we have proved the general case for $n \in N$. $\square$

Let us now see how to deal with estimates for higher order commutators. We consider the case $n = 2$ in detail. Let $T : \bar{A} \to \bar{B}$, and for $f \in \bar{A}_{\theta,q;J}$ let us set

$$u(t) = (TD_j(t)f - D_j(t)Tf)$$

then,

$$C_2(Tf) = \frac{1}{2}\int_0^\infty u(t)\,(\log t)^2\,\frac{dt}{t} - \int_0^\infty \tilde{u}(t)\log t\,\frac{dt}{t}$$

with

$$\int_0^\infty \tilde{u}(t)\frac{dt}{t} = \int_0^\infty u(t)\log t\,\frac{dt}{t}, \int_0^\infty u(t)\frac{dt}{t} = 0$$

and

$$\Phi_{\theta,q}(J(t,\tilde{u}(t);\bar{A})) \le c\,\|f\|_{\bar{A}_{\theta,q;J}}$$

Thus $C_2(Tf)$ can be rewritten as

$$C_2(Tf) = \int_0^\infty \left(-\int_0^t u(s)\frac{ds}{s} - \tilde{u}(t)\right)\log t\,\frac{dt}{t}$$

Now, since

$$\int_0^\infty \left(-\int_0^t u(s)\frac{ds}{s} - \tilde{u}(t)\right)\frac{dt}{t} = \int_0^\infty u(t)\log t\,\frac{dt}{t} - \int_0^\infty u(t)\log t\,\frac{dt}{t} = 0$$

Theorem 2 implies that

$$\|C_2(Tf)\|_{\bar{B}_{\theta,q;J}} \le c\Phi_{\theta,q}(J(t,\int_0^t u(s)\frac{ds}{s};\bar{A})) + \Phi_{\theta,q}(J(t,\tilde{u}(t);\bar{A}))$$

but by hypothesis

$$\Phi_{\theta,q}(J(t,\tilde{u}(t);\bar{A})) \le c\,\|f\|_{\bar{A}_{\theta,q;J}}$$

and another application of Theorem 2 gives

$$\Phi_{\theta,q}(J(t,\int_0^t u(s)\frac{ds}{s};\bar{A})) \le c\,\|f\|_{\bar{A}_{\theta,q;J}}$$

At an earlier stage in the development of the theory (cf. [10], [11]) considerable effort was spent characterizing the domain and range spaces associated with the operators Ω. The proofs of those characterizations were in some sense because not only were the proofs complicated but they required methods disjoint from the proof of the commutator theorems. A different approach resolving this issue was given in [25]. From our new perspective we will show that properly formulated these results are due to suitable cancellations. Recall that we define the domain spaces for the operator Ω as follows

$$Dom_{\theta,q;J}\Omega(\bar{H}) = \{a \in \bar{H}_{\theta,q;J} : \Omega a \in \bar{H}_{\theta,q;J}\} \tag{14}$$

It is shown in the papers quoted above that

$$Dom_{\theta,q}\Omega(\bar{H}) = \bar{H}_{\theta,q,\log;J}$$

The proofs in [10], [11], and more recently in [5] are indirect. Direct proofs are given for weighted L^p spaces and then appeal is made to the fact that general real interpolation spaces can be obtained as partial retracts of such spaces. Let us now give a new proof of this result which emphasizes the rôle of cancellation.

We start by observing an equivalent way to describe the domain spaces associated with Ω: $a \in Dom_{\theta,q}\Omega(\bar{H}) \iff a$ can be represented by $a = \int_0^\infty u(s)\frac{ds}{s}$, with $\Phi_{\theta,q}(J(t,u(t);\bar{H})) < \infty$, and, moreover,

$$\int_0^\infty u(s) \log s \frac{ds}{s} = 0.$$

The simple argument is given in [5]. Once we know this (14) is a consequence of the following

Theorem 4 *Suppose that*

$$a = \int_0^\infty u(s)\frac{ds}{s}$$

with

$$\int_0^\infty u(s) \log s \frac{ds}{s} = 0; \Phi_{\theta,q}(J(t,u(t);\bar{H})) < \infty$$

Then,

$$a \in \bar{H}_{\theta,q,\log;J}$$

and there exists an absolute constant $c > 0$, such that

$$\|a\|_{\bar{H}_{\theta,q,\log;J}} \le c\Phi_{\theta,q}(J(t,u(t);\bar{H}))$$

Proof.   Again the method of proof is to exploit the available cancellation to obtain a more favorable estimation of the size of a. The only "new" ingredient is that we require a version of Hardy's inequality for the $\Phi_{\theta,q,\log^\beta}$ norms. Indeed, by integration by parts, using the cancellation, let us write

$$a = \int_0^\infty u(s) \log s \frac{1}{\log s} \frac{ds}{s} = \int_0^\infty \left(\int_0^t u(s) \log s \frac{ds}{s} \right) \frac{1}{(\log t)^2} \frac{dt}{t}$$

Then, as above,

$$J(t, \tfrac{1}{(\log t)^2} \int_0^t u(s) \log s \tfrac{ds}{s}; \bar{H}) \le \tfrac{1}{(\log t)^2} \int_0^t J(s,u(s);\bar{H}) |\log s| \tfrac{ds}{s} + t\tfrac{1}{(\log t)^2} \int_t^\infty J(s,u(s);\bar{H}) |\log s| \tfrac{ds}{s^2}$$

Now, the Hardy type estimates (cf. [1])

$$\Phi_{\theta,q,\log} \left(\frac{1}{(\log t)^2} \int_0^t J(s,u(s);\bar{H}) |\log s| \frac{ds}{s} \right) \le c\Phi_{\theta,q}(J(s,u(s);\bar{H}))$$

$$\Phi_{\theta,q,\log} \left(t\frac{1}{(\log t)^2} \int_t^\infty J(s,u(s);\bar{H}) |\log s| \frac{ds}{s^2} \right) \le c\Phi_{\theta,q}(J(s,u(s);\bar{H}))$$

show that

$$\Phi_{\theta,q,\log}\left(J\left(t,\frac{1}{(\log t)^2}\int_0^t u(s)\log s\frac{ds}{s};\bar H\right)\right)\le c\Phi_{\theta,q}(J(s,u(s);\bar H))$$

and we are done. $\square$

We consider now individual commutation theorems with the higher order Ω_n operators. We shall prove that (cf. [21])

$$\|[T,\Omega_n]f\|_{\bar B_{\theta,q,J}}\le c\|f\|_{\bar A_{\theta,q,\log^{n-1};J}}\tag{15}$$

To see what is needed to prove (15), observe that

$$[T,\Omega_n]f=\frac{1}{n!}\int_0^\infty\left(T(D_J(t)(f))-D_J(t)(Tf)\right)(\log t)^n\frac{dt}{t}$$

with the cancellation

$$\int_0^\infty\left(T(D_J(t)(f))-D_J(t)(Tf)\right)\frac{dt}{t}=0$$

Consequently the result follows from our next

Theorem 5 *Suppose that*

$$a=\int_0^\infty u(s)(\log s)^n\frac{ds}{s}$$

with

$$\int_0^\infty u(s)\frac{ds}{s}=0,\ \Phi_{\theta,q,\log^{n-1}}(J(t,u(t);\bar H))<\infty$$

then, $a\in\bar H_{\theta,q;J}$, and moreover, there exists an absolute constant $c>0$, such that

$$\|a\|_{\bar H_{\theta,q;J}}\le c\Phi_{\theta,q,\log^{n-1}}(J(t,u(t);\bar H))$$

Proof. We can only integrate by parts once and we arrive to the representation

$$a=\frac{1}{n}\int_0^\infty\left(\int_0^t u(s)\frac{ds}{s}\right)(\log t)^{n-1}\frac{dt}{t}$$

we estimate in the usual fashion and arrive to

$$|\log t|^{n-1}J(t,\int_0^t u(s)\tfrac{ds}{s};\bar H)\le c\,|\log t|^{n-1}\int_0^t J(s,u(s);\bar H)\tfrac{ds}{s}+$$
$$c\,|\log t|^{n-1}t\int_t^\infty J(s,u(s);\bar H)\tfrac{ds}{s^2}$$

Now,

$$\Phi_{\theta,q}(|\log t|^{n-1}J(t,\int_0^t u(s)\tfrac{ds}{s};\bar H))\le c\Phi_{\theta,q,\log^{n-1}}(\int_0^t J(s,u(s);\bar H)\tfrac{ds}{s})+$$
$$c\Phi_{\theta,q,\log^{n-1}}(t\int_t^\infty J(s,u(s);\bar H)\tfrac{ds}{s^2})$$

$$\le c\Phi_{\theta,q,\log^{n-1}}(J(t,u(t);\bar H))$$

as we wished to show. $\square$

References

[1] C. Bennett and K. Rudnick, *On Lorentz-Zygmund spaces*, Diss. Math. **175** (1980), 1-72.

[2] J. Bergh and J. Löfström, *Interpolation Spaces: An introduction*, Springer-Verlag, Berlin, Heidelberg and New York, 1976.

[3] A.P. Calderón, *Intermediate spaces and interpolation, the complex method*, Studia Math. **24** (1964), 113-190.

[4] M. Carro, J. Cerda, M. Milman, and J. Soria, *Schechter methods of interpolation and commutators*, submitted

[5] M. Carro, J. Cerda, and J. Soria, *A unified theory of commutators for interpolation methods*, preprint.

[6] R. Coifman and L. Grafakos, *Hardy space estimates for multilinear operators I*, Rev. Mat. Iberoamericana **8** (1992), 45-67.

[7] R. Coifman, P. L. Lions, Y. Meyer and S. Semmes, *Compacité par compensation et éspaces de Hardy*, C.R. Acad. Sci. Paris **309** (1989), 945-949.

[8] —————, *Compensated compactness and Hardy spaces*, Jour. Math. Pures et Appl. **72** (1993), 247-286.

[9] R. Coifman, R. Rochberg, and G. Weiss, *Factorization theorems for Hardy spaces in several variables*, Ann. Math. **103** (1976), 611-635.

[10] M. Cwikel, B. Jawerth, and M. Milman, *The domain spaces of quasilogarithmic operators*, Trans. Amer. Math. Soc. **317** (1990), 599-609.

[11] M. Cwikel, B. Jawerth, M. Milman, and R. Rochberg, *Differential estimates and commutators in interpolation theory*, "Analysis at Urbana II", London Math. Soc., Cambridge Univ. Press 1989, pp. 170-220.

[12] Fan Ming and S. Kaijser, *Complex interpolation with derivatives of analytic functions*, preprint, Uppsala, 1992.

[13] T. Iwaniec and A. Lutoborski, *Integral estimates for null Lagrangians*, preprint.

[14] T. Iwaniec, and C. Sbordone, *Weak minima of variational integrals*, preprint, University of Naples, 1992.

[15] B. Jawerth, R. Rochberg, and G. Weiss, *Commutator and other second order estimates in real interpolation theory*, Ark. Mat. **24** (1986), 191-219.

[16] N. Kalton, *Nonlinear commutators in interpolation theory*, Mem. Amer. Soc. **373,** Amer. Math. Soc., Providence, RI. 1988.

[17] N. Kalton, *Differential of complex interpolation processes for Köthe function spaces*, Trans. Amer. Math. Soc. **333** (1992), 479-529.

[18] N. Kalton, *Trace class operators and commutators*, Journal of Functional Analysis **86** (1989), 41-74.

[19] C. Li, A. McIntosh, K. Zhang, *Higher Integrability and Reverse Hölder Inequalities*, preprint, Macquarie University, 1993.

[20] M. Milman, *A commutator theorem with applications*, Coll. Math. **44** (1993), 201-210.

[21] M. Milman, *Higher order commutator in the real method of interpolation*, submitted

[22] M. Milman, *Extrapolation and Optimal Decompositions with Applications to Analysis*, Lecture Notes in Mathematics **1580**, Srpinger-Verlag, New York, 1994.

[23] M. Milman, *Inequalities for Jacobians: Interpolation methods*, Proc. 2nd Latin American Analysis Seminar, Rev. Mat. Colombiana **27** (1993), 67-81, Errata, to appear.

[24] M. Milman, *Integrability of the Jacobian of orientation preserving maps: interpolation tecniques*, Comptes Rendus Acad. Sc. Paris **317** (1993), 539-543.

[25] M. Milman and T. Schonbek, *Second order estimates in interpolation theory and applications*, Proc. Amer. math. Soc. **110** (1990), 961-969.

[26] S. Müller, *Higher integrability of determinants and weak convergence in L^1*, J. Reine Angew Math. **412** (1990), 20-34.

[27] R. Rochberg, *Higher order estimates in complex interpolation theory*, Pacific Journal, to appear.

[28] R. Rochberg and G. Weiss, *Derivatives of analytic families of Banach spaces*, Ann. Math. **118** (1983), 315-347.

[29] M. Schechter, *Complex interpolation*, Comp. Math. **18** (1967), 117-147.

Department of Mathematics, Florida Atlantic University, Fl 33431
E-mail address: milman@acc.fau.edu
Department of Mathematics, Washington University, St Louis, Missouri 63130
E-mail address: rr@artsci.wustl.edu

Contemporary Mathematics
Volume **189**, 1995

AN EXTENSION OF NAIMARK'S DILATION THEOREM TO KREIN SPACES

FERNANDO PELAEZ

ABSTRACT. An extension of the classical Naimark's dilation theorem to the general Krein space setting is given. This allows us to give a new proof of the fact that every bounded operator on a Krein space has minimal unitary dilations.

1. INTRODUCTION AND NOTATION

Let T be a bounded operator in a Hilbert space $\mathcal{H}$. An operator U in a Hilbert space $\tilde{\mathcal{H}}$ is called a unitary dilation of T if U is unitary, $\tilde{\mathcal{H}}$ contains $\mathcal{H}$ as a closed subspace and the (strong) dilation property holds: $T^n = P_{\mathcal{H}}^{\tilde{\mathcal{H}}} U^n \mid_{\mathcal{H}}$, $n = 1, 2, 3, \ldots$ where $P_{\mathcal{H}}^{\tilde{\mathcal{H}}}$ is the orthogonal projection from $\tilde{\mathcal{H}}$ onto $\mathcal{H}$. If T has a unitary dilation then T is a contraction ($\parallel T \parallel \leq 1$). The converse is also true and is the content of a well known theorem due to Sz.-Nagy [11], every contractive operator in a Hilbert space has a unitary dilation; moreover, this dilation can be chosen to be minimal, that is, not admitting a decomposition into the direct sum of two unitary operators such that one of them is also a dilation of T. Such a minimal dilation is essentially unique (two minimal unitary dilations of T are always isomorphic) and it is called *the* minimal unitary dilation of T. Nagy's theorem is the starting point of a rich theory [12]. If T is a contraction in a Hilbert space $\mathcal{H}$ then the operator-valued function γ defined on the group of integers Z by $\gamma(n) = T^n$ if $n \geq 0$ and $\gamma(n) = (T^*)^{-n}$ if $n < 0$ is a function of positive type ([12] I.8.1), that is, every function $h : Z \to \mathcal{H}$ with finite support verifies $\sum_{n,m \in Z} \langle \gamma(n-m)h(n), h(m) \rangle \geq 0$.

An important result concerning unitary dilations of operator-valued functions of positive type defined on a general group is the classical Naimark's theorem [10]. It establishes that if $\mathcal{H}$ is a Hilbert space, Γ a group with neutral element e and $\gamma : \Gamma \to \mathcal{L}(\mathcal{H})$ a function of positive type ($\sum_{t,s \in \Gamma} \langle \gamma(s^{-1}t) h(t), h(s) \rangle \geq 0$ for every function $h : \Gamma \to \mathcal{H}$ with finite support) with $\gamma(e) = I$ (the identity operator in H) then there exist a Hilbert space $\tilde{\mathcal{H}}$ that contains $\mathcal{H}$ as a closed subspace and a group of unitary operators $\{ U(t) \ / \ t \in \Gamma \}$ in $\tilde{\mathcal{H}}$ such that $\gamma(t) = P_{\mathcal{H}}^{\tilde{\mathcal{H}}} U(t) \mid_{\mathcal{H}}$ for every $t \in \Gamma$. Such a dilation can be chosen to be minimal and essential uniqueness of the minimal unitary dilation is also obtained. Therefore, Nagy's theorem follows from a straightforward application of the theorem of Naimark. A similar approach works for the unitary dilation of continuous contractive semigroups ([12] I.8.2).

1991 *Mathematics Subject Classification.* 47B50, 47A20.

This paper is in final form and no version of it will be submitted for publication elsewhere

In [6], Davis, by a modification of the methods of [12], obtained the corresponding Nagy's result for an arbitrary operator. Of course, the conclusion differs by the fact that the dilation acts in a Krein space as a "J-unitary" operator (that is, it preserves the indefinite Krein space inner product). He also gave a continuous version for semigroups of operators in a Hilbert space ([7]).

Arocena characterized those Hilbert space operator-valued functions in a general group which have unitary dilations to a Krein space which contains the original Hilbert space as a positive regular subspace ([2], Theorem IV.1). As Nagy's theorem follows from that of Naimark's, Marcantognini showed ([9]) that David's result can be deduced from Arocena's theorem.

Now, let T be a bounded operator in a Krein space $\mathcal{K}$. A unitary dilation of T is a bounded and bijective operator U defined in a Krein space $\tilde{\mathcal{H}}$ containing $\mathcal{K}$ as a regular subspace such that U is isometric (in the sense that preserves the indefinite product of $\tilde{\mathcal{H}}$) and such that $T^n = P_{\mathcal{K}}^{\tilde{\mathcal{H}}} U^n |_{\mathcal{K}}$ and $(T^*)^n = P_{\mathcal{K}}^{\tilde{\mathcal{H}}} U^{-n} |_K$, $n = 1, 2, 3, \ldots$. The notion of minimality is the same as in the Hilbert space setting. The dilation properties of Krein space operators are similar to those in the Hilbert space case. However, two minimal unitary dilations of a given operator in a Krein space are not necessarily isomorphic. This is due to the fact that a densely defined Krein space isometry may fail to have a continuous extension. We refer the reader to [8] for the constructions of isometric and unitary dilations of a given operator in a Krein space and for the treatment of essential uniqueness conditions.

Our aim is to give a Krein space version of Arocena's theorem (Theorem IV.1 in [2]) for Krein space operator-valued functions (Theorem 2.1 below). This allows us (following the ideas in [9]) to give a "Naimark type proof" of the fact that every bounded operator in a Krein space has minimal unitary dilations.

Familiarity with operator theory on Krein spaces is presumed and we refer the reader to [1], [3], [4] and [8] for more details. Most of the notation adopted here is the same as in [8]. If $(\mathcal{K}, [.,.])$ is a Krein space ($[.,.]$ is the indefinite product on $\mathcal{K}$) and J is any fundamental symmetry on it, we write $\langle .,. \rangle_J$ for the Hilbert-space inner product induced by J (i.e $\langle x, y \rangle_J = [Jx, y]$) and $\mathcal{K}_J$ for the vector space $\mathcal{K}$ endowed with the product $\langle .,. \rangle_J$.

The topological notions correspond to the norm induced by any of the J-inner products, which are all equivalents. All the other notions are to be taken in their Krein space versions.

By a regular subspace of a Krein space $\mathcal{K}$ we mean a closed subspace M of $\mathcal{K}$ which is a Krein space in the inner product inherited from $\mathcal{K}$. This is equivalent to say that M is orthocomplemented. If M is a regular subspace of $\mathcal{K}$ we denote by $P_M^{\mathcal{K}}$ the orthogonal projection from $\mathcal{K}$ onto M.

If $\{M_t\}_{t \in \Gamma}$ is a family of subsets of the Krein space $\mathcal{K}$ then the symbol $\bigvee_{t \in \Gamma} M_t$ denotes the closed linear span of these sets.

By $\mathcal{L}(\mathcal{H}, \mathcal{K})$ we denote the set of all everywhere defined and bounded (continuous) operators (linear operators) from the Krein space $\mathcal{H}$ to the Krein space $\mathcal{K}$. We use $\mathcal{L}(\mathcal{K})$ for $\mathcal{L}(\mathcal{K}, \mathcal{K})$.

The adjoint of an operator $A \in \mathcal{L}(\mathcal{H}, \mathcal{K})$ is denoted by A^* and is the unique operator in $\mathcal{L}(\mathcal{K}, \mathcal{H})$ such that $[Ah, k]_{\mathcal{K}} = [h, A^* k]_{\mathcal{H}}$ holds for every $h \in \mathcal{H}$ and $k \in \mathcal{K}$. The Hilbert space adjoint is denoted by $A^{\times}$. An operator $V \in \mathcal{L}(\mathcal{H}, \mathcal{K})$ is an isometry if $[Vh, Vh'] = [h, h'] \ \forall \ h, h' \in \mathcal{H}$. An isomorphism between $\mathcal{H}$ and $\mathcal{K}$

is an isometry from $\mathcal{H}$ onto $\mathcal{K}$. We also fix the following notation:
Let $(\mathcal{K}, [.,.])$ be a Krein space and let Γ be a group with neutral element e. We denote by $E = E(\Gamma, \mathcal{K})$ the set of the functions $h : \Gamma \longrightarrow \mathcal{K}$ with finite support. If $h \in E$ and $u \in \Gamma$ then h_u is the function in E defined by $h_u(t) = h(u^{-1}t)$, $t \in \Gamma$. If x is a vector in $\mathcal{K}$ then $\delta_{t,x}$ denotes the function in E with support equals $\{t\}$ such that $\delta_{t,x}(t) = x$. If $\gamma : \Gamma \to \mathcal{L}(\mathcal{K})$ is a function, then B_γ denotes the complex-valued form on Γ defined by

$$B_\gamma(h, h') = \sum_{t,s \in \Gamma} [\, \gamma(s^{-1}t)\, h(t)\, ,\, h'(s)\,], \qquad h, h' \in E.$$

Analogously, if $k : \Gamma \times \Gamma \to \mathcal{L}(K)$ is a kernel, then B_k denotes the complex-valued form on Γ defined by

$$B_k(h, h') = \sum_{t,s \in \Gamma} [\, k(t,s)\, h(t)\, ,\, h'(s)\,], \qquad h, h' \in E.$$

2. UNITARY DILATIONS OF KREIN SPACE OPERATOR-VALUED FUNCTIONS ON A GROUP

Let $(\mathcal{K}, [.,.])$ be a Krein space, Γ a group with neutral element e and $\gamma : \Gamma \to \mathcal{L}(\mathcal{K})$ a hermitian function (i.e $\gamma(t^{-1}) = \gamma(t)^*$) such that $\gamma(e) = I_\mathcal{K}$ (the identity operator in $\mathcal{K}$) and let us suppose that there exists a minimal unitary dilation of γ in a Krein space $\tilde{\mathcal{H}}$. By this we mean that there exist a Krein space $\tilde{\mathcal{H}}$ that contains a regular subspace $\mathcal{H}$ isomorphic to $\mathcal{K}$ and a representation $\{\, U(t) \,/\, t \in \Gamma \,\} \subset \mathcal{L}(\tilde{\mathcal{H}})$ of Γ by means of unitary operators on $\tilde{\mathcal{H}}$ such that:

(i) $\gamma(t) = \tau^{-1} P_\mathcal{H}^{\tilde{\mathcal{H}}} U(t) \tau$, $\forall t \in \Gamma$ ($\tau : \mathcal{K} \longrightarrow \mathcal{H}$ is an isomorphism);

(ii) $\tilde{\mathcal{H}} = \bigvee_{t \in \Gamma} U(t) \mathcal{H}$.

We are going to study the properties of the form B_γ. For every function $h \in E$ we have:

$$B_\gamma(h, h) = \sum_{t,s \in \Gamma} \left[\, \tau^{-1} P_\mathcal{H}^{\tilde{\mathcal{H}}} U(s^{-1}t)\tau h(t)\, ,\, h(s)\, \right]_\mathcal{K}$$

$$= \sum_{t,s \in \Gamma} \left[\, P_\mathcal{H}^{\tilde{\mathcal{H}}} U(s)^* U(t)\tau h(t)\, ,\, \tau h(s)\, \right]_\mathcal{H} = \sum_{t,s \in \Gamma} [\, U(s)^* U(t)\tau h(t)\, ,\, \tau h(s)\,]_{\tilde{\mathcal{H}}}$$

$$= \sum_{t,s \in \Gamma} [\, U(t)\tau h(t)\, ,\, U(s)\tau h(s)\,]_{\tilde{\mathcal{H}}} = \left[\, \sum_{t \in \Gamma} U(t)\tau h(t)\, ,\, \sum_{t \in \Gamma} U(t)\tau h(t)\, \right]_{\tilde{\mathcal{H}}}$$

If J' is a fundamental symmetry on $\tilde{\mathcal{H}}$ such that $J' \mathcal{H} \subset \mathcal{H}$ then

$$|\, B_\gamma(h, h)\, | \leq \left\langle\, \sum_{t \in \Gamma} U(t)\tau h(t)\, ,\, \sum_{t \in \Gamma} U(t)\tau h(t)\, \right\rangle_{J'} =$$

$$= \sum_{t,s \in \Gamma} \langle\, U(t)\tau h(t)\, ,\, U(s)\tau h(s)\, \rangle_{J'} = \sum_{t,s \in \Gamma} [\, J'U(t)\tau h(t)\, ,\, U(s)\tau h(s)\,]_{\tilde{\mathcal{H}}}$$

$$= \sum_{t,s \in \Gamma} [\, U(s)^* J'U(t)\tau h(t)\, ,\, \tau h(s)\,]_{\tilde{\mathcal{H}}} = \sum_{t,s \in \Gamma} \left[\, P_\mathcal{H}^{\tilde{\mathcal{H}}} U(s)^* J'U(t)\tau h(t)\, ,\, \tau h(s)\, \right]_\mathcal{H}$$

$$= \sum_{t,s \in \Gamma} \Big[\tau^{-1} P_{\mathcal{H}}^{\tilde{\mathcal{H}}} U(s)^* J' U(t) \tau \, h(t) \,,\, h(s) \Big]_{\mathcal{K}}$$

If we define the $\mathcal{L}(\mathcal{K})$-valued kernel $k : \Gamma \times \Gamma \longrightarrow \mathcal{L}(\mathcal{K})$ by:

(2.1) $$k(t,s) \,=\, \tau^{-1} \, P_{\mathcal{H}}^{\tilde{\mathcal{H}}} \, U(s)^* \, J' \, U(t) \, \tau.$$

then we already showed that k majorizes γ in the sense that

$$\mid B_\gamma(h,h) \mid \,\leq\, B_k(h,h), \qquad h \in E.$$

It is immediately checked that k is hermitian (i.e $k(t,s)^* = k(s,t)$, $\forall t, s \in \Gamma$) and that $k(e,e)$ is a fundamental symmetry on $\mathcal{K}$. Although k is not necessarily translation invariant, we have

$$B_k(h_u, h_u) \,=\, \sum_{t,s \in \Gamma} \Big[\tau^{-1} P_{\mathcal{H}}^{\tilde{\mathcal{H}}} U(s)^* J' U(t) \tau h(u^{-1}t) \,,\, h(u^{-1}s) \Big]_{\mathcal{K}} =$$

$$\sum_{t,s \in \Gamma} \Big[P_{\mathcal{H}}^{\tilde{\mathcal{H}}} U(us)^* J' U(ut) \tau h(t) \,,\, \tau h(s) \Big]_{\mathcal{H}} = \sum_{t,s \in \Gamma} [U(us)^* J' U(ut) \tau h(t) \,,\, \tau h(s)]_{\tilde{\mathcal{H}}}$$

$$= \sum_{t \in \Gamma} \langle \, U(u)U(t)\tau h(t) \,,\, U(u)U(s)\tau h(s) \, \rangle_{J'}$$

$$= \Big\langle \sum_{t \in \Gamma} U(u)U(t)\tau h(t) \,,\, \sum_{t \in \Gamma} U(u)U(t)\tau h(t) \Big\rangle_{J'}$$

$$\leq \parallel U(u) \parallel^2 \parallel \sum_{t \in \Gamma} U(t) \, \tau \, h(t) \parallel^2 \,=\, \parallel U(u) \parallel^2 \, B_k(h,h)$$

for every function $h \in E$ and $u \in \Gamma$, and then, the kernel k satisfies the bounding condition:

$$\mid B_k(h_u, h_u) \mid \,\leq\, \parallel U(u) \parallel^2 \, B_k(h,h), \qquad h \in E, \; u \in \Gamma.$$

Now, let us assume that the minimality condition (ii) holds. Let $h \in E$ such that $B_k(h,h) \neq 0$ and set $f = \sum_{t \in \Gamma} U(t) \tau h(t)$. Since $\parallel f \parallel_{J'}^2 := \langle f, f \rangle_{J'} = B_k(h,h)$ it follows that $f \neq 0$ and then $[f, J'f]_{\tilde{\mathcal{H}}} = \langle f, f \rangle_{J'} = \parallel f \parallel_{J'}^2 = \parallel f \parallel \parallel J'f \parallel > \frac{1}{2} \parallel f \parallel \parallel J'f \parallel$. By (ii) we can approximate $J'f$ by a vector g of the form $g = \sum_{t \in \Gamma} U(t) \tau h'(t)$ for some $h' \in E$ such that $[f, g]_{\tilde{\mathcal{H}}} > \frac{1}{2} \parallel f \parallel \parallel g \parallel$. Since $[f, g]_{\tilde{\mathcal{H}}} = B_\gamma(h, h')$, $\parallel f \parallel_{J'}^2 = B_k(h,h)$ and $\parallel g \parallel_{J'}^2 = B_k(h',h')$ we have proved the following

$\forall \, h \in E$ with $B_k(h,h) \neq 0$, $\exists h' \in E$ with $B_k(h',h') \neq 0$ such that

$$\mid B_\gamma(h,h') \mid \geq B_k(h,h)^{\frac{1}{2}} \, B_k(h',h')^{\frac{1}{2}}.$$

We can also consider some continuity properties; if Γ is a topological group and if $t, \bar{t}, s, \bar{s} \in \Gamma$ and $x, y \in \mathcal{K}$ then it is immediate to check that:

$$\mid [\, (k(t,s) - k(\bar{t},\bar{s})) \, x \,,\, y \,]_{\mathcal{K}} \mid =$$

$$\mid \langle \, (U(t) - U(\bar{t}))\tau x \,,\, U(s)\tau y \, \rangle_{J'} + \langle \, U(\bar{t})\tau x \,,\, (U(s) - U(\bar{s}))\tau y \, \rangle_{J'} \mid$$

$$\leq \parallel (U(t) - U(\bar{t}))\tau x \parallel_{J'} \parallel U(s)\tau y \parallel_{J'} + \parallel U(\bar{t})\tau x \parallel_{J'} \parallel (U(s) - U(\bar{s}))\tau y \parallel_{J'}$$

It follows that if $\{\, U(t) \,/\, t \in \Gamma \,\}$ is a strongly continuous group of bounded operators then $k(t,s)$ is continuous, as a function of two variables, with respect to the weak topology of operators.

The properties of the kernel k that we have already shown are not only necessary but also sufficient to assure the existence of a dilation :

Theorem 2.1. *Let* $(\,\mathcal{K}\,,\,[.,.]\,)$ *be a Krein space,* Γ *a group with neutral element* e *and* $\gamma : \Gamma \longrightarrow \mathcal{L}(\mathcal{K})$ *a hermitian function such that* $\gamma(e) = I_{\mathcal{K}}$. *Then the following conditions (I) and (II) are equivalent:*

(I) There exist a Krein space $\tilde{\mathcal{H}}$ *that contains a regular subspace* $\mathcal{H}$ *isomorphic to* $\mathcal{K}$ *and a representation* $\{\, U(t) \,/\, t \in \Gamma \,\} \subset \mathcal{L}(\tilde{\mathcal{H}})$ *of* Γ *by means of unitary operators on* $\tilde{\mathcal{H}}$ *such that:*

    (i) $\gamma(t) = \tau^{-1} P_{\mathcal{H}}^{\tilde{\mathcal{H}}} U(t)\,\tau, \quad \forall\, t \in \Gamma \qquad (\; \tau : \mathcal{K} \longrightarrow \mathcal{H}$ *is an isomorphism* $);$

    (ii) $\tilde{\mathcal{H}} = \bigvee_{t \in \Gamma} U(t)\,\mathcal{H}.$

(II) There exists a kernel $k : \Gamma \times \Gamma \longrightarrow \mathcal{L}(\mathcal{K})$ *such that:*

    (a) $J := k(e,e)$ *is a fundamental symmetry on* $\mathcal{K}$.
    (b) *There exists* $\lambda > 0$ *such that* $\mid B_\gamma(h,h) \mid \,\leq\, \lambda\, B_k(h,h), \quad \forall\, h \in E$.
    (c) *There exists a function* $\rho : \Gamma \to [0, +\infty)$ *such that*
       $\mid B_k(h_u, h_u) \mid \,\leq\, \rho(u)\, B_k(h,h), \quad \forall\, h \in E, \; \forall\, u \in \Gamma$.
    (d) *There exists* $\delta > 0$ *such that* $\forall\, h \in E$ *with* $B_k(h,h) \neq 0$, $\exists\, h' \in E$ *with*
       $B_k(h',h') \neq 0$ *such that* $\mid B_\gamma(h,h') \mid \,\geq\, \delta\, B_k(h,h)^{\frac{1}{2}}\, B_k(h',h')^{\frac{1}{2}}$.

Proof. Assume that (II) holds.

From the positive form B_k, by a standard construction, we can obtain a Hilbert space $(\tilde{\mathcal{H}}, \langle .,. \rangle)$ and an operator $\tau : \mathcal{K} \longrightarrow \tilde{\mathcal{H}}$ such that

$$(2.2) \qquad \langle\, \tau x\,,\, \tau y\,\rangle_{\tilde{\mathcal{H}}} \,=\, \langle\, x\,,\, y\,\rangle_J, \qquad x, y \in \mathcal{K}.$$

Indeed, if $E_0 := \{\, h \in E \,/\, B_k(h,h) = 0 \,\}$, $E' := E/E_0$ and $\pi : E \to E'$ is the canonical projection, then, setting $\langle\, \pi h\,,\, \pi h'\,\rangle := B_k(h,h')$ $(h, h' \in E)$, we obtain a scalar product in the quotient E', so the corresponding completion $\tilde{\mathcal{H}}$ is a Hilbert space and π can be considered as a map from E onto a dense subspace of $\tilde{\mathcal{H}}$. The norm induced by $\langle .,. \rangle$ will be denoted by $\|\ \|$. Let $\tau : \mathcal{K} \to \tilde{\mathcal{H}}$ be the map defined by $\tau\, x = \pi\, \delta_{e,x}$. We have: $\langle\, \tau x\,,\, \tau y\,\rangle_{\tilde{\mathcal{H}}} = \langle\, \pi\, \delta_{e,x}\,,\, \pi\, \delta_{e,y}\,\rangle_{\tilde{\mathcal{H}}} = B_k(\delta_{e,x}, \delta_{e,y}) = $
$= [k(e,e)x, y] = [Jx, y] = \langle x, y \rangle_J, \; \forall x, y \in \mathcal{K}$.

We introduce the "group of translations" in the following way: for each $t \in \Gamma$ consider the operator $U(t) : E' \to E'$ defined by $U(t)\pi h = \pi h_t$. Since condition (b) holds, we obtain $\langle U(t)\pi h, U(t)\pi h \rangle_{E'} = \langle \pi h_t, \pi h_t \rangle_{E'} = B_k(h_t, h_t) \leq \rho(t) B_k(h,h) = $
$= \rho(t)\,\langle\, \pi h\,,\, \pi h\,\rangle_{E'}, \; \forall h \in E$. This allows us to extend $U(t)$ to a bounded operator $U(t) : \tilde{\mathcal{H}} \longrightarrow \tilde{\mathcal{H}}$ such that $\| U(t) \| \leq \sqrt{\rho(t)}$. It is easy to see that $\{\, U(t) \,/\, t \in \Gamma \,\} \subset$ $\mathcal{L}(\tilde{\mathcal{H}})$ is a group of operators (i.e $U(ts) = U(t)U(s)\; \forall t, s \in \Gamma$, and $U(e) = I$). Since any $h \in E$ is of the form $h = \sum_{i=1}^{n} \delta_{t_i, x_i}$ then any element of E' is of the form $\sum_{i=1}^{n} \pi\, \delta_{t_i, x_i} = \sum_{i=1}^{n} U(t_i)\pi\delta_{e,x_i} = \sum_{i=1}^{n} U(t_i)\tau x_i$. It follows that our construction has the minimality property (ii), that is, $\tilde{\mathcal{H}} = \bigvee_{t \in \Gamma} U(t)\,\mathcal{H}$ where $\mathcal{H} := \tau\,\mathcal{K}$.

In this step we are going to introduce an indefinite scalar product in the vectorial space $\tilde{\mathcal{H}}$ using the form B_γ. Condition (b) implies that (see [4], Lemma IV.1.1) $\mid B_\gamma(h,h') \mid \,\leq\, 2\,\lambda\, B_k(h,h)\, B_k(h',h'), \; \forall\, h, h' \in E$. It follows that $(\pi\, h, \pi\, h') :=$

$B_\gamma(h, h')$ determines a "well-defined" indefinite scalar product in the quotient E' which can also be extended to all $\tilde{\mathcal{H}}$ in such a way that

$$(2.3) \qquad |\,(\,f\,,\,g\,)\,| \le 2\,\lambda\,\|\,f\,\|\,\|\,g\,\|, \qquad f,g \in \tilde{\mathcal{H}}.$$

If $x, y \in \mathcal{K}$ then $(\tau x, \tau y)_{\tilde{\mathcal{H}}} = (\pi \delta_{e,x}, \pi \delta_{e,y})_{\tilde{\mathcal{H}}} = B_\gamma(\delta_{e,x}, \delta_{e,y}) = [x, y]$. So

$$(2.4) \qquad (\,\tau x\,,\,\tau y\,)_{\tilde{\mathcal{H}}} = [\,x\,,\,y\,], \qquad x, y \in \mathcal{K}.$$

We had already shown that for each $t \in \Gamma$, $U(t)$ is a continuous operator in $\tilde{\mathcal{H}}$. Moreover, since B_γ is an invariant traslation form, we have $(U(t)\pi h, U(t)\pi h) = (\pi h_t, \pi h_t) = B_\gamma(h_t, h_t) = B_\gamma(h, h) = (\pi h, \pi h)$, $\forall h \in E$. It follows that $U(t)$ preserves the indefinite product:

$$(2.5) \qquad (\,U(t)\,f\,,\,U(t)\,f\,) = (\,f\,,\,f\,), \qquad f \in \tilde{\mathcal{H}}.$$

According to (2.3), the sesquilinear form $(.,.)$ is bounded. Then there exists an operator $G \in \mathcal{L}(\tilde{\mathcal{H}})$, $G^\times = G$, such that $(f, g) = \langle Gf, g \rangle_{\tilde{\mathcal{H}}}$, $\forall f, g \in \tilde{\mathcal{H}}$. It is known, see [4], Theorem 1.3, that $(\tilde{\mathcal{H}}, (.,.))$ is a Krein space and that the norm induced by any of its fundamental symmetries is equivalent to the norm $\|\ \|$ if and only if the operator G is boundly invertible. It is easy to see that our last hypothesis (d) ensures that condition on G. By (2.2) and (2.4) the operator τ is a continuous isometry in $\tilde{\mathcal{H}}$. It follows that $\mathcal{H} = \tau \mathcal{K}$ is a regular subspace of $\tilde{\mathcal{H}}$.

Finally, if $t \in \Gamma$, $x, y \in \mathcal{K}$ we have $[\gamma(t)x, y] = B_\gamma(\delta_{t,x}, \delta_{e,y}) = (\pi \delta_{t,x}, \pi \delta_{e,y})_{\tilde{\mathcal{H}}}$
$= (U(t)\pi\delta_{e,x}, \pi\delta_{e,y})_{\tilde{\mathcal{H}}} = (U(t)\tau x, \tau y)_{\tilde{\mathcal{H}}} = (P_{\mathcal{H}}^{\tilde{\mathcal{H}}} U(t)\tau x, \tau y)_{\mathcal{H}} = [\tau^{-1} P_{\mathcal{H}}^{\tilde{\mathcal{H}}} U(t)\tau x, y]$
and then

$$\gamma(t) = \tau^{-1}\, P_{\mathcal{H}}^{\tilde{\mathcal{H}}}\, U(t)\, \tau, \qquad t \in \Gamma. \qquad \square$$

Remarks

(1) If $(\mathcal{K}, [.,.])$ is a Hilbert space then we obtain Arocena's result. In addition, if γ is a function of positive type, we can take $k(t, s) = \gamma(s^{-1}t)$, $\tilde{\mathcal{H}}$ becames a Hilbert space and Theorem 2.1 reduces to Naimark's theorem.

(2) If $\{U'(t)/t \in \Gamma\} \subset \mathcal{L}(\tilde{\mathcal{H}}')$ is another minimal unitary dilation of γ then, in a routine way, we obtain a well defined and densely defined isometry V on $\tilde{\mathcal{H}}$ having dense range in $\tilde{\mathcal{H}}'$ such that $V U(t)\tau h(t) = U'(t)\tau' h(t), t \in \Gamma$.

(3) If Γ is a topological group, ρ is a continuous function and k is continuous as a function of two variables with respect to the weak topology of operators, then, it is to see that $\{U(t) / t \in \Gamma\}$ is strongly continuous.

3. DILATION OF A BOUNDED GENERAL OPERATOR

Fix a Krein space $(\mathcal{K}, [.,.])$ and an operator $T \in \mathcal{L}(\mathcal{K})$. We define, on the group of integers Z, the $\mathcal{L}(\mathcal{K})$-valued function

$$\gamma(n) = \begin{cases} T^n, & \text{if } n \ge 0 \\ (T^*)^{-n}, & \text{if } n < 0 \end{cases}$$

and we consider the form

$$B_\gamma : E \times E \longrightarrow C \ / \ B_\gamma(h, h') = \sum_{n,m \in Z} [\,\gamma(n-m)\,h(n)\,,\,h'(m)\,]$$

where, in this case, E is the set of the sequences $h = (h(n))_{n \in Z}$ with finite support. Following the ideas in [9], we are going to construct a kernel k which satisfies the conditions (a), (b), (c) and (d) of Theorem 2.1. The well-known change of variables:

$$g(n) = \begin{cases} \sum_{k=n}^{\infty} T^{k-n} h(k), \; if \; n > 0 \\ \sum_{k=-\infty}^{-1} (T^*)^{-k} h(k) + h(0) + \sum_{k=1}^{\infty} T^k h(k), \; if \; n = 0 \\ \sum_{k=-\infty}^{n} (T^*)^{n-k} h(k), \; if \; n < 0 \end{cases}$$

determines a bijection $h = (h(n))_{n \in Z} \mapsto g = (g(n))_{n \in Z}$ from E onto E whose inverse is given by

$$h(n) = \begin{cases} g(n) - T\, g(n+1)\,, \quad n > 0 \\ g(0) - T\, g(1) - T^*\, g(-1)\,, \quad n = 0 \\ g(n) - T^*\, g(n-1)\,, \quad n < 0 \end{cases}$$

and such that:

$$B_\gamma(h, h') = \sum_{n=-\infty}^{\infty} \langle\, \mathcal{A}(n,n)\, g(n)\,,\, g'(n)\, \rangle_J\,, \qquad h, h' \in E.$$

where $\mathcal{A}$ is the matrix defined by

$$\mathcal{A}(n,m) = \begin{cases} J\, A_T\,, \; if \; n = m > 0 \\ J\, A_{T^*}\,, \; if \; n = m < 0 \\ J\,, \; if \; n = m = 0 \\ 0\,, \; otherwise. \end{cases}$$

where A_T denotes the operator $A_T = I - T^*T$. We have $\mathcal{A}(n,n)^\times = \mathcal{A}(n,n)$. The espectral theorem for self-adjoints operators on Hilbert space allows us to write

$$\mathcal{A}(n,n) = \int_{-\infty}^{\infty} \lambda\, dE_n(\lambda)$$

where $E_n(.)$ is a spectral measure on $\mathcal{K}_J$. As it is known, the following operators plays an important role

$$\mid \mathcal{A}(n,n) \mid = \int_{-\infty}^{\infty} \mid \lambda \mid\, dE_n(\lambda), \quad sgn\, \mathcal{A}(n,n) = \int_{-\infty}^{\infty} (\, sgn\, \lambda\,)\, dE_n(\lambda),$$

$$\mid \mathcal{A}(n,n) \mid^{1/2} = \int_{-\infty}^{\infty} \mid \lambda \mid^{1/2}\, dE_n(\lambda).$$

Let us observe that $|\mathcal{A}(n,n)| = |A_T|$ (the positive square root of $(A_T)^\times A_T$), if $n > 0$; $|\mathcal{A}(n,n)| = |A_{T^*}|$ (the positive square root of $(A_{T^*})^\times A_{T^*}$), if $n < 0$ and $|\mathcal{A}(n,n)| = |J| = I$ if $n = 0$.

If $\prec, \succ$ denotes the product in E given by $\prec h, h' \succ := \sum_{k=1}^{\infty} \langle\, h(k)\,,\, h'(k)\, \rangle_J$ and $\mathcal{C}$ is the matrix of the application $h = (h(n))_{n \in Z} \mapsto g = (g(n))_{n \in Z}$ then we have $\mid B_\gamma(h,h) \mid \leq \sum_{-\infty}^{\infty} \langle\, | \mathcal{A}(n,n) |\, g(n)\,,\, g'(n)\, \rangle_J = \prec |\mathcal{A}|\, g, g \succ = \prec (\mathcal{C}^* |\mathcal{A}| \mathcal{C})\, h,\, h \succ = \sum_{n,m \in Z} \langle\, (\mathcal{C}^* |\mathcal{A}| \mathcal{C})(m,n)\, h(n)\,,\, h(m)\, \rangle_J = \sum_{n,m \in Z} [\, J\, (\mathcal{C}^* |\mathcal{A}| \mathcal{C})(m,n)\, h(n)\,,\, h(m)\,].$ It follows that

$$\mid B_\gamma(h,h) \mid \leq B_k(h,h), \qquad h \in E$$

with $k : Z \times Z \to \mathcal{L}(\mathcal{K})$ the operator-valued kernel defined by

(3.6) $$k(n,m) = J\,(\,C^* \mid \mathcal{A} \mid C\,)\,(m,n), \qquad m,n \in Z.$$

By construction, the kernel k satisfies the condition (b) of Theorem 2.1. Computing the product of matrices we obtain

$$k(n,m) = \begin{cases} T^{-n} J (T^*)^{-m} + \sum_{j=n\vee m}^{-1} T^{j-n} J \mid A_{T^*} \mid (T^*)^{j-m}, \ if \ n,m \leq -1 \\ T^{-n} J T^m, \ if \ n \leq 0, \ m \geq 0 \\ (T^*)^n J (T^*)^{-m}, \ if \ n \geq 0, \ m \leq 0 \\ (T^*)^n J T^m + \sum_{j=1}^{n \vee m} (T^*)^{n-j} J \mid A_T \mid (T^*)^{m-j}, \ if \ n,m \geq 1 \end{cases}$$

It follows that $k(0,0) = J$ and $k(n,m)^* = k(m,n)$, $\forall\, n,m \in Z$. If $h \in E$ is such that $B_k(h,h) \neq 0$ and $g = Ch$, defining $g' \in E$ by $g'(n) = (\,sgn\,\mathcal{A}(n,n)\,)\,g(n)$, $\forall n \in Z$, we obtain $B_k(h',h') = \prec \mid \mathcal{A} \mid g',\, g' \succ = B_k(h,h)$. Then $\mid B_\gamma(h,h') \mid = \sum_{n=-\infty}^{\infty} \langle\,\mathcal{A}(n,n)\,g(n),\,(sgn\mathcal{A}(n,n))\,g(n)\,\rangle_J = B_k(h,h) = B_k(h,h)^{1/2} B_k(h',h')^{1/2}$. So the condition (d) of the theorem is also satisfied.

Now, let $h \in E$, $g = C\,h$ and $\bar{g} = C\,h_1$ (h_1 is the function in E such that $h_1(n) = h(n-1)$, $n \in Z$). We have:

$$\bar{g}(n) = \begin{cases} g(n-1) & if \ n < 0 \\ T\,g(0) + A_{T^*}\,g(-1) & if \ n = 0 \\ g(0) - T^*\,g(-1) & if \ n = 1 \\ g(n-1) & if \ n > 1 \end{cases}$$

The vector space E can be viewed as a subspace of $l^2(Z, \mathcal{K}_J) := \{x : Z \to \mathcal{K} \,/\, \sum_{-\infty}^{\infty} \parallel x(n) \parallel_J^2 < \infty\}$. If $E_{|\mathcal{A}|}$ is the vector space E equipped with the $|\mathcal{A}|$-semidefinite form $\langle\,h,\,h'\,\rangle_{|\mathcal{A}|} = \prec |\mathcal{A}|\,h,\,h' \succ$, then $E_{|\mathcal{A}|}$ can be regarded as a semidefinite inner product space whose topology admits that of $l^2(Z, \mathcal{K}_J)$ as a majorant. It follows that the linear map $g \mapsto \bar{g}$ is $|\mathcal{A}|$-continuous. Thus, there exists a constant $\alpha > 0$ (which depends only on T) such that $B_k(h_1,h_1) = \,= \prec |\mathcal{A}|\,\bar{g},\,\bar{g} \succ \leq \alpha \prec |\mathcal{A}|\,g,\,g \succ \leq \alpha\,B_k(h,h)$. Analogously, there exists a constant $\beta > 0$ such that $B_k(h_{-1},h_{-1}) \leq \beta\,B_k(h,h)$. Setting $\rho = max\,\{\alpha,\beta\}$ it follows that $B_k(h_u,h_u) \leq \rho^{|u|}\,B_k(h,h)$, $\forall\,h \in E$, $\forall u \in \Gamma$. Thus Theorem 2.1 allows us to ensure the existence of a minimal unitary dilation of T.

Let us recall the so called *canonical* minimal unitary dilation of T. Define the operators

$$D_T = \mid J\,A_T \mid^{1/2} = \mid J - JT^*T \mid^{1/2} = \mid J - T^{\times} JT \mid^{1/2}.$$

$$J_T = sgn\,(J\,A_T) = sgn\,(J - JT^*T) = sgn\,(J - T^{\times} JT).$$

D_T is referred to as a defect operator of T. The corresponding defect space is

$$\mathcal{D}_T = \overline{D_T\,\mathcal{K}}$$

The defect space may be viewed as a Krein space in the inner product

$$[\,x,\,y\,]_{\mathcal{D}_T} = \langle\,J(T)\,x,\,y\,\rangle_J, \qquad x,y \in \mathcal{D}_T.$$

We may in a similar manner define $D_{T^{\times}}$, $J_{T^{\times}}$ and $\mathcal{D}_{T^{\times}}$. If we consider the Krein space

$$\tilde{\mathcal{H}}_c := \ldots\ldots \oplus \mathcal{D}_{T^{\times}} \oplus \mathcal{D}_{T^{\times}} \oplus \mathcal{K} \oplus \mathcal{D}_T \oplus \mathcal{D}_T \oplus \ldots\ldots$$

and chose

$$J_c := \ldots \oplus J_{T^\times} \oplus J_{T^\times} \oplus J \oplus J_T \oplus J_T \oplus \ldots$$

as a fundamental symmetry on $\tilde{\mathcal{H}}_c$ then (see [5], 1.5 and [8], Theorem 3.1.6) there exists an operator $L \in \mathcal{L}(\mathcal{D}_{T^\times}, \mathcal{D}_T)$ such that the operator $U_c : \tilde{\mathcal{H}}_c \to \tilde{\mathcal{H}}_c$ defined by

$$U_c = \begin{pmatrix}
& & & & \cdots & & & & \\
\cdots & I & 0 & 0 & 0 & 0 & 0 & 0 & \cdots \\
\cdots & 0 & I & 0 & 0 & 0 & 0 & 0 & \cdots \\
\cdots & 0 & 0 & D_{T^\times} & T & 0 & 0 & 0 & \cdots \\
\cdots & 0 & 0 & L & D_T & 0 & 0 & 0 & \cdots \\
\cdots & 0 & 0 & 0 & 0 & I & 0 & 0 & \cdots \\
\cdots & 0 & 0 & 0 & 0 & 0 & I & 0 & \cdots \\
& & & & \cdots & & & &
\end{pmatrix}$$

is a minimal unitary dilation of T wich is called the canonical dilation of T. This dilation gives rise to a kernel k_c wich verifies the conditions (a) throught (d) of the Theorem 2.1 and wich can be obtained using (2.1). The computation of (2.1) yields

$$k_c(n,m) \;=\; P_{\mathcal{K}}^{\tilde{\mathcal{H}}_c} \, (U^*)^m \, J_c \, U^n \, |_{\mathcal{K}} \;=$$

$$= \begin{cases}
T^{-n} J (T^*)^{-m} + \sum_{j=n \vee m}^{-1} T^{j-n} D_{T^\times} J_{T^\times} (D_{T^\times})^* (T^*)^{j-m}, & \text{if } n,m \leq -1 \\
T^{-n} J T^m, & \text{if } n \leq 0, \ m \geq 0 \\
(T^*)^n J (T^*)^{-m}, & \text{if } n \geq 0, \ m \leq 0 \\
(T^*)^n J T^m + \sum_{j=1}^{n \vee m} (T^*)^{n-j} (D_T)^* J_T D_T (T^*)^{m-j}, & \text{if } n,m \geq 1
\end{cases}$$

and then k_c equals the kernel k given by (3.6) since the following identities holds

$$(D_T)^* J_T D_T = J D_T^\times D_T = J \mid J A_T \mid = J \mid A_T \mid$$

$$D_{T^\times} J_{T^\times} (D_{T^\times})^* = D_{T^\times} D_{T^\times} J = \mid A_{T^*} J \mid \; J = J \mid A_{T^*} \mid J \; J = J \mid A_{T^*} \mid$$

It follows that our proof of the existence of at least one minimal unitary dilation of the operator T based on the Theorem 2.1 and using the kernel (3.6) gives rise to the canonical dilation of T.

REFERENCES

1. T. Ando, *Linear operators on Krein spaces*, Hokkaido University , Research Institute of Applied Electricity, Division of Applied Mathematics, Sapporo, 1979.
2. R. Arocena, *Scattering functions, Fourier transforms of measures, realization of linear systems and dilations of operators to Krein spaces: a unified approach* , Publ. Math. d'Orsay 85-02 (1985), 1-55.
3. T. Azizov and I. Iokhvidov, *Linear Operators in Spaces with Indefinite Metric*, English transl., Wiley, New York, 1989.
4. J. Bognár, *Indefinite Inner Product Spaces*, Springer, Berlin-New York, 1974.
5. T. Constantinescu and A. Gheondea, *Minimal signature in lifting of operators I*, J. Operator Theory 22, 1989, 345-367.
6. Ch. Davis, *J-unitary dilation of a general operator*, Acta Sci. Math. (Szeged), 31 (1970), 75-86.
7. Ch. Davis, *Dilation of uniformly continuous semigroups*, Rev. Roumaine Math. Pures Appl., 15 (1970), 975-983.
8. M. Dritschel and J. Rovnyak, *Extension theorems for contraction operators on Krein spaces*, Oper. Theory: Adv. Appl., Vol. 47, Birkhauser, Basel-Boston, (1990), 221-305.

9. S. Marcantognini, *A new proof of a Theorem of Davis*, Acta Científica Venezolana, Vol. 37, No.1 (1986), 102-103.

10. M.A. Naimark, *Positive definite operator functions on a commutative group*, Bulletin (Izvestiya) Acad. Sci. URSS (ser math.), 7 (1943), 237-244. (Russian, with English summary).

11. B. Sz.- Nagy, *Sur les contractions de l'espace de Hilbert*, Acta. Sci. Math., 15 (1953), 87-92.

12. B. Sz.- Nagy and C. Foias, *Harmonic Analysis of Operators on Hilbert Space*, North Holland, New York, (1970).

CENTRO DE MATEMATICA, FACULTAD DE CIENCIAS, UNIVERSIDAD DE LA REPUBLICA, EDUARDO ACEVEDO 1139, MONTEVIDEO, URUGUAY Y DEPARTAMENTO DE MÉTODOS CUANTITATIVOS, FACULTAD DE CIENCIAS ECONÓMICAS Y ADMINISTRACIÓN, MONTEVIDEO, URUGUAY

E-mail address: FPELAEZ@cmat.edu.uy

Contemporary Mathematics
Volume **189**, 1995

Approximation by analytic operator-valued functions

V.V. Peller[*]

Abstract

In [PY1] it was proved that any matrix-valued function Φ in $H^\infty + C$ has a unique superoptimal approximation by bounded analytic functions, i.e., there exists a unique function Q in H^∞ which minimizes not only the supremum of $\|\Phi(\zeta) - Q(\zeta)\|$, $\zeta \in \mathbb{T}$, but also the suprema of all further singular values of the matrices $\Phi(\zeta) - Q(\zeta)$, $\zeta \in \mathbb{T}$. In this paper we obtain a similar result for operator-valued functions and study properties of such approximations.

Introduction

A classical result ([Kh], see also [AAK]) claims that if φ is a scalar function on the unit circle $\mathbb{T}$, which belongs to the class $H^\infty + C$, then there exists a unique function $f \in H^\infty$ such that

$$\operatorname{dist}_{L^\infty(\mathbb{T})}\{\varphi, H^\infty\} = \|\varphi - f\|_{L^\infty(\mathbb{T})}.$$

[*]1991 *Mathematics Subject Classification.* Primary 47B35, 30D50.
The author was supported by a grant from NSF

However in the matrix-valued case it is easy to see that the corresponding statement does not hold (see [PY1] for more details). In this case $\|\varphi - f\|_\infty$ means by definition

$$\sup_{\zeta \in \mathbb{T}} \|\varphi(\zeta) - f(\zeta)\|,$$

where the norm is the operator norm on the finite-dimensional Hilbert space $\mathbb{C}^n$. However it was shown in [PY1] that if we minimize not only the L^∞-norm of $\|\Phi(\zeta) - Q(\zeta)\|$, $\zeta \in \mathbb{T}$, but also the L^∞-norms of all further singular values, then we get uniqueness.

Let us state the result more precisely. Put

$$\Omega_0 = \{Q \in H^\infty : Q \text{ minimizes } \operatorname{ess\,sup}_{\zeta \in \mathbb{T}} \|\Phi(\zeta) - Q(\zeta)\|\};$$

$$\Omega_j = \{Q \in \Omega_{j-1} : Q \text{ minimizes } \operatorname{ess\,sup}_{\zeta \in \mathbb{T}} s_j(\Phi(\zeta) - Q(\zeta))\}.$$

Here s_j is the jth singular value, i.e., the distance to the set of operators (matrices) of rank at most j. Put

$$t_j \stackrel{\text{def}}{=} \operatorname{ess\,sup}_{\zeta \in \mathbb{T}} s_j(\Phi(\zeta) - Q(\zeta)), \quad Q \in \Omega_j.$$

It was proved in [PY1] that if Φ is an $n \times m$ matrix function in $H^\infty + C$, then the set $\Omega_{\min\{n,m\}-1}$ consists of one function, which is called the superoptimal approximation of Φ by H^∞ functions. It was also proved in [PY1] that if Q is the superoptimal approximation of Φ, then for all j the singular values $s_j(\Phi(\zeta) - Q(\zeta))$ are constant almost everywhere on $\mathbb{T}$ and coincide with the above numbers t_j. The t_j are called the *superoptimal singular values*.

The method of [PY1] is based on certain special factorizations (thematic factorizations) and on the analyticity property of minors of unitary-valued completions of inner matrix columns.

Later Treil [T] gave a different approach to this problem which also allowed him to cover the case of operator-valued matrices (the case of infinite dimensional matrix-valued functions).

In this paper we treat the case of infinite-dimensional matrix-valued functions by using the approach based on thematic factorizations [PY1].

However in the infinite-dimensional situation the property of analyticity of minors of inner column functions does not help and we apply another method to get a thematic factorization.

To define a superoptimal approximation in the infinite-dimensional case we have to consider the infinite sequence of the sets Ω_j defined above. An operator-valued function Q in H^∞ is called a *superoptimal approximation* if $Q \in \cap_{j \geq 0} \Omega_j$.

Recall that given a function $\Phi \in L^\infty(\mathcal{B})$ ($\mathcal{B}$ is the space of bounded linear operators from a Hilbert space $\mathcal{H}$ to a Hilbert space $\mathcal{K}$), the *Toeplitz operator* T_Φ is defined on the Hardy class $H^2(\mathcal{H})$ by

$$T_\Phi f = \mathbb{P}_+ \Phi f, \quad f \in H^2(\mathcal{H}),$$

where $\mathbb{P}_+$ is the orthogonal projection on $L^2(\mathcal{K})$ onto $H^2(\mathcal{K})$. The *Hankel operator* H_Φ is defined on $H^2(\mathcal{H})$ by

$$H_\Phi f = \mathbb{P}_- \Phi f, \quad f \in H^2(\mathcal{H}),$$

where $\mathbb{P}_- = I - \mathbb{P}_+$.

It is well known that $\|H_\Phi\| = \mathrm{dist}\,_{L^\infty}\{\Phi, H^\infty(\mathcal{B})\}$ and H_Φ is compact if and only if $\Phi \in H^\infty(\mathcal{B}) + C(\mathcal{C})$, where $\mathcal{C}$ is the space of compact operators on Hilbert space (these are operator-valued analogs of Nehari's and Hartman's theorems, see [Pa]). As in the finite-dimensional case we are going to prove the uniqueness of a superoptimal approximation under the assumption that H_Φ is compact.

Recall several notions which will be used throughout the paper. Given a matrix function Φ (of finite or infinite size), we denote by Φ^t its transpose. A bounded analytic matrix function Φ in $\mathbb{D}$ is called *inner* if $\Phi(\zeta)$ is isometric for almost all $\zeta \in \mathbb{T}$. It is called *outer* if the operator of multiplication by Φ on the corresponding Hardy class H^2 of vector functions has dense range. Finally Φ is called *co-outer* if Φ^t is outer. We refer the reader to [H], [SF] for more details.

In Section 1 we construct unitary completions of inner column functions which have certain important properties. They are similar to the ones constructed in [PY1] in the finite-dimensional case. We introduce the class of the so-called thematic matrix functions and we prove that a Toeplitz operator whose symbol is thematic must have trivial kernel and dense range.

In Section 2 using the results of Section 1 we construct certain special factorizations which are similar to the thematic factorizations constructed in [PY1] in the finite-dimensional case. However the method of construction of such factorizations developed in [PY1] uses the fact that all minors of thematic matrices on the first column are in H^∞. This does not work in the infinite-dimensional case. The construction given in Section 2 is based on the fact that the Hankel operator whose symbol is a thematic function, naturally related to the problem, is invertible.

In Section 3 we prove an inequality between the superoptimal singular values and the singular values of the Hankel operator which is similar to the finite-dimensional case (see [PY2]). The proof is based on a finite-dimensional approximation which reduces the case to the finite-dimensional one.

Finally, in Section 4 we use the inequality proved in Section 3 to prove the uniqueness of a superoptimal approximation for functions in $H^\infty(\mathcal{B}) + C(\mathcal{C})$ and study some properties of superoptimal approximation.

I am grateful to H.Bercovici and V.I.Vasyunin for helpful discussions.

1. Unitary completions of inner columns

Let $\varphi = \{\varphi_j\}_{j\geq 0}$ be an inner column. Consider the space $L \overset{\text{def}}{=}$ $\operatorname{Ker} T_{\varphi^t}$. Clearly it is invariant under multiplication by z. So by the Beurling-Lax theorem (see [N]) $L = \Theta H^2(\mathcal{K})$ ($\mathcal{K} = \ell^2$ or $\mathcal{K} = \mathbb{C}^m$, $m \in \mathbb{Z}_+$) for some inner function Θ. We need the following property of Θ whose proof is suggested by Vasyunin.

Lemma 1.1. *Let $\varphi = \{\varphi_j\}_{j\geq 0}$ be an inner column and let $L = \operatorname{Ker} T_{\varphi^t} = \Theta H^2(\mathcal{K})$. Then $\dim \operatorname{Ker} \Theta^*(\zeta) = 1$ for almost all $\zeta \in \mathbb{T}$.*

Proof. Let us first show that $\dim \operatorname{Ker} \Theta^*(\zeta) \geq 1$. Consider the block matrix

$$U = \begin{pmatrix} \bar{\varphi} & \Theta \end{pmatrix}.$$

It is easy to see that $U(\zeta)$ is isometric a.e. on $\mathbb{T}$. It follows that the columns of $\Theta(\zeta)$ are orthogonal to $\bar{\varphi}(\zeta)$ a.e. on $\mathbb{T}$. Therefore $\bar{\varphi}(\zeta) \in \operatorname{Ker} \Theta^*(\zeta)$ a.e., which proves that $\dim \operatorname{Ker} \Theta^*(\zeta) \geq 1$.

To show that $\dim \operatorname{Ker} \Theta^*(\zeta) \leq 1$ we assume without loss of generality that $\varphi_0 \neq \mathbb{O}$. Consider the matrix

$$
G = \begin{pmatrix}
-\varphi_1 & -\varphi_2 & -\varphi_3 & \cdots \\
\varphi_0 & 0 & 0 & \cdots \\
0 & \varphi_0 & 0 & \cdots \\
0 & 0 & \varphi_0 & \cdots \\
\vdots & \vdots & \vdots & \ddots
\end{pmatrix}.
$$

It is easy to see that $GH^2(\ell^2) \subset L = \Theta H^2(\mathcal{K})$. The proof will be completed if we show that $\dim \operatorname{Ker} G^t(\zeta) \leq 1$ a.e. on $\mathbb{T}$. Assume that $c = \{c_j\}_{j \geq 0} \in \operatorname{Ker} G^t(\zeta)$. We have

$$
\begin{pmatrix}
-\varphi_1(\zeta) & \varphi_0(\zeta) & 0 & 0 & \cdots \\
-\varphi_2(\zeta) & 0 & \varphi_0(\zeta) & 0 & \cdots \\
-\varphi_3(\zeta) & 0 & 0 & \varphi_0(\zeta) & \cdots \\
\vdots & \vdots & \vdots & \vdots & \ddots
\end{pmatrix}
\begin{pmatrix}
c_0 \\
c_1 \\
c_2 \\
\vdots
\end{pmatrix}
=
\begin{pmatrix}
0 \\
0 \\
0 \\
\vdots
\end{pmatrix}. \tag{1.1}
$$

If $\varphi_0(\zeta) \neq 0$, we have from (1.1)

$$
c_j = \frac{\varphi_j(\zeta)}{\varphi_0(\zeta)} c_0, \quad j \geq 1,
$$

which proves that the c_j with $j \geq 1$ are uniquely determined by c_0 and so $\dim \operatorname{Ker} G^t(\zeta) \leq 1$. ∎

Lemma 1.2. *Under the hypothesis of Lemma 1.1 the function* Θ *is co-outer.*

Proof. Assume the contrary. Then $\Theta^t = \Upsilon F$, where F is an outer function and Υ is an inner function . It follows that $\Theta = F^t \Upsilon^t$. Multiplying this equality by $(\Upsilon^t)^*$ on the right, we find that $F^t = \Theta(\Upsilon^t)^*$. Since $(\Upsilon^t)^*(\zeta)$ is isometric a.e. on $\mathbb{T}$, it follows that F^t is inner.

Consider the space $F^t H^2(\ell^2)$. Let us show that it is contained in $\mathrm{Ker}\, T_{\varphi^t}$. Note first that

$$\Theta(\zeta)c \perp \bar{\varphi}(\zeta) \quad \text{a.e. on } \mathbb{T} \text{ for any } c \in \ell^2. \qquad (1.2)$$

Indeed this follows from the fact that $\Theta g \subset \mathrm{Ker}\, T_{\varphi^t}$, where $g(\zeta) \equiv c$. Let now $f \in H^2(\ell^2)$. We have

$$(F^t(\zeta)f(\zeta), \bar{\varphi}(\zeta))_{\ell^2} = \left(\Theta(\zeta)(\Upsilon^t)^*(\zeta)f(\zeta), \bar{\varphi}(\zeta) \right)_{\ell^2} = 0$$

by (1.2).

It follows that

$$F^t H^2(\ell^2) = \Theta(\Upsilon^t)^* H^2(\ell^2) \subset \Theta H^2(\ell^2).$$

Multiplying the last inclusion on the left by Θ^*, we obtain

$$(\Upsilon^t)^* H^2(\ell^2) \subset H^2(\ell^2).$$

Clearly this implies that Υ^t is a constant unitary matrix. ∎

Remark. Lemmas 1.1 and 1.2 allow us to obtain unitary-valued completions to inner columns which have important properties. Namely, let $\varphi = \{\varphi_j\}_{j \geq 0}$ be an inner column. Then there exists a unitary-valued matrix $\begin{pmatrix} \varphi & \Theta \end{pmatrix}$ such that Θ is inner and co-outer. We need the following procedure to construct such a completion. As before $L \overset{\text{def}}{=} \mathrm{Ker}\, T_{\varphi^t} = \Theta H^2(\ell^2)$. Then the orthogonal projection P_L onto L is given by $P_L f = \Theta \mathbb{P}_+ \Theta^* f$, $f \in H^2(\ell^2)$. Let C be a constant vector function. Then $P_L C = \Theta \mathbb{P}_+ \Theta^* C = \Theta \Theta^*(0)C$. Clearly $\overline{P_L C}$ is pointwise orthogonal to φ. Since $\Theta(\zeta)$ is isometric on $\mathbb{T}$, it follows that for any constant vector functions C_1 and C_2

$$(P_L C_1, P_L C_2)_{H^2(\ell^2)} = (P_L C_1(\zeta), P_L C_2(\zeta))_{\ell^2} \quad \text{a.e.}$$

It is easy to see that a constant function C is in $\mathrm{Ker}\, P_L$ if and only if $C = \gamma\varphi(0)$, $\gamma \in \mathbb{C}$. Since Θ is co-outer, it follows that $\Theta^*(0)$ has dense range in ℓ^2. Therefore to construct a desired unitary-valued completion we can choose constant functions C_j, $j \geq 0$, such that the $\Theta^*(0)C_j$ form an orthonormal basis in ℓ^2. Let U be a unitary operator on ℓ^2 such that

$\{U\Theta^*(0)C_j\}_{j\geq 0}$ is the standard orthonormal basis in ℓ^2. Let $\Xi \stackrel{\text{def}}{=} \Theta U^*$. Clearly Ξ is an inner function and $\Xi H^2(\ell^2) = \Theta H^2(\ell^2) = L$. It is easy to see now that

$$\Xi = \left(\begin{array}{cccc} P_L C_0 & P_L C_1 & P_L C_2 & \cdots \end{array} \right)$$

i.e. the columns of Ξ are the $P_L C_j$, $j \geq 0$.

Definition. A matrix function is called *thematic* if it is unitary-valued and has the form $\left(\begin{array}{cc} \varphi & \Xi \end{array} \right)$, where φ is a column function which is inner and co-outer and Ξ is a matrix function which is also inner and co-outer.

The following theorem reveals an important property of thematic functions.

Theorem 1.3. *Let Ω be a thematic matrix function of the form $\Omega = \left(\begin{array}{cc} \varphi & \Xi \end{array} \right)$, where φ is a co-outer inner column and Ξ is a co-outer inner function. Then the Toeplitz operator T_Ω on $H^2(\ell^2)$ has trivial kernel and dense range.*

Proof. Let us first show that $\operatorname{Ker} T_\Omega = \{\mathbb{O}\}$. Let $f \in \operatorname{Ker} T_\Omega$. It follows that

$$\Omega f = \left(\begin{array}{c} \varphi_0 f_0 + \sum_{k\geq 0} \overline{\xi_{0k}} f_k \\ \varphi_1 f_0 + \sum_{k\geq 0} \overline{\xi_{1k}} f_k \\ \vdots \end{array} \right) \in H^2_-(\ell^2), \qquad (1.3)$$

where $f = \{f_j\}_{j\geq 0}$, $\varphi = \{\varphi_j\}_{j\geq 0}$, and $\Xi = \{\xi_{jk}\}_{j\geq 0, k\geq 1}$.

Since Ω is unitary-valued, it follows from (1.3) that

$$\sum_{j\geq 0} \overline{\varphi_j} \left(\varphi_j + \sum_{k\geq 0} \overline{\xi_{jk}} f_k \right) = \sum_{j\geq 0} |\varphi_j|^2 f_0 = f_0 \in H^2_-,$$

which implies that $f_0 = \mathbb{O}$ and so $\bar{\Xi}\{f_k\}_{k\geq 1} \in H^2_-(\ell^2)$.

Since Ξ^t is an outer function, for each $j \geq 1$ there exists a sequence $\{g_n\}_{n\geq 0}$ in $H^2(\ell^2)$ such that $\Xi^t g_n$ converges to $\boldsymbol{e}_j$, where $\boldsymbol{e}_j$ is the function identically equal to the e_j. Therefore

$$g_n^* \Xi \{f_k\}_{k\geq 1} \to f_j, \quad j \geq 1,$$

and so $f_j \in H^2_-$, $j \geq 1$, which implies that $f_j = \mathbb{O}$.

Let us now prove that $\operatorname{Ker} T_{\Omega^*} = \{\mathbb{O}\}$. Let $f = \{f_k\}_{k \geq 0} \in \operatorname{Ker} T_{\Omega^*}$. We have

$$\begin{pmatrix} \varphi^* \\ \Xi^t \end{pmatrix} f = \begin{pmatrix} \varphi^* f \\ \Xi^t f \end{pmatrix} \in H^2_-(\ell^2). \tag{1.4}$$

Hence $\Xi^t f \in H^2_-(\ell^2)$. Since both Ξ^t and f are analytic, it follows that $\Xi^t f = \mathbb{O}$. Therefore the column $\{f_j(\zeta)\}_{j \geq 0}$ is orthogonal in ℓ^2 to the vectors $\gamma_k(\zeta) \overset{\text{def}}{=} \{\overline{\xi_{jk}}(\zeta)\}_{k \geq 0}$, $j \geq 1$, for almost all $\zeta \in \mathbb{T}$. Since the vectors $\varphi(\zeta)$ and $\gamma_k(\zeta)$, $j \geq 1$, form an orthonormal basis in ℓ^2, we have $f(\zeta) = \tau(\zeta)\varphi(z)$, $\zeta \in \mathbb{T}$, for some $\tau \in L^2$. It follows from (1.4) that

$$\varphi^* f = \tau \in H^2_-. \tag{1.5}$$

Now we can use the fact that φ is co-outer and so there exist sequences of functions $\{\lambda_j^{(n)}\}_{n \geq 1}$, $j \geq 0$, in H^2 such that for each n only finitely many $\lambda_j^{(n)}$ can be nonzero and $\sum_{j \geq 0} \lambda_j^{(n)} \varphi_j \to \mathbf{1}$ as $n \to \infty$. Therefore by (1.5), $\sum_{j \geq 0} \lambda_j^{(n)} f_j \to \tau$ which contradicts the fact that $\tau \in H^2_-$. ∎

Remark. It can also be shown that under the hypotheses of Theorem 1.3 the operator T_{Ω^t} has trivial kernel and dense range.

Theorem 1.4. *Let V be a thematic matrix. Then the operators $H_V^* H_V$ and $H_{V^*}^* H_{V^*}$ are unitarily equivalent.*

Proof. By Theorem 1.3 the Toeplitz operator T_V on $H^2(\ell^2)$ has trivial kernel and dense range. In [PK] it was shown that if v is a scalar unimodular function such that the Toeplitz operator T_v on H^2 has trivial kernel and dense range, then the operators $H_v^* H_v$ and $H_{v^*}^* H_{v^*}$ are unitarily equivalent. The proof given in [PK] also works for operator-valued functions which proves the result. ∎

Corollary 1.5. *Let $V = \begin{pmatrix} v & \Xi \end{pmatrix}$ be a thematic matrix function. Suppose that the Hankel operator $H_{\bar{v}}$ is compact. Then the Toeplitz operator T_V is invertible.*

Proof. Since $H_{\bar{v}}$ is compact, it follows that H_{V^*} is compact and by Theorem 1.4, H_V is compact. We have

$$T_V^* T_V = I - H_V^* H_V \quad \text{and} \quad T_{V^*}^* T_{V^*} = I - H_{V^*}^* H_{V^*},$$

and so T_V is Fredholm. The result follows now from Theorem 1.3. ■

The same reasoning proves that under the hypotheses of Corollary 1.5 the operator T_{V^t} is invertible (see the Remark after Theorem 1.3).

2. Factorizations

In this section we obtain certain factorizations of $\Phi - Q$ where $Q \in \Omega_m$ which are similar to the so-called thematic factorizations introduced in [PY1]. Then we shall prove an important inequality between the superoptimal singular values and the Hankel singular values that was established in [PY2] in the matrix (finite-dimensional) case.

The construction of such factorizations is similar to the finite-dimensional case (see [PY1]). We assume that $\Phi \notin H^\infty(\mathcal{B})$. Otherwise the situation is trivial. Let $v \in H^2(\ell^2)$ be a maximizing vector of H_Φ (such a vector exists since H_Φ is compact) and let $H_\Phi v = t_0 w$. Then as in [PY1] for the finite-dimensional case the columns v and $\bar{z}\bar{w}$ admit the following factorizations

$$ v = v_{(i)} \vartheta_1 h, \qquad \bar{z}\bar{w} = w_{(i)} \vartheta_2 h, $$

where the columns $v_{(i)}$ and $w_{(i)}$ are inner and co-outer h is a scalar outer function, and ϑ_1 and ϑ_2 are scalar inner functions. By Lemma 1.2 we can find unitary-valued completions of the form

$$ V = \left(\begin{array}{cc} v_{(i)} & \bar{\alpha} \end{array} \right), \qquad W = \left(\begin{array}{cc} w_{(i)} & \bar{\beta} \end{array} \right)^t, $$

where α and β are inner and co-outer. Put $u_0 \stackrel{\text{def}}{=} \bar{z}\bar{\vartheta}_1\bar{\vartheta}_2\bar{h}/h$.

Let now Q_0 be an arbitrary best approximation of Φ by bounded analytic functions, that is $\|\Phi - Q_0\|_\infty = \|H_\Phi\|$. Exactly as in [PY1] one can prove that

$$ W(\Phi - Q)V = \left(\begin{array}{cc} t_0 u_0 & \mathbb{O} \\ \mathbb{O} & \Phi_1 \end{array} \right), \qquad (2.1) $$

where Φ_1 is a function in $L^\infty(\mathcal{B})$, $\|\Phi_1\| \leq t_0$. One can prove by exactly the same method as in [PY1] in the finite matrix case that a function

$Q \in H^{\infty}(\mathcal{B})$ is a best approximation to Φ if and only if there exists a function $Q_1 \in H^{\infty}(\mathcal{B})$ such that

$$W(\Phi - Q)V = \begin{pmatrix} t_0 u_0 & \mathbb{O} \\ \mathbb{O} & \Phi_1 - Q_1 \end{pmatrix} \qquad (2.2)$$

and $\|\Phi_1 - Q_1\|_{\infty} \leq t_0$.

The factorization (2.1) reduces the problem of finding a superoptimal approximation to Φ to the corresponding problem for Φ_1. Namely Q is a superoptimal approximation to Φ if and only if Q_1 is a superoptimal approximation to Φ_1.

To repeat this procedure we have to prove that $\Phi_1 \in H^{\infty}(\mathcal{B}) + C(\mathcal{C})$. To prove this in the finite-dimensional case in [PY1] the following property of thematic matrices was used: all 2×2 minors of a thematic matrix on the first column are in H^{∞}. This implied that all entries of Φ_1 are in $H^{\infty} + C$.

However in the infinite-dimensional case even if we prove that all entries of Φ_1 are in $H^{\infty} + C$, this would not imply the compactness of H_{Φ_1}. So we are going to use another method which is based on Theorem 1.4.

Theorem 2.1. *Let $\Phi \in H^{\infty}(\mathcal{B}) + C(\mathcal{C})$ and suppose that (2.1) holds. Then $\Phi_1 \in H^{\infty}(\mathcal{B}) + C(\mathcal{C})$.*

Proof. Let us first prove that $u_0 \in QC$, and the Hankel operators $H_{v_{(i)}^*}$ and $H_{w_{(i)}^*}$ are compact. The proof is similar to the proof given in [PY1] in the finite-dimensional case. It follows from (2.2) that

$$w_{(i)}^t(\Phi - Q)v_{(i)} = t_0 u_0.$$

Since $w_{(i)}, v_{(i)} \in H^{\infty}$ and $\Phi - Q \in H^{\infty}(\mathcal{B}) + C(\mathcal{C})$, it follows that $u_0 \in H^{\infty} + C$. Since u_0 has the form $u_0 = \vartheta \bar{h}/h$, where ϑ is an inner function and h is an outer function in H^2, it is easy to see that the Toeplitz operator T_{u_0} on H^2 has dense range (see [PK]). This implies that $u_0 \in QC$ (see [PK]).

It is also easy to see from (2.2) that

$$\overline{u_0} w_{(i)}^t(\Phi - Q) = t_0 v_{(i)}^*.$$

It follows that $H_{v_{(i)}^*}$ is compact. To prove that $H_{w_{(i)}^*}$ is compact, we can consider the transposed matrix $(\Phi - Q)^t$ and interchange the roles of $v_{(i)}$ and $w_{(i)}$.

Let us show that $\Phi_1 \in H^\infty(\mathcal{B}) + C(\mathcal{C})$. Since $H_{v_{(i)}^*}$ is compact, it follows that H_{V^*} is compact. Then $H_{V^*}^* H_{V^*}$ is compact and by Theorem 1.4 the operator $H_V^* H_V$ must also be compact which implies that H_V is compact. Similarly H_W is compact. Therefore all three factors on the left-hand side of (2.1) belong to $H^\infty(\mathcal{B}) + C(\mathcal{C})$ and so does their product, which implies that $\Phi_1 \in H^\infty(\mathcal{B}) + C(\mathcal{C})$. ∎

If $\Phi_1 \in H^\infty(\mathcal{B})$, then $Q_1 = \Phi_1$ is the unique superoptimal approximation to Φ_1 and we stop the process. If $\Phi_1 \notin H^\infty(\mathcal{B})$ (which means that $t_1 \neq 0$), we take an arbitrary best approximation Q_1 and factorize $\Phi_1 - Q_1$ as we have done for $\Phi - Q_0$. If the process does not stop at the mth step, we obtain the following factorization for $\Phi - Q$ where Q is a function in Ω_{m-1}

$$\Phi - Q = \qquad\qquad (2.3)$$

$$W_0^* \begin{pmatrix} 1 & \mathbb{O} \\ \mathbb{O} & W_1^* \end{pmatrix} \cdots \begin{pmatrix} I_{m-1} & \mathbb{O} \\ \mathbb{O} & W_{m-1}^* \end{pmatrix} D \begin{pmatrix} I_{m-1} & \mathbb{O} \\ \mathbb{O} & V_{m-1}^* \end{pmatrix} \cdots \begin{pmatrix} 1 & \mathbb{O} \\ \mathbb{O} & V_1^* \end{pmatrix} V_0^*,$$

where

$$D = \begin{pmatrix} t_0 u_0 & 0 & \cdots & 0 & 0 \\ 0 & t_1 u_1 & \cdots & 0 & 0 \\ \vdots & \vdots & \ddots & \vdots & \vdots \\ 0 & 0 & \cdots & t_{m-1} u_{m-1} & 0 \\ 0 & 0 & \cdots & 0 & \Phi_m \end{pmatrix},$$

$\Phi_m \in H^\infty(\mathcal{B}) + C(\mathcal{C})$ and $\|\Phi_m\|_\infty \leq t_{m-1}$. Now we can obtain the following description of Ω_m: an H^∞ function Q belongs to Ω_m if and only if there exists a function Q_m such that it is a best approximation of Φ_m and $\Phi - Q$ is equal to the right-hand side of (2.3) with

$$D = \begin{pmatrix} t_0 u_0 & 0 & \cdots & 0 & 0 \\ 0 & t_1 u_1 & \cdots & 0 & 0 \\ \vdots & \vdots & \ddots & \vdots & \vdots \\ 0 & 0 & \cdots & t_{m-1} u_{m-1} & 0 \\ 0 & 0 & \cdots & 0 & \Phi_m - Q_m \end{pmatrix}.$$

If the sequence $\{t_m\}_{m \geq 0}$ does not vanish, we can continue this process indefinitely long and we obtain an infinite sequence of factorizations

(2.3). With such a sequence we can associate the sequence of *indices* $\{k_m\}_{m \geq 0}$, $k_m \overset{\text{def}}{=} -\text{wind}\, u_m$ (since $u_m \in QC$, the winding number is well-defined as $\text{wind}\, u_m = -\text{ind}\, T_{u_m}$; it is negative because u_m has the form $u_m = \bar{z}\bar{\vartheta}\bar{h}/h$, where ϑ is an inner function and h is an outer function in H^2).

Consider the *extended t-sequence*

$$t_0, \cdots, t_0, t_1, \cdots, t_1, t_2, \cdots,$$

where t_j is repeated k_j times. Denote the terms of this extended t-sequence by $\{\tilde{t}_j\}_{j \geq 0}$ (i.e. $\tilde{t}_0 = \cdots = \tilde{t}_{k_0 - 1} = t_0$, $\tilde{t}_{k_0} = \cdots = \tilde{t}_{k_0 + k_1 - 1} = t_1$, etc). If $t_m = 0$ for some m, the index k_m is not defined but we can assume that $k_m = 1$ in this case.

The extended t-sequence was defined in the finite-dimensional case in [PY1]. It was shown in [PY1] that the indices k_j do depend on the choice of factorizations. However in [PY2] it was shown that the sums of the indices that correspond to equal superoptimal singular values t_j are determined by the function Φ. In other words the extended t-sequence is uniquely determined by the function Φ.

I conjecture that the same is true in the infinite-dimensional case.

3. Superoptimal and Hankel singular values

In [PY2] the following inequality between the Hankel singular values and the superoptimal singular values was proved in the finite-dimensional case: $\tilde{t}_j \leq s_j(H_\Phi)$. In this section we prove the same inequality in the infinite-dimensional case.

Theorem 3.1. *Let $\Phi \in H^\infty(\mathcal{B}) + C(\mathcal{C})$. Then*

$$\tilde{t}_j \leq s_j(H_\Phi), \quad j \geq 0.$$

Corollary 3.2. *Under the hypotheses of Theorem 3.1*

$$t_j \leq s_j(H_\Phi), \quad j \geq 0.$$

Proof of Theorem 3.1. Let $n \in \mathbb{Z}_+$. Denote by P_n the orthogonal projection on ℓ^2 onto the linear span of the basis vectors $e_0, \cdots, e_n$, where $\{e_k\}_{k \geq 0}$ is the standard orthonormal basis of ℓ^2. Let us fix $m \geq 0$ and find $j \in \mathbb{Z}_+$ such that $\tilde{t}_j = t_m$. Consider the matrix D in the factorization (2.3). Let $D_n = P_n D P_n$, $n \geq m$. Then D_n has the form

$$D_n = \begin{pmatrix} t_0 u_0 & 0 & \cdots & 0 & 0 \\ 0 & t_1 u_1 & \cdots & 0 & 0 \\ \vdots & \vdots & \ddots & \vdots & \vdots \\ 0 & 0 & \cdots & t_{m-1} u_{m-1} & 0 \\ 0 & 0 & \cdots & 0 & \Phi^{(n)} \end{pmatrix},$$

Since $D \in H^\infty(\mathcal{B}) + C(\mathcal{C})$, it follows that $\|H_{\Phi^{(n)}} - H_{\Phi_m}\| \to 0$ and so $\|H_{\Phi^{(n)}}\| \to \|H_{\Phi_m}\| = t_m$ as $n \to \infty$. Let

$$\Psi^{(n)} \stackrel{\text{def}}{=}$$

$$W_0^* \begin{pmatrix} 1 & \mathbb{O} \\ \mathbb{O} & W_1^* \end{pmatrix} \cdots \begin{pmatrix} I_m & \mathbb{O} \\ \mathbb{O} & W_m^* \end{pmatrix} D_n \begin{pmatrix} I_m & \mathbb{O} \\ \mathbb{O} & V_m^* \end{pmatrix} \cdots \begin{pmatrix} 1 & \mathbb{O} \\ \mathbb{O} & V_1^* \end{pmatrix} V_0^*. \quad (3.1)$$

Since all matrix functions on the right-hand side of (3.1) belong to $H^\infty(\mathcal{B}) + C(\mathcal{C})$, it follows that $\|H_\Phi - H_{\Psi^{(n)}}\| \to 0$ and so $t_j(H_{\Psi^{(n)}}) \to t_j(H_\Phi)$. Therefore it is sufficient to show that $\|H_{\Phi^{(n)}}\| \leq s_j(H_{\Psi^{(n)}})$. Let us now fix $n \in \mathbb{Z}_+$.

Consider the matrix $\begin{pmatrix} I_m & \mathbb{O} \\ \mathbb{O} & W_m^* \end{pmatrix}$. If we keep its first $n+1$ columns and change the remaining columns arbitrarily, the product on the right-hand side of (3.1) does not change. Let $W_m^* = \begin{pmatrix} \bar{w} & \beta \end{pmatrix}$, where both w and β are inner and co-outer. As we observed in the remark after Lemma 1.2, there constants $\{C_r\}_{r \geq 1}$ such that the columns of β are $P_L C_r$, $r \geq 1$, where P_L is the orthogonal projection onto the subspace $L = \text{Ker} \, T_{w^t}$.

We need the following fact.

Lemma 3.3. *Let w be a column function which is inner and co-outer. Suppose that the Hankel operator $H_{\bar{w}}$ is compact. Then the Toeplitz operator T_{w^t} maps $H^2(\ell^2)$ onto H^2.*

Let us first complete the proof of Theorem 3.1. As we have mentioned in the proof of Theorem 2.1, the operator $H_{\bar{w}}$ is compact. So it follows from Lemma 3.3 that $T|L^{\perp}$ is an invertible operator from $L^{\perp}$ onto H^2 which implies that the operator $(T_{w^t})^*T_{w^t}$ is invertible on $L^{\perp}$. So 0 is an isolated point of the spectrum of $(T_{w^t})^*T_{w^t}$. Clearly, P_L is the spectral projection of the selfadjoint operator $(T_{w^t})^*T_{w^t}$ onto L which can be written down with the help of the Dunford-Riesz integral formula. Let $P^{(M)}$ be the truncation projection onto span $\{e_d : 0 \le d \le M\}$. Let w_M be a column function which is inner and co-outer and such that $P^{(M-1)}w_M = P^{(M-1)}w$ and $w_M \in$ Range $P^{(M)}$. Let $L_M = \operatorname{Ker} T_{w_M^t}$ and let P_{L_M} be the orthogonal projection onto L_M. It follows from the Dunford-Riesz formula that $\|P_L - P_{L_M}\| \to 0$ as $M \to \infty$. It is easy to see that $P^{(M)}$ and P_{L_M} commute. Let K be the size of Φ_M and suppose that $M > K$. Consider the columns $P^{(M)}P_{L_M}C_r$, $1 \le r \le K$. Clearly $\|P_L C_r - P^{(M)}P_{L_M}C_r\| \to 0$, $1 \le r \le K$.

Choose now such a large number M such that the numbers $\|P_L C_r - P^{(M)}P_{L_M}C_r\|$ are small, $1 \le r \le K$. It follows that the numbers $|(\bar{w}_M, P^{(M)}P_{L_M}C_r)|$, $1 \le r \le K$, and $|(P^{(M)}P_{L_M}C_r, P^{(M)}P_{L_M}C_s)|$, $1 \le r, s \le K$, $r \ne s$, are small, and the norms of $P^{(M)}P_{L_M}C_r$, $1 \le r \le K$, are close to 1. Using the Gram–Schmidt orthogonalization with the vectors functions $\bar{w}_M, P^{(M)}P_{L_M}C_1, \cdots, P^{(M)}P_{L_M}C_r$, we can construct vector functions $D_1, \cdots, D_r$ such that $\bar{w}_M, D_1, \cdots, D_r$ is an orthonormal family (both in $H^2(\ell^2)$ and pointwise). We can now complete this family to an orthonormal family $\bar{w}_M, D_1, \cdots, D_r, \cdots, D_M$ (in $H^2(\ell^2)$ and pointwise). Put now

$$(W_m^{(M)})^* \stackrel{\text{def}}{=} \begin{pmatrix} I_m & \mathbb{O} & \mathbb{O} \\ \mathbb{O} & W^* & \mathbb{O} \\ \mathbb{O} & \mathbb{O} & I \end{pmatrix},$$

where

$$W^* = \begin{pmatrix} \bar{w}_M & D_1 & \cdots & D_M \end{pmatrix}$$

and I is the infinite-dimensional identity matrix.

It is easy to see that if we replace W_m by $W_m^{(M)}$ on the right-hand side of (3.1), the norm of the right-hand side will not change much. Then we can repeat the same procedure with all W_r, $r \le m$, and finally we can do the same with all V_r, $r \leq m$. Clearly the result now follows from the finite-dimensional case (see [PY2]). ∎

Proof of Lemma 3.3. Clearly, it is sufficient to show that $T_{w^t} T_{\bar{w}}$ is onto. We have

$$T_{w^t} T_{\bar{w}} = T_{w^t \bar{w}} - H_{\bar{w}}^* H_{\bar{w}} = I - H_{\bar{w}}^* H_{\bar{w}}. \tag{3.2}$$

Since $H_{\bar{w}}$ is compact, the operator on the right-hand side of (3.2) is Fredholm and has zero index. So the result will follow if we prove that $\operatorname{Ker} H_{\bar{w}} = \{\mathbb{O}\}$.

Let $w = \{w_s\}_{s \geq 0}$. Suppose that $f \in \operatorname{Ker} H_{\bar{w}}$. Then $\bar{w}_s f \in H_-^2$ for any $j \in \mathbb{Z}_+$. Since w is co-outer, it follows that $f \in H_-^2$, which completes the proof. ∎

4. Properties of superoptimal approximations

In this section we shall show that for a function $\Phi \in H^\infty(\mathcal{B}) + C(\mathcal{C})$ there exists a unique superoptimal approximation by bounded analytic functions. The existence of such an approximation follows easily from a compactness argument. So we only need to prove uniqueness. We shall also prove that if $\Phi \in H^\infty(\mathcal{B}) + C(\mathcal{C})$ and Q is its superoptimal approximation, then $Q^* \in H^\infty(\mathcal{B}) + C(\mathcal{C})$.

Theorem 4.1. *Let* $\Phi \in H^\infty(\mathcal{B}) + C(\mathcal{C})$. *Then there is a unique superoptimal approximation by functions in* $H^\infty(\mathcal{B})$.

Proof. Let Q_1 and Q_2 be superoptimal approximation of Φ. Then each of them admits a representation of the form (2.3). It follows that $\|Q_1 - Q_2\| \leq 2t_m$ for every $m \in \mathbb{Z}_+$. Since $t_m \to 0$ by Corollary 3.2, we have $Q_1 = Q_2$. ∎

Theorem 4.2. *Let* $\Phi \in H^\infty(\mathcal{B}) + C(\mathcal{C})$ *and let* Q *be its superoptimal approximation by functions in* $H^\infty(\mathcal{B})$. *Then the singular values* $s_j(\Phi(\zeta) - Q(\zeta))$, $j \in \mathbb{Z}_+$, *are constant almost everywhere on* $\mathbb{T}$.

The result follows immediately from (2.3).

In [PY1] it was proved that the operator of superoptimal approximation in the finite-dimensional case has certain hereditary properties, that is under certain assumptions on a function space X we prove that if a matrix function Φ belongs to X, then its superoptimal approximation Q also belongs to X. In particular it was shown in [PY1] that this is true if $X = QC \overset{\text{def}}{=} H^\infty + C \cap \overline{H^\infty + C}$. The following result is an analogue of that result in the infinite-dimensional case.

Theorem 4.3. *Let $\Phi \in H^\infty(\mathcal{B}) + C(\mathcal{C})$ and let Q be the superoptimal approximation of Q by functions in $H^\infty(\mathcal{B})$. Then $Q^* \in H^\infty(\mathcal{B}) + C(\mathcal{C})$.*

Proof. We have to prove that the Hankel operator H_{Q^*} is compact. Put

$$R_m = W_0^* \begin{pmatrix} 1 & \mathbb{O} \\ \mathbb{O} & W_1^* \end{pmatrix} \cdots \begin{pmatrix} I_m & \mathbb{O} \\ \mathbb{O} & W_m^* \end{pmatrix} D^{(m)} \begin{pmatrix} I_m & \mathbb{O} \\ \mathbb{O} & V_m^* \end{pmatrix} \cdots \begin{pmatrix} 1 & \mathbb{O} \\ \mathbb{O} & V_1^* \end{pmatrix} V_0^*,$$

where

$$D^{(m)} = \begin{pmatrix} t_0 u_0 & 0 & \cdots & 0 & 0 \\ 0 & t_1 u_1 & \cdots & 0 & 0 \\ \vdots & \vdots & \ddots & \vdots & \vdots \\ 0 & 0 & \cdots & t_{m-1} u_{m-1} & 0 \\ 0 & 0 & \cdots & 0 & \mathbb{O} \end{pmatrix},$$

Since $H_{\bar{v}_d}$ and $H_{\bar{v}_d}$ are compact for $d \leq m$, it follows from Theorem 1.4 that the operator $H_{R_m^*}$ is compact. Clearly, $\|\Phi - Q - R_m\| \leq t_m$. Hence by Corollary 3.2 $\|\Phi - Q - R_m\| \to 0$, which proves the result. ∎

Problem. Let $0 < p < \infty$. Suppose that the Hankel operator H_Φ belongs to the Schatten–von Neumann class $\boldsymbol{S}_p$ (see [P], where such Hankel operators are described in terms of Φ). Let Q be the superoptimal approximation of Φ by functions in $H^\infty(\mathcal{B})$. Is it true that $H_{Q^*} \in \boldsymbol{S}_p$?

Note that in the finite dimensional case the corresponding assertion holds (see [PY1]).

We conclude the paper by the following result which gives an infinite factorization of of $\Phi - Q$.

Theorem 4.4. *Let* $\Phi \in H^{\infty}(\mathcal{B}) + C(\mathcal{C})$ *and let* Q *be the superoptimal approximation of* Φ *by functions in* $H^{\infty}(\mathcal{B})$. *Then* $\Phi - Q$ *admits a factorization of the form*

$$W_0^* \begin{pmatrix} 1 & \mathbb{O} \\ \mathbb{O} & W_1^* \end{pmatrix} \begin{pmatrix} I_2 & \mathbb{O} \\ \mathbb{O} & W_2^* \end{pmatrix} \cdots D \cdots \begin{pmatrix} I_2 & \mathbb{O} \\ \mathbb{O} & V_2^* \end{pmatrix} \begin{pmatrix} 1 & \mathbb{O} \\ \mathbb{O} & V_1^* \end{pmatrix} V_0^*, \quad (4.1)$$

where

$$D = \begin{pmatrix} t_0 u_0 & 0 & 0 & \cdots \\ 0 & t_1 u_1 & 0 & \cdots \\ 0 & 0 & t_2 u_2 & \cdots \\ \vdots & \vdots & \vdots & \ddots \end{pmatrix},$$

both infinite products on the right-hand side of (4.1) being convergent in the strong operator topology.

The result follows immediately from (2.3) and Corollary 3.2.

References

[AAK] V.M. Adamyan, D.Z. Arov, and M.G. Krein, *On infinite Hankel matrices and generalized problems of Carathéodory-Fejér and F. Riesz*, Funktsional. Anal. i Prilozhen. **2:1** (1968), 1-19.

[H] H. Helson, *Lectures on invariant subspaces*, Acad. Press, NY, 1964.

[Kh] S. Khavinson, *On some extremal problems of the theory of analytic functions*, Uchen. Zapiski Mosk. Universiteta, Matem. **144**:4 (1951), 133-143. English transl.: Amer Math. Soc. Translations (2) **32** (1963), 139-154.

[N] N.K. Nikol'skii, *Treatise on the shift operator*, Springer-Verlag, Berlin–Heidelberg–New York–Tokyo, 1986.

[Pa] L.B. Page, *Bounded and compact vectorial Hankel operators*, Trans. Amer. Math. Soc. **64** (1970), 529-539.

[P] V.V. Peller, *Vectorial Hankel operators, commutators and related operators of the Schatten–von Neumann class* $\mathfrak{S}_p$, Int. Eq. Op. Theory **5** (1982), 244-272.

[PKh] V.V. Peller and S.V. Khrushchev, *Hankel operators, best approx-imation and stationary Gaussian processes*, Uspekhi Mat. Nauk **37:1** (1982), 53-124. English Transl.: Russian Math. Surveys **37:1** (1982), 61-144.

[PY1] V.V. Peller and N.J. Young, *Superoptimal analytic approxima-tions of matrix functions*, J. Funct. Anal. **120** (1994), 300-343.

[PY2] V.V. Peller and N.J. Young, *Superoptimal singular values and indices of matrix functions*, Int. Eq. Op. Theory **20** (1994), 350-363.

[SF] B. Sz.-Nagy and C. Foias, *Analyse harmonique des opérateurs de l'espace de Hilbert*, Akadémiai Kiadó, Budapest, 1967.

[T] S.R. Treil, On superoptimal approximation by analytic and mero-morphic matrix-valued functions, to appear in J. Funct. Anal.

Kansas State University, Manhattan, KS 66506
E-mail address: peller@math.ksu.edu

Contemporary Mathematics
Volume **189**, 1995

Character sums and explicit estimates for L-functions

JUAN C. PERAL

ABSTRACT. We prove in this paper, by the use of a suitable Fourier expansion, sharp bounds for certain Dirichlet character sums leading to explicit upper estimates for L-functions in points which relate to the value of the corresponding class number.

1. Introduction.

Assume that χ is a primitive, non-principal, Dirichlet character, modulo q, and consider the expression given by:

$$S_\chi(f) = \sum_{a=1}^{q-1} \chi(a) f(\frac{a}{q})$$

where f is a function defined in $[0,1]$. Sums of this kind, for suitable f, appear frequently in number theory, for example when studying the distribution of quadratic residues, exponential sums with multiplicative coefficients, or certain important values of L-fuctions such as the ones related to the class number $h(-q)$.

The case of f being the function $f(t) = t^m$ has been treated by several authors; Ayoub, Chowla and Walum [2], and Fine [5], study the sign of these sums in relation to the size of the class number $h(-q)$ for χ being the Legendre symbol and q being a prime number such that $q \equiv 3\,(4)$. In [18] Williams studied a more general kind of sums, proving estimates such as:

$$S_\chi(m) = \sum_{a=1}^{q-1} \chi(a)(\frac{a}{q})^m = O(q^{\frac{1}{2}} \ln q).$$

In [17] Toyoizumi gives explicit bounds for the constants implicit in the O sign of Williams' result proving, by the use of the generalized Bernoulli numbers, the following result:

1991 *Mathematics Subject Classification.* Primary 11L40, 11M20; Secondary 42A05.

Theorem.

(a) *Assume $\chi(-1) = 1$. Then:*

$$| S_\chi(m) | \le \frac{2\zeta(2)e^{2\pi}m!}{(2\pi)^{m+1}}q^{\frac{1}{2}}.$$

(b) *Assume $\chi(-1) = -1$. Then:*

$$| S_\chi(m) | \le (\frac{2\zeta(3)e^{2\pi}m!}{(2\pi)^{m+1}} + \frac{| L(1,\chi) |}{\pi})q^{\frac{1}{2}}.$$

Slightly different sums were used by Rosser [15] in order to investigate the non vanishing of $L(s,\chi)$ for $s \in [0,1]$ in the case of q a small modulo.

The aim of this work is to give an expression for the general sums $S_\chi(f)$, from which a variety of consequences will follow. The representation will be achieved by the use of the gaussian sums together with an adequate Fourier expansion. Among the consequences we will prove upper bounds for the case $S_\chi(m)$, improving the result in [17] quoted before. In fact our estimates are sharp in a sense specified later on. As an application we will also prove explicit inequalities for $|L(1,\chi)|$, in the case of χ being a primitive non-principal Dirichlet character, such as the ones considered in Kanemitsu [8], for even and odd characters, and Hua [7], or Louboutin [9] for the case of even characters.

2. Statements of the results.

We will enunciate in this section the main results. The representation of the sums $S_\chi(f)$ aluded to, is included in the following lemma, where we assume that the derivative of the function f is integrable:

Lemma.

(a) *Assume χ and f as before, and $\chi(-1) = 1$. Then:*

$$\sum_{a=1}^{a=q} \chi(a)f(\frac{a}{q}) = -\frac{\sqrt{q}}{\pi\tau_q(\chi)} \int_0^1 \sum \overline{\chi(k)}\frac{\sin 2\pi kt}{k} f'(t)dt.$$

(b) *Assume $\chi(-1) = -1$. Then:*

$$\sum_{a=1}^{a=q} \chi(a)f(\frac{a}{q}) + (f(1) - f(0))\frac{\sqrt{q}}{i\pi\tau_q(\chi)}L(\chi,1)$$

$$= -\frac{\sqrt{q}}{i\pi\tau_q(\chi)} \int_0^1 \sum \overline{\chi(k)}\frac{\cos 2\pi kt}{k} f'(t)dt.$$

As a consequence of the lemma we will greatly improve the result given in [17], in fact we will prove:

THEOREM 1.

(a) *Assume that* $\chi(-1) = 1$. *Then:*

$$\mid S_\chi(m) \mid \leq q^{\frac{1}{2}} \left(\frac{m-1}{2(m+1)} \right)$$

(b) *Assume that* $\chi(-1) = -1$. *Then:*

$$\mid S_\chi(m) + \frac{q^{\frac{1}{2}}}{\pi i \tau_q(\chi)} L(\overline{\chi}, 1) \mid \leq q^{\frac{1}{2}} \left(\frac{m}{\pi} \int_0^1 \ln \frac{1}{2\sin(\pi t)} t^{m-1} dt \right)$$

That the results of the preceding theorem are sharp is the content of the following corollary:

COROLLARY 1. *Let* χ *be the Legendre symbol. Then, for fixed natural m and any $\epsilon > 0$, we have:*

(a) *There are infinitely many primes* $q \equiv 1(4)$ *such that:*

$$S_\chi(m) \geq q^{\frac{1}{2}} \left(\frac{m-1}{2(m+1)} - \epsilon \right)$$

(b) *There are infinitely many primes* $q \equiv 3(4)$ *such that*

$$S_\chi(m) \geq q^{\frac{1}{2}} \left(\frac{m}{\pi} \int_0^1 \ln \frac{1}{2\sin(\pi t)} t^{m-1} dt - \frac{L(\chi, 1)}{\pi} - \epsilon \right)$$

Another result is included in the:

COROLLARY 2.

Let $q \equiv 3\,(4)$ *be a prime number, and m a non-negative integer. Then for the Legendre symbol we have:*

$$\frac{1}{q^m} \sum_{a=1}^{q-1} (\frac{a}{q}) a^m + h(-q) \leq \frac{\sqrt{q}}{\pi} \{\ln(m) + O(1)\}$$

THEOREM 2.

(a) *Assume that* χ *is a primitive, odd, non-principal Dirichlet character, modulo q, then:*

$$|L(1, \chi)| \leq \frac{1}{2} \ln q + 1 + \gamma - \ln 2$$

(b) *Assume that* χ *is a primitive, even, non-principal Dirichlet character, modulo q, then:*

$$|L(1, \chi)| \leq \frac{1}{2} \ln q + 1 - \ln 2$$

Here γ is the Euler constant.

3. Proofs of the results

For the proofs of the results we will use a partial summation and the gaussian sums jointly with the expression of the incomplete sums of a character by means of an adequate Fourier developement. For any χ as the given, let τ_q be

$$\tau_q(\chi) = \frac{1}{\sqrt{q}} \sum_{n=1}^{k} \chi(n) e^{\frac{2\pi i n}{q}} ;$$

this number is known to have absolute value one. We are going to use the finite Fourier expansion for primitive and non-principal characters which reads as follows:

$$\chi(n) = \frac{1}{\sqrt{q}} \tau_q(\chi) \sum_{r=1}^{q} \overline{\chi}(r) e^{\frac{-2\pi i r n}{q}} .$$

The fact $\sum_{m=1}^{q} \chi(m) = 0$ will be used too. See Apostol [1] for these expansions. The exact value of $\tau_q(\chi)$ is known for q any prime number and χ the Legendre symbol.

3.1. Proof of the lemma.

We will prove some auxiliary results, the first is just a partial sumation valid for χ a non-principal Dirichlet character and f a function defined in $[0, 1]$ together with its derivative, which is assumed to be integrable. We have

$$\sum_{a=1}^{q-1} \chi(a) f(\frac{a}{q}) = -\int_0^1 \left(\sum_{a=1}^{[qt]} \chi(a) \right) f'(t) dt$$

In fact the inner sum is constant in the intervals $(\frac{r}{q}; \frac{r+1}{q})$ so we have:

$$-\int_0^1 \left(\sum_{a=1}^{[qt]} \chi(a) \right) f'(t) dt = -\sum_{r=1}^{q-1} \int_{\frac{r}{q}}^{\frac{r+1}{q}} \left(\sum_{a=1}^{r} \chi(a) \right) f'(t) dt$$

$$= -\sum_{r=1}^{q-1} \left(\sum_{a=1}^{r} \chi(a) \left(f(\frac{r+1}{q}) - f(\frac{r}{q}) \right) \right)$$

$$= -\sum_{a=1}^{q-1} \chi(a) \left(\sum_{r=a}^{q-1} f(\frac{r+1}{q}) - f(\frac{r}{q})^m \right)$$

$$= -\sum_{a=1}^{q-1} \chi(a) \left(f(1) - f(\frac{a}{q}) \right) = 0 + \sum_{a=1}^{q-1} \chi(a) f(\frac{a}{q})$$

Now we will use a Fourier expansion in order to get a suitable expression for the inner sum in the integral of the preceeding formula. We will get:

$$\sum_{a=1}^{[\lambda q]} \chi(a) = \frac{\sqrt{q}}{\pi \overline{\tau}_q(\chi)} \sum_{n=1}^{\infty} \overline{\chi}(n) \frac{\sin 2\pi n \lambda}{n}$$

for the case $\chi(-1) = 1$, $\lambda \in [0,1)$, and $\lambda \neq \frac{r}{q}$, and

$$\sum_{a=1}^{[\lambda q]} \chi(a) = \frac{\sqrt{q}}{i\pi \overline{\tau_q(\chi)}} L(\overline{\chi}, 1) - \frac{\sqrt{q}}{i\pi \overline{\tau_q(\chi)}} \sum_{n=1}^{\infty} \overline{\chi}(n) \frac{\cos 2\pi n\lambda}{n}$$

for the case $\chi(-1) = -1$ and $\lambda \in [0,1)$, $\lambda \neq \frac{r}{q}$.

These kind of expansions were first used by Polya [14]. For any $\lambda \in [0,1)$ take $f_\lambda(x)$ the characteristic function of the interval $[0,\lambda)$ and periodic of period one on the real numbers. Its Fourier development is given by

$$f_\lambda(x) = \lambda + \frac{1}{\pi} \sum_{\substack{n=-\infty \\ n \neq 0}}^{\infty} \frac{e^{2\pi i n x}}{n} \left(\frac{\sin 2\pi n\lambda}{2} + \frac{1 - \cos 2\pi n\lambda}{2i} \right)$$

which is convergent to the function except for $x = 0$ and $x = \lambda$.

If we take the values of $f_\lambda(\frac{a}{q})$ multipied by $\chi(a)$, and add from $a = 1$ to $a = q - 1$, we get:

$$\sum_{a=1}^{q-1} \chi(a) f_\lambda\left(\frac{a}{q}\right) = \sum_{a=1}^{[\lambda q]} \chi(a)$$

$$= \lambda \sum_{a=1}^{q-1} \chi(a) + \frac{1}{\pi} \sum_{\substack{n=-\infty \\ n \neq 0}}^{\infty} \left(\frac{\sin 2\pi n\lambda}{2n} + \frac{1 - \cos 2\pi n\lambda}{2in} \right) \overline{\sum_{a=1}^{q-1} \overline{\chi}(a) e^{\frac{-2\pi i n a}{q}}}$$

$$= 0 + \frac{\sqrt{q}}{\pi} \sum_{\substack{n=-\infty \\ n \neq 0}}^{\infty} \left(\frac{\sin 2\pi n\lambda}{2n} + \frac{1 - \cos 2\pi n\lambda}{2in} \right) \frac{\overline{\chi}(n)}{\overline{\tau_q(\chi)}}$$

where we have used the finite Fourier expansion for primitive Dirichlet characters quoted before, and where the overline means complex conjugation. Grouping together the terms with n and $-n$ we finally get:

$$\sum_{a=1}^{[\lambda q]} \chi(a) = \frac{\sqrt{q}}{\pi \overline{\tau_q(\chi)}} \sum_{n=1}^{\infty} (\overline{\chi}(n) + \overline{\chi}(-n)) \frac{\sin 2\pi n\lambda}{2n}$$

$$+ \frac{\sqrt{q}}{\pi \overline{\tau_q(\chi)}} \sum_{n=1}^{\infty} (\overline{\chi}(n) - \overline{\chi}(-n)) \frac{1 - \cos 2\pi n\lambda}{2in}$$

In the first case we have $\chi(-1) = 1$ so $\overline{\chi}(n) = \overline{\chi}(-n)$ and the second series vanishes. In the second case $\chi(-1) = -1$ so $\overline{\chi}(n) = -\overline{\chi}(-n)$ and hence the first series vanishes in this case. So we have proved the validity of the second and third formulas. The importance of the role played by the condition $\chi(-1) = 1$ or -1 becomes clear in this result.

The lemma is now a consequence of the preceeding formulas. Let us prove part b). In this case we have, by (1) and (3):

$$\sum_{a=1}^{q-1} \chi(a) f(\frac{a}{q}) = -\int_0^1 \sum_{a=1}^{[tq]} \chi(a) f'(t) dt$$

$$= -\int_0^1 \{\frac{\sqrt{q}}{i\pi\overline{\tau_q}(\chi)} L(\overline{\chi},1) - \frac{\sqrt{q}}{i\pi\overline{\tau_q}(\chi)} \sum_{n=1}^{\infty} \overline{\chi}(n) \frac{\cos 2\pi nt}{n}\} f'(t) dt$$

$$= -\{f(1) - f(0)\} \frac{\sqrt{q}}{i\pi\overline{\tau_q}(\chi)} L(\overline{\chi},1) + \frac{\sqrt{q}}{i\pi\overline{\tau_q}(\chi)} \int_0^1 \sum_{n=1}^{\infty} \overline{\chi}(n) \frac{\cos 2\pi nt}{n} f'(t) dt$$

which proves this part of the lemma, the other being identical.

3.2. Proof of Theorem 1.

We state now two well known facts concerning Fourier coefficients of decreasing (or convex) functions. Let f be defined in $[0,1]$ integrable and decreasing (increasing). Then:

$$I_n = \int_0^1 f(t) \sin 2\pi nt \, dt \geq 0 \, (\leq 0).$$

In fact we have:

$$I_n = \int_0^1 f(t) \sin 2\pi nt \, dt = \sum_{k=0}^{n-1} \int_{\frac{k}{n}}^{\frac{k+1}{n}} f(t) \sin 2\pi nt \, dt$$

$$= \sum_{k=0}^{n-1} \int_0^{\frac{1}{n}} f(t + \frac{k}{n}) \sin 2\pi nt \, dt = \sum_{k=0}^{n-1} \int_0^{\frac{1}{2n}} \{f(t + \frac{k}{n}) - f(\frac{k+1}{n} - t)\} \sin 2\pi nt \, dt$$

where we have used that $\sin 2\pi n(\frac{1}{n} - t) = -\sin 2\pi nt$. Now observe that if f is decreasing then the function under the last integral is positive, so $I_n \geq 0$. The same argument gives the result in the case of f increasing.

Assume now that f is integrable and convex (concave) in $[0,1]$. Then:

$$J_n = \int_0^1 f(t) \cos 2\pi nt \, dt \geq 0 \, (\leq 0).$$

Indeed, by the definition we have:

$$J_n = \int_0^1 f(t) \cos 2\pi nt dt = \sum_{k=0}^{n-1} \int_{\frac{k}{n}}^{\frac{k+1}{n}} f(t) \cos 2\pi nt \, dt$$

$$= \sum_{k=0}^{n-1} \int_0^{\frac{1}{n}} f(t + \frac{k}{n}) \cos 2\pi nt \, dt$$

$$= \sum_{k=0}^{n-1} \int_0^{\frac{1}{4n}} \{f(t+\frac{k}{n}) + f(\frac{k+1}{n} - t) - f(\frac{2k+1}{2n} - t) - f(\frac{2k+1}{2n} + t)\} \cos 2\pi nt \, dt$$

where we have used the symmetries of cos. Now if f is convex, the function under the last integral is again positive, and so is J_n. The same argument gives the result in the other case.

Let us now prove part a) of Theorem 1. If we use the lemma we get

$$S_\chi(m) = -m \int_0^1 \left(\sum_{a=1}^{[qt]} \chi(a) \right) t^{m-1} dt = \frac{-m\sqrt{q}}{\pi \overline{\tau_q}(\chi)} \int_0^1 \left(\sum_{n=1}^\infty \overline{\chi}(n) \frac{\sin 2\pi n t}{n} \right) t^{m-1} dt$$

From this we have

$$\mid S_\chi(m) \mid \le \frac{\sqrt{q}}{\pi} m \sum_{n=1}^\infty \frac{1}{n} \mid I_n \mid$$

and using (4) we get that I_n is negative for all n so we obtain:

$$\mid \frac{S_\chi(m)}{q^{\frac{1}{2}}} \mid \le \frac{1}{\pi} m \sum_{n=1}^\infty \frac{1}{n} \mid I_n \mid = -\frac{1}{\pi} m \int_0^1 \sum_1^\infty \frac{\sin 2\pi n t}{n} t^{m-1} dt$$

and taking into account that

$$\frac{1}{\pi} \sum_1^\infty \frac{\sin 2\pi n t}{n} = \frac{1}{2} - t$$

and that

$$m \int_0^1 (t - \frac{1}{2}) t^{m-1} dt = \frac{m-1}{2(m+1)}$$

we get the first part of Theorem 1.

If we assume now that $\chi(-1) = -1$ and use the lemma again we get

$$S_\chi(m) = -m \int_0^1 \left(\sum_{a=1}^{[qt]} \chi(a) \right) t^{m-1} dt$$

$$= -\frac{\sqrt{q}}{\pi i \overline{\tau_q}(\chi)} L(\overline{\chi}, 1) + \frac{m\sqrt{q}}{\pi i \overline{\tau_q}(\chi)} \int_0^1 \left(\sum_{n=1}^\infty \overline{\chi}(n) \frac{\cos 2\pi n t}{n} \right) t^{m-1} dt$$

In order to prove the second part of Theorem 1, we make the same reasoning as for the first part, taking into account that now J_n is positive, as proved before, and finally using the fact that

$$\sum_{n=1}^\infty \frac{\cos 2\pi n t}{n} = -\ln(2 \sin \pi t)$$

for $t \in (0, 1)$.

3.3. Proof of Corollary 1.

We are going to use an argument as the one used by Montgomery [10]. We will prove part a), the proof of part b) being similar. Let us take $P = 4 \prod_{q \le N} q$ where q is prime. Then using quadratic reciprocity, we can find an a such that if $p \equiv a(P)$ then $\left(\frac{q}{p}\right) = +1$ for all the primes numbers $q \leq N$. Now for those $p \equiv a(P)$ we have:

$$\frac{1}{p^{m+\frac{1}{2}}} \sum_{a=1}^{p-1} \left(\frac{a}{p}\right) a^m = \frac{m}{\pi} \int_0^1 \sum_{n=1}^{\infty} \left(\frac{n}{p}\right) \frac{\sin 2\pi nt}{n} (1-t)^{m-1} dt$$

$$= \frac{m}{\pi} \int_0^1 \left(\sum_{n=1}^{\infty} \frac{\sin 2\pi nt}{n} - \sum_{n=N+1}^{\infty} \frac{\sin 2\pi nt}{n} + \sum_{n=N+1}^{\infty} \left(\frac{n}{p}\right) \frac{\sin 2\pi nt}{n} \right) (1-t)^{m-1} dt$$

due to the fact that for any $n \leq N$, $\left(\frac{n}{p}\right) = +1$, so we have:

$$\frac{1}{p^{m+\frac{1}{2}}} \sum_{a=1}^{p-1} \left(\frac{a}{p}\right) a^m \geq \frac{m}{\pi} \int_0^1 \pi(\frac{1}{2} - t)(1-t)^{m-1} dt - \frac{2m}{\pi} \sum_{N+1}^{\infty} \frac{|I_n|}{n}$$

$$= \frac{m-1}{2(m+1)} - \frac{2m}{\pi} \sum_{N+1}^{\infty} \frac{|I_n|}{n} \geq \frac{m-1}{2(m+1)} - \frac{m}{\pi^2} \sum_{N+1}^{\infty} \frac{1}{n^2} \geq \frac{m-1}{2(m+1)} - \frac{m}{\pi^2} \frac{1}{N}$$

by using the bound $|I_n| \leq \frac{1}{2\pi n}$ which is easily proved. So it is enough to take N such that $N > [\frac{m}{\epsilon \pi^2}]$.

3.4. Proof of Corollary 2.
Assume that m is a non-negative integer. Then:

$$H(m) = -\int_0^1 \ln(t)(1-t)^m dt = \frac{1}{m+1}\{1 + \frac{1}{2} + \frac{1}{3} + ... + \frac{1}{m+1}\}$$

In fact performing an integration by parts we get;

$$H(m) = -m \int_0^1 t(1-t)^{m-1} \ln(t) dt + \int_0^1 t(1-t)^{m-1} dt$$

$$= m \int_0^1 (1-t)^m \ln(t) dt - m \int_0^1 (1-t)^{m-1} \ln(t) dt + \frac{1}{m+1}$$

so the integral $H(m)$ satisfies the recurrence relation:

$$(m+1)H(m) = mH(m-1) + \frac{1}{m+1}$$

which together with $H(0) = 1$ implies by induction the validity of the equality given for $H(m)$.

Now in order to prove Corollary 2 it is enough to apply the theorem from which we know that the upper bound is given by the following integral:

$$\frac{\sqrt{q}}{\pi} m \int_0^1 \ln(\frac{1}{2\sin \pi t}) t^{m-1}$$

This integral can be estimated in terms of m as follows. By the well known product expansion for the function sine given by the formula:

$$\frac{2\sin \pi t}{2\pi t} = \prod_{k=1}^{\infty} (1 - \frac{t^2}{k^2})$$

valid for any non-integer t we get the equality:

$$\int_0^1 \ln\left(\frac{1}{2\sin\pi t}\right)t^{m-1} = -\int_0^1 \ln(1-t)t^{m-1}dt$$

$$-\int_0^1 \ln(2\pi t)t^{m-1}dt - \int_0^1 \ln(1+t)t^{m-1}dt + \int_0^1 \sum_{k=2}^{\infty} \ln\left(1+\frac{t^2}{k^2-t^2}\right)t^{m-1}dt$$

and using (6) we see that the main term in this equality is the first integral whose contribution is:

$$\frac{1}{m}\left\{1+\frac{1}{2}+\frac{1}{3}+...+\frac{1}{m}\right\}$$

In fact all the other terms are of the order of magnitude $O(m^{-1})$; the second because the integral there is exactly $\frac{1}{m^2}$, the third by a direct integration, and the last integral is easily dominated because we have for $t \in [0,1]$ that

$$\sum_{k=2}^{\infty} \ln\left(1+\frac{t^2}{k^2-t^2}\right) \leq \frac{4}{3}\left(\frac{\pi^2}{6}-1\right)t^2$$

from which the assertion follows.

Observe that with $m \equiv \sqrt{q}$ and taking into account that one can directly estimate the sum

$$\frac{1}{q^m}\sum_{a=1}^{q-1}\left(\frac{a}{q}\right)a^m$$

by $O(\sqrt{q})$ by a trivial comparision with an integral, we get the bound

$$h(-q) \leq \frac{\sqrt{q}}{2\pi}\ln q + O(\sqrt{q})$$

whitout using, at least in a direct way, the Polya-Vinogradov inequality in order to get the extra factor $\frac{1}{2}$ for the trivial estimate $\frac{\sqrt{q}}{\pi}\ln q$.

Note that a direct bound of the order

$$\frac{1}{q^m}\sum_{a=1}^{q-1}\left(\frac{a}{q}\right)a^m = O(\sqrt{q})$$

for m of the order of smaller powers of q would greatly improve the known bounds for $h(-q)$.

3.5. Proof of Theorem 2.

The interest of the estimate in Theorem 2 lies in the fact that it is explicit. Of course the estimates of Burgess [3,4], Pintz [13], and Stephens [16] give better bounds by a factor of one half in the main term, but these estimates are not effective. The result of this theorem should be compared with the results of Hua [7], and the improvement of Louboutin [9], whose proofs are valid for even characters. The paper of Kanemitsu [8] also deals with these kind of estimates.

The proof of Theorem 2 involves the same ideas that the proof of Corollary 2 jointly with a more careful estimate of the integrals giving the second order term. In fact in order to prove part a) of Theorem 2 it is enough to apply Theorem

1, from which we know that the upper bound, as in the case of Corollary 2, is
given by the integral:

$$\frac{\sqrt{q}}{\pi}m\int_0^1 \ln(\frac{1}{2\sin\pi t})t^{m-1}$$

Now arguing as before we have the following inequality:

$$|L(1,\chi)| \le \{1 + \frac{1}{2} + \frac{1}{3} + \ldots + \frac{1}{m}\} + \frac{1}{m} - \ln(2\pi)$$

$$+ m\int_0^1 \sum_{k=2}^{\infty} \ln(1 + \frac{t^2}{k^2 - t^2})t^{m-1}dt - m\int_0^1 \ln(1+t)t^{m-1}dt + \frac{\pi}{q^{\frac{1}{2}}}|S_\chi(m)|$$

Now a trivial estimate gives

$$|S_\chi(m)| \le \frac{q}{m+1}$$

and taking into account that

$$\int_0^1 \sum_{k=2}^{\infty} \ln(1 + \frac{t^2}{k^2 - t^2})t^{m-1}dt \le \int_0^1 \sum_{k=2}^{\infty} \ln\frac{k^2}{k^2 - 1}t^{m-1}dt = \frac{\ln 2}{m}$$

we get that the joint contribution of the last two integrals is at most $\frac{1}{m}$, and now
taking $m = [\pi\sqrt{q}]$ and collecting all the terms we get the assertion of part a).

For the proof of the second part of the theorem we take the sum $\sum_1^{H-1} S_\chi(m)$
which due to the fact that in this case $\chi(-1) = 1$ is easily seen to be equal to:

$$q\sum_{a=1}^{a=q-1} \frac{\chi(a)}{a} - q\sum_{a=1}^{a=q-1} \chi(a)\frac{(1 - \frac{a}{q})^H}{a}.$$

So we obtain:

$$|\sum_{a=1}^{a=q-1} \frac{\chi(a)}{a}| \le \frac{1}{q}|\sum_1^{H-1} S_\chi(m)| + \sum_{a=1}^{a=q-1} \frac{(1 - \frac{a}{q})^H}{a}$$

now we apply the Theorem 1 and get $|\sum_1^{H-1} S_\chi(m)| \le \frac{H\sqrt{q}}{2}$ and on the other
hand one has:

$$\sum_{a=1}^{a=q-1} \frac{(1 - \frac{a}{q})^H}{a} \le \sum_{a=1}^{\infty} \frac{e^{\frac{-Ha}{q}}}{a} = \ln(\frac{e^{\frac{H}{q}}}{e^{\frac{H}{q}} - 1}) \le \frac{H}{q} + \ln\frac{q}{H}$$

and by choosing $H = [2\sqrt{q}]$ we get part b) of Theorem 2.

4. Remarks

1. For a fixed m, the size of the qs for which we are near of reaching the
best constant in the Corollary 1, are at least of the order of magnitude e^m, so
even though the Theorem 1 can not be improved in general, it seems possible
to get some improvement if we fix the relative sizes of m and q, this should
be achieved if we use in the integral expression for the sum $S_\chi(m)$ the results
about the behaviour of small segments of the incomplete sums for χ and the

Polya-Vinogradov inequality. The results of Burgess [3,4] already quoted, and the work of Montgomery and Vaughan [11] deal with these kind of sums.

2. A more precise study of the integrals I_n, J_n is also possible in this particular case of f being the function $f(t) = t^m$.

REFERENCES

1. T. M. Apostol, *Introduction to Analytic Number Theory*, Springer-Verlag, New York, 1976.
2. R. Ayoub, S. Chowla, H. Walum, *On sums involving Quadratic characters*, Journal London Math. Soc. **42** (1967), 152–154.
3. D. A. Burgess, *On character sums and L-series II*, Proc. London Math. Soc. **13** (1963), 524–536.
4. D. A. Burgess, *Estimating $L_\chi(1)$*, Norske Vid. Selsk. Forh. **39** (1966), 101–108.
5. N. J. Fine, *On a question of Ayoub, Chowla and Walum concernig character sums.*, Illinois J. Math. **14** (1970), 88–90.
6. A. Hildebrand, *On the constant in the Polya-Vinogradov inequality*, Canadian Math. Bull. **31 (3)** (1988), 347–352.
7. L. K. Hua, *Introduction to Number theory*, Springer Verlag, 1982.
8. S. Kanemitsu, *On same bounds for the value of dirichlet L-functions at s=1*, Mem. Fac. Sci. Kyushu **31** (1977), 15–23.
9. S. Louboutin, *Majorations explicites de $|L(\chi, 1)|$*, C.R.A.S. **316. Serie 1** (1993), 11–14.
10. H. Montgomery, *Distribution questions concernig a character sum*, Colloquia Mathematica Societatis Janos Bolyai **13** (1974), 195–203.
11. H. Montgomery and R. C. Vaughan, *Exponential sums with multiplicative coefficients*, Inventiones Math. **43** (1977), 69–82.
12. J. C. Peral, *On the sums $S - \chi(m)$.*, Acta Arithmetica. To appear.
13. J. Pintz, *Elementary methods in the theory of the L-function VII*, Acta Arithmetica **32** (1977), 397–406.
14. G. Polya, *Uber die Verteilung der quadratischen Reste and Nicht-reste*, Gottingen Nachrichten (1918), 21–29.
15. J. B. Rosser, *Real roots of Dirichlet characters*, Bulletin of AMS. **55** (1949), 906–913.
16. P. J. Stephens, *Optimizing the size of $L(\chi, 1)$*, Proc. London Math. Soc. **24** (1972), 1–14.
17. M. Toyoizumi, *On certain character sums*, Acta Arithmetica **55** (1990), 229–232.
18. K. S. Williams, *A class of character sums*, Journal London Math. Soc. **46 6yr 1971**, 67–72.

DEPARTAMENTO DE MATEMÁTICAS, FACULTAD DE CIENCIAS, UNIVERSIDAD DEL PAÍS VASCO, APARTADO 644, 48080 BILBAO, SPAIN

Contemporary Mathematics
Volume **189**, 1995

On the resolvent of the dyadic paraproduct, and a nonlinear operation on RH_p weights

MARÍA CRISTINA PEREYRA

ABSTRACT. The existence of a bounded inverse of $(I - \Pi_b)$ on L^p (Π_b is the dyadic paraproduct) does not imply the same for $(I - \lambda\Pi_b)$, $-1 \leq \lambda < 1$ (we present a counterexample); but it guarantees the existence of $1 < p_o$ such that there exist a bounded inverse in L^{p_o} for every $-1 \leq \lambda \leq 1$. This is equivalent to showing that the RH_p^d class of weights is not preserved under certain nonlinear operation involving λ, but if $\omega \in RH_p^d$ then there exists $1 < p_o$ such that the transformed weight $\omega_\lambda \in RH_{p_o}^d$ for all $-1 \leq \lambda \leq 1$.

1. Introduction

The necessary and sufficient conditions for the existence of a solution $f \in L^p(\mathbf{R})$ of the equation:

(1) $$(I - \Pi_b)f = g, \quad \|f\|_p \leq C\|g\|_p;$$

are known (see [**8**]). Here I is the identity operator, Π_b is the dyadic paraproduct (see (3) for the precise definition), associated to the function b of bounded mean oscillation (BMO), and g is a function in $L^p(\mathbf{R})$.

We are interested in studying this equation under a simple perturbation:

(2) $$(I - \lambda\Pi_b)f = g, \quad -1 \leq \lambda \leq 1.$$

It is known, by spectral theory, that as soon as (1) is solvable, then so is (2) in a neighborhood of $\lambda = 1$; clearly the same is true near $\lambda = 0$. What

---

1991 *Mathematics Subject Classification*. Primary 42c; Secondary 47b.
Key words and phrases. RH_p weights, dyadic paraproduct.
The author was supported by NSF grant #DMS 9304580

about $0 < \lambda < 1$? Does invertibility of $(I - \Pi_b)$ in L^p guarantees invertibility of $(I - \lambda\Pi_b)$ for $0 < \lambda < 1$?

The answer to this question is no. Counterexamples were constructed for $-1 < \lambda < 0$ (see [8]) it was not clear then what the answer was for $0 < \lambda < 1$.

The questions we are asking can be rephrased in terms of preserving reverse Hölder p (RH_p) weights (see [6]) under certain nonlinear operation involving λ (see §3 for definitions and precise statements).

THEOREM 1.0.1. There exist a doubling weight $\omega \in RH_p^d$, $1 < p < \infty$, and $0 < \lambda < 1$, such that $\omega_\lambda \notin RH_p^d$ (ω_λ will be defined later).

The weight ω_λ is given by an infinite product where λ appears in each factor (see (5)). Inserting $\lambda = 1$ we get back the weight ω we started with. Unfolding the product, you get $\omega_\lambda = 1 + \lambda b + ...$ where $b \in BMO$. At first sight we are tempted to say that ω_λ plays a role much like the more traditional $\omega^\lambda = e^{\lambda b}$ does. Theorem 1.0.1 makes clear the substantial differences between them, since if $\omega \in RH_p^d$ then $\omega^\lambda \in RH_p^d$ for all $0 \le \lambda \le 1$ (this is just a trivial application of Hölder's inequality!).

To the weight ω given by Theorem 1.0.1 we can associate a function $b \in BMO$, such that to ω_λ corresponds the function $\lambda b \in BMO$, and $(I - \Pi_b)$ is invertible in L^p, but $(I - \lambda\Pi_b) = (I - \Pi_{\lambda b})$ is not.

Nevertheless we will show that:

THEOREM 1.0.2. There exists $p_0 > 1$ such that if $(I - \Pi_b)$ is invertible in L^p, then $(I - \lambda\Pi_b)$ is invertible in L^{p_0} for all $-1 \le \lambda \le 1$.

Notation and basic definitions are in §2. In §3 we recall a dyadic characterization of weights, the correspondence we mentioned between weights and functions in BMO becomes clear. In §4 we recall the necessary and sufficient conditions for inverting $(I - \lambda\Pi_b)$ in L^p, and we prove Theorem 1.0.2. In the last section we construct the counterexample.

Throughout this paper C will denote a constant that might change from line to line.

We would like to express here our inmense gratitude to Mischa Cotlar for a life dedicated to people and mathematics, thanks Mischa for being.

2. Preliminaries

2.1. Dyadic intervals and Haar basis. We will work on $\mathbf{R}$ but everything holds in $\mathbf{R}^n$ (see [8]).

Let us denote by $\mathcal{D}$ the family of all dyadic intervals in $\mathbf{R}$, i.e. intervals of the form $(j2^{-k}, (j+1)2^{-k}]$, $j, k \in \mathbf{Z}$. $\mathcal{D}_k$ denotes the kth generation of $\mathcal{D}$, consisting of those dyadic intervals of length 2^{-k}. Given any interval J, $\mathcal{D}(J)$ denotes the family of dyadic subintervals of J; $\mathcal{D}_n(J) = \{I \in \mathcal{D}(J) : |I| = 2^{-n}|J|\}$, $|I|$ denotes the length of the interval I. Given an interval J we will denote the right and left halves respectively by J_r and J_l; they are the elements of $\mathcal{D}_1(J)$.

The *Haar function* associated to an interval I is given by:

$$h_I = \frac{1}{|I|^{1/2}}(\chi_{I_r}(x) - \chi_{I_l}(x)),$$

here χ_I denotes the characteristic function of the interval I.

The Haar functions indexed on the dyadics, $\{h_I\}_{I \in \mathcal{D}}$, form a basis of $L^2(\mathbf{R})$.

2.2. Expectation and Difference operators. We define the expectation and difference operators for locally integrable functions by:

$$
\begin{aligned}
E_k f(x) &= \frac{1}{|I|} \int_I f(t)dt = m_I f, \quad x \in I \in \mathcal{D}, \\
\Delta_k f(x) &= E_{k+1} f(x) - E_k f(x).
\end{aligned}
$$

As operators defined on L^p, it is clear that $E_k = \sum_{j<k} \Delta_j$, and that $\sum_j \Delta_j =$ identity operator. It is not hard to check that:

$$\Delta_k f(x) = \sum_{I \in \mathcal{D}_k} \langle f, h_I \rangle h_I(x),$$

here $\langle .,. \rangle$ denotes the standard inner product in L^2. This proves that the Haar system is complete.

2.3. Dyadic paraproducts and BMO. A locally integrable function b is in the space of *bounded mean oscillation*, BMO, if

$$\frac{1}{|J|} \int_J |b(x) - m_J b|^2 \le C|J|,$$

for every interval J, recall that $m_J b = \frac{1}{|J|} \int_J b$. This is equivalent to the *Carleson condition*

$$b \in BMO \iff \sum_{I \in \mathcal{D}(J)} b_I^2 \le C|J| \quad \forall J,$$

where here $b_I = \langle b, h_I \rangle$. (It is just an application of Plancherel's Theorem for orthonormal basis.)

Define formally the *dyadic paraproduct* associated to a function $b \in BMO$ by:

(3) $$\Pi_b f = \sum_k E_k f \Delta_k b.$$

The dyadic paraproduct is a bilinear operator known to be bounded in L^p, more precisely,

$$\|\Pi_b f\|_p \le C\|b\|_{BMO}\|f\|_p.$$

See [**7**], or [**2**], for more about BMO, paraproducts and related subjects.

2.4. Weights. A *doubling weight* ω is a positive locally integrable function such that $\omega(2I) \leq C\omega(I)$ for all intervals I. (We are using the notations $\omega(I) = \int_I \omega$, and $2I$ is an interval concentric to I and with double length.)

A weight ω is in A_∞ if given $\epsilon > 0$, there exists $\delta > 0$ such that for any interval I, $E \subset I$, such that $|E| < \delta|I|$, then $\omega(E) \leq \epsilon\omega(I)$. (Eg: $\omega(x) = |x|^\alpha$ for $-1 < \alpha$.)

Every A_∞ weight is doubling, but the converse is not true (see [4],[9]). There are equivalent definitions of A_∞, for example:

A weight ω is in A_∞ if for every interval I

$$\frac{1}{|I|} \int_I \omega \leq C \exp\left(\frac{1}{|I|} \int_I \ln \omega\right).$$

The smallest of such C's is called the A_∞ constant of the weight ω.

See [6] for the general theory of weights.

A weight ω is in RH_p (reverse Hölder p) if for every interval I

$$\left(\frac{1}{|I|} \int_I \omega^p\right)^{1/p} \leq C \frac{1}{|I|} \int_I \omega.$$

The smallest of such C's is called the RH_p constant of the weight ω.

(Eg: $\omega(x) = |x|^\alpha$ for $\alpha > -1/p$.)

The main properties of these classes of weights are the following:

(a) if $\omega \in RH_p$ then $\omega \in RH_{p+\epsilon}$ for some $\epsilon > 0$;

(b) if $p < q$ then $RH_q \subset RH_p$;

(c) $A_\infty = \cup_{p>1} RH_p$.

Property (b) is a trivial consequence of Hölder's inequality. Property (a) is a classical result of Gehring (see [5], [6]). And property (c) can be found in [6], Thm. 2.11.

Remark 1: The A_∞ constant of a weight ω forces a lower bound p_o, $p_o > 1$, on the range of p's such that ω is not in RH_p. This can be seen in the proof of property (c) (see [5], [6]).

There is a classical correspondence between A_∞ weights and functions in BMO. Namely, if $\omega \in A_\infty$ then $b = \log \omega \in BMO$. Conversely if, $b \in BMO$ and has sufficiently small BMO norm, then $e^b \in A_\infty$. We will consider a different correspondence in the next section, first introduced in [3].

3. Dyadic characterization of weights

In this section we will restrict our attention to functions defined on the unit interval $I_o = [0,1]$.

Let ω be a weight defined on I_o, such that $E_o\omega = m_{I_o}\omega = 1$. By the Lebesgue differentiation theorem, $\lim_{k\to\infty} E_k\omega(x) = \omega(x)$ for a.e. x; we can then write the telescoping product:

$$\omega(x) = \prod_{k=0}^{\infty} \frac{E_{k+1}\omega(x)}{E_k\omega(x)} = \prod_{k=0}^{\infty} \left(1 + \frac{\Delta_k\omega(x)}{E_k\omega(x)}\right).$$

Let us define the function b, at least formally, by

(4) $$b = \sum_{k=0}^{\infty} \Delta_k b, \quad \Delta_k b = \frac{\Delta_k\omega}{E_k\omega}.$$

Still at the formal level, given a locally integrable function b, we can write:

(5) $$\omega = \prod_{k=0}^{\infty}(1 + \Delta_k b).$$

The partial products are always defined and, in case of convergence, they correspond to the expectation of ω at level k, i.e.:

(6) $$E_k\omega = \prod_{j=0}^{k-1}(1 + \Delta_j b).$$

DEFINITION 3.0.1. A locally integrable function b is of RH_p^d-type if

(7) $$\sum_{I\in\mathcal{D}(J)} m_I^p\omega\, b_I^2 \leq C m_J^p\omega\, |J|, \quad \forall J \in \mathcal{D}(I_o);$$

here $m_I\omega = \prod_{I\subset I'\in\mathcal{D}(I_o)}(1 + b_{I'}h_{I'}(x_I))$, where $x_I \in I$.

Under certain conditions, the formal equations (4) and (5) make sense. Properties of the weight ω can be read off properties on the corresponding function b, and viceversa. We have the following *dictionary*:

  (i) if $|\Delta_k b| < 1$ then the partial products in (6) converge weakly to a positive measure.
  (ii) $|\Delta_k b| < 1 - \epsilon$ for all $k \geq 0$ if and only if ω is a dyadic doubling weight.
  (iii) $b \in BMO^d$ if and only if $\omega \in A_\infty^d$. The A_∞ constant of ω depends only on the BMO norm of b.
  (iv) b is of RH_p^d-type if and only if $\omega \in RH_p^d$.

Remark 2: In this setting all conditions are *dyadic*, i.e. they hold on dyadic intervals, the superscript d indicates that (e.g. BMO^d, A_∞^d, etc).

The results (i)-(iii) appeared first in [**3**]. This characterization of the dyadic RH_p classes is due to S. Buckley (see [**1**]).

Remark 3: For the dyadic theory to resemble the classical theory (in particular if we want properties (a)-(c) to hold), we have to assume that the weight ω is dyadic doubling (i.e. $\omega(\tilde{I}) \leq C\omega(I)$, where $\tilde{I}$ is the dyadic parent of I); or equivalently that the corresponding b satisfies $|\Delta_k b| < 1 - \epsilon$ (by (ii)).

4. Inverting $(I - \lambda \Pi_b)$

The following results are known (see [8]). We will state the theorems for functions defined on the unit interval $I_o = [0, 1]$, in that case we must assume that the functions have mean value zero on I_o. The results are true for functions in $L^p(\mathbf{R})$. Let $L_o^p(I_o) = \{f \in L^p(I_o) : \int_{I_o} f = E_0 f = 0\}$.

When writing spaces of functions or classes of weights we will sometimes "forget" the domain of definition I_o, (e.g. L_o^p, RH_p^d, where it should read $L_o^p(I_o)$, $RH_p^d(I_o)$.)

THEOREM 4.0.1. Given a locally integrable function b such that $|\Delta_k b| < 1 - \epsilon$ for all $k \geq 0$, the operator $(I - \Pi_b)$ has a bounded inverse in $L_o^p(I_o)$ if and only if b is of RH_p^d-type. Moreover, we have an explicit formula for the inverse operator

$$(I - \Pi_b)^{-1} g(x) = \sum_{k=0}^{\infty} \Delta_k g(x) \prod_{j > k} (1 + \Delta_j b(x)).$$

Let $\omega = \prod_{j=0}^{\infty} (1 + \Delta_j b)$. We can write, by (6),

$$\prod_{j > k} (1 + \Delta_j b(x)) = \frac{\omega(x)}{E_k \omega(x)(1 + \Delta_k b(x))}.$$

Given a doubling weight ω, define formally the operator

$$P_\omega g(x) = \sum_{k=0}^{\infty} \frac{\omega(x) \, \Delta_k g(x)}{E_k \omega(x) \, (1 + \Delta_k b(x))},$$

(recall that $\Delta_k b = \Delta_k \omega / E_k \omega$).

THEOREM 4.0.2. P_ω is a well defined and bounded operator in $L^p(I_o)$ if and only if $\omega \in RH_p^d(I_o)$.

The operator P_ω is an example of the *multiplier operator*:

$$Tf(x) = \sum_{k=0}^{\infty} \omega_k(x) \Delta_k f(x).$$

Clearly if the multipliers are constant functions, $\omega_k = a_k$, then T is bounded in L^p if and only if the sequence of a_k's is bounded. The general conditions on the sequence of multipliers that will guarantee boundedness of T are not known. The necessary and sufficient conditions are known for a few particular cases (see [8]).

PROOF. [Sketch of the proof of Theorem 4.0.1] Suppose that f is a solution of the equation $f = g + \Pi_b f$. It is not hard to check that $\Delta_k f = \Delta_k g + E_k f \Delta_k b$, using the properties of the expectation and difference operators, and the definition of the dyadic paraproduct. Recall that $\Delta_k = E_{k+1} - E_k$, we obtain then the *recurrence equation*

$$E_{k+1}f = \Delta_k g + (1 + \Delta_k b)E_k f.$$

Solving the recurrence, and passing to the limit in L^p (using Theorem 4.0.2) we get that $f = P_\omega g$. $\square$

The paraproduct is a bilinear operation, in particular $\lambda\Pi_b = \Pi_{\lambda b}$. Therefore, by the previous theorems, questions about the invertibility of $(I - \lambda\Pi_b)$ are reduced to questions about the weight

$$(8) \qquad\qquad \omega_\lambda = \prod_{k=0}^{\infty}(1 + \lambda\Delta_k b),$$

corresponding to the function λb.

Given a doubling weight $\omega \in RH_p^d$, $b \in BMO$ corresponding to ω as described in the previous section, the operator $(I - \Pi_b)$ is invertible in L^p. By spectral theory (the resolvent is an open set of $\mathbf{C}$), we know that $(I - \lambda\Pi_b)$ will be invertible in neighborhoods of $\lambda = 1$ and $\lambda = 0$. Is this last statement true for $-1 < \lambda < 1$?

This question can be translated into a question about weights.

Does multiplication by $-1 < \lambda < 1$ on the b side preserves RH_p^d weights? Graphically:

$$b \longleftrightarrow \omega \in RH_p^d$$
$$\lambda b \longleftrightarrow \omega_\lambda \in RH_p^d \quad ?$$

The answer is negative. For $-1 < \lambda < 0$ counterexamples were constructed in [8]. We will present a counterexample for $0 < \lambda < 1$ in the last section.

If we replace RH_p^d by doubling A_∞^d, then the statement is true, more precisely:

LEMMA 4.0.3. *Given a doubling weight* $\omega \in A_\infty^d$ *then* ω_λ *is a doubling* A_∞^d *weight for every* $-1 \leq \lambda \leq 1$. *Moreover the* A_∞ *constants are uniformly bounded.*

PROOF. ω is a doubling A_∞^d weight $\Longleftrightarrow b \in BMO$ and $|\Delta_k b| < 1 - \epsilon$ (properties (ii) and (iii)) $\Rightarrow \lambda b \in BMO$, and, since $-1 \leq \lambda \leq 1$, certainly $|\Delta_k \lambda b| = |\lambda\Delta_k b| < 1 - \epsilon$, and $\|\lambda b\|_{BMO} \leq \|b\|_{BMO} \Longleftrightarrow \omega_\lambda$ is a doubling A_∞^d weight with A_∞ constant depending only on $\|b\|_{BMO}$. $\square$

Nevertheless it is true that:

THEOREM 4.0.4. *Given a doubling weight* $\omega \in RH_p^d$, *then there exists* $p_o > 1$ *such that* $\omega_\lambda \in RH_{p_o}^d$ *for all* $-1 \leq \lambda \leq 1$.

This implies that:

THEOREM 4.0.5. Let b be a locally integrable function such that $|\Delta_k b| < 1 - \epsilon$ for all k. If $(I - \Pi_b)$ has a bounded inverse in L^p then there exists $p_o > 1$ such that $(I - \lambda \Pi_b)$ is invertible in L^{p_o} for all $-1 \leq \lambda \leq 1$.

PROOF. If $|\Delta_k b| < 1 - \epsilon$ and $(I - \Pi_b)$ has a bounded inverse in L^p then, by Theorem 4.0.1, b is of RH_p^d-type $\Longleftrightarrow$ the corresponding ω is doubling and in RH_p^d (by (ii) and (iv)) $\Rightarrow$ there exists $p_o > 1$ such that $\omega_\lambda \in RH_{p_o}^d$ for all $-1 \leq \lambda \leq 1$ (by Theorem 4.0.4), and ω_λ is certainly doubling (by Lemma 4.0.3) $\Longleftrightarrow$ λb is of RH_p^d-type and $|\Delta_k \lambda b| < 1 - \epsilon$, and using once more Theorem 4.0.1 we conclude that $(I - \lambda \Pi_b)$ is invertible in L^{p_o} for all $-1 \leq \lambda \leq 1$. $\square$

PROOF. [proof of Theorem 4.0.4] Given a doubling weight $\omega \in RH_p^d$ then $\omega \in A_\infty^d$ (by property (b)). By Lemma 4.0.3 it is also true that for $-1 \leq \lambda \leq 1$, ω_λ are doubling A_∞^d weights, with A_∞ constants uniformly bounded. That implies (see Remark 1 in page 4) the existence of $p_o > 1$ so that $\omega_\lambda \in RH_p^d$ for all $p > p_o$ and for all $-1 \leq \lambda \leq 1$. $\square$

5. Counterexample

THEOREM 5.0.6. There exist a doubling dyadic weight ω on $[0,1]$, $1 < p < \infty$, and $0 < \lambda < 1$, such that $\omega \in RH_p^d$ but $\omega_\lambda \notin RH_p^d$.

The proof of this theorem will follow easily from the next two lemmas.

LEMMA 5.0.7. There exists a one-parameter family of weights ω^t, $1/2 \leq t \leq 1$, with the following properties:
1) $\int_0^1 \omega^t = 1$ for all $1/2 \leq t \leq 1$,
2) ω^t is doubling for all $1/2 \leq t \leq 1$,
3) $\omega^t \in RH_p^d([0,1])$ if and only if $1 < p < p_t$, where $p_t = \ln 8 / \ln[8t^2(t-1)]$ for $1/2 \leq t < (1+\sqrt{5})/4$, and $p_t = \infty$ for $(1+\sqrt{5})/4 \leq t \leq 1$.

The weights ω^t constructed in the previous lemma have the property that their structure is preserved under the operation $(\omega^t)_\lambda$, in particular we can keep track of the RH_p^d classes that they belong to. More precisely:

LEMMA 5.0.8. $(\omega^t)_\lambda = \omega^{t_\lambda}$, where $t_\lambda = 1/2 + \lambda(t - 1/2)$, for all $1/2 \leq t \leq 1$.

PROOF. [Proof of Theorem 5.0.6] Choose $(1+\sqrt{5})/4 \leq t_o \leq 1$ (eg. $t_o = 7/8$). Set $\omega = \omega^{t_o}$. By Lemma 5.0.7, ω is doubling and $\omega \in RH_p^d$ for all $1 < p < \infty$. In particular, $\omega \in RH_{p_o}^d$ for $p_o = p_{(t_o)_\lambda} > 1$, where $(t_o)_\lambda = 1/2 + \lambda(t_o - 1/2) > 1/2$, and we choose $\lambda > 0$ such that

$$(t_o)_\lambda < (1+\sqrt{5})/4$$

(for $t_o = 7/8$, and $(t_o)_\lambda = 2/3$ we get $\lambda = 4/9 < 1$).
Combining lemmas 5.0.8 and 5.0.7, we get that $\omega_\lambda = (\omega^{t_o})_\lambda = \omega^{(t_o)_\lambda}$ is doubling and belongs to RH_p^d only for $1 < p < p_{(t_o)_\lambda}$; in particular it does not

belong to $RH_{p_o}^d$ for $p_o = p_{(t_o)_\lambda}$ (in our example $p_o = p_{2/3} = \ln 8/\ln(32/27) > 1$). $\square$

PROOF. [Proof of Lemma 5.0.7]

Fix $1/2 \leq t \leq 1$.

Let $I_k = (2^{-k}, 2^{-k+1}]$ for $k = 1, 2, \ldots$. Clearly $I_o = (0,1] = \cup_{k=1}^\infty I_k$. Define the step function

$$\omega^t(x) = \sum_{k=1}^\infty c_k(t)\chi_{I_k}(x),$$

where if we let $s = 1 - t$, then

$$c_k(t) = \begin{cases} 2^k(t^2s)^n s & k = 3n+1 \\ 2^k(t^2s)^n ts & k = 3n+2 \\ 2^k(t^2s)^n t^3 & k = 3n+3 \end{cases}$$

Remark: The numbers t and s represent the proportion of the mass of ω on a given interval I_k^* that we have distributed among its children I_{k+1}, I_{k+1}^*; where $I_k^* = \cup_{j>k} I_j$ (the sibling of I_k).

1) Just computing and since $s + t = 1$, we get

$$\int_0^1 \omega^t = \sum_{k=1}^\infty c_k(t)|I_k| = (s + ts + t^3)\sum_{n=0}^\infty (t^2s)^n = \frac{s + ts + t^3}{1 - t^2s} = 1$$

2) We want to show that ω^t is a dyadic doubling weight. We must check that the mass in any dyadic interval I is comparable with that of its parent $\tilde{I}$, i.e. $\omega^t(\tilde{I}) \leq C\omega^t(I)$, for all $I \in \mathcal{D}(I_o)$. It is enough to consider only those intervals $\tilde{I}_k = I_k \cup I_k^*$. Because of the scale invariance it is enough to consider only the case $k = 1$, $\tilde{I}_1 = I_o = I_1 \cup I_1^*$, $I_1^* = \cup_{j\geq 2} I_j$. By 1) $\omega^t(\tilde{I}_1) = 1$, by definition $\omega^t(I_1) = s = 1 - t \leq 1/2$. It follows that $\omega^t(I_1^*) = t \geq 1/2$. Clearly $\omega^t(\tilde{I}) \leq s^{-1}\omega(I)$.

3) We want to show that $\omega^t \in RH_p^d$ for $1 < p < p_t$. Scaling once more, it is enough to check that for $p < p_t$

$$\int_0^1 (\omega^t)^p \leq C\left(\int_0^1 \omega^t\right)^p = C.$$

Now $(\omega^t)^p(x) = \sum_{k+1}^\infty c_k^p \chi_{I_k}(x)$, hence

$$\int_0^1 (\omega^t)^p(x)dx = \sum_{k+1}^\infty c_k^p 2^{-k}$$

$$= ((2s)^p + (2^2ts)^p + (2^3t^3)^p)\sum_{n=0}^\infty \left[\frac{(2^3t^2s)^p}{2^3}\right]^n.$$

This series converges if and only if

(9)
$$\frac{(2^3t^2s)^p}{2^3} < 1.$$

Recall that $s = 1 - t$. Set $f(t) = 8t^2(1 - t)$. It is a straightforward calculation to check that for $(1 + \sqrt{5})/4 \leq t \leq 1$ then $f(t) \leq 1$. Hence in this range of t's (9) holds for every $1 < p$, i.e. $p_t = \infty$. For $1/2 \leq t < (1 + \sqrt{5})/4$ we have that $f(t) > 1$. Therefore (9) holds only if and only if $p < p_t = \ln 8/\ln(8t^2(1 - t))$.

This finishes the proof of the lemma. $\square$

PROOF. [Proof of Lemma 5.0.8] We can use the results in §3 to write:

$$\omega^t = \prod_{k=0}^{\infty}(1 + \Delta_k b^t),$$

$$\omega_\lambda^t = \prod_{k=0}^{\infty}(1 + \lambda\Delta_k b^t).$$

We can find explicitly b^t. Recall that

$$\Delta_k b^t = \frac{\Delta_k \omega^t}{E_k \omega^t} = \frac{E_{k+1}\omega^t - E_k \omega^t}{E_k \omega^t}.$$

By the definition of ω^t, it is clear that $\Delta_k b^t(x)$ is not 0 only for those $x \in I_k^* = \cup_{j>k}I_j$, and in that case, $|\Delta_k b^t(x)| = t - s$ (by the remark in the proof of Lemma 5.0.7). This implies that for $0 < \lambda < 1$

$$|\lambda\Delta_k b^t(x)| = \begin{cases} \lambda(t - s) & x \in I_k^* \\ 0 & \text{otherwise} \end{cases}$$

We can write $\lambda(t - s) = t_\lambda - s_\lambda$ where $t_\lambda = 1/2 + \lambda(t - 1/2)$ and $s_\lambda = 1/2 - \lambda(t - 1/2)$; clearly $t_\lambda + s_\lambda = 1$. The structure of the weight ω_λ^t is exactly the same as the structure of the initial weight ω^t, except that we now replace t by t_λ. $\square$

More refined versions of this counterexample will appear elsewhere. A slightly more delicate construction will allow us to construct examples where the index p_o in Theorem 5.0.6 can be made as close to one as we want (in our example the worst $p_o = p_{2/3}$).

REFERENCES

1. S. BUCKLEY, *Summation conditions on weights*. Michigan Math. J. 40 #1, 153-170, 1993.
2. M. CHRIST, *Lectures on singular integral operators*. Regional Conferences Series in Math, AMS # 77, 1990.
3. R. FEFFERMAN, C. KENIG, J. PIPHER, *The theory of weights and the Dirichlet problem for elliptic equations*. Annals of Math. # 134, 65-124, 1991.
4. C. FEFFERMAN, B. MUCKENHOUPT, *Two non-equivalent conditions for weight functions*. Proc. AMS, vol 45, # 1, 1974.
5. F.W. GEHRING, *The L^p integrability of the partial derivatives of a quasiconformal mapping*. Acta Math. #130, 265-277, 1973.
6. J. GARCIA-CUERVA, J.L. RUBIO de FRANCIA, *Weighted norm inequalities and related topics*. North Holland, 1985.
7. Y. MEYER, *Ondelettes et Opérateurs II*. Herman 1990.

8. M.C. PEREYRA, *On the resolvent of dyadic paraproducts.* To appear in Revista Matemática Iberoamericana, vol 10 # 3, 1994.
9. J.O. STRÖMBERG, *Non-equivalence between two kinds of conditions on weight functions.* Proc. Sym. in Pure Math., vol XXX, part I, 1979.
10. L. WARD, *Personal communication.* 1994.

DEPARTMENT OF MATHEMATICS, PRINCETON UNIVERSITY, PRINCETON, NJ 08544
Current address: Department of Mathematics, IAS, Princeton, NJ 08540
E-mail address: crisp@math.princeton.edu

Contemporary Mathematics
Volume **189**, 1995

On an analogue of the Wold decomposition for a π- semi-unitary operator and its model representation

V. A. STRAUSS

1. Introduction

The present papers deals with operator theory in indefinite inner product spaces. There are some monographs (see, for instance [**1**], [**2**]) which are devoted to this area of problems and contain corresponding terminological and bibliographical information. We shall consider the different aspects of generalizations for π-semi-unitary operators of the well-known Wold decomposition for semi-unitary operators. After that we shall describe a model representation of functional type for π-semi-unitary operators.

2. Matrix representation of a π- semi-unitary operator

First of all let us introduce the main notation which will be used in the present paper. Let $\mathcal{H}$ be a separable space Π_κ i.e. a complex Hilbert space with a usual scalar product $(\cdot, \cdot)$ and an indefinite scalar product $[\cdot, \cdot]$, defined by $[x, y] = (Jx, y)$, where $J = P_+ - P_-$, $P_+ + P_- = I$, P_+ and P_- are orthoprojections in Hilbert sense, $\kappa = min \, dim\{P_+\mathcal{H}, P_-\mathcal{H}\}$, $0 < \kappa < \infty$. Usually J is called a fundamental symmetry. Without loss of generality we may consider the case $\kappa = dim P_-\mathcal{H}$. The symbols corresponding to the form $[\cdot, \cdot]$ will be given in square brackets and will be equiped by prefix "π-". For inctance we shall denote by symbol $[\perp]$ a π-orthogonality and by symbol $\perp$ an orthogonality in a Hilbert sense. As usualy $\mathbf{C}$ is the complex plane, $T \subset C$ is the unit circle, $\sigma(C)$ and $\sigma_p(C)$ are respectively the spectrum and the point spectrum (set of eigen-values) of an operator C. In this paper we consider linear operators only.

Let U be a π-semi-unitary operator, i.e. a domain of operator U is $\mathcal{H}$, $U\mathcal{H} \subset \mathcal{H}$ and for all $x, y \in \mathcal{H}$ we have $[Ux, Uy] = [x, y]$. Everywhere below we shall assume $U\mathcal{H} \neq \mathcal{H}$.

1991 *Mathematics Subject Classification*. Primary 47B50.

We call the subspace $\mathcal{L} = (U\mathcal{H})^{[\perp]}$ the defect subspace of U. The defect subspace of U is a wandering subspace for U, but a wandering subspace is not uniquely determined by U in Π_κ (see [8], Example 6.4). By virtue of above conditions the subspace $\mathcal{L}$ is not equal 0 or $\mathcal{H}$ and it is positive. In fact, as far as U is semi-unitary, the subspace $UP_-\mathcal{H}$ is a κ-dimensional negative subspace and hence it is a maximal negative subspace (let us remind a negative subspace is named maximal negative if it isn't a proper subset of other negative subspace). So the subspace $(UP_-\mathcal{H})^{[\perp]}$ is positive. It remains to note that $\mathcal{L} \subset (UP_-\mathcal{H})^{[\perp]}$.

Let us introduce some subspaces connected with U. Let $\mathcal{H}_s$ be a closed linear span of a set of subspaces of the form $U^m\mathcal{L}$, $m = 0, 1, \ldots$. It is clear that $\mathcal{H}_s$ is a non-negative subspace with generally speaking isotropic part. If $\mathcal{H}_s$ is a positive subspace then $U|_{\mathcal{H}_s}$ is a usual operator without any singularities connected with an indefinite metric, therefore everywhere below we assume that $\mathcal{H}_s \cap \mathcal{H}_s^{[\perp]} \neq \{0\}$. Note that a regular case was studied in [8] and [9]. We introduce the subspace $\mathcal{H}_r = \mathcal{H}_s^{[\perp]}$ also.

Proposition 1. $U\mathcal{H}_r = \mathcal{H}_r$.

In order to prove this proposition we first mention the following theorem.

Theorem 1. Let $\mathcal{H}$ be a space of the form

$$\mathcal{H} = \mathcal{H}_0 \oplus \mathcal{H}_1 \oplus \mathcal{H}_2 \oplus \mathcal{H}_3, \tag{1}$$

where $\mathcal{H}_0$ and $\mathcal{H}_1$ have a same finite dimension and let a fundamental symmetry J have a representation of the form

$$J = \begin{pmatrix} 0 & V^{-1} & 0 & 0 \\ V & 0 & 0 & 0 \\ 0 & 0 & I_2 & 0 \\ 0 & 0 & 0 & J_3 \end{pmatrix}$$

with respect to the decomposition (1), where $V : \mathcal{H}_0 \to \mathcal{H}_1$ is an isometric operator, I_2 is an identical operator in $\mathcal{H}_2$ and J_3 is a fundamental symmetry of the inner product $[\cdot, \cdot]$ restricted on the space $\mathcal{H}_3$. Let the operator U admit the representation

$$U = \begin{pmatrix} U_{00} & 0 & 0 & 0 \\ U_{10} & U_{11} & U_{12} & U_{13} \\ U_{20} & 0 & U_{22} & 0 \\ U_{30} & 0 & 0 & U_{33} \end{pmatrix} \tag{2}$$

with respect to the decomposition (1). Then the operator U is a π-semi-unitary operator iff the following conditions are fulfilled

(3)

$a)$ $V^{-1}U_{11}^*V_{00} = I_0$;

$b)$ $U_{22}^*U_{22} = I_2$;

$c)$ $U_{33}^*J_3U_{33} = J_3$;

$d)$ $U_{20} = -U_{22}U_{12}^*VU_{00}$;

$e)$ $U_{10} = U_{11}V\{-0.5(U_{20}^*U_{20} + U_{30}^*J_3U_{30}) + iA\}$;

$f)$ $U_{30} = -U_{33}J_3U_{13}^*VU_{00}$;

$g)$ $\mathcal{L} = \mathcal{H}_2 \ominus U_{22}\mathcal{H}_2$;

$h)$ $\mathcal{H}_3 = U_{33}\mathcal{H}_3$;

where I_0 and A are respectively identical and selfadjoint operators acting on $\mathcal{H}_0$.

The proof of this theorem is omitted so far as it is easy and can be reduced to the direct comparison betweem the definition of π-semi-unitary operator and the condition (3). Furthermore the similar results concerning the decomposition of a contraction on Π_κ or on Krein space are contained in [3],[4] and our theorem may be considered as their simple spesification in the case of π- semi-unitary operator.

In order to prove Proposition 1 let us denote $\mathcal{H}_1=\mathcal{H}_r \cap \mathcal{H}_s$, $\mathcal{H}_0 = J\mathcal{H}_1$, $\mathcal{H}_2 = \mathcal{H}_s \ominus \mathcal{H}_1$, $\mathcal{H}_3 = \{\mathcal{H}_s \oplus \mathcal{H}_0\}^{[\perp]}$.

Here J is a fundamental symmetry mentioned at beginning of this section and it is connected with a Hilbert scalar product on $\mathcal{H}$. On the other hand there are different topologically equvalent ways for the definition on $\mathcal{H}$ a Hilbert scalar product together with a corresponding fundamental symmetry. The concrete choice is not essential in this theory, hence without loss of generality we can suppose that the equality $(\cdot,\cdot)=[\cdot,\cdot]$ take place in $\mathcal{H}_2$ and $\mathcal{H}_3 \perp \{\mathcal{H}_s \oplus \mathcal{H}_0\}^{[\perp]}$ (note that in this connection it isn't necessary to change the subspace $\mathcal{H}_0$). In this case the decomposition $\mathcal{H} = \mathcal{H}_0 \dotplus \mathcal{H}_1 \dotplus \mathcal{H}_2 \dotplus \mathcal{H}_3$, the operators J and U satisfy the conditions of Theorem 1. Below the choice of $(\cdot,\cdot)$ will be the same.

Now the correctness of this proposition immediately follows from the matrix representation (2) and property (3).

3. Wold decomposition and related topics

Further consideration of properties of π-semi-unitary operator is assosiated with a Wold decomposition [12] or, more exactly, with its indefinite analogue. Let us remind that the subspace $\mathcal{M} \subset \mathcal{H}$ is named regular if $\mathcal{H} = \mathcal{M} \oplus \mathcal{M}^{[\perp]}$.

Definition 1. An operator B acting on $\mathcal{H}$ is called completely π-nonunitary if there is no regular invariant (for B) subspace $\mathcal{H}' \neq \{0\}$ such that the operator $B|_{\mathcal{H}'}$ is π-unitary ($\mathcal{H}'$ may be a definite subspace).

Definition 2. An indefinite Wold decomposition of U is the decomposition of the form

(4)
$$\mathcal{H}_N[+]\mathcal{H}_U,$$

where $\mathcal{H}_N$ and $\mathcal{H}_U$ are invariant subspaces of the operator U, while $U|_{\mathcal{H}_U}$ is a completely π-nonunitary operator and if $\mathcal{H}_U \neq \{0\}$ then $U|_{\mathcal{H}_U}$ is a π-unitary operator.

We must note that these definitions are not generally accepted. For instance the operator in Example 6.5 from [8] is completely π-nonunitary and has an indefinite Wold decomposition ($\mathcal{H}_U = \{0\}$) in the sense of Definition 2, but it does not have a Wold decomposition in the sense of [8].

Proposition 2. For the elements of decomposition (4) (if it exists) there are the following inclusions

$$\mathcal{H}_U \subset \mathcal{H}_r, \ \mathcal{H}_s \subset \mathcal{H}_N.$$

In fact by virtue of the defect subspace we have $\mathcal{H}_U[\perp]\mathcal{L}$ which gives $\mathcal{H}_s[\perp]\mathcal{H}_U$, the rest is trivial.

Below we shall give two examples which show that in contrast to the definite case in the case of Π_κ the problems of of existence and uniqueness of decomposition (4) have a negative solution in general.

Let $U_r = U|_{\mathcal{H}_r}$.

Lemma. Let $\{\lambda_j\}$ be the set of critical poins of the spectral function of the operator U_{33} (see [5], [6]) and let the condition

(5) $$\Delta \cap \sigma(U_{11}) = \emptyset$$

be fullfilled for the arc $\Delta = \{\xi : \ \xi = e^{it}, \ t \in (\alpha, \beta)\}$, where $e^{i\alpha}, e^{i\beta} \notin \{\lambda_j\}$. If $\sigma(U_{33}) \cap \Delta \neq \emptyset$, there exists a π-orthogonal (=π-selfadjoint) projection $E(\Delta)$ such that

a) $E(\Delta)U = UE(\Delta)$;
b) $\sigma(U|_{E(\Delta)\mathcal{H}}) \subset \bar{\Delta}$;
c) $E(\Delta)\mathcal{H} \subset \mathcal{H}_r$;
d) $\sigma(U_{r|_{(I-E(\Delta))\mathcal{H}_r}}) \subset C \setminus \Delta$.

Proof. According to the decomposition $\mathcal{H}_r = \mathcal{H}_1 \oplus \mathcal{H}_3$ we have

$$U_r = \begin{pmatrix} U_{11} & U_{13} \\ 0 & U_{33} \end{pmatrix},$$

hence there exists the following representation for the resolvent of the operator U_r

(6) $$R_\lambda(U_r) = \begin{pmatrix} R_\lambda(U_{11}) & -R_\lambda(U_{11})U_{13}R_\lambda(U_{33}) \\ 0 & R_\lambda(U_{33}) \end{pmatrix}.$$

According to this representation and using the scheme given for instance in [10] and [7] we can construct the projection $E^r(\Delta)$ (during the process of the construction one uses an improper Riesz integral on the curve lying symmetrically with respect to $\mathbf{T}$) into an invariant subspace of the operator U_r. This subspace corresponds to the part of spectrum of operator U_r concentrated on the arc Δ

without $Ker(U_r - e^{i\alpha}I)$ and $Ker(U_r - e^{i\beta}I)$. By virtue of condition (5) the operator $E^r(\Delta)$ has in turn the representation

$$E^r(\Delta) = \begin{pmatrix} 0 & E_{13} \\ 0 & E_{33}(\Delta) \end{pmatrix},$$

where $E_{33}(\Delta)$ is a π--orthogonal projection corresponding to the spectral function of U_{33}. It should be noted that in our case $E_{33}(T)\mathcal{H}_3$ may be not equal to $\mathcal{H}_3$, since $\sigma(U_{33}) \setminus T \neq \emptyset$ in general. ¿From the form of $E^r(\Delta)$ we have that

(7)
$$\left. \begin{array}{l} \sigma(U_r|_{(E^r(\Delta)\mathcal{H}_r)}) \subset \bar{\Delta}; \\ \sigma(U_r|_{(I-E^r(\Delta))\mathcal{H}_r}) \subset C \setminus \Delta; \end{array} \right\}$$

and

(8)
$$\left. \begin{array}{l} E_{13}(\Delta)E_{33}(\Delta) = E_{13}(\Delta); \\ E_{33}^*(\Delta) = J_3 E_{33(\Delta)} J_3; \\ E_{33}^2(\Delta) = E_{33}(\Delta). \end{array} \right\}$$

Moreover since $E^r(\Delta)$ and U_r commute with each other then

(9)
$$\left. \begin{array}{l} U_{11}E_{13}(\Delta) + U_{33}E_{33} = E_{13}(\Delta)U_{33}; \\ E_{33}(\Delta)U_{33} = U_{33}E_{33}(\Delta). \end{array} \right\}$$

Now let's define $E(\Delta)$ on all $\mathcal{H}$ by the matrix representation

$$E(\Delta) = \begin{pmatrix} 0 & 0 & 0 & 0 \\ E_{13}(\Delta)J_3 E_{13}^*(\Delta)V & 0 & 0 & E_{13}(\Delta) \\ 0 & 0 & 0 & 0 \\ E_{33}(\Delta)J_3 E_{13}^*(\Delta)V & 0 & 0 & E_{33}(\Delta) \end{pmatrix}$$

corresponding to the decomposition (1). The direct verification with due regard for (8) shows that $E(\Delta)$ is a π-orthogonal projection and its image is the same as the image of $E^r(\Delta)$. By virtue of conditions (7) the proof would be finished if we show that $E(\Delta)U = UE(\Delta)$. These operators commute if and only if the conditions (9) and the following conditions

(10)
$$\left. \begin{array}{l} U_{11}E_{13}(\Delta)J_3 E_{13}^*(\Delta)V + U_{13}E_{33}(\Delta)J_3 E_{13}^*(\Delta)V = \\ = E_{13}(\Delta)J_3 E_{13}^*(\Delta)VU_{00} + E_{13}(\Delta)U_{30}; \\ U_{33}E_{33}(\Delta)J_3 E_{13}^*(\Delta)V = \\ = E_{33}(\Delta)J_3 E_{13}^*(\Delta)VU_{00} + E_{33}(\Delta)U_{30}; \end{array} \right\}$$

are satisfied simultaneously. We omit the further proof because the condition (10) can be obtained by means of a simple transformation of the set of the conditions (3), (8) and (9).

Remark. From the construction of $E(\Delta)$ it is follows that for arcs Δ_1 and Δ_2 which satisfy (5) we have

$$E(\Delta_1)E(\Delta_2) = E(\Delta_2)E(\Delta_1) = E(\Delta_1 \cap \Delta_2).$$

Proposition 3. Let an arc Δ satisfy (5) and let $E(\Delta)$ be the corresponding π-orthogonal projection. If for the operator U there exists the decomposition (4) then

$$(11) \qquad\qquad E(\Delta)\mathcal{H} \subset \mathcal{H}_U.$$

In fact let P_N and P_U be the π-orthogonal projections into $\mathcal{H}_N$ and $\mathcal{H}_U$ respectively. Since $\mathcal{H}_U \subset \mathcal{H}_r$ the operator P_U can be regarded also as an operator acting on $\mathcal{H}_r$. Moreover P_U commutes with the operators U_r and U_r^{-1}, the operator $E^r(\Delta)$ belongs to a weak closure of the algebra generated by U_r and U_r^{-1}, therefore on $\mathcal{H}_r$ we have $E^r(\Delta)P_U = P_U E^r(\Delta)$. Thus $P_U E(\Delta) = P_U E^r(\Delta)E(\Delta) = E^r(\Delta)P_U E(\Delta) = E(\Delta)P_U E(\Delta)$. Since P_U and $E(\Delta)$ are π-orthogonal projections then $P_U E(\Delta) = E(\Delta)P_U$ and therefore $P_N E(\Delta) = E(\Delta)P_N$. If we assume that $E(\Delta)P_N \neq 0$ then we obtain that in $\mathcal{H}_N$ there exists the regular subspace (the image of the operator $E(\Delta)P_N$) invariant with respect to U, moreover $U|_{E(\Delta)P_N \mathcal{H}}$ acts as a π-unitary operator. This result contradits the definition of the decomposition (4). Thus $E(\Delta)P_N = 0$ and the conclusion (11) is proved.

Denote by $\tilde{\mathcal{H}}_r$ the closed linear span of the subspaces $E(\Delta)\mathcal{H}$, where Δ runs through the set of all arcs satisfying the conditions (5) and $E(\Delta)$ is the projection introduced in the Lemma.

Proposition 4. In order that an operator U has the decomposition (4) it is necessary that

$$(12) \qquad\qquad \tilde{\mathcal{H}}_r \cap \mathcal{H}_s = \{0\}.$$

The correctness of Proposition 4 follows immediately from Proposition 2 and Proposition 3. On the basis of this proposition it is easy to construct an example of π-semi-unitary operator without the decomposition (4). Such an example is contained in [11]. The description of the mentioned example has the small and simply eliminable inaccuracy (one of the coefficient is skipped), but here we shall give an another example illustrating the insufficiency of the conditions (12) for the existence of the decomposition (4).

Example 1. Let a combination of systems $\{e_m\}_{-\infty}^\infty$ and $\{g_j\}_1^7$ be an orthogonal basis in $\mathcal{H}$ and let operators J and U be determined by the following conditions

$$Jg_j = g_{6-j},\ j = 1, 2, \ldots, 5;\ Jg_6 = g_7,\ Jg_7 = g_6;$$
$$Je_m = e_m,\ m = 0, \pm 1, \pm 2, \ldots;$$
$$Ue_m = \lambda_m e_m + (\lambda_m - 1)g_7,\ m = -1, -2, \ldots;$$
$$Ue_0 = e_1 + g_4 + g_7;\ Ue_m = e_{m+1},\ m = 1, 2, \ldots;$$
$$Ug_1 = g_1 + 2ig_2 - 2g_3 - 2ig_4 - 2ig_7 - 2ie_1;$$
$$Ug_2 = g_2 + 2ig_3 - 2.5g_4 - ig_5 - g_7 - e_1;$$
$$Ug_3 = g_3 + 2ig_4 - 2g_5;\ Ug_4 = g_4 + 2ig_5;\ Ug_5 = g_5;$$
$$Ug_6 = g_6 - 0.5(1 + \textstyle\sum_{m=-1}^{-\infty} |\lambda_m - 1|^2)g_7 - e_1 - \sum_{m=-1}^{-\infty}(1 - \lambda_m)e_m;$$
$$Ug_7 = g_7;$$

where $\lambda_m = exp(i/m)$, $m = -1, -2, \dots$. Direct verification shows that U is a π-semi-unitary operator, the defect subspace of U is one-dimensional subspace and it is generated e_0, $\mathcal{H}_s$ is equal to the closed linear span of $g_4 + g_7$, g_5 and $\{e_m\}_0^\infty$, $\tilde{\mathcal{H}}_r$ is formed by g_7 and $\{e_m\}_{-1}^{-\infty}$.

Suppose that for U there exists the decomposition (4). In this case $\mathcal{H}_N$ is formed as a linear span of $\mathcal{H}_s$ and of certain vectors which are π-orthogonal to $\tilde{\mathcal{H}}_r$ and orthogonal in Hilbert sense to $\mathcal{H}_s$. There are four linear independent vectors of such kind only, i.e. g_1, g_2, g_3 and $g_7 - g_4$. As far as $\mathcal{H}_N$ is a nondegerated subspace at least two vectors in the form $x = g_1 + \alpha_1 g_3 + \beta_1(g_7 - g_4)$ and $y = g_2 + \alpha_2 g_3 + \beta_2(g_7 - g_4)$ must belong to $\mathcal{H}_N$. Since $U\mathcal{H}_N \subset \mathcal{H}_N$ then

$$(U - I)^2 y = -4g_4 - (9i + 4\alpha_2)g_5 + e_1 - e_2 \in \mathcal{H}_N$$

and consequently $g_4 \in \mathcal{H}_N$. On the other hand $(g_4 + g_7) \in \mathcal{H}_N$, moreover at the same time $g_7 \in \tilde{\mathcal{H}}_r$. This leads to a contradiction.

Theorem 2. For the given operator U there exist two finite nonintersecting systems of complex numbers $\{\lambda_j\}$ and $\{\mu_m\}$ from T (each of these systems may be the empty set) and the homomorphism E which maps the lattice L of the subsets from T, where L is generated by the arcs of the form

$$\Delta = \{\xi : \xi = e^{it}, t \in (\alpha; \beta)\}, e^{i\alpha}, e^{i\beta} \notin \{\lambda_j\} \cup \{\mu_m\}, \Delta \cap \{\mu_m\} = \emptyset$$

and possibly by certain subsets from $\{\mu_m\}$, into the commutative lattice of the π-ortogonal projections such that for every $X \in L$

(13)

a) $E(X)U = UE(X)$;

b) if $E(X) \neq 0$, then the operator $U|_{E(X)\mathcal{H}}$ is π − unitary and $\sigma(U|_{E(X)\mathcal{H}}) \subset \bar{X}$;

c) $E(X)\mathcal{H} \subset \mathcal{H}_r$;

d) $\sigma(U|_{\hat{\mathcal{H}}_r^{[\perp]} \cap \mathcal{H}_r}) \cap T \subset \sigma(U_{11})$, where $\hat{\mathcal{H}}_r$ is the closure linear span of the subspaces in the form of $qE(X)\mathcal{H}$;

e) the codimension of $\hat{\mathcal{H}}_r$ is finite with respect to $\mathcal{H}_r$.

Proof. In order to construct the lattice L and the homomorphism E satisfying the conditions (13a-d) it is sufficient to use the Lemma assuming that $\{\mu_m\} = \sigma(U_{11}) \cap T$ and $\{\lambda_j\}$ is equal to the set of the critical points of the spectral function of U_{33} which are not belong to $\{\mu_m\}$. The condition (13e) is fulfilled after this procedure only if the codimension of $\tilde{\mathcal{H}}_r$ with respect to $\mathcal{H}_r$ is finite. In such case $\hat{\mathcal{H}}_r = \tilde{\mathcal{H}}_r$ and the proof is finished, therefore below we shall consider the opposite case only.

Let $\mathcal{H}_r' = \tilde{\mathcal{H}}_r^{[\perp]} \cap \mathcal{H}_r$. Then $\mathcal{H}_r'$ is an infinite-dimensional invariant subspace of the operator U and according to the Lemma we have $\sigma(U|_{\mathcal{H}_r'}) \subset \sigma(U_{11}) \cup (\sigma(U_{33}) \setminus T)$. In virtue of the structure of the operator U_r all points of $\sigma(U|_{\mathcal{H}_r'})$ are isolated and $\sigma(U|_{\mathcal{H}_r'}) \subset \sigma_p(U)$, moreover the dimension of the subspace corresponding to the spectrum $\sigma(U|_{\mathcal{H}'_r}) \setminus T$ is a finite value and therefore we can neglect it. Let $\mathcal{N}_\nu$ be a root subspace for $\nu \in \sigma(U|_{\mathcal{H}'_r}) \cap T$ and $\dim \mathcal{N}_\nu = \infty$. Since $\mathcal{H}_1$ is a finite-dimensional subspace and U_{33} is a π-unitary operator on $\mathcal{H}_3$ there exists a regular eigen-subspace $\mathcal{M}_\nu$ of the operator U with finite codimension with respect to $\mathcal{N}_\nu$. Now let us denote by P_ν the π-orthogonal projection into $\mathcal{M}_\nu$ and let $\{\nu\} \in L$, $E(\{\nu\}) = P_\nu$. In fact this step completes the proof of Theorem 2 since the codimension of the linear span of $\tilde{\mathcal{H}}_r$ and $\mathcal{H}'_r$ is a finite value with respect to $\mathcal{H}_r$.

Corollary 1. $\mathcal{H}_r$ must be a finite-dimensional subspace in case U is a completely π-nonunitary operator.

Corollary 2. If $\sigma_p(U|_{\mathcal{H}_s}) \subset C \setminus T$ then there exists the decomposition (4) for the operator U.

In fact in this case $\{\mu\} = \emptyset$ and therefore $T \in L$.

Corollary 3. An operator U has the decomposition (4) if and only if there exists a regular subspace $\mathcal{H}' \subset \mathcal{H}_r$ such that it is an invariant subspace with respect to U and it has a finite codimension with respect to $\mathcal{H}_r$.

Proof. If the decomposition (4) exists for U then according to the Corollary 1 the subspace $\mathcal{H}_s$ has a finite codimension with respect to the subspace $\mathcal{H}_N$ and it is easy to see that a codimension $\mathcal{H}_U$ with respect to $\mathcal{H}_r$ is equal to a codimension $\mathcal{H}_s$ with respect to $\mathcal{H}_N$. Since $\mathcal{H}_U$ is a finite-dimensional subspace then the necessity of our theorem is proved. Now let $\mathcal{H}'$ be a regular subspace, $\mathcal{H}' \subset \mathcal{H}_r$, the subspace $\mathcal{H}'$ have a finite codimension with respect to $\mathcal{H}_r$ and $U\mathcal{H}' \subset \mathcal{H}'$. From the listed conditions and the Proposition 1 we have $U\mathcal{H}' = \mathcal{H}'$. Now let us consider the operator $U|_{\mathcal{H}'^{[\perp]}}$. If this operator is completely π-nonunitary then the proof is finished, therefore let us assume the contrary. In this case there exists a regular subspace $\mathcal{H}'' \subset \mathcal{H}'^{[\perp]}$ such that $U\mathcal{H}'' = \mathcal{H}''$. By virtue of the Proposition 1 we have $\mathcal{H}''[\perp]\mathcal{H}_s$. As far as a codimension $\mathcal{H}_s$ with respect to $\mathcal{H}'^{[\perp]}$ is equal to a codimension $\mathcal{H}'$ with respect to $\mathcal{H}_r$, without loss of generality we can assume that if the regular subspace $\mathcal{H}''' \subset \mathcal{H}'^{[\perp]}$ is such that $\mathcal{H}'' \subset \mathcal{H}'''$ and $U\mathcal{H}''' = \mathcal{H}'''$, then $\mathcal{H}'' = \mathcal{H}'''$. Now we can set $\mathcal{H}_U = \mathcal{H}'[+]\mathcal{H}''$.

4. The problem of ambiguity for generalized Wold decomposition

It is wellknown that elements of Wold decomposition are uniquely determined in the case of Hilbert space. In Pontrjagin space a situation isn't so simple.

Theorem 3. Let an operator U have the decomposition (4). This decomposition is uniquely determined if and only if $\sigma_p(U|_{\mathcal{H}_U}) \cap \sigma_p(U|_{\mathcal{H}_N}) = \emptyset$.

Proof. Necessity. Let us assume there exists the point $\lambda \in C$, such that $\lambda \in \sigma_p(U|_{\mathcal{H}_U}) \cap \sigma_p(U|_{\mathcal{H}_N})$. Let us show that in this case $\sigma(U|_{\mathcal{H}_U}) \cap \sigma(U_{11}) \neq \emptyset$. In fact if $\lambda \in \sigma_p(U|_{\mathcal{H}_N})$, we have either $\lambda \in \sigma(U|_{11})$ or $1/\bar\lambda \in \sigma(U_{11})$, as otherwise we can separate away from finite-dimensional invariant subspace $\mathcal{H}_r \cap \mathcal{H}_N$ with the help of Riesz projector the regular invariant subspace correponding to the pair of (perhaps coinsident) eigenvalues λ and $1/\bar\lambda$ (we can use, for instance, for this purpose the matrix representation of the operator $U|_{\mathcal{H}_N \cap \mathcal{H}_r}$ analogous to (6)).

Further, without loss of generality we may assume $\mathcal{H}_U \perp \mathcal{H}_N$. In this case the subspaces $\mathcal{H}_U$ and $\mathcal{H}_N$ are invariant not only for U but also for J. Let us fix such orthogonal basis in the space $\mathcal{H}_1$ that the operator U_{11} has the Jordan normal form with respect to this basis and let e be the fixed eigenvector from this basis corresponding the eigenvalue $\lambda \in \sigma_p(U|_{\mathcal{H}_U}) \cap \sigma(U_{11})$. Denote $g = Je_1 \,(= V^{-1}e)$. By virtue of the choice of canonical symmetry J and the relation (3a) we have $g \in \mathcal{H}_N$ and

$$(14) \qquad\qquad Ug = (1/\bar{\lambda})g + z,$$

where $z \in \{g\}^\perp \cap \mathcal{H}_N = \{e\}^\perp \cap \mathcal{H}_N$.

Now take some vector $h \in \mathcal{H}_U$, corresponding to the eigenvalue $1/\bar{\lambda}$ (recall that for π-unitary operator $U|_{\mathcal{H}_U}$ the condition $\lambda \in \sigma_p(U|_{\mathcal{H}_U})$ implies $(1/\bar{\lambda} \in \sigma_p(U|_{\mathcal{H}_U})$ and introduce the new subspace $\mathcal{H}'_N$ by the formula

$$(15) \qquad\qquad \mathcal{H}'_N = (\{e\}^\perp \cap \mathcal{H}_N) \oplus \{\xi(g+h)\}_{\xi \in C}.$$

By virtue of the choice of h and the equality (14) $\mathcal{H}_N{}'$ is a nondegenerate and invariant (with respect to U) subspace and $U|_{\mathcal{H}'_N}$ is a completely π-nonunitary operator. In fact, otherwise there exists the invariant with respect to U regular subspace $\mathcal{H}' \subset \mathcal{H}'_N$, such that $U|_{\mathcal{H}'}$ is a π-unitary operator. In this case $\mathcal{H}'[\perp]\mathcal{H}_s$ and since $\mathcal{H}' \in \{e\}^{[\perp]} \cap \mathcal{H}'_N$. Simultaneously by virtue of (15) $\{e\}^{[\perp]} \cap \mathcal{H}'_N = \{e\}^{[\perp]} \cap \mathcal{H}_N$ since $\mathcal{H}' \subset \mathcal{H}_N$. It's a contradiction! For the completion the proof of necessity it remains to show that if $\mathcal{H}'_U = \mathcal{H}'^{[\perp]}_N$, then $U\mathcal{H}'_U = \mathcal{H}'_U$. Actually it is obviously since $U\mathcal{H}'_U[\perp]U\mathcal{H}'_N$ and by the construction $U\mathcal{H}'_N[+]\mathcal{H}'_N$. Thus we have presented the second decomposition of the form (4).

Now let us go to the proof of sufficiency. We use the scheme of proof by contradiction for this purpose again, i.e. assume

$$(16) \qquad\qquad \sigma_p(U|_{\mathcal{H}_U}) \cap \sigma_p(U|_{\mathcal{H}_N}) = \emptyset$$

and simultaneously there exist the decomposition (4) and the decomposition $\mathcal{H}'_N[+]\mathcal{H}'_U$ of the same type. Note that for operators $U|_{\mathcal{H}_N}$ and there are the the represetations by the form (2) this the properties (3). It is easy to show (see the above reasoning) that $\sigma_p(U|_{\mathcal{H}_N}) \subset \sigma(U_{11}) \cup \sigma(U_{00})$ and $\sigma_p(U|_{\mathcal{H}_N}{}') \subset \sigma(U_{11}) \cup \sigma(U_{00})$, hence there exists a polynom $Q(\xi)$ such that its set of roots belongs to $\sigma(U_{11}) \cup \sigma(U_{00})$ and

$$(17) \qquad\qquad Q(U)\mathcal{H}_N \subset \mathcal{H}_s, \ Q(U)\mathcal{H}'_N \subset \mathcal{H}_s.$$

Let us denote $\mathcal{H}' = CLin\{\mathcal{H}_N, \mathcal{H}'_N\}$. Then $\mathcal{H}'$ is an invariant subspace of the operator U and by virtue of (17) $Q(U)\mathcal{H}' \subset \mathcal{H}_s$, therefore

$$(18) \qquad\qquad \sigma_p(U|_{\mathcal{H}'}) \subset \sigma(U_{11}) \cup \sigma(U_{00}).$$

Without loss of generality we can assume $\mathcal{H}'_N \not\subset \mathcal{H}_N$. Consider the subspace $P_U \mathcal{H}'$, where P_U is a π-orthoprojection on the space $\mathcal{H}_U$. From $\mathcal{H}_N \subset \mathcal{H}'$, $P_U \mathcal{H}_N = \{0\}$, $\mathcal{H}_N \neq \mathcal{H}'$ we have $P_U \mathcal{H}' \neq \{0\}$, $P_U \mathcal{H}' \subset \mathcal{H}'$ and $P_U \mathcal{H}' \subset \mathcal{H}_U$. By virtue of (18) we receive $\sigma_p(U|_{\mathcal{H}_U}) \cap (\sigma(U_{11}) \cup \sigma(U_{00})) \neq \emptyset$. As the set $\sigma_p(U|_{\mathcal{H}_U})$ is symmetric with respect to T, $\sigma_p(U|_{\mathcal{H}_U}) \cap \sigma(U_{11}) \neq \emptyset$. It contradicts (16).

5. Closing remarks

The existence of generalized Wold decomposition for π-semi-unitary operator facilitates a construction it's model representation since in this case it is possible to consider two separate problems of this construction for π-unitary and π-semi-unitary completely π-nonunitary operators respectively. Below we describe a model representation for π-unitary operator.

Let us consider again a separable Pontrjagin space $\mathcal{H}$ and a π-unitary operator U. For the simplicity we shall suppose that U has a unitary spectrum only. Let E_λ be a spectral function of U with the domain $[0,2\pi] \setminus \Lambda$, where Λ is a (finite) set of critical points, $0, 2\pi \notin \Lambda$. We shall consider the case $\Lambda \neq \emptyset$ only. Let us denote by $\tilde{\mathcal{H}}$ the closed linear span of the subspaces $E(\Delta)\mathcal{H}$, where Δ runs all set of segments $\Delta \subset [0,2\pi] \setminus \Lambda$. We shall consider the case $\mathcal{H}_1 = \tilde{\mathcal{H}} \cap \tilde{\mathcal{H}}^{[\perp]} \neq \{0\}$ and denote $\tilde{U} = U|_{\tilde{\mathcal{H}}}$.

Now let us consider a separable (probably finite dimensional) Hilbert space $\mathcal{E}$ with a fixed orthonormal basis $\{e_j\}_1^\alpha$, a non-decreasing continuous on the left acting on the segment $[0,2\pi]$ function $\sigma(t)$ and generated by such function the Lebegue measure μ_σ. After this let us consider a system $\{h_j(t)\}_{j=1}^\alpha$ of μ_σ-measurable scalar functions such that

a) for all $t \in [0; 2\pi]$ and $j = 1, 2, \ldots, \alpha$ $h_j(t) = 0$ or $h_j(t) = 1$;

b) if $h_j(t) = 1$ and $j > 1$ then $h_{j-1}(t) = 1$;

c) for all $j = 1, 2, \ldots, \alpha$ $\mu_\sigma\{t : h_j(t) = 1\} > 0$.

Now the space $L^2_{\vec{\sigma}}(\mathcal{E})$ is a Hilbert space of vector-valued functions $f(t) = \sum_{j=1}^\alpha f_j(t) e_j$, where for all $j = 1, 2, \ldots, \alpha$ we have $f_j(t) = f_j(t) h_j(t)$ and

$$\int_0^{2\pi} \|f(t)\|^2_{\mathcal{E}} d\sigma(t) < \infty .$$

The scalar product on $L^2_{\vec{\sigma}}(\mathcal{E})$ is naturaly defined.

Further, let $\{\tilde{g}_1(t), \tilde{g}_2(t), \ldots, \tilde{g}_n(t)\}$ be a set of vector-valued functions of the same type as functions from the space $L^2_{\vec{\sigma}}(\mathcal{E})$ but

$$\int_0^{2\pi} \|\tilde{g}_j(t)\|^2_{\mathcal{E}} d\sigma(t) = \infty ,$$

and what is more the set $\{\tilde{g}_1(t), \tilde{g}_2(t), \ldots, \tilde{g}_n(t)\}$ is linear independent modulo $L^2_{\vec{\sigma}}(\mathcal{E})$. We shall denote the linear span of $\{\tilde{g}_1(t), \tilde{g}_2(t), \ldots, \tilde{g}_n(t)\}$ and $L^2_{\vec{\sigma}}(\mathcal{E})$ by $\tilde{L}^2_{\vec{\sigma}}(\mathcal{E})$ and we shall consider $\tilde{L}^2_{\vec{\sigma}}(\mathcal{E})$ as a Hilbert space where functions from the set $\{\tilde{g}_1(t), \tilde{g}_2(t), \ldots, \tilde{g}_n(t)\}$ are mutually orthogonal, orthogonal with respect to $L^2_{\tilde{\sigma}}(\mathcal{E})$ and have unit norm.

Theorem 4. There exist corresponding to above conditions a Hilbert functional space $L^2_{\tilde{\sigma}}(\mathcal{E})$, a set of functions $\{\tilde{g}_1(t), \tilde{g}_2(t), \ldots, \tilde{g}_n(t)\}$, a continuous and invertible continuous operator $W \colon \tilde{L}^2_{\tilde{\sigma}}(\mathcal{E}) \to \tilde{\mathcal{H}}$, such that the space $\tilde{L}^2_{\tilde{\sigma}}(\mathcal{E})$ is invariant with respect to the operator e^{-iT} of multiplication by function e^{-it} and

$$\tilde{U} = W(e^{-iT})^* W^{-1} \, ,$$

where "*" is the symbol of conjugation in Hilbert sense.

We omit the proof of this theorem because it repeats practically the analogous proof for π-selfadjoint operator which is contained in [13] (see [14] also). The another approach to this problem was described in [15].

A model representation for π-semi-unitary completely π-nonunitary operator U is described in [16] for the case when $\mathcal{H}_s$ is maximal nonnegative suspace and $\sigma_p(U) \subset T$.

REFERENCES

1. Azizov,T.Ya.; Iohvidov,I.S. Foundation of the theory of linear operators in spaces with an indefinite metric, Moscow, Nauka, 1985, 362 p. (Russian)
2. Bognar, J. Indefinite inner product spaces, Springer-Verlag, Berlin-Heidelberg-New York, 1974.
3. Bruinsma, P; Dijksma, A.; de Snoo, H.S.V. Unitary dilatation of contraction in Π_κ-spaces, in OT: Advances and Applications, 28 (1988), Birkhaüser Verlag, pp. 27-42.
4. Constantinescu, T.; Gheondea, A., On unitary dilatations and characteristic functions in indefinite inner product spaces, in OT: Advances and Applications, 24 (1987), Birkhaüser Verlag, pp. 87-102.
5. Iohvidov,I.S.; Krein, M.G.; Langer, H., Introduction to the spectral theory of linear operators in space with an indefinite metric, Akademie-Verlag, Berlin, 1982.
6. Krein M.G.; Langer, G.K. (=Langer, H.), On the spectral function of a self-adjoint operator in a space with indefinite metric, Dokl.Akad.Nauk SSSR 152(1963), No 1, pp. 39-42. (Russian)
7. Langer, H., Spectral functions of definitizable operators in Krein space, in Lectures Notes in Math.,948 (1982), Springer-Verlag, Berlin-Heidlberg-New York, pp. 1-46.
8. McEnnis, B.W., Shifts on indefinite inner product spaces, PacificJ.Math., 81 (1979), pp. 113-130.
9. McEnnis, B.W., Shifts on indefinite inner product spaces II, Report, Ohio State University.
10. Strauss, V.A., On the theory of self-adjoint operators in Banach spaces with Hermitian form, Sibirsky Matem.Zhurnal, XIX(1978), No 3, pp. 685-692.(Russian)

11. Strauss, V.A., On the analog of Wold decomposition for a π-semi-unitary operator, Uspehi Matem.Nauk 43(1988), No 1, pp. 185-186. (Russian)

12. Sz.-Nagy, B; Foias, C., Harmonic analysis of operators in Hilbert space, Amsterdam-Budapest, 1970.

13. Strauss, V.A., Models of π-selfadjoint operator, Functional Analysis. Spectral theory, Interinstitutional collectional of scientific works, Ul'yanovsk. Gos. Ped. Inst., Ul'yanovsk, 1984, pp.123-133. (Russian)

14. V.A.Strauss, Functional representation of the algebra generated by the self-adjoint operator in Pontryagin space, Funktsionalyi analiz i ego prilozheniya (Functional analysis and its applications), Moscow, 1986, V.20, No 1, pp.91-92. (Russian)

15. P.Jonas, H.Langer, B.Textorius, Models and unitary equivalence of cyclic selfadjoint operators in Pontrjagin spaces, Operator Theory: Advances and Aplications, Birkh. Verlag, Basel, 1992, vol.59, pp. 252-284.

16. V.A.Strauss, On certain properties of π- isometric operator generated by shift, Functional Analysis. Spectral Theory,Interinstitutional collectional of scientific works, Ul'yanovsk. Gos. Ped. Inst., Ul'yanovsk, 1985, pp.137-147. (Russian)

DEPARTMENT OF APPLIED MATHEMATICS, TECHNICAL STATE UNIVERSITY OF CHELYABINSK, LENIN AVENUE 76, CHELYABINSK 454080, RUSSIA
E-mail address: str@math.tu-chel.ac.ru

Contemporary Mathematics
Volume **189**, 1995

Maximal Estimates for Oscillatory Integrals with Concave Phase

BJÖRN G. WALTHER

ABSTRACT. Define a family of linear operators by

$$[S_t f]\hat{}(\xi) = \exp(it|\xi|^a)\,\hat{f}(\xi),$$

$a > 0$, $f \in \mathcal{S}(\mathbf{R}^n)$, and its maximal operator by

$$[S^* f](x) = \sup_{0 \le t \le 1} |[S_t f](x)|.$$

The results in this paper concern the case $0 < a < 1$. For $n = 1$ we prove that S^* is bounded $H^s(\mathbf{R}) \supset \mathcal{S}(\mathbf{R}) \to L^2(-A, A)$ if $s > a/4$, and that $a/4$ cannot be replaced by a smaller number.

1. Introduction, The Main Theorem

1.1 Notation. For x and ξ in $\mathbf{R}^n$ we let $x\xi = x_1\xi_1 + \ldots + x_n\xi_n$. If a is a real positive number and if f is in the Schwartz class $\mathcal{S}(\mathbf{R}^n)$ we define a family of linear operators and its associated maximal operator by

(1.1 a) $\qquad [S_t f](x) = \dfrac{1}{(2\pi)^n} \displaystyle\int_{\mathbf{R}^n} e^{i(x\xi + t|\xi|^a)}\,\widehat{f}(\xi)\,d\xi, \quad t \ge 0,$

(1.1 b) $\qquad [S^* f](x) = \sup_{0 \le t \le 1} |[S_t f](x)|.$

1991 *Mathematics Subject Classification.* 42B25, 42B08, 42A45.

Key words and phrases. Oscillatory integrals, Almost Everywhere Convergence, Maximal Functions, Summability of Fourier Integrals.

The material presented here is part of research made under supervision of Professor Per Sjölin, Royal Institute of Technology, Stockholm, to whom the author wants to express his gratitude for patience, guidance and support. We would also like to thank the referee and Professor Allan Gut for valuable comments.

This paper is in final form and no version of it will be submitted for publication elsewhere.

Here $\widehat{f}$ is the Fourier transform of f,

$$\widehat{f}(\xi) = \int_{\mathbf{R}^n} e^{-ix\xi} f(x)\, dx.$$

We also introduce fractional Sobolev spaces

$$H^s(\mathbf{R}^n) = \{f \in \mathcal{S}'(\mathbf{R}^n) : \|f\|_{H^s(\mathbf{R}^n)} < \infty\}$$

where

$$\|f\|^2_{H^s(\mathbf{R}^n)} = \int_{\mathbf{R}^n} (1 + |\xi|^2)^s |\widehat{f}(\xi)|^2 d\xi.$$

We shall prove the following theorem.

1.2 THEOREM. *Let* $0 < a < 1$.

(a) *There is a number* C_A *independent of* $f \in \mathcal{S}(\mathbf{R})$ *such that the inequality*

$$\int_{-A}^{A} [S^* f\,](x)^2 dx \le C_A \|f\|^2_{H^s(\mathbf{R})}$$

holds if $s > a/4$.

(b) *Conversely, assume that such a number exists. Then* $s \ge a/4$.

1.3 Discussion. For $a = 2$ let $u(x,t) = [S_t f\,](x)$. At least for $f \in \mathcal{S}(\mathbf{R}^n)$ it is easy to check that

$$\begin{cases} \Delta_x u = i\partial_t u \\ u(x,0) = f(x) \end{cases}$$

where Δ_x is the Laplace operator with respect to the space variable x. This means that u is a solution to the time-dependent Schrödinger equation without potential. After allowing a more general class of initial values f and after making the values of $u(x,t)$ precise the problem of convergence of $u(x,t)$ to $f(x)$ as $t \to 0$ is natural to study as a counterpart to convergence problems for other partial differential equations in mathematical physics.

For $a \ne 2$ one can study the properties of S_t as $t \to 0$ and view this as a study of a summability method for Fourier integrals. Related to this convergence problem is the following problem regarding S^*.

1.4 PROBLEM. *For which values of* s *is there a number* C_A *independent of* $f \in \mathcal{S}(\mathbf{R}^n)$ *such that the inequality*

(1.2) $$\int_{|x| \le A} [S^* f\,](x)^2 dx \le C_A \|f\|^2_{H^s(\mathbf{R}^n)}$$

holds?

This problem has been solved in full in a one-dimensional special case.

1.5 THEOREM. (Carleson [4], Dahlberg, Kenig [6], Kenig, Ruiz [9], Sjölin [11]) *Let* $a > 1$ *and* $n = 1$. *Then there is a number* C_A *independent of* $f \in \mathcal{S}(\mathbf{R}^n)$ *such that the inequality* (1.2) *holds if and only if* $s \ge 1/4$.

2. A Linearization and its Related Smoothing Inequality

2.1 Notation. We first generalize the definition of S^* as follows. For any bounded measurable function m on $\mathbf{R}^n \times \mathbf{R}$ and for $f \in \mathcal{S}(\mathbf{R}^n)$ we define

$$[S_m^* f](x) = \sup_{t \in \mathbf{R}} | \int_{\mathbf{R}^n} e^{ix\xi} m(\xi, t) \widehat{f}(\xi) \, d\xi|.$$

2.2 THEOREM. (Soljanik*) *Assume that $\xi \mapsto m(\xi, t)$ is a radial function for each t. Then there is a number C_A independent of $f \in \mathcal{S}(\mathbf{R}^n)$ such that*

$$\int_{|x| \leq A} [S_m^* f](x)^2 dx \leq C_A \|f\|_{H^s(\mathbf{R}^n)}^2$$

holds if $s > 1/2$.

In the proof of this theorem Soljanik uses polar coordinates and an estimate for the Fourier transform of a measure supported by the unit sphere. (See Hörmander [**7, theorem 7.1.26 p. 173**].)

2.3 Discussion on the Local Smoothing Inequality. In this subsection constants denoted by C may be different at each occurrence.

Sjölin [**11, theorem 2 p. 700**] states and proves theorem 2.2 in the special case $m(\xi, t) = \exp(it|\xi|^a)$, $a > 1$. His proof is more elaborate than Soljanik's but interesting in its own right, since he derives and uses a so called local smoothing inequality which we here shall analyze further. (For other results on local smoothing see Ben-Artzi, Devinatz [**1**] and Kenig, Ponce, Vega [**8**].)

Choose smooth cut-off functions χ_n and χ_1 on $\mathbf{R}^n$ and $\mathbf{R}$ respectively. For $f \in \mathcal{S}(\mathbf{R}^n)$ define

$$(Sf)[x](t) = \chi_n(x)\chi_1(t)[S_t f](x),$$

$$\|Sf\|_{L^2(\mathbf{R}^n, H^s(\mathbf{R}))}^2 = \int_{\mathbf{R}^n} \|(Sf)[x]\|_{H^s(\mathbf{R})}^2 \, dx,$$

$$\|Sf\|_{L^2(\mathbf{R}^n, L^\infty(\mathbf{R}))}^2 = \int_{\mathbf{R}^n} \|(Sf)[x]\|_{L^\infty(\mathbf{R})}^2 \, dx.$$

The local smoothing inequality valid for $a > 1$ is

$$(2.1) \qquad \|Sf\|_{L^2(\mathbf{R}^n, H^0(\mathbf{R}))} \leq C\|f\|_{H^{(1-a)/2}(\mathbf{R}^n)}.$$

Differentiation with respect to t yields

$$(2.2) \qquad \|Sf\|_{L^2(\mathbf{R}^n, H^1(\mathbf{R}))} \leq C\|f\|_{H^{(1+a)/2}(\mathbf{R}^n)}$$

and interpolating between (2.1) and (2.2) (see Bergh, Löfström [**2, theorem 5.1.2 p. 107 and 6.4.5 (7) p. 153**]) leads to

$$\|Sf\|_{L^2(\mathbf{R}^n, H^s(\mathbf{R}))} \leq C\|f\|_{H^{(1-a)/2+as}(\mathbf{R}^n)}, \quad 0 < s < 1.$$

---

*Letter communication April 1990

Using a standard argument for Bessel potentials in one dimension we arrive at

$$(2.3) \qquad \|Sf\|_{L^2(\mathbf{R}^n, L^\infty(\mathbf{R}))} \le C\|f\|_{H^{(1-a)/2+as}(\mathbf{R}^n)}, \quad s > 1/2.$$

Due to the compactness of supports in the definition of S the boundedness property in (2.3) implies the boundedness property of S^* sought for. (Cf. [11, pp. 704-706].)

For the case $0 < a \le 1$ we can do a little better. Instead of deriving a smoothing inequality we apply Parseval's formula and get

$$\|Sf\|_{L^2(\mathbf{R}^n, H^0(\mathbf{R}))} \le C\|f\|_{H^0(\mathbf{R}^n)}.$$

Applying time differentiation, interpolation and the argument for Bessel potentials as above yields

$$\|Sf\|_{L^2(\mathbf{R}^n, L^\infty(\mathbf{R}))} \le C\|f\|_{H^{sa}(\mathbf{R}^n)}, \quad s > 1/2.$$

Hence we have the following observation.

2.4 OBSERVATION. (Cf. Cowling [5].) *Let* $0 < a \le 1$. *Then there is a number* C_A *independent of* $f \in \mathcal{S}(\mathbf{R}^n)$ *such that the inequality*

$$(2.4) \qquad \int_{|x| \le A} [S^* f](x)^2 dx \le C_A\|f\|^2_{H^s(\mathbf{R}^n)}$$

holds if $s > a/2$.

2.5 Remark. Let the interval $J_{a,n}$ be defined by

$$s \in J_{a,n} \quad \text{if and only if} \quad (2.4) \text{ holds for some } C_A \text{ independent of } f \in \mathcal{S}(\mathbf{R}^n).$$

$s_{a,n} = \inf J_{a,n}$ is called the *critical regularity*. Theorem 1.2 tells us that $s_{a,1} = a/4$ for $0 < a < 1$. On the other hand $s_{a,1} = 1/4$ for $a > 1$ according to theorem 1.5.

Let $n \ge 2$. Observation 2.4 tells us that $s_{a,n} \le a/2$ for $0 < a < 1$. This is in contrast with the case $a > 1$ where $s_{a,n} \ge 1/4$ according to Vega [14]. The estimate $s_{a,n} \le 1/2$, $a > 1$ is due to Sjölin [11] and Vega [14]. (See also Ben-Artzi, Devinatz [1].) Recently Bourgain [3] has proved that $s_{2,2} < 1/2$.

3. Another Linearization and its Boundedness, Proof of 1.2 (a)

3.1 Notation. In this section we shall use functions χ and ψ such that $\chi \in \mathcal{C}_0^\infty(\mathbf{R})$ is even, non-negative and equal to 1 on a neighbourhood of 0 and $\psi = 1 - \chi$. From these functions we obtain three families of functions as follows: for each positive integer N and M set

$$\chi_N(\xi) = \chi(\frac{\xi}{N}), \quad \psi_M(\xi) = \psi(M\xi), \quad \rho_{NM}(\xi) = \chi_N(\xi)\psi_M(\xi).$$

3.2 The Linear Operator R_t. Let t be a measurable function on $\mathbf{R}^n$ with $0 < t < 1$ and choose a smooth real-valued cut-off function φ on $\mathbf{R}^n$. Following Sjölin [**11, p. 707**] we study

$$(3.1) \qquad [R_t f](x) = \varphi(x) \int_{\mathbf{R}^n} e^{i(x\xi + t(x)|\xi|^a)} |\xi|^{-s} \widehat{f}(\xi)\, d\xi, \quad f \in \mathcal{S}(\mathbf{R}^n).$$

If R_t is a t-uniformly bounded operator on $L^2(\mathbf{R}^n)$ it follows that the inequality

$$\int_{|x| \leq A} [S^* f](x)^2 dx \leq C_A \|f\|_{H^s(\mathbf{R}^n)}^2$$

holds for some number C_A independent of f. Part (a) of theorem 1.2 therefore follows by proving t-uniform $L^2(\mathbf{R}^n)$-boundedness for R_t in a one-dimensional special case. To derive this boundedness we need some information about the inverse Fourier transform of

$$(3.2) \qquad m(\xi) = e^{\pm i|\xi|^a} |\xi|^{-2s}, \quad \xi \in \mathbf{R}, \quad 0 < a < 4s < \min\{2a, 1\}.$$

Write $m = \chi m + \psi m$ and let K_χ and K_ψ be the inverse Fourier transforms of χm and ψm respectively.

3.3 LEMMA. K_χ is bounded and

$$K_\chi(x) = \mathcal{O}(|x|^{c(\chi)}), \ |x| \to \infty, \quad c(\chi) = -1 + 2s.$$

3.4 Lemma. (Miyachi [**10, proposition 5.1 (i) p. 289**]) K_ψ decays fast at infinity and

$$K_\psi(x) = \mathcal{O}(|x|^{c(\psi)}), \ |x| \to 0, \quad c(\psi) = \frac{4s - 2 + a}{2 - 2a}.$$

3.5 Remark. Let ν be one of the functions χ and ψ. The conditions on a and s in (3.2) give $0 < c(\nu) + 1 < 1$.

3.6 LEMMA. Let $f(x) = |x|^\alpha$, $x \in \mathbf{R}^n$, $\alpha > -n$, and let $g \in \mathcal{C}(\mathbf{R}^n)$ decay fast at infinity. Then

$$(f * g)(x) = \mathcal{O}(f(x)), \quad |x| \to \infty.$$

The proof is omitted.

3.7 *Proof of 3.3:* Since $-2s > -1$, the integral

$$\int_{\mathbf{R}} e^{i(x\xi \pm |\xi|^a)} \chi(\xi) |\xi|^{-2s}\, d\xi$$

is absolutely convergent. Hence K_χ is bounded (and continuous). To derive the asymptotic estimate we write

$$2\pi K_\chi(x) = \lim_{M \to \infty} \int_{\mathbf{R}} e^{ix\xi}(e^{i|\xi|^a} - 1)[\chi\psi_M](\xi)\, |\xi|^{-2s}\, d\xi + \int_{\mathbf{R}} e^{ix\xi}\chi(\xi)\, |\xi|^{-2s}\, d\xi$$

$$= \lim_{M \to \infty} G_M(x) + H(x).$$

490 BJÖRN G. WALTHER

By Taylor's formula and integration by parts

$$|G_M(x)| \leq \frac{C}{|x|} \left(\int_{\mathbf{R}} [\chi\psi_M](\xi)\, |\xi|^{a-2s-1}\, d\xi + \int_{\mathbf{R}} |[\chi\psi_M]'(\xi)|\, |\xi|^{a-2s}\, d\xi \right),$$

where C is independent of M. The first integral remains bounded as $M \to \infty$, since $a - 2s > 0$. To bound the second integral independent of M we notice again that $a - 2s > 0$ and also that $|[\chi\psi_M]'(\xi)|$ is like two approximative units whose support approach 0 as $M \to \infty$.

To handle H we use a well-known formula for Fourier transforms of homogeneous distributions (see e.g. Stein [**13, lemma 1 (a) p. 117**]) and lemma 3.6. We get

$$\int_{\mathbf{R}} e^{ix\xi}\chi(\xi)|\xi|^{-2s}\, d\xi = \gamma_s \int_{\mathbf{R}} \widehat{\chi}(y-x)|y|^{-1+2s}\, dy = \mathcal{O}(|x|^{-1+2s}), \quad |x| \to \infty$$

for some constant γ_s. We can now conclude that

$$K_\chi(x) = \mathcal{O}(|x|^{-1}) + \mathcal{O}(|x|^{-1+2s}) = \mathcal{O}(|x|^{-1+2s}), \quad |x| \to \infty,$$

and the lemma is proved.

3.8 THEOREM. *The formula* (3.1) *defines a t-uniformly bounded operator* R_t *on* $L^2(\mathbf{R})$ *if* $0 < a < 1$ *and* $s > a/4$.

PROOF. Functions f and g appearing in the integrals below belong to $\mathcal{S}(\mathbf{R})$. Constants denoted by C (sometimes with subscripts) may be different at each occurrence.

It will be enough to prove the statement of the theorem when

$$a < 4s < \min\{2a, 1\}.$$

Set

$$[R_N f](x) = \varphi(x) \int_{\mathbf{R}} e^{i(x\xi + t(x)|\xi|^a)} \chi_N(\xi)|\xi|^{-s}\widehat{f}(\xi)\, d\xi,$$

$$[R_{NM} f](x) = \varphi(x) \int_{\mathbf{R}} e^{i(x\xi + t(x)|\xi|^a)} \rho_{NM}(\xi)|\xi|^{-s}\widehat{f}(\xi)\, d\xi.$$

Here the integrations are performed over compact sets; the $L^2(\mathbf{R})$-boundedness for each R_N and R_{NM} can easily be verified. Computations of the adjoints show that

$$[R_N^* g](x) = \iint_{\mathbf{R}^2} e^{ix\xi}\varphi(y)e^{-i(y\xi + t(y)|\xi|^a)}\chi_N(\xi)|\xi|^{-s}g(y)\, dy d\xi,$$

$$[R_{NM}^* g](x) = \iint_{\mathbf{R}^2} e^{ix\xi}\varphi(y)e^{-i(y\xi + t(y)|\xi|^a)}\rho_{NM}(\xi)\,|\xi|^{-s}g(y)\, dy d\xi.$$

We will prove that the operators $R_N^*, N = 1, 2, \ldots$ are bounded on $L^2(\mathbf{R})$ uniformly with respect to N and t. Then this will hold for the operators $R_N, N =$

$1, 2, \ldots$ too. Since

$$[R_t f](x) = \lim_{N \to \infty} [R_N f](x), \quad f \in \mathcal{S}(\mathbf{R}),$$

we can by Fatou's lemma conclude that R_t is bounded on $L^2(\mathbf{R})$ and that the bound is independent of t.

A computation involving Fubini's theorem (cf. [**11, p. 708**]) shows that

$$\int_{-A}^{A} |[R_{NM}^* g](x)|^2 dx =$$

$$\int_{-A}^{A} \left[\iint_{\mathbf{R}^2} \left\{ \iint_{\mathbf{R}^2} e^{ix\xi} \varphi(y) e^{-i(y\xi + t(y)|\xi|^a)} \rho_{NM}(\xi)|\xi|^{-s} \right. \right.$$

$$e^{-ix\eta} \varphi(z) e^{i(z\eta + t(z)|\eta|^a)} \rho_{NM}(\eta)|\eta|^{-s} \, d\xi d\eta \Bigg\} \, g(y)\overline{g(z)} \, dy dz \Bigg] \, dx =$$

$$\iint_{\mathbf{R}^2} \varphi(y) \left[\int_{\mathbf{R}} e^{-i((y-z)\xi + t(y)|\xi|^a)} \rho_{NM}(\xi)|\xi|^{-s} \right.$$

$$\left. \left\{ \int_{-A}^{A} e^{i(x-z)\xi} \widehat{p}(z, x - z) \, dx \right\} \, d\xi \right] g(y)\overline{g(z)} \, dy dz.$$

Here $\widehat{p}(z, x - z)$ denotes the Fourier transform of the function

$$\eta \mapsto \varphi(z) e^{it(z)|\eta|^a} \rho_{NM}(\eta)|\eta|^{-s}$$

evaluated at the point $x - z$. Passing A to infinity and using Fatou's lemma and Lebesgue's theorem on dominated convergence yields

$$\int_{\mathbf{R}} |[R_N^* g](x)|^2 dx \leq \liminf_{M \to \infty} \int_{\mathbf{R}} |[R_{NM}^* g](x)|^2 \, dx$$

$$= 2\pi \iint_{\mathbf{R}^2} K_N(y, z) g(y)\overline{g(z)} \, dy dz,$$

where

(3.3) $\qquad K_N(y, z) = \varphi(y)\varphi(z) \int_{\mathbf{R}} e^{-i((y-z)\xi + (t(y)-t(z))|\xi|^a)} \chi_N(\xi)^2 |\xi|^{-2s} \, d\xi.$

We shall prove that there is a number C independent of y, z and N such that

(3.4) $\qquad |K_N(y, z)| \leq C|\varphi(y)\,\varphi(z)|(|y - z|^{c(\psi)} + |y - z|^{c(\chi)}).$

Assume that this has been done. For $0 < \beta < 1$ let I_β be the Riesz potential of order β,

$$[I_\beta f](x) = \int_{\mathbf{R}} |x - y|^{-1+\beta} f(y) \, dy.$$

(See [**13, pp. 117, 119-120**].) We want to estimate the double integral

(3.5) $\qquad \iint_{\mathbf{R}^2} |K_N(y, z)| \, |g(y)g(z)| \, dy dz.$

Let ν be one of the functions χ and ψ. Choose exponents p and q such that

$$0 < \frac{1}{q} = \frac{1}{p} - (c(\nu) + 1) \le \frac{1}{2} \le \frac{1}{p} < 1.$$

This is possible, since by remark 3.5

$$\max\{\frac{1}{2}, c(\nu) + 1\} < \min\{1, c(\nu) + \frac{3}{2}\}.$$

After replacing $K_N(y, z)$ in (3.5) by a term from the right hand side of (3.4) we apply Hölder's inequality and the inequality of Hardy, Littlewood and Sobolev and get

$$C \iint_{\mathbf{R}^2} |\varphi(y)\varphi(z)||y - z|^{c(\nu)} |g(y)g(z)| \, dydz$$

$$= C \int_{\mathbf{R}} [\, I_{c(\nu)+1}(|\varphi g|)\,](y) \, |\varphi(y)g(y)| \, dy$$

$$\le C \|I_{c(\nu)+1}(|\varphi g|)\|_{L^q(\mathbf{R})} \|\varphi g\|_{L^{q^*}(\mathbf{R})}$$

$$\le CC_\nu \|\varphi g\|_{L^p(\mathbf{R})} \|\varphi g\|_{L^{q^*}(\mathbf{R})}$$

(3.6) $$\le CC_{\nu,\varphi} \|g\|_{L^2(\mathbf{R})}^2,$$

where we have used the compact support of φ in the last inequality. (q^* is the dual exponent of q.) From (3.5) and (3.6) we can now conclude that

$$\int_{\mathbf{R}} |[R_N^* g](x)|^2 dx \le 2\pi \iint_{\mathbf{R}^2} K_N(y, z)g(y)\overline{g(z)} \, dydz \le C(C_{\chi,\varphi} + C_{\psi,\varphi})\|g\|_{L^2(\mathbf{R})}^2,$$

where the constant in the last inequality is independent of N, i.e. the operators R_N^* are uniformly bounded as desired.

Proof of the estimate (3.4): For $t(y) - t(z) \ne 0$ we set

$$\eta = |t(y) - t(z)|^{1/a}\xi, \quad v = -|t(y) - t(z)|^{-1/a}(y - z), \quad L = N|t(y) - t(z)|^{1/a}.$$

By a change of variables in (3.3)

$$K_N(y, z) = \varphi(y)\varphi(z)|t(y) - t(z)|^{(2s-1)/a} \int_{\mathbf{R}} e^{iv\eta} m(\eta)\chi(\frac{\eta}{L})^2 \, d\eta.$$

We define $K_{N,\nu}(y, z)$ by replacing m by νm in this formula for $K_N(y, z)$. (As before ν denotes either ψ or χ.) By Lemma 3.3 and 3.4 and a change of variables we get

$$|K_{N,\nu}(y, z)| = |\varphi(y)\varphi(z)||t(y) - t(z)|^{(2s-1)/a}| \int_{\mathbf{R}} K_\nu(u)L\widehat{\chi^2}(Lu - Lv) \, du|$$

$$\le C_\nu |\varphi(y)\varphi(z)||t(y) - t(z)|^{(2s-1)/a} \int_{\mathbf{R}} |Lu|^{c(\nu)}|\widehat{\chi^2}(Lu - Lv)|L \, du L^{-c(\nu)}$$

$$\le C_\nu |\varphi(y)\varphi(z)||t(y) - t(z)|^{(2s-1)/a}|Lv|^{c(\nu)} L^{-c(\nu)}$$

$$= C_\nu |\varphi(y)\varphi(z)||t(y) - t(z)|^{(2s-1-c(\nu))/a}|y - z|^{c(\nu)}.$$

To get the last inequality we have also used Lemma 3.6. In the case $\nu = \chi$ the exponent of $|t(y) - t(z)|$ is 0. In the case $\nu = \psi$ it is $(1 - 4s)/(2 - 2a)$, which is positive since $s < 1/4$ and $a < 1$. We can now write

$$|K_N(y, z)| \leq |K_{N,\chi}(y, z)| + |K_{N,\psi}(y, z)|$$
$$\leq C|\varphi(y)\varphi(z)| \, (|y - z|^{c(\chi)} + |y - z|^{c(\psi)}),$$

where $C = C_\chi + C_\psi 2^{(1-4s)/(2-2a)}$ does not depend on y, z or N.

Now we have proved the estimate (3.4) in the case $t(y) \neq t(z)$. Since the integral

$$\int_{\mathbf{R}} e^{-i(y-z)\xi + \varepsilon|\xi|^a)} \chi_N(\xi)^2 |\xi|^{-2s} \, d\xi$$

is continuous with respect to ε this estimate is valid also in the case $t(y) = t(z)$.

We have proved the estimate (3.4) and the proof of our theorem is finished.

4. Proof of 1.2 (b)

4.1 Introduction. In section 3 we proved part (a) of theorem 1.2 by studying a linear operator naturally associated to S^*. Here we shall prove part (b) of the same theorem by exhibiting a family $\{f_N\}_{N \geq 1}$ for which $\|S^* f_N\|_{L^2(-A,A)}$ increases faster than $\|f_N\|_{H^s(\mathbf{R})}$ as $N \to \infty$, provided $s < a/4$. This family is taken over from Sjölin [12] where it is used in the study of global estimates of $S^* f$, $a > 1$.

4.2 *Proof of* 1.2 (b): Choose $\varphi \in C_0^\infty(\mathbf{R})$ with $\operatorname{supp} \varphi \subset \,]-1, 1[$. Define f_N by $\widehat{f_N}(\xi) = \varphi(N^{a/2-1}\xi + N^{a/2})$. Then

$$\|f_N\|_{H^s(\mathbf{R})}^2 \leq \mathcal{O}(N^{2s+1-a/2}), \quad N \to \infty.$$

Next we compute $[S_t f_N](x)$. By a change of variable

$$[S_t f_N](x) = \frac{N^{1-a/2}}{2\pi} \int_{\mathbf{R}} e^{ix(N^{1-a/2}\xi - N)} e^{it|N^{1-a/2}\xi - N|^a} \varphi(\xi) \, d\xi$$
$$= \frac{N^{1-a/2}}{2\pi} \int_{\mathbf{R}} e^{iF(x,\xi)} \varphi(\xi) \, d\xi.$$

Applying Taylor's formula to the function $\eta \mapsto (1 - \eta)^a$ yields

$$F(x, \xi) = tN^a - Nx + (N^{1-a/2}x - atN^{a/2})\xi + \frac{a(a-1)t}{2}\xi^2 + \mathcal{O}(tN^{-a/2}).$$

Choose $t = xN^{1-a}/a$. Then

$$|[S_t f_N](x)| = \frac{N^{1-a/2}}{2\pi} \left| \int_{\mathbf{R}} e^{iG(x,\xi)} \varphi(\xi) \, d\xi \right|,$$

where

$$G(x, \xi) = \frac{(a-1)x}{2N^{a-1}}\xi^2 + \mathcal{O}(xN^{1-3a/2}).$$

To control the space variable we choose $\varepsilon > 0$ and restrict x to

$$M_\varepsilon = \{x' \in \mathbf{R} : (1 - \varepsilon)N^{a-1} \le x' \le N^{a-1}\}.$$

Then G simplifies to

$$G(x, \xi) = \frac{a-1}{2}\xi^2 + \mathcal{O}(\varepsilon), \quad \varepsilon \to 0.$$

At this stage we may choose φ such that

$$\int_{\mathbf{R}} e^{i(a-1)\xi^2/2}\varphi(\xi)\,d\xi \ne 0.$$

If ε is small enough it follows that

$$\inf_{x \in M_\varepsilon} |\int_{\mathbf{R}} e^{iG(x,\xi)}\varphi(\xi)d\xi| = c > 0.$$

Hence, if $N^{a-1} \le A$, then

$$\int_{-A}^{A} [S^*f_N](x)^2dx \ge \int_{M_\varepsilon} [S^*f_N](x)^2\,dx \ge (\frac{N^{1-a/2}c}{2\pi})^2 \varepsilon N^{a-1} = \frac{\varepsilon c^2}{4\pi^2}N$$

Now, if the inequality in 1.2 (a) holds, then

$$N \le C\,N^{2s+1-a/2}, \quad N \to \infty,$$

where C depends on φ, ε, c and A but not on N. This forces us to choose $s \ge a/4$.

REFERENCES

1. M. Ben-Artzi, A. Devinatz, *Local Smoothing and Convergence Properties of Schrödinger Type Equations*, J. Func. Anal. **101** (1991), 231–254.
2. J. Bergh, J. Löfström, *Interpolation Spaces*, Springer-Verlag, Berlin, Heidelberg and New York, 1976.
3. J. Bourgain, *A remark on Schrödinger Operators*, Israel J. Math. **77** (1992), 1–16.
4. L. Carleson, *Some Problems in Harmonic Analysis related to Statistical Mechanics*, Euclidean Harmonic Analysis (Proceedings of Seminars Held at the University of Maryland, 1979) (Benedetto, J.J., ed.), Lecture Notes in Math., vol. 779, Springer-Verlag, Berlin, Heidelberg and New York, 1980, pp. 5–45.
5. M. G. Cowling, *Pointwise Behaviour of Solutions to Schrödinger Equations*, Harmonic Analysis (Proceedings of a Conference Held in Cortona, Italy, July 1 – 9, 1982) (Mauceri, G.; Ricci, F.; Weiss, G., eds.), Lecture Notes in Math., vol. 992, Springer-Verlag, Berlin, Heidelberg, New York and Tokyo, 1983, pp. 83–90.
6. B.E.J. Dahlberg, C.E. Kenig, *A note on the Almost Everywhere behaviour of Solutions to the Schrödinger equation*, Harmonic Analysis (Proceedings of a Conference Held at the University of Minnesota, Minneapolis, April 20 – 30, 1981) (Ricci, F.; Weiss, G., eds.), Lecture Notes in Math., vol. 908, Springer-Verlag, Berlin, Heidelberg and New York, 1982, pp. 205–209.
7. L. Hörmander, *The Analysis of Linear Partial Differential Operators I*, Springer-Verlag, Berlin, Heidelberg, New York and Tokyo, 1983.
8. C.E. Kenig, G. Ponce, L. Vega, *Oscillatory Integrals and Regularity of Dispersive Equations*, Indiana Univ. Math. J. **40** (1991), 33–69.
9. C.E. Kenig, A. Ruiz, *A Strong Type (2, 2) Esimate for a Maximal Operator associated to the Schrödinger equation*, Trans. Amer. Math. Soc. **280** (1983), 239–246.
10. A. Miyachi, *On some Singular Fourier Multipliers*, J. Fac. Sci. Univ. Tokyo Sect. IA Math. **28** (1981), 267–315.

11. P. Sjölin, *Regularity of Solutions to the Schrödinger Equation*, Duke Math. J. **55** (1987), 699–715.
12. P. Sjölin, *Global Maximal Estimates for Solutions to the Schrödinger Equation*, Studia Math. **110** (1994), 105–114.
13. E.M. Stein, *Singular Integrals and Differentiability Properties of Functions*, Princeton University Press, Princeton, New Jersey, 1970.
14. L. Vega, *Schrödinger Equations: Pointwise Convergence to the Initial Data*, Proc. Amer. Math. Soc. **102** (1988), 874–878.

DEPARTMENT OF MATHEMATICS, UPPSALA UNIVERSITY, S – 751 06 UPPSALA, SWEDEN

E-mail address: walther@math.uu.se

Contemporary Mathematics
Volume **189**, 1995

Algebras of Multilinear Forms on Groups

GUANGHUA ZHAO AND BERTRAM M. SCHREIBER

Dedicated to Mischa Cotlar on the occasion of his eightieth birthday

ABSTRACT. For locally compact groups G_i, $i = 1, 2, \cdots, n$, let $CB(G_1, \cdots, G_n)$ denote the Banach space of completely bounded multilinear forms on $C_0(G_1) \times \cdots \times C_0(G_n)$ in the completely bounded norm. $CB(G_1, \cdots, G_n)$ has the structure of a Banach $*$-algebra under a multiplication and adjoint operation which agree with the convolution structure on the measure algebra $M(G_1 \times \cdots \times G_n)$. If the G_i are all abelian, $CB(G_1, \cdots, G_n)$ carries a naturally defined Fourier transform which generalizes the Fourier-Stieltjes transform on measure algebras. Various other aspects of $CB(G_1, \cdots, G_n)$ are investigated.

1. Introduction

For locally compact spaces X_i, $i = 1, 2, \cdots, n$, let $MM(X_1, X_2, \cdots, X_n)$ denote the space of all bounded n-linear forms $u : C_0(X_1) \times C_0(X_2) \times \cdots \times C_0(X_n) \to C$. Equipped with the usual norm on n-linear forms, $MM(X_1, X_2, \cdots, X_n)$ is a Banach space. Its elements are called *multimeasures* on $X_1 \times X_2 \times \cdots \times X_n$, *bimeasures* when $n = 2$, and *trimeasures* if $n = 3$.

In [**8**], [**9**], [**10**], [**11**] the second author and others showed that if G_1 and G_2 are locally compact groups, then $MM(G_1, G_2)$ is a Banach $*$-algebra consistent with the convolution structure of the measure algebra $M(G_1 \times G_2)$, and they studied the structure of $MM(G_1, G_2)$ as a normed algebra. The arguments in the papers cited above, however, depend heavily on the classical inequality of Grothendieck, which is invalid when $n \geq 3$. In fact, it has been shown that it is impossible to introduce an algebra structure on $MM(G_1, G_2, \cdots, G_n)$ consistent with that of $M(G_1 \times G_2 \times \cdots \times G_n)$ when $n \geq 3$. (See [**11**, Theorem 6].)

Nevertheless, for locally compact groups $G_1, G_2, \cdots, G_n$ there is an appropriate normed subspace of $MM(G_1, G_2, \cdots, G_n)$ on which one can introduce a Banach-algebra structure, namely the Banach space $CB(G_1, G_2, \cdots, G_n)$ of

1991 *Mathematics Subject Classification.* Primary 43A10; Secondary 46L05.

all *completely bounded n*-linear forms. In fact, it follows from Remark 2.3 below that this is the most appropriate setting for an extension of the results on $MM(G_1, G_2)$. As we shall see below, there is a convolution and an adjoint operation on the space $CB(G_1, G_2, \cdots, G_n)$ which generalize the convolution structure of the measure algebra of $G_1 \times G_2 \times \cdots \times G_n$ and which will make $CB(G_1, G_2, \cdots, G_n)$ into a Banach $*$-algebra. In the case of LCA (locally compact abelian) groups, we shall also define and study a Fourier transform for completely bounded n- linear forms which generalizes the Fourier-Stieltjes transform of measures.

Some of the material herein has been modified from the doctoral dissertation of the first author submitted to Wayne State University in July, 1990. The fact that $CB(G_1, G_2, \cdots, G_n)$ is a Banach algebra (perhaps not the explicit formulations (10) and (11) below for the multiplication and adjoint, however) can, in fact, be extracted from more recent abstract results in the theory of operator spaces, for instance [**2**], [**6**], [**7**]. Nevertheless, given the widespread interest in Harmonic Analysis and convolutions, it seems appropriate to develop this context explicitly. We hope that this will stimulate further research into the harmonic analysis of multilinear forms on groups.

Throughout the paper the inner product on a Hilbert space will be denoted by $(\cdot|\cdot)$. For Hilbert spaces H and K, $L(H, K)$ denotes the space of all bounded linear operators from H to K, and $L(H) = L(H, H)$. Recall that the σ-weak (operator) topology on $L(H)$ is the locally convex topology given by the family of seminorms

$$p(T) = \left| \sum_{n=1}^{\infty} (T\xi_n | \eta_n) \right|, \quad \sum_{n=1}^{\infty} \|\xi_n\|^2 < \infty, \quad \sum_{n=1}^{\infty} \|\eta_n\|^2 < \infty.$$

X, Y, X_i will always denote locally compact Hausdorff spaces, $\mathcal{A}, \mathcal{B}, \mathcal{A}_i$ denote C^*-algebras, and G, H, G_i are locally compact groups. $\mathcal{L}^\infty(X)$ and $C_0(X)$ are the spaces of bounded functions on X which are, respectively, Borel locally measurable and continuous with limit zero at infinity. As usual, $C_0(X)^*$ is identified with $M(X)$, the space of bounded regular Borel measures on X.

If $\mathcal{A}$ is a C^*-algebra, $M_k(\mathcal{A})$ denotes the algebra of all $k \times k$ matrices with entries in $\mathcal{A}$. Recall that we may identify the bidual $\mathcal{A}^{**}$ of $\mathcal{A}$ with the enveloping von Neumann algebra $\mathcal{N}$ of $\mathcal{A}$, the double commutant $\mathcal{A}''$ when $\mathcal{A}$ is represented as an algebra of operators. Under this identification, the weak* topology on $\mathcal{A}^{**}$ becomes the σ-weak topology on $\mathcal{N}$. For the fundamental facts about C^*-algebras that we will need, we refer the reader to [**12**],[**16**].

For $i = 1, 2, \cdots, n$, let $\mathcal{A}_i$ be a C^*- algebra and $\theta_i : \mathcal{A}_i \to L(H_i)$ be a representation. Then $\theta_1 \otimes \cdots \otimes \theta_n$ denotes the unique representation of the (spatial) tensor product C^*-algebra $\mathcal{A}_1 \otimes \cdots \otimes \mathcal{A}_n$ on the tensor-product Hilbert space $H_1 \otimes \cdots \otimes H_n$ satisfying

$$(1) \qquad (\theta_1 \otimes \cdots \otimes \theta_n)(a_1 \otimes \cdots \otimes a_n) = \theta_1(a_1) \otimes \cdots \otimes \theta_n(a_n).$$

For each i there is a unique extension of θ_i to $\mathcal{A}_i''$ which is continuous with respect to the weak* topology on $\mathcal{A}_i''$ and the σ- weak topology on $L(H_i)$, namely θ_i^{**}. We shall denote these representations θ_i^{**} by θ_i also. $\theta_1 \otimes \cdots \otimes \theta_n$ will thus be extended to $(\mathcal{A}_1 \otimes \cdots \otimes \mathcal{A}_n)''$. To clarify this extension, let us observe the following.

Lemma 1.1. *Let $\{T_\alpha\}$ be a net in $L(H)$ and $S \in L(K)$. If $\{T_\alpha\}$ converges σ-weakly to zero in $L(H)$, then $\{T_\alpha \otimes S\}$ converges σ-weakly to zero in $L(H \otimes K)$.*

Proof. Choose two sequences $\{\xi_i\}$ and $\{\eta_i\}$ in the Hilbert-space tensor product $H \otimes K$ such that $\sum_1^\infty \|\xi_i\|^2 < \infty$ and $\sum_1^\infty \|\eta_i\|^2 < \infty$. If $\{e_{1\lambda}\}_{\lambda \in \Lambda}$ is an orthonormal basis of H and $\{e_{2\omega}\}_{\omega \in \Omega}$ is an orthonormal basis of K, then $\{e_{1\lambda} \otimes e_{2\omega}\}_{(\lambda,\omega) \in \Lambda \times \Omega}$ is an orthonormal basis of $H \otimes K$. Thus we can easily write

$$\xi_i = \sum_{j=1}^\infty x_{ij} \otimes y_{ij} \quad \text{and} \quad \eta_i = \sum_{k=1}^\infty x'_{ik} \otimes y'_{ik}$$

with $x_{ij}, x'_{ij} \in H$, $y_{ik}, y'_{ik} \in K$, and $\{x'_{ik}\}_{k=1}^\infty$, $\{y_{ij}\}_{j=1}^\infty$ being orthonormal sets in H and K, respectively. Then $\|\xi_i\|^2 = \sum_{j=1}^\infty \|x_{ij}\|^2$ and $\|\eta_i\|^2 = \sum_{k=1}^\infty \|y'_{ik}\|^2$. Hence $\sum_{i=1}^\infty \sum_{j=1}^\infty \|x_{ij}\|^2 < \infty$ and $\sum_{i=1}^\infty \sum_{k=1}^\infty \|y'_{ik}\|^2 < \infty$. Note that

$$
\begin{aligned}
\sum_{i=1}^\infty ((T_\alpha \otimes S)\xi_i | \eta_i) &= \sum_{i=1}^\infty \left(\sum_{j=1}^\infty T_\alpha x_{ij} \otimes S y_{ij} \Big| \sum_{k=1}^\infty x'_{ik} \otimes y'_{ik} \right) \\
&= \sum_{i=1}^\infty \sum_{j=1}^\infty \sum_{k=1}^\infty (T_\alpha x_{ij} | x'_{ik})(S y_{ij} | y'_{ik}) \\
&= \sum_{i=1}^\infty \sum_{j=1}^\infty \left(T_\alpha x_{ij} \Big| \sum_{k=1}^\infty (y'_{ik} | S y_{ij}) x'_{ik} \right).
\end{aligned}
$$

Let $z_{ij} = \sum_{k=1}^\infty (y'_{ik} | S y_{ij}) x'_{ik}$, then

$$
\begin{aligned}
\sum_{i=1}^\infty \sum_{j=1}^\infty \|z_{ij}\|^2 &= \sum_{i=1}^\infty \sum_{j=1}^\infty \sum_{k=1}^\infty |(y'_{ik} | S y_{ij})|^2 \\
&= \sum_{i=1}^\infty \sum_{k=1}^\infty \sum_{j=1}^\infty |(S^* y'_{ik} | y_{ij})|^2 \\
&\leq \sum_{i=1}^\infty \sum_{k=1}^\infty \|S^* y'_{ik}\|^2 \\
&\leq \|S\|^2 \sum_{i=1}^\infty \sum_{k=1}^\infty \|y'_{ik}\|^2 \\
&= \|S\|^2 \sum_{i=1}^\infty \|\eta_i\|^2 < \infty.
\end{aligned}
$$

By assumption,

$$\left| \sum_{i=1}^{\infty} \sum_{j=1}^{\infty} (T_\alpha x_{ij} | z_{ij}) \right| \to 0.$$

It follows that

$$\left| \sum_{i=1}^{\infty} (T_\alpha \otimes S\xi_i | \eta_i) \right| \to 0. \quad \square$$

Corollary 1.2. *Let $\mathcal{A}_i$ be a C^*- algebra and $\theta_i : \mathcal{A}_i \to L(H_i)$ be a representation, $i = 1, 2, \cdots, n$. Then*

$$\mathcal{A}_1'' \otimes \cdots \otimes \mathcal{A}_n'' \subseteq (\mathcal{A}_1 \otimes \cdots \otimes \mathcal{A}_n)'',$$

and (1) holds for all $a_i \in \mathcal{A}_i''$, $i = 1, 2, \cdots, n$.

Proof. Clearly it suffices to consider the case $n = 2$. Let $a_1 \in \mathcal{A}_1''$ and $a_2 \in \mathcal{A}_2$, and choose a net $\{a_1^\alpha\} \in \mathcal{A}_1$ converging weak* to a_1. Then by Lemma 1.1, $a_1^\alpha \otimes a_2$ converges weak* to $a_1 \otimes a_2$, so $a_1 \otimes a_2 \in (\mathcal{A}_1 \otimes \mathcal{A}_2)''$. Since $\theta_1 \otimes \theta_2$ is weak*-to-σ-weak continuous,

$$(\theta_1 \otimes \theta_2)(a_1^\alpha \otimes a_2) \to (\theta_1 \otimes \theta_2)(a_1 \otimes a_2)$$

σ-weakly. But $(\theta_1 \otimes \theta_2)(a_1^\alpha \otimes a_2) = \theta_1(a_1^\alpha) \otimes \theta_2(a_2)$. By Lemma 1.1 again,

$$\theta_1(a_1^\alpha) \otimes \theta_2(a_2) \to \theta_1(a_1) \otimes \theta_2(a_2)$$

σ-weakly. Therefore,

$$(\theta_1 \otimes \theta_2)(a_1 \otimes a_2) = \theta_1(a_1) \otimes \theta_2(a_2).$$

Now choose $\{a_2^\alpha\} \in \mathcal{A}_2$ converging weak* to a_2 and argue as above to complete the proof. $\square$

Considering $C_0(X)$ as a commutative C^*-algebra and embedding $\mathcal{L}^\infty(X)$ in $C_0(X)^{**}$ via integration, we may view $\mathcal{L}^\infty(X)$ as a subspace of $C_0(X)^{**}$. In fact, as is well known, $\mathcal{L}^\infty(X)$ is a C^*-subalgebra of $C_0(X)^{**}$. By the weak* topology of $\mathcal{L}^\infty(X)$ we shall mean the topology inherited from the weak* topology on $C_0(X)^{**}$. Recall that the C^*-algebra $C_0(X) \otimes C_0(Y)$ is canonically isomorphic to $C_0(X \times Y)$. Corollary 1.2 implies that if θ and π are representations of $C_0(X)$ and $C_0(Y)$ on H and K, respectively, then

$$(\theta \otimes \pi)(f \otimes g) = \theta(f) \otimes \pi(g)$$

for all $f \in \mathcal{L}^\infty(X)$ and $g \in \mathcal{L}^\infty(Y)$.

The following lemma is obvious. Since it is needed in several places, we include it here for reference.

Lemma 1.3. *If $T : X \to Y$ is a continuous map, then the map $f \to f \circ T$ from $\mathcal{L}^\infty(Y)$ to $\mathcal{L}^\infty(X)$ is weak* continuous.*

2. Completely Bounded Multilinear Forms

Completely bounded multilinear operators and completely bounded norms were first introduced in [5].

Definition 2.1. Let $\mathcal{A}_1, \mathcal{A}_2, \cdots, \mathcal{A}_n$ and $\mathcal{A}$ be C^*-algebras, and let $u : \mathcal{A}_1 \times \mathcal{A}_2 \times \cdots \times \mathcal{A}_n \to \mathcal{A}$ be an n-linear operator. For each $k \geq 1$, the n-linear operator $u_k : M_k(\mathcal{A}_1) \times M_k(\mathcal{A}_2) \times \cdots \times M_k(\mathcal{A}_n) \to M_k(\mathcal{A})$ is defined as follows. If $u(a_1, \cdots, a_n) = v_1(a_1)v_2(a_2)\cdots v_n(a_n)$ for some linear operators $v_i : \mathcal{A}_i \to \mathcal{A}$, then the entrywise extension of each v_i to $M_k(\mathcal{A}_i)$ induces an obvious extension u_k of u via multiplication in $M_k(\mathcal{A})$. Motivated by this case, one defines u_k for any u by

$$(2) \qquad u_k(A_1, A_2, \cdots, A_n) = \left(\sum_{r,s,\cdots,t} u(a_{1ir}, a_{2rs}, \cdots, a_{ntj}) \right)$$

for all $A_l = (a_{lij}) \in M_k(\mathcal{A}_l)$ $(1 \leq l \leq n)$. The operator u is said to be *completely bounded* with completely bounded norm $\|u\|_{cb}$ if

$$(3) \qquad \|u\|_{cb} = \sup\{\|u_k\| : k \geq 1\}$$

is finite.

The following theorem is a special case of the Christensen-Sinclair representation theorem for completely bounded multilinear operators [5]. (See also [14],[1],[2].)

Theorem 2.2. *A complex-valued n-linear form u on $\mathcal{A}_1 \times \mathcal{A}_2 \times \cdots \times \mathcal{A}_n$ is completely bounded if and only if there are Hilbert spaces $H_1, H_2, \cdots, H_n$, operators $U_i \in L(H_{i+1}, H_i)$ for $i = 1, 2, \cdots, n-1$, two vectors $\xi \in H_n$ and $\eta \in H_1$, and representations $\theta_i : \mathcal{A}_i \to L(H_i)$ for $i = 1, 2, \cdots, n$ such that*

$$(4) \qquad u(a_1, a_2, \cdots, a_n) = (\theta_1(a_1)U_1\theta_2(a_2)U_2 \cdots \theta_n(a_n)\xi | \eta)$$

for $a_i \in \mathcal{A}_i$, $i = 1, 2, \cdots, n$. Moreover, we may choose ξ, η and U_i such that $\|\xi\| = \|\eta\| = 1$ and $\|u\|_{cb} = \|U_1\|\|U_2\| \cdots \|U_n\|$.

Actually every completely bounded n-linear form has a representation of a simpler type [17], [4, Cor. 3.2], namely

$$(5) \qquad u(a_1, a_2, \cdots, a_n) = (\theta_1(a_1)\theta_2(a_2) \cdots \theta_n(a_n)\xi | \xi')$$

where all the representations θ_i act on the same Hilbert space K, $\xi, \xi' \in K$, and $\|u\|_{cb} = \|\xi\|\|\xi'\|$. We will use (4) and (5) alternatively.

Whenever a completely bounded n-linear form u is represented as in (4) or (5), the representations θ_i $(1 \leq i \leq n)$ are said to be associated with u.

Remark 2.3. [5] For a *bilinear* form u on $C_0(X) \times C_0(Y)$, the representation (4) is equivalent to the Grothendieck inequality, as is easily seen.

Definition 2.4. For C^*-algebras $\mathcal{A}_1, \mathcal{A}_2, \cdots, \mathcal{A}_n$, let $CB(\mathcal{A}_1, \cdots, \mathcal{A}_n)$ be the space of all completely bounded n-linear forms on $\mathcal{A}_1 \times \cdots \times \mathcal{A}_n$, equipped with the completely bounded norm. It is known that $CB(\mathcal{A}_1, \cdots, \mathcal{A}_n)$ is the dual space of $\mathcal{A}_1 \otimes_h \mathcal{A}_2 \otimes_h \cdots \otimes_h \mathcal{A}_n$, the Haagerup tensor product of the indicated algebras. For a proof, see [**14**, Theorem 3.1] and [**13**, Prop. 3.7]. In particular, $CB(\mathcal{A}_1, \cdots, \mathcal{A}_n)$ is a Banach space in the completely bounded norm. When $\mathcal{A}_i = C_0(X_i)$, $i = 1, \cdots, n$, we shall denote the space $CB(\mathcal{A}_1, \mathcal{A}_2, \cdots, \mathcal{A}_n)$ by $CB(X_1, X_2, \cdots, X_n)$.

For expositional convenience, we shall just consider the case $n = 3$. The extension of our results to higher n is just a matter of notation.

Each measure $\mu \in M(X_1 \times X_2 \times X_3)$ corresponds to a trilinear form u_μ on $C_0(X_1) \times C_0(X_2) \times C_0(X_3)$, namely

$$u_\mu(f_1, f_2, f_3) = \int_{X_1 \times X_2 \times X_3} f_1(x_1)f_2(x_2)f_3(x_3)d\mu(x_1, x_2, x_3).$$

The form u_μ is completely bounded, since

$$u_\mu(f_1, f_2, f_3) = (\theta_1(f_1)\theta_2(f_2)\theta_3(f_3)g|1),$$

where $\theta_1(f_1), \theta_2(f_2)$, and $\theta_3(f_3)$ are the multiplications by $f_1 \otimes 1 \otimes 1, 1 \otimes f_2 \otimes 1$, and $1 \otimes 1 \otimes f_3$, respectively, on $L^2(X_1 \times X_2 \times X_3, |\mu|)$, and $g \in \mathcal{L}^\infty(X_1 \times X_2 \times X_3)$ is such that $d\mu = gd|\mu|$. The map $\mu \to u_\mu$ is obviously linear. It is injective since the span of the elements of the form $f_1(x_1)f_2(x_2)f_3(x_3)$ with $f_i \in C_0(X_i)$ is dense in $C_0(X_1 \times X_2 \times X_3)$. It is also continuous because $\|u_\mu\|_{cb} \le \|g\|_2\|1\|_2 = \|\mu\|$. Identifying μ with u_μ, we can view $M(X_1 \times X_2 \times X_3) \subseteq CB(X_1, X_2, X_3)$.

It is well known that, in general, $M(X_1 \times X_2 \times X_3)$ is not dense in $MM(X_1, X_2, X_3)$. Our next theorem shows that this is still true if we replace $MM(X_1, X_2, X_3)$ by $CB(X_1, X_2, X_3)$.

Theorem 2.5. *If any two of the spaces X_1, X_2, X_3 contain nonvoid perfect sets, then $M(X_1 \times X_2 \times X_3)$ is not dense in $CB(X_1, X_2, X_3)$.*

Proof. Suppose X_1 and X_2 contain nonvoid perfect sets. We shall show that if $M(X_1 \times X_2 \times X_3)$ is dense in $CB(X_1, X_2, X_3)$, then $M(X_1 \times X_2)$ is dense in $MM(X_1, X_2)$. Indeed, if $\epsilon > 0$ and $u \in MM(X_1, X_2)$, then by Remark 2.3,

$$u(f_1, f_2) = (\theta_1(f_1)U_1\theta_2(f_2)\xi|\eta),$$

where θ_1, θ_2 are representations of $C_0(X_1)$ and $C_0(X_2)$ on some Hilbert spaces H_1 and H_2, respectively, $U_1 \in L(H_2, H_1)$, $\xi \in H_2$, and $\eta \in H_1$. Choose $x_{3,0}$ in X_3. Let θ_3 be the representation of $C_0(X_3)$ on $H_3 = H_2$ defined by $\theta_3(f_3) = f_3(x_{3,0})I$. Set

$$u'(f_1, f_2, f_3) = (\theta_1(f_1)U_1\theta_2(f_2)I\theta_3(f_3)\xi|\eta);$$

then $u' \in CB(X_1, X_2, X_3)$. If $M(X_1 \times X_2 \times X_3)$ is dense in $CB(X_1, X_2, X_3)$, then there exists a measure $\mu' \in M(X_1 \times X_2 \times X_3)$ such that in the trimeasure

norm,
$$\|\mu' - u'\| \leq \|\mu' - u'\|_{cb} < \epsilon.$$
Now choose $f_{3,0} \in C_0(X_3)$ such that $f_{3,0}(x_{3,0}) = 1 = \|f_{3,0}\|$. Let μ be the measure on $X_1 \times X_2$ given by

$$\int_{X_1 \times X_2} f(x_1, x_2) d\mu(x_1, x_2) = \int_{X_1 \times X_2 \times X_3} f(x_1, x_2) f_{3,0}(x_3) d\mu'(x_1, x_2, x_3)$$

for $f \in C_0(X_1 \times X_2)$. Since $u'(f_1, f_2, f_{3,0}) = u(f_1, f_2)$, we have

$$\begin{aligned}
\|\mu - u\| &= \sup\{|\mu(f_1, f_2) - u(f_1, f_2)| : \|f_1\|_\infty \leq 1, \|f_2\|_\infty \leq 1\} \\
&= \sup\{|\mu'(f_1, f_2, f_{3,0}) - u'(f_1, f_2, f_{3,0})| : \|f_1\|_\infty \leq 1, \|f_2\|_\infty \leq 1\} \\
&\leq \|\mu' - u'\| < \epsilon
\end{aligned}$$

Therefore $M(X_1 \times X_2)$ is dense in $MM(X_1, X_2)$. But [**11**, Theorem 3] says this is not the case, and the proof is complete. $\square$

We now extend the elements of $CB(X_1, X_2, X_3)$ to completely bounded trilinear forms on $C_0(X_1)^{**} \times C_0(X_2)^{**} \times C_0(X_3)^{**}$.

Proposition 2.6. *Let* $u \in CB(\mathcal{A}_1, \mathcal{A}_2, \mathcal{A}_3)$. *There is a unique, separately weak***-continuous, completely bounded, trilinear form* $\tilde{u}$ *on* $\mathcal{A}_1^{**} \times \mathcal{A}_2^{**} \times \mathcal{A}_3^{**}$ *which extends* u *and satisfies* $\|u\|_{cb} = \|\tilde{u}\|_{cb}$.

Proof. Let u be represented as in (4) (with $n = 3$) so that $\|u\|_{cb} = \|U_1\| \|U_2\|$. Extend each θ_i to $\mathcal{A}_i^{**}$ as above. The right-hand side of (4) extends to $\mathcal{A}_1^{**} \times \mathcal{A}_2^{**} \times \mathcal{A}_3^{**}$, defining a trilinear form $\tilde{u}$ on that space. By Theorem 2.2, $\tilde{u}$ is completely bounded with $\|u\|_{cb} = \|\tilde{u}\|_{cb}$. The separate weak* continuity of $\tilde{u}$ follows from the weak*-to-σ-weak continuity of θ_i, and the uniqueness is clear, since $\mathcal{A}_i$ is weak* dense in $\mathcal{A}_i^{**}$, $i = 1, 2, 3$. $\square$

Corollary 2.7. *Each* $u \in CB(X_1, X_2, X_3)$ *extends isometrically to a (unique) separately weak***- continuous, completely bounded, trilinear form on* $\mathcal{L}^\infty(X_1) \times \mathcal{L}^\infty(X_2) \times \mathcal{L}^\infty(X_3)$ *(which will also be denoted by* u*).*

Theorem 2.8. *If* $u \in CB(\mathcal{A}_1, \mathcal{A}_2, \mathcal{A}_3)$ *and* $v \in CB(\mathcal{B}_1, \mathcal{B}_2, \mathcal{B}_3)$, *then there exists a unique* $u \otimes v \in CB(\mathcal{A}_1 \otimes \mathcal{B}_1, \mathcal{A}_2 \otimes \mathcal{B}_2, \mathcal{A}_3 \otimes \mathcal{B}_3)$ *satisfying*

$$(6) \qquad (u \otimes v)(a_1 \otimes b_1, a_2 \otimes b_2, a_3 \otimes b_3) = u(a_1, a_2, a_3) v(b_1, b_2, b_3).$$

Moreover, (6) holds for $a_i \in \mathcal{A}_i^{**}$ *and* $b_i \in \mathcal{B}_i^{**}$, $i = 1, 2, 3$, *and*

$$\|u \otimes v\|_{cb} \leq \|u\|_{cb} \|v\|_{cb}.$$

Proof. Let

$$(7) \qquad u(a_1, a_2, a_3) = (\theta_1(a_1) \theta_2(a_2) \theta_3(a_3) \xi | \xi')$$
$$(8) \qquad v(b_1, b_2, b_3) = (\pi_1(b_1) \pi_2(b_2) \pi_3(b_3) \eta | \eta')$$

with $\|u\|_{cb} = \|\xi\|\|\xi'\|$ and $\|v\|_{cb} = \|\eta\|\|\eta'\|$, where for $i = 1, 2, 3$, θ_i is a representation of $\mathcal{A}_i$ on H, π_i is a representation of $\mathcal{B}_i$ on K, $\xi, \xi' \in H$, and $\eta, \eta' \in K$. For $x_i \in \mathcal{A}_i \otimes \mathcal{B}_i$, $i = 1, 2, 3$, define

$$(9) \quad \begin{aligned} (u \otimes v)(x_1, x_2, x_3) = \\ ((\theta_1 \otimes \pi_1)(x_1)(\theta_2 \otimes \pi_2)(x_2)(\theta_3 \otimes \pi_3)(x_3)(\xi \otimes \eta)|\xi' \otimes \eta'). \end{aligned}$$

Then $u \otimes v \in CB(\mathcal{A}_1 \otimes \mathcal{B}_1, \mathcal{A}_2 \otimes \mathcal{B}_2, \mathcal{A}_3 \otimes \mathcal{B}_3)$, and $\|u \otimes v\|_{cb} \leq \|\xi \otimes \eta\|\|\xi' \otimes \eta'\| = \|\xi\|\|\eta\|\|\xi'\|\|\eta'\| = \|u\|_{cb}\|v\|_{cb}$. Furthermore,

$$\begin{aligned} (u \otimes v)(a_1 \otimes b_1, a_2 \otimes b_2, a_3 \otimes b_3) \\ = ((\theta_1(a_1) \otimes \pi_1(b_1))(\theta_2(a_2) \otimes \pi_2(b_2))(\theta_3(a_3) \otimes \pi_3(b_3))(\xi \otimes \eta)|\xi' \otimes \eta') \\ = (\theta_1(a_1)\theta_2(a_2)\theta_3(a_3)\xi|\xi')(\pi_1(b_1)\pi_2(b_2)\pi_3(b_3)\eta|\eta') \\ = u(a_1, a_2, a_3)v(b_1, b_2, b_3). \end{aligned}$$

The uniqueness is clear, and the extension to second duals follows from Corollary 1.2. $\square$

Remark 2.9. For $\mu \in M(X_1 \times X_2 \times X_3)$ and $\nu \in M(Y_1 \times Y_2 \times Y_3)$, we have $u_\mu \otimes u_\nu = u_{\mu \times \nu}$.

3. Banach *-Algebras of Completely Bounded Multilinear Forms

Let us introduce a multiplication and an adjoint operation on $CB(G_1, G_2, G_3)$. Recall that for $\mu, \nu \in M(G)$, $\mu * \nu$ is defined by the formula $\mu * \nu(f) = \mu \times \nu(Mf)$, $f \in C_0(G)$, where $Mf(x, y) = f(xy)$. Of course, we have implicitly extended $\mu \times \nu$ from $C_0(G \times G)$ to $\mathcal{L}^\infty(G \times G)$. We shall use the same approach to define a convolution in $CB(G_1, G_2, G_3)$.

Definition 3.1. If G is a locally compact group and f is a function on G, set

$$Mf(x, y) = f(xy), \quad \check{f}(x) = f(x^{-1}), \quad f^*(x) = \overline{f(x^{-1})}.$$

For $u, v \in CB(G_1, G_2, G_3)$, and $f_i \in C_0(X_i)$, $i = 1, 2, 3$, define

$$(10) \quad (u * v)(f_1, f_2, f_3) = (u \otimes v)(Mf_1, Mf_2, Mf_3),$$

$$(11) \quad u^*(f_1, f_2, f_3) = \overline{u(f_1^*, f_2^*, f_3^*)}.$$

If θ is a representation of $C_0(G)$ on H we denote by $\tilde{\theta}$ the representation defined by $\tilde{\theta}(f) = [\theta(f^*)]^*$.

Remarks 3.2. (i) If u, v are represented as in (7) and (8), then according to (9) and (10), $u * v$ has a representation

$$(12) \quad \begin{aligned} (u * v)(f_1, f_2, f_3) = \\ ((\theta_1 \otimes \pi_1)(Mf_1)(\theta_2 \otimes \pi_2)(Mf_2)(\theta_3 \otimes \pi_3)(Mf_3)(\xi \otimes \eta)|\xi' \otimes \eta'). \end{aligned}$$

Since M is a $*$-homomorphism from $C_0(G_i)$ to $\mathcal{L}^\infty(G_i \times G_i)$, the above representation shows that $u * v \in CB(G_1, G_2, G_3)$ with associated representations $(\theta_i \otimes \pi_i)M$, $i = 1, 2, 3$. Moreover,

$$\|u * v\|_{cb} \leq \|\xi \otimes \eta\| \|\xi' \otimes \eta'\| = \|\xi\| \|\eta\| \|\xi'\| \|\eta'\| = \|u\|_{cb} \|v\|_{cb}.$$

(ii) To see $u^* \in CB(G_1, G_2, G_3)$, let $A_l = (f_{lij}) \in M_k(C_0(G_l))$ and $\tilde{A}_l = (f_{lij}^*), l = 1, 2, 3$. Then

$$
\begin{aligned}
\|(u^*)_k(A_1, A_2, A_3)\| &= \left\| \left(\sum_{r,s} u^*(f_{1ir}, f_{2rs}, f_{3sj}) \right) \right\| \\
&= \left\| \left(\sum_{r,s} \overline{u(f_{1ir}^*, f_{2rs}^*, f_{3sj}^*)} \right) \right\| \\
&= \left\| \left(\sum_{r,s} u(f_{1ir}^*, f_{2rs}^*, f_{3sj}^*) \right) \right\| \\
&= \left\| u_k(\tilde{A}_1, \tilde{A}_2, \tilde{A}_3) \right\|.
\end{aligned}
$$

Since $\|A_l\| = \sup_{x \in G_l} \|(f_{lij}(x))\| = \sup_{x \in G_l} \|(\bar{f}_{lij}(x))\| = \sup_{x \in G_l} \left\| \left(f_{lij}^*(x) \right) \right\|$
$= \|\tilde{A}_l\|$, $\|(u^*)_k\| = \|u_k\|$, and hence $u^* \in CB(G_1, G_2, G_3)$ with $\|u^*\|_{cb} = \|u\|_{cb}$.

(iii) Equations (10) and (11) hold for $f_i \in \mathcal{L}^\infty(G_i)$. Indeed, by Lemma 1.3, the map $f \to \check{f}$, and thus the map $f \to f^*$, and the map M are weak* continuous.

Lemma 3.3. *Let θ, π be two representations of $C_0(G)$ on H and K, respectively. Then $(\theta \otimes \pi)(Mf) = (\tilde{\pi} \otimes \tilde{\theta})(M\check{f})$ for $f \in C_0(G)$ (identifying $H \otimes K$ with $K \otimes H$).*

Proof. Define $T : G \times G \to G \times G$ by $T(x, y) = (y^{-1}, x^{-1})$. For $f, g \in C_0(G)$,

$$
\begin{aligned}
(\theta \otimes \pi)(f \otimes g) &= \theta(f) \otimes \pi(g) = \pi(g) \otimes \theta(f) \\
&= \tilde{\pi}(\check{g}) \otimes \tilde{\theta}(\check{f}) = (\tilde{\pi} \otimes \tilde{\theta})(\check{g} \otimes \check{f}) \\
&= (\tilde{\pi} \otimes \tilde{\theta})((f \otimes g) \circ T).
\end{aligned}
$$

Since $\theta \otimes \pi$ and $\tilde{\pi} \otimes \tilde{\theta}$ are continuous with respect to the weak* topology on $C_0(G \times G)^{**}$ and the σ-weak topology on $L(H \otimes K)$, by Lemma 1.3, $(\theta \otimes \pi)(h) = (\tilde{\pi} \otimes \tilde{\theta})(h \circ T)$ for all $h \in \mathcal{L}^\infty(X \times Y)$. Since $(Mf) \circ T = M\check{f}$, the lemma follows. $\square$

Corollary 3.4. *For $u, v \in CB(G_1, G_2, G_3)$, $(u * v)^* = v^* * u^*$.*

Proof. Let u, v be represented as in (7) and (8), respectively. Then

$$u^*(f_1, f_2, f_3) = \overline{u(f_1^*, f_2^*, f_3^*)} = \left(\tilde{\theta}_3(f_3) \tilde{\theta}_2(f_2) \tilde{\theta}_1(f_1) \xi' | \xi \right),$$

$$v^*(f_1, f_2, f_3) = (\tilde{\pi}_3(f_3) \tilde{\pi}_2(f_2) \tilde{\pi}_1(f_1) \eta' | \eta).$$

Note that these are not standard representations for u^* and v^* because of the order of the $*$-representations. Nevertheless,

$$(v^* \otimes u^*)(\Phi_1, \Phi_2, \Phi_3) =$$
$$\left((\tilde{\pi}_3 \otimes \tilde{\theta}_3)(\Phi_3)(\tilde{\pi}_2 \otimes \tilde{\theta}_2)(\Phi_2)(\tilde{\pi}_1 \otimes \tilde{\theta}_1)(\Phi_1)(\eta' \otimes \xi')|\eta \otimes \xi \right)$$

for all $\Phi_i \in C_0(G_i \times G_i)^{**}$ because both sides are separately weak* continuous and equal when $\Phi_i = f_i \otimes g_i$ with $f_i, g_i \in C_0(G_i)$, $i = 1, 2, 3$. So

$$(v^* * u^*)(f_1, f_2, f_3) =$$
$$\left((\tilde{\pi}_3 \otimes \tilde{\theta}_3)(Mf_3)(\tilde{\pi}_2 \otimes \tilde{\theta}_2)(Mf_2)(\tilde{\pi}_1 \otimes \tilde{\theta}_1)(Mf_1)(\eta' \otimes \xi')|\eta \otimes \xi \right).$$

On the other hand,

$$(u * v)^*(f_1, f_2, f_3) = \overline{(u * v)(f_1^*, f_2^*, f_3^*)}$$
$$= \overline{((\theta_3 \otimes \pi_3)(\overline{Mf_3^*})(\theta_2 \otimes \pi_2)(\overline{Mf_2^*})(\theta_1 \otimes \pi_1)(\overline{Mf_1^*})(\xi' \otimes \eta')|\xi \otimes \eta)}$$
$$= ((\theta_3 \otimes \pi_3)(M\check{f}_3)(\theta_2 \otimes \pi_2)(M\check{f}_2)(\theta_1 \otimes \pi_1)(M\check{f}_1)(\xi' \otimes \eta')|\xi \otimes \eta).$$

Now the equation $(u * v)^* = v^* * u^*$ follows from lemma 3.3. $\square$

Lemma 3.5. *Let θ, π, δ be representations of $C_0(G)$ on H, K, L respectively. Then for $f \in C_0(G)$,*

$$\{\theta \otimes [(\pi \otimes \delta)M]\}Mf = \{[(\theta \otimes \pi)M] \otimes \delta\}Mf.$$

Proof. Let $T_1, T_2 : G \times G \times G \to G \times G$ be defined by $T_1(x, y, z) = (xy, z)$ and $T_2(x, y, z) = (x, yz)$, respectively. Note that Corollary 1.2 gives

$$\{\theta \otimes [(\pi \otimes \delta)M]\}(f \otimes g) = \theta(f) \otimes (\pi \otimes \delta)(Mg)$$
$$= [\theta \otimes (\pi \otimes \delta)](f \otimes Mg)$$
$$= [\theta \otimes (\pi \otimes \delta)]((f \otimes g) \circ T_2).$$

Thus, as in the proof of Lemma 3.3, for all $h \in \mathcal{L}^\infty(G \times G)$,

$$\{\theta \otimes [(\pi \otimes \delta)M]\}(h) = [\theta \otimes (\pi \otimes \delta)](h \circ T_2).$$

A similar argument shows that

$$\{[(\theta \otimes \pi)M] \otimes \delta\}(h) = [(\theta \otimes \pi) \otimes \delta](h \circ T_1).$$

Our lemma now follows from the identity $(Mf) \circ T_1 = (Mf) \circ T_2.\square$

Corollary 3.6. *The multiplication defined by (10) is associative.*

Proof. For $u, v, w \in CB(G_1, G_2, G_3)$, let u, v be represented as in (7) and (8), respectively, and let

$$w(f_1, f_2, f_3) = (\delta_1(f_1)\delta_2(f_2)\delta_3(f_3)\zeta|\zeta'),$$

where δ_i is a representation of $C_0(G_i)$, $i = 1, 2, 3$, on some Hilbert space L, and $\zeta, \zeta' \in L$. By (12), the associated representations of $u * v$ are $(\theta_i \otimes \pi_i)M$, $i = 1, 2, 3$, and those of $v * w$ are $(\pi_i \otimes \delta_i)M$, $i = 1, 2, 3$. Thus for $f_i \in C_0(G_i)$, $i = 1, 2, 3$,

$$
\begin{aligned}
((u * v) * w)(f_1, f_2, f_3) &= ((u * v) \otimes w)(Mf_1, Mf_2, Mf_3) \\
&= (U((\xi \otimes \eta) \otimes \zeta)|(\xi' \otimes \eta') \otimes \zeta'),
\end{aligned}
$$

where

$$
U = \{[(\theta_1 \otimes \pi_1)M] \otimes \delta_1\}(Mf_1)\{[(\theta_2 \otimes \pi_2)M] \otimes \delta_2\}(Mf_2)\{[(\theta_3 \otimes \pi_3)M] \otimes \delta_3\}(Mf_3).
$$

On the other hand,

$$
\begin{aligned}
(u * (v * w))(f_1, f_2, f_3) &= (u \otimes (v * w))(Mf_1, Mf_2, Mf_3) \\
&= (V(\xi \otimes (\eta \otimes \zeta))|\xi' \otimes (\eta' \otimes \zeta')),
\end{aligned}
$$

where

$$
V = \{\theta_1 \otimes [(\pi_1 \otimes \delta_1)M]\}(Mf_1)\{\theta_2 \otimes [(\pi_2 \otimes \delta_2)M]\}(Mf_2)\{\theta_3 \otimes [(\pi_3 \otimes \delta_3)M]\}(Mf_3).
$$

By Lemma 3.5, $U = V$ and hence $(u * v) * w = u * (v * w)$. $\square$

Theorem 3.7. *The multiplication* (10) *and the adjoint operation* (11) *define a unital Banach $*$-algebra structure with isometric involution on $CB(G_1, G_2, G_3)$ which extends the $*$-algebra structure of $M(G_1 \times G_2 \times G_3)$.*

Proof. If $u, v \in CB(G_1, G_2, G_3)$, we have shown in Remarks 3.2 that $u * v, u^* \in CB(G_1, G_2, G_3)$, $\|u^*\|_{cb} = \|u\|_{cb}$, and $\|u * v\|_{cb} \leq \|u\|_{cb}\|v\|_{cb}$. It is easy to check, given Corollary 3.4, that $u \to u^*$ is an adjoint operation in $CB(G_1, G_2, G_3)$. If $\mu \in M(G_1 \times G_2 \times G_3)$, then

$$
\begin{aligned}
u_\mu^*(f_1, f_2, f_3) &= \overline{u_\mu(f_1^*, f_2^*, f_3^*)} \\
&= \overline{\int_{G_1 \times G_2 \times G_3} f_1^*(x_1)f_2^*(x_2)f_3^*(x_3)d\mu(x_1, x_2, x_3)} \\
&= \int_{G_1 \times G_2 \times G_3} f_1(x_1)f_2(x_2)f_3(x_3)d\mu^*(x_1, x_2, x_3) \\
&= u_{\mu^*}(f_1, f_2, f_3).
\end{aligned}
$$

Since Theorem 2.8 gives immediately that

$$
(u_1 + u_2) \otimes v = u_1 \otimes v + u_2 \otimes v,
$$

$$
u \otimes (v_1 + v_2) = u \otimes v_1 + u \otimes v_2,
$$

it follows that the multiplication is distributive. This, along with Corollary 3.6, shows that (10) defines a multiplication on $CB(G_1, G_2, G_3)$.

Let $e \in G$ be the identity and $\theta_e(f) = f(e)$, $f \in C_0(G)$. Define $T : G \to G \times G$ by $T(x) = (x, e)$. If θ is a representation of $C_0(G)$ on H, then for $f, g \in C_0(G)$,

$$(\theta \otimes \theta_e)(f \otimes g) \;=\; \theta(f) \otimes g(e) = g(e)\theta(f)$$
$$=\; \theta(g(e)f) = \theta((f \otimes g) \circ T).$$

Thus by Lemma 1.3, $(\theta \otimes \theta_e)(h) = \theta(h \circ T)$, $h \in \mathcal{L}^\infty(G \times G)$. In particular,

(13) $(\theta \otimes \theta_e)(Mf) = \theta(Mf \circ T)$, $f, g \in C_0(G)$.

Let $e = (e_1, e_2, e_3)$, where e_i is the identity of G_i, and let u_e be the trilinear form induced by the point mass δ_e, that is,

$$u_e(f_1, f_2, f_3) = f_1(e_1)f_2(e_2)f_3(e_3).$$

Let us show that u_e is an identity of $CB(G_1, G_2, G_3)$. We may write the above formula as

$$u_e(f_1, f_2, f_3) = (\theta_{e_1}(f_1)\theta_{e_2}(f_2)\theta_{e_3}(f_3)1|1).$$

Applying (13),

$$(u * u_e)(f_1, f_2, f_3) = (u \otimes u_e)(Mf_1, Mf_2, Mf_3)$$
$$=\; ((\theta_1 \otimes \theta_{e_1})(Mf_1)(\theta_2 \otimes \theta_{e_2})(Mf_2)(\theta_3 \otimes \theta_{e_3})(Mf_3)(\xi \otimes 1)|\xi' \otimes 1)$$
$$=\; (\theta_1(f_1)\theta_2(f_2)\theta_3(f_3)\xi|\xi')$$
$$=\; u(f_1, f_2, f_3).$$

Similarly, $u_e * u = u$.

Finally,

$$u_{\mu*\nu}(f_1, f_2, f_3) \;=\; \int_{G_1 \times G_2 \times G_3} f_1(x_1)f_2(x_2)f_3(x_3)\, d\mu * \nu(x_1, x_2, x_3)$$
$$=\; \int_{G_1^2 \times G_2^2 \times G_3^2} f_1(x_1y_1)f_2(x_2y_2)f_3(x_3y_3)d(\mu \times \nu)(x_1, \cdots, y_3)$$
$$=\; u_{\mu \times \nu}(Mf_1, Mf_2, Mf_3)$$
$$=\; (u_\mu \otimes u_\nu)(Mf_1, Mf_2, Mf_3)$$
$$=\; (u_\mu * u_\nu)(f_1, f_2, f_3),$$

and the proof is complete. $\square$

Theorem 3.8. $CB(G_1, G_2, G_3)$ *is commutative if and only if* $G_1, G_2,$ *and* G_3 *are all abelian.*

Proof. Let the G_i all be abelian. To see $u * v = v * u$, it follows from (12) that we need only show that for G abelian and θ and π representations of $C_0(G)$ (and identifying, as usual, $H \otimes K$ with $K \otimes H$), we have

$$(\theta \otimes \pi)(Mf) = (\pi \otimes \theta)(Mf).$$

But this follows easily by observing that $(\theta \otimes \pi)(f \otimes g) = (\pi \otimes \theta)(g \otimes f)$ for $f, g \in C_0(G)$, and that in our case the map $T(x, y) = (y, x)$ satisfies $Mf \circ T = Mf$ and arguing as in Lemma 3.3.

Conversely, if $CB(G_1, G_2, G_3)$ is commutative, then $M(G_1 \times G_2 \times G_3)$, as a subalgebra of $CB(G_1, G_2, G_3)$, is commutative. Hence $G_1 \times G_2 \times G_3$ is abelian. $\square$

Theorem 3.9. *Let H_i be a closed, normal subgroup of G_i and $\phi_i : G_i \to G_i/H_i$ be the quotient map, $i = 1, 2, 3$. For $u \in CB(G_1, G_2, G_3)$, define $\sigma(u) \in CB(G_1/H_1, G_2/H_2, G_3/H_3)$ by*

$$\sigma(u)(f_1, f_2, f_3) = u(f_1 \circ \phi_1, f_2 \circ \phi_2, f_3 \circ \phi_3), \qquad f_i \in C_0(G_i/H_i).$$

Then σ is a $*$-algebra homomorphism with $\|\sigma(u)\|_{cb} \leq \|u\|_{cb}$.

Proof. The map $\delta_i : C_0(G_i/H_i) \to C(G_i)$ defined by $\delta_i(f_i) = f_i \circ \phi_i$ is obviously a $*$-homomorphism. For $f_i \in C_0(G_i/H_i)$, $i = 1, 2, 3$, Proposition 2.6 allows us to write

$$
\begin{aligned}
\sigma(u)(f_1, f_2, f_3) &= u(f_1 \circ \phi_1, f_2 \circ \phi_2, f_3 \circ \phi_3) \\
&= (\theta_1(f_1 \circ \phi_1)\theta_2(f_2 \circ \phi_2)\theta_3(f_3 \circ \phi_3)\xi|\xi') \\
&= ((\theta_1\delta_1)(f_1)(\theta_2\delta_2)(f_2)(\theta_3\delta_3)(f_3)\xi|\xi').
\end{aligned}
$$

Thus $\sigma(u) \in CB(G_1/H_1, G_2/H_2, G_3/H_3)$, and

$$\|\sigma(u)\|_{cb} \leq \|u\|_{cb}.$$

By Remark 3.2 (iii), we have that

$$
\begin{aligned}
\sigma(u^*)(f_1, f_2, f_3) &= u^*(f_1 \circ \phi_1, f_2 \circ \phi_2, f_3 \circ \phi_3) \\
&= \overline{u((f_1 \circ \phi_1)^*, (f_2 \circ \phi_2)^*, (f_3 \circ \phi_3)^*)} \\
&= \overline{u(f_1^* \circ \phi_1, f_2^* \circ \phi_2, f_3^* \circ \phi_3)} \\
&= \overline{\sigma(u)(f_1^*, f_2^*, f_3^*)} \\
&= \sigma(u)^*(f_1, f_2, f_3).
\end{aligned}
$$

Since σ is obviously linear, to complete the proof we just need to show that

$$\sigma(u * v) = \sigma(u) * \sigma(v), \qquad u, v \in CB(G_1, G_2, G_3).$$

Now by Remark 3.2 (iii),

$$
\begin{aligned}
\sigma(u * v)(f_1, f_2, f_3) &= (u * v)(f_1 \circ \phi_1, f_2 \circ \phi_2, f_3 \circ \phi_3) \\
&= (u \otimes v)(M(f_1 \circ \phi_1), M(f_2 \circ \phi_2), M(f_3 \circ \phi_3)) \\
&= (U(\xi \otimes \eta)|\xi' \otimes \eta'),
\end{aligned}
$$

where $U = (\theta_1 \otimes \pi_1)[M(f_1 \circ \phi_1)](\theta_2 \otimes \pi_2)[M(f_2 \circ \phi_2)](\theta_3 \otimes \pi_3)[M(f_3 \circ \phi_3)]$, and

$$
\begin{aligned}
(\sigma(u) * \sigma(v))(f_1, f_2, f_3) &= (\sigma(u) \otimes \sigma(v))(Mf_1, Mf_2, Mf_3) \\
&= (V(\xi \otimes \eta)|\xi' \otimes \eta'),
\end{aligned}
$$

where $V = (\theta_1\delta_1 \otimes \pi_1\delta_1)(Mf_1)(\theta_2\delta_2 \otimes \pi_2\delta_2)(Mf_2)(\theta_3\delta_3 \otimes \pi_3\delta_3)(Mf_3)$.

For $g_i, h_i \in C_0(G_i/H_i)$,

$$
\begin{aligned}
(\theta_i\delta_i \otimes \pi_i\delta_i)(g_i \otimes h_i) &= \theta_i\delta_i(g_i) \otimes \pi_i\delta_i(h_i) \\
&= \theta_i(g_i \circ \phi_i) \otimes \pi_i(h_i \circ \phi_i) \\
&= (\theta_i \otimes \pi_i)((g_i \circ \phi_i) \otimes (h_i \circ \phi_i)).
\end{aligned}
$$

Hence, as before, $(\theta_i\delta_i \otimes \pi_i\delta_i)(Mf_i) = (\theta_i \otimes \pi_i)[M(f_i \circ \phi_i)]$, $i = 1, 2, 3$. So $U = V$, and hence $\sigma(u * v) = \sigma(u) * \sigma(v)$. $\square$

4. Fourier Transforms of Completely Bounded Multilinear Forms

Definition 4.1. Let G_i be a LCA group with character group Γ_i, $i = 1, 2, 3$, and let $u \in CB(G_1, G_2, G_3)$. We define the *Fourier transform* $\hat{u}$ of u by the formula

$$\hat{u}(\gamma_1, \gamma_2, \gamma_3) = u(\bar\gamma_1, \bar\gamma_2, \bar\gamma_3), \quad \gamma_i \in \Gamma_i, i = 1, 2, 3.$$

It is obvious from Proposition 2.6 that $\|\hat{u}\|_\infty \leq \|u\|_{cb}$, $u \in CB(G_1, G_2, G_3)$. It is also clear that $\hat{u}$ determines u uniquely, because the trigonometric polynomials are weak* dense in $C_0(G_i)^{**}$ and u is separately weak* continuous.

If $\mu \in M(G_1 \times G_2 \times G_3)$, then clearly $\hat{u}_\mu(\gamma_1, \gamma_2, \gamma_3) = \hat\mu(\gamma_1, \gamma_2, \gamma_3)$.

Theorem 4.2. *If G_i, $i = 1, 2, 3$ are LCA groups and $u, v \in CB(G_1, G_2, G_3)$, then*

$$\widehat{u * v} = \hat{u}\hat{v} \ \text{ and } \ \widehat{u^*} = \bar{\hat{u}}.$$

Proof. For $\gamma_i \in \Gamma_i$, $i = 1, 2, 3$,

$$
\begin{aligned}
\widehat{u * v}(\gamma_1, \gamma_2, \gamma_3) &= (u * v)(\bar\gamma_1, \bar\gamma_2, \bar\gamma_3) \\
&= (u \otimes v)(M\bar\gamma_1, M\bar\gamma_2, M\bar\gamma_3) \\
&= (u \otimes v)(\bar\gamma_1 \otimes \bar\gamma_1, \bar\gamma_2 \otimes \bar\gamma_2, \bar\gamma_3 \otimes \bar\gamma_3) \\
&= u(\bar\gamma_1, \bar\gamma_2, \bar\gamma_3)v(\bar\gamma_1, \bar\gamma_2, \bar\gamma_3) \\
&= \hat{u}(\gamma_1, \gamma_2, \gamma_3)\hat{v}(\gamma_1, \gamma_2, \gamma_3),
\end{aligned}
$$

and

$$
\begin{aligned}
\widehat{u^*}(\gamma_1, \gamma_2, \gamma_3) &= u^*(\bar\gamma_1, \bar\gamma_2, \bar\gamma_3) \\
&= \overline{u(\bar\gamma_1{}^*, \bar\gamma_2{}^*, \bar\gamma_3{}^*)} \\
&= \overline{u(\bar\gamma_1, \bar\gamma_2, \bar\gamma_3)} \\
&= \overline{\hat{u}(\gamma_1, \gamma_2, \gamma_3)}. \ \square
\end{aligned}
$$

We conclude with a remark which compares our work with Ylinen's work in [**18**].

Remark 4.3. For locally compact groups $G_1, G_2, \cdots, G_n$, Ylinen [18] studied completely bounded n-linear forms $\Phi : C^*(G_1) \times \cdots \times C^*(G_n) \to C$ and their Fourier transforms, where $C^*(G_i)$ is the group C^*-algebra of G_i. A convolution was introduced on the space of all such completely bounded n-linear forms via the Fourier transform. It turned out that the space equipped with the completely bounded norm is always a *commutative* unital Banach algebra with respect to this convolution [18, Sec. 6]. It is obvious that the space we study in this paper is quite different from the one studied in [18]. If the G_i are all LCA groups, however, then the two spaces do coincide, only with the roles of G_i and Γ_i interchanged, since $C^*(G_i) = C_0(\Gamma_i)$ in this case. Moreover, since Fourier transforms for abelian G_i in [18] were also defined via the unique separately weak* continuous extensions to $C^*(G_1)^{**} \times \cdots \times C^*(G_n)^{**}$, it is easy to see that the two definitions of Fourier transform are the same.

REFERENCES

1. D.P. Blecher, *Tensor products of Operator spaces II*, Canad. J. Math. **44** (1992), 75-90.
2. D.P. Blecher and R.R. Smith, *The dual of the Haagerup tensor product*, J. Lond. Math. Soc. (2) **45** (1992), 126-144.
3. N. Bourbaki, *Topological Vector Spaces, Chapters 1-5*, Springer-Verlag, Berlin/Heidelberg/New York, 1987.
4. E. Christensen, E.G. Effros, and A.M. Sinclair, *Completely bounded multilinear maps and C^* algebraic cohomology*, Inv. Math. **90** (1987), 279-296.
5. E. Christensen and A.M. Sinclair, *Representations of completely bounded multilinear operators*, J. Funct. Anal. **72** (1987), 151-181.
6. E.G. Effros, J. Kraus, and Z.-J. Ruan, *On two quantized tensor products*, Operator Algebras, Mathematical Physics, and Low Dimensional Topology (R. Herman and B. Tanbay, eds.), Research Notes in Math., vol. 5, Peters, Wellesley, Mass., 1993, pp. 125-145.
7. E.G. Effros and Z.-J. Ruan, *Operator convolution algebras: an approach to quantum groups*, preprint, 1991.
8. J.E. Gilbert, T. Ito, and B.M. Schreiber, *Bimeasure algebras on locally compact groups*, J. Funct. Anal. **64** (1985), 134- 162.
9. C.C. Graham and B.M. Schreiber, *Bimeasure algebras on LCA groups*, Pacific J. Math. **115** (1984), 91-127.
10. ——, *Sets of interpolation for Fourier transforms of bimeasures*, Colloq. Math. **51** (1987), 149-154.
11. ——, *Projections in spaces of bimeasures*, Canad. Math. Bull. **31** (1) (1988), 19-25.
12. G.J. Murphy, *C^*-Algebras and Operator Theory*, Academic Press, San Diego, 1990.
13. V.I. Paulsen, *Completely Bounded Maps and Dilations*, Research Notes in Mathematics Series, Vol. 146, Pitman, London, 1986.
14. V.I. Paulsen and R.R. Smith, *Multilinear maps and tensor norms on operator systems*, J. Funct. Anal. **73** (1987), 258- 276.
15. S. Saeki, *Tensor products of $C(X)$-spaces and their conjugate spaces*, J. Math. Soc. Japan **28** (1976), 33-47.
16. M. Takesaki, *Theory of Operator Algebras I*, Springer-Verlag, Berlin/Heidelberg/New York, 1979.
17. K. Ylinen, *Representing completely bounded multilinear operators*, preprint, 1986.
18. K. Ylinen, *Noncommutative Fourier transforms of bounded bilinear forms and completely bounded multilinear operators*, J. Funct. Anal. **79** (1988), 144-165.

DEPARTMENT OF MATHEMATICS AND COMPUTER SCIENCE, FAYETTEVILLE STATE UNIVERSITY, FAYETTEVILLE, NC 28301
E-mail address: Internet: GZhao@Hazel.FSUFAY.Edu

DEPARTMENT OF MATHEMATICS, WAYNE STATE UNIVERSITY, DETROIT, MI 48202
E-mail address: Internet: BertS@Math.Wayne.Edu

Other Titles in This Series

(*Continued from the front of this publication*)

(See the AMS catalog for earlier titles)